Biological Activity of Natural Secondary Metabolite Products

Biological Activity of Natural Secondary Metabolite Products

Special Issue Editor

Toshio Morikawa

MDPI • Basel • Beijing • Wuhan • Barcelona • Belgrade

MDPI

Special Issue Editor
Toshio Morikawa
Kindai University
Japan

Editorial Office
MDPI
St. Alban-Anlage 66
Basel, Switzerland

This is a reprint of articles from the Special Issue published online in the open access journal *International Journal of Molecular Sciences* (ISSN 1422-0067) from 2017 to 2018 (available at: http://www.mdpi.com/journal/ijms/special_issues/natural_secondary_metabolite)

For citation purposes, cite each article independently as indicated on the article page online and as indicated below:

LastName, A.A.; LastName, B.B.; LastName, C.C. Article Title. *Journal Name* **Year**, *Article Number*, Page Range.

ISBN 978-3-03897-174-0 (Pbk)
ISBN 978-3-03897-175-7 (PDF)

Contents

About the Special Issue Editor

Toshio Morikawa is Professor at the Pharmaceutical Research and Technology Institute, Kindai University, Japan. He was born in Kyoto Prefecture, Japan in 1972 and received his Ph.D. under the supervision of Professor Masayuki Yoshikawa at Kyoto Pharmaceutical University in 2002. In 2001, he started his academic career at Kyoto Pharmaceutical University as an Assistant Professor. He became a Lecturer in 2005, an Associate Professor in 2010, and a Professor in 2015 at the Pharmaceutical Research and Technology Institute, Kindai University. He received The JSP Award for Young Scientists in 2005 and The JSP Award for Scientific Contributions in 2018. His current research program focuses on the search for bioactive constituents from natural resources and the development of new functional foods for the prevention and improvement of lifestyle diseases. He has published over 200 papers in peer reviewed journals and is currently serving on the editorial board of the Journal of Natural Medicines, Traditional and Kampo Medicine, and the Japanese Journal of Food Chemistry and Safety. He has also served twice as a Guest Editor for *IJMS*.

Preface to "Biological Activity of Natural Secondary Metabolite Products"

Natural secondary metabolite products that are isolated from plants, animals, microorganisms, etc., are classified as polyketides, isoprenoids, aromatics (phenylpropanoids), alkaloids, etc. Their chemical diversity and variety of biological activities have attracted the attention of chemists, biochemists, biologists, etc. The Special Issue on the "Biological Activity of Natural Secondary Metabolite Products" is intended to present biologically active natural products as candidates and/or leads for pharmaceuticals, dietary supplements, functional foods, cosmetics, food additives, etc. The research fields covered in this Special Issue of IJMS include the chemistry of natural products, phytochemistry, pharmacognosy, food chemistry, bioorganic synthetic chemistry, chemical biology, molecular biology, molecular pharmacology, and other research fields related to bioactive natural secondary metabolite products.

Toshio Morikawa
Special Issue Editor

International Journal of
Molecular Sciences

MDPI

Article

Tapirira guianensis Aubl. Extracts Inhibit Proliferation and Migration of Oral Cancer Cells Lines

Renato José Silva-Oliveira [1], Gabriela Francine Lopes [2], Luiz Fernando Camargos [3],
Ana Maciel Ribeiro [4], Fábio Vieira dos Santos [3], Richele Priscila Severino [5],
Vanessa Gisele Pasqualotto Severino [5], Ana Paula Terezan [5], Ralph Gruppi Thomé [6],
Hélio Batista dos Santos [6], Rui Manuel Reis [1,7,8] and Rosy Iara Maciel de Azambuja Ribeiro [3,*]

[1] Molecular Oncology Research Center, Barretos Cancer Hospital, Barretos 14784-400, Brazil; renatokjso@gmail.com (R.J.S.-O.); ruireis.hcb@gmail.com (R.M.R.)
[2] Laboratory of Experimental Pathology, Federal University of São João del Rei—CCO/UFSJ, Divinópolis 35501-296, Brazil; gabrielafrancine_lp@hotmail.com
[3] Laboratory of Mutagenesis, Federal University of São João del Rei—CCO/UFSJ, Divinópolis 35501-296, Brazil; luizfcamargos@yahoo.com.br (L.F.C.); fabiosantos@ufsj.edu.br (F.V.d.S.)
[4] Medical School, Federal University of Minas Gerais—UFMG, Belo Horizonte 31270-901, Brazil; anamacielr@gmail.com
[5] Special Academic Unit of Physics and Chemistry, Federal University of Goiás, Catalão 75704-020, Brazil; richeleps@ufg.br (R.P.S.); vanessa.pasqualotto@gmail.com (V.G.P.S.); apterezan@hotmail.com (A.P.T.)
[6] Laboratory of Tissue Processing, Federal University of São João del Rei—CCO/UFSJ, Divinópolis 35501-296, Brazil; ralphthome@gmail.com (R.G.T.); hbsufsj@gmail.com (H.B.d.S.)
[7] Life and Health Sciences Research Institute (ICVS), Health Sciences School, University of Minho, Braga 4710-057, Portugal
[8] 3ICVS/3B's-PT Government Associate Laboratory, Braga 4710-057, Portugal
* Correspondence: rosy@ufsj.edu.br; Tel.: +55-37-3221-1610; Fax: +55-37-3221-1614

Academic Editor: Toshio Morikawa
Received: 1 September 2016; Accepted: 12 October 2016; Published: 8 November 2016

Abstract: Cancer of the head and neck is a group of upper aerodigestive tract neoplasms in which aggressive treatments may cause harmful side effects to the patient. In the last decade, investigations on natural compounds have been particularly successful in the field of anticancer drug research. Our aim is to evaluate the antitumor effect of *Tapirira guianensis* Aubl. extracts on a panel of head and neck squamous cell carcinoma (HNSCC) cell lines. Analysis of secondary metabolites classes in fractions of *T. guianensis* was performed using Nuclear Magnetic Resonance (NMR). Mutagenicity effect was evaluated by Ames mutagenicity assay. The cytotoxic effect, and migration and invasion inhibition were measured. Additionally, the expression level of apoptosis-related molecules (PARP, Caspases 3, and Fas) and MMP-2 was detected using Western blot. Heterogeneous cytotoxicity response was observed for all fractions, which showed migration inhibition, reduced matrix degradation, and decreased cell invasion ability. Expression levels of MMP-2 decreased in all fractions, and particularly in the hexane fraction. Furthermore, overexpression of FAS and caspase-3, and increase of cleaved PARP indicates possible apoptosis extrinsic pathway activation. Antiproliferative activity of *T. guianensis* extract in HNSCC cells lines suggests the possibility of developing an anticancer agent or an additive with synergic activities associated with conventional anticancer therapy.

Keywords: HNSCC; cytotoxic activity; alkaloids; apoptosis markers

1. Introduction

Cancer of the head and neck is a group of upper aerodigestive system neoplasms that corresponds to the seventh most common cancer worldwide [1]. They are aggressive tumors and can often be invasive and metastatic [2]. This pathology is a condition that continues to be diagnosed in an advanced stage with low survival rate [3].

The prevalence of oral cancer in Brazil is high, especially among men in their fifth decade of life [4]. According to INCA (Brazilian National Cancer Institute), approximately 300,000 new cases of oral cancer were diagnosed in 2012 and it was responsible for 145,000 deaths. The five-year survival rate of these patients is 50%–60% [4–6]. Unfortunately, even aggressive treatments, such as surgery, radiotherapy and chemotherapy are not curative, and cause severe co-morbidities. To overcome these limitations, the phytomedicine, which uses therapeutic agents derived from plants as an adjuvant treatment in combination with surgery and or radiotherapy, can constitute potential options [7,8].

Some plants are known to contribute significantly within the traditional medicine of developing countries [9,10]. In the last decade, research on natural compounds has been particularly successful in the field of anticancer drug research. Examples of anticancer agents developed from higher plants (also known as vascular plants) are the alkaloids vinblastine and vincristine, which were both obtained from the Madagascar periwinkle (*Catharanthus roseus*) [11,12]. Brazil is home of the greatest biodiversity, representing more than 20% of the total number of species on Earth [13]. This fact, together with the need for developing new drugs, justifies the incentive for research programs in investigation of potential medicinal plants.

Tapirira guianensis Aubl., a member of the Anacardiaceae family, is popularly known in Brazil as "tatapiririca", "cedroí" or "Pau-pombo". It is widely distributed throughout the Brazilian territory, and its leaves and bark are used by folk medicine to treat dermatitis, syphilis and as a cleanser, due to its antibacterial and antifungal activity [14]. Preliminary study demonstrated that the $CHCl_3$ extract and two isolated compounds of the seeds of *T. guianensis* displayed cytotoxicity activity against human prostate cancer [15], however the cytotoxic action of the leaves constituents remain unknown.

Studies have shown the presence of secondary compounds of the families of tannins, coumarins, flavones, flavonols, flavanones, saponins, steroids and alkaloids in *T. guianensis* leaves [16,17]. The presence of hidrobenzofuranides norisoprenoids that are long chain metabolites with suggested anti-tumor activity was also identified in the leaves [17]. Longatti et al. found that crude extract and the hydroalcohol, chloroform and ethyl acetate fractions showed significant inhibition of matrix metalloproteinase 2 (MMP-2) [16].

Despite these findings, the knowledge of *T. guianensis* anti-neoplastic activity and its molecular mechanisms is still scarce. Therefore, in this study, we aim to evaluate the effect of this species in head and neck tumor cell lines.

2. Results

2.1. Nuclear Magnetic Resonance (NMR) Analysis of T. guianensis

The identification by ^{1}H NMR of secondary metabolites classes present on the crude extract and fractions obtained by liquid–liquid extraction from *T. guianensis* was based on their chemical shifts (ppm) features.

The NMR spectrum of the crude extract (C3) presented signals of chemical shifts characteristics of alkaloids, coumarins and flavonoids in aromatic region of 6.0–7.5 ppm, hydrogen bonded to carbon sp^2 and/or neighbors heteroatoms in region of 3.4–5.3 ppm for alkaloids and coumarins and sugars (saponin and/or glycosylated flavonoids) and hydrogen bonded to carbon sp^3 in region of 0.7–2.4 ppm for saponin and steroids. The ethyl acetate fraction (C1) showed signals of chemical shifts typical of coumarins and flavonoids in aromatic region of 6.0–7.3 ppm, hydrogen bonded to carbon sp^2 and/or neighbors heteroatoms in region of 3.0–5.8 ppm for coumarins and glycosylated flavonoids, and hydrogen bonded to carbon sp^3 in region of 0.5–1.5 ppm for steroids. The hydroalcoholic fraction

(C2), on the other hand, exhibited signals of chemical shifts characteristics of alkaloids, coumarins and flavonoids in aromatic region of 6.2–8.2 ppm, hydrogen bonded to carbon sp^2 and/or neighbors heteroatoms in region of 3.2–5.4 ppm for alkaloids, coumarins and sugars (saponin and/or glycosylated flavonoids) and hydrogen bonded to carbon sp^3 in region of 0.8–2.3 ppm for saponin and steroids. Lastly, the hexane fraction (C4) displayed signals of chemical shifts typical of alkaloids in aromatic region of 6.0–7.3 ppm and hydrogen bonded to carbon sp^2 and/or neighbors heteroatoms in region of 3.0–5.8 ppm, a hydrogen bonded to carbon sp^3 in region of 0.5–1.5 ppm for steroids.

2.2. Ames Mutagenicity Assay

All evaluated doses of *T. guaianensis* hydroalcoholic extract induced a significant increase in revertants frequency employing the TA98 strain of *Salmonella typhimurium* (Table 1). The Ames test mutagenic index (MI) values observed ranged between 3.37 (to lowest dose) and 14.84 (to highest dose), indicating a high mutation index of this extract. However, in the TA100 strain without S9, the mutagenicity was not observed. None of the assessed doses induced significant alterations in revertants frequency to this strain when compared with negative control group dimethyl sulfoxide (DMSO) and all MI values were lower than 2. Nevertheless, in studies with TA100 with metabolic activation (+S9), the analyzed extract was mutagenic, but at a lower level, suggesting the presence of indirect mutagens that induce base substitutions in DNA structure.

Table 1. Revertant/plate, standard deviation and mutagenicity index (MI) in the strains TA98 and TA100 of *S. typhimurium* after treatment with different doses of the extract from *T. guianensis*, without (−S9) and with (+S9) metabolic activation.

Treatment	TA98				TA100			
	−S9		+S9		−S9		+S9	
	Mean ± SD	MI	Mean ± SD	MI	Mean ± SD	MI	Mean ± SD	MI
NC	16.7 ± 6.0	–	41.7 ± 5.69	–	121.7 ± 18.2	–	97.00 ± 40.85	–
Extract (µg/mL)	–	–	–	–	–	–	–	–
4.25	56.0 ± 11.0 *	3.3	140.3 ± 87.5 **	3.4	130.0 ± 37.2	1.1	198.7 ± 24.5 *	2.0
8.67	67.3 ± 16.3 *	4.0	180.7 ± 42.2 *	4.3	118.0 ± 74.6	1.0	232.0 ± 8.9 *	2.4
16.68	136.0 ± 16.7 *	8.1	309.7 ± 74.6 *	7.4	68.7 ± 33.1	0.6	250.7 ± 35.2 *	2.6
24.09	186.0 ± 29.9 *	11.1	423.7 ± 60.5 *	10.2	110.7 ± 11.5	0.9	306.3 ± 31.2 *	3.2
30.97	246.3 ± 43.7 *	14.7	315.0 ± 10.5 *	7.6	119.7 ± 29.0	1.0	308.3 ± 41.4 *	3.2
PC	426.3 ± 40.3 *	–	2839.7 ± 515.3 *	–	471.0 ± 194.2 *	–	1257.3 ± 159.3 *	–

Mean ± SD: Revertants Mean Frequency per plate ± Standard Deviation; MI: Mutagenicity Index; NC: Negative Control (DMSO − 80 µL/plate); PC: Positive Control 4-nitro-o-phenylenediamine (10 µg/plate − TA98 (−S9)), Methylmethane sulfonate (260 µg/plate − TA100 (−S9)) or 2-Aminoanthracene (5 µg/plate − TA98 (+S9) and TA100 (+S9)); * $p < 0.05$ and MI > 2; ** MI > 2.

2.3. Efficacy of T. guianensis Samples in Head and Neck Squamous Cell Carcinoma Cell Lines

The half-maximal inhibitory concentration (IC_{50}) was determined to assess the cytotoxicity of *T. guianensis* samples in head and neck tumor cell lines (Table 2) and a representation of the proliferation and survival curves of the head and neck tumor cell lines is depicted in Figure S1. The ethyl acetate fraction (C1) showed IC_{50} values between 28.0 ± 2.9 and 244.0 ± 4.8 µg/mL. The hydroalcoholic fraction (C2) exhibited low values 14.0 ± 2.0 and 15.0 ± 2 µg/mL for SCC14 and HN13 cell lines, respectively. The crude extract of *T. guianensis* (C3) display IC_{50} values between 45.0 ± 4.0 and 349.0 ± 5.6 µg/mL. The hexane fraction (C4) showed an IC_{50} values varying from 58.0 ± 1.9 to 592.0 ± 20 µg/mL. We observed that the SCC25 cell lines exhibited a resistant phenotype for all compounds with IC_{50} values varying from 240 ± 26.0 to 592.0 ± 2.0 µg/mL (Table 2).

Table 2. Classification and IC_{50} value response to different fractions of the *T. guianensis* extract in HNSCC cell lines.

Cell Line	Anatomic Site	IC_{50} Value (Mean $\pm$ DP) mg/mL			
–	–	C1	C2	C3	C4
SCC14	Hypopharynx	0.029 ± 0.0007	0.014 ± 0.002	0.058 ± 0.002	0.074 ± 0.002
SCC25	Oral Cavity	0.244 ± 0.048	0.240 ± 0.026	0.349 ± 0.056	0.592 ± 0.020
Fadu	Hypopharynx	0.050 ± 0.002	0.023 ± 0.002	0.056 ± 0.003	0.186 ± 0.018
HN13	Tongue	0.028 ± 0.0002	0.015 ± 0.002	0.045 ± 0.004	0.058 ± 0.001

We further evaluated the growth inhibition (GI) classification and showed a moderate sensitivity (MS) and highly sensitive (HS) for ethyl acetate (C1) and hydroalcoholic fractions (C2) in SCC14, Fadu and HN13 cell lines, respectively (Table 3). The crude extract (C3) and hexane fraction (C4) showed a moderate sensitivity in SCC14, Fadu and HN13 cell lines (Table 3). As expected, the SCC25 cell line displayed a resistant (R) phenotype for all compounds (Table 3; Figure 1). Based on these results, the most sensitive cell line (HN13) was chosen for the subsequent cellular characterization.

Table 3. Growth inhibition to different fractions of the *T. guianensis* extract in HSCC cell lines.

Cell Line	GI Classification at 0.05 mg/mL (%)			
–	C1	C2	C3	C4
SCC14	MS (58.3%)	HS (63.3%)	MS (55.3%)	MS (46.7%)
SCC25	R (25.3%)	R (28%)	R (25%)	R (0%)
Fadu	MS (47%)	HS (66.3%)	MS (50.3%)	MS (47.3%)
HN13	MS (58.3%)	HS (66.3%)	MS (55.3%)	MS (46.6%)

HS: highly sensitive; MS: moderate sensitive and R: resistant.

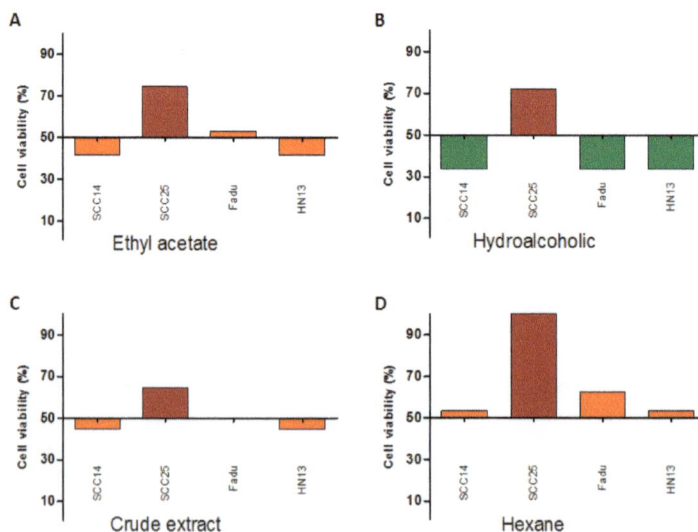

Figure 1. Cytotoxicity profile of SCC14, SCC25, Fadu, HN13 head and neck tumor cell lines, exposed to the different fractions of the *T. guianensis* extract: (**A**) Ethyl acetate (C1); (**B**) Hydroalcoholic (C2); (**C**) Crude extract (C3); and (**D**) Hexane (C4). Bars represent the cell viability at 50 µg/mL. Bars represent the GI score classification. Green (HS: highly sensitive); Orange (MS: moderate sensitive) and Red (R: resistant).

2.4. Wound Healing Migration Assay

In the HN13 cell line, we observed that after 24 h, 60.20% ± 9.35% of the control (DMSO) wound was closed (Figure 2A,B). All fractions and extract exhibited a significant reduction of wound closure, with the hydroalcoholic (C2) and crude extract (C3), showing the highest closure rate, namely 30.21% ± 5.81% and 19.54% ± 2.03%, respectively ($p < 0.01$) (Figure 2A,B).

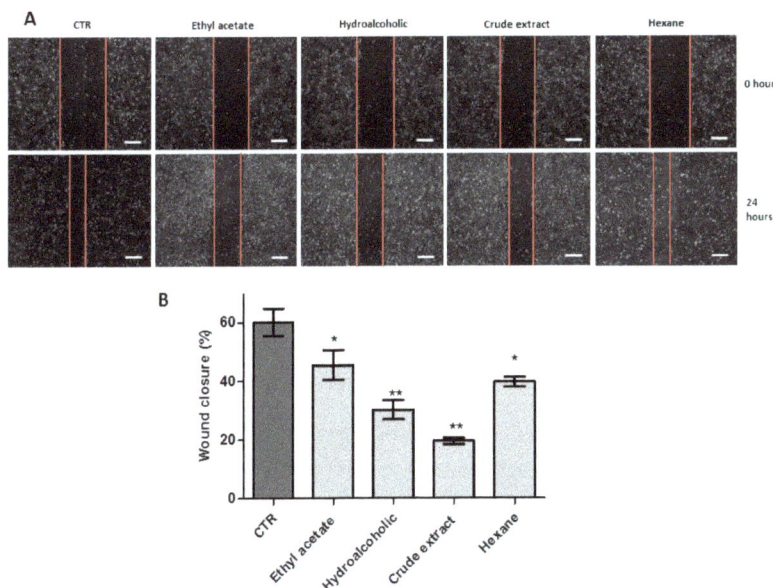

Figure 2. (**A**) Representative images of wound healing assay of HN13 cell line after 24 h, exposed to ethyl acetate fraction (25 µg/mL), hydroalcoholic fraction (10 µg/mL), crude extract (50 µg/mL) and hexane fraction (60 µg/mL). The red lines represent the distance between both edges of the wound; Scale bars, 200 µm; (**B**) bars represent the relative migration expressed as the means ± SD for different compounds. All experiments were repeated for three times. * $p < 0.05$; ** $p < 0.01$.

2.5. Invasion Assay

We further addressed whether *T. guianensis* extract and fractions could impair cell migration and invasion in vitro (Figure 3). Using matrigel invasion assay, all compounds reduced the invasion of HN13 cell lines, however the C4 fraction showed greater effect on cell invasiveness ability by matrix degradation ($p < 0.01$) (Figure 3A,D). We previously observed the migration inhibition induced by the hydroalcoholic (C2) and crude extract (C3) by wound healing assay. Additionally, we assess migration potential using empty inserts (without Matrigel), where we found a significant reduction of the migratory potential exposed to the hydroalcoholic (C2), crude extract (C3) and hexane (C4) fractions. Finally, no significant differences were detected between C2 and C4 (Figure 3B,E). In order to interrogate whether the invasion capability was associated with metalloproteinase activity, we analyzed MMP-2 protein expression by Western blot that showed the presence of the MMP-2 (72 kDa) band only in the C1 treatment (Figure 3C).

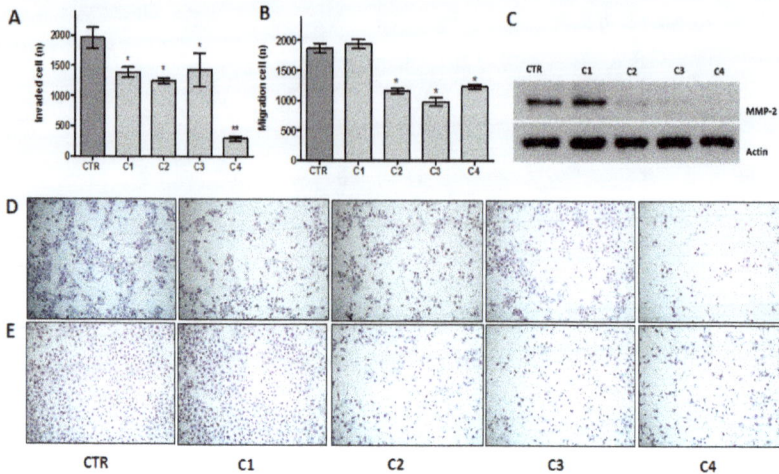

Figure 3. Effect of crude extract (C3) and fractions (C1, C2 and C4) of *T. guianensis* on HN13 cellular migration and invasion. (**A**) Invasion was measured at 24 h by Matrigel invasion assay, the results were expressed in relation to the DMSO control (considered as 100% of invasion) as the mean percentage of invasion ± SD; (**B**) analysis of index cell migration measured at 24 h by trans-well, the results are expressed as the means ± SD; (**C**) western blot analysis using the antibodies of anti-human MMP-2. The cells lines were incubated with compounds for 24 h; (**D,E**) representative images (at ×40 magnification) of migration and invasion assays, respectively. * $p < 0.05$; ** $p < 0.01$.

2.6. Expression of Apoptosis Markers (PARP, Caspase 3, and Fas)

To elucidate whether crude extract and fractions induce cancer cell death, HN13 cells were treated for 24 h, and the expression of total and cleaved PARP, caspase 3, FAS and tumor necrosis factor receptor 1 (TNFR1) proteins was evaluated by Western blot (Figure 4). All compounds showed a significant increase of cleaved PARP compare to control (DMSO), being more significant for the crude extract (C3) and hexane fraction (C4) (Figure 4B). Accordingly, we found increased levels of caspase 3 in HN13 cell lines treated with crude extract (C3) and hexane fraction (C4) (Figure 4C). Interestingly, we found significant increased levels of FAS in the C3 and C4 fraction (Figure 4D), and only the C4 fraction showed increasing levels of TNFR1 (Figure 4E).

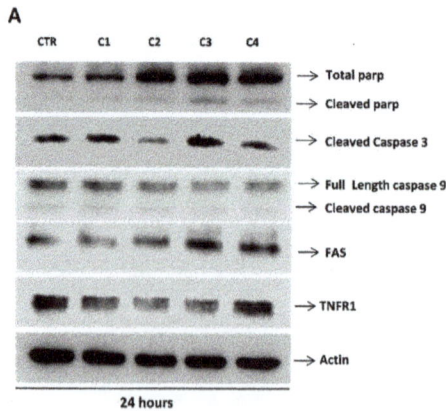

Figure 4. *Cont.*

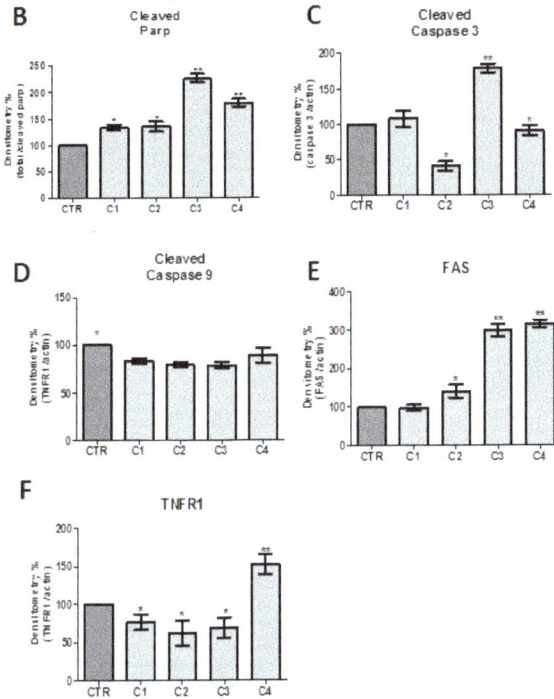

Figure 4. Effect of crude extract (C3) and fractions (C1, C2 and C4) of *T. guianensis* on intracellular signaling pathway activation in HN13 cells. HN13 cell lines were incubated with compounds (IC$_{50}$ values) for 24 h. (**A**) Western blot, using the antibodies of anti-human PARP (Full Length: 116 kDa and cleaved: 89 kDa), Caspase 3 (Full Length: 35 kDa), Caspase 9 (cleaved: 19 kDa) Fas (40 kDa) and TNFR1 (55 kDa) and Actin (45 kDa); (**B–F**) densitometric analysis of Western blot data of the four proteins. Bars results are expressed as the means $\pm$ SD. * $p < 0.05$; ** $p < 0.01$.

3. Discussion

In the present study, we have reported the antineoplastic potential of the crude extract and fraction of *T. guianensis* that showed a significant cytotoxicity effect in a panel of HNSCC cell lines. These data were further supported by results from migration and invasion assay, which demonstrated a significant reduction in migration index and cellular invasion. Moreover, the analysis of key cell death players indicates an activation of the extrinsic apoptotic pathway.

A variety of studies conducted by the NCI (National Cancer Institute), showed that more than 1000 different phytochemicals have cancer-preventive activity [18] Brazil contains the richest flora worldwide, and the great majority of it is unexplored in terms of therapeutic proprieties.

We initiated this study by assessing the mutagenicity of *T. guianensis* crude extract, using the Ames assay. The hydroalcoholic extract was mutagenic in TA98 strain without and with metabolic activation. High values of mutagenic index were observed in these conditions, indicating that compounds present in the extract are able to induce gene mutations. Nevertheless, in TA100 strain of *S. typhimurium*, the extract was mutagenic only after metabolic activation (+S9). According to Mortelmans and Zeiger (2000), the strain TA98 identifies agents able to induce gene mutations by deletions or additions of nucleotides on DNA structure, resulting in frameshift mutations [19]. Besides, TA100 strain detects compounds that cause base substitution in DNA. Considering these findings, it was identified that the crude extract presents in its phytochemical constitution agents able to induce frameshift mutations directly or indirectly. However, studies performed in TA100 indicated that substitution of bases was induced by the extract only after its modification by hepatic enzymes, present in the S9 fraction.

Several studies have shown that some flavonoids can be potent mutagenic agents in different biological systems [20,21]. In accordance with Procházková et al. (2011), the antioxidant or pro-oxidant effects of flavonoids are related with the concentration during exposition and particular molecular characteristics [22]. The values indicated in these tests may suggest the mechanisms of action of the mutagenic activity evidenced in the extracts. However, the concentrations of growth inhibition of the different cell lines do not relate with mutagenic assay.

A previous study of seeds of *T. guianensis* isolated two cyclic compounds which were broadly active against a panel of human cancer cell lines [12]. Based on putative anti-tumoral effect of *T. guianensis*, we tested its anti-proliferative action in a panel of four head and neck cancer cell lines Briefly, we showed that all fractions caused a moderate (SCC14 and Fadu) or high (HN13) cytotoxicity, with exception of the SCC25 cell line, which showed a resistant profile. We observed the classification and response to different fractions from extract of *T. guianensis* in head and neck tumors cell lines and verified that hydroalcoholic fraction showed the best results with low IC_{50} values (14.0 ± 2.0 and 15.0 ± 2 µg/mL). The NMR analyses showed signals of chemical shifts characteristics of alkaloids, coumarins and flavonoids in hydroalcoholic fraction (C2). Flavonoid has been identified in crude extract of leaves of the *T. guianensis*. In our study, rutin and quercetin were the flavonoids identified, the latter present in greater quantities. These results are in agreement with previous report that detected these components [16]. Moreover, in vitro studies have found the anticancer effects of flavonoids in the NPC nasopharyngeal carcinoma cell line [23]; K562 leukemia cell line [24,25]; 184-B5/HER normal breast cell line [26] and MCF-7 breast cancer cell line [27]. Furthermore, in vivo approaches indicated that flavonoid compounds, similar to hydrobenzofuranoids obtained from *T. guianensis*, exhibit similar biological activity [26,27]. Therefore, we suggest that the presence of flavonoids compounds within the extract used in this study can be related to cytotoxic effects.

Cell motility is considered to be an important determinant of the metastatic potential of cancer cells and can constitute promising treatment strategy [26,28]. Therefore, we used different experiments to evaluate the effects of *T. guianensis* extract and fractions on the motility, migration and invasion. We detected significant results in wound closure of the HN13 cell line exposed to all compounds. Importantly, the hydroalcohol (C2), crude extract (C3) and hexane (C4) compounds, which showed high levels of flavonoids such as quercetin, showed reduced invasion on the transwell migration assay. Previous in vitro study showed that suppression of the HGF/c-Met signaling pathway by quercetin, contributes to the anti-metastatic action and inhibit the cellular migration in melanoma cell lines [29] and also inhibited the Wnt pathway in teratocarcinoma cell line [28].

We also showed that all compound inhibited the invasion in matrigel assay. MMPs are known ECM-degrading enzymes and theirs activity are associated with tumor invasiveness [30] and associated with tumorigenesis, progression and prognosis of head and neck tumor [31–33]. Herein, we observed that hydroalcohol (C2), crude extract (C3), and hexane (C4) fractions induced MMP-2 downregulation. These results are in line with previous work from our group that found that extracts of *T. guianensis* inhibited matrix metalloproteinases [16].

Our findings suggest that several classes of secondary metabolites of C2, C3 and especially C4 fraction may be responsible for the inhibition of cellular proliferation. Total alkaloids found in these compounds can be related with reduced invasion. A recent study showed that total alkaloids from *Rubus alceifolius* Poir inhibited the enzyme activity and reduced the expression of MMP-2 and MMP-9 [34]. Alkaloids are secondary compounds that exhibit anti-tumor activity with broad spectrum of action and they deserve further and deeper studies. Other studies have shown that alkaloids derived from other plants also present antitumor activity, such as opium alkalois (papaverine, noscapine and narceine) that induces apoptosis in cancer cell lines [35], capsaicin (an alkaloid isolated from the chili pepper) that modulates cell cycle progression and induces apoptosis in human KB cell line (epidermoid carcinoma) through mitochondrial membrane permeabilization and caspase activation, suggesting an antineoplastic activity [36]. We believe that our results offer important perspectives on cancer targeted therapy, especially in the control of metastasis mechanisms, exhibit in vitro for C2, C3 and C4 compounds.

Apoptosis pathway was assessed by Western blot, where we detected an overexpression of FAS, caspase-3, TNFR1 and increased level of cleaved PARP in C3 and C4, which indicates a possible extrinsic pathway of apoptosis. Fas-mediated apoptosis, ligands of Fas leads to strong caspase-8 activation at the DISC, thereby activating other caspases including caspase-3 in the absence of mitochondrial involvement [37]. Despite the absence of moderate anti-neoplastic effect of the C3 and C4 fraction, these fractions have secondary metabolites that activate an extrinsic activation mechanism mediated by FAS and TNFR1 increased. Nevertheless, the mechanisms in which *T. guianensis* participates into the complex signal pathways to achieve its anticancer role need further investigation, such as monitoring of other proteins involved in extrinsic apoptotic signaling

4. Materials and Methods

4.1. Plant Material and Extraction

Samples of the *T. guianensis* were collected within the Cerrado region of Minas Gerais, Brazil (Latitude S18°58'08'' and Longitude W49°27'54''). After identification, a voucher specimen was registered (143407) in the Herbarium of the Department of Botany, of the Federal University of Minas Gerais, in Belo Horizonte (Brazil).

Dried and powdered leaves (100 g) of *T. guianensis* were extracted by maceration (70% hydroalcoholic solution, 15 days), which was filtered and afterwards lyophilized, resulting in 2.111 g of crude extract. The extract was dissolved in CH_3CH_2OH/H_2O (7:3) and successively extracted with hexane (C_6H_{14}), chloroform ($CHCl_3$) and ethyl acetate ($C_4H_8O_2$), resulting in 1.1059 g (4677%), 0.3904 g (20.04%) and 0.2321 g (11.92%), respectively, and the hydroalcoholic residue was 0.2363 g (12.13%) [12]. We further used the fractions: ethyl acetate (C1), hidroalcoholic (C2), crude extract (C3) and hexane (C4).

4.2. Nuclear Magnetic Resonance (NMR) Analysis of T. guianensis

The extract and fractions from *T. guianensis* were prepared with deutered solvents for nuclear magnetic resonance (NMR) analysis. The ^1H NMR spectra were obtained using Bruker DRX-400MHz spectrometer (Bruker, Billerica, MA, USA) with chloroform-d ($CDCl_3$) and methanol-d4 (CD_3OD) using TMS as internal standard.

4.3. Ames Mutagenicity Assay

The *Salmonella typhimurium* mutagenicity assay was performed using the plate incorporation protocol with the strains TA100 and TA98 of *Salmonella typhimurium* [38]. Five different concentrations of the crude extract obtained from the leaves of *T. guianensis* were evaluated in this assay: 21.68, 16.26, 10.84, 5.42 and 2.71 mg/plate. The selection of the concentrations was based on the solubility limit of the samples in DMSO and on their toxicity in the TA98 and TA100 [38].

Toxicity was apparent both as a reduction in the number of his + revertants and as an alteration in the auxotrophic background (i.e., background lawn). The maximum volume of DMSO employed in the studies was 80 µL per plate.

The various concentrations of extract tested were added to 500 µL of buffer pH 7.4, 100 µL of bacterial culture and 2 mL of top agar. After agitation, the mixtures were poured on to plates containing minimum agar. The plates were incubated at 37 °C for 48 h and the his+ revertant colonies were manually counted. All experiments were performed in triplicate. The standard mutagens used as positive controls in experiments were 4-nitro-*o*-phenylenediamine (10 µg/plate) for TA98 and methylmethane sulfonate (260 µg/plate). DMSO (solvent) served as the negative control (80 µL/plate).

4.4. Preparation of Standard Solutions

Each sample was accurately weighed so that 0.1 g of each could be separately re-dissolved in 1 mL of DMSO. From these solutions, serial dilutions were made to obtain lower concentrations (5, 10, 25, 50, 75, 100, and 200 µg/mL) and then, were frozen at −20 °C, for later use.

4.5. Cell Lines and Cell Culture

The head and neck cell lines SCC14, SCC25, Fadu and HN13 (kindly provided by Rui Manuel Reis, Barretos Cancer Hospital) were used to determine the cytotoxic effect of different *T. guianensis* extract and fractions [39]. All cell lines were initially authenticated by short tandem repeat (STR) DNA typing according to the International Reference Standard for Authentication of Human Cell Lines using a panel of 8 (D5S818, D13S317, D7S820, D16S539, vWA, TH01, TPOX and CSF1P0) STR loci plus gender determination amelogenin (AMEL), using the fluorescent labeling primers as reported and tested for mycoplasma contamination [40]. The cell lines were maintained in Dulbecco's modified Eagle's medium (DMEM) (Sigma Aldrich, St. Louis, MO, USA), containing 10% fetal bovine serum (FBS) (Gibco-Life Technologies, Grand Island, NY, USA), 1% penicillin/streptomycin (Sigma Aldrich). Cells were incubated in a humidified atmosphere of 95% air and 5% CO_2 at 37 °C.

4.6. Cell Viability Assay

The cell viability was assessed by MTT assay (Promega, Madison, WI, USA) as previous described [39,41]. To determine the IC_{50} values, 5×10^3 cell lines/well were seeded in a 96 well plate and incubated at increased concentrations of different extract fractions of *T. guianensis*, under reduced FBS concentration (0.5%) for 72 h. The results were expressed as a percentage relatively to control cells (DMSO treatment). The IC_{50} values were calculated by nonlinear regression analysis using GraphPad Prism software (5.01 version, GraphPad Software, Inc., La Jolla, CA, USA). The growth inhibition (GI) was detected at fixed concentration of the 50 µg/mL, and cell lines were scored as highly sensitive (HS) with GI > 60%, moderately sensitive (MS) with GI between 40% and 60% and resistant with GI < 40% as previously described [39,42]. All the assays were done in triplicate and repeated at least three times.

4.7. Wound Healing Migration Assay

To assess the potential effect of *T. guianensis* fractions and crude extract in inhibition of cell migration, HN13 cell line was plated in 6 well plates (2.5×10^5) in DMEM 10% FBS + 1% P/S and allowed to adhere overnight. After reaching 95% of confluence, mono-layer cells were washed with PBS, scraped with a plastic 200 µL pipette tip. The cell line was incubated with fixed concentrations of each fraction in DMEM 0.5% FBS + 1% P/S. Images of the wound were captured after 0 and 24 h by Olympus IX71 optical microscope (Olympus Optical CO. Hamburg, Germany) and the relative migration distance was measured by the following formula:

$$\text{percentage of wound closure } (\%) = \frac{100 \times (A - B)}{A}$$

A, the width of cell wounds before incubation, and *B* the width of cell wounds after incubation [43].

4.8. Matrigel Invasion Assay

The invasion potential of *T. guianensis* fractions and crude extract was evaluated by BD BioCoat Matrigel invasion chambers Kit (BD Biosciences, San Jose, CA, USA), following manufacturer instructions and as previously described [44]. HN13 cell line (2.5×10^4) cells were plated in the Matrigel-coated 24-well transwell inserts, containing DMEM (free-serum) and 2 µg/mL of the different compounds, as a chemo-attractant was used DMEM 10% FBS. Additionally, control inserts (without Matrigel), were used for measuring the migratory index of HN13 cell line exposed to different compounds.

HN13 cell line was allowed to invade and migrate for 24 h. Then, the insert membrane was fixed with methanol iced and stained with hematoxylin/eosin. The membranes were photographed at 40× magnification microscope and counted. The results were expressed in relation to the DMSO control (considered as 100% of invasion) as the mean percentage of invasion ± SD.

4.9. Western Blot

To assess apoptosis and MMP-2, HN13 cell line was cultivated in 6 well plates and after 90% plate confluence, cells were scraped in buffer lyses that contained 50 mM Tris (pH 7.6–8), 150 mM NaCl, 5 mM EDTA, 1 mM Na_3VO_4, 10 mM NaF, 10 mM $Na_4P_2O_7$, 1% NP-40, and protease cocktail inhibitors. The cellular lysate was centrifuged at 13,000 rpm for 15 min in 5 °C and total protein of the supernatant was quantified by Bradford method. Briefly, 20 µg of proteins from lysates were resolved by 10% SDS-PAGE and transferred to nitrocelulose membranes in TransBlot Turbo transfer (Bio-Rad Laboratories, Hercules, CA, USA) and incubated in 5% nonfat dried milk in TBS-T for 1 h at room temperature before primary antibody overnight incubation with PARP total/cleaved (#9532, Cell Signaling Technology, Danvers, MA, USA), FAS (#4233, Cell Signaling Technology), CASPASE-3 (#14220, Cell Signaling Technology), CASPASE-9 (#9505, Cell Signaling Technology), TNFR1(#3736, Cell Signaling Technology), β-actin (#3700, Cell Signaling Technology) at 1:100 dilution, and at 1:500 dilution for MMP-2 (sc-58386-Santa Cruz Biotechnology, Dallas, TX, USA). After washing with TBS-T, membranes were incubated with the secondary antibody anti-rabbit (#7074, Cell Signaling Technology) at dilution 1:5000. Immune detection was done with ECL Western Blotting Detection Reagents in automatic ImageQuant mini LAS4000 (GE Healthcare Life Sciences, Pittsburgh, PA, USA). Densitometric data from Western blots were performed with ImageJ software (NIH-Scion Corporation, Bethesda, MD, USA). Caspase-3, FAS and TNFR1 values were normalized to β-actin levels and PARP cleaved to total PARP status. All experiments were carried out in triplicate.

4.10. Statistical Analysis

For Proliferation assay the results were expressed as the means ± SD. The results for Matrigel invasion assay were expressed in relation to the DMSO control (invaded cells) as the mean number of cells ± SD. Single comparisons between the different conditions studied were done using Student's *t* test, and differences between groups were tested using two-way analysis of variance. Statistical analysis was done using GraphPad Prism version 5.01 (5.01 version, GraphPad Software, Inc.). The level of significance in all the statistical analyses was set at $p \leq 0.05$. The results obtained were evaluated employing the statistical software Salanal and adopting the Bernstein et al. (1982) [45]. The data of Ames mutagenicity assay were assessed by analysis of variance (ANOVA) followed by a linear-regression. Furthermore, the mutagenic index (MI), which is the average number of revertants per plate divided by the average number of revertants per plate with the negative (solvent) control, was also calculated for each concentration. A sample was considered mutagenic when a dose–response relationship and a two-fold increase in the number of mutants (MI $\geq$ 2) with at least one concentration were observed.

5. Conclusions

The present study showed cytotoxic activity of *T. guianensis* extract on oral cancer cells lines, as well as an ability to inhibit tumor migration and invasion, constituting a putative anticancer agent, alone or in combination with classic chemotherapy and radiotherapy approaches. Nevertheless, further studies are needed to identify the mechanisms by which *T. guianensis* extract acts as inhibitors of cell proliferation.

Supplementary Materials: Supplementary materials can be found at www.mdpi.com/1422-0067/17/11/1839/s1.

Acknowledgments: The authors acknowledge CNPq, FAPEMIG (APQ-00953-12), Ministério da Ciência, Tecnologia e Inovação—MCTI, and Fundo Nacional de Saúde—FNS (1302/13), FINEP (MCTI/FINEP/MS/SCTIE/DECIT-01/2013—FPXII-BIOPLAT) and CAPES (Brazilian agencies) for the financial support of this study. We thank Arali Aparecida Costa Araujo for her support in collecting of *T. guianensis* and João Máximo De Siqueira (UFSJ) and Paulo Cezar Vieira (UFSCar) for their support in the phytochemical analysis.

Author Contributions: Renato José Silva-Oliveira carried out the studies of cell line and cell culture, proliferation assay, wound healing migration assay, matrigel invasion assay, Western blot analysis, and statistical analysis. He also participated of data acquisition and its interpretation and performed the statistical analysis. Gabriela Francine Lopes has been responsible for the preparation of extracts, carried out the studies of cell culture and cell viability assay. Ana Maciel Ribeiro carried out the Statistical Analysis and has been involved in drafting the manuscript. Fábio Vieira dos Santos and Luiz Fernando Camargos carried out the Ames mutagenicity assay and interpretation of data. Ralph Gruppi Thomé and Hélio Batista dos Santos helped to draft the manuscript and have been involved in revising critically the manuscript. Ana Paula Terezan, Richele Priscila Severino and Vanessa Gisele Pasqualotto Severino participated in acquisition of nuclear magnetic resonance data of *T. guianensis*, analysis and interpretation of these data, and have been involved in revising it critically for important intellectual content. Rui Manuel Reis participated in the design of the study, and has made substantial contributions to analysis and interpretation of data and helped to draft the manuscript. Rosy Iara Maciel de Azambuja Ribeiro conceived the study, participated in its design and coordination, and has made substantial contributions to analysis and interpretation of data and drafted the manuscript. All authors read and approved the final manuscript.

Conflicts of Interest: The authors declared no potential conflicts of interest with respect to the research authorship and/or publication of this article and no competing financial interests in relation to the work described.

References

1. Jemal, A.; Siegel, R.; Xu, J.; Ward, E. Cancer statistics, 2010. *CA Cancer J. Clin.* **2010**, *60*, 277–300. [CrossRef] [PubMed]
2. Suh, Y.; Amelio, I.; Guerrero Urbano, T.; Tavassoli, M. Clinical update on cancer: Molecular oncology of head and neck cancer. *Cell Death Dis.* **2014**, *5*, e1018. [CrossRef] [PubMed]
3. Hoffmann, T.K. Systemic therapy strategies for head-neck carcinomas: Current status. *Laryngorhinootologie* **2012**, *91*, 123–143.
4. Instituto Nacional de Câncer, Estimativa 2014: Incidência de câncer no brasil. Rio de Janeiro, R.J. Available online: http://www.inca.gov.br/estimativa/2014 (accessed on 23 October 2015).
5. Bonfante, G.M.D.S.; Machado, C.J.; Souza, P.E.A.D.; Andrade, E.I.G.; Acurcio, F.D.A.; Cherchiglia, M.L. Specific 5-year oral cancer survival and associated factors in cancer outpatients in the Brazilian Unified National Health System. *Cadernos de Saúde Pública* **2014**, *30*, 983–997. [CrossRef] [PubMed]
6. Shah, J.P. Cervical lymph node metastases—Diagnostic, therapeutic, and prognostic implications. *Oncology* **1990**, *4*, 61–69. [PubMed]
7. Shaikh, R.; Pund, M.; Dawane, A.; Iliyas, S. Evaluation of anticancer, antioxidant, and possible anti-inflammatory properties of selected medicinal plants used in Indian traditional medication. *J. Tradit. Complement. Med.* **2014**, *4*, 253–257. [CrossRef] [PubMed]
8. Majid, M.Z.; Zaini, Z.M.; Razak, F.A. Apoptosis-inducing effect of three medicinal plants on oral cancer cells KB and ORL-48. *Sci. World J.* **2014**, *2014*, 125353–125361. [CrossRef] [PubMed]
9. Pitchai, D.; Roy, A.; Ignatius, C. In vitro evaluation of anticancer potentials of lupeol isolated from *Elephantopus scaber* L. on MCF-7 cell line. *J. Adv. Pharm. Technol. Res.* **2014**, *5*, 179–184. [CrossRef] [PubMed]
10. Kim, J.W.; Amin, A.R.; Shin, D.M. Chemoprevention of head and neck cancer with green tea polyphenols. *Cancer Prev. Res.* **2010**, *3*, 900–909. [CrossRef] [PubMed]
11. Bhavana, S.M.; Lakshmi, C.R. Oral oncoprevention by phytochemicals—A systematic review disclosing the therapeutic dilemma. *Adv. Pharm. Bull.* **2014**, *4*, 413–420. [PubMed]
12. Safarzadeh, E.; Sandoghchian Shotorbani, S.; Baradaran, B. Herbal medicine as inducers of apoptosis in cancer treatment. *Adv. Pharm. Bull.* **2014**, *4*, 421–427. [PubMed]
13. Brazil, Ministry of the Environment. Brazilian biodiversity. Available online: http://www.mma.gov.br/biodiversidade/biodiversidade-brasileira (accessed on 23 October 2015).
14. Roumy, V.; Fabre, N.; Portet, B.; Bourdy, G.; Acebey, L.; Vigor, C.; Valentin, A.; Moulis, C. Four anti-protozoal and anti-bacterial compounds from Tapirira guianensis. *Phytochemistry* **2009**, *70*, 305–311. [CrossRef] [PubMed]
15. David, J.M.; Chavez, J.P.; Chai, H.B.; Pezzuto, J.M.; Cordell, G.A. Two new cytotoxic compounds from Tapirira guianensis. *J. Nat. Prod.* **1998**, *61*, 287–289. [CrossRef] [PubMed]

16. Longatti, T.R.; Cenzi, G.; Lima, L.A.R.S.; Oliveira, R.J.S.; Oliveira, V.N.; Da Silva, S.L.; Ribeiro, R.I.M.A. Inhibition of gelatinases by vegetable extracts of the species *Tapirira guianensis* (stick pigeon). *Br. J. Pharm. Res.* **2011**, *1*, 133–140. [CrossRef]

17. Correia, S.D.J.; David, J.M.; Silva, E.P.D.; David, J.P.; Lopes, L.M.X.; Guedes, M.L.S. Flavonoids, norisoprenoids and other terpenes from leaves of Tapirira guianensis. *Química Nova* **2008**, *31*, 2056–2059. [CrossRef]

18. Surh, Y.J. Cancer chemoprevention with dietary phytochemicals. *Nat. Rev. Cancer* **2003**, *3*, 768–780. [CrossRef] [PubMed]

19. Mortelmans, K.; Zeiger, E. The Ames Salmonella/microsome mutagenicity assay. *Mutat. Res.* **2000**, *455*, 29–60. [CrossRef]

20. Duthie, S.J.; Johnson, W.; Dobson, V.L. The effect of dietary flavonoids on DNA damage (strand breaks and oxidised pyrimdines) and growth in human cells. *Mutat. Res.* **1997**, *390*, 141–151. [CrossRef]

21. Santos, F.V.; Tubaldini, F.R.; Colus, I.M.; Andreo, M.A.; Bauab, T.M.; Leite, C.Q.; Vilegas, W.; Varanda, E.A. Mutagenicity of Mouriri pusa Gardner and Mouriri elliptica Martius. *Food Chem. Toxicol.* **2008**, *46*, 2721–2727. [CrossRef] [PubMed]

22. Prochazkova, D.; Bousova, I.; Wilhelmova, N. Antioxidant and prooxidant properties of flavonoids. *Fitoterapia* **2011**, *82*, 513–523. [CrossRef] [PubMed]

23. Fang, C.Y.; Wu, C.C.; Hsu, H.Y.; Chuang, H.Y.; Huang, S.Y.; Tsai, C.H.; Chang, Y.; Tsao, G.S.; Chen, C.L.; Chen, J.Y. EGCG inhibits proliferation, invasiveness and tumor growth by up-regulation of adhesion molecules, suppression of gelatinases activity, and induction of apoptosis in nasopharyngeal carcinoma cells. *Int. J. Mol. Sci.* **2015**, *16*, 2530–2558. [CrossRef] [PubMed]

24. Zhang, D.; Zhuang, Y.; Pan, J.; Wang, H.; Li, H.; Yu, Y.; Wang, D. Investigation of effects and mechanisms of total flavonoids of Astragalus and calycosin on human erythroleukemia cells. *Oxid. Med. Cell. Longev.* **2012**, *2012*, 209843–209848. [CrossRef] [PubMed]

25. Costa, P.M.D. Antitumor Potential of Hydrobenzofuranoids Isolated from the Leaves of tapirira guianensis (anacardiaceae). Available online: http://www.repositorio.ufc.br/bitstream/riufc/2614/1/2006_dis_pmcosta.pdf (accessed on 23 October 2015).

26. Du, Y.; Feng, J.; Wang, R.; Zhang, H.; Liu, J. Effects of flavonoids from *Potamogeton crispus* L. on proliferation, migration, and invasion of human ovarian cancer cells. *PLoS ONE* **2015**, *10*, e0130685. [CrossRef] [PubMed]

27. Hayashi, A.; Gillen, A.C.; Lott, J.R. Effects of daily oral administration of quercetin chalcone and modified citrus pectin on implanted colon-25 tumor growth in Balb-c mice. *Altern. Med. Rev.* **2000**, *5*, 546–552. [PubMed]

28. Cao, H.H.; Cheng, C.Y.; Su, T.; Fu, X.Q.; Guo, H.; Li, T.; Tse, A.K.; Kwan, H.Y.; Yu, H.; Yu, Z.L. Quercetin inhibits HGF/c-met signaling and HGF-stimulated melanoma cell migration and invasion. *Mol. Cancer* **2015**, *14*, 103–115. [CrossRef] [PubMed]

29. Mojsin, M.; Vicentic, J.M.; Schwirtlich, M.; Topalovic, V.; Stevanovic, M. Quercetin reduces pluripotency, migration and adhesion of human teratocarcinoma cell line NT2/D1 by inhibiting Wnt/β-catenin signaling. *Food Funct.* **2014**, *5*, 2564–2573. [CrossRef] [PubMed]

30. Yadav, L.; Puri, N.; Rastogi, V.; Satpute, P.; Ahmad, R.; Kaur, G. Matrix metalloproteinases and cancer—Roles in threat and therapy. *Asian Pac. J. Cancer Prev.* **2014**, *15*, 1085–1091. [CrossRef] [PubMed]

31. Wang, D.; Muller, S.; Amin, A.R.; Huang, D.; Su, L.; Hu, Z.; Rahman, M.A.; Nannapaneni, S.; Koenig, L.; Chen, Z.; et al. The pivotal role of integrin β1 in metastasis of head and neck squamous cell carcinoma. *Clin. Cancer Res.* **2012**, *18*, 4589–4599. [CrossRef] [PubMed]

32. Liu, R.R.; Li, M.D.; Li, T.; Tan, Y.; Zhang, M.; Chen, J.C. Matrix metalloproteinase 2 (MMP2) protein expression and laryngeal cancer prognosis: A meta analysis. *Int. J. Clin. Exp. Med.* **2015**, *8*, 2261–2266. [PubMed]

33. Jafarian, A.H.; Vazife Mostaan, L.; Mohammadian Roshan, N.; Khazaeni, K.; Parsazad, S.; Gilan, H. Relationship between the expression of matrix metalloproteinase and clinicopathologic features in oral squamous cell carcinoma. *Iran. J. Otorhinolaryngol.* **2015**, *27*, 219–223. [PubMed]

34. Zhao, J.; Liu, L.; Wan, Y.; Zhang, Y.; Zhuang, Q.; Zhong, X.; Hong, Z.; Peng, J. Inhibition of hepatocellular carcinoma by total alkaloids of rubus alceifolius poir involves suppression of hedgehog signaling. *Integr. Cancer Ther.* **2015**, *14*, 394–401. [CrossRef] [PubMed]

35. Afzali, M.; Ghaeli, P.; Khanavi, M.; Parsa, M.; Montazeri, H.; Ghahremani, M.H.; Ostad, S.N. Non-addictive opium alkaloids selectively induce apoptosis in cancer cells compared to normal cells. *Daru* **2015**, *23*, 16–24. [CrossRef] [PubMed]
36. Lin, C.H.; Lu, W.C.; Wang, C.W.; Chan, Y.C.; Chen, M.K. Capsaicin induces cell cycle arrest and apoptosis in human kb cancer cells. *BMC Complement. Altern. Med.* **2013**, *13*, 46–55. [CrossRef] [PubMed]
37. Petak, I.; Houghton, J.A. Shared pathways: Death receptors and cytotoxic drugs in cancer therapy. *Pathol. Oncol. Res.* **2001**, *7*, 95–106. [CrossRef] [PubMed]
38. Maron, D.M.; Ames, B.N. Revised methods for the salmonella mutagenicity test. *Mutat. Res.* **1983**, *113*, 173–215. [CrossRef]
39. Silva-Oliveira, R.J.; Silva, V.A.; Martinho, O.; Cruvinel-Carloni, A.; Melendez, M.E.; Rosa, M.N.; de Paula, F.E.; de Souza Viana, L.; Carvalho, A.L.; Reis, R.M. Cytotoxicity of allitinib, an irreversible anti-EGFR agent, in a large panel of human cancer-derived cell lines: Kras mutation status as a predictive biomarker. *Cell. Oncol.* **2016**, *39*, 253–263. [CrossRef] [PubMed]
40. Dirks, W.G.; Faehnrich, S.; Estella, I.A.; Drexler, H.G. Short tandem repeat DNA typing provides an international reference standard for authentication of human cell lines. *ALTEX* **2005**, *22*, 103–109. [PubMed]
41. Martinho, O.; Silva-Oliveira, R.; Miranda-Goncalves, V.; Clara, C.; Almeida, J.R.; Carvalho, A.L.; Barata, J.T.; Reis, R.M. In vitro and in vivo analysis of RTK inhibitor efficacy and identification of its novel targets in glioblastomas. *Transl. Oncol.* **2013**, *6*, 187–196. [CrossRef] [PubMed]
42. Konecny, G.E.; Glas, R.; Dering, J.; Manivong, K.; Qi, J.; Finn, R.S.; Yang, G.R.; Hong, K.L.; Ginther, C.; Winterhoff, B.; et al. Activity of the multikinase inhibitor dasatinib against ovarian cancer cells. *Br. J. Cancer* **2009**, *101*, 1699–1708. [CrossRef] [PubMed]
43. Martinho, O.; Zucca, L.E.; Reis, R.M. AXL as a modulator of sunitinib response in glioblastoma cell lines. *Exp. Cell Res.* **2015**, *332*, 1–10. [CrossRef] [PubMed]
44. Moniz, S.; Martinho, O.; Pinto, F.; Sousa, B.; Loureiro, C.; Oliveira, M.J.; Moita, L.F.; Honavar, M.; Pinheiro, C.; Pires, M.; et al. Loss of WNK2 expression by promoter gene methylation occurs in adult gliomas and triggers rac1-mediated tumour cell invasiveness. *Hum. Mol. Genet.* **2013**, *22*, 84–95. [CrossRef] [PubMed]
45. Bernstein, L.; Kaldor, J.; McCann, J.; Pike, M.C. An empirical approach to the statistical analysis of mutagenesis data from the salmonella test. *Mutat. Res.* **1982**, *97*, 267–281. [CrossRef]

International Journal of
Molecular Sciences

MDPI

Article

Berberine Suppresses Cyclin D1 Expression through Proteasomal Degradation in Human Hepatoma Cells

Ning Wang [1], Xuanbin Wang [2,3], Hor-Yue Tan [1], Sha Li [1], Chi Man Tsang [4], Sai-Wah Tsao [4] and Yibin Feng [1,2,3,*]

1 School of Chinese Medicine, The University of Hong Kong, 10 Sassoon Road, Pokfulam, Hong Kong, China; ckwang@hku.hk (N.W.); hoeytan@hku.hk (H.-Y.T.); u3003781@connect.hku.hk (S.L.)
2 Laboratory of Chinese Herbal Pharmacology, Oncology Center, Renmin Hospital, Hubei University of Medicine, Shiyan 442000, China; wangxb@hbmu.edu.cn
3 Hubei Key Laboratory of Wudang Local Chinese Medicine Research and School of Pharmacy, Hubei University of Medicine, Shiyan 442000, China
4 Department of Anatomy, The University of Hong Kong, 21 Sassoon Road, Pokfulam, Hong Kong, China; annatsan@hku.hk (C.M.T.); gswtsao@hku.hk (S.-W.T.)
* Correspondence: yfeng@hku.hk; Tel.: +852-258-90431; Fax: +852-216-84259

Academic Editor: Toshio Morikawa
Received: 27 September 2016; Accepted: 9 November 2016; Published: 15 November 2016

Abstract: The aim of this study is to explore the underlying mechanism on berberine-induced Cyclin D1 degradation in human hepatic carcinoma. We observed that berberine could suppress both in vitro and in vivo expression of Cyclin D1 in hepatoma cells. Berberine exhibits dose- and time-dependent inhibition on Cyclin D1 expression in human hepatoma cell HepG2. Berberine increases the phosphorylation of Cyclin D1 at Thr286 site and potentiates Cyclin D1 nuclear export to cytoplasm for proteasomal degradation. In addition, berberine recruits the Skp, Cullin, F-box containing complex-β-Transducin Repeat Containing Protein (SCF$^{\beta\text{-TrCP}}$) complex to facilitate Cyclin D1 ubiquitin-proteasome dependent proteolysis. Knockdown of β-TrCP blocks Cyclin D1 turnover induced by berberine; blocking the protein degradation induced by berberine in HepG2 cells increases tumor cell resistance to berberine. Our results shed light on berberine's potential as an anti-tumor agent for clinical cancer therapy.

Keywords: berberine; Cyclin D1; ubiquitinated-dependent proteolysis; β-TrCP; tumor growth inhibition

1. Introduction

Overexpression of Cyclin D1 in various human cancers is regarded as a key mechanism underlying tumor angiogenesis, progression, and metastasis [1–6]. Cyclin D1 overexpression is also found to enhance cancer cells' resistance to chemotherapeutic agents [7]. Disruption of Cyclin D1 proteolysis is one of the major mechanisms that cancer cells accumulate Cyclin D1 [8]. In particular, it was noticed that Cyclin D1 was overexpressed in hepatocellular carcinoma (HCC) and was associated with aggressive forms of HCC [9,10]. Chronic overexpression of Cyclin D1 in transgenic mice with HCC was also observed [11].

Berberine is a natural product belonging to the group of isoquinoline alkaloids that are present in many medical plants. The anti-tumor action of berberine was extensively reported, in which berberine was shown to modulate several different signal transductions to induce tumor cell cycle redistribution and apoptosis, and to inhibit tumor cell migration [12,13]. Several studies revealed that inhibitory effect of berberine on Cyclin D1 expression in various cancer cell lines including neuroblastoma SK-N-SH & SK-N-MC cells [14], human epidermoid carcinoma A431 cells [15], human prostate carcinoma LNCap, DU145 & PC-3 cells [16], human leukemia cells HL-60 [17], and pulmonary giant cell carcinoma

PG cells [11], indicating that Cyclin D1 may be a potential target for berberine in cancer therapy. However, the exact mechanism of Cyclin D1 inhibition in berberine-treated cancer cells has not been well documented. A recent study reveals that berberine suppresses the activity of the AP-1 signaling pathway and decreases the binding of transcription factors to the Cyclin D1 AP-1 motif, indicating that transcriptional inhibition of Cyclin D1 may be involved in berberine's anti-tumor effect [18]. It is interesting to examine whether the inhibitory action of berberine on Cyclin D1 expression in liver cancer cells shares the same mechanism and to figure out the exact machinery that undergoes Cyclin D1 suppression in human hepatoma cells exposed to berberine.

In this study, the underlying mechanism of Cyclin D1 suppression by berberine in human hepatoma cells was examined. It was observed that berberine could suppress both in vitro and in vivo expression of Cyclin D1 in hepatoma cells. Dose- and time-dependent Cyclin D1 inhibition is observed in HepG2 cells exposed to berberine; and the rapid ablation of Cyclin D1 induced by 6 h berberine treatment is found independent of transcriptional inhibition. We found Cyclin D1 undergoes ubiquitinated degradation in berberine-treated HepG2 cells, and phosphorylation at Thr-286 site of Cyclin D1 is required for berberine-driven Cyclin D1 degradation. The β-transducin repeat-containing protein (β-TrCP) recruitment as E3 ligases by berberine are induced when Cyclin D1 proteolyzes. Genetic depletion of β-TrCP attenuates berberine's inductive action on Cyclin D1 degradation as well as berberine's anti-tumor effect. Our results indicate that involvement of β-TrCP as E3 ligase in Cyclin D1 ubiquitination-dependent proteolysis is the mechanism in berberine's inhibitory action on Cyclin D1 expression in HepG2 cells, and contributes partially to the anti-tumor action of berberine. This sheds light on berberine's potential in the agent list for liver cancer therapy.

2. Results

2.1. Berberine Suppresses In Vitro and In Vivo Cyclin D1 Expression in Hepatoma Cells

It was extensively reported by our previous studies that berberine could suppress both in vitro and in vivo growth of HCC [19–21]. Consistently, we observed reduced expression of Cyclin D1 in hepatoma cells with berberine treatment (Figure 1A). While berberine significantly reduced proliferation of xenografted hepatoma, the expression of Cyclin D1 in hepatoma xenograft was in parallel inhibited (Figure 1B). These observations confirmed the property of berberine in suppressing in vitro and in vivo expression of Cyclin D1 in hepatoma. To further profile the action of berberine, we systemically examined Cyclin D1 expression in berberine-treated HepG2 cells. HepG2 cells with 6 h exposure to berberine exhibit significant dose-dependent reduction of Cyclin D1 expression (Figure 1C). Time-dependent manner of Cyclin D1 expression inhibition was also observed in HepG2 cells exposed to100 μM berberine (Figure 1C). Six hour exposure of 100 μM berberine to HepG2 cells was unable to carry out any alteration on the cell phase distribution, indicating that the rapid suppression on Cyclin D1 is not attributed to the cycle arrest induction by berberine in HepG2 cells (Figure 1D). This observation indicates that Cyclin D1 inhibition may occur prior to cell cycle change and can cause redistribution of cell cycle phases. Our findings reveal that berberine could rapidly inhibit Cyclin D1 expression in time- and dose-dependent manner but is independent on cell cycle.

2.2. Berberine Triggers Post-Translational Suppression on Cyclin D1 Expression

A previous study reveals that berberine suppresses the activity of the AP-1 signaling pathway and decreases the binding of transcription factors to the Cyclin D1 AP-1 motif, indicating that transcriptional inhibition of Cyclin D1 may be involved in the anti-tumor effect of berberine [11]. To determine if inhibition of berberine on Cyclin D1 expression in hepatoma cells undergoes the same mechanism, we issued a quantitative real-time polymerase chain reaction (qPCR) analysis to quantify the Cyclin D1 mRNA transcripts in HepG2 cells exposed to berberine. Interestingly, we found that either 6 or 12 h exposure to berberine could not suppress the transcripts level of Cyclin D1, however, the protein expression was significantly inhibited (Figure 2A). To further examine if Cyclin

D1 suppression by berberine in HepG2 cells undergoes at a post-transcriptional level, we analyzed the protein expression in HepG2 cells with or without 100 μM berberine intervention in the presence of cycloheximide, a translation and protein synthesis inhibitor. We found that 100 μM berberine could shorten the half-life of Cyclin D1 protein in the presence of 150 μg/mL cycloheximide (Figure 2B). This action is further confirmed by the observation that presence of 20 nM MG-132, a proteasome inhibitor, is able to completely block the Cyclin D1 ablation induced by berberine exposure in HepG2 cells (Figure 2C). Our results show that berberine could induce a rapid post-translational degradation of Cyclin D1 in HepG2 cells.

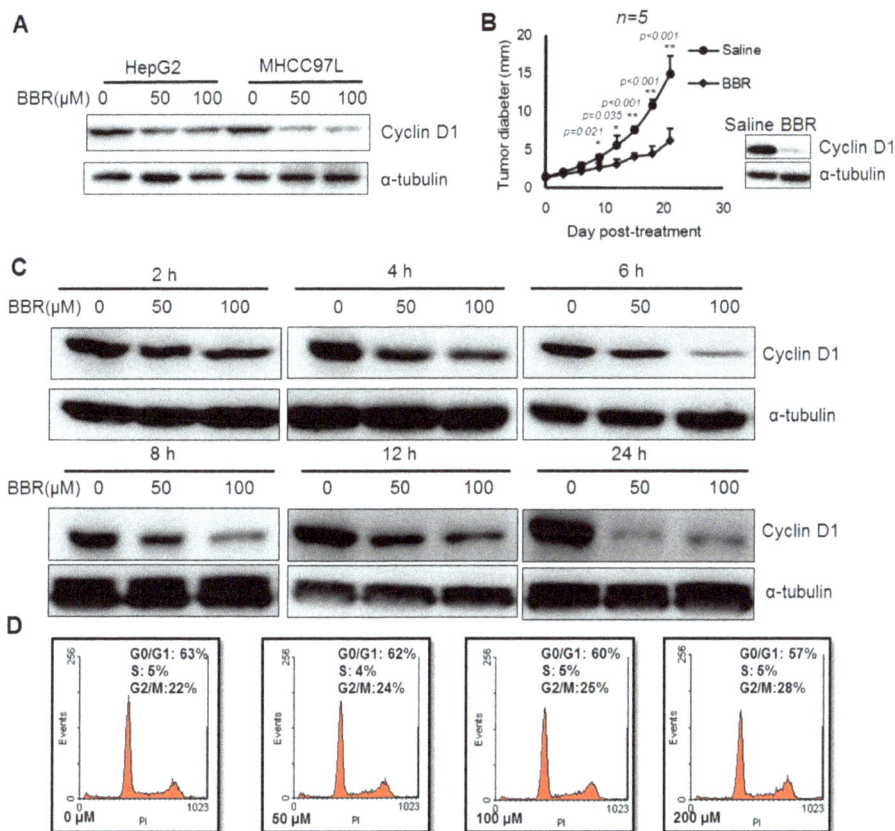

Figure 1. Berberine suppresses Cyclin D1 expression in hepatoma cells. (**A**) HepG2 and MHCC97L cells were treated with 100 μM berberine for 24 h. the expression of Cyclin D1 was inhibited; (**B**) Xenograft model was established as described and treatment of berberine can lead to reduced tumor size as well as Cyclin D1 expression; (**C**) Upon 6 h exposure of 100 μM berberine, the expression of Cyclin D1 was potently repressed. Cyclin D1 was detected by immunoblotting with β-actin as internal control; (**D**) HepG2 cells were treated with berberine at different doses for 6 h and then subject to cell cycle analysis. No significant cell cycle phase redistribution was observed. * $p < 0.05$,** $p < 0.01$.

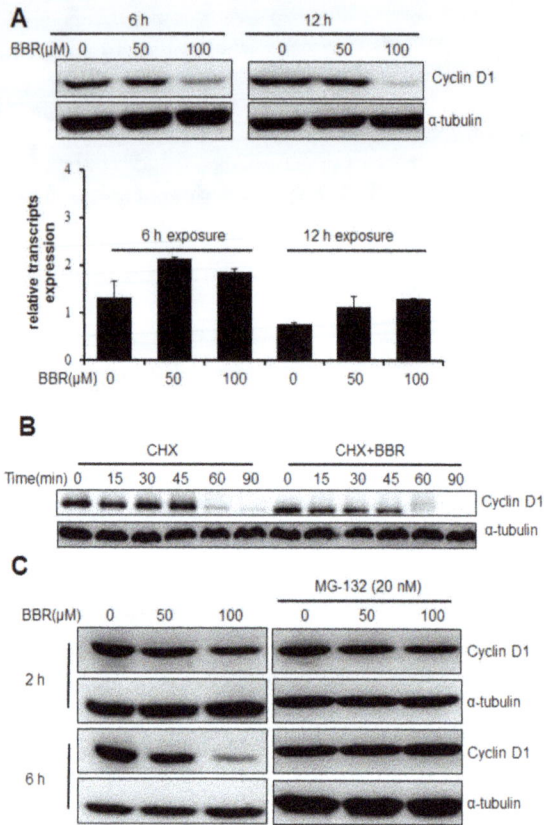

Figure 2. Berberine inhibits Cyclin D1 expression in HepG2 cells via post-translational control. (**A**) qPCR was used to detected the mRNA transcript of Cyclin D1 with GAPDH as internal control. No mRNA changed while Cyclin D1 protein was reduced by berberine; (**B**) Cells were treated with berberine in the presence of 150 μg/ml Cycloheximide. Reduced half-life in berberine-treated cells were found; (**C**) Cells were treated with berberine in the presence of 20 nM MG-132. Cyclin D1 was detected by immunoblotting with α-tubulin as internal control.

2.3. Berberine Promotes Cyclin D1 Ubiquitination in HepG2 Cells and Facilitates β-TrCP Binding

A direct evidence of berberine-induced Cyclin D1 unbiquitination in HepG2 cells was observed (Figure 3A). The endogenous expressing Cyclin D1 in HepG2 cells with berberine treatment in the presence of 20 nM MG-132 was immunoprecipitated using specific antibody against Cyclin D1 and analyzed using antibody against ubiquitin. Increased ubiquitinated Cyclin D1 was found in a dose-dependent manner, indicating that berberine could promote the ubiquitination of endogenous Cyclin D1. We observed that one of the F-box proteins, β-TrCP, could be triggered to bind to the skp1-cullin-F-box (SCF) protein complex of Cyclin D1 upon berberine exposure (Figure 3B). As recently reported, β-TrCP could serve as an E3 ligase and be incorporated in the SCF complex-facilitating ubiquitination dependent Cyclin D1 proteolysis [22]. To figure out the direct evidence of the involvement of β-TrCP in berberine-induced Cyclin D1 ablation, we used specific siRNA against human *BTRC* gene to block its expression in HepG2 cells. Partial genetic deletion of β-TrCP in HepG2 cells attenuates berberine's action on Cyclin D1 expression (Figure 3C). Our results may indicate that β-TrCP serves as the particular E3 ligase in berberine-driving Cyclin D1 proteolysis in HepG2 cells.

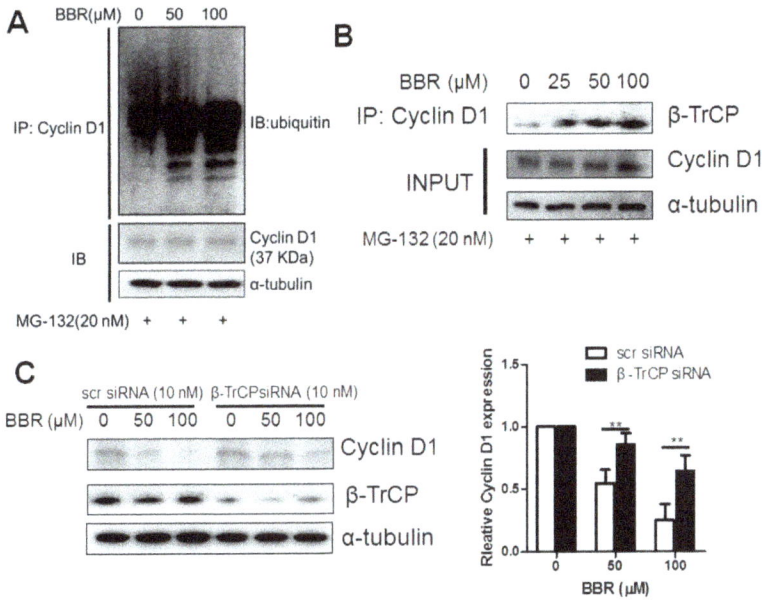

Figure 3. Berberine induces Cyclin D1 ubiquitination and recruits β-TrCPas an E3 ligase. (**A**) Cells were treated by berberine for 6 h in the presence of MG-132 (20 nM). Ubiquitinated Cyclin D1 was precipitated with antibody against Cyclin D1 and detected with ubiquitin antibody; (**B**) Cells were treated with berberine for 6 h in the presence of MG-132 (20 nM). Ubiquitinated Cyclin D1 was precipitated with antibody against Cyclin D1 and β-TrCP was detected with β-TrCP antibody; (**C**) shows that genetic knockdown of β-TrCP attenuates berberine's effect on Cyclin D1 degradation. (+ means presence of the chemicals), ** $p < 0.01$.

2.4. Berberine Promotes Cyclin D1 Phosphorylation and Nuclear Export in HepG2 Cells

Previous studies reported that Cyclin D1 turnover was mediated by ubiquitin-dependent proteasomal degradation and dependent on T286 (the threonine 286) phosphorylation [23]. However, it was observed that certain mutations stabilized Cyclin D1 but did not affect its polyubiquitylation, which could prove that the regulation of Cyclin D1 degradation may be ubiquitin-independent [24]. Identifying if the berberine-induced Cyclin D1 degradation in HepG2 is dependent on the phosphorylation on its T286 site, we first examined if berberine could promote the Cyclin D1 phosphorylation in HepG2 cells. Western blot analysis indicates that berberine-facilitated Cyclin D1 repression in HepG2 cells was accompanied with increases in Thr-286 phophorylation in the presence of MG132, the proteasome inhibitor (Figure 4A), and the effect of berberine in triggering Cyclin D1 phosphorylation in HepG2 cells is in dose- and time-dependent manner. This indicates that phosphorylation of Cyclin D1 at the T286 site may be involved in its degradation induced by berberine. Since the ubiquitination process of Cyclin D1 is conducted in cytoplasm, the nuclear export is necessary for berberine-facilitated Cyclin D1 degradation. Both immunofluorescence and immunoblotting analysis exhibit that berberine could reduce the nuclear localization of Cyclin D1 in HepG2 cells (Figure 4B,C). These data suggest that the ability of berberine to promote phosphorylation dependent nuclear transport and ubiquitination of Cyclin D1 plays an integral role in its subsequent degradation.

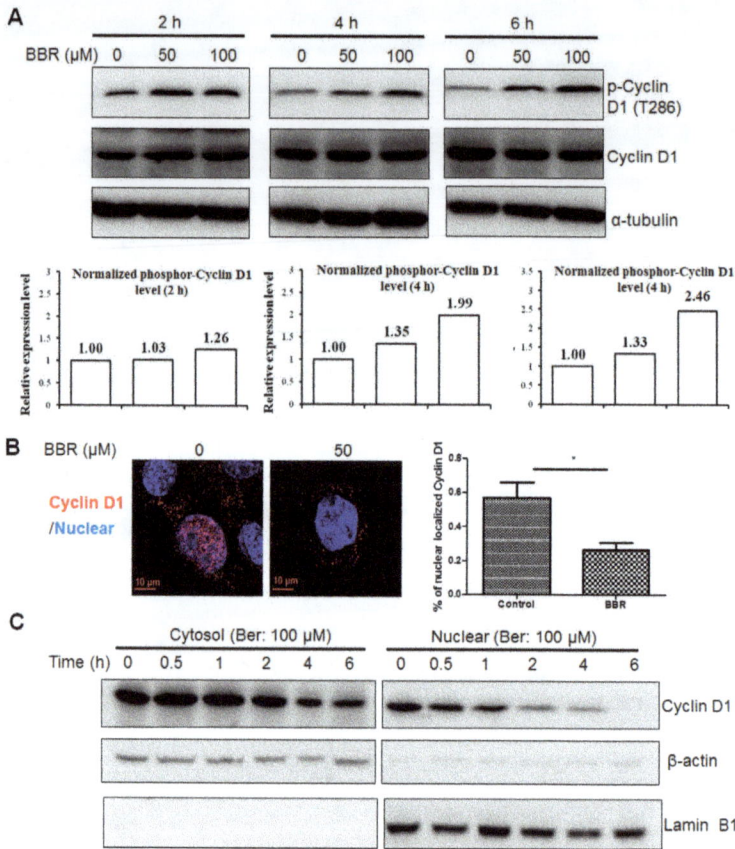

Figure 4. Berberine induces Cyclin D1 phosphorylation at T286 site and its nuclear export in HepG2 cells. (**A**) Cells were treated with berberine in the presence of MG132. The expression of phosphor-Cyclin D1 was normalized by total Cyclin D1 to avoid fluctuation induced by dynamic degradation of Cyclin D1; (**B**) Cells were treated with berberine for 6 h and fixed. Cyclin D1 was stained (Red) and DAPI was used to stain the nucleus; (**C**) Cells were treated with berberine and cytosolic and nuclear fractions were collected. β-actin and Lamin B1 were used as internal controls, respectively. * $p < 0.05$.

2.5. Berberine-Induced Cyclin D1 Degradation Is T286 Phosphorylation Dependent

In order to determine if phosphorylation of Cyclin D1 at T286 site is required for its degradation induced by berberine in HepG2 cells, we transfected pcDNA plasmid encoding either HA-tagged Cyclin D1 (wild-type, wt) or HA-tagged Cyclin D1 T286A mutant (mut) into HepG2 cells which were then exposed to berberine for 6 h. Immunoblotting analysis shows that the wild-type exogenous Cyclin D1 undergoes rapid degradation in the presence of berberine while mutant Cyclin D1 remains intact (Figure 5A). Since previous study reports that β-TrCP recruitment requires T286 phosphorylation of Cyclin D1, we issued that the recruitment of protein complex including β-TrCP to Cyclin D1 should be observed in cells transfected with wt Cyclin D1 but rather mut Cyclin D1. The protein complex in HepG2 cells transfected with pcDNA3 plasmid encoding either wt HA-Cyclin D1 or mut HA-Cyclin D1 T286A was precipitated by HA antibody and β-TrCP was detected by immunoblotting. Recruitment of β-TrCP was observed in HepG2 cells transfected with wt Cyclin D1 plasmid but not in cells with mut Cyclin D1 transfection when exposing to berberine, suggesting that T286 phosphorylation is required for the recruitment of β-TrCP as E3 ligase for the ubiquitination of Cyclin D1 driven by berberine (Figure 5B). This may indicate that berberine-induced Cyclin D1 ablation is T286-depdendent. To further identify the contribution of Cyclin D1 ablation in berberine's anti-tumor action, respectively, the plasmid encoding either HA-Cyclin D1 wt or HA-Cyclin D1 T286A were transfected into HepG2 cells followed by berberine treatment and WST-1 assay was used to detect the cell response to berberine. We found that cells with expression of mut Cyclin D1 show more resistance to berberine's effect than cells with wt Cyclin D1 transfection (Figure 5C). This indicates that cells that could not undergo T286 phosphorylation-mediated protein degradation when exposed to berberine are more likely to survive upon berberine treatment. These results exhibit that berberine induced Cyclin D1 degradation partially contributes to berberine's effect.

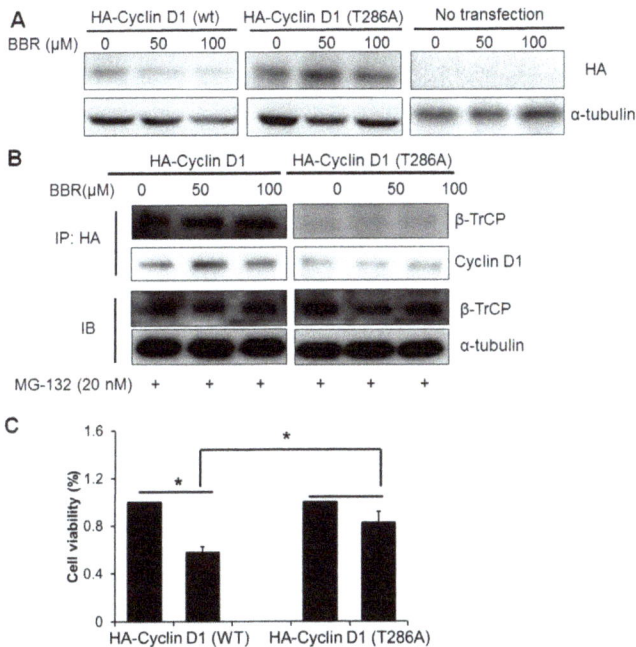

Figure 5. T286 phosphorylation is required from Cyclin D1 ubiquitin-proteasomal degradation induced by berberine. (**A**) Cells expressing HA-tagged wt and mutant Cyclin D1 was treated with berberine for 6 h. Expression of exogenous Cyclin D1 was blotted with hemagglutinin (HA) antibody; (**B**) Cells expressing HA-tagged wt and mutant Cyclin D1 was treated with berberine for 6 h. The protein complex was precipitated with HA antibody and precipitated β-TrCP and Cyclin D1 were detected. INPUT level of β-TrCP was detected as control; (**C**) Cells expressing HA-tagged wt and mutant Cyclin D1 was treated with berberine for 24 h. Cell viability was detected by 3-(4,5-dimethylthiazol-2-yl)-2,5-diphenyltetrazolium bromide (MTT) assay (+ means presence of the chemicals). * $p < 0.05$.

3. Discussion

In HCC, Cyclin D1 was found overexpressed and associated with aggressive forms of HCC [9,16]. Therefore, targeting Cyclin D1 by small molecule agents may be a therapeutically relevant strategy for the treatment of Cyclin D1-overexpressing HCC [22]. As a natural product with a long history and being intensively focused on its anti-tumor activity, berberine was reported to suppress Cyclin D1 expression in various human cancer cell lines, however, few of studies reported the underlying mechanism on Cyclin D1 inhibition action of berberine. From a translational perspective, understanding how berberine-facilitated Cyclin D1 inhibition is an important and integral step in drug discovery. In our study, we found a rapid suppression action of berberine on Cyclin D1 expression in human hepatoma cells HepG2, and berberine promotes an ubiquitination-dependent proteolysis of Cyclin D1 in HepG2 cells. This kind of effect of berberine is dependent on Cyclin D1's phosphorylation at the T286 site. Some previous studies show that berberine could upregulate the AMP-kinase and MAPK p42/p44 [25,26]. Phosphorylation of the related signaling by berberine may be responsible for its various biological functions, and our finding shows Cyclin D1 phosphorylation by berberine may be related to Cyclin D1 degradation in tumor cells. These findings suggest that the ubiquitin-proteasome signal pathway involves as a novel mechanism in Cyclin D1 ablation induced by berberine in HepG2 cells.

It was noticed that berberine can suppress the expression of Cyclin D1 in different hepatoma cell lines including HepG2 and MHCC97L. As well, Cyclin D1 was potently inhibited in berberine-treated hepatoma xenograft. The detailed mechanism of berberine in suppressing Cyclin D1 was elaborated in a particular cell line HepG2. The origin of HepG2 remains to be controversial though there are a plenty of studies that regarded it as a cell line of hepatocellular carcinoma. However, it was recently shown that HepG2 cells share more genetic similarity with hepatoblastoma but not hepatocellular carcinoma [27]. An increasing number of HCC cell lines has been developed and was used in the study of liver cancer, however, not all the cell lines have a correlation with the clinical features of liver tumor. Chen et al. compared the genomic data of tumor samples from clinical setting and that of commonly used HCC cell lines, and found that around half of cell lines have poor correlation in genetic features with human tumor samples. Fortunately, the four commonly used hepatoma cell lines, HepG2, Huh7, Hep3B, and PLC/PRF/5 exhibited high correlation to the tumors [28]. In our findings, the post-transcriptional mechanism of berberine-induced Cyclin D1 degradation was proven in one of clinically correlated cell line HepG2. The significance of this study may be increased with this mechanism being validated in other hepatoma cell lines.

The ubiquitin-proteasome dependent proteolysis is the important system in the control of protein degradation in cells [29]. The ubiquitination system is consisted of ubiquitin-activating enzymes (E1), ubiquitin-conjugating enzymes (E2), and ubiquitin-protein ligases (E3) [30,31], among which E3 is the specific enzyme for each degraded protein. For proteins controlling cell cycle, the Skp1-Cullin-F-box (SCF) complex is the particular E3 ligase for its ubiquitination [32]. β-TrCP is one of the F-box proteins that contain a protein structural motif of approximately 50 amino acids mediating protein-protein interactions. β-TrCP is linked closely to cancer for its activity in the degradation of IκBα and β-catenin [33]. Cyclin D1 was also reported as the substrate of β-TrCP in tumor cells under glucose starvation or particular anti-tumor agent treatment. Increased interaction between β-TrCP and Cyclin D1 was shown to promote Cyclin D1 protelysis in LNCap cells with exposure of peroxisome proliferator-activated receptor-γ (PPARγ) agonist STG28 and thereby contributed to its anti-tumor activity [22]. In our study, we observed that berberine, a natural product with wide spectrum of anti-tumor activity, could promote the recruitment of SCF protein complex and Cyclin D1 in HepG2 cells and facilitate Cyclin D1 proteolysis. We found that Cyclin D1 expression inhibition by berberine is dependent on ubiquitination pathway, and the particular F-box protein β-TrCP is involved. Knockdown of β-TrCP expression attenuates the Cyclin D1 turnover induced by berberine in HepG2 cells in a dose-dependent manner, indicating that β-TrCP plays a key role in berberine's action. Moreover, genetic deletion of β-TrCP partially increases the viability of HepG2 cells with exposure of berberine, revealing that Cyclin D1 degradation induced by berberine may contribute partially to its anti-tumor activity. The overall scheme of the mechanism underlying berberine's action on Cyclin D1 degradation is shown in Figure 6. We found that long-termed treatment of berberine increases its potency in suppressing tumor cell growth as well as in potentiating Cyclin D1 turnover. Our findings in this study indicate berberine's potential as an anti-tumor agent with clear mechanism in inducing Cyclin D1 degradation.

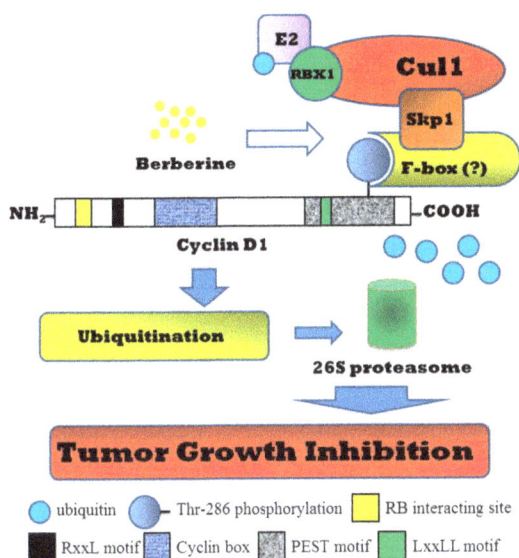

Figure 6. The overall scheme on the mechanism underlying berberine's control on Cyclin D1 degradation in HepG2 cells.

4. Materials and Methods

4.1. Chemicals and Plasmids

Berberine chloride, protein synthesis inhibitor cycloheximide and proteasome inhibitor MG-132 were purchased from Sigma-aldrich (St. Louis, MO, USA). Plasmid pcDNA3 Cyclin D1-HA (Plasmid 11181) and pcDNA3 Cyclin D1-HA (T286A, Plasmid 11182) were kindly provided by Bruce Zetter (Harvard Medical School, deposited by Addgene, Cambridge, MA, USA); plasmid pcDNA3 HA-ubiquitin (Plasmid 18712) was provided by Edward Yeh (The University of Texas-Houston Health Science Center, deposited by Addgene).

4.2. Cell Line and Cell Culture

The human hepatoma cell line HepG2 was obtained from American Type Culture Collection (ATCC, Manassas, VA, USA). MHCC97L cells were kindly gifted by Man Kwan from Department of Surgery, The University of Hong Kong (Hong Kong, China). Cells were maintained in the high glucose Dulbecco's Modified Eagle Medium (DMEM, Invitrogen, Carlsbad, CA, USA) supplemented with 10% FBS (Invitrogen), and incubated in a humidified atmosphere containing 5% CO_2 at 37 °C.

4.3. Xenograft Model

The protocol for animal study was approved by the Committee on the Use of Live Animals in Teaching and Research (CULATR) of the University of Hong Kong (code: 2441-11). Animal was housed in Laboratory Animal Centre of The University of Hong Kong with humane care. Four-week-old female BALB/c nude mice received 1×10^6 MHCC97L cells by subcutaneous injection at the right flank. One week after injection, mice were randomized into two groups. The treatment group of mice received intraperitoneal injection of berberine (10 mg/kg/2 days) while mice in control group received the same volume of saline buffer. Treatment lasted three weeks and at the end of study, mice were sacrificed by overdose of pentobarbital (200 mg/kg) and tumor was dissected out for analysis.

4.4. Real-Time Quantitative Polymease Chain Reaction

Total RNA was extracted and purified using RNeasy Mini Kit (Qiagen, Hilden, Germany) following the manufacturer's instruction. Reverse-transcription reaction was performed using QuantiTech Reverse Transcription Kit (Qiagen) to prepare cDNA samples. The quantitative real-time PCR (qRT-PCR) was conducted by QuantiTect SYBR Green PCR Kit (Qiagen) with 1 μM primers for *CCND1* (right: 5′-GACCTCCTCCTCGCACTTCT-3′; left: 5′-GAAGATCGTCGCCACCTG-3′; Invitrogen, USA) on LightCycler 480 real-time PCR system (Roche, Basel, Switzerland). The expression of *GAPDH* was used as endogenous control (right: 5′-GCCCAATACGACCAAATCC-3′; left: 5′-GCTAGGGACGGCCTGAAG-3′ Invitrogen, USA) for the normalization of gene expression of *CCND1*.

4.5. Cell Cycle Analysis

HepG2 cells exposed to berberine (0, 50, 100 μM) for 6 h were collected and fixed in ice-cold 70% ethanol overnight. Cells were then centrifuged for 5 min at 1500 rpm at room temperature. Ethanol was discarded and cell pellet was re-suspended in PBS containing propidium iodide (5 μg/mL) and RNase A (50 units/mL). Cell cycle phase distribution were examined by flow cytometer (Epics XL, Beckman Coulter, Brea, CA, USA) and analyzed by Winmidi V2.9 program.

4.6. Immunofluoscence

HepG2 cells were seeded in 10 mm cover slip and incubated overnight. Then cells were treated with berberine (0, 50, 100 μM) for 6 h. Cells were fixed in 4% paraformaldehyde for 1 h and then penetrated in 0.1% Triton-X100 for 15 min. Cells were blocked in 5% normal goat serum in PBS overnight at 4 °C followed by incubation with Cyclin D1 primary antibody (1:50) overnight at 4 °C. After washing, the bound primary antibody was detected using Texas Red goat anti-rabbit antibody (Santa Cruz, 1:200) at room temperature for 2 h. The nuclear counterstaining was performed using a 4,6-diamidino-2-phenylindole-containing mounting medium (Invitrogen) before examination. Images were taken using confocal microscope (Carl Zeiss, Oberkochen, Germany, 400 magnification, CCD camera).

4.7. Subcellular Fractionation

Cells were lysed with cold hypotonic buffer (10 mM Hepes, 10 mM KCl, 0.1 mM EDTA, 0.4% NP-40, 0.05 mM DTT) containing protease inhibitor cocktail (Roche) for 5 min and then supernatant (Cytoplasmic fraction) was collected by centrifugation at 14,000× g 4 °C. The residue was then extracted with nuclear extraction buffer (20 mM Hepes, 400 mM NaCl, 1 mM EDTA, 0.05 mM DTT, in the presence of protease inhibitor cocktail) on ice for 30 min, followed by centrifugation at 14,000× g for 10 min at 4 °C. Supernannt was collected as nuclear fraction. Both cytoplasmic and nuclear fraction was separated and immunoblotted with β-actin and LAMIN B1 as control, respectively [22].

4.8. Immunoblotting

Protein was isolated on SDS-PAGE and then transferred to polyvinylidene fluoride membrane (PVDF, Biorad, Hercules, CA, USA). The membrane was then blocked with 5% BSA overnight at 4 °C, followed by incubation with respective primary antibodies overnight at 4 °C. After washing, the membrane was then incubated with appropriate secondary antibody (Abcam, Cambridge, UK) at room temperature for 2 h. Image was captured using a chemiluminenescence imaging system (Bio-rad, Biorad) with ECL advanced kit (GE Healthcare, Little Chalfont, UK) as substrate.

4.9. Co-Immunoprecipitation Assay

Cells were treated with berberine in the presence of MG-132 for 6 h. Collected cell pellets were extracted using NP-40 lysis buffer (Invitrogen) supplemented with cocktail protease inhibitor (Roche)

Int. J. Mol. Sci. **2016**, *17*, 1899

for 5 min on ice followed by centrifuging at 14,000 rpm at 4 °C for 10 min. The supernatant was collected and aliquoted. Co-immunoprecipitation assay was performed using Dynabeads® protein G kit (Invitrogen) following manufacturer's instruction. Briefly, each 1.5 mg of magnetic beads were transferred to a 1.5 mL microcentrifuge tube and separated on the magnet (Millipore, Billerica, MA, USA) to remove the supernatant. Diluted antibodies were bound by incubating with the beads for 10 min at room temperature with rotation. The beads were collected by placing the tube on the magnet and removing the supernatant. The cell lysate was then incubated with antibody-bound beads for 10 min at room temperature with rotation and then discarded. After washing, the bound protein was eluted by incubating the beads with 20 μL elution buffer for 2 min at room temperature with rotation. The supernatant was collected and the eluted proteins were denatured and analyzed by immunoblotting.

4.10. RNA Interference

HepG2 cells were seeded in DMEM medium supplemented with 10% FBS and 1% antibiotics with 70% confluence. 24 h before transfection, medium was discarded and replaced with serum- and antibiotic-free DMEM medium. Transfection was carried out using Lipofectamine 2000 (Invitrogen) according to manufacturer's instruction. 10 μg of siRNA against human β-TrCP (sc-37178, Santa Cruz, CA, USA) was transfected. The cells were supplemented with DMEM medium with 10% FBS and 1% antibiotics 6 h after transfection. Treatment of berberine was conducted within 48 h after transfection.

4.11. Statistical Analysis

All experiments were conducted in triplicate except particular notice. Results were analyzed using student *t*-test and expressed as mean $\pm$ SD.

5. Conclusions

In conclusion, we observed that berberine exhibits dose- and time-dependent inhibition on Cyclin D1 expression in human hepatoma cells. Berberine increases the phosphorylation of Cyclin D1 at Thr286 site, and recruits the SCFβ-TrCP complex to facilitate Cyclin D1 ubiquitin-proteasome dependent proteolysis. In addition, berberine potentiates Cyclin D1 nuclear export to cytoplasm for proteasomal degradation. Knockdown of β-TrCP blocks Cyclin D1 turnover induced by berberine; blocking the protein degradation induced by berberine in HepG2 cells increases tumor cell resistance to berberine. Our results shed light on berberine's potential as an anti-tumor agent for clinical cancer therapy.

Acknowledgments: The study was financially supported by grants from the research council of the University of Hong Kong (Project Codes: 104001764, 10400699, 104002320, 104002889, 104003422), Wong's Donation on Modern Oncology of Chinese Medicine (Project code: 200006276), Gala Family Trust (Project Code: 200007008), Government-Matching Grant Scheme (Project Code: 207060411), National Natural Science Foundation of China (Project Code: 81302808), the Open Project of Hubei Key Laboratory of Wudang Local Chinese Medicine Research, Hubei University of Medicine (Grant No. WDCM001), the Young Scientist Innovation Team Project of Hubei Colleges (Grant No.T201510) and the Research Grant Committee (RGC) of Hong Kong SAR of China (RGC General Research Fund, Project Code: 766211). The authors would like to express thanks to Keith Wong, Cindy Lee, Alex Shek, and the Faculty Core Facility for their technical support.

Author Contributions: Ning Wang and Yibin Feng conceived and designed the experiments; Ning Wang, Hor-Yue Tan, and Sha Li performed the experiments; Yibin Feng, Xuanbin Wang, and Chi Man Tsang analyzed the data; Sai-Wah Tsao contributed reagents/materials/analysis tools; Ning Wang and Yibin Feng wrote the paper.

Conflicts of Interest: The authors declare no conflict of interest.

References

1. Chung, D.C. Cyclin D1 in human neuroendocrine: Tumorigenesis. *Ann. N. Y. Acad. Sci.* **2004**, *1014*, 209–217. [CrossRef] [PubMed]
2. Fu, M.; Wang, C.; Li, Z.; Sakamaki, T.; Pestell, R.G. Minireview: Cyclin D1: Normal and abnormal functions. *Endocrinology* **2004**, *145*, 5439–5447. [CrossRef] [PubMed]
3. Stacey, D.W. Cyclin D1 serves as a cell cycle regulatory switch in actively proliferating cells. *Curr. Opin. Cell Biol.* **2003**, *15*, 158–163. [CrossRef]
4. Weinstein, I.B. Cancer. Addiction to oncogenes—The achilles heal of cancer. *Science* **2002**, *297*, 63–64. [CrossRef] [PubMed]
5. Diehl, J.A. Cycling to cancer with cyclin D1. *Cancer Biol. Ther.* **2002**, *1*, 226–231. [CrossRef] [PubMed]
6. Wang, C.; Li, Z.; Fu, M.; Bouras, T.; Pestell, R.G. Signal transduction mediated by cyclin D1: From mitogens to cell proliferation: A molecular target with therapeutic potential. *Cancer Treat. Res* **2004**, *119*, 217–237. [PubMed]
7. Freemantle, S.J.; Liu, X.; Feng, Q.; Galimberti, F.; Blumen, S.; Sekula, D.; Kitareewan, S.; Dragnev, K.H.; Dmitrovsky, E. Cyclin degradation for cancer therapy and chemoprevention. *J. Cell. Biochem.* **2007**, *102*, 869–877. [CrossRef] [PubMed]
8. Lin, D.I.; Barbash, O.; Kumar, K.G.; Weber, J.D.; Harper, J.W.; Klein-Szanto, A.J.; Rustgi, A.; Fuchs, S.Y.; Diehl, J.A. Phosphorylation-dependent ubiquitination of cyclin D1 by the SCF (FBX4-αB crystallin) complex. *Mol. Cell* **2006**, *24*, 355–366. [CrossRef] [PubMed]
9. Choi, M.S.; Yuk, D.Y.; Oh, J.H.; Jung, H.Y.; Han, S.B.; Moon, D.C.; Hong, J.T. Berberine inhibits human neuroblastoma cell growth through induction of p53-dependent apoptosis. *Anticancer Res.* **2008**, *28*, 3777–3784. [PubMed]
10. Mantena, S.K.; Sharma, S.D.; Katiyar, S.K. Berberine inhibits growth, induces G1 arrest and apoptosis in human epidermoid carcinoma A431 cells by regulating Cdki-Cdk-cyclin cascade, disruption of mitochondrial membrane potential and cleavage of caspase 3 and PARP. *Carcinogenesis* **2006**, *27*, 2018–2027. [CrossRef] [PubMed]
11. Luo, Y.; Hao, Y.; Shi, T.P.; Deng, W.W.; Li, N. Berberine inhibits cyclin D1 expression via suppressed binding of AP-1 transcription factors to CCND1 AP-1 motif. *Acta Pharmacol. Sin.* **2008**, *29*, 628–633. [CrossRef] [PubMed]
12. Tang, J.; Feng, Y.; Tsao, S.; Wang, N.; Curtain, R.; Wang, Y. Berberine and coptidis rhizoma as novel antineoplastic agents: A review of traditional use and biomedical investigations. *J. Ethnopharmacol.* **2009**, *126*, 5–17. [CrossRef] [PubMed]
13. Tsang, C.M.; Lau, E.P.; Di, K.; Cheung, P.Y.; Hau, P.M.; Ching, Y.P.; Wong, Y.C.; Cheung, A.L.; Wan, T.S.; Tong, Y.; et al. Berberine inhibits Rho GTPases and cell migration at low doses but induces G2 arrest and apoptosis at high doses in human cancer cells. *Int. J. Mol. Med.* **2009**, *24*, 131–138. [PubMed]
14. Liang, K.W.; Yin, S.C.; Ting, C.T.; Lin, S.J.; Hsueh, C.M.; Chen, C.Y.; Hsu, S.L. Berberine inhibits platelet-derived growth factor-induced growth and migration partly through an AMPK-dependent pathway in vascular smooth muscle cells. *Eur. J. Pharmacol.* **2008**, *590*, 343–354. [CrossRef] [PubMed]
15. Liang, K.W.; Ting, C.T.; Yin, S.C.; Chen, Y.T.; Lin, S.J.; Liao, J.K.; Hsu, S.L. Berberine suppresses MEK/ERK-dependent Egr-1 signaling pathway and inhibits vascular smooth muscle cell regrowth after in vitro mechanical injury. *Biochem. Pharmacol.* **2006**, *71*, 806–817. [CrossRef] [PubMed]
16. Mantena, S.K.; Sharma, S.D.; Katiyar, S.K. Berberine, a natural product, induces G1-phase cell cycle arrest and caspase-3-dependent apoptosis in human prostate carcinoma cells. *Mol. Cancer Ther.* **2006**, *5*, 296–308. [CrossRef] [PubMed]
17. Khan, M.; Giessrigl, B.; Vonach, C.; Madlener, S.; Prinz, S.; Herbaceck, I.; Holzl, C.; Bauer, S.; Viola, K.; Mikulits, W.; et al. Berberine and a *Berberis lycium* extract inactivate Cdc25A and induce α-tubulin acetylation that correlate with HL-60 cell cycle inhibition and apoptosis. *Mutat. Res.* **2010**, *683*, 123–130. [CrossRef] [PubMed]
18. Santra, M.K.; Wajapeyee, N.; Green, M.R. F-box protein FBXO31 mediates cyclin D1 degradation to induce G1 arrest after DNA damage. *Nature* **2009**, *459*, 722–725. [CrossRef] [PubMed]

19. Tsang, C.M.; Cheung, K.C.; Cheung, Y.C.; Man, K.; Lui, V.W.; Tsao, S.W.; Feng, Y. Berberine suppresses Id-1 expression and inhibits the growth and development of lung metastases in hepatocellular carcinoma. *Biochim. Biophys. Acta* **2015**, *1852*, 541–551. [CrossRef] [PubMed]

20. Wang, N.; Zhu, M.; Wang, X.; Tan, H.Y.; Tsao, S.W.; Feng, Y. Berberine-induced tumor suppressor p53 upregulation gets involved in the regulatory network of MIR-23a in hepatocellular carcinoma. *Biochim. Biophys. Acta* **2014**, *1839*, 849–857. [CrossRef] [PubMed]

21. Wang, N.; Feng, Y.; Zhu, M.; Tsang, C.M.; Man, K.; Tong, Y.; Tsao, S.W. Berberine induces autophagic cell death and mitochondrial apoptosis in liver cancer cells: The cellular mechanism. *J. Cell. Biochem.* **2010**, *111*, 1426–1436. [CrossRef] [PubMed]

22. Wei, S.; Yang, H.C.; Chuang, H.C.; Yang, J.; Kulp, S.K.; Lu, P.J.; Lai, M.D.; Chen, C.S. A novel mechanism by which thiazolidinediones facilitate the proteasomal degradation of cyclin D1 in cancer cells. *J. Biol. Chem.* **2008**, *283*, 26759–26770. [CrossRef] [PubMed]

23. Diehl, J.A.; Cheng, M.; Roussel, M.F.; Sherr, C.J. Glycogen synthase kinase-3β regulates cyclin D1 proteolysis and subcellular localization. *Genes Dev.* **1998**, *12*, 3499–3511. [CrossRef] [PubMed]

24. Yu, Z.K.; Gervais, J.L.; Zhang, H. Human CUL-1 associates with the SKP1/SKP2 complex and regulates p21(CIP1/WAF1) and cyclin D proteins. *Proc. Natl. Acad. Sci. USA* **1998**, *95*, 11324–11329. [CrossRef] [PubMed]

25. Turner, N.; Li, J.Y.; Gosby, A.; To, S.W.; Cheng, Z.; Miyoshi, H.; Taketo, M.M.; Cooney, G.J.; Kraegen, E.W.; James, D.E.; et al. Berberine and its more biologically available derivative, dihydroberberine, inhibit mitochondrial respiratory complex I: A mechanism for the action of berberine to activate amp-activated protein kinase and improve insulin action. *Diabetes* **2008**, *57*, 1414–1418. [CrossRef] [PubMed]

26. Cui, G.; Qin, X.; Zhang, Y.; Gong, Z.; Ge, B.; Zang, Y.Q. Berberine differentially modulates the activities of ERK, p38 MAPK, and JNK to suppress TH17 and TH1 T cell differentiation in type 1 diabetic mice. *J. Biol. Chem.* **2009**, *284*, 28420–28429. [CrossRef] [PubMed]

27. Lopez-Terrada, D.; Cheung, S.W.; Finegold, M.J.; Knowles, B.B. Hep G2 is a hepatoblastoma-derived cell line. *Hum. Pathol.* **2009**, *40*, 1512–1515. [CrossRef] [PubMed]

28. Chen, B.; Sirota, M.; Fan-Minogue, H.; Hadley, D.; Butte, A.J. Relating hepatocellular carcinoma tumor samples and cell lines using gene expression data in translational research. *BMC Med Genom.* **2015**, *8*, S5. [CrossRef] [PubMed]

29. Mo, Z.; Zu, X.; Xie, Z.; Li, W.; Ning, H.; Jiang, Y.; Xu, W. Antitumor effect of F-PBF (β-TrCP)-induced targeted PTTG1 degradation in HeLa cells. *J. Biotechnol.* **2009**, *139*, 6–11. [CrossRef] [PubMed]

30. Hershko, A.; Ciechanover, A. The ubiquitin system. *Annu. Rev. Biochem.* **1998**, *67*, 425–479. [CrossRef] [PubMed]

31. Pickart, C.M. Mechanisms underlying ubiquitination. *Annu. Rev. Biochem.* **2001**, *70*, 503–533. [CrossRef] [PubMed]

32. Frescas, D.; Pagano, M. Deregulated proteolysis by the F-box proteins SKP2 and β-TrCP: Tipping the scales of cancer. *Nat. Rev. Cancer* **2008**, *8*, 438–449. [CrossRef] [PubMed]

33. Li, X.; Liu, J.; Gao, T. β-TrCP-mediated ubiquitination and degradation of PHLPP1 are negatively regulated by Akt. *Mol. Cell. Biol.* **2009**, *29*, 6192–6205. [CrossRef] [PubMed]

International Journal of
Molecular Sciences

MDPI

Article

The Antiproliferative Effect of Chakasaponins I and II, Floratheasaponin A, and Epigallocatechin 3-O-Gallate Isolated from *Camellia sinensis* on Human Digestive Tract Carcinoma Cell Lines

Niichiro Kitagawa [1,2,†], Toshio Morikawa [1,3,*,†], Chiaki Motai [1,2], Kiyofumi Ninomiya [1,3], Shuhei Okugawa [1,2], Ayaka Nishida [1], Masayuki Yoshikawa [1] and Osamu Muraoka [1,3,*]

[1]   Pharmaceutical Research and Technology Institute, Kindai University, 3-4-1 Kowakae, Higashi-osaka, Osaka 577-8502, Japan; kitagawa@koshiroseiyaku.co.jp (N.K.); motai@koshiroseiyaku.co.jp (C.M.); ninomiya@phar.kindai.ac.jp (K.N.); okugawa_kameokanousan@koshiroseiyaku.co.jp (S.O.); 1311610173m@kindai.ac.jp (A.N.); m-yoshikawa@leto.eonet.ne.jp (M.Y.)
[2]   Koshiro Company Ltd., 2-5-8 Doshomachi, Chuo-ku, Osaka 541-0045, Japan
[3]   Antiaging Center, Kindai University, 3-4-1 Kowakae, Higashi-osaka, Osaka 577-8502, Japan
*   Correspondence: morikawa@kindai.ac.jp (T.M.); muraoka@phar.kindai.ac.jp (O.M.); Tel.: +81-6-4307-4306 (T.M.); +81-6-4307-4015 (O.M.); Fax: +81-6-6729-3577 (T.M.); +81-6-6721-2505 (O.M.)
†   These authors contributed equally to this work.

Academic Editor: Gopinadhan Paliyath
Received: 20 October 2016; Accepted: 17 November 2016; Published: 26 November 2016

Abstract: Acylated oleanane-type triterpene saponins, namely chakasaponins I (**1**) and II (**2**), floratheasaponin A (**3**), and their analogs, together with catechins—including (−)-epigallocatechin 3-O-gallate (**4**), flavonoids, and caffeine—have been isolated as characteristic functional constituents from the extracts of "tea flower", the flower buds of *Camellia sinensis* (Theaceae), which have common components with that of the leaf part. These isolates exhibited antiproliferative activities against human digestive tract carcinoma HSC-2, HSC-4, MKN-45, and Caco-2 cells. The antiproliferative activities of the saponins (**1–3**, IC_{50} = 4.4–14.1, 6.2–18.2, 4.5–17.3, and 19.3–40.6 μM, respectively) were more potent than those of catechins, flavonoids, and caffeine. To characterize the mechanisms of action of principal saponin constituents **1–3**, a flow cytometric analysis using annexin-V/7-aminoactinomycin D (7-AAD) double staining in HSC-2 cells was performed. The percentage of apoptotic cells increased in a concentration-dependent manner. DNA fragmentation and caspase-3/7 activation were also detected after 48 h. These results suggested that antiproliferative activities of **1–3** induce apoptotic cell death via activation of caspase-3/7.

Keywords: chakasaponin; floratheasaponin; (−)-epigallocatechin 3-O-gallate; anti-proliferative activity; tea flower; *Camellia sinensis*; apoptosis

---

1. Introduction

Saponins, which comprises a triterpene or steroid aglycone with oligosugar chains, are a large, structurally diverse group of bioactive natural products that are widely distributed in higher plants and marine organisms, such as starfish and sea cucumbers [1–6]. Saponins have been reported to possess a number of important bioactive properties, such as expectorant, anti-inflammatory, vasoprotective, hypocholesterolemic, immunomodulatory, hypoglycemic, cytotoxic, molluscicidal, antifungal, and antiparasitic activities [1,3,5,6]. Recently, we reported the identification and biological properties of saponin constituents from "tea flower", the flower buds of *Camellia sinensis* (L.) O. Kuntze (Theaceae) [7–11]; the major saponin constituents were chakasaponins I (**1**) and II (**2**) and floratheasaponin A (**3**) (Figure 1). "Tea flower" has been used as food-garnishing agent in some Japanese dishes (e.g., botebote-cha in Shimane prefecture) or in drinks in some rural areas (e.g., batabata-cha in Niigata prefecture). With regard to the biofunctions of saponin constituents in "tea flower", antihyperlipidemic, antihyperglycemic, gastroprotective, antiobesity, antiallergic, and pancreatic lipase and amyloid β aggregation-inhibitory effects have been reported [11]. A variety of health foods and beverages made from "tea flower" have been developed in Japan, Taiwan, and neighboring Asian countries based on the abovementioned evidence. In the current study, we investigated the antiproliferative activities of the active constituents in "tea flower", namely saponins, catechins (including (–)-epigallocatechin 3-*O*-gallate (**4**)), flavonoids, and caffeine, which have components common with those of the leaf part ("green tea") [12]. The antiproliferative effects against human digestive tract carcinoma HSC-2, HSC-4, MKN-45, and Caco-2 cells were examined.

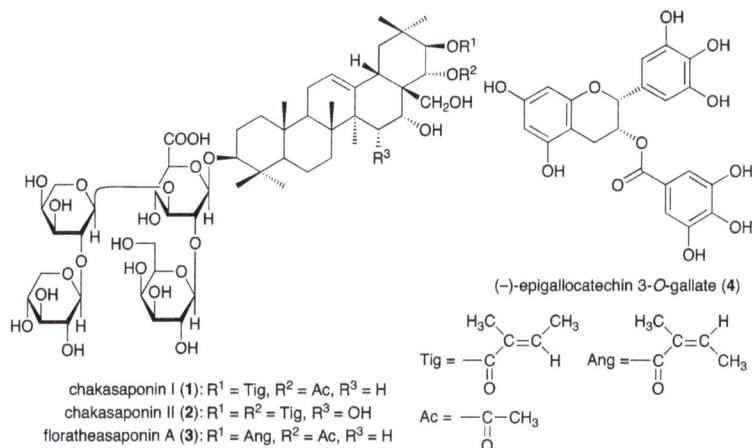

chakasaponin I (**1**): R¹ = Tig, R² = Ac, R³ = H
chakasaponin II (**2**): R¹ = R² = Tig, R³ = OH
floratheasaponin A (**3**): R¹ = Ang, R² = Ac, R³ = H

(–)-epigallocatechin 3-*O*-gallate (**4**)

Figure 1. Structures of chakasaponins I (**1**) and II (**2**), floratheasaponin A (**3**), and (–)-epigallocatechin 3-*O*-gallate (**4**).

2. Results and Discussion

2.1. Antiproliferative Activities of Constituents Isolated from "Tea Flower" against Human Gastric Carcinoma HSC-2, HSC-4, MKN-45, and Caco-2 Cells

Digestive tract carcinoma refers to a group malignancy located in the oral cavity, pharynx, larynx, esophagus, stomach, and large intestines and is the most common type of cancer worldwide [13]. In our previous studies of bioactive components from natural medicines isolated from the flowers of *Bellis perennis*, we observed antiproliferative activities of acylated oleanane-type triterpene saponins, perennisaponins A–T, against HSC-2, HSC-4, and MKN-45 cells [14]. The structures of perennisaponins resemble that of saponin constituents isolated from *C. sinensis*, and thus, similar activities were expected. Two analytical protocols have been developed with respect to nine acylated oleanane-type triterpene saponins, including chakasaponins I (**1**), II (**2**), and III, floratheasaponin A (**3**), and 15 polyphenols, including (+)-catechin, (–)-epicatechin, (–)-epigallocatechin, (–)-epicatechin 3-*O*-gallate, (–)-epigallocatechin 3-*O*-gallate (**4**), kaempferol, kaempferol 3-*O*-β-D-glucopyranoside, kaempferol 3-*O*-β-D-galactopyranoside, kaempferol 3-*O*-β-D-glucopyranosyl-(1→3)-α-L-rhamnopyranosyl-(1→6)-β-D-glucopyranoside, kaempferol 3-*O*-β-D-glucopyranosyl-(1→3)-α-L-rhamnopyranosyl-(1→6)-β-D-galactopyranoside, quercetin, quercetin 3-*O*-β-D-glucopyranoside, quercetin 3-*O*-β-D-galactopyranoside, rutin, and caffeine (Figure S1) [8–10]. Previously, the antiproliferative effects of dietary catechins and flavonoids from "green tea", such as compound **4** and quercetin derivatives, against HSC-2, MKN-45, and Caco-2 cells have been reported [15–18]. To identify the active principles, the inhibitory activities of abovementioned "tea flower" constituents against human gastric carcinoma HSC-2, HSC-4, MKN-45, and Caco-2 cells were evaluated. As shown in Table 1, the saponin constituents (**1–3**) show antiproliferative activities, with IC_{50} values of 4.4–14.1 μM against HSC-2, 6.2–18.2 μM against HSC-4, 4.5–17.3 μM against MKN-45, and 19.3–40.6 μM against Caco-2 cells. The results are as follows: the IC_{50} values of chakasaponin I (**1**) against HSC-2, HSC-4, MKN-45, and Caco-2 cells were equal to 4.6, 17.5, 16.8, and 30.2 μM, respectively, and those of floratheasaponin A (**3**) were equal to 4.4, 6.2, 4.5, and 19.3 μM, respectively. The compounds having a common theasapogenol B (=barringtogenol C) moiety as an aglycone show relatively strong activities. These activities are higher than those of chakasaponin II (**2**), which has an IC_{50} value of 14.1, 18.2, 17.3, and 40.6 μM against HSC-2, HSC-4, MKN-45, and Caco-2 cells, respectively, and those of chakasaponin III, which has an IC_{50} value of 19.4, 22.1, 21.1, and 52.2 μM, respectively, except for the antiproliferative activity against MKN-45 cells. These results indicate that the presence of the 16β-hydroxy moiety in the aglycone part reduces the activity. Conversely, the polyphenol constituents (–)-epigallocatechin and (–)-epigallocatechin 3-*O*-gallate (**4**) show an antiproliferative activity against HSC-2 cells with IC_{50} values equal to 54.6 and 28.3 μM, respectively, whereas (–)-epigallocatechin, compound **4**, quercetin, and quercetin 3-*O*-β-D-glucopyranoside have IC_{50} values of 23.8, 27.2, 77.2, and 67.3 μM, respectively, against HSC-4 cells (Table S1). The concentration dependencies of the antiproliferative activity of compounds **1–4** against HSC-2 cells are observed in a range of 3–100 μM for compounds **1, 2**, and **4** and a range of 0.3–10 μM for compound **3**, as shown in Figure S2.

Table 1. Antiproliferative effects of compounds **1–4** from "tea flower" on human digestive tract carcinoma HSC-2, HSC-4, MKN-45, and Caco-2 cells.

Treatment	IC_{50} (μM) [a]			
	HSC-2	HSC-4	MKN-45	Caco-2
Chakasaponin I (**1**)	4.6	17.5	16.8	30.2
Chakasaponin II (**2**)	14.1	18.2	17.3	40.6
Floratheasaponin A (**3**)	4.4	6.2	4.5	19.3
(–)-Epigallocatechin 3-*O*-gallate (**4**)	28.3	27.2	>100 (70.3)	>100 (71.8)
5-FU	>100 (55.5)	>100 (77.3)	>100 (52.4)	>100 (80.2)
Cisplatin	14.5	7.0	circa 100	>100 (69.5)
Doxorubicin	0.040	0.18	0.12	>100 (80.9)
Camptothecin	0.020	0.089	0.10	>100 (63.7)
Taxol	0.0012	0.0059	0.15	>100 (61.5)

Each value represents the mean ± S.E.M. (*n* = 4); [a] values in parentheses present percent of cell viability at 100 μM; commercial 5-FU (5-fluorouracil), cisplatin, doxorubicin, and camptothecin were purchased from Wako Pure Chemical Industries, Ltd. (Osaka, Japan) and taxol was from Tocris Bioscience (Bristol, UK).

2.2. Effects of Cell Cycle Distribution in HSC-2 Cells

Apoptosis and cell cycle dysfunction are closely associated biochemical processes, and any disturbance in cell cycle progression may lead to apoptotic cell death [19]. The effects of cell cycle distribution were analyzed to determine the mechanism associated with the growth inhibitory effect of chakasaponins I (**1**) and II (**2**), floratheasaponin A (**3**), and (–)-epigallocatechin 3-*O*-gallate (**4**) on HSC-2 cells [20,21]. The cell distribution in G0/G1, S, and G2/M phases shown in blue, red, and green areas, respectively, were determined after a 48 h incubation (Figure 2). As shown in Table 2, compounds **1–3** significantly induce populations of the cell distribution in S and G2/M phases, but reduce that of the G0/G1 phase. However, compound **4** does not affect the cell cycle distribution at the effective concentration. These results imply that compounds **1–3** induced the cell cycle arrest at the G2/M phase.

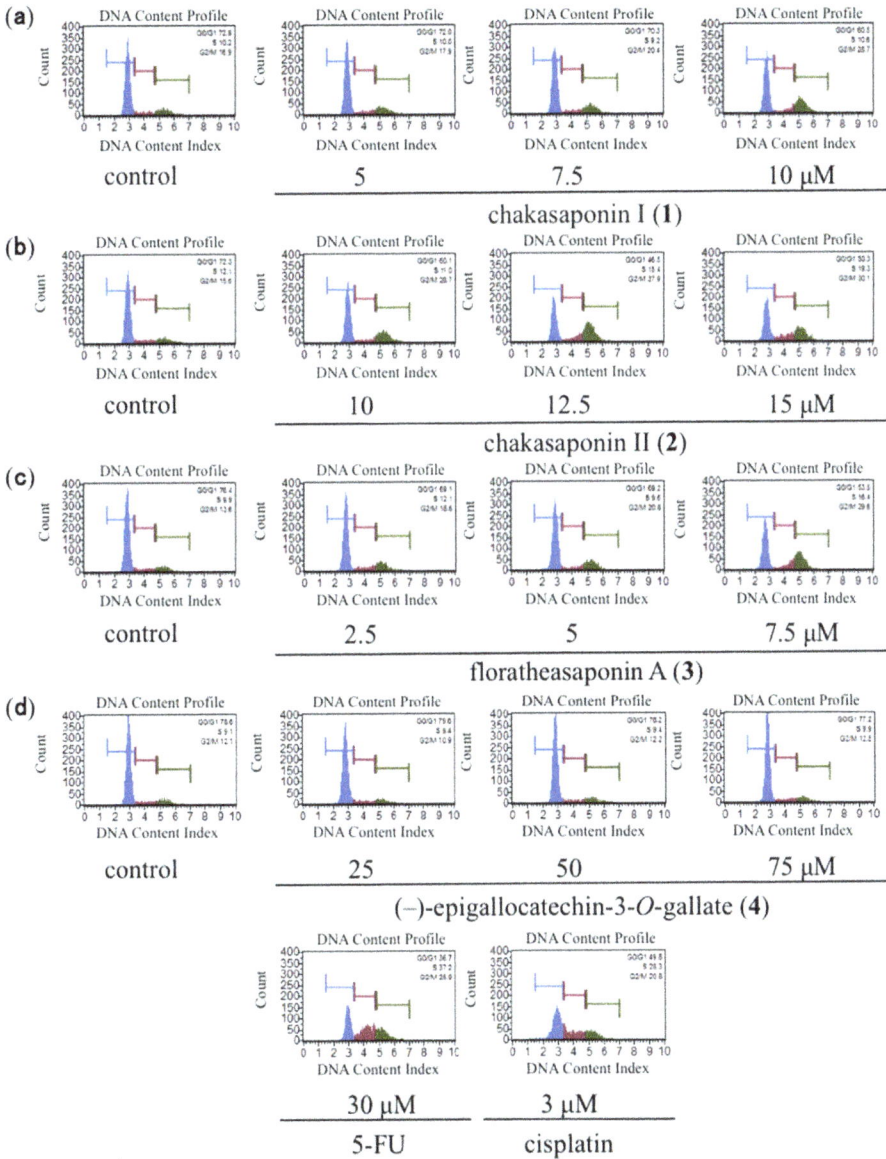

Figure 2. Effects of **1–4**, 5-FU, and cisplatin on the cell cycle distribution of HSC-2 cells. Cell cycle distribution was measured by Muse® Cell Analyzer using a Muse Cell Cycle Kit (Merck Millipore, Darmstadt, Germany); HSC-2 cells were treated with (**a**) 5, 7.5, and 10 µM of **1**; (**b**) 10, 12.5, and 15 µM of **2**; (**c**) 2.5, 5, and 7.5 µM of **3**; (**d**) 25, 50, and 75 µM of **4**, 30 µM of 5-FU, and 3 µM of cisplatin for 48 h; the data represent the mean percentages ± SD of total apoptosis (*n* = 3); commercial 5-FU and cisplatin were purchased from Wako Pure Chemical Industries, Ltd. (Osaka, Japan).

Table 2. Effects of 1–4, 5-FU, and cisplatin on cell cycle distribution of HSC-2 cells for 48 h.

Treatment	Concentration (μM)	Cell Cycle (%) [a]		
		G0/G1 Phase	S Phase	G2/M Phase
Control	–	73.5 ± 0.8	10.8 ± 0.4	15.6 ± 0.7
Chakasaponin I (1)	5	71.6 ± 0.4	10.2 ± 0.4	18.0 ± 0.1
–	7.5	69.1 ± 1.4 *	10.6 ± 0.8	20.1 ± 0.8 **
–	10	59.9 ± 0.2 **	11.2 ± 1.0	28.7 ± 0.9 **
Control	–	72.4 ± 1.1	11.6 ± 0.4	16.0 ± 0.7
Chakasaponin II (2)	10	60.3 ± 0.1 **	10.5 ± 0.4	28.9 ± 0.3 **
–	12.5	44.1 ± 1.3 **	18.8 ± 2.4 **	36.8 ± 1.3 **
–	15	49.2 ± 0.8 **	20.0 ± 0.3 **	30.5 ± 0.6 **
Control	–	76.5 ± 1.4	9.8 ± 0.6	13.5 ± 0.8
Floratheasaponin A (3)	2.5	69.4 ± 1.0 *	11.8 ± 0.1	18.5 ± 0.9 *
–	5	67.6 ± 1.7 **	10.3 ± 0.7	21.7 ± 0.9 **
–	7.5	53.1 ± 0.8 **	15.9 ± 1.8 **	30.3 ± 1.1 **
Control	–	78.5 ± 1.2	9.4 ± 0.5	11.8 ± 0.7
(–)-Epigallocatechin 3-O-gallate (4)	25	78.8 ± 1.1	9.7 ± 0.5	11.2 ± 0.6
–	50	78.6 ± 1.2	9.0 ± 0.5	12.0 ± 0.6
–	75	76.8 ± 0.5	10.1 ± 0.1	12.7 ± 0.4
5-FU	30	36.4 ± 0.3 **	37.9 ± 0.9 **	25.4 ± 0.6 **
Cisplatin	3	50.7 ± 0.7 **	27.9 ± 0.5 **	20.0 ± 0.3 **

Each value represents the mean ± S.E.M. (n = 3); [a] cell cycle distribution was measured by Muse® Cell Analyzer using a Muse Cell Cycle Assay Kit (Merck Millipore); asterisks denote significant differences from the control group, * $p < 0.05$, ** $p < 0.01$; commercial 5-FU and cisplatin were purchased from Wako Pure Chemical Industries, Ltd. (Osaka, Japan).

2.3. Quantification of Apoptotic Cell Death Using Annexin-V Binding Assay in HSC-2 Cells

Apoptosis plays an important role in the homeostatic maintenance of the tissue by selectively eliminating excessive cells. On the other hand, the induction of apoptosis of carcinoma cells is also recognized to be useful in cancer treatment, since cytotoxic agents used in chemotherapy of leukemia and solid tumors are known to cause apoptosis in target cells. Thus, the induction of apoptosis of cancer cells may be useful in cancer treatment [22]. To determine the apoptosis-inducing effects of compounds 1–4 against HSC-2 cells, staining with annexin-V/7-aminoactinomycin D (7-AAD) as a marker of early and late apoptotic events was performed (Figure 3) [23]. As shown in Table 3, the percentage of total apoptosis of compounds 1–3 increase in a concentration-dependent manner. However, compound 4 does not induce apoptotic cell death by this annexin-V-binding assay.

2.4. Evaluation of Apoptotic Morphological Changes in HSC-2 Cells

Representative morphological features of apoptosis in HSC-2 cells were examined under an inverted light fluorescence microscope using 4′,6-diamidino-2-phenylindole dihydrochloride (DAPI) as a staining agent [24]. As shown in Figure 4, the cell shrinkage caused by compounds 1–3 was mediated partially through apoptosis induction as nuclear chromatin condensation. Nuclei of the cells treated with compound 4 showed the obvious changes only slightly, while evident nuclear condensation was observed in cells treated with compounds 1–3.

2.5. DNA Fragmentation in HSC-2 Cells

DNA fragmentation in HSC-2 cells treated with compounds 1–4 were examined. Apoptotic cells display condensed chromatin and fragmented nuclei, but nonapoptotic cells maintain their structure [24–26]. As shown in Figure 5, DNA ladder formation, which is indicative of apoptosis, is observed in HSC-2 cells treated with compounds 1–3 for 48 h at effective concentrations. This fragmentation shows the same pattern as those treated with actinomycin D at a concentration of 0.1 μM.

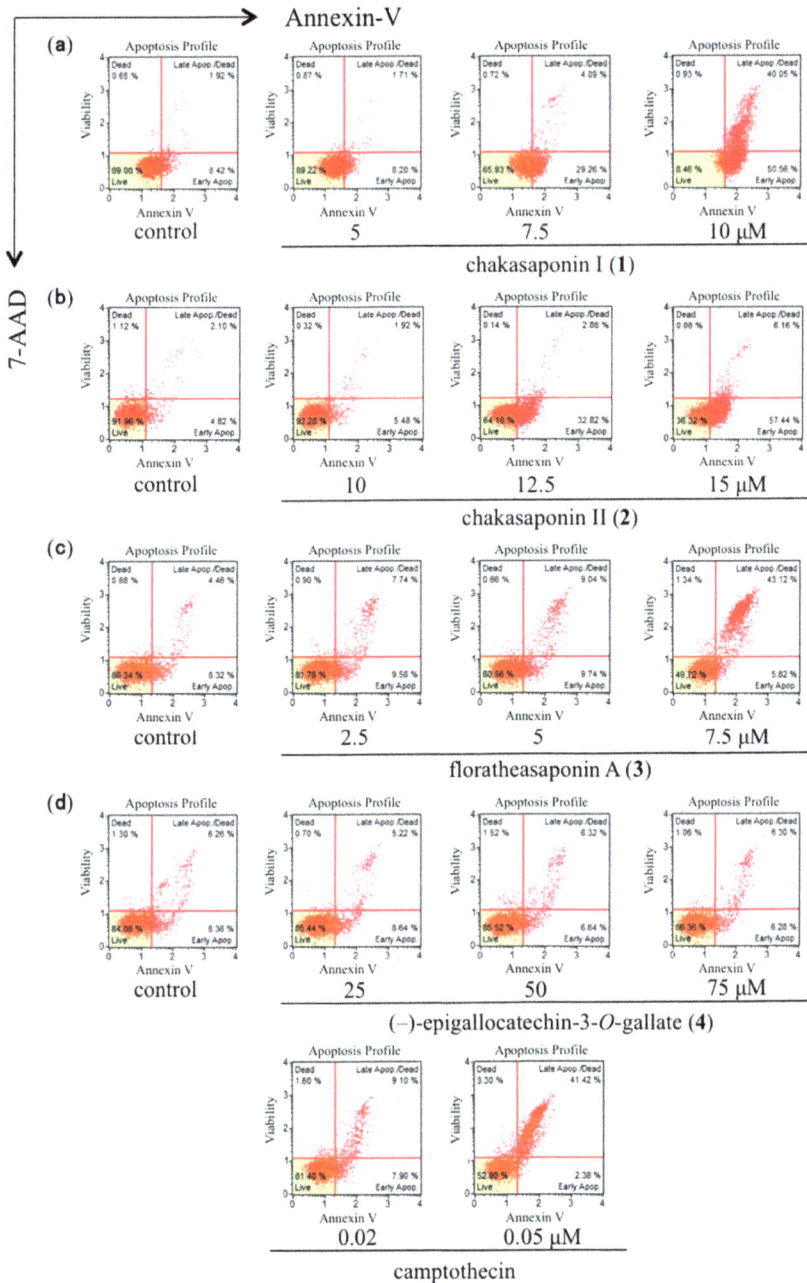

Figure 3. Effects of **1**–**4** and camptothecin on the apoptosis of HSC-2 cells. Annexin-V-binding was measured by Muse® Cell Analyzer using a Muse Annexin-V and Deal Cell Kit (Merck Millipore); HSC-2 cells were treated with (**a**) 5, 7.5, and 10 μM of **1**; (**b**) 10, 12.5, and 15 μM of **2**; (**c**) 2.5, 5, and 7.5 μM of **3**; (**d**) 25, 50, and 75 μM of **4** and 0.02 and 0.05 μM of camptothecin for 24 h; the data represent the mean percentages ± SD of total apoptosis (*n* = 3); commercial camptothecin was purchased from Wako Pure Chemical Industries, Ltd. (Osaka, Japan).

Table 3. Effects of **1–4** and camptothecin on the apoptosis of HSC-2 cells for 24 h.

Treatment	Concentration (μM)	Total Apoptotic Cells (%) [a]
Control	–	8.9 ± 1.5
Chakasaponin I (**1**)	5	10.0 ± 0.4
–	7.5	35.6 ± 3.0 **
–	10	89.7 ± 0.4 **
Control	–	6.3 ± 0.4
Chakasaponin II (**2**)	10	7.0 ± 0.3
–	12.5	35.7 ± 0.4 **
–	15	64.2 ± 1.0 **
Control	–	12.9 ± 0.4
Floratheasaponin A (**3**)	2.5	17.5 ± 1.0
–	5	18.7 ± 0.3 *
–	7.5	50.1 ± 1.9 **
Control	–	13.9 ± 0.5
(–)-Epigallocatechin 3-*O*-gallate (**4**)	25	14.4 ± 0.6
–	50	13.1 ± 0.1
–	75	12.7 ± 0.1
Camptothecin	0.02	16.9 ± 0.2 **
–	0.05	43.9 ± 0.7 **

Each value represents the mean ± S.E.M. (n = 3); [a] cell cycle distribution was measured by Muse® Cell Analyzer using a Muse Annexin-V and Deal Cell Kit (Merck Millipore); asterisks denote significant differences from the control group, * $p < 0.05$, ** $p < 0.01$; commercial camptothecin was purchased from Wako Pure Chemical Industries, Ltd. (Osaka, Japan).

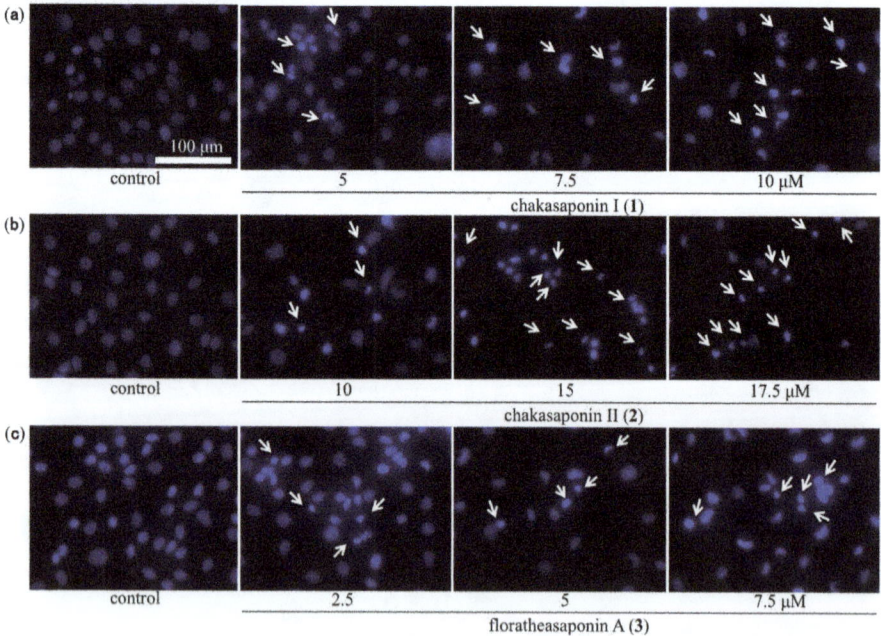

(a) control | 5 | 7.5 | 10 μM
chakasaponin I (**1**)

(b) control | 10 | 15 | 17.5 μM
chakasaponin II (**2**)

(c) control | 2.5 | 5 | 7.5 μM
floratheasaponin A (**3**)

Figure 4. *Cont.*

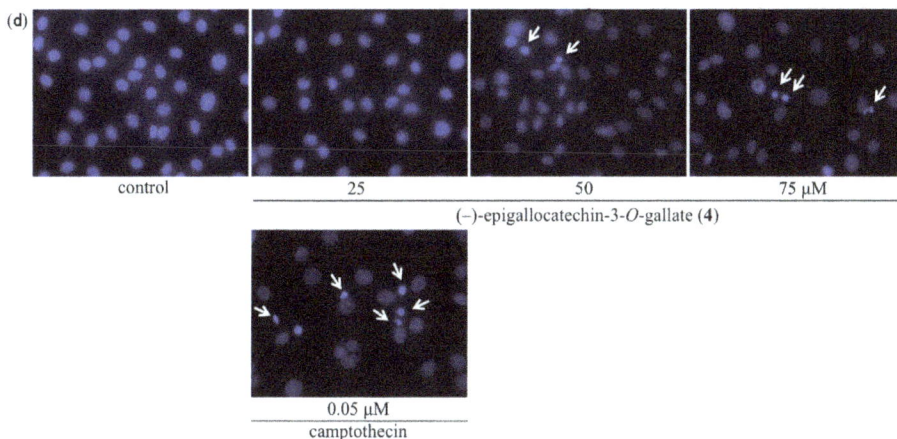

Figure 4. Morphological analysis of HSC-2 cells treated with **1–4** and camptothecin. Morphology of representative fields of HSC-2 cells stained with 4′,6-diamidino-2-phenylindole dihydrochloride (DAPI) after treatment with (**a**) 5, 7.5, and 10 μM of **1**; (**b**) 10, 15, and 17.5 μM of **2**; (**c**) 2.5, 5, and 7.5 μM of **3**; (**d**) 25, 50, and 75 μM of **4** and 0.05 μM of camptothecin for 48 h; the cells indicated by *arrows* represent fragmented and condensed nuclear chromatins; commercial camptothecin was purchased from Wako Pure Chemical Industries, Ltd. (Osaka, Japan).

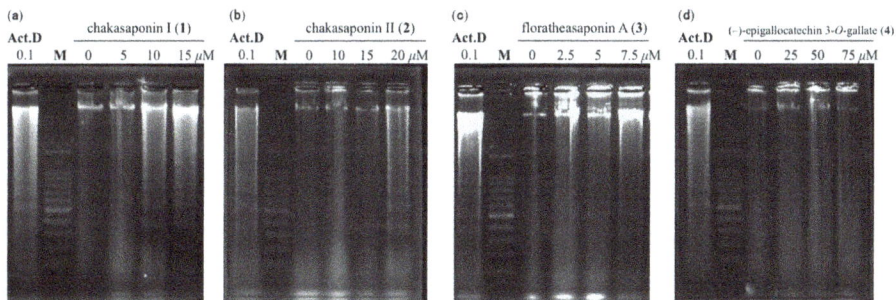

Figure 5. DNA fragmentation in HSC-2 cells treated with **1–4** and actinomycin D. Representative DNA fragmentation of HSC-2 treated with (**a**) 5, 10, and 15 μM of **1**; (**b**) 10, 15, and 20 μM of **2**; (**c**) 2.5, 5, and 7.5 μM of **3**; (**d**) 25, 50, and 75 μM of **4**, and (Act.D) 0.1 μM of actinomycin D for 48 h; (M) represents a marker (100 bp DNA ladder); commercial actinomycin D was purchased from Wako Pure Chemical Industries, Ltd. (Osaka, Japan).

2.6. Effects of Caspase-3/7 in HSC-2 Cells

Caspase play a central role in the apoptotic signaling. During apoptosis, caspase activity contributes to the degradation of DNA and leads to further disruption of cellular components, resulting in alterations of cell morphology [22,27–29]. To confirm that caspases are involved the enzyme activity of caspase-3/7 in HSC-2 cells after coculture with compounds **1–4**, the activities of caspase-3/7 were measured by Muse® Cell Analyzer (Figure 6). As shown in Table 4, the activation of caspase-3/7 by compounds **1–3** was found to occur in a concentration-dependent manner. These results suggest that the antiproliferative effects of acylated saponin constituents isolated from "tea flower" (**1–3**) against HSC-2 involve apoptotic cell death via activation of caspase-3/7. The efficacies of these saponins were found to be stronger than that of compound **4**, the most abundant polyphenol constituent in "green tea".

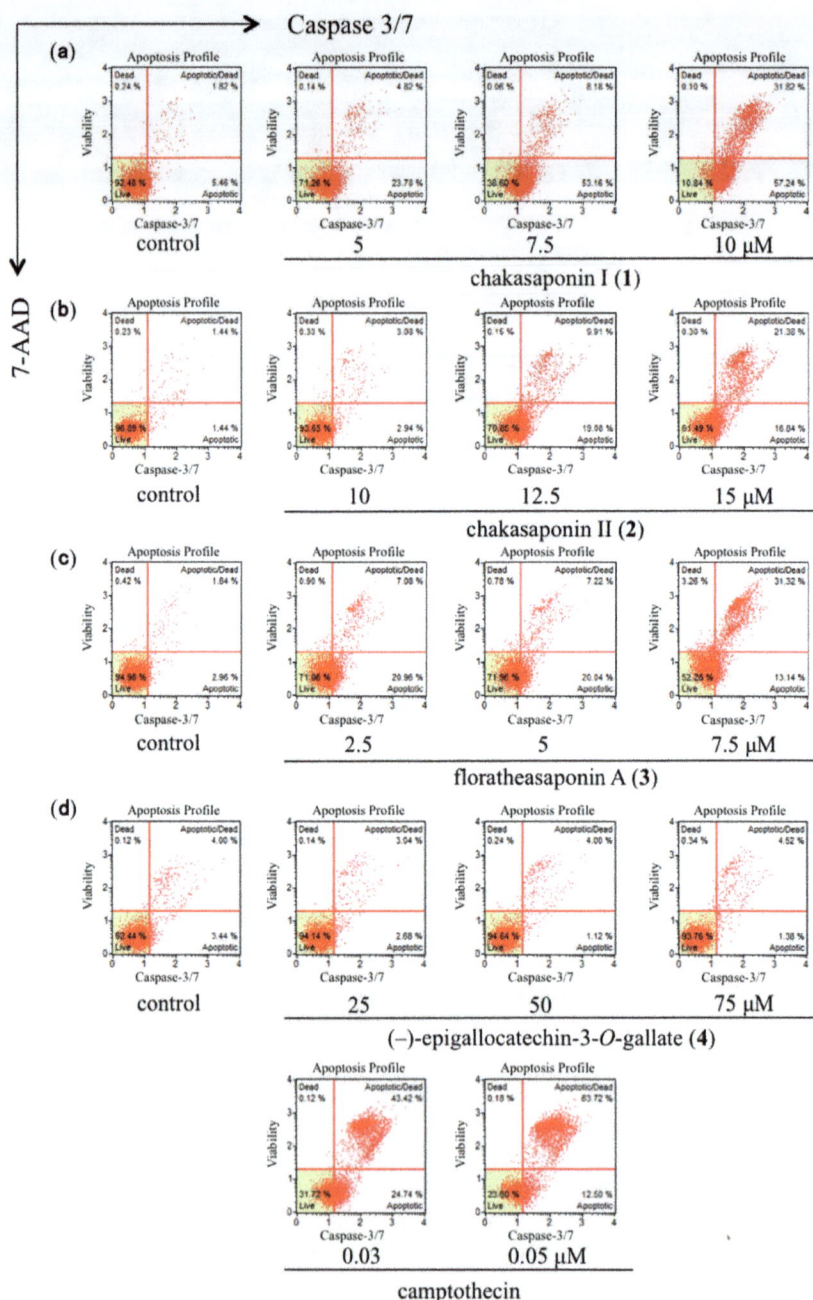

Figure 6. Effects of **1–4** and camptothecin on caspase-3/7 in HSC-2 cells. Activities of caspase-3/7 were measured by Muse® Cell Analyzer using a Muse Caspase-3/7 Kit (Merck Millipore); HSC-2 cells were treated with (**a**) 5, 7.5, and 10 μM of **1**; (**b**) 10, 12.5, and 15 μM of **2**; (**c**) 2.5, 5, and 7.5 μM of **3**; (**d**) 25, 50, and 75 μM of **4** and 0.03 and 0.05 μM of camptothecin for 24 h; the data represent the mean percentages ± SD of total apoptosis (*n* = 3); commercial camptothecin was purchased from Wako Pure Chemical Industries, Ltd. (Osaka, Japan).

Table 4. Effects of **1–4** and camptothecin on caspase-3/7 in HSC-2 cells.

Treatment	Concentration (μM)	Total Apoptotic Cells (%) [a]
Control	–	7.1 ± 0.4
Chakasaponin I (**1**)	5	25.9 ± 3.8 **
–	7.5	59.6 ± 6.3 **
–	10	90.2 ± 0.9 **
Control	–	2.8 ± 0.1
Chakasaponin II (**2**)	10	5.9 ± 0.1
–	12.5	35.6 ± 1.5 **
–	15	50.9 ± 9.7 **
Control	–	4.2 ± 0.5
Floratheasaponin A (**3**)	2.5	27.9 ± 0.6 **
–	5	27.9 ± 1.0 **
–	7.5	45.3 ± 0.6 **
Control	–	7.2 ± 0.4
(–)-Epigallocatechin 3-*O*-gallate (**4**)	25	5.9 ± 0.2
–	50	4.9 ± 0.4
–	75	5.9 ± 0.1
Camptothecin	0.03	67.2 ± 1.2 **
–	0.05	75.0 ± 2.1 **

Each value represents the mean $\pm$ S.E.M. ($n = 3$); [a] cell cycle distribution was measured by Muse® Cell Analyzer using a Muse Caspase-3/7 Kit (Merck Millipore); asterisks denote significant differences from the control group, ** $p < 0.01$; commercial camptothecin was purchased from Wako Pure Chemical Industries, Ltd. (Osaka, Japan).

3. Materials and Methods

3.1. Chemicals Constituents from "Tea Flower"

In previous studies, chakasaponins I (**1**, 0.497%), II (**2**, 0.153%), and III (0.194%), floratheasaponin A (**3**, 0.038%), (–)-epicatechin 3-*O*-gallate (0.015%), kaempferol (0.00080%), kaempferol 3-*O*-β-D-glucopyranoside (0.021%), kaempferol 3-*O*-β-D-galactopyranoside (0.011%), kaempferol 3-*O*-rutinoside (0.051%, 0.042%), kaempferol 3-*O*-β-D-glucopyranosyl-(1→3)-α-L-rhamnopyranosyl-(1→6)-β-D-glucopyranoside (0.409%), kaempferol 3-*O*-β-D-glucopyranosyl-(1→3)-α-L-rhamnopyranosyl-(1→6)-β-D-galactopyranoside (0.047%), quercetin 3-*O*-β-D-glucopyranoside (0.0063%), quercetin 3-*O*-β-D-galactopyranoside (0.0032%), and rutin (0.021%), were obtained from the methanol extract form the dried flower buds of *C. sinensis* collected in Fujian province, China (CSS-F1) [9]. The other chemicals, (–)-epigallocatechin 3-*O*-gallate (**4**), (+)-catechin, (–)-epicatechin, (–)-epigallocatechin, (–)-epicatechin 3-*O*-gallate, and quercetin were purchased from Funakoshi Co., Ltd. (Tokyo, Japan) and caffeine was from Nakalai Tesque Inc. (Kyoto, Japan) (Figure S1).

3.2. Reagents

Fetal bovine serum (FBS) was purchased from Life Technologies (Carlsbad, CA, USA); minimum essential medium (MEM) and RPMI 1640 medium were from Sigma-Aldrich Chemical (St. Louis, MO, USA); other chemicals were from Wako Pure Chemical Industries, Co., Ltd. (Osaka, Japan). The 96-well microplate was purchased from Sumitomo Bakelite Co., Ltd. (Tokyo, Japan).

3.3. Cell Viability Assays

HSC-2 (RCB1945), HSC-4 (RCB1902), MKN-45 (RCB1001), and Caco-2 cells (RCB0988) were obtained from RIKEN Bio Resource Center (Thukuba, Japan). Cells were cultured in MEM (for HSC-2 and Caco-2) or RPMI 1640 medium (for HSC-4 and MKN-45) containing FBS (10% for HSC-2, HSC-4, and MKN-45; 20% for Caco-2), 0.1 mM nonessential amino acids (for Caco-2), penicillin G (100 U/mL), and streptomycin (100 μg/mL) at 37 °C under a 5% CO_2 atmosphere. The cells were inoculated into a 96-well tissue culture plate (HSC-2 and HSC-4: 3×10^3; MKN-45: 7.5×10^3 cells/well; Caco-2: 2×10^4 cells/well in 100 μL/well). After 24 h incubation, 100 μL/well of medium containing a test sample was added. After 48 h incubation, cell viability was assessed using the 3-(4,5-dimethylthiazol-2-yl)-2,5-diphenyltetrazolium bromide (MTT) colorimetric assay. In this assay, 20 μL of MTT (5 mg/mL in phosphate-buffered saline (PBS(−)) solution was added to the medium. After 4 h incubation, the medium was removed, and 100 μL of isopropanol containing 0.04 M HCl was added to dissolve the formazan produced in the cells. The optical density (OD) of the formazan solution was measured using a microplate reader at 570 nm (reference: 655 nm). Inhibition (%) was obtained by the following formula and the IC_{50} was determined graphically.

$$\text{Inhibition (\%)} = ((\text{O.D. (sample)} - \text{O.D. (control)}) / \text{O.D. (control)}) \times 100$$

Each test compound was dissolved in dimethylsulfoxide (DMSO), and this solution was added to the medium (final DMSO concentration: 0.5%). Cisplatin, 5-fluorouracil (5-FU), doxorubicin, camptothecin, and taxol were used as reference compounds.

3.4. Cell Cycle Analysis

The cell cycle distribution analysis was measured by Muse® Cell Analyzer (Merck-Millipore, Darmstadt, Germany) using a Muse Cell Cycle Assay Kit (Merck-Millipore) according to the manufacturer's instructions. Briefly, HSC-2 cells were inoculated into a 6-well tissue culture plate (2×10^5 cells/1.5 mL/well) and cultured in MEM containing 10% FBS, penicillin G (100 U/mL), and streptomycin (100 μg/mL) at 37 °C under a 5% CO_2 atmosphere. After 24 h incubation, 500 μL/well of medium containing the test sample was added. After 48 h treatment, the cells were harvested by trypsinization, and suspended in 300 μL of PBS(−). The cells were added in 700 μL of ice-cold ethanol and incubated overnight at −20 °C. The fixed cells were collected by centrifugation at $300 \times g$ for 5 min and washed twice with 250 μL of PBS(−), then suspended in 200 μL of cell cycle reagent and incubated for 30 min in dark [30]. Each test compound was dissolved in DMSO, and the solution was added to the medium (final DMSO concentration: 0.5%). 5-FU was used as a reference compound.

3.5. Annexin-V/7-AAD Assay

Apoptosis was measured by a Muse Annexin-V and Dead Cell Kit (Merck-Millipore) according to the manufacturer's instructions. Briefly, HSC-2 cells were inoculated into a 6-well tissue culture plate (2×10^5 cells/1.5 mL/well). After 24 h incubation, 500 μL/well of medium containing the test sample was added. After 24 h treatment, the cells were harvested by trypsinization and a 100 μL cell suspension was labeled for 20 min in the dark with the same volume of Muse Annexin-V and Dead Cell Reagent. Subsequently, quantitative detection of annexin-V/7-AAD-positive cells was performed using the Muse® Cell Analyzer. Cells stained with annexin-V only were defined as early apoptotic, while annexin-V and 7-AAD double-stained cells were defined as late apoptotic [14]. Each test compound was dissolved in DMSO, and the solution was added to the medium (final DMSO concentration: 0.5%). Camptothecin was used as a reference compound.

3.6. DAPI Staining for Morphological Analysis

HSC-2 cells (2×10^5 cells/2 mL/well) were seeded onto coverslips in a 6-well tissue culture plate and cultured in MEM containing 10% FBS, penicillin G (100 U/mL), and streptomycin (100 μg/mL) at 37 °C under 5% CO_2 atmosphere. After 24 h incubation, the medium was replaced with 2 mL fresh medium per well containing the test sample; then, the cells were cultured for 48 h. Next, the medium was removed and washed twice with PBS(−) and fixed with 4% paraformaldehyde phosphate buffer solution (pH 7.4, Wako Pure Chemical Industries, Ltd., Osaka, Japan). The cells were permeabilized by 0.2% Triton X-100 in PBS(−), stained with DAPI (1 μg/mL in PBS(−)), and were observed by fluorescence microscopy (EVOS® FL Cell Imaging System, Thermo Fisher Scientific, Waltham, MA, USA) [14]. Each test compound was dissolved in DMSO, and the solution was added to the medium (final DMSO concentration: 0.5%). Camptothecin was used as a reference compound.

3.7. Agarose Gel Electrophoresis for the Detection of DNA Fragmentation

HSC-2 cells were inoculated into a 6-well tissue culture plate (2×10^5 cells/1.5 mL/well). After 24 h incubation, 500 μL/well of medium containing the test sample was added. After 48 h treatment, the cells were harvested and suspended in a solution containing 50 mM Tris-HCl (pH 8.0), 150 mM NaCl, 10 mM ethylenediaminetetraacetic acid (EDTA), and 0.5% sodium dodecyl sulfate (SDS) at room temperature for 30 min. The lysates were incubated with RNase A (100 μg/mL) for 1 h at 37 °C, then proteinase K (500 μg/mL) was added and incubated at 50 °C for 2 h. DNA was extracted twice with an equal volume of 25:24:1 (*v*/*v*/*v*) phenol:chloroform:isoamyl alcohol and was purified by ethanol precipitation. The DNA was subjected to electrophoresis on 2.0% agarose gels and stained by ethidium bromide. The DNA bands were detected under UV illumination [24]. Each test compound was dissolved in DMSO, and the solution was added to the medium (final DMSO concentration: 0.5%). Actinomycin D was used as a reference compound.

3.8. Caspase-3/7 Assay

Apoptotic status based on caspase-3/7 activation was measured by a Muse Caspase-3/7 Kit (Merck-Millipore). Briefly, HSC-2 cells were inoculated into a 6-well tissue culture plate (2×10^5 cells/1.5 mL/well). After 24 h incubation, 500 μL/well of medium containing the test sample was added. After 48 h treatment, the cells were harvested by trypsinization and stained according to the manufacturer's instructions. Subsequently, quantitative detection of caspase-3/7-positive cells was performed using the Muse® Cell Analyzer [31–33]. Each test compound was dissolved in DMSO, and the solution was added to the medium (final DMSO concentration: 0.5%). Camptothecin was used as a reference compound.

3.9. Statistics

Values are expressed as means ± S.E.M. Significant differences were calculated using Dunnett's test. Probability (*p*) values less than 0.05 were considered significant.

4. Conclusions

Acylated oleanane-type triterpene saponins obtained from the flower buds of *C. sinensis* (tea flower), namely chakasaponins I (**1**) and II (**2**) and floratheasaponin A (**3**), and the most abundant polyphenol constituent in "green tea", (−)-epigallocatechin 3-*O*-gallate (**4**), showed antiproliferative activities against human digestive tract carcinoma HSC-2, HSC-4, MKN-45, and Caco-2 cells. These activities of the saponins (**1–3**, IC_{50} = 4.4–14.1, 6.2–18.2, 4.5–17.3, and 19.3–40.6 μM, respectively) were more potent than those of **4** (IC_{50} = 28.3, 27.2, >100, and >100 μM, respectively). In our previous study, we have demonstrated that a similar acylated oleanane-type triterpene saponin obtained from the flowers of *Bellis perennis* (Asteraceae), perennisaponin O, showed a relatively strong activity against HSC-2, HSC-4, and MKN-45 cells (IC_{50} = 11.2, 14.3, and 6.9 μM, respectively). Furthermore, the mechanism of action of perennisaponin O against HSC-2 was found to involve apoptotic cell death [14]. To clarify the mechanisms of action of compounds **1–3** against HSC-2 cells, effects of cell cycle distribution were analyzed at each effective concentration. The present study demonstrated a significant inhibition of cell proliferation by compounds **1–3** with cell cycle arrest occurring at the G2/M phase in a concentration-dependent manner. These saponins (compounds **1–3**) efficiently induce apoptosis in HSC-2 cells, as demonstrated by annexin V/7-AAD assay, morphological changes, and DNA fragmentation. Furthermore, the apoptotic cell death triggered by compounds **1–3** in HCS-2 cells was dependent on the activation of caspase-3/7. The efficacy of these saponins (**1–3**) as apoptosis inducers was found to be higher than that of compound **4**. Recently, compound **4** has received great attention in cancer research related to cancer preventive effects [34–36], the synergistic enhancement by the combination with different anticancer drugs [37,38], and clinical applications [39–41]. It is well recognized that compound **4** is able to bind to multiple molecular targets, thus, it can affect a range of signaling pathways, resulting in growth inhibition, apoptosis or suppressions of invasion, angiogenesis, and metastasis [42]. Ability of compound **4** to induce cell death in cancer cells is considered a key mechanism related to its anticancer function [43]. There are conflicting reports of apoptotic and nonapoptotic cell death induced by compound **4**, and the exact molecular mechanisms have not been fully elucidated yet. However, most of the previous reports have concluded that this compound induces caspase-mediated apoptosis in various tumor cells via the mitochondrial pathway [44,45] or via the death receptor [46–48]. Conversely, several reports demonstrated the involvement of nonapoptotic cell death, such as caspase-independent necrosis-like cell death [49] or reactive oxygen species (ROS)-mediated lysosomal membrane permeabilization [50]. The detailed mechanism of action of the antiproliferative activity of compound **4** needs further studies. In conclusion, we herein described that the saponin constituents of "tea flower" compounds **1–3** showed some antiproliferative activities against HSC-2 cells by the apoptotic pathway via caspase-3/7 activation. On the basis of the abovementioned evidence, these saponins can potentially be useful for the treatment and/or prevention of the digestive tract cancer. Further investigations are recommended.

Supplementary Materials: Supplementary materials can be found at www.mdpi.com/1422-0067/17/12/1979/s1.

Acknowledgments: This work was supported by the MEXT-Supported Program for the Strategic Research Foundation at Private Universities, 2014–2018, Japan (S1411037, Toshio Morikawa), as well as JSPS KAKENHI, Japan, Grant Numbers 15K08008 (Toshio Morikawa), 15K08009 (Kiyofumi Ninomiya), and 16K08313 (Osamu Muraoka).

Author Contributions: Niichiro Kitagawa, Toshio Morikawa, Kiyofumi Ninomiya, Masayuki Yoshikawa, and Osamu Muraoka conceived and designed the experiments; Niichiro Kitagawa, Toshio Morikawa, Chiaki Motai, Kiyofumi Ninomiya, Shuhei Okugawa, and Ayaka Nishida performed the experiments; Niichiro Kitagawa, Toshio Morikawa, and Chiaki Motai analyzed the data; Niichiro Kitagawa, Toshio Maroikawa, Masayuki Yoshikawa, and Osamu Muraoka contributed the materials; Niichiro Kitagawa, Toshio Maroikawa, and Osamu Muraoka wrote the paper.

Conflicts of Interest: The authors declare no conflict of interest.

References

1. Sparg, S.G.; Light, M.E.; van Staden, J. Biological activities and distribution of plant saponins. *J. Ethnopharmacol.* **2004**, *94*, 219–243. [CrossRef] [PubMed]
2. Vincken, J.-P.; Heng, L.; de Groot, A.; Gruppen, H. Saponins, classification and occurrence in the plant kingdom. *Phytochemistry* **2007**, *68*, 275–297. [CrossRef] [PubMed]
3. Podolak, I.; Galanty, A.; Sobolewska, D. Saponins as cytotoxic agents: A review. *Phytochem. Rev.* **2010**, *9*, 425–474. [CrossRef] [PubMed]
4. Dinda, B.; Debnath, S.; Mohanta, B.C.; Harigaya, Y. Naturally occurring triterpenoid saponins. *Chem. Biodivers.* **2010**, *7*, 2327–2580. [CrossRef] [PubMed]
5. Osbourn, A.; Goss, R.J.M.; Field, R.A. The saponins—Polar isoprenoids with important and diverse biological activities. *Nat. Prod. Rep.* **2011**, *28*, 1261–1268. [CrossRef] [PubMed]
6. Faizal, A.; Geelen, D. Saponins and their role in biological processes in plants. *Phytochem. Rev.* **2013**, *12*, 877–893. [CrossRef]
7. Yoshikawa, M.; Morikawa, T.; Yamamoto, K.; Kato, Y.; Nagatomo, A.; Matsuda, H. Floratheasaponins A–C, acylated oleanane-type triterpene oligoglycosides with anti-hyperlipidemic activities from flowers of the tea plant (*Camellia sinensis*). *J. Nat. Prod.* **2005**, *68*, 1360–1365. [CrossRef] [PubMed]
8. Morikawa, T.; Miyake, S.; Miki, Y.; Ninomiya, K.; Yoshikawa, M.; Muraoka, O. Quantitative analysis of acylated oleanane-type triterpene saponins, chakasaponins I–III and floratheasaponins A–F, in the flower buds of *Camellia sinensis* from different regional origins. *J. Nat. Med.* **2012**, *66*, 608–613. [CrossRef] [PubMed]
9. Morikawa, T.; Ninomiya, K.; Miyake, S.; Miki, Y.; Okamoto, M.; Yoshikawa, M.; Muraoka, O. Flavonol glycosides with lipid accumulation inhibitory activity and simultaneous quantitative analysis of 15 polyphenols and caffeine in the flower buds of *Camellia sinensis* from different regions by LCMS. *Food Chem.* **2013**, *140*, 353–360. [CrossRef] [PubMed]
10. Morikawa, T.; Lee, I.J.; Okugawa, S.; Miyake, S.; Miki, Y.; Ninomiya, K.; Kitagawa, N.; Yoshikawa, M.; Muraoka, O. Quantitative analysis of catechin, flavonoid, and saponin constituents in "tea flower", the flower buds of *Camellia sinensis*, from different regions in Taiwan. *Nat. Prod. Commun.* **2013**, *8*, 1553–1557. [PubMed]

11. Matsuda, H.; Nakamura, S.; Morikawa, T.; Muraoka, O.; Yoshikawa, M. New biofunctional effects of the flower buds of *Camellia sinensis* and its bioactive acylated oleanane-type triterpene oligoglycosides. *J. Nat. Med.* **2016**, *70*, 689–701. [CrossRef] [PubMed]

12. Morikawa, T.; Nakamura, S.; Kato, Y.; Muraoka, O.; Matsuda, H.; Yoshikawa, M. Bioactive saponins and glycosides. XXVIII. New triterpene saponins, foliatheasaponins I, II, III, IV, and V, from tencha (the leaves of *Camellia sinensis*). *Chem. Pharm. Bull.* **2007**, *55*, 293–298. [CrossRef] [PubMed]

13. Zhang, X.; Zhang, Y.; Gu, D.; Cao, C.; Zhang, Q.; Xu, Z.; Gong, Y.; Chen, J.; Tang, C. Increased risk of developing digestive tract cancer in subjects carrying the *PLCE1* r32274223 A > G polymorphism: Evidence from a meta-analysis. *PLoS ONE* **2013**, *8*, e76425.

14. Ninomiya, K.; Motai, C.; Nishida, E.; Kitagawa, N.; Yoshihara, K.; Hayakawa, T.; Muraoka, O.; Li, X.; Nakamura, S.; Yoshikawa, M.; et al. Acylated oleanane-type triterpene saponins from the flowers of *Bellis perennis* show anti-proliferative activities against human digestive tract carcinoma cell lines. *J. Nat. Med.* **2016**, *70*, 435–451. [CrossRef] [PubMed]

15. Weisburg, J.H.; Weissman, D.B.; Sedaghat, T.; Babich, H. In vitro cytotoxicity of epigallocatechin gallate and tea extracts to cancerous and normal cells from the human oral cavity. *Basic Clin. Pharmacol. Toxicol.* **2004**, *95*, 191–200. [CrossRef] [PubMed]

16. Babich, H.; Krupka, M.E.; Nissim, H.A.; Zuckerbraun, H.L. Differential in vitro cytotoxicity of (–)-epicatechin gallate (ECG) to cancer and normal cells from the human oral cavity. *Toxicol. In Vitro* **2005**, *19*, 231–242. [CrossRef] [PubMed]

17. Ran, Z.-H.; Xu, Q.; Tong, J.-L.; Xiao, S.-D. Apoptotic effect of epigallocatechin-3-gallate on the human gastric cancer cell line MKN45 via activation of the mitochondrial pathway. *World J. Gastroenterol.* **2007**, *13*, 4255–4259. [CrossRef] [PubMed]

18. Delgado, L.; Fernandes, I.; González-Manzano, S.; de Freitas, V.; Mateus, N.; Santos-Buelga, C. Anti-proliferative effects of quercetin and catechin metabolites. *Food Funct.* **2014**, *5*, 797–803. [CrossRef] [PubMed]

19. Elmore, S. Apoptosis: A review of programmed cell death. *Toxicol. Pathol.* **2007**, *35*, 495–516. [CrossRef] [PubMed]

20. V G M, N.; Atmakur, H.; Katragadda, S.B.; Devabakthuni, B.; Kota, A.; S, C.K.; Kuncha, M.; M V P S, V.V.; Kulkarni, P.; Janaswamy, M.R.; et al. Antioxidant, hepatoprotective and cytotoxic effects of icetexanes isolated from stem-bark of *Premna tomentosa*. *Phytomedicine* **2014**, *21*, 497–505. [CrossRef] [PubMed]

21. Moskot, M.; Gabig-Ciminska, M.; Jakóbkiewicz-Banecka, J.; Wesierska, M.; Bochenska, K.; Wegrzyn, G. Cell cycle is disturbed in mucopolysaccharidosis type II fibroblasts, and can be improved by genistein. *Gene* **2016**, *585*, 100–103. [CrossRef] [PubMed]

22. Matsuda, H.; Yoshida, K.; Miyagawa, K.; Nemoto, Y.; Asao, Y.; Yoshikawa, M. Nuphar alkaloids with immediately apoptosis-inducing activity from *Nuphar pumilum* and their structural requirements for the activity. *Bioorg. Med. Chem.* **2006**, *16*, 1567–1573. [CrossRef] [PubMed]

23. Kloesch, B.; Becker, T.; Dietersdorfer, E.; Kiener, H.; Steiner, G. Anti-inflammatory and apoptotic effects of the polyphenol curcumin on human fibroblast-like synoviocytes. *Int. Immunopharmacol.* **2013**, *15*, 400–405. [CrossRef] [PubMed]

24. Zhao, Q.; Huo, X.-C.; Sun, F.-D.; Dong, R.-Q. Polyphenol-rich extract of *Salvia chinensis* exhibits anticancer activity in different cancer cell lines, and induces cell cycle arrest at the G_0/G_1-phase, apoptosis and loss of mitochondrial membrane potential in pancreatic cancer cells. *Mol. Med. Rep.* **2015**, *12*, 4843–4850. [PubMed]

25. Adrian, C.; Martin, S.J. The mitochondrial apoptosome: A killer unleashed by the cytochrome seas. *Trends Biochem. Sci.* **2001**, *26*, 390–397. [CrossRef]

26. Kim, R.; Emi, M.; Tanabe, K. Caspase-dependent and independent cell death pathways after DNA damage. *Oncol. Rep.* **2005**, *14*, 595–599. [CrossRef] [PubMed]

27. Matsuda, H.; Akaki, J.; Nakamura, S.; Okazaki, Y.; Kojima, H.; Tamesada, M.; Yoshikawa, M. Apoptosis-inducing effects of sterols from the dried powder of cultured mycelium of *Cordyceps sinensis*. *Chem. Pharm. Bull.* **2009**, *57*, 411–414. [CrossRef] [PubMed]

28. Miyamoto, D.; Endo, N.; Oku, N.; Arima, Y.; Suzuki, T.; Suzuki, Y. β-Thujaplicin zinc chelate induces apoptosis in mouse high metastatic melanoma B16BL6 cells. *Biol. Pharm. Bull.* **1998**, *21*, 1258–1262. [CrossRef] [PubMed]

29. Wang, J.; Yuan, L.; Xiao, H.; Xiao, C.; Wang, Y.; Liu, X. Momordin Ic induces HepG2 cell apoptosis through MAPK and PI3K/Akt-mediated mitochondrial pathways. *Apoptosis* **2013**, *18*, 751–765. [CrossRef] [PubMed]
30. Al Dhaheri, Y.; Attoub, S.; Ramadan, G.; Arafat, K.; Bajbouj, K.; Karuvantevida, N.; AbuQamar, S.; Eid, A.; Iratni, R. Carnosol induces ROS-mediated beclin 1-independent autophagy and apoptosis in triple negative breast cancer. *PLoS ONE* **2014**, *9*, e109630. [CrossRef] [PubMed]
31. D'Alessandro, R.; Refolo, M.G.; Lippolis, C.; Giannuzzi, G.; Carella, N.; Messa, C.; Cavallini, A.; Carr, B.I. Antagonism of Sorafenib and Regorafenib actions by platelet factors in hepatocellular carcinoma cell lines. *BMC Cancer* **2014**, *14*, 351–360. [CrossRef] [PubMed]
32. Carr, B.I.; Cavallini, A.; D'Alessandro, R.; Refolo, M.G.; Lippolis, C.; Mazzocca, A.; Messa, C. Platelet extracts induce growth, migration and invasion in human hepatocellular carcinoma in vitro. *BMC Cancer* **2014**, *14*, 43–52. [CrossRef] [PubMed]
33. D'Alassandro, R.; Refolo, M.G.; Lippolis, C.; Carella, N.; Messa, C.; Cavallini, A.; Carr, B.I. Modulation of regorafenib effects on HCC cell lines by epidermal growth factor. *Cancer Chemother. Pharmacol.* **2015**, *75*, 1237–1245. [CrossRef] [PubMed]
34. Fujiki, H.; Suganuma, M. Green tea and cancer prevention. *Proc. Jpn. Acad.* **2002**, *78*, 263–270. [CrossRef]
35. Fujiki, H.; Suganuma, M.; Okabe, S.; Sueoka, E.; Sueoka, N.; Fujimoto, N.; Goto, Y.; Matsuyama, S.; Imai, K.; Nakachi, K. Cancer prevention with green tea and monitoring by a new biomarker, hnRNP B1. *Mutat. Res.* **2001**, *480–481*, 229–304. [CrossRef]
36. Fujiki, H.; Sueoka, E.; Watanabe, T.; Suganuma, M. Primary cancer prevention by green tea, and tertiary cancer prevention by the combination of green tea catechins and anticancer compounds. *J. Cancer Prev.* **2015**, *20*, 1–4. [CrossRef] [PubMed]
37. Suganuma, M.; Saha, A.; Fujiki, H. New cancer treatment strategy using combination of green tea catechins and anticancer drugs. *Cancer Sci.* **2011**, *102*, 317–323. [CrossRef] [PubMed]
38. Fujiki, H.; Sueoka, E.; Watanabe, T.; Suganuma, M. Synergistic enhancement of anticancer effects on numerous human cancer cell lines treated with the combination of EGCG, other green tea catechins, and anticancer compounds. *J. Cancer Res. Clin. Oncol.* **2015**, *141*, 1511–1522. [CrossRef] [PubMed]
39. Fujiki, H.; Suganuma, M.; Okabe, S.; Sueoka, N.; Komori, A.; Sueoka, E.; Kozu, T.; Tada, Y.; Suga, K.; Imai, K.; et al. Cancer inhibition by green tea. *Mutat. Res.* **1998**, *402*, 307–310. [CrossRef]
40. Singh, B.N.; Shankar, S.; Srivastava, R.K. Green tea catechin, epigallocatechin-3-gallate (EGCG): Mechanisms, perspectives and clinical applications. *Biochem. Pharmacol.* **2011**, *82*, 1807–1821. [CrossRef] [PubMed]
41. Lecumberri, E.; Dupertuis, Y.M.; Miralbell, R.; Pichard, C. Green tea polyphenol epigallocatechin-3-gallate (EGCG) as adjuvant in cancer therapy. *Clin. Nutr.* **2013**, *32*, 894–903. [CrossRef] [PubMed]
42. Khan, N.; Mukhtar, H. Multitargeted therapy of cancer by green tea polyphenols. *Cancer Lett.* **2008**, *269*, 269–280. [CrossRef] [PubMed]
43. Ahmed, N.; Feyes, D.K.; Nieminen, A.L.; Agarwal, R.; Mukhtar, H. Green tea constituent epigallocatechin-3-gallate and induction of apoptosis and cell cycle arrest in human carcinoma cells. *J. Natl. Cancer Inst.* **1997**, *89*, 1881–1886. [CrossRef]
44. Lee, J.-H.; Jeong, Y.-J.; Lee, S.-W.; Kim, D.; Oh, S.-J.; Lim, H.-S.; Oh, H.-K.; Kim, S.-H.; Kim, W.-J.; Jung, J.-Y. EGCG induces apoptosis in human laryngeal epidermoid carcinoma Hep2 cells via mitochondria with the release of apoptosis-inducing factor and endonuclease G. *Cancer Lett.* **2010**, *290*, 68–75. [CrossRef] [PubMed]
45. Qanungo, S.; Das, M.; Haldar, S.; Basu, A. Epigallocatechin-3-gallate induces mitochondrial membrane depolarization and caspase-dependent apoptosis in pancreatic cancer cells. *Carcinogenesis* **2005**, *26*, 958–967. [CrossRef] [PubMed]
46. Hayakawa, S.; Saeki, K.; Sazuka, M.; Suzuki, Y.; Shoji, Y.; Nakamura, Y.; Ohta, T.; Kaji, K.; Yuo, A.; Isemura, M. Apoptosis induction by epigallocatechin gallate involves its binding to Fas. *Biochem. Biophys. Res. Commun.* **2001**, *285*, 1102–1106. [CrossRef] [PubMed]
47. Lim, Y.C.; Cha, Y.Y. Epigallocatechin-3-gallate induces growth inhibition and apoptosis of human anaplastic thyroid carcinoma cells through suppression of EGFR/ERK pathway and cyclin B1/CDK1 complex. *J. Surg. Oncol.* **2011**, *104*, 776–780. [CrossRef] [PubMed]
48. Onoda, C.; Kuribayashi, K.; Nirasawa, S.; Tsuji, N.; Tanaka, M.; Kobayashi, D.; Watanabe, N. (−)-Epigallocatechin-3-gallate induces apoptosis in gastric cancer cell lines by down-regulating survivin expression. *Int. J. Oncol.* **2011**, *38*, 1403–1408. [PubMed]

49. Iwasaki, R.; Ito, K.; Ishida, T.; Hamanoue, M.; Adachi, S.; Watanabe, T.; Sato, Y. Catechin, green tea component, causes caspase-independent necrosis-like cell death in chronic myelogenous leukemia. *Cancer Sci.* **2009**, *100*, 349–356. [CrossRef] [PubMed]

50. Zhang, Y.; Yang, N.-D.; Zhou, F.; Shen, T.; Duan, T.; Zhou, J.; Shi, Y.; Zhu, X.-Q.; Shen, H.-M. (–)-Epigallocatechin-3-gallate induces non-apoptotic cell death in human cancer cells via ROS-mediated lysosomal membrane permeabilization. *PLoS ONE* **2012**, *7*, e46749. [CrossRef] [PubMed]

International Journal of
Molecular Sciences

MDPI

Article

Hepatoprotective Effect of *Cuscuta campestris* Yunck. Whole Plant on Carbon Tetrachloride Induced Chronic Liver Injury in Mice

Wen-Huang Peng [1,†], Yi-Wen Chen [1,†], Meng-Shiou Lee [1], Wen-Te Chang [1], Jen-Chieh Tsai [2], Ying-Chih Lin [3] and Ming-Kuem Lin [1,4,*]

[1] Department of Chinese Pharmaceutical Sciences and Chinese Medicine Resources, College of Biopharmaceutical and Food Sciences, China Medical University, 91 Hsueh-Shih Road, Taichung 40402, Taiwan; whpeng@mail.cmu.edu.tw (W.-H.P.); kathy@hsbf.com.tw (Y.-W.C.); leemengshiou@mail.cmu.edu.tw (M.-S.L.); wtchang@mail.cmu.edu.tw (W.-T.C.)
[2] Department of Health and Nutrition Biotechnology, College of Health Science, Asia University, 500 Liufeng Rd., Wufeng, Taichung 41354, Taiwan; jenchieh@mail.dyu.edu.tw
[3] Department of Optometry, Jen-Teh Junior College of Medicine, Nursing and Management, 79-9 Sha-Luen Hu Xi-Zhou Li Hou-Loung Town, Miaoli 356, Taiwan; u462013@yahoo.com.tw
[4] Graduate Institute of Biotechnology, National Chung Hsing University, 145 Xingda Rd., South Dist., Taichung 402, Taiwan
*   Correspondence: linmk@mail.cmu.edu.tw; Tel.: +886-4-2205-3366 (ext. 5212); Fax: +886-4-2207-8083
†   These authors contributed equally to this work.

Academic Editor: Toshio Morikawa
Received: 11 October 2016; Accepted: 1 December 2016; Published: 7 December 2016

Abstract: Cuscuta seeds and whole plant have been used to nourish the liver and kidney. This study was aimed to investigate the hepatoprotective activity of the ethanol extract of *Cuscuta campestris* Yunck. whole plant (CC_{EtOH}). The hepatoprotective effect of CC_{EtOH} (20, 100 and 500 mg/kg) was evaluated on carbon tetrachloride (CCl_4)-induced chronic liver injury. Serum alanine aminotransferase, aspartate aminotransferase, triglyceride and cholesterol were measured and the fibrosis was histologically examined. CC_{EtOH} exhibited a significant inhibition of the increase of serum alanine aminotransferase, aspartate aminotransferase, triglyceride and cholesterol. Histological analyses showed that fibrosis of liver induced by CCl_4 were significantly reduced by CC_{EtOH}. In addition, 20, 100 and 500 mg/kg of the extract decreased the level of malondialdehyde (MDA) and enhanced the activities of anti-oxidative enzymes including superoxide dismutase (SOD), glutathione peroxidase (GPx) and glutathione reductase (GRd) in the liver. We demonstrate that the hepatoprotective mechanisms of CC_{EtOH} were likely to be associated to the decrease in MDA level by increasing the activities of antioxidant enzymes such as SOD, GPx and GRd. In addition, our findings provide evidence that *C. campestris* Yunck. whole plant possesses a hepatoprotective activity to ameliorate chronic liver injury.

Keywords: *Cuscuta campestris*; hepatoprotective effect; carbon tetrachloride; fibrosis; antioxidant effect

1. Introduction

Cuscuta seed, Cuscutae Semen or Tu-Si-Zi, which has been widely used to nourish the liver and kidney, mainly refer to the seeds of *Cuscuta chinensis* Lam. Some phytochemical and pharmacological studies have reported the beneficial activities of Cuscuta seeds [1]. For example, Cuscuta seeds have activities to improve defective kidneys [2], prevent liver against damage [3] and alleviate inflammation/pain [4]. Crude polysaccharides from Cuscuta seeds have an immunostimulating activity [5]. Alongside with the use of the seeds, the Cuscuta whole plant has also been recorded in the famous book "Shen Nong's Herbal" and some ancient medicinal books to treat dermatosis. In addition, it has been used as a folk medicine to treat adiposity or as a substitute for the Cuscuta seeds. However, no pharmaceutical study has been reported yet. In this study, *C. campestris* Yunck. whole plant was locally collected and used for the first time to examine its pharmaceutical activity.

Liver tissue injury can be caused by the ingestion of chemicals or drugs or by infection through virus infiltration [6]. Among them, carbon tetrachloride (CCl_4) is commonly used to study hepatotoxicity in animal models [7,8]. CCl_4 can be metabolized into the highly reactive trichloromethyl radical [9] and then trigger lipid peroxidation [10]. Therefore, blocking the lipid peroxidation can protect liver against CCl_4-induced injury [11,12]. In this study, the hepatoprotective activity of the ethanol extract of *C. campestris* whole plant (CC_{EtOH}) was investigated on CCl_4-induced chronic liver injury in mice. Once liver damage has occurred, liver marker enzymes (alanine aminotransferase (ALT), and aspartate aminotransferase (AST)) and lipid profile (total triglyceride and cholesterol) will be increased [13]. Therefore, the levels of serum ALT, AST, cholesterol and triglyceride were measured in this study. In addition, liver biopsies were performed for examining the pathological changes. To elucidate the underlying mechanism of the hepatoprotective activity, the levels of malondialdehyde (MDA) and the activities of anti-oxidative enzymes (superoxide dismutase (SOD), glutathione peroxidase (GPx), and glutathione reductase (GRd)) in liver were also measured. Silymarin was examined as the positive control as it is a promising agent for liver protection. Our findings in this study provide evidence that *C. campestris* Yunck. whole plant possesses a hepatoprotective activity and the underlying mechanism is likely to be associated to the increase of the anti-oxidation by increasing the activities of antioxidant enzymes such as SOD, GPx and GRd.

2. Results and Discussion

2.1. Effect of CC_{EtOH} on CCl_4-Induced Hepatotoxicity

First of all, the hepatotoxicity effect of CCl_4 and the protective effect of CC_{EtOH} were examined using the serum of CCl_4-induced mice. As shown in Figure 1, the CCl_4 group exhibited significant increases of serum ALT, AST, triglyceride and cholesterol. However, these increases were obviously inhibited by treatment with CC_{EtOH} (20, 100 and 500 mg/kg) and silymarin (200 mg/kg). In addition, the inhibitions were in a dose-dependent manner. These results clearly suggested that CC_{EtOH} possess protective properties against CCl_4-induced liver injury.

2.2. Histological Analyses

The results of hematoxylin and eosin histological analyses showed that CCl_4 induced histological changes including increased hepatic cells cloudy swelling, cytoplasmic vacuolization, lymphocytes infiltration, hepatocellular and necrosis (Figure 2C,D) when compared to the control group (Figure 2A,B). The liver damages were reduced by treatment with CC_{EtOH} (20, 100 and 500 mg/kg) (Figure 2G–L). Since CCl_4 induced fibrosis, Sirius Red staining was conducted and the score of liver fibrosis were examined. The results showed that the levels of inflammation and fibrosis are significantly decreased by treatment with CC_{EtOH} (100 and 500 mg/kg) (Figures 2 and 3 and Table 1). Histological examinations showed that treatment with CC_{EtOH} significantly prevents CCl_4-induced liver injury.

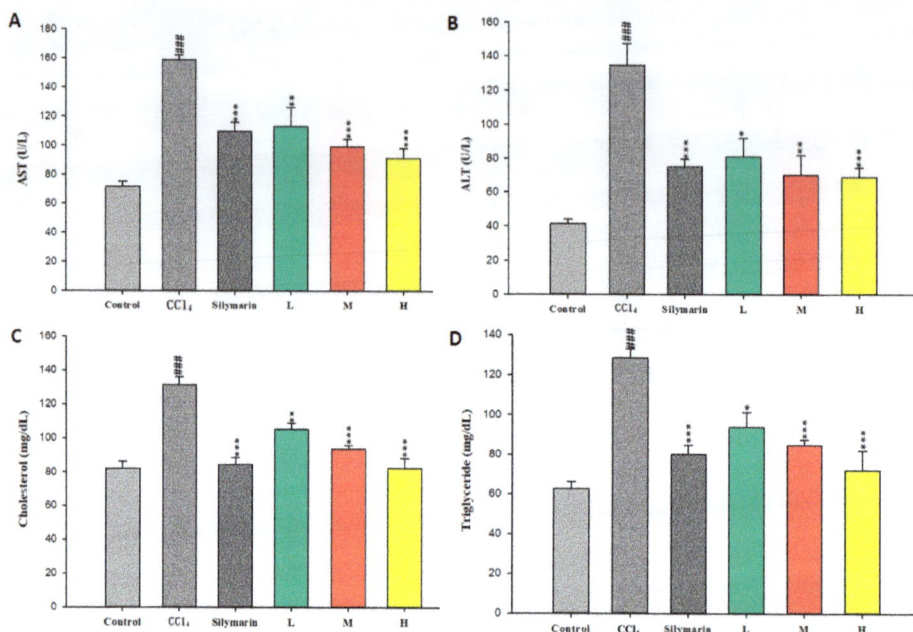

Figure 1. The effects of silymarin and the ethanol extracts of *Cuscuta campestris* whole plant at low (L), middle (M) and high (H) concentrations (20, 100 and 500 mg/kg, respectively) on serum: aspartate aminotransferase (AST) (**A**); and alanine aminotransferase (ALT) (**B**) activities; and cholesterol (**C**); and triglyceride (**D**) levels in mice treated with CCl$_4$. Values are mean ± SEM (n = 10). # indicates significant difference from the control group (### $p < 0.001$). * indicates significant difference from the CCl$_4$ group (* $p < 0.05$, ** $p < 0.01$ and *** $p < 0.001$).

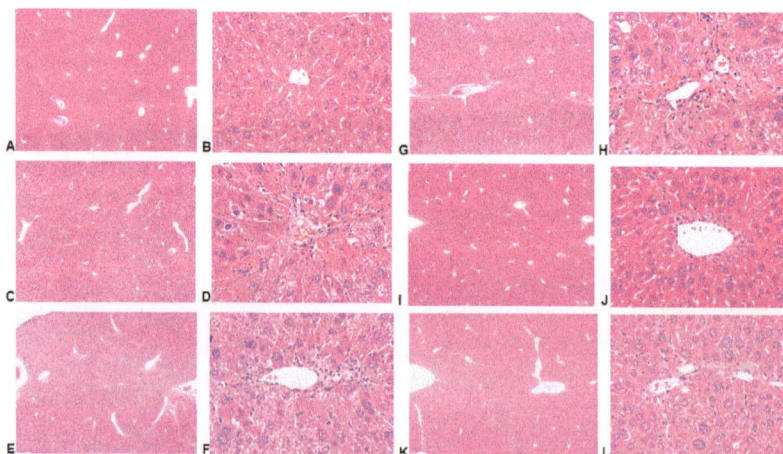

Figure 2. Hepatic histological analyses of the effects of silymarin and the ethanol extracts of *Cuscuta campestris* whole plant (CC$_{EtOH}$) on CCl$_4$-induced liver damage in mice using H&E staining (40× (**A,C,E,G,I,K**) and 200× (**B,D,F,H,J,L**) magnification): (**A,B**) control group; (**C,D**) animals treated with CCl$_4$; (**E,F**) animals treated with silymarin (200 mg/kg) and CCl$_4$; and (**G–L**) animals treated with CC$_{EtOH}$ (20, 100 and 500 mg/kg) and CCl$_4$, respectively.

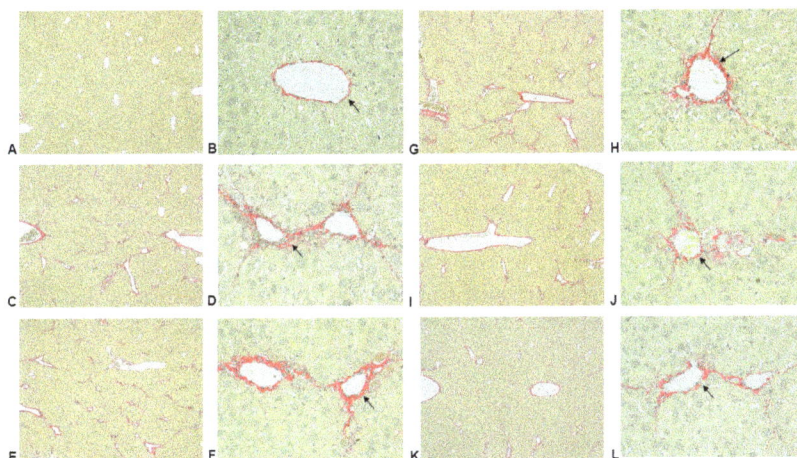

Figure 3. Hepatic histological analyses of the effects of silymarin and the ethanol extracts of *Cuscuta campestris* whole plant (CC$_{EtOH}$) on CCl$_4$-induced liver damage in mice using Sirius Red staining ($40\times$ (**A,C,E,G,I,K**) and $200\times$ (**B,D,F,H,J**) magnification): (**A,B**) control group; (**C,D**) animals treated with CCl$_4$; (**E,F**) animals treated with silymarin (200 mg/kg) and CCl$_4$; and (**G–L**) animals treated with CC$_{EtOH}$ (20, 100 and 500 mg/kg) and CCl$_4$, respectively. Arrows indicate the fibrosis.

Table 1. Quantitative the protective effects of silymarin and CC$_{EtOH}$ on CCl$_4$-induced hepatic fibrosis based on histological analyses using Sirius Red staining [1].

Group	Histopathologic Score of Liver Fibrosis	
	Observation [2]	Image (%) [3]
Normal	0	0.7 ± 0.3
CCl$_4$	1.6 ± 0.5	2.9 ± 0.8
Silymarin/CCl$_4$	1.4 ± 0.5	1.5 ± 0.5 *
CC$_{EtOH}$ 20 mg/kg/CCl$_4$	1.4 ± 0.5	2.1 ± 0.9
CC$_{EtOH}$ 100 mg/kg/CCl$_4$	1.2 ± 0.4	2.3 ± 0.7
CC$_{EtOH}$ 500 mg/kg/CCl$_4$	1.1 ± 0.5 *	1.9 ± 0.6 *

[1] Hepatic fibrosis was scored 0–4 according to the method of Ruwart et al. [14] as mentioned in the Materials and Methods; [2] The scores were obtained by the following calculation: the sum of the number per grade of affected mice/the total number of examined mice (n = 9–10); [3] The final Sirius Red positive area (%) was calculated by Image-Plus and was divided the sum of the number per SR positive area (%) of affected mice by the total number of examined mice (n = 9–10). * Statistically significant difference between CCl$_4$ group and drug-treated groups at $p < 0.05$.

2.3. *Effect of CC$_{EtOH}$ on MDA Level*

As MDA level is usually used to elucidate the level of lipid peroxidation in liver, the effects of CC$_{EtOH}$ on CCl$_4$-induced MDA production were examined. As shown in Figure 4, the level of MDA in the CCl$_4$ group was dramatically increased ($p < 0.001$) compared with the control group, however, the levels of MDA were significantly reduced by treatment with CC$_{EtOH}$ (20, 100 and 500 mg/kg) ($p < 0.001$) and silymarin (200 mg/kg) ($p < 0.001$) compared with the CCl$_4$ group. The results suggested that the CCl$_4$-induced hepatic lipid peroxidation is reduced by CC$_{EtOH}$.

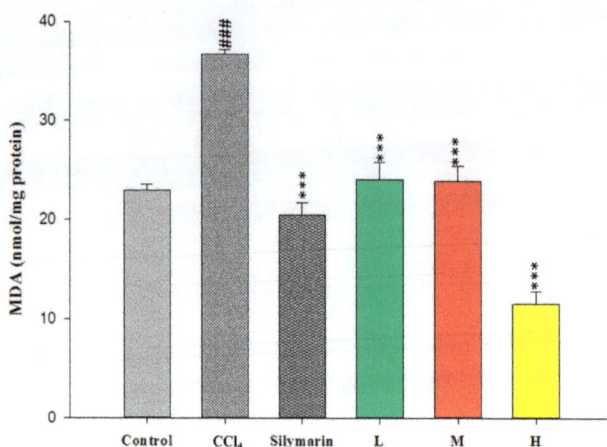

Figure 4. The effects of silymarin and the ethanol extracts of *Cuscuta campestris* whole plant on malondialdehyde (MDA) content in mice treated with CCl$_4$. Values are mean ± SEM (*n* = 10). $^{#}$ indicates significant difference from the control group ($^{###}$ *p* < 0.001). * indicates significant difference from the CCl$_4$ group (*** *p* < 0.001).

2.4. Effect of CC$_{EtOH}$ on Antioxidant Enzymatic Activities

To evaluate the antioxidant effects of CC$_{EtOH}$, SOD, GPx and GRd were measured in the liver. The activities of these hepatic enzymes in the CCl$_4$ group were dramatically decreased compared with the control group (Figures 5–7). However, treatment with CC$_{EtOH}$ at the three doses and silymarin significantly increased the levels of SOD, GPx and GRd activities. The results suggested that the inhibitory effect of CCl$_4$ on these hepatic enzymes was reversed by CC$_{EtOH}$.

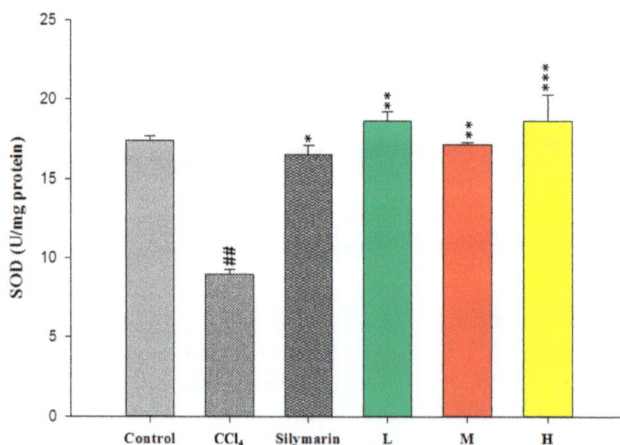

Figure 5. The effects of silymarin and the ethanol extracts of *Cuscuta campestris* whole plant on superoxide dismutase (SOD) activity in mice treated with CCl$_4$. Values are mean ± SEM (*n* = 10). $^{#}$ indicates significant difference from the control group ($^{##}$ *p* < 0.01). * indicates significant difference from the CCl$_4$ group (* *p* < 0.05, ** *p* < 0.01 and *** *p* < 0.001).

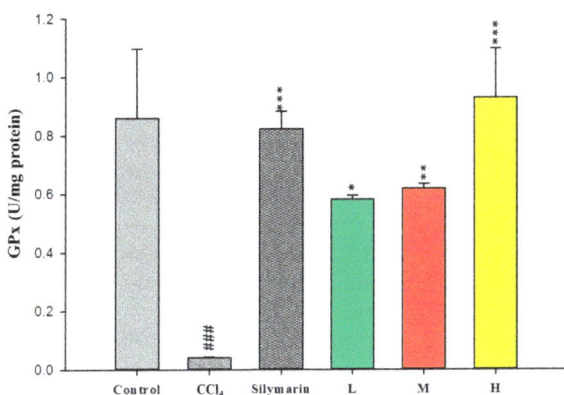

Figure 6. The effects of silymarin and the ethanol extracts of *Cuscuta campestris* whole plant on glutathione peroxidase (GPx) activity in mice treated with CCl$_4$. Values are mean ± SEM (n = 10). [#] indicates significant difference from the control group ([###] $p < 0.001$). * indicates significant difference from the CCl$_4$ group (* $p < 0.05$, ** $p < 0.01$ and *** $p < 0.001$).

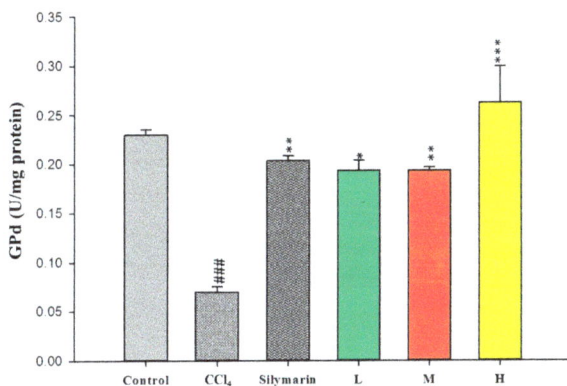

Figure 7. The effects of silymarin and the ethanol extracts of *Cuscuta campestris* whole plant on glutathione reductase (GRd) activity in mice treated with CCl$_4$. Values are mean ± SEM (n = 10). [#] indicates significant difference from the control group ([###] $p < 0.001$). * indicates significant difference from the CCl$_4$ group (* $p < 0.05$, ** $p < 0.01$ and *** $p < 0.001$).

2.5. Phytochemical Analysis of CC$_{EtOH}$

As shown in Figure S1, the HPLC chromatograms of CC$_{EtOH}$ and the standards showed that peaks at the retention times of 17.5 and 24.5 min were hyperoside and quercetin, respectively. Two peaks showing at the retention times of 17.5 and 24.5 min were detected in the CC$_{EtOH}$ chromatogram. The results revealed that hyperoside, quercetin and their glycosides are present in CC$_{EtOH}$.

Cuscuta whole plant has also been used as a folk medicinal material to treat adiposity or as a substitute for the Cuscuta seeds, which have been widely used to nourish the liver and kidney in Chinese medicine. In this study, we demonstrated in the first time that the whole plant of *C. campestris* exhibits a hepatoprotective activity.

CCl$_4$ has been commonly employed for the evaluation of hepatoprotective activity of different kinds of herbal extracts and drugs [8,15]. CCl$_4$ is thought to be transformed into trichloromethyl radicals, which are hepatotoxic metabolites. These radicals are able to react with sulfhydryl groups of glutathione (GSH) and protein. In addition, they can trigger protein oxidation and lipid peroxidation, which result in hepatocellular damage [7,12,16]. In this study, CCl_4 was used to induce chronic hepatic injury and our results revealed that the injury was significantly reduced by CC_{EtOH}, clearly demonstrating that *Cuscuta campestris* possesses hepatoprotective activity.

Previous studies have shown that hepatic damage increases AST and ALT activities in the hepatocytes [16] and the levels of ALT, AST, triglyceride and cholesterol in serum are increased by administering CCl_4 to mice [13,17,18]. In this study, serum ALT, AST, triglyceride and cholesterol levels were increased after CCl_4 administration and these increases were all significantly decreased by treatment with CC_{EtOH} at three concentrations (Figure 1). In addition, histological analyses showed that hepatic cell injury induced by CCl_4 was accompanied by fibrosis and such injury was attenuated by CC_{EtOH} (Figure 3). The quantitative histopathologic score of the fibrosis of hepatocytes showed that the fibrosis levels were significantly decreased by CC_{EtOH} (100 and 500 mg/kg; Table 1). These results indicated that CC_{EtOH} can prevent liver against fibrosis.

Lipid peroxidation has been shown to be an important cause of CCl_4-induced liver injury [10]. Malondialdehyde (MDA) is the end product of the lipid peroxidation and thus commonly used as an indicator of the CCl_4-induced liver injury [19]. In this study, the increased hepatic MDA levels induced by CCl_4 were significantly decreased by treatment of CCl_4 (Figure 4). Therefore, these results indicated that CCl_4 can protect the liver against CCl_4-induced injury through inhibiting MDA production. Superoxide dismutase (SOD), glutathione peroxidase (GPx) and glutathione reductase (GRd) are anti-oxidative enzymes which are easily inactivated by reactive oxygen species (ROS) and lipid peroxides which are caused by CCl_4 [16]. In this study, the activities of SOD, GPx and GRd from the CCl_4-induced injury livers were measured. The results showed that their activities were promoted by treatment with CC_{EtOH} (Figures 5–7), suggesting that CC_{EtOH} is able to reduce ROS production by increasing hepatic anti-oxidative enzymes activities and thus prevent the development of CCl_4-induced liver damage. To confirm the anti-oxidative activity of the extract used in this study, the catechin-equivalent phenolics and quercetin-equivalent flavonoid concentrations of the extract were examined and determined as 58.61 ± 0.8 and 15.032 ± 1.3 mg/g CC_{EtOH}, respectively. In addition, 1,1-Diphenyl-2-picrylhydrazyl (DPPH) scavenging of the extract was examined. The catechin equivalent DPPH scavenging capability was determined as 25.19 ± 0.54 mg/g CC_{EtOH} and the IC_{50} of CC_{EtOH} for DPPH scavenging is approximately 1.71 mg/mL. These results support that CC_{EtOH} containing hyperoside and quercetin has a capability to increase the anti-oxidant systems in liver. Moreover, the inhibitory effect on MDA production was also likely due to the increase in SOD, GPx and GRd activities.

Phytochemical analyses by HPLC showed that the major compounds in CC_{EtOH} are hyperoside, quercetin and flavonoid glycosides (Figure S1). Although hyperoside and quercetin have been detected in both of seeds and whole plant of *C. campestris*, their amounts and the other flavonoids are different [4,20]. Hyperoside have been shown to increase the level of heme oxygenase-1, an important enzyme in antioxidant defense systems to reduce oxidative stress [21]. Quercetin has shown a high antioxidant activity by reducing the production of reactive oxygen species and nitric oxide [20,22]. The methanol extract of the seeds of *Cuscuta chinensis* containing hyperoside, quercetin and kaempferol have been reported to increase the activities of SOD, GPx and GRd in the liver [4]. Hyperoside and quercetin have been shown to exhibit hepatoprotective effect against CCl_4-induced liver injury [23,24]. Therefore, hyperoside and quercetin can be the major active constituents in CC_{EtOH} which contribute to the hepatoprotective effect.

3. Materials and Methods

3.1. Plant Materials and Preparation of Plant Extract

Cuscuta campestris Yunck. grown on *Bidens pilosa* var. radiata was collected from Miaoli County, Taiwan. They were authenticated by Ming-Kuem Lin and Wen-Huang Peng in several aspects, including the morphology of its flowers and the chemical compositions of its seeds [20]. The whole plants of *C. campestris* Yunck. were dried in a circulating air oven, and then ground. The powder (1.05 kg) was extracted with 75% ethanol three times. The filtrates were collected and concentrated with a rotary evaporator under reduced pressure. The concentrated extract was then lyophilized and weighted. The yield ratio of CC_{EtOH} (91 g) was 8.7% (w/w). The extract was stored in $-20\,^{\circ}C$ before the experiments.

3.2. Chemicals

Silymarin, quercetin and kampferol were purchased from Sigma-Aldrich Chemical Co. (Saint Louis, MO, USA). Carboxymethylcellulose (CMC) and carbon tetrachloride (CCl_4) was purchased from Merck Co. (Munchen, Germany). CCl_4 was dissolved into olive oil as a 40% (v/v) solution. All other reagents used were of analytical grades (Merck Co., Munchen, Germany).

3.3. Experimental Animals

ICR male mice (18–22 g) were purchased from BioLASCO Taiwan Co., Ltd. (Taipei, Taiwan). These mice were maintained in standard cages with a 12-h:12-h light-dark cycle, relative humidity 55% $\pm$ 5%, and 22 $\pm$ 1 $^{\circ}C$ for seven days before the experiment. Food and water ad libitum were supplied by following the NIH Guide for the Care and Use of Laboratory Animals. The experimental protocol conducted in this study has been approved by the Institutional Animal Care and Use Committee, China Medical University (104-70-N; 18 December 2014).

3.4. Experimental Design of CCl_4-Induced Hepatotoxicity

Sixty experimental mice were randomly separated into 6 groups. For the control group and the CCl_4 group, mice were orally administered 1% carboxymethyl cellulose (CMC). For the silymarin group, mice were orally administered silymarin (200 mg/kg in 1% CMC). For the CC_{EtOH} groups, mice were orally administered CC_{EtOH} (20, 100 and 500 mg/kg in 1% CMC). Theses oral administrations were conducted using a feeding tube with 100 μL/10 g Body Weight every day for 9 consecutive weeks. After one week of the administration of the silymarin and experimental drugs, CCl_4 (40 μL/kg BW, 40% in olive oil) was started to inject intraperitoneally into all mice except for mice in the control group at one hour before the administration of the experimental drugs for every 3.5 days (twice a week) and 8 consecutive weeks. Thus, there were sixteen times of CCl_4 treatment for the five CCl_4 treated groups. The control mice received an equivalent volume of olive oil. One week after the last administration of the experimental drugs, the mice were sacrificed under anesthesia and their blood was collected for evaluation of the biochemical parameters (AST, ALT, triglyceride and cholesterol). Their liver tissues were obtained for MDA assay, histological analysis and antioxidant enzymatic activity measurements.

3.5. Serum Biochemistry

The serum was obtained as descripted previously [4]. Serum ALT, AST, triglyceride and cholesterol were measured using spectrophotometric diagnostic kits (Roche, Berlin, Germany).

3.6. Histological Analysis

Histological Analysis was performed according to the method of the previous report by staining with hematoxylin and eosin [8,11] and with Sirius Red [25], and then observed under light microscopy (Olympus, Tokyo, Japan). For quantitative scoring of the hepatic fibrosis, the values were used according to the published method [14], as the following: none: normal liver, score 0; slight: increase of collagen without formation of septa, score 1; mild: septa do not connect with each other and incomplete septa formation from portal tract to central vein, score 2; moderate: septa interconnecting completely but thin (incomplete cirrhosis), score 3; and remarkable: with thick septa (complete cirrhosis), score 4.

3.7. MDA Level as Well as Antioxidant Enzymatic Activity Measurement

MDA level was determined as descripted previously using the thiobarbituric acid reacting substance method [26]. SOD, GPx and GRd enzymatic activities were determined according to the published methods [27–29]. MDA, SOD, GPx and GRd assay kits were purchased from Randox Laboratory Ltd. (Antrim, UK).

3.8. Statistical Analyses

All data were shown as mean ± SEM. SPSS statistics software program was used to do the statistical data analyses. One-way ANOVA followed by Scheffe's multiple range test was used to perform the statistical analyses. For the histological analyses, non-parametric Kruskal–Wallis test followed by the Mann–Whitney U-test was used to carry out the statistical analyses. The criterion for statistical significance was $p < 0.05$.

3.9. Phytochemical Analysis of CC_{EtOH} by HPLC

The HPLC profile of CC_{EtOH} was determined and compared with the standard (hyperoside and quercetin), which was conducted as described previously [20]. Quantification was performed by comparing the sample peak with the corresponding standard compound.

4. Conclusions

The present study clearly elucidated that CC_{EtOH} exhibited a hepatoprotective activity against CCl_4-induced chronic liver injury in mice. The underlying mechanisms were likely the decreasing in MDA level through enhancing the activities of hepatic anti-oxidative enzymes such as GPx, GRd and SOD, and thereby the significant decrease of serum ALT, AST, triglyceride and cholesterol. In addition, fibrosis of liver was significantly reduced by CC_{EtOH}. Therefore, *C. campestris* can be developed into pharmacological agents to prevent some liver disorders.

Supplementary Materials: Supplementary materials can be found at www.mdpi.com/1422-0067/17/12/2056/s1.

Acknowledgments: This work is supported by grants from National Science Council (MOST 105-2320-B-039-012-MY3 and MOHW103-CMAP-M-114-122408) and China Medical University (CMU104-TC-02). We would like to express our thanks to Tun-Jen Cheng for his help with the anti-oxidation experiment and Jiunn-Wang Liao from National Chung Hsing University for his help with the histological analyses.

Author Contributions: Conceived and designed the experiments: Wen-Huang Peng, and Ming-Kuem Lin. Performed the experiments: Yi-Wen Chen and Jen-Chieh Tsai. Analyzed the data: Meng-Shiou Lee, Wen-Te Chang, and Ming-Kuem Lin. Contributed reagents/materials/analysis tools: Ying-Chih Lin. Wrote the paper: Wen-Huang Peng and Ming-Kuem Lin.

Conflicts of Interest: The authors declare no conflict of interest.

References

1. Donnapee, S.; Li, J.; Yang, X.; Ge, A.H.; Donkor, P.O.; Gao, X.M.; Chang, Y.X. *Cuscuta chinensis* Lam.: A systematic review on ethnopharmacology, phytochemistry and pharmacology of an important traditional herbal medicine. *J. Ethnopharmacol.* **2014**, *157*, 292–308. [CrossRef] [PubMed]
2. Yang, J.; Wang, Y.; Bao, Y.; Guo, J. The total flavones from *Semen cuscutae* reverse the reduction of testosterone level and the expression of androgen receptor gene in kidney-yang deficient mice. *J. Ethnopharmacol.* **2008**, *119*, 166–171. [CrossRef] [PubMed]
3. Yen, F.L.; Wu, T.H.; Lin, L.T.; Lin, C.C. Hepatoprotective and antioxidant effects of *Cuscuta chinensis* against acetaminophen-induced hepatotoxicity in rats. *J. Ethnopharmacol.* **2007**, *111*, 123–128. [CrossRef] [PubMed]
4. Liao, J.C.; Chang, W.T.; Lee, M.S.; Chiu, Y.J.; Chao, W.K.; Lin, Y.C.; Lin, M.K.; Peng, W.H. Antinociceptive and anti-inflammatory activities of *Cuscuta chinensis* seeds in mice. *Am. J. Chin. Med.* **2014**, *42*, 223–242. [CrossRef] [PubMed]
5. Bao, X.; Wang, Z.; Fang, J.; Li, X. Structural features of an immunostimulating and antioxidant acidic polysaccharide from the seeds of *Cuscuta chinensis*. *Planta Med.* **2002**, *68*, 237–243. [CrossRef] [PubMed]

6. Lee, S.H.; Heo, S.I.; Li, L.; Lee, M.J.; Wang, M.H. Antioxidant and hepatoprotective activities of *Cirsium setidens* NAKAI against CCl₄-induced liver damage. *Am. J. Chin. Med.* **2008**, *36*, 107–114. [CrossRef] [PubMed]

7. Wang, T.; Sun, N.L.; Zhang, W.D.; Li, H.L.; Lu, G.C.; Yuan, B.J.; Jiang, H.; She, J.H.; Zhang, C. Protective effect of dehydrocavidine on carbon tetrachloride-induced acute hepatotoxicity in rats. *J. Ethnopharmacol.* **2008**, *117*, 300–308. [CrossRef] [PubMed]

8. Tsai, J.C.; Peng, W.H.; Chiu, T.H.; Huang, S.C.; Huang, T.H.; Lai, S.C.; Lai, Z.R.; Lee, C.Y. Hepatoprotective effect of *Scoparia dulcis* on carbon tetrachloride induced acute liver injury in mice. *Am. J. Chin. Med.* **2010**, *38*, 761–775. [CrossRef] [PubMed]

9. Brattin, W.J.; Glende, E.A., Jr.; Recknagel, R.O. Pathological mechanisms in carbon tetrachloride hepatotoxicity. *J. Free Radic. Biol. Med.* **1985**, *1*, 27–38. [CrossRef]

10. Basu, S. Carbon tetrachloride-induced lipid peroxidation: Eicosanoid formation and their regulation by antioxidant nutrients. *Toxicology* **2003**, *189*, 113–127. [CrossRef]

11. Lee, C.Y.; Peng, W.H.; Cheng, H.Y.; Chen, F.N.; Lai, M.T.; Chiu, T.H. Hepatoprotective effect of *Phyllanthus* in Taiwan on acute liver damage induced by carbon tetrachloride. *Am. J. Chin. Med.* **2006**, *34*, 471–482. [CrossRef] [PubMed]

12. Weber, L.W.; Boll, M.; Stampfl, A. Hepatotoxicity and mechanism of action of haloalkanes: Carbon tetrachloride as a toxicological model. *Crit. Rev. Toxicol.* **2003**, *33*, 105–136. [CrossRef] [PubMed]

13. Adewole, S.O.; Salako, A.A.; Doherty, O.W.; Naicker, T. Effect of melatonin on carbon tetrachloride-induced kidney injury in Wistar rats. *Afr. J. Biomed. Res.* **2007**, *10*, 153–164. [CrossRef]

14. Ruwart, M.J.; Wilkinson, K.F.; Rush, B.D.; Vidmar, T.J.; Peters, K.M.; Henley, K.S.; Appelman, H.D.; Kim, K.Y.; Schuppan, D.; Hahn, E.G. The integrated value of serum procollagen III peptide over time predicts hepatic hydroxyproline content and stainable collagen in a model of dietary cirrhosis in the rat. *Hepatology* **1989**, *10*, 801–806. [CrossRef] [PubMed]

15. Manibusan, M.K.; Odin, M.; Eastmond, D.A. Postulated carbon tetrachloride mode of action: A review. *J. Environ. Sci. Health C Environ. Carcinog. Ecotoxicol. Rev.* **2007**, *25*, 185–209. [CrossRef] [PubMed]

16. Recknagel, R.O.; Glende, E.A., Jr.; Dolak, J.A.; Waller, R.L. Mechanisms of carbon tetrachloride toxicity. *Pharmacol. Ther.* **1989**, *43*, 139–154. [CrossRef]

17. Lee, C.H.; Park, S.W.; Kim, Y.S.; Kang, S.S.; Kim, J.A.; Lee, S.H.; Lee, S.M. Protective mechanism of glycyrrhizin on acute liver injury induced by carbon tetrachloride in mice. *Biol. Pharm. Bull.* **2007**, *30*, 1898–1904. [CrossRef] [PubMed]

18. Tseng, S.H.; Chien, T.Y.; Tzeng, C.F.; Lin, Y.H.; Wu, C.H.; Wang, C.C. Prevention of hepatic oxidative injury by Xiao-Chen-Chi-Tang in mice. *J. Ethnopharmacol.* **2007**, *111*, 232–239. [CrossRef] [PubMed]

19. Hung, M.Y.; Fu, T.Y.; Shih, P.H.; Lee, C.P.; Yen, G.C. Du-Zhong (*Eucommia ulmoides* Oliv.) leaves inhibit CCl₄-induced hepatic damage in rats. *Food Chem. Toxicol.* **2006**, *44*, 1424–1431. [CrossRef] [PubMed]

20. Lee, M.S.; Chen, C.J.; Wan, L.; Koizumi, A.; Chang, W.T.; Yang, M.J.; Lin, W.H.; Tsai, F.J.; Lin, M.K. Quercetin is increased in heat-processed *Cuscuta campestris* seeds, which enhances the seed's anti-inflammatory and anti-cancer activities. *Process Biochem.* **2011**, *46*, 2248–2254. [CrossRef]

21. Park, J.Y.; Han, X.; Piao, M.J.; Oh, M.C.; Fernando, P.M.; Kang, K.A.; Ryu, Y.S.; Jung, U.; Kim, I.G.; Hyun, J.W. Hyperoside induces endogenous antioxidant system to alleviate oxidative stress. *J. Cancer Prev.* **2016**, *21*, 41–47. [CrossRef] [PubMed]

22. Zhang, M.; Swarts, S.G.; Yin, L.; Liu, C.; Tian, Y.; Cao, Y.; Swarts, M.; Yang, S.; Zhang, S.B.; Zhang, K.; et al. Antioxidant properties of quercetin. *Adv. Exp. Med. Biol.* **2011**, *701*, 283–289. [PubMed]

23. Choi, J.H.; Kim, D.W.; Yun, N.; Choi, J.S.; Islam, M.N.; Kim, Y.S.; Lee, S.M. Protective effects of hyperoside against carbon tetrachloride-induced liver damage in mice. *J. Nat. Prod.* **2011**, *74*, 1055–1060. [CrossRef] [PubMed]

24. Domitrović, R.; Jakovac, H.; Vasiljev Marchesi, V.; Vladimir-Knežević, S.; Cvijanović, O.; Tadić, Z.; Romić, Z.; Rahelić, D. Differential hepatoprotective mechanisms of rutin and quercetin in CCl₄-intoxicated BALB/cN mice. *Acta Pharmacol. Sin.* **2012**, *33*, 1260–1270. [CrossRef] [PubMed]

25. Junqueira, L.C.; Bignolas, G.; Brentani, R.R. Picrosirius staining plus polarization microscopy, a specific method for collagen detection in tissue sections. *Histochem. J.* **1979**, *11*, 447–455. [CrossRef] [PubMed]

26. Draper, H.H.; Hadley, M. Malondialdehyde determination as index of lipid peroxidation. *Methods Enzymol.* **1990**, *186*, 421–431. [PubMed]

Int. J. Mol. Sci. **2016**, *17*, 2056

27. Misra, H.P.; Fridovich, I. The role of superoxide anion in the autooxidation of epinephrine and a simple assay for superoxide dismutase. *J. Biol. Chem.* **1972**, *247*, 3170–3175. [PubMed]
28. Carlberg, I.; Mannervik, B. Glutathione reductase. *Methods Enzymol.* **1984**, *113*, 484–490.
29. Flohe, L.; Gunzler, W.A. Assays of glutathione peroxidase. *Methods Enzymol.* **1984**, *105*, 114–121. [PubMed]

International Journal of
Molecular Sciences

MDPI

Article

Chemical Composition and Antioxidant Activity of *Euterpe oleracea* Roots and Leaflets

Christel Brunschwig [†], Louis-Jérôme Leba [†], Mona Saout, Karine Martial, Didier Bereau and Jean-Charles Robinson *

UMR QUALITROP, Université de Guyane, Campus Universitaire de Troubiran, P.O. Box 792, 97337 Cayenne Cedex, French Guiana, France; christel.brunschwig@gmail.com (C.B.); louis-jerome.leba@univ-guyane.fr (L.-J.L.); mona.saout@gmail.com (M.S.); karine.martial@univ-guyane.fr (K.M.); didier.bereau@univ-guyane.fr (D.B.)
* Correspondence: jean-charles.robinson@univ-guyane.fr; Tel.: +594-594-29-99-46
† These authors contributed equally to this work.

Academic Editor: Toshio Morikawa
Received: 27 October 2016; Accepted: 12 December 2016; Published: 29 December 2016

Abstract: *Euterpe oleracea* (açaí) is a palm tree well known for the high antioxidant activity of its berries used as dietary supplements. Little is known about the biological activity and the composition of its vegetative organs. The objective of this study was to investigate the antioxidant activity of root and leaflet extracts of *Euterpe oleracea* (*E. oleracea*) and characterize their phytochemicals. *E. oleracea* roots and leaflets extracts were screened in different chemical antioxidant assays (DPPH—2,2-diphenyl-1-picrylhydrazyl, FRAP—ferric feducing antioxidant power, and ORAC—oxygen radical absorbance capacity), in a DNA nicking assay and in a cellular antioxidant activity assay. Their polyphenolic profiles were determined by UV and LC-MS/MS. *E. oleracea* leaflets had higher antioxidant activity than *E. oleracea* berries, and leaflets of *Oenocarpus bacaba* and *Oenocarpus bataua*, as well as similar antioxidant activity to green tea. *E. oleracea* leaflet extracts were more complex than root extracts, with fourteen compounds, including caffeoylquinic acids and C-glycosyl derivatives of apigenin and luteolin. In the roots, six caffeoylquinic and caffeoylshikimic acids were identified. Qualitative compositions of *E. oleracea*, *Oenocarpus bacaba* and *Oenocarpus bataua* leaflets were quite similar, whereas the quantitative compositions were quite different. These results provide new prospects for the valorization of roots and leaflets of *E. oleracea* in the pharmaceutical, food or cosmetic industry, as they are currently by-products of the açaí industry.

Keywords: *Euterpe oleracea*; açaí; antioxidant activity

1. Introduction

Euterpe oleracea (*E. oleracea*), known as açaí worldwide and wassaye in French Guiana, is a palm tree native from the Amazonian rainforest, which gives purple berries with high potential currently used as dietary food supplements or in cosmetics thanks to their high antioxidant activity. Traditionally, açaí berries are consumed as a beverage called "açaí vino" in Brazil. Since the mid 2000s, many scientific studies focused on *E. oleracea* berries due to its biological activities including anti-proliferative, anti-inflammatory, antioxidant, and cardioprotective properties. *E. oleracea* was also found to be extremely rich in antioxidant compounds including phenolic acids, flavonoids and anthocyanins [1,2].

While açaí berries are mainly used as foodstuffs, *E. oleracea* vegetative organs (such as leaves and roots) are traditionally used in medicine, curing snake bites [3], diabetes, kidney and liver pains, fever, anemia, as well as arthritis, or used as hemostatic [4,5]. Roots of *Euterpe precatoria*, a close species to *E. oleracea*, also showed anti-inflammatory [6] and antioxidant properties [7]. The vegetative organs of *E. oleracea* are considered as waste from the açaí palm heart industry and could be potential sources of phytochemicals with cosmetic, pharmaceutical or nutritional applications. Currently, there is interest in valorizing polyphenols-containing food wastes such as grape skins, olive mill waste water, citrus peels, seed wastes or berry leaves [8,9] as there is a growing demand for polyphenols for pharmaceutical, cosmetic and food industries, especially to treat cardiovascular, and skin diseases, cancer, aging, and diet issues. Some studies have already been carried out on açaí seeds, another waste from the açaí industry. Antioxidant activity [10] and anti-nociceptive activity have been demonstrated [11], showing that this waste could find applications in the food or pharmaceutical industries. Therefore, there is also a real value in enhancing the knowledge of the phytochemical composition and pharmacological activities of açaí roots and leaves. A previous study showed that roots and leaflets of two other Amazonian palm trees (*Oenocarpus bacaba* and *Oenocarpus bataua*) had high antioxidant activity, underlining both *Oenocarpus* vegetative organs as good sources of antioxidant compounds [12]. In this paper, we therefore investigated the antioxidant activity of root and leaflet crude extracts of *E. oleracea* using different chemical antioxidant assays (DPPH—2,2-diphenyl-1-picrylhydrazyl, FRAP—ferric reducing antioxidant power, and ORAC—oxygen radical absorbance capacity), a DNA nicking assay and a cellular antioxidant activity assay, while LC-MS/MS was performed to elucidate the structures of the bioactive compounds.

2. Results and Discussion

2.1. Antioxidant Activity

2.1.1. Antioxidant Activity in the Chemical Assays (DPPH, FRAP, and ORAC)

Antioxidant properties of root and leaflet extracts from *E. oleracea* were investigated using different chemical assays (DPPH, FRAP and ORAC) having different modes of action: hydrogen transfer (ORAC), electron transfer (FRAP) or a mixed mode (DPPH) [13]. In all antioxidant tests, best results were obtained for acetone extracts independently of the organ used. Leaflet extracts were more active than root extracts, independently of the solvent used.

In the DPPH assay, values for *E. oleracea* leaflet extracts ranged from 480 to 990 μmol Trolox Eq/g Dry Matter (μmol TEq/g DM). DPPH activity of *E. oleracea* acetone leaflet extracts were at least two-fold higher than *E. oleracea* and *O. bataua* berries [14], two-fold higher than *Oenocarpus* leaflets acetone extracts [12], and slightly lower than green tea (1200 μmol TEq/g DM) (Table 1).

In the FRAP assay, all values for leaflet extracts ranged from 1000 to 1400 μmol Fe(II)Eq/g DM. They were two-fold higher than palm berries [14–16], at least two-fold higher than antioxidant foodstuffs like lettuce [17], but lower than green tea extracts (1900–2700 μmol Fe(II)Eq/g DM) (Table 1). When comparing leaflet acetone extracts, *E. oleracea* FRAP activity was two-fold higher than *O. bataua* and similar to *O. bacaba* [12].

The ORAC assay, considered as one of the most relevant amongst antioxidant chemical assays, gave ORAC values from 1600 to 2200 μmol TEq/g DM, for leaflet extracts. They were at least three-fold more active than palm berries in the ORAC assay [14], at least 1.5-fold more active than the best *Oenocarpus* leaflet extracts [12] and almost equivalent to green tea with 2400 μmol TEq/g DM (Table 1). Besides, even if *E. oleracea* roots were less active than *E. oleracea* leaflets in all the antioxidant tests, their activities were as high or higher than those of *E. oleracea* and *O. bataua* berries, as illustrated by ORAC results (300–1300 μmol TEq/g DM) (Table 1).

Root and leaflet extracts from *E. oleracea* were active across the board in the different chemical antioxidant assays (DPPH, FRAP, and ORAC) showing that they act according to different mechanisms of action such as electron and/or hydrogen transfer. Root and leaflet extracts were at least as active as palm berries, considered as super fruits. Leaflet extracts were the most antioxidant extracts, almost equivalent to green tea leaves, which have one of the strongest recorded antioxidant activity.

Table 1. Antioxidant activity and total phenolic content (TPC) of root and leaflet extracts of *Euterpe oleracea*.

Extract Name (Plant/Part/Solvent)	TPC (µg GAEq/mg DM) *	DPPH (µmol TEq/g DM) *	FRAP (µmol Fe(II) Eq/g DM) *	ORAC (µmol TEq/g DM) *	EC$_{50}$ in NHDF (µg/mL)	CAA in NHDF (µmole QEq/g DM)
WRW	14 ± 2 [d]	105 ± 5 [c]	310 ± 129 [e]	302 ± 124 [d]	30 ± 7	10 ± 2
WRA	58 ± 9 [b,c]	471 ± 101 [b]	769 ± 137 [c,d]	1259 ± 320 [c]	14 ± 2	29 ± 4
WRM	28 ± 4 [c,d]	237 ± 72 [c]	449 ± 111 [d,e]	686 ± 186 [d]	16 ± 1	14 ± 1
WLW	62 ± 7 [a,b]	479 ± 39 [b,c]	965 ± 178 [b,c]	1643 ± 385 [b,c]	29 ± 9	45 ± 12
WLA	84 ± 7 [a]	991 ± 81 [a]	1381 ± 376 [a]	2229 ± 484 [a]	10 ± 1	208 ± 23
WLM	76 ± 7 [a,b]	660 ± 72 [b]	1194 ± 180 [a,b]	2177 ± 611 [a,b]	16 ± 2	87 ± 12
TLW [e]	100 ± 9	748 ± 52	1911 ± 82	1348 ± 45	-	-
TLA [e]	126 ± 2	1185 ± 34	2686 ± 107	2167 ± 19	-	-
TLM [e]	98 ± 8	1045 ± 69	2677 ± 142	2366 ± 31	-	-
ObtLA [f]	50	460	730	1200	14	100
ObcLA [f]	60	540	1030	1560	24	50
ObcBA [g]	30	240	200	170	-	-
WBA [g]	40	240	130	430	-	-

* Averages with the same letter [a–d] within columns are not significantly different (*p* < 0.05) using Fisher's Least Significant Difference test; W: Wassaye (*E. oleracea*); Obt: *O. bataua*; Obc: *O. bacaba*; R: roots; L: leaflets; B: berries; T: Green Tea leaves; W: water; A: acetone/water 70/30; M: methanol/water 70/30; CAA: cellular antioxidant activity; NHDF: normal human dermal fibroblasts; GAEq: Gallic acid equivalent; TEq: Trolox equivalent; QEq: Quercetin equivalent; DM: dry matter; TPC: total phenolic content; EC$_{50}$: median effective concentration; CAA: cellular antioxidant activity; DPPH: 2,2-Diphenyl-1-picrylhydrazyl; FRAP: ferric reducing antioxidant power; ORAC: oxygen radical absorbance capacity; [e] data from Leba et al., 2014 [16]; [f] data from Leba et al., 2016 [12]; [g] data from Rezaire et al., [14].

2.1.2. Antioxidant Activity in the DNA Nicking Assay

Despite optimization of the DNA nicking assay conditions, the sensitivity using methanol extracts was low [16], therefore only aqueous and acetone extracts of *E. oleracea* were assessed in the DNA nicking assay. All root extracts of *E. oleracea* were found to be active in this assay. The aqueous and acetone root extracts had an antioxidant effect at all concentrations tested; they protected form I from degradation and reduced the formation of linear nicked form III in a similar way to Trolox at 1 mg/mL (Figure 1). Aqueous leaflet extracts had an antioxidant effect in the DNA nicking assay at all concentrations tested (1 mg/mL and 10 mg/mL), while acetone leaflet extracts had an antioxidant effect at 10 mg/mL and a prooxidant effect at a lower concentration of 1 mg/mL. The overall high potency of the root and leaflet extracts of *E. oleracea* in the DNA nicking assay encouraged the evaluation of their antioxidant activity in an assay physiologically more relevant, i.e., a cell-based antioxidant assay.

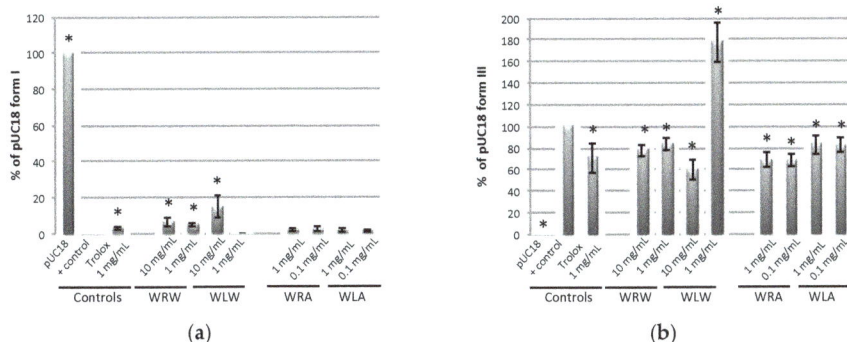

Figure 1. Antioxidant activity of *Euterpe oleracea* extracts in the DNA nicking assay indicating: (a) protection of supercoiled form I; and (b) formation of nicked linear form III; W: Wassaye (*E. oleracea*); R: roots; L: leaflets; W: water; A: acetone/water 70/30; (* *p* < 0.05; *n* = 3).

2.1.3. Cytotoxicity and Cellular Antioxidant Activity (CAA)

As root and leaflet extracts of *E. oleracea* were not cytotoxic to normal human dermal fibroblasts (NHDF) at 100 µg/mL in the MTT (3-(4,5-dimethylthiazol-2-yl)-2,5-diphenyltetrazolium bromide) assay (Table S1), they were tested at this concentration in the Cellular Antioxidant Activity (CAA) assay.

Root and leaflet extracts from *E. oleracea* were quite active in the cellular antioxidant assay with the median effective concentration (EC_{50}) ranging from 10 to 30 µg/mL (Table 1). Most of the palm extracts were more active than fruits including kiwis, blueberries, carrots and strawberries, which had a high antioxidant activity (EC_{50} from 50 to 200 µg/mL) in a similar cell-based assay [18].

Leaflet extracts of *E. oleracea* were more active in the CAA assay (45–208 µmol Quercetin Eq/g DM) than root extracts (10–30 µmol Quercetin Eq/g DM) and acetone extracts were more active than methanol extracts. Activities for *E. oleracea* were quite similar to *Oenocarpus* roots (10–50 µmol Quercetin Eq/g DM), whereas *E. oleracea* leaflets were at least two-fold superior than *Oenocarpus* leaflets (7–100 µmol Quercetin Eq/g DM) [12]. Interestingly, there was a very high correlation between the ORA activity and the cell-based activity (Pearson's r correlation coefficients of 0.976) (Figure S1), which are both based on the ROO· radical scavenging capacity of the extracts. These correlations were also found in the work of Girard-Lalancette et al., 2009 and Leba et al., 2016 [12,18]. The high potency of extracts in the ORAC assay translated well in the CAA assay, meaning that compounds were able to act in a cellular environment, either by crossing membrane cells or by directly scavenging radicals outside the cells. These results encourage investigating the chemical composition of *E. oleracea* root and leaflet extracts.

2.2. Total Phenolic Content (TPC)

The Total Phenolic Content (TPC) of root and leaflet extracts from *E. oleracea* ranged from 14 to 84 µg Gallic acid equivalent/mg Dry Matter (µg GAEq/mg DM) (Table 1).

Best results were found with acetone independently of the organ used. The highest values were found for the leaflet extracts of *E. oleracea*, with 84 µg GAEq/mg DM, which was two-fold higher than the TPC of palm berries including *E. oleracea*, *O. bataua* [14] and *O. bacaba* [15], known to be antioxidants. The TPC of *E. oleracea* roots were globally similar to that of *Oenocarpus* roots extracts [12], whereas *E. oleracea* leaflets were richer in polyphenols than *Oenocarpus* leaflets [12]. Moreover, the TPC of *E. oleracea* leaflets was almost similar to the TPC of green tea leaves (100–125 µg GAEq/mg), which has one of the strongest recorded antioxidant activities. There was a high correlation between the total polyphenolic content (TPC) and the antioxidant activity in the different assays ($p < 0.05$) with high Pearson's r correlation coefficients between TPC and DPPH (0.945), TPC and FRAP (0.983), TPC and ORAC (0.986), and TPC and CAA activity (0.968) (Figure S1).

2.3. Identification of Compounds in Root and Leaflet Extracts by LC-MS/MS

The UV chromatograms recorded at 320 nm showed that both *E. oleracea* root and leaflet extracts (Figure 2) contained polyphenols, whose structures (Figure 3) were then determined by LC-MS/MS. Leaflet extracts showed more complex polyphenol profiles (fourteen compounds) than root extracts (six compounds), while some peaks (**1**, **2**, and **3**) were common to both extracts.

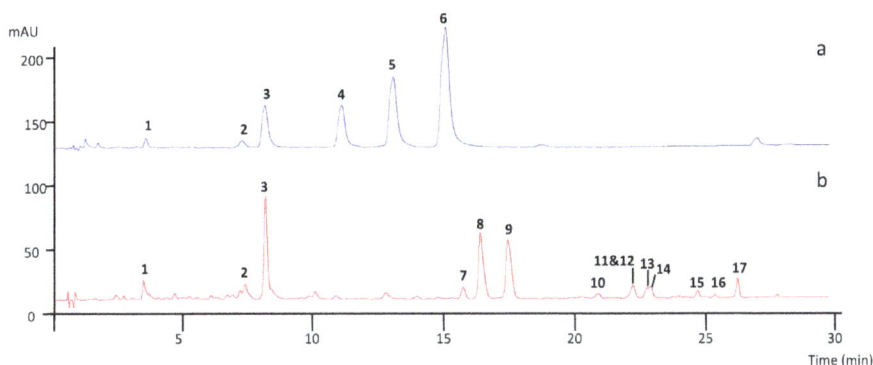

Figure 2. Representative chromatograms of: (**a**) root extracts; and (**b**) leaflet extracts of *Euterpe oleracea* at λ = 320 nm; Kinetex PFP column 100 × 4.6 mm, 2.6 μm; (**1**) 3-CQA; (**2**) 4-CQA; (**3**) 5-CQA; (**4**) 4-CSA; (**5**) 5-CSA; (**6**) CSA; (**7**) 6,8-di-C-hexosyl apigenin; (**8,9**) 6,8-di-C-hexosyl apigenin sulfate; (**10,14,15**) 6-C-hexosyl-8-C-pentosyl apigenin isomers; (**12,16**) 6-C-pentosyl-8-C-hexosyl apigenin isomer; (**11**) 8-C-glucosyl luteolin; (**13**) 6-C-glucosyl luteolin; and (**17**) 6-C-glucosyl apigenin; CQA: caffeoylquinic acid; CSA: caffeoylshikimic acid.

Hydroxycinnamic acids

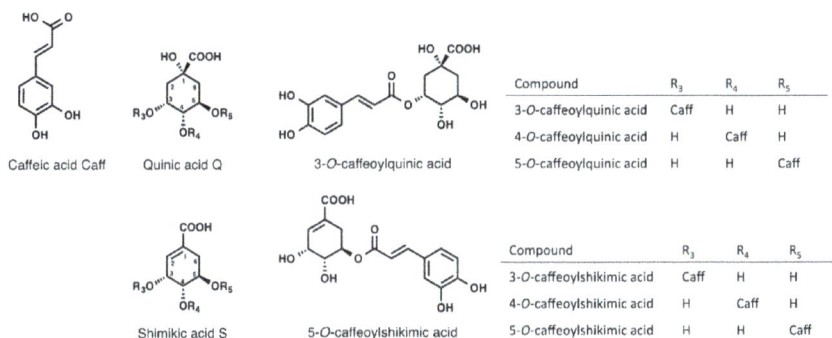

Flavones

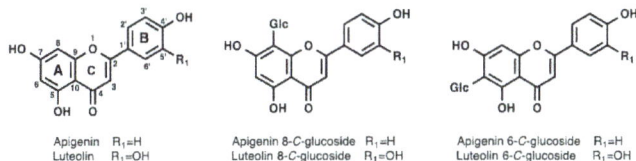

Figure 3. Structure of main components of root and leaflet extracts of *Euterpe oleracea*; Glc: glucose.

2.3.1. Characterization of Caffeoylquinic Derivatives (Mr = 354) in Root and Leaflet Extracts

In the root and leaflet extracts, three peaks (**1**, **2**, and **3**) at retention time (t_R) 4.2, 7.2 and 9.1 min yielded $[M - H]^-$ ions at m/z 353 and $[M + Na]^+$ at m/z 377 and showed UV spectra (at λ = 298 nm and 320 nm) characteristic of caffeoylquinic acids (CQA, Mr = 354).

In positive mode, the ions at m/z 163 and 145, produced by the loss of the quinic acid moiety [19] were not conclusive of the structure of the CQA. The structure of CQA isomers was assigned according to their diagnostic ions at m/z 191, 179 and 173 in the negative mode [20]. Compound **2** (t_R 7.2 min) was assigned to 4-CQA, having a base peak at m/z 173 in the MS^2 spectrum. Compounds **1** (t_R 4.2 min) and **3** (t_R 9.1 min), both having m/z 191 as a base peak were distinguished by the intense ion at m/z 179 in the MS^2 spectrum, characteristic of 3-CQA for compound **1**, whereas compound **3** was assigned to 5-CQA (Table 2). The retention order of the CQA isomers (**3** < **4** < **5**) was similar to the study of Regos and Treutter, 2010 [21] using a similar PFP column.

2.3.2. Characterization of Caffeoylshikimic Derivatives (Mr = 336) in Root Extracts

In the root extracts, three peaks (**4**, **5**, **6**) at t_R 12.2, 14.2 and 16.2 min yielded [M − H]$^-$ ions at m/z 335, [M + Na]$^+$ at m/z 359, and showed UV spectra (λ = 298 nm and 320 nm) corresponding to caffeoylshikimic acids (CSA, Mr = 336). In positive mode, the ions at m/z 163 and 145 were characteristic of the caffeoyl moiety [19]. The structure of CSA was tentatively assigned according to their diagnostic ions at m/z 317, 291, 179, 161 and 135 in negative mode [22–24] (Table 2). The fragmentation pattern of compound **4** with an intense ion at m/z 161 was conclusive of a 4-CSA. The absence of the ion at m/z 161 in the MS^2 spectrum of compound **5** and the presence of ions at m/z 317 and m/z 291 was indicative of a 5-CSA isomer. The fragmentation pattern of compound **6** was close to that of a 4-CSA isomer but was not conclusive of the structure.

2.3.3. Characterization of Apigenin Derivatives (Mr = 432, Mr = 594, Mr = 674, and Mr = 564) in Leaflet Extracts

The UV spectra of eleven compounds (**7–17**) in the leaflet extracts of *E. oleracea* showing a λ_{max} at 340 nm were indicative of flavone derivatives. Compound **17** (Mr = 432) gave [M−H]$^-$ at m/z 431 and [M + H]$^+$ at m/z 433. Characteristic losses of 90, 120 and 150 Da were diagnostic of hexose residues. The fragment ions at m/z 311 (aglycone + 42) and m/z 341 (aglycone + 72) in negative mode clearly indicate 6-C-glycosyl apigenin [25], whose identity was confirmed with a synthetic standard. Nine compounds, i.e., compounds **7** (Mr = 594), **8**, **9** (Mr = 674) **10**, **12**, **14**, **15**, and **16** (Mr = 564), showed the characteristic MS^2 fragmentation of di-C-glycosyl flavones, the occurrence of fragment ions at m/z 353 (aglycone + 83) and 383 (aglycone + 113) being indicative of apigenin derivatives [25]. Compound **7** (Mr = 594) yielded [M − H]$^-$ at m/z 593, with a fragmentation pattern in negative mode characteristic of hexose derivatives (losses of 90 Da and 120 Da) and was assigned to a 6,8-di-C-hexosyl apigenin. Compounds **8** and **9** (Mr = 674) yielded [M − H]$^-$ at m/z 673 and [M + K]$^+$ at 713. The MS^2 spectrum showed a first loss of 80 Da, characteristic of sulfates, followed by successive losses of 90 Da, 120 Da and 150 Da, characteristics of hexose residues in both positive and negative modes. Compounds **8** and **9** were tentatively assigned to 6,8-di-C-hexosyl apigenin sulfates. The lack of a hypsochromic shift in the band I of **8** and **9** compared to compound **7** suggest the sulfation occurred on position **7** or on the sugar moieties [26]. Six peaks (**10**, **12**, **14**, **15**, **16**) (Mr = 564) at t_R 21.2, 22.3, 23.2, 24.9 and 25.6 min yielded [M − H]$^-$ ions at m/z 563 in the negative mode. Their MS^2 spectrum showed losses of 60 Da, 90 Da, 120 Da and 150 Da, characteristic of pentose and hexose residues. They were assigned as 6,8-di-C-pentosyl-hexosyl apigenin derivatives. The [M − H-60]$^-$ ion resulting from the fragmentation of the pentose moiety was more intense in **12** (100%) and **16** (25%) than in other compounds, indicating that the pentose moiety was on position 6, which fragments preferentially over sugars on position 8 [25]. Therefore, compounds **12** and **16** were identified as 6-C-pentosyl-8-C-hexosyl apigenin isomers. The others compounds (**10**, **14**, and **15**) were assigned as 6-C-hexosyl-8-C-pentosyl apigenin isomers.

Table 2. Identification of main components of root and leaflet extracts of *Euterpe oleracea*.

N	t_R (min)	UV λ_{max} (nm)	Negative Mode MS	Negative Mode MS² (%)	Positive Mode MS	Positive Mode MS² (%)	Tentative Identity	Abbreviation	Extracts
1	4.2	238, 295sh, 321	353 [M − H]⁻	MS²[353]: 191 (100), 179 (25)	377 [M + Na]⁺	377/353/163/145	3-Caffeoylquinic acid	3-CQA	Roots, leaflets
2	7.9	237, 286sh, 322	353 [M − H]⁻	MS²[353]: 173 (100)	377 [M + Na]⁺	377/353/163/145	4-Caffeoylquinic acid	4-CQA	Roots, leaflets
3	9.1	239, 295sh, 323	353 [M − H]⁻	MS²[353]: 191 (100)	377 [M + Na]⁺	377/353/163/145	5-Caffeoylquinic acid	5-CQA	Roots, leaflets
4	12.2	239, 295sh, 325	335 [M − H]⁻	MS²[335]: 291 (100), 179 (60), 161 (80), 135 (75)	359 [M + Na]⁺	359/163/145	4-Caffeoylshikimic acid	4-CSA	Roots
5	14.2	239, 295sh, 325	335 [M − H]⁻	MS²[335]: 317 (100), 291 (100), 179 (50)	359 [M + Na]⁺	359/163/145	5-Caffeoylshikimic acid	5-CSA	Roots
6	16.2	239, 295sh, 325	335 [M − H]⁻	MS²[335]: 317 (5), 291 (5), 179 (75), 161 (100), 135 (80)	359 [M + Na]⁺	359/163/145	Caffeoylshikimic acid	CSA	Roots
7	16.1	239, 270, 335	593 [M − H]⁻	MS²[593]: 503 (20), 473 (25), 383 (50), 353 (50)	595 [M + Na]⁺	595/577/457/427/317	6,8-di-C-hexosyl apigenin	Di-Glc-Api	Leaflets
8	16.8	239, 270, 335	673 [M − H]⁻	MS²[673]: 593 (100), 575 (10), 503 (10), 473 (15), 413 (5), 383 (10), 353 (10); MS³[593]: 503 (25), 473 (10), 383(50), 353 (100); MS³[673→593]: 473(100), 413(20), 383(60), 353(100); MS³[673→473]: 383 (100), 353 (100)	713 [M + Na]⁺	MS²[713]: 633 (100), 593 (25), 543 (25), 513 (25), 423 (10), 393 (10), 351 (10); MS³[713→633]: 573 (100), 543 (100), 513 (100), 483 (20), 423 (75), 393 (20), 351 (50)	6,8-di-C-hexosyl apigenin sulfate	Di-Glc-Api Sulf	Leaflets
9	17.8	239, 270, 335	673 [M − H]⁻	MS²[673]: 593 (100), 503 (10), 473 (20), 383 (10), 353 (10); MS²[393]: 473 (60), 383 (30), 353 (100); MS³[673→593]: 473 (100), 383 (50), 353 (100); MS³[673→473]: 383 (100), 353 (100)	713 [M + K]⁺	MS²[713]: 633 (100), 593 (25), 543 (25), 513 (75), 483 (25), 423 (30), 393 (30), 363 (10), 351 (10); MS³[713→633]: 573 (20), 543 (100), 513 (70), 483 (5), 423 (50), 393 (20), 381 (50), 351 (25)	6,8-di-C-hexosyl apigenin sulfate	Di-Glc-Api Sulf	Leaflets
10	21.2	238, 272, 339	563 [M − H]⁻	MS²[563]: 473 (10), 443 (30), 383 (80), 353 (80)	-	-	6-C-hexosyl-8-C-pentosyl apigenin isomer		Leaflets
11	22.1	240, 270, 342	447 [M − H]⁻	447/429/411/357/327/299	449 [M + H]⁺	449/413/383	8-C-glycosyl luteolin (orientin)*		Leaflets
12	22.3	238, 272, 339	563 [M − H]⁻	MS²[563]: 503 (100), 473 (10), 443 (20), 413 (75), 383 (50), 353 (10)	-	-	6-C-pentosyl-8-C-hexosyl apigenin isomer		Leaflets
13	22.9	241, 270, 346	447 [M − H]⁻	447/429/411/357/327/299	449 [M + H]⁺	MS²[449]: 431 (10), 413 (80), 395 (20), 383 (50), 353 (60), 329 (10), 299 (100)	6-C-glycosyl luteolin (isoorientin)*		Leaflets
14	23.2	238, 272, 340	563 [M − H]⁻	MS²[563]: 473 (75), 443 (75), 383 (25), 353 (100)	-	-	6-C-hexosyl-8-C-pentosyl apigenin isomer		Leaflets
15	24.9	241, 271, 336	563 [M − H]⁻	MS²[563]: 443 (100), 383 (30), 353 (50), 323 (20)	-	-	6-C-hexosyl-8-C-pentosyl apigenin isomer		Leaflets
16	25.6	238, 277, 335	563 [M − H]⁻	MS²[563]: 503 (25), 473 (10), 383 (10), 353 (100)	-	-	6-C-pentosyl-8-C-hexosyl apigenin isomer		Leaflets
17	26.4	241, 270, 337	431 [M − H]⁻	431/413/395/341/311/283	433 [M + H]⁺	MS²[433]: 397 (75), 379 (30), 367 (100), 337 (50), 313 (40), 295 (20), 283 (50)	6-C-glycosyl apigenin (isovitexin)*		Leaflets

* Structure confirmed using standard compounds.

2.3.4. Characterization of Luteolin Derivatives (Mr = 448)

Compounds **11** and **13** (*Mr* = 432) gave [M − H]⁻ at *m/z* 447 and [M + H]⁺ at *m/z* 449. Characteristics losses of 90 Da and 120 Da were diagnostics of hexose residues. The fragments ions at *m/z* 327 (aglycone + 42) and *m/z* 357 (aglycone + 72) in the negative mode clearly indicated *C*-glycosyl luteolin derivatives [25]. Based on retention order and injection of synthetic standards, **11** was identified as 8-*C*-glycosyl luteolin and **13** as 6-*C*-glycosyl luteolin.

2.3.5. Quantification of Chemical Compounds in Root and Leaflet Extracts

Root extracts of *E. oleracea* were mainly characterized by the presence of hydroxycinnamic acids HCA, namely three caffeoylquinic acids (CQA) and three caffeoylshikimic acids (CSA) (Figure 4). Hydroxycinnamic acids are known to display biological properties such as antiviral, anti-inflammatory and antioxidant activity [27]. They also play important physiological role of defense against pathogens and disease resistance [27], which is essential for vulnerable vegetative organs such as roots. The CSA content was higher than the CQA content in the root extracts, (2.5 mg/g DM against 0.5 mg/g DM for the most concentrated extract), i.e., 70%–85% of all HCA (Figure 4). 5-CQA was the major compound amongst CQA, while each of the three CSA was found in almost similar amounts. The qualitative composition of *E. oleracea* roots was similar to that of *O. bataua* and *O. bacaba* [12]. However, the quantitative compositions of *O. bacaba* were different with a global two-fold lower content in compound compared to *E. oleracea*.

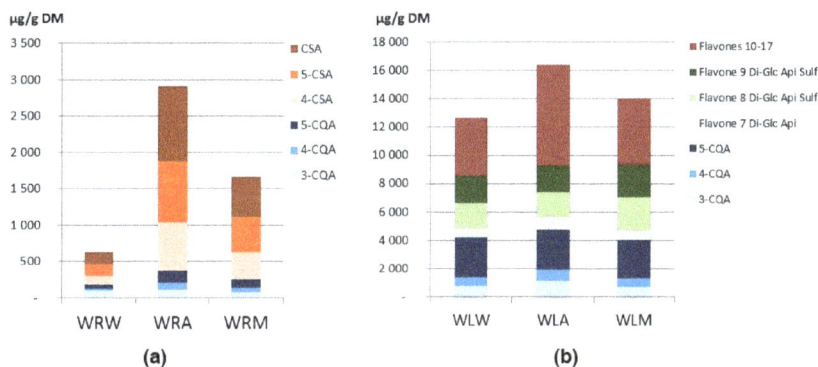

Figure 4. Chemical composition of: (**a**) root extracts; and (**b**) leaflet extracts of *Euterpe oleracea*; W: Wassaye (*E. oleracea*); R: roots; L: leaflets; W: water; A: acetone/water 70/30; M: methanol/water 70/30; CQA: caffeoylquinic acid; CSA: caffeoylshikimic acid (for identification of compounds **7–17**, see Table 2); Di-Glc-Api: 6,8-di-*C*-hexosyl apigenin; Di-Glc-Api Sulf: 6,8-di-*C*-hexosyl apigenin sulfate; *n* = 3 biological replicates.

Leaflet extracts were characterized by two main compound families: three caffeoylquinic acids, as in the root extracts, and eleven flavones (Figure 4). 5-CQA was the main CQA in the leaflet extracts (40%–70% of total CQA). Higher contents of CQA (up to 4.5 mg/g DM) were found in the leaflet extracts compared to the root extracts. Leaflet flavones were quantified mainly as *C*-monoglycosyl and *C*-diglycosyl derivatives of apigenin, and to a lesser extent as *C*-glycosyl derivatives of luteolin. Di-*C*-glycosyl derivatives of apigenin (compounds **7–9**) were the major compounds in leaflet extracts of *E. oleracea*, accounting for almost 50% of all *C*-glycosylflavones, the sulfated flavones **8** and **9** being the major ones with about 2 mg/g DM. Apigenin *C*-glycosides are phytochemical markers specific to a few *Arecaceae* species, while *C*-glycosyl luteolin derivatives are more common amongst palm tree species [28]. More recently, *C*-glycosyl derivatives of apigenin, close to those found in this work, were identified in another Amazonian palm tree species, *Mauritia flexuosa* [29]. The presence of sulfated flavones in the leaflet extracts of *E. oleracea* is interesting as natural sulfated flavonoids inspire the design of smaller bioactive compounds with sulfate groups [30]. *E. oleracea* leaflets composition was qualitatively similar to that of *Oenocarpus* leaflets [12] but the quantitative composition was conspicuously different (maximum 1500 and 5000 µg/g DM for *O. bataua* and *O. bacaba*, respectively, against 16000 µg/g DM for *E. oleracea*). It was interesting to observe that these three palm tree species (*E. oleracea*, *O. bacaba* and *O. bataua*) have the same qualitative composition of their roots or of their leaflets but have quite different antioxidant activities. Indeed, *E. oleracea* leaflets were two fold more active than *O. bacaba* and *O. bataua* leaflets. The difference of activity could come from the quantity of polyphenolic compounds in the extracts, rather than the qualitative composition. These results interrogate the fact that there might be a particular regulation of key enzymes in the biosynthetic pathway of polyphenols in *E. oleracea* leaflets. To validate this hypothesis, numerous molecular and biochemical studies will be required.

2.4. Relationship between the Chemical Composition and the Antioxidant Activity

Principal Component Analysis (PCA) was carried out using the data from the chemical antioxidant assays (DPPH, FRAP, ORAC) and the quantification of compounds in root and leaflet extracts of *E. oleracea*. The antioxidant activity of the roots was linked to the CSA and CQA contents (negative side of PC1) (Figure 5). There was no clear differentiation of the methanol and acetone root extracts according to the PCA, as they had very similar antioxidant activity and chemical composition.

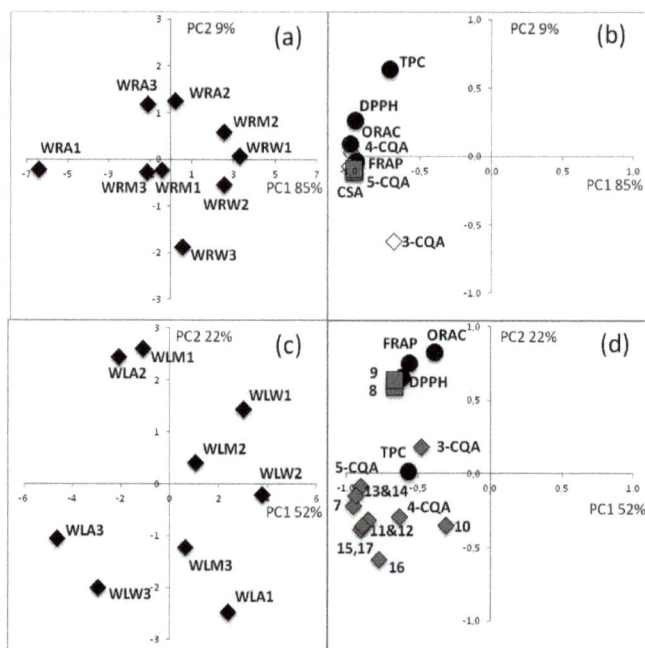

Figure 5. Principal Component Analysis (PCA) plots for antioxidant activity and chemical composition of: root extracts (**a,b**); and leaflet extracts (**c,d**) of *Euterpe oleracea*; W: Wassaye (*E. oleracea*); R: roots; L: leaflets; W: water; A: acetone/water: 70/30; M: methanol/water: 70/30; CQA: caffeoylquinic acid; CSA: caffeoylshikimic acid, (for identification of compounds **7–17**, see Table 2); DPPH: 2,2-Diphenyl-1-picrylhydrazyl; FRAP: ferric reducing antioxidant power; ORAC: oxygen radical absorbance capacity.

For the leaflet extracts, the most active extracts in all the chemical assays were on the positive side of PC2, which was correlated with high contents of flavones **8** and **9** (Figure 5). These results translated quite well to the CAA assay, where the leaflet extracts of *E. oleracea* were very active. The PCA highlighted a correlation between the antioxidant activity of *E. oleracea* leaflet extracts and the presence of di-C-glycosyl apigenin sulfates (**8** and **9**). Indeed, hydroxyl substituted flavones such as apigenin or luteolin derivatives are known to have high antioxidant activity due to their structures [31].

3. Materials and Methods

3.1. Chemicals and Reagents

Solvents used for extraction and LC-MS analysis were of HPLC grade, obtained from Carlo Erba Reagents (Val de Reuil, France). Standards of 5-*O*-caffeoylquinic acid, orientin (>99%), isoorientin (>99%), and isovitexin (>99%) were obtained from Extrasynthèse (Genay, France). Gallic acid-1-hydrate (99%) was purchased from Panreac Quimica (Barcelona, Spain).

3.2. Plant Materials

Roots and leaflets of three specimens of *E. oleracea* were harvested on February 2013, in Macouria, French Guiana. Samples were washed, cut, freeze-dried and grinded immediately after collection. Dried matter was then stored at $-20\ ^\circ$C to limit degradation until extraction.

3.3. Extraction

Extractions were performed following the method previously published [12]. Three solvents were used for extraction of roots and leaflets from *E. oleracea* and green tea leaves: water (W), methanol/water 70/30 v/v (M) and acetone/water 70/30 v/v (A). The combination of acetone or methanol with water is commonly used to extract polyphenols [32]. Ultrasounds were used to assist the extraction of polyphenols of leaflets and roots from *E. oleracea*, as they are known to be a green alternative to conventional extraction methods and give optimized extraction efficiency and reduced extraction time [33]. Preliminary studies using ultrasound extraction were carried out to determine that four successive extractions gave optimized extraction efficiency.

3.4. Total Phenolics by Folin–Ciocalteu

Total Phenolic Content (TPC) was determined using the Folin–Ciocalteu method as previously described [12].

3.5. Chemical Antioxidant Assays (DPPH, FRAP, ORAC)

Chemical antioxidant properties were assessed using DPPH, FRAP and ORAC assays as described in our previous paper [12]. These assays are based on different antioxidant mode of action like electron transfer (FRAP), hydrogen transfer (ORAC) or a mixed mechanism implying both electron and hydrogen transfer (DPPH) [13].

3.6. Cellular Assay

3.6.1. Cell Culture

Normal Human Dermal Fibroblasts (NHDF) were purchased from PromoCell (Heidelberg, Germany), cultured in growth medium RPMI GlutaMAX™ (89%) (Thermo Fischer Scientific, Waltham, MA, USA), supplemented with 5% FBS (Fetal Bovine Serum), 1% antibiotics, 5% CO_2 and maintained at 37 $^\circ$C. NHDF cells were used to determine the cytotoxicity of *E. oleracea* extracts and to assess the antioxidant activity of the extracts in a cell-based assay described in Section 3.6.3.

3.6.2. Cytotoxicity Assay

Cytotoxicity of the extracts in NHDF was measured using the MTT (3-(4,5-dimethylthiazol-2 -yl)-2,5-diphenyltetrazolium bromide) method [34]. NHDF cells were incubated for 24 h at 37 °C, in presence or absence of *E. oleracea* extracts at 100, 200, 300, 400 and 500 µg/mL. The cells were washed twice with 500 µL of a phosphate buffer solution (PBS), then incubated at 37 °C for 3 h with MTT at 0.5 mg/mL. MTT was removed using 1 mL of DMSO and the plate was kept in the dark at room temperature during 30 min before recording the absorbance at 595 nm using a plate-reader (Dynex, Magellan Biosciences, Tampa, FL, USA) to detect viable cells. A decrease of the absorbance by more than 20% was set as the limit for cytotoxicity.

3.6.3. Cellular Antioxidant Activity (CAA) Assay

The cell-based assay was done according to the method of Wolfe and Liu [35], with some modifications. Briefly, NHDF cells were seeded for 24 h on a 96-well microplate at a density of 11×10^4 cells by well in 100 µL of RPMI growth medium. The growth medium was removed and the wells were washed with PBS. Wells were treated in triplicates with 100 µL of extracts at four concentrations (10, 25, 50 and 100 µg/mL), not cytotoxic to the NHDF cells, plus 50 µM 2′,7′-dichlorodihydrofluorescein diacetate (DCFH-DA) dissolved in the treatment medium and kept in the dark for 1 h. Control wells (cells treated with 50 µM of DCFH-DA, without extract), and blank wells (cells treated only with growth medium) were included. Then, the wells were drained from the treatment medium and 100 µL of 2,2′-Azobis(2-amidinopropane) dihydrochloride (AAPH) at 250 µM was added in each well (except in blank wells). Immediately after the AAPH addition, the plate was placed into a plate-reader (Dynex, Magellan Biosciences) at 37 °C and the fluorescence was recorded for 30 min ($\lambda_{excitation}$ 485 nm, $\lambda_{emission}$ 538 nm). After blank subtraction from the fluorescence readings, the area under the curve of fluorescence versus time was integrated to calculate the CAA value at each extracts concentration tested as follows: CAA unit: $100 - (\int SA / \int CA) \times 100$ where $\int SA$ is the integrated area under the sample fluorescence versus time curve and $\int CA$ is the integrated area from the control curve. The median effective concentration EC_{50} (µg/mL), was determined for *E. oleracea* extracts from the median effect plot of log (fa/fu) versus log (concentration), where fa is the fraction affected and fu is the fraction unaffected by the treatment. Quercetin was used as a standard in the CAA assay and the cellular antioxidant activity of *E. oleracea* extracts was then expressed in µmol Quercetin equivalents/g of dried matter (µmol QEq/g DM).

3.6.4. DNA Nicking Assay

The DNA nicking assay was performed using the pUC18 plasmid. All conditions required to analyze aqueous and organic extracts were previously optimized [16]. Four microliters of *E. oleracea* extracts (max 7% acetone) at various concentrations were added to 4 µL of the Fenton reaction mixture, containing: plasmid DNA (150 µg/µL), phosphate buffer (50 mM, pH 7.4), H_2O_2 (30 mM), $FeSO_4$ (2 mM/8 mM) and EDTA-Na$_2$ (3.75 mM/15 mM) for aqueous and acetone extracts respectively. The final volume was adjusted to 24 µL using distilled water and incubated for 20 min and 15 min at 37 °C for aqueous and acetone extracts respectively. Two microliters of loading dye was added to the incubated mixture, and 10 µL were loaded onto 1% (w/v) agarose gel. After electrophoresis in TAE buffer (0.04 M tris-acetate and 1 mM EDTA, pH 7.4) using a DNA subcell (Bio-Rad Laboratories, Hercules, CA, USA), the agarose gel was stained with ethidium bromide for 15 min and DNA bands were analyzed using a Bio-Rad Gel Doc™ XR (Bio-Rad Laboratories). The prooxidant and antioxidant activity of the extracts was assessed using band intensity (NIH Image J) of form I (supercoiled DNA) and form III (nicked linear DNA) compared to the appropriate controls. DNA incubated with and without Fenton's reagent was used as positive and negative control, respectively. Trolox was used as a control of DNA protection, at a concentration of 0.01 and 0.1 mg/mL for aqueous and acetone extracts, respectively.

3.6.5. Analysis by LC-MS/MS

LC-MS/MS analysis was performed on an ion trap (500-MS Varian, Palo Alto, CA, USA) equipped with an electrospray ionization source and coupled to a Pro Star LC system (Agilent Technologies, Santa Clara, CA, USA) and a Pro Star 335 PDA (Agilent). Extracts were filtered using 0.2 μm PTFE filters. Twenty microliters of samples were injected onto a Kinetex PFP column, 100 × 4.6 mm, 2.6 μm (Agilent) using a gradient of 1% formic acid (A) and acetonitrile (B) at 25 °C. The gradient was as follows: 5%–10% B in 10 min, 10%–20% in 10 min, 20% held for 10 min, 20%–100% in 5 min, 100% held for 5 min, and returned to initial conditions for 9 min to reequilibrate the column. The flow rate was 1 mL/min. The total run time per sample was 60 min.

Analyses were first performed using Full scan mode (m/z 120–750), both in negative mode and positive mode to identify the molecular ions and then in TurboDDS™ mode (Data Dependent Scanning, Varian) to acquire MS^2 spectra. Helium was used as the damping and collision gas at 0.8 mL/min. The operation parameters were as follows: nebulizer gas pressure 50 psi, drying gas pressure 25 psi, drying gas temperature 350 °C, needle voltage −5000 V/5000 V, sprayshield voltage −600 V, spray chamber 50 °C, capillary voltage 100 V/−70 V (negative ionization mode/positive ionization mode).

The identification of compounds was performed according to their MS^2 fragmentation [15–17,26,27], using standards when available. Quantification was carried out with UV detection. 5-CQA and 6-C-glycosyl apigenin (isovitexin) were used as calibration standards (λ_{max} = 320 nm) to quantify hydroxycinnamic acids (caffeoylquinic acids—CQA and caffeoylshikimic acids—CSA) and flavones, respectively. Gallic acid was used as the internal standard at λ_{max} = 280 nm.

3.6.6. Statistical Data Analysis

Results of antioxidant activity were presented as the mean of three biological replicates and standard error (SE) or standard error of the mean (SEM). Comparisons between extracts in the TPC, ORAC, FRAP, DPPH and DNA nicking assays were performed using an ANOVA followed by multiple comparisons using Fisher's Least Significant Difference test ($p < 0.05$). Principal component analysis (PCA) was performed on the phytochemical composition and chemical antioxidant activity (DPPH, FRAP, and ORAC) of roots and leaflet extracts from *E. oleracea* using the Statistica program (Statsoft, Paris, France).

4. Conclusions

Root and leaflet extracts from *E. oleracea* were active across the board in different chemical antioxidant assays (DPPH, FRAP, ORAC, DNA nicking) and biological assay (CAA). The high antioxidant activity of the root and leaflet extracts of *E. oleracea* was respectively correlated to the presence of hydroxycinnamic acids and apigenin C-glycosides. The high antioxidant activity of *E. oleracea* leaflets seemed to come from their highest content of polyphenols, suggesting a particular regulation of the polyphenols biosynthetic pathway. This study shows for the first time that *Euterpe oleracea* roots and leaflets, which are currently by-products of the palm heart industry, could, along with the berries, be valorized as a new non-cytotoxic source of antioxidants containing hydroxycinnamic acids and flavonoids for pharmaceutical, nutraceutical or cosmetic applications.

Supplementary Materials: Supplementary materials can be found at www.mdpi.com/1422-0067/18/1/61/s1.

Acknowledgments: We thank the European Regional Development Fund (ERDF/FEDER, presage 30574), "la Région Guyane" (French Guiana Region), the French Ministry of Higher Education and Research (MESR) and the "Centre National d'Etudes Spatiales" (CNES) for financial support of this research project (Palmazon Project, Valorization of Amazonian palm trees from French Guiana, 2010). The University of French Guiana is gratefully acknowledged for support. We also thank Mathew Njoroge from the University of Cape Town, for his work in proof reading this manuscript.

Author Contributions: Christel Brunschwig performed the ORAC assay, identified and quantified the compounds by LC-MS/MS, performed statistical analysis, and wrote the manuscript; Louis-Jérôme Leba performed the extractions, the DPPH assay, the Folin-Ciocalteu assay, performed statistical analysis, and wrote the manuscript; Mona Saout and Karine Martial helped in performing the experiments; Didier Bereau managed, edited and revised the manuscript; and Jean-Charles Robinson managed, designed and revised the manuscript. The final version of the manuscript has been read and accepted by all the authors.

Conflicts of Interest: The authors declare no conflict of interest.

Abbreviations

DPPH	2,2-Diphenyl-1-picrylhydrazyl
FRAP	Ferric Reducing Antioxidant Power
ORAC	Oxygen Radical Absorbance Capacity
CAA	Cellular Antioxidant Activity
TPC	Total Polyphenol Content
GAEq	Gallic acid equivalent
TEq	Trolox equivalent
QEq	Quercetin equivalent
DM	Dry Matter
PCA	Principal Component Analysis
NHDF	Normal Human Dermal Fibroblasts
HCA	Hydroxycinnamic acid
CQA	Caffeoylquinic acid
CSA	Caffeoylshikimic acid
MTT	3-(4,5-Dimethylthiazol-2-yl)-2,5-diphenyltetrazolium bromide

References

1. Yamaguchi, K.K.D.L.; Pereira, L.F.R.; Lamarão, C.V.; Lima, E.S.; da Veiga, V.F., Jr. Amazon açai: Chemistry and biological activities: A review. *Food Chem.* **2015**, *179*, 137–151. [CrossRef] [PubMed]
2. Heinrich, M.; Dhanji, T.; Casselman, I. Açai (*Euterpe oleracea* Mart.)—A phytochemical and pharmacological assessment of the species' health claims. *Phytochem. Lett.* **2011**, *4*, 10–21. [CrossRef]
3. Grenand, P.; Moretti, C.; Jacquemin, H.; Prévost, M.-F. *Pharmacopées traditionnelles en Guyane*; IRD Editions: Paris, France, 2004.
4. Bourdy, G.; deWalt, S.J.; Chávez de Michel, L.R.; Roca, A.; Deharo, E.; Muñoz, V.; Balderrama, L.; Quenevo, C.; Gimenez, A. Medicinal plants uses of the Tacana, an Amazonian Bolivian ethnic group. *J. Ethnopharmacol.* **2000**, *70*, 87–109. [CrossRef]
5. Hajdu, Z.; Hohmann, J. An ethnopharmacological survey of the traditional medicine utilized in the community of Porvenir, Bajo Paraguá Indian Reservation, Bolivia. *J. Ethnopharmacol.* **2012**, *139*, 838–857. [CrossRef] [PubMed]
6. Deharo, E.; Baelmans, R.; Gimenez, A.; Quenevo, C.; Bourdy, G. In vitro immunomodulatory activity of plants used by the Tacana ethnic group in Bolivia. *Phytomedicine* **2004**, *11*, 516–522. [CrossRef] [PubMed]
7. Galotta, A.L.Q.A.; Boaventura, M.A.D.; Lima, L.A.R.S. Antioxidant and cytotoxic activities of "Açaí" (*Euterpe precatoria* mart.). *Quim. Nova* **2008**, *31*, 1427–1430. [CrossRef]
8. Schieber, A.; Stintzing, F.; Carle, R. By-products of plant food processing as a source of functional compounds—recent developments. *Trends Food Sci. Technol.* **2001**, *12*, 401–413. [CrossRef]
9. Ferlemi, A.-V.; Lamari, F.N. Berry Leaves: An Alternative Source of Bioactive Natural Products of Nutritional and Medicinal Value. *Antioxidants* **2016**, *5*, 17. [CrossRef] [PubMed]
10. Barros, L.; Calhelha, R.C.; Queiroz, M.J.R.P.; Santos-Buelga, C.; Santos, E.A.; Regis, W.C.B.; Ferreira, I.C.F.R. The powerful in vitro bioactivity of *Euterpe oleracea* Mart. seeds and related phenolic compounds. *Ind. Crops Prod.* **2015**, *76*, 318–322. [CrossRef]
11. Sudo, R.T.; Neto, M.L.; Monteiro, C.E.S.; Amaral, R.V.; Resende, A.C.; Souza, P.J.C.; Zapata-Sudo, G.; Moura, R. Antinociceptive effects of hydroalcoholic extract from *Euterpe oleracea* Mart. (Açaí) in a rodent model of acute and neuropathic pain. *BMC Complement. Altern. Med.* **2015**, *15*, 208. [CrossRef] [PubMed]

12. Leba, L.-J.; Brunschwig, C.; Saout, M.; Martial, K.; Bereau, D.; Robinson, J.-C. *Oenocarpus bacaba* and *Oenocarpus bataua* Leaflets and Roots: A New Source of Antioxidant Compounds. *Int. J. Mol. Sci.* **2016**, *17*, 1014. [CrossRef] [PubMed]

13. Prior, R.L.; Wu, X.; Schaich, K. Standardized methods for the determination of antioxidant capacity and phenolics in foods and dietary supplements. *J. Agric. Food Chem.* **2005**, *53*, 4290–4302. [CrossRef] [PubMed]

14. Rezaire, A.; Robinson, J.-C.; Bereau, D.; Verbaere, A.; Sommerer, N.; Khan, M.K.; Durand, P.; Prost, E.; Fils-Lycaon, B. Amazonian palm *Oenocarpus bataua* ("patawa"): Chemical and biological antioxidant activity–Phytochemical composition. *Food Chem.* **2014**, *149*, 62–70. [CrossRef] [PubMed]

15. Abadio Finco, F.D.B.; Kammerer, D.R.; Carle, R.; Tseng, W.-H.; Böser, S.; Graeve, L. Antioxidant activity and characterization of phenolic compounds from bacaba (*Oenocarpus bacaba* Mart.) fruit by HPLC-DAD-MS(n). *J. Agric. Food Chem.* **2012**, *60*, 7665–7673. [CrossRef] [PubMed]

16. Leba, L.-J.; Brunschwig, C.; Saout, M.; Martial, K.; Vulcain, E.; Bereau, D.; Robinson, J.-C. Optimization of a DNA nicking assay to evaluate *Oenocarpus bataua* and *Camellia sinensis* antioxidant capacity. *Int. J. Mol. Sci.* **2014**, *15*, 18023–18039. [CrossRef] [PubMed]

17. Tiveron, A.P.; Melo, P.S.; Bergamaschi, K.B.; Vieira, T.M.F.S.; Regitano-d'Arce, M.A.B.; Alencar, S.M. Antioxidant activity of Brazilian vegetables and its relation with phenolic composition. *Int. J. Mol. Sci.* **2012**, *13*, 8943–8957. [CrossRef] [PubMed]

18. Girard-Lalancette, K.; Pichette, A.; Legault, J. Sensitive cell-based assay using DCFH oxidation for the determination of pro- and antioxidant properties of compounds and mixtures: Analysis of fruit and vegetable juices. *Food Chem.* **2009**, *115*, 720–726. [CrossRef]

19. Fang, N.; Yu, S.; Prior, R.L. LC/MS/MS characterization of phenolic constituents in dried plums. *J. Agric. Food Chem.* **2002**, *50*, 3579–3585. [CrossRef] [PubMed]

20. Clifford, M.N.; Johnston, K.L.; Knight, S.; Kuhnert, N. Hierarchical scheme for LC-MSn identification of chlorogenic acids. *J. Agric. Food Chem.* **2003**, *51*, 2900–2911. [CrossRef] [PubMed]

21. Regos, I.; Treutter, D. Optimization of a high-performance liquid chromatography method for the analysis of complex polyphenol mixtures and application for sainfoin extracts (*Onobrychis viciifolia*). *J. Chromatogr. A* **2010**, *1217*, 6169–6177. [CrossRef] [PubMed]

22. Hammouda, H.; Kalthoum Chérif, J.; Trabelsi-Ayadi, M.; Baron, A.; Guyot, S. Detailed polyphenol and tannin composition and its variability in Tunisian dates (*Phoenix dactylifera* L.) at different maturity stages. *J. Agric. Food Chem.* **2013**, *61*, 3252–3263. [CrossRef] [PubMed]

23. Jaiswal, R.; Sovdat, T.; Vivan, F.; Kuhnert, N. Profiling and characterization by LC-MSn of the chlorogenic acids and hydroxycinnamoylshikimate esters in maté (Ilex paraguariensis). *J. Agric. Food Chem.* **2010**, *58*, 5471–5484. [CrossRef] [PubMed]

24. Parveen, I.; Threadgill, M.D.; Hauck, B.; Donnison, I.; Winters, A. Isolation, identification and quantitation of hydroxycinnamic acid conjugates, potential platform chemicals, in the leaves and stems of *Miscanthus ×giganteus* using LC-ESI-MS(n). *Phytochemistry* **2011**, *72*, 2376–2384. [CrossRef] [PubMed]

25. Ferreres, F.; Silva, B.M.; Andrade, P.B.; Seabra, R.M.; Ferreira, M.A. Approach to the study of C-glycosyl flavones by ion trap HPLC-PAD-ESI/MS/MS: Application to seeds of quince (*Cydonia oblonga*). *Phytochem. Anal.* **2003**, *14*, 352–359. [CrossRef] [PubMed]

26. Barron, D.; Varin, L.; Ibrahim, R.K.; Harborne, J.B.; Williams, C.A. Sulphated flavonoids–an update. *Phytochemistry* **1988**, *27*, 2375–2395. [CrossRef]

27. Karaköse, H.; Jaiswal, R.; Kuhnert, N. Characterization and quantification of hydroxycinnamate derivatives in *Stevia rebaudiana* leaves by LC-MS(n). *J. Agric. Food Chem.* **2011**, *59*, 10143–10150. [CrossRef] [PubMed]

28. Williams, C.A.; Harborne, J.B.; Clifford, H.T. Negatively charged flavones and tricin as chemosystematic markers in the palmae. *Phytochemistry* **1973**, *12*, 2417–2430. [CrossRef]

29. De Oliveira, D.M.; Siqueira, E.P.; Nunes, Y.R.F.; Cota, B.B. Flavonoids from leaves of *Mauritia flexuosa*. *Braz. J. Pharmacogn.* **2013**, *23*, 614–620. [CrossRef]

30. Correia-da-Silva, M.; Sousa, E.; Pinto, M.M.M. Emerging sulfated flavonoids and other polyphenols as drugs: Nature as an inspiration. *Med. Res. Rev.* **2014**, *34*, 223–279. [CrossRef] [PubMed]

31. Heim, K.E.; Tagliaferro, A.R.; Bobilya, D.J. Flavonoid antioxidants: Chemistry, metabolism and structure-activity relationships. *J. Nutr. Biochem.* **2002**, *13*, 572–584. [CrossRef]

32. Ignat, I.; Volf, I.; Popa, V.I. A critical review of methods for characterisation of polyphenolic compounds in Fruits and Vegetables. *Food Chem.* **2011**, *126*, 1821–1835. [CrossRef] [PubMed]
33. Barba, F.J.; Zhu, Z.; Koubaa, M.; Sant'Ana, A.S.; Orlien, V. Green alternative methods for the extraction of antioxidant bioactive compounds from winery wastes and by-products: A review. *Trends Food Sci. Tech.* **2016**, *46*, 96–109. [CrossRef]
34. Mosmann, T. Rapid colorimetric assay for cellular growth and survival: Application to proliferation and cytotoxicity assays. *J. Immunol. Methods* **1983**, *65*, 55–63. [CrossRef]
35. Wolfe, K.L.; Rui, H.L. Cellular antioxidant activity (CAA) assay for assessing antioxidants, foods, and dietary supplements. *J. Agric. Food Chem.* **2007**, *55*, 8896–8907. [CrossRef] [PubMed]

International Journal of
Molecular Sciences

MDPI

Article

New Abietane and Kaurane Type Diterpenoids from the Stems of *Tripterygium regelii*

Dongsheng Fan [1], Shuangyan Zhou [2], Zhiyuan Zheng [1], Guo-Yuan Zhu [1], Xiaojun Yao [1,2], Ming-Rong Yang [1], Zhi-Hong Jiang [1] and Li-Ping Bai [1,*]

[1] State Key Laboratory of Quality Research in Chinese Medicine, Macau Institute for Applied Research in Medicine and Health, Macau University of Science and Technology, Taipa, Macau 999078, China; fandongsheng1985@sina.com (D.F.); zyzheng@must.edu.mo (Z.Z.); gyzhu@must.edu.mo (G.-Y.Z.); xjyao@must.edu.mo (X.Y.); mryang@must.edu.mo (M.-R.Y.); zhjiang@must.edu.mo (Z.-H.J.)
[2] State Key Laboratory of Applied Organic Chemistry and Department of Chemistry, Lanzhou University, Lanzhou 730000, China; zhoushy13@lzu.edu.cn
* Correspondence: lpbai@must.edu.mo; Tel.: +853-8897-2403; Fax: +853-2888-0091

Academic Editor: Jianhua Zhu
Received: 10 December 2016; Accepted: 6 January 2017; Published: 13 January 2017

Abstract: Eleven new abietane type (**1-11**), and one new kaurane (**12**), diterpenes, together with eleven known compounds (**13–23**), were isolated and identified from the stems of *Tripterygium regelii*, which has been used as a traditional folk Chinese medicine for the treatment of rheumatoid arthritis in China. The structures of new compounds were characterized by means of the interpretation of high-resolution electrospray ionization mass spectrometry (HRESIMS), extensive nuclear magnetic resonance (NMR) spectroscopic data and comparisons of their experimental CD spectra with calculated electronic circular dichroism (ECD) spectra. Compound **1** is the first abietane type diterpene with an 18→1 lactone ring. Compound **19** was isolated from the plants of the *Tripterygium* genus for the first time, and compounds **14–17** were isolated from *T. regelii* for the first time. Triregelin I (**9**) showed significant cytotoxicity against A2780 and HepG2 with IC_{50} values of 5.88 and 11.74 µM, respectively. It was found that this compound was inactive against MCF-7 cells. The discovery of these twelve new diterpenes not only provided information on chemical substances of *T. regelii*, but also contributed to the chemical diversity of natural terpenoids.

Keywords: *Tripterygium regelii*; diterpenoids; cytotoxicity

1. Introduction

Diterpenes are naturally-occurring 20-carbon terpenoids that display a wide array of potentially useful biological effects. Abietanes are a large group of diterpenoids, which have been isolated from a variety of terrestrial plants, such as families of Araucariaceae, Cupressaceae, Phyllocladaceae, Pinaceae, Podocarpaceae, Asteraceae, Celastraceae, Hydrocharitaceae, and Lamiaceae, etc. [1,2]. Furthermore, this class of diterpenes have been found from fungal species [2]. So far, it has been reported that some abietane type diterpenes displayed a broad spectrum of promising biological activities including anticancer [3], cytotoxic [4,5], antiviral [6–9], anti-inflammatory [10], and anti-oxidant [9] effects, and so on. For example, tanshinone IIA was regarded as a potent cytotoxic compound for human leukemia cells [11]. Carnosol has been found to possess favorable anticancer and chemo-preventive effects [12]. Triptolide is a promising lead compound to treat inflammatory, immunological and cancerous diseases [13]. Recently, it was reported that miltirone is an inhibitor of P-glycoprotein [14].

As a part of ongoing research work on bioactive constituents from *Tripterygium regelii* [15–17], the methanolic extract of the stems of *T. regelii* was further investigated, leading to the isolation and characterization of twenty three diterpenoids, including eleven new abietane (**1−11**) and one new kaurane type (**12**) diterpenes, as well as eleven known abietane compounds (**13**−**23**) (Figure 1). Herein, this paper reports the isolation and structural elucidation of these new diterpenes, as well as cytotoxic evaluation of seventeen diterpenes on three cancer cell lines.

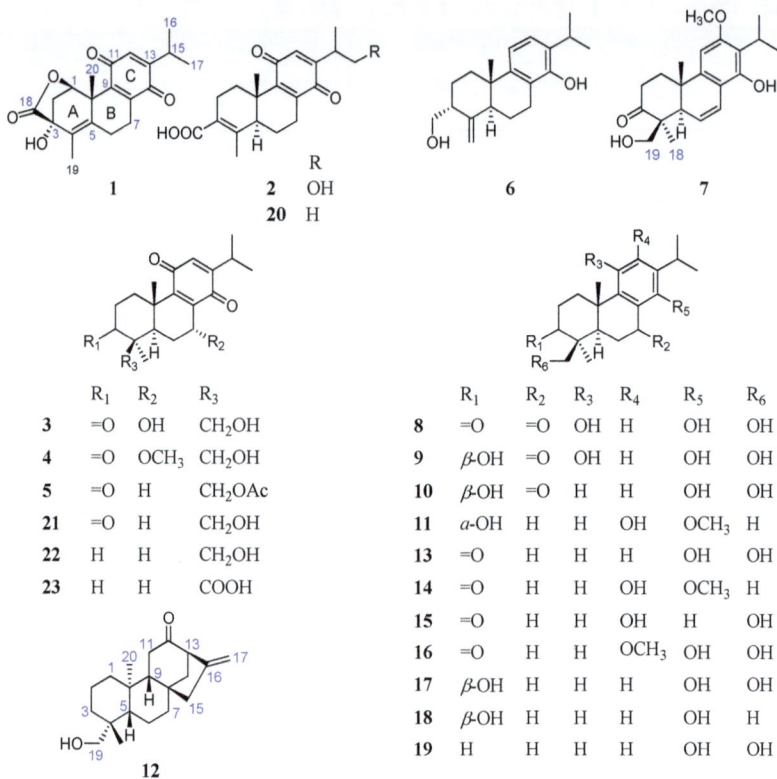

	R
2	OH
20	H

	R_1	R_2	R_3
3	=O	OH	CH_2OH
4	=O	OCH_3	CH_2OH
5	=O	H	CH_2OAc
21	=O	H	CH_2OH
22	H	H	CH_2OH
23	H	H	COOH

	R_1	R_2	R_3	R_4	R_5	R_6
8	=O	=O	OH	H	OH	OH
9	β-OH	=O	OH	H	OH	OH
10	β-OH	=O	H	H	OH	OH
11	a-OH	H	H	OH	OCH_3	H
13	=O	H	H	H	OH	OH
14	=O	H	H	OH	OCH_3	H
15	=O	H	H	OH	H	OH
16	=O	H	H	OCH_3	OH	OH
17	β-OH	H	H	H	OH	OH
18	β-OH	H	H	H	OH	H
19	H	H	H	H	OH	OH

12

Figure 1. The chemical structures of compounds **1**−**23**.

2. Results and Discussion

Compound **1** was obtained as a yellow amorphous powder with a molecular formula of $C_{20}H_{22}O_5$, which was determined by a protonated molecular ion at *m/z* 343.1546 [M + H]$^+$ (calcd for $C_{20}H_{23}O_5$, 343.1540) in its high-resolution electrospray ionization mass spectrometry (HRESIMS), indicating 10 degrees of unsaturation. IR spectrum of **1** showed a lactone carbonyl band at 1732 cm^{-1} and benzoquinone bands at 1680 and 1601 cm^{-1}. The UV spectrum of **1** exhibited an absorption maximum at 260 nm, which is characteristic of a *p*-benzoquinone. The ^{1}H nuclear magnetic resonance (NMR) spectroscopic data (Table 1) exhibited the characteristic signals for a benzoquinone proton (δ_H 6.41 (1H, d, *J* = 1.2 Hz, H-12)), an oxygenated methine (δ_H 5.86 (1H, d, *J* = 6.0 Hz, H-1)), an isopropyl moiety including a methine (δ_H 3.00 (1H, sept d, *J* = 6.6, 1.2 Hz, H-15)) and two secondary methyls (δ_H 1.12 and 1.11 (each 3H, d, *J* = 6.6 Hz, H$_3$-16 and H$_3$-17)), and two tertiary methyls (δ_H 1.85 and 1.54 (each 3H, s, H$_3$-19 and H$_3$-20)). The ^{13}C NMR spectroscopic data (Table 2) displayed resonances for 20 carbons, which were confirmed by distortionless enhancement by polarization transfer (DEPT) and heteronuclear single quantum coherence (HSQC) experiments to be an ester carbonyl carbon (δ_C 177.3), a trisubstituted *p*-benzoquinone (δ_C 187.3 (C-11), 186.8 (C-14), 153.9 (C-13), 146.0 (C-8), 144.6 (C-9) and 131.6 (C-12)), a tetrasubstituted double bond (δ_C 132.6 (C-5) and 130.2 (C-4)), two aliphatic quaternary carbons (including an oxygenated one), two methines (including an oxygenated one), three methylenes, and four methyl groups. These spectroscopic data (Tables 1 and 2) suggested that compound **1** is an abietane type diterpene with a *p*-benzoquinone C-ring [18,19], structurally similar to the known triptoquinone A (**20**) [18], an 18(4→3)-*abeo*-abietane quinone type diterpene, except for the A-ring. The $\Delta^{4,5}$ double bond was inferred from the HMBC correlations from H$_2$-2, H$_2$-6 and H$_3$-19 to C-4 (δ_C 130.2), from H-7 (δ_H 2.90), H$_3$-19 and H$_3$-20 to C-5 (δ_C 132.6). The oxygenated methine (δ_H 5.86; δ_C 78.4) was assigned to C-1 based on the HMBC correlations from H-1 proton (δ_H 5.86) to C-2 (δ_C 39.1), C-3 (δ_C 74.3), C-5 (δ_C 132.6), C-9 (δ_C 144.6), C-10 (δ_C 44.7), and C-20 (δ_C 23.7). The key HMBC correlation from H-1 (δ_H 5.86) to C-18 (δ_C 177.3) suggested a lactone formed between C-1 and C-18, accounting for the remaining one degree of unsaturation. Hydroxylation of C-3 was inferred from the HMBC correlations from H-1 (δ_H 5.86), H$_2$-2 (δ_H 2.36 and 1.96) and H$_3$-19 (δ_H 1.85) to C-3 (δ_C 74.3). It was deduced that the proton at C-1 and the hydroxyl group at C-3 should be a *cis* relationship due to the lactone between C-1 and C-3. Therefore, the proposed structure of **1** was established as a lactone derivative of triptoquinone A bearing 5*S*, 10*S* absolute configuration by X-ray crystallographic analysis [18] (Figure 2).

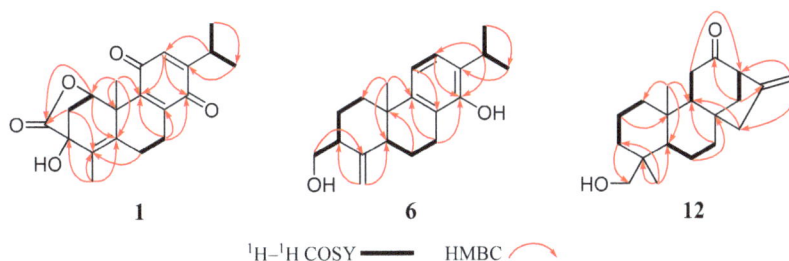

Figure 2. The ^{1}H–^{1}H COSY and key HMBC correlations of compounds **1**, **6**, and **12**.

However, the relative configuration of the substituents at the C-1 and C-3 could not be assigned by nuclear Overhauser effect spectroscopy (NOESY) experiment, owing to the fact that no any key NOE effects were observed (Figure 3). Hence, electron circular dichroism (ECD) calculations were conducted to determine the absolute configuration of compound **1** by time-dependent density functional theory (TDDFT) with the B3LYP/DGDZVP method [20,21]. The calculated ECD of (1*R*, 3*R*)-**1** matched well with the experimental CD spectrum (Figure 4A) of **1**. Therefore, compound **1** was determined as proposed, and given the trivial name of triregelin A.

Int. J. Mol. Sci. 2017, 18, 147

Table 1. ^{1}H NMR (600 MHz) spectroscopic data for compounds 1–5 and 12.

Position	δ_H (J in Hz)					
	1 [a]	2 [a]	3 [a]	4 [a]	5 [a]	12 [a]
1	5.86, d (6.0)	1.45, m [c] / 2.77, m [c]	1.89, ddd (14.4, 10.2, 5.4) / 2.80, ddd (14.4, 9.0, 6.0) [c]	1.90, m / 2.75, m [c]	1.65, m / 2.77, m [c]	0.79, td (13.2, 3.6) / 1.79, m [c]
2	1.96, d (10.8) / 2.36, m [c]	2.46, m / 2.55, m	2.46, m [c] / 2.73, ddd (16.2, 10.2, 6.0) [c]	2.45, m [c] / 2.75, m [c]	2.52, ddd (15.6, 6.6, 3.6) / 3.03, ddd (15.6, 7.2, 3.6) [c]	1.43, dt (13.8, 3.6) / 1.55, m [c]
3	-	-	-	-	-	0.95, dd (13.8, 4.2) / 1.79, m [c]
5	2.16, m	2.22, m [c]	2.46, d (13.8) [c]	2.47, d (13.8) [c]	1.78, dd (12.6, 1.8)	1.00, m [c]
6	2.76, dd (13.2, 6.0)	1.50, m [c] / 2.24, m [c]	1.65, td (13.8, 4.2) / 1.96, br d (13.8)	1.41, td (13.8, 3.0) / 2.02, dt (13.8, 1.8)	1.59, m / 1.96, br d, d (13.2, 7.2)	1.36, qd (12.6, 3.6) / 1.72, m [c]
7	2.37, m [c] / 2.90, dd (19.8, 6.0)	2.39, ddd (18.6, 11.4, 7.2) / 2.80, m	4.81, br s	4.39, dd (3.0, 1.8)	2.34, ddd (18.6, 12.0, 7.2) / 2.81, m [c]	1.62, dd (13.8, 4.8) / 1.72, m [c]
9	-	-	-	-	-	1.58, m [c]
11	-	-	-	-	-	-
12	6.41, d (1.2)	6.47, s	6.44, s	6.42, d (1.2)	6.37, d (1.2)	2.24, d (17.0) / 2.53, dd (17.0, 9.6)
13	-	-	-	-	-	-
14	-	-	-	-	-	3.21, d (4.8)
15	3.00, sept d (6.6, 1.2)	3.12, m	3.02, sept (7.0)	3.04, sept (7.2, 1.2)	3.00, d (7.2) [c]	1.51, dd (12.6, 4.8) / 2.40, d (12.6)
16	1.12, d (6.6)	1.17, d (7.2)	1.14, d (7.0)	1.12, d (7.2)	1.10, d (7.2)	2.36, s
17	1.11, d (6.6)	3.67, d (7.2)	1.13, d (7.0)	1.13, d (7.2)	1.11, d (7.2)	4.87, s / 4.99, s
18	-	-	1.37, s	1.36, s	1.22, s	0.98, s
19	1.85, s	2.11, s	3.47, t (10.8) / 4.02, dd (10.8, 2.4)	3.44, d (12.0) / 4.05, dd (12.0)	4.56, d (12.0) / 4.08, d (12.0)	3.68, dd (10.1, 4.0) / 3.44, dd (10.1, 4.0)
20	1.54, s	1.18, s	1.24, s	1.22, s	1.44, s	0.85, s
OH-7	-	-	2.77, s [c]	3.26, br s	-	-
OH-19	-	-	3.15, dd (10.8, 2.4)	-	-	1.09, br s
OMe-7	-	-	-	3.50, s	-	-
OAc-19	-	-	-	-	2.03, s	-

[a] Measured in $CDCl_3$; [c] Overlapping signal was assigned from ^{1}H–^{1}H COSY, HSQC and HMBC experiments. The signals of br, s, d, t, q, sept and m represent broad, singlet, doublet, triplet, quartet, septet and multiplet splitting patterns of protons, respectively.

Table 2. ^{13}C NMR (150 MHz) spectroscopic data for compounds 1–12.

Position	δ_C, Type											
	1[a]	2[a]	3[a]	4[a]	5[a]	6[a]	7[a]	8[a]	9[a]	10[a]	11[b]	12[a]
1	78.4, CH	31.8, CH_2	34.1, CH_2	34.1, CH_2	34.7, CH_2	38.2, CH_2	35.1, CH_2	35.1, CH_2	34.5, CH_2	36.2, CH_2	32.8, CH_2	39.7, CH_2
2	39.1, CH_2	24.6, CH_2	34.2, CH_2	34.2, CH_2	34.9, CH_2	27.4, CH_2	35.3, CH_2	34.8, CH_2	28.3, CH_2	28.1, CH_2	27.0, CH_2	17.8, CH_2
3	74.3, C	147.9, C	220.4, C	220.8, C	212.4, C	46.1, CH	214.8, C	219.0, C	79.7, CH	79.9, CH	74.8, CH	35.5, CH_2
4	130.2, C	124.5, C	49.9, C	49.7, C	51.3, C	150.8, C	53.0, C	50.3, C	42.8, C	42.3, C	38.3, C	38.6, C
5	132.6, C	47.3, CH	45.2, CH	44.9, CH	52.9, CH	47.9, CH	51.5, CH	49.4, CH	49.6, CH	48.8, CH	44.0, CH	56.2, CH
6	22.2, CH_2	18.7, CH_2	26.1, CH_2	22.5, CH_2	18.5, CH_2	20.8, CH_2	124.0, CH	35.6, CH_2	35.5, CH_2	35.7, CH_2	18.9, CH_2	20.2, CH_2
7	27.0, CH_2	25.2, CH_2	61.9, CH	69.8, CH	26.0, CH_2	23.7, CH_2	122.1, CH	204.3, C	205.6, C	205.2, C	25.1, CH_2	39.4, CH_2
8	146.0, C	142.6, C	140.9, C	139.3, C	142.8, C	120.7, C	113.6, C	114.9, C	115.1, C	114.2, C	119.7, C	44.1, C
9	144.6, C	149.0, C	148.7, C	148.6, C	148.0, C	145.8, C	145.2, C	133.1, C	134.7, C	153.1, C	150.0, C	57.6, CH
10	44.7, C	36.6, C	37.5, C	37.3, C	37.6, C	39.7, C	37.7, C	38.3, C	39.5, C	37.5, C	37.9, C	39.2, C
11	187.3, C	187.4, C	187.8, C	187.9, C	187.5, C	117.6, CH	98.4, CH	144.5, C	144.2, C	113.6, CH	108.8, CH	35.9, CH_2
12	131.6, CH	134.0, CH	132.4, CH	131.8, CH	132.0, CH	123.3, CH	158.3, C	123.8, CH	123.9, CH	133.6, CH	156.5, C	211.5, C
13	153.9, C	149.0, C	153.6, C	154.0, C	153.3, C	130.3, C	119.6, C	136.6, C	136.3, C	134.9, C	125.1, C	60.7, CH
14	186.8, C	188.1, C	188.7, C	186.4, C	187.4, C	150.3, C	150.3, C	155.4, C	155.7, C	160.7, C	156.9, C	39.4, CH_2
15	26.6, CH	34.5, CH	26.4, CH	26.5, CH	26.4, CH	26.9, CH	24.3, CH	26.0, CH	26.0, CH	26.1, CH	26.1, CH	48.2, CH
16	21.3, CH_3	15.4, CH_3	21.3, CH_3	21.3, CH_3	21.3, CH_3	22.6, CH_3	20.9, CH_3	22.1, CH_3	22.1, CH_3	22.1, CH_3	21.6, CH_3	148.8, C
17	21.4, CH_3	66.6, CH_2	21.3, CH_3	21.4, CH_3	21.3, CH_3	22.8, CH_3	20.9, CH_3	22.2, CH_3	22.2, CH_3	22.3, CH_3	21.7, CH_3	107.8, CH_2
18	177.3, C	173.7, C	22.3, CH_3	22.3, CH_3	21.8, CH_3	64.7, CH_2	19.7, CH_3	22.6, CH_3	22.4, CH_3	22.0, CH_3	29.1, CH_3	26.9, CH_3
19	11.8, CH_3	18.5, CH_3	65.7, CH_2	65.8, CH_2	65.7, CH_2	104.5, CH_2	65.9, CH_2	65.5, CH_2	63.7, CH_2	63.7, CH_2	22.4, CH_3	65.4, CH_2
20	23.7, CH_3	19.2, CH_3	19.7, CH_3	20.0, CH_3	20.2, CH_3	22.5, CH_3	20.3, CH_3	18.3, CH_3	18.3, CH_3	24.2, CH_3	25.2, CH_3	16.4, CH_3
OMe-7	–	–	–	–	–	–	55.7, CH_3	–	–	–	–	–
OMe-12	–	–	–	–	–	–	–	–	–	–	–	–
OMe-14	–	–	–	57.9, CH_3	–	–	–	–	–	–	60.5, CH_3	–
OAc-19	–	–	–	–	20.9, CH_3 / 170.8, C	–	–	–	–	–	–	–

[a] Measured in CDCl$_3$; [b] Measured in pyridine-d_5.

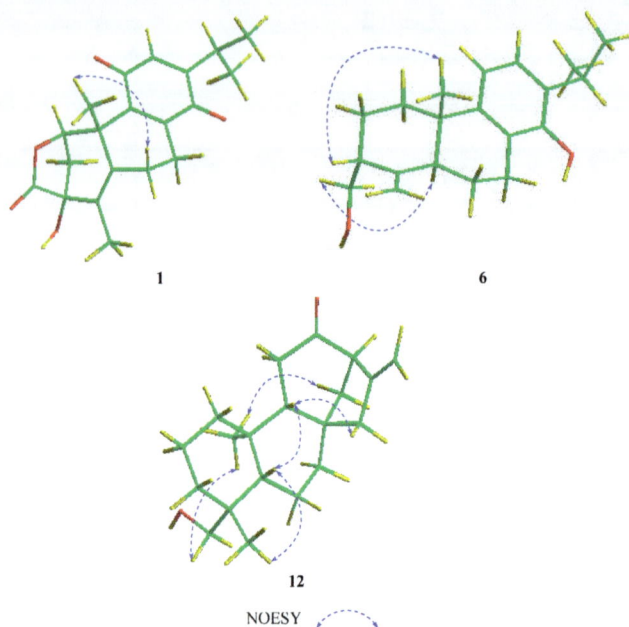

Figure 3. The selected NOESY correlations of compounds **1**, **6**, and **12**. The red, yellow and green atoms represent oxygens, hydrogens and carbons, respectively.

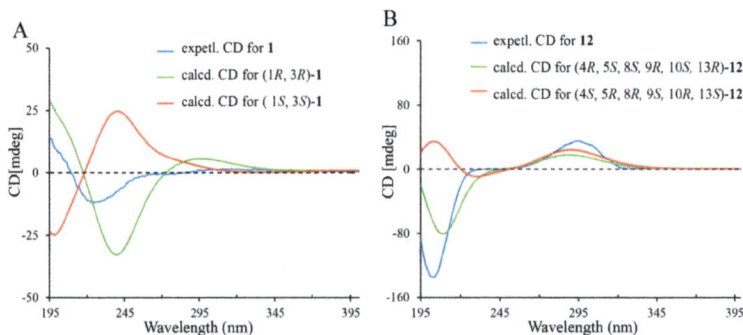

Figure 4. Experimental and calculated CD spectra of compounds **1** (**A**) and **12** (**B**).

Compound **2** was assigned as a molecular formula of $C_{20}H_{24}O_5$ based on a deprotonated molecular ion at *m/z* 343.1560 [M − H]$^-$ (calcd for $C_{20}H_{23}O_5$, 343.1551) in its HRESIMS. IR spectrum of **2** exhibited a conjugated carboxylic acid band at 1688 cm^{-1} and benzoquinone band at 1649 cm^{-1}. The ^{1}H and ^{13}C NMR spectroscopic data (Tables 1 and 2) of **2** were closely analogous to those of triptoquinone A (**20**) [18], except for the absence of a secondary methyl group and the presence of a hydroxymethyl group (δ_H 3.67 (2H, d, *J* = 7.2 Hz, H$_2$-17); δ_C 66.6). The hydroxymethyl group was allocated to be at C-15, as deduced by ^{1}H–^{1}H COSY correlation of H-15/H$_2$-17 and HMBC correlations from the hydroxymethyl protons to C-13 (δ_C 149.0) and C-15 (δ_C 34.5). Therefore, compound **2** was characterized and given a trivial name of triregelin B.

Compound **3** showed a molecular formula of $C_{20}H_{26}O_5$, as established from an [M + H]$^+$ ion at *m/z* 347.1866 (calcd $C_{20}H_{27}O_5$, 347.1853) in the HRESIMS. Analysis of the NMR spectroscopic data

(Tables 1 and 2) indicated that **3** was structurally related to triptoquinone B (**21**) [18] except for the absence of the C-7 methylene in triptoquinone B and the presence of an additional hydroxyl proton (δ_H 2.77) and an oxygenated methine (δ_H 4.81, δ_C 61.9) in **3**. These data suggested hydroxylation of C-7 in **3**, which was supported by HMBC correlation from the hydroxyl proton (δ_H 2.77) to C-7 (δ_C 61.9). The α-orientation of the hydroxyl group at C-7 was deduced from the NOESY correlation of H-7/H$_3$-20. Thus, compound **3** was identified and named triregelin C.

Compound **4** gave a molecular formula of $C_{21}H_{28}O_5$, as deduced from an [M + H]$^+$ ion at *m/z* 361.2008 (calcd $C_{21}H_{29}O_5$, 361.2010), 14.0142 atomic mass units (amu) more than that of **3** in the HRESIMS. The ^1H and ^{13}C NMR spectroscopic data (Tables 1 and 2) of **4** were closely similar to those of **3**, except for the appearance of a methoxyl group. The methoxyl group was assigned at C-7, as evidenced from the observed HMBC correlation from the methoxyl protons (δ_H 3.50) to C-7 (δ_C 69.8). Thus, compound **4** was characterized and named triregelin D.

Compound **5** showed a molecular formula of $C_{22}H_{28}O_5$ on the basis of a protonated molecular ion at *m/z* 373.1998 [M + H]$^+$ (calcd $C_{22}H_{29}O_5$, 373.2010) in its HRESIMS, 42.0106 amu more than that of triptoquinone B (**21**) [18]. The 1D NMR spectroscopic data (Tables 1 and 2) of **5** were analogous to those of triptoquinone B (**21**) [18], except for the presence of an acetyl group (δ_H 2.03; δ_C 170.8, 20.9). The acetyl group was allocated to C-19, as evidenced from the HMBC correlations from H$_2$-19 (δ_H 4.56, 4.08) to the carbonyl carbon (δ_C 170.8) of the acetyl group. Therefore, compound **5** was determined and named triregelin E.

Compound **6** had a molecular formula of $C_{20}H_{28}O_2$ deduced from a protonated molecular ion at *m/z* 301.2162 [M + H]$^+$ (calcd for $C_{20}H_{29}O_2$, 301.2162) in the HRESIMS. IR spectrum of **6** displayed a double bond band at 1626 cm^{-1}, and aromatic ring bands at 1580 and 1424 cm^{-1}. The ^1H NMR data (Table 3) showed the characteristic signals for two coupled aromatic protons (δ_H 7.04 and 6.09 (1H each, d, *J* = 7.8 Hz)), two singlet vinylic protons (δ_H 4.86 and 4.77), an oxygenated methylene (δ_H 3.95 and 3.70), an isopropyl moiety (δ_H 3.15, 1.26 and 1.25), and a tertiary methyl group (δ_H 0.99). The ^{13}C NMR data (Table 2) displayed resonances for 20 carbons, which were ascribed to a tetrasubstituted benzene ring, an exocyclic double bond, an aliphatic quaternary carbon, three methines, five methylenes (including an oxygenated one) and three methyl groups. The ^1H and ^{13}C NMR spectroscopic data of **6** were similar to those of triptobenzene P [22], an 18 (4→3)-*abeo*-abietane diterpene previously isolated from *T. wilfordii*, except for the following two differences. One difference is the replacement of the methoxyl group at C-12 in triptobenzene P by a hydrogen in **6**, which was supported by ^1H–^1H COSY correlation of H-11/H-12, and HMBC correlations from H-12 (δ_H 7.04) to C-9 (δ_C 145.8) and C-15 (δ_C 26.9). The other difference is the downfield shift of C-14 (δ_C 150.3) in **6** relative to that (δ_C 123.8) in triptobenzene P, indicating hydroxylation of C-14 in **6**. Thus, the planar structure of **6** was established as 12-demethoxy-14-hydroxy-triptobenzene P, which was confirmed by the ^1H–^1H COSY and HMBC data (Figure 2). The NOE correlations of H-5α/H$_2$-18 and H$_3$-20β/H-3 indicated α-orientation of CH$_2$-18 (Figure 3). Thus, compound **6** was defined and named triregelin F.

Compound **7** gave a molecular formula of $C_{21}H_{28}O_4$, as established from an [M + H]$^+$ ion at *m/z* 345.2055 (calcd for $C_{21}H_{29}O_4$, 345.2060) in the HRESIMS. IR spectrum of **7** exhibited a carbonyl band at 1703 cm^{-1}, and aromatic ring bands at 1604, 1566 and 1455 cm^{-1}. The ^1H and ^{13}C NMR spectroscopic data (Tables 2 and 3) of **7** were very similar to those of triptobenzene A (**13**) [23], except for the absence of two methylene groups and an aromatic proton, and the presence of a double bond (δ_H 6.82 (1H, dd, *J* = 10.2, 3.0 Hz, H-7); 122.1 and 5.86 (1H, dd, *J* = 10.2, 3.0 Hz, H-6); δ_C 124.0), and a methoxyl group (δ_H 3.80 (3H,s, OCH$_3$-12); δ_C 55.7). The double bond was assigned at between C-6 and C-7, which was supported by HMBC correlations from H-6 (δ_H 5.86) to C-4 (δ_C 53.0) and C-10 (δ_C 37.7), and from H-7 (δ_H 6.82) to C-9 (δ_C 145.2) and C-14 (δ_C 150.3). The methoxyl group was located at C-12, as deduced from the HMBC correlation from the methoxyl protons (δ_H 3.80) to C-12 (δ_C 158.3). The key NOE correlations of H-5α/H$_3$-18 and H$_3$-20/H$_2$-19 were observed in the NOESY spectrum. Accordingly, compound **7** was elucidated as illustrated in Figure 1, and named triregelin G.

Table 3. ¹H NMR (600 MHz) spectroscopic data for compounds **6**–**11**.

Position	δH (J in Hz)					
	6[a]	7[a]	8[a]	9[a]	10[a]	11[b]
1	1.63, td (13.2, 4.2) / 2.32, dt (13.2, 4.2)	2.14, td (13.2, 5.4) / 2.48, ddd (12.6, 6.0, 3.0)	2.07, ddd (16.2, 9.6, 4.8) / 3.31, m[c]	1.50, td (13.8, 3.6) / 3.34, dt (13.8, 3.6)	1.70, td (13.8, 4.2) / 2.36, dt (13.8, 3.0)	1.97, dt (12.6, 3.6) / 2.38, td (13.2, 3.6)
2	1.43, qd (13.0, 4.2) / 1.92, m[c]	2.61, ddd (15.6, 5.4, 3.0) / 2.83, ddd (15.6, 13.2, 6.0)	2.54, ddd (15.6, 8.4, 7.2) / 2.74, m[c]	1.89, m[c] / 2.02, m	1.98, m / 2.04, m	1.87, m[c] / 2.10, tt (14.4, 3.6)
3	2.20, m[c]	-	-	3.56, dd (11.4, 3.6)	3.55, dd (11.4, 3.6)	3.68, q (3.6)
5	2.17, d (12.6)[c]	2.69, t (3.0)	2.64, d (15.0)[c]	1.91, dd (14.4, 2.4)[c]	1.94, dd (14.4, 3.6)	2.22, dd (12.6, 2.4)
6	1.81, qd (12.6, 6.0) / 1.95, m[c]	5.86, dd (10.2, 3.0)	2.63, d (16.2) / 2.72, m[c]	2.64, dd (16.8, 14.4) / 2.74, dd (16.8, 2.4)	2.67, dd (18.0, 14.4) / 2.80, dd (18.0, 3.6)	1.88, m[c] / 1.72, m[c]
7	2.61, ddd (16.4, 12.6, 7.2) / 2.87, dd (16.4, 6.0)	6.82, dd (10.2, 3.0)	-	-	-	2.81, ddd (16.2, 11.4, 7.8) / 3.13, dd (16.2, 6.6)
11	6.90, d (7.8)	6.34, s	6.83, s	6.77, s	6.74, d (7.8)	7.07, s
12	7.04, d (7.8)	-	-	-	7.36, d (7.8)	-
15	3.15, sept (7.2)	3.44, sept (7.2)	3.32, sept (6.6)[c]	3.30, sept (6.6)	3.32, sept (6.6)	3.76, sept (7.2)
16	1.26, d (7.2)	1.33, d (7.2)	1.20, d (6.6)	1.18, d (6.6)	1.22, d (6.6)	1.72, d (7.2)[c]
17	1.25, d (7.2)	1.33, d (7.2)	2.21, d (6.6)	1.20, d (6.6)	1.20, d (6.6)	1.68, d (7.2)
18	3.70, dd (10.8, 6.0) / 3.95, dd (10.8, 6.0)	1.27, s	1.35, s	1.30, s	1.29, s	1.27, s
19	4.77, s / 4.86, s	3.84, d (12.0, 4.8) / 4.14, d (12.0)	3.54, d (11.4) / 4.06, d (11.4)	3.42, dd (11.4, 9.0) / 4.35, d (11.4)	3.50, dd (11.4, 7.8) / 4.36, d (11.4)	0.96, s
20	0.99, s	1.21, s	1.42, s	1.35, s	1.19, s	1.25, s
OH-3	-	4.90, s	4.62, s	4.47, s	2.56, s	5.69, d (3.6)
OH-11	-	-	-	-	-	-
OH-12	-	-	-	-	-	-
OH-14	-	-	12.79, s	12.96, s	13.07, s	10.80, s
OH-19	-	1.77, br s	2.96, s	2.80, br d (9.0)	2.83, br d (7.8)	-
OMe-12	-	-	-	-	-	-
OMe-14	-	3.80, s	-	-	-	3.72, s

[a] Measured in CDCl₃; [b] Measured in pyridine-d₅; [c] Overlapping signal was assigned from ¹H-¹H COSY, HSQC, and HMBC experiments. The signals of br, s, d, t, q, sept and m represent broad, singlet, doublet, triplet, quartet, septet and multiplet splitting patterns of protons, respectively.

Compound **8** had a molecular formula of $C_{20}H_{26}O_5$, according to an $[M + H]^+$ ion at m/z 347.1840 $[M + H]^+$ (calcd for $C_{20}H_{27}O_5$, 347.1853). The ^1H and ^{13}C NMR spectroscopic data (Tables 2 and 3) of **8** were closely related to those of triptobenzene A (**13**) [23]. However, one of the key differences was the replacement of the methylene at C-7 in triptobenzene A by a keto carbonyl carbon (δ_C 204.3) in **8**, as evidenced from HMBC correlation from H-5 (δ_H 2.64) to C-7. The other difference was the absence of a doublet aromatic proton and the presence of an additional hydroxyl proton (δ_H 4.62) together with the downfield shift of C-11 (δ_C 144.5) in **8** compared to that in triptobenzene A, which suggested hydroxylation of C-11 in **8**. Therefore, compound **8** was assigned and named triregelin H.

Compound **9** had a molecular formula of $C_{20}H_{28}O_5$ based on a protonated molecular ion at m/z 349.2005 $[M + H]^+$ (calcd for $C_{20}H_{29}O_5$, 349.2010), with 2.0161 amu more than that of **8** in the HRESIMS. The ^{13}C NMR spectroscopic data (Table 2) of **9** were closely comparable to those of **8**, except for the absence of the C-3 keto carbonyl in **8**, and the presence of an oxygenated methine (δ_H 3.56 (1H, dd, J = 11.4, 3.6 Hz, H-3); δ_C 79.7) in **9**. These suggested that the C-3 keto carbonyl group in **8** was reduced to be a hydroxyl group in **9**. The hydroxyl group at C-3 was β-oriented, as deduced from the NOESY correlations of H-3/H-5α and H-3/H₃-18. Therefore, compound **9** was identified and named triregelin I.

Compound **10** displayed a molecular formula of $C_{20}H_{28}O_4$ established by a protonated molecular ion at m/z 333.2053 $[M + H]^+$ (calcd for $C_{20}H_{29}O_4$, 333.2060), revealing 15.9948 amu less than that in **9** in the HRESIMS. The ^1H and ^{13}C NMR spectroscopic data (Tables 2 and 3) of **10** were very similar to those of **9** except for the presence of an extra doublet aromatic proton (δ_H 6.74, H-11) and the upfield shift of C-11 (δ_C 113.6) relative to that (δ_C 144.2) in **9**. These data revealed the dehydroxylation of C-11 in **10**, which was further supported by ^1H–^1H COSY correlation of H-11/H-12, and HMBC correlations from H-11 to C-8 (δ_C 114.2), C-10 (δ_C 37.5), and C-13 (δ_C 134.9). Hence, compound **10** was elucidated and named triregelin J.

Compound **11** showed a molecular formula of $C_{21}H_{32}O_3$, as deduced from a protonated molecular ion at m/z 333.2426 $[M + H]^+$ (calcd for $C_{21}H_{33}O_3$, 333.2424) in the HRESIMS. Comparison of the NMR spectroscopic data (Tables 2 and 3) of **11** with neotriptonoterpene (**14**) [24] showed that both compounds were structurally comparable, except for the absence of the C-3 keto carbonyl in neotriptonoterpene (**14**) and the presence of an extra oxygenated methine (δ_H 3.68; δ_C 74.8) in **11**. These suggested that the C-3 keto carbonyl group in neotriptonoterpene (**14**) was reduced to be a hydroxyl group in **11**. The C-3 hydroxyl group was α-oriented, as inferred from the coupling constant ($J_{2,3}$ = 3.6 Hz) and the NOESY correlation between H-3 and H₃-19. Accordingly, the compound **11** was characterized and named triregelin K.

Compound **12**, white amorphous power, had a molecular formula of $C_{20}H_{30}O_2$, as deduced from an $[M + H]^+$ ion at m/z 303.2322 (calcd for $C_{20}H_{31}O_2$, 303.2319) in the HRESIMS. IR spectrum of **12** exhibited a strong carbonyl band at 1710 cm⁻¹. The ^1H NMR spectrum (Table 1) exhibited the characteristic signals for a vinylic group (δ_H 4.99 and 4.87), an oxygenated methylene (δ_H 3.68 and 3.44), a hydroxyl group (δ_H 1.09), and two tertiary methyls (δ_H 0.98 and 0.85). The ^{13}C NMR and DEPT spectra (Table 2) showed 20 carbon signals including a carbonyl group, an exocyclic double bond, three quaternary carbons, there methines, nine methylenes (including an oxygenated one) and two methyl groups. All the above NMR data indicated that **12** was a kaurane type diterpenoid, and structurally similar to (−)-*ent*-kaur-16-en-19-ol [25–27]. The distinct difference was that the C-12 methylene in (−)-*ent*-kaur-16-en-19-ol was oxidized to be a keto carbonyl group in **12**, as deduced from the downfield shift of C-12 (δ_C 211.5), and the HMBC correlations from H-9 (δ_C 1.58) and H₂-14 (δ_H 2.40, 1.51) to C-12. Finally, the planar structure of **12** was confirmed on the basis of the ^1H–^1H COSY and HMBC experiments (Figure 2). In the NOESY spectrum, the correlations of H₃-20/H₂-19 and H₃-20/H₂-14 indicated that these protons were in the same face. In the same way, the other key NOE cross peaks of H-5/H-9 and H-9/H₂-15 were also observed (Figure 3), suggesting H-5, H-9, and H₂-15 were in the other face. However, **12** displayed a positive specific rotation ($[\alpha]_D^{21}$ +50.86 (*c* 0.50, MeOH)) in contrast to the negative one reported for (−)-*ent*-kaur-16-en-19-ol [27]. ECD curves for the two possible stereo-structures (4*R*, 5*S*, 8*S*, 9*R*, 10*S*, 13*R*-**12** and 4*S*, 5*R*, 8*R*, 9*S*, 10*R*, 13*S*-**12**) were, therefore, calculated to determine the absolute configuration of **12**. As illustrated in Figure 4B, the calculated profile of 4*R*, 5*S*, 8*S*, 9*R*, 10*S*, and 13*R*-**12** were in good agreement with the experimental CD spectrum of **12**. Therefore, compound **12** was identified and named triregelin L.

The compounds **3**, **6**, **7**, and **9** were also selected to calculate their ECD data in order to further confirm their absolute configurations. As the results, their experimental CD spectra showed similar CD pattern to the calculated ones of (4*S*, 5*R*, 7*R*, 10*S*)-**3**, (3*R*, 5*S*, 10*S*)-**6**, (4*S*, 5*R*, 10*S*)-**7**, and (3*S*, 4*S*, 5*R*, 10*S*)-**9**, respectively (Figure S1). The HRMS, UV, IR, NMR and CD spectra (Figures S2–S121) of twelve new compounds were shown in supplementary materials.

In addition, eleven known abietanes were also isolated from the stems of *T. regelii*, including triptobenzene A (**13**) [23], neotriptonoterpene (**14**) [24], triptobenzene M (**15**) [28], wilforol F (**16**) [29], triptobenzene J (**17**) [30], triptobenzene B (**18**) [23], abieta-8, 11, 13-triene-14, 19-diol (**19**) [31], triptoquinone A (**20**), triptoquinone B (**21**), triptoquinone D (**22**) and triptoquinone F (**23**) [18]. These compounds were identified by comparison of their spectroscopic (1D NMR and specific rotation) and HRMS data with those reported in the literature. Compound **1** is the first abietane type diterpene with an 18→1 lactone ring. The discovery of the above twelve new diterpenes contributed to the chemical diversity of natural terpenoids.

Cytotoxic effects of seventeen diterpenes (**2**, **7–11**, **13–23**) were evaluated against three cancer cell lines of A2780, HepG2 and MCF-7. As the results show (Table 4), compound **9** displayed cytotoxicity against A2780, HepG2, and MCF-7 cells with IC_{50} values of 5.88, 11.74, and 46.40 μM, respectively. Compound **11** showed solely cytotoxic effect on MCF-7 cell with an IC_{50} value of 26.70 μM. Compound **14** exhibited weak cytotoxic activity on A2780, HepG2, and MCF-7 cells with IC_{50} values of 65.80, 35.45, and 64.80 μM, respectively.

Table 4. Cytotoxic effects of diterpenes on three cancer cell lines of A2780, HepG2, and MCF-7.

Compounds *	IC_{50} (μM) against A2780	IC_{50} (μM) against HepG2	IC_{50} (μM) against MCF-7
9	5.88 ± 2.22	11.74 ± 1.92	46.40 ± 3.54
11	>100	>100	26.70 ± 5.57
14	65.80 ± 21.53	35.45 ± 8.23	64.80 ± 24.90
taxol	0.006 ± 0.001	0.003 ± 0.0002	0.005 ± 0.001

* Seventeen compounds (**2**, **7–11**, **13–23**) were evaluated for cytotoxic effects against three cancer cell lines; IC_{50} values for other tested compounds were larger than 100 μM on three cancer cells.

3. Materials and Methods

3.1. General Experimental Procedures

Optical rotations were obtained using a Rudolph Research Analytical Autopol I automatic polarimeter (Rudolph Research Analytical, Hackettstown, NJ, USA). IR spectra were measured on an Agilent Cary 600 series FT-IR spectrometer (KBr) (Agilent, Santa Clara, CA, USA). Ultraviolet (UV) spectra were recorded on a Beckman Coulter DU® 800 spectrophotometer (Beckman Coulter, Fullerton, CA, USA). HRMS spectra were carried out on an Agilent 6230 electrospray ionization (ESI) time-of-flight (TOF) mass spectrometer (Agilent, Santa Clara, CA, USA). Nuclear magnetic resonance (NMR) spectra were measured on a Bruker Ascend 600 NMR spectrometer at 600 MHz for ¹H NMR and 150 MHz for ¹³C NMR (Bruker, Zurich, Switzerland). Chemical shifts were expressed in δ (ppm) with tetramethylsilane (TMS) as an internal reference, and coupling constants (*J*) were reported in hertz (Hz). Circular dichroism spectra were measured on a Jasco J1500 CD spectrometer (Jasco Corporation, Tokyo, Japan). Medium pressure liquid chromatography (MPLC) was conducted on a Sepacore Flash Chromatography System (Buchi, Flawil, Switzerland) by employing a flash column (460 mm × 36 mm, i.d., Buchi) packed with Bondapak Waters ODS (40–63 μm, Waters, Milford, MA, USA). Preparative high performance liquid chromatography (HPLC) was carried out on a Waters

Xbridge Prep C$_8$ column (10 mm × 250 mm, 5 μm) by utilizing a Waters liquid chromatography system equipped with 1525 Binary HPLC Pump and 2489 UV/Visible detector (Waters, Milford, MA, USA). Semi-preparative HPLC was done on a Waters Xbridge Prep C$_{18}$ column (10 mm × 250 mm, 5 μm) by using an Agilent 1100 liquid chromatography system coupled with a quaternary pump and a diode array detector (DAD) (Agilent, Santa Clara, CA, USA). Column chromatography was conducted on silica gel (40−60 μm, Grace, Columbia, MD, USA) and Bondapak Waters ODS (40–63 μm, Waters). Thin layer chromatographies (TLCs) were performed on pre-coated silica gel 60 F$_{254}$ plates and TLC silica gel 60 RP-18 F$_{254S}$ plates (200 μm thick, Merck KGaA, Darmstadt, Germany), which were used to monitor fractions. Spots on the TLC were visualized by UV light (254 nm) or heating after spraying with 5% H$_2$SO$_4$ in ethanol.

3.2. Plant Material

The stems of *T. regelii* used in this study were collected from Changbai Mountain in Jilin province, China, in October 2012. The plant was authenticated by Liang Xu, Liaoning University of Traditional Chinese Medicine (Dalian, China). A voucher specimen (No. MUST-TR201210) has been deposited at State Key Laboratory of Quality Research in Chinese Medicine, Macau University of Science and Technology.

3.3. Extraction and Isolation

The air-dried stems of *T. regelii* (8.0 kg) were powdered, and extracted three times with methanol (64 L) under ultrasonic-assisted extraction at room temperature for 1 h. The methanol extract was evaporated under reduced pressure to yield a dark brown residue, which was then suspended in H$_2$O, and successively partitioned with *n*-hexane, ethyl acetate (EtOAc) and *n*-butanol. Then, the EtOAc-soluble extract (150.0 g) was fractionated over a silica gel column using a gradient system of petroleum ether (PE)-acetone (100:0–35:65, *v/v*) to provide thirteen fractions (Fr.1–Fr.13). Fraction 5 was chromatographed over an ODS column using a gradient system of MeOH–H$_2$O (50:50–100:0, *v/v*) to yield eight fractions (Fr.5-1–Fr.5-8). Fractions 5-2 (110.0 mg) and 5-4 (130.0 mg) were further separated on silica gel columns eluted with PE–EtOAc (95:5–55:45, *v/v*) to afford compounds **22** (15.0 mg) and **23** (30.0 mg), respectively. Similarly, fraction 7 (5.0 g) was subjected to an ODS column with a gradient system of MeOH–H$_2$O (40:60–90:10, *v/v*) to give nine fractions (Fr.7-1–Fr.7-9). Compounds **20** (50.0 mg) and **21** (30.0 mg) were obtained by silica gel columns separation using PE–EtOAc (90:10–30:70, *v/v*) as eluting solvents from fractions 7-2 (150.0 mg) and 7-4 (100.0 mg), respectively. Fraction 8 (5.4 g) was subjected to an ODS column using a gradient system of MeOH–H$_2$O (35:65–80:20, *v/v*) to afford ten fractions (Fr.8-1–Fr.8-10). The fraction 8-3 (2.5 g) was fractionated over a silica gel column, with a gradient elution by PE–EtOAc (90:10–63:35, *v/v*), to produce seven fractions (Fr.8-3-1–Fr.8-3-7). Fraction 8-3-4 (40.5 mg) was purified by semi-preparative HPLC using CH$_3$CN–H$_2$O (52:48, *v/v*) as mobile phase to afford compounds **5** (0.6 mg), **12** (0.7 mg) and **18** (1.0 mg). Fraction 8-3-5 (200.6 mg) was subjected to semi-preparative HPLC using CH$_3$CN-H$_2$O (58:42, *v/v*) as solvent system to give compounds **6** (1.9 mg) and **11** (2.0 mg), as well as subfraction 8-3-5-2. Compounds **14** (1.6 mg) and **19** (1.0 mg) were obtained by preparative HPLC with an isocratic elution of CH$_3$CN-H$_2$O (55:45, *v/v*) from the subfraction 8-3-5-2 (65.7 mg). Fraction 11 (5.5 g) was chromatographed over an ODS column, eluted with MeOH–H$_2$O (30:70-100:0, *v/v*), to afford sixteen fractions (Fr.11-1–Fr.11-16). Compound **3** (5.0 mg) was isolated by preparative HPLC eluting with a MeOH–H$_2$O (35:65, *v/v*) solvent system from fraction 11-4 (26.9 mg). Fraction 11-6 (49.5 mg) was separated by semi-preparative HPLC using CH$_3$CN–H$_2$O (35:65, v/v) as mobile phase to yield compound **2** (5.0 mg). Fraction 11-7 (261.1 mg) was subjected to a silica gel column with a gradient elution of PE–EtOAc (20:80–10:90, *v/v*) and purified by semi-preparative HPLC using CH$_3$CN–H$_2$O (40:60, *v/v*) as eluting solvent to afford compounds **8** (1.2 mg), **4** (2.0 mg), **13** (1.6 mg) and fraction 11-7-3. Compounds **7** (0.6 mg) and **16** (3.1 mg) were purified by semi-preparative HPLC with a CH$_3$CN–H$_2$O (42:58, *v/v*) solvent system from fraction 11-7-3 (56.9 mg). Fraction 11-8 (284.3 mg) was isolated by preparative HPLC using MeOH−H$_2$O (46:54, v/v) as mobile phase to yield compound **1** (0.59 mg) and five fractions (Fr.11-8-1−Fr.11-8-5). Fraction 11-8-2 (42.5 mg) was subjected to semi-preparative HPLC with a CH$_3$CN−H$_2$O (47:53, v/v) solvent system to give compound **15** (2.0 mg). Fraction 11-8-3 (30.3 mg) was isolated by semi-preparative HPLC using CH$_3$CN−H$_2$O (45:55, v/v) as eluting solvent to yield compounds **10** (1.6 mg) and **17** (2.1 mg). Fraction 12 (9.0 g) was separated by MPLC using a gradient system of MeOH–H$_2$O (5:95–100:0, 50 mL/min) to obtain six fractions (Fr.12-1–Fr.12-6). Fraction 12-5 was chromatographed over a silica gel column using CHCl$_3$–MeOH (100:0–90:10, v/v) as solvent system, and then purified by semi-preparative HPLC using CH$_3$CN–H$_2$O (39:61, v/v) as mobile phase to give compound **9** (1.2 mg).

3.4. Structural Characterization

Triregelin A (**1**): yellow amorphous powder; $[\alpha]_D^{21}$ + 3.8 (c 0.50, MeOH); IR (KBr) ν_{max}: 3444, 2925, 2854, 1732, 1601, 1455, 1377, 1260, 1167, 1086, 1013, 957, 893, 803, 756, 667 cm^{-1}; UV (MeOH) λ_{max} (log ε) 260 (2.35) nm; CD (c 2.92 × 10^{-3} mol/L, MeOH) λ_{max} ($\Delta\varepsilon$) 225 (−1.22), 311 (+0.16); ^{1}H NMR (CDCl$_3$, 600 MHz) and ^{13}C NMR (CDCl$_3$, 150 MHz) data, see Tables 1 and 2; HRESIMS m/z 343.1546 [M + H]$^+$ (calcd for C$_{20}$H$_{23}$O$_5$, 343.1540).

Triregelin B (**2**): yellow amorphous powder; $[\alpha]_D^{21}$ + 75.5 (c 1.00, MeOH); IR (KBr) ν_{max}: 3421, 2970, 2937, 2881, 1688, 1649, 1436, 1375, 1248, 1104, 1030, 977, 905, 798, 659, 599 cm^{-1}; UV (MeOH) λ_{max} (log ε) 258 (3.17) nm; CD (c 1.46 × 10^{-3} mol/L, MeOH) λ_{max} ($\Delta\varepsilon$) 269 (+7.24), 355 (+0.35), 476 (−0.54); ^{1}H NMR (CDCl$_3$, 600 MHz) and ^{13}C NMR (CDCl$_3$, 150 MHz) data, see Tables 1 and 2; HRESIMS m/z 343.1560 [M − H]$^-$ (calcd for C$_{20}$H$_{23}$O$_5$, 343.1551).

Triregelin C (**3**): yellow, amorphous powder; $[\alpha]_D^{21}$ − 34.0 (c 1.00, MeOH); IR (KBr) ν_{max}: 3437, 2966, 2936, 2876, 1700, 1650, 1463, 1430, 1383, 1294, 1235, 1165,1107, 1044, 974, 933, 900, 813, 741 cm^{-1}; UV (MeOH) λ_{max} (log ε) 256 (3.42) nm; CD (c 1.45 × 10^{-3} mol/L, MeOH) λ_{max} ($\Delta\varepsilon$) 283 (+2.21), 359 (−0.25), 478 (−0.32); ^{1}H NMR (CDCl$_3$, 600 MHz) and ^{13}C NMR (CDCl$_3$, 150 MHz,) data, see Tables 1 and 2; HRESIMS m/z 347.1866 [M + H]$^+$ (calcd for C$_{20}$H$_{27}$O$_5$, 347.1853).

Triregelin D (**4**): yellow, amorphous powder; $[\alpha]_D^{21}$ − 9.3 (c 1.00, MeOH); IR (KBr) ν_{max}: 3435, 2964, 2932, 2877, 1688, 1652, 1606, 1462, 1426, 1383, 1293, 1234, 1192, 1085, 1039, 910, 856 cm^{-1}; UV (MeOH) λ_{max} (log ε) 255 (3.19) nm; CD (c 1.39 × 10^{-3} mol/L, MeOH) λ_{max} ($\Delta\varepsilon$) 271 (+3.05), 357 (−0.22), 481 (−0.61); ^{1}H NMR (CDCl$_3$, 600 MHz) and ^{13}C NMR (CDCl$_3$, 150 MHz) data, see Tables 1 and 2; HRESIMS m/z 361.2008 [M + H]$^+$ (calcd for C$_{21}$H$_{29}$O$_5$, 361.2010).

Triregelin E (**5**): yellow, amorphous powder; $[\alpha]_D^{21}$ + 17.0 (c 0.50, MeOH); IR (KBr) ν_{max}: 3455, 2962, 2930, 2874, 1744, 1713, 1650, 1604, 1464, 1384, 1294, 1233, 1104, 1042, 906, 802, 757, 666, 603 cm^{-1}; UV (MeOH) λ_{max} (log ε) 257 (2.75) nm; CD (c 1.34 × 10^{-3} mol/L, MeOH) λ_{max} ($\Delta\varepsilon$) 261 (+3.35), 349 (+0.33), 474 (−0.30); ^{1}H NMR (CDCl$_3$, 600 MHz) and ^{13}C NMR (CDCl$_3$, 150 MHz) data, see Tables 1 and 2; HRESIMS m/z 373.1998 [M + H]$^+$ (calcd for C$_{22}$H$_{29}$O$_5$, 373.2010).

Triregelin F (**6**): yellow, amorphous powder; $[\alpha]_D^{21}$+169.3 (c 1.00, MeOH); IR (KBr) ν_{max}: 3437, 2964, 2872, 1626, 1424, 1250, 1160, 1115, 1059, 896, 815, 705 cm^{-1}; UV (MeOH) λ_{max} (log ε) 223 (3.03), 270 (2.35) nm; CD (c 1.67 × 10^{-3} mol/L, MeOH) λ_{max} ($\Delta\varepsilon$) 209 (+5.51), 217 (sh) (+3.41), 265 (+1.02); ^{1}H NMR (CDCl$_3$, 600 MHz) and ^{13}C NMR (CDCl$_3$, 150 MHz) data, see Tables 2 and 3; HRESIMS m/z 301.2162 [M + H]$^+$(calcd for C$_{20}$H$_{29}$O$_2$, 301.2162).

Triregelin G (**7**): yellow, amorphous powder; $[\alpha]_D^{21}$ − 46.9 (c 0.50, MeOH); IR (KBr) ν_{max}: 3382, 2958, 2929, 2872, 1703, 1604, 1566, 1455, 1417, 1378, 1312, 1261, 1224, 1137, 1107, 1056, 802, 756, 667 cm^{-1}; UV (MeOH) λ_{max} (log ε) 233 (3.23), 279 (2.37), 312 (2.45) nm; CD (c 1.45 × 10^{-3} mol/L, MeOH) λ_{max} ($\Delta\varepsilon$) 239 (−2.50), 286 (+1.08), 311 (−1.02), 392 (+0.31); ^{1}H NMR (CDCl$_3$, 600 MHz) and ^{13}C NMR (CDCl$_3$, 150 MHz) data, see Tables 2 and 3; HRESIMS m/z 345.2055 [M + H]$^+$ (calcd for C$_{21}$H$_{29}$O$_4$, 345.2060).

Triregelin H (**8**): yellow, amorphous powder; $[\alpha]_D^{21}$ + 56.4 (c 1.00, MeOH); IR (KBr) ν_{max}: 3398, 2961, 2926, 2872, 1697, 1624, 1429, 1382, 1349, 1301, 1233, 1162, 1107, 1040, 964, 894, 801, 754 cm^{-1}; UV (MeOH) λ_{max} (log ε) 238 (2.82), 270 (2.38) nm; CD (c 1.45 × 10^{-3} mol/L, MeOH) λ_{max} ($\Delta\varepsilon$) 233

(+2.77), 270 (−0.27), 310 (−1.08), 377 (+1.25); ^{1}H NMR (CDCl$_3$, 600 MHz) and ^{13}C NMR (CDCl$_3$, 150 MHz) data, see Tables 2 and 3; HRESIMS m/z 347.1840 [M + H]$^+$ (calcd for C$_{20}$H$_{27}$O$_5$, 347.1853).

Triregelin I (**9**): white, amorphous powder; $[\alpha]_D^{21}$ + 29.4 (*c* 0.50, MeOH); IR (KBr) ν_{max}: 3396, 2959, 2924, 2854, 1714, 1592, 1428, 1348, 1260, 1168, 1114, 1028, 970, 800, 755, 709 cm^{-1}; UV (MeOH) λ_{max} (log ε) 235 (3.32), 270 (3.22), 380 (2.95) nm; CD (*c* 1.44 × 10^{-3} mol/L, MeOH) λ_{max} ($\Delta\varepsilon$) 206 (+2.23), 235 (+0.88), 271 (−0.65), 311 (−0.61), 375 (+0.96); ^{1}H NMR (CDCl$_3$, 600 MHz) and ^{13}C NMR (CDCl$_3$, 150 MHz) data, see Tables 2 and 3; HRESIMS m/z 349.2005 [M + H]$^+$ (calcd for C$_{20}$H$_{29}$O$_5$, 349.2010).

Triregelin J (**10**): white, amorphous powder; $[\alpha]_D^{21}$ + 1.4 (*c* 1.00, MeOH); IR (KBr) ν_{max}: 3364, 2962, 2937, 2870, 1622, 1558, 1455, 1427, 1381, 1347, 1251, 1212, 1160, 1113, 1080, 1037, 981, 914, 821, 757, 711, 663, 582 cm^{-1}; UV (MeOH) λ_{max} (log ε) 216 (3.40), 266 (2.33), 343 (2.42) nm; CD (*c* 1.51 × 10^{-3} mol/L, MeOH) λ_{max} ($\Delta\varepsilon$) 218 (+1.82), 230 (+1.34), 266 (−3.14), 333 (+1.23), 346 (+1.71); ^{1}H NMR (CDCl$_3$, 600 MHz) and ^{13}C NMR (CDCl$_3$, 150 MHz) data, see Tables 2 and 3; HRESIMS m/z 333.2053 [M + H]$^+$ (calcd for C$_{20}$H$_{29}$O$_4$, 333.2060).

Triregelin K (**11**): white, amorphous powder; $[\alpha]_D^{21}$ + 28.1 (*c* 0.25, MeOH); IR (KBr) ν_{max}: 3396, 3210, 2926, 2865, 1737, 1607, 1581, 1441, 1412, 1373, 1331, 1308, 1268, 1206, 1101, 1041, 942, 926, 857, 800, 756, 695, 660 cm^{-1}; UV (MeOH) λ_{max} (log ε) 227 (2.82), 283 (2.36) nm; CD (*c* 1.51 × 10^{-3} mol/L, MeOH) λ_{max} ($\Delta\varepsilon$) 228 (+0.97); ^{1}H NMR (pyridine-d_5, 600 MHz) and ^{13}C NMR (pyridine-d_5, 150 MHz) data, see Tables 2 and 3; HRESIMS m/z 333.2426 [M + H]$^+$ (calcd for C$_{21}$H$_{33}$O$_3$, 333.2424).

Triregelin L (**12**): white, amorphous powder; $[\alpha]_D^{21}$ + 50.86 (*c* 0.50, MeOH); IR (KBr) ν_{max}: 3475, 3072, 2961, 2924, 2867, 1710, 1655, 1607, 1510, 1445, 1415, 1369, 1261, 1089, 1028, 882, 801, 701, 665 cm^{-1}; UV (MeOH) λ_{max} (log ε) 203 (3.15) nm; CD (*c* 1.66 × 10^{-1} mol/L, MeOH) λ_{max} ($\Delta\varepsilon$) 203 (−24.65), 295 (+6.40); ^{1}H NMR (CDCl$_3$, 600 MHz) and ^{13}C NMR (CDCl$_3$, 150 MHz) data, see Tables 1 and 2; HRESIMS m/z 303.2322 [M + H]$^+$ (calcd for C$_{20}$H$_{31}$O$_2$, 303.2319).

3.5. Calculation Methods of Electronic Circular Dichroism (ECD) Spectra

The Gaussian 09 software package [32] was used to conduct all of the ECD calculations. The molecule geometries of molecules were firstly optimized at the level of B3LYP/6-31G (d, p) and the output geometries were subsequently employed to perform ECD calculations using time-dependent density functional theory (TDDFT) with the method of B3LYP/DGDZVP [20,21] since this method usually offers desirable outcomes [33]. The model of polarizable continuum was utilized to simulate the solvation effect in the calculations of circular dichroism. The experimental condition was simulated by using methanol as the solvent. The absolute configurations of all compounds were defined by comparing the calculated ECD curves with the experimental spectra.

3.6. Cytotoxicity of Diterpenes against Three Cancer Cell Lines

The A2780 (ovarian carcinoma) cell line was obtained from the KeyGEN biotech (Nanjing, China). HepG2 (hepatocellular carcinoma) and MCF-7 (human breast cancer) cell lines were purchased from the American Type Culture Collection. All of the cell lines were cultured in Dulbecco's modified Eagle medium (DMEM) (Invitrogen) supplemented with 10% (*v/v*) heat-inactivated fetal bovine serum (FBS) (Invitrogen), 100 U/mL penicillin, and 100 μg/mL streptomycin (Invitrogen) in a humidified atmosphere of 5% CO$_2$/95% air at 37 °C. Briefly, cells were seeded in 96-well plates in triplicate at a density of 2 × 10^3 cells/well (100 μL) and cultured at 37 °C in a 5% CO$_2$ humidified atmosphere for 24 h. Then, the cells were treated with fresh culture medium containing various concentrations of tested compounds and incubated at 37 °C under a humidified atmosphere of 5% CO$_2$/95% air for another 72 h. After that, the supernatant in each well was discarded and the cells were washed by phosphate-buffered saline (PBS) to avoid the possible effect of culture medium and tested compounds on the following MTT (3-(4,5-dimethylthiazol-2-yl)-2,5-diphenyltetrazolium bromide) assay. Subsequently, cells were incubated for 4 h at 37 °C in culture medium containing a final concentration of 0.5 mg/mL MTT (100 μL). The formed formazan crystals were dissolved in DMSO (100 μL) after removing the supernatant in each well. A microplate reader (Infinite 200 PRO,

Int. J. Mol. Sci. **2017**, *18*, 147

Tecan, Männedorf, Switzerland) was employed to determine the absorbance of each well at 570 nm. GraphPad Prism 6 software (Prism 6.0, GraphPad Software, Inc., La Jolla, CA, USA) was used to calculate the IC_{50} values (concentration that suppresses 50% of cell growth) of all tested compounds. All assays were performed in triplicate in three independent experiments. Data was expressed as mean $\pm$ SD ($n = 3$).

4. Conclusions

To sum up, 23 diterpenoids were isolated from the Chinese herbal medicine *T. regelii*, including eleven new abietane, and one new kaurane, diterpenes. Importantly, triregelin A (**1**) represents the first abietane diterpene bearing an 18→1 lactone ring. Triregelin I (**9**) exhibited significant cytotoxic effects on A2780 and HepG2 cancer cells with IC_{50} values of 5.88 μM and 11.74 μM, respectively, and was found inactive against MCF-7 cancer cells. Triregelin K (**11**) displayed a weak cytotoxic effect on MCF-7 cell with an IC_{50} value of 26.70 μM.

Supplementary Materials: Supplementary materials can be found at www.mdpi.com/1422-0067/18/1/147/s1.

Acknowledgments: This research was supported financially by Macao Science and Technology Development Fund, Macao Special Administrative Region (Grant No. 056/2013/A2, and 063/2011/A3).

Author Contributions: Li-Ping Bai and Zhi-Hong Jiang conceived and designed the experiments; Dongsheng Fan and Zhiyuan Zheng performed the experiments; Shuangyan Zhou and Xiaojun Yao conducted the calculations of ECD. Dongsheng Fan and Zhiyuan Zheng analyzed the data; Guo-Yuan Zhu guided the structural elucidations; Ming-Rong Yang performed the measurement of all IR spectra; Li-Ping Bai and Dongsheng Fan wrote the paper; Zhi-Hong Jiang revised the manuscript.

Conflicts of Interest: The authors declare no conflict of interest.

Abbreviations

HRESIMS	High resolution electrospray ionization mass spectrometry
CD	Circular dichroism
UV	Ultraviolet visible
IR	Infrared
NMR	Nuclear magnetic resonance
DEPT	Distortionless enhancement by polarization transfer
HSQC	Heteronuclear single quantum coherence
HMBC	Heteronuclear multiple bond correlation
^{1}H–^{1}H COSY	Proton–proton correlation spectroscopy
NOESY	Nuclear Overhauser effect spectroscopy

References

1. Rodríguez, B. ^{1}H and ^{13}C NMR spectral assignments of some natural abietane diterpenoids. *Magn. Reson. Chem.* **2003**, *41*, 741–746. [CrossRef]

2. González, M.A. Aromatic abietane diterpenoids: Their biological activity and synthesis. *Nat. Prod. Rep.* **2015**, *32*, 684–704. [CrossRef] [PubMed]

3. Yang, X.W.; Feng, L.; Li, S.M.; Liu, X.H.; Li, Y.L.; Wu, L.; Shen, Y.H.; Tian, J.M.; Zhang, X.; Liu, X.R. Isolation, structure, and bioactivities of abiesadines A–Y, 25 new diterpenes from Abies georgei orr. *Bioorg. Med. Chem.* **2010**, *18*, 744–754. [CrossRef] [PubMed]

4. Burmistrova, O.; Simões, M.F.T.; Rijo, P.; Quintana, J.; Bermejo, J.; Estévez, F. Antiproliferative activity of abietane diterpenoids against human tumor cells. *J. Nat. Prod.* **2013**, *76*, 1413–1423. [CrossRef] [PubMed]

5. Kafil, V.; Eskandani, M.; Omidi, Y.; Nazemiyeh, H.; Barar, J. Abietane diterpenoid of Salvia sahendica boiss and buhse potently inhibits MCF-7 breast carcinoma cells by suppression of the PI3K/AKT pathway. *RSC Adv.* **2015**, *5*, 18041–18050. [CrossRef]

6. Zhang, G.J.; Li, Y.H.; Jiang, J.D.; Yu, S.S.; Qu, J.; Ma, S.G.; Liu, Y.B.; Yu, D.Q. Anti-coxsackie virus b diterpenes from the roots of *Illicium jiadifengpi*. *Tetrahedron* **2013**, *69*, 1017–1023. [CrossRef]

7. Zhang, G.J.; Li, Y.H.; Jiang, J.D.; Yu, S.S.; Wang, X.J.; Zhuang, P.Y.; Zhang, Y.; Qu, J.; Ma, S.G.; Li, Y. Diterpenes and sesquiterpenes with anti-coxsackie virus B3 activity from the stems of *Illicium jiadifengpi*. *Tetrahedron* **2014**, *70*, 4494–4499. [CrossRef]

8. González, M.A.; Zaragozá, R.N.J. Semisynthesis of the antiviral abietane diterpenoid jiadifenoic acid C from callitrisic acid (4-epidehydroabietic acid) isolated from sandarac resin. *J. Nat. Prod.* **2014**, *77*, 2114–2117. [CrossRef] [PubMed]

9. Wang, Y.D.; Zhang, G.J.; Qu, J.; Li, Y.H.; Jiang, J.D.; Liu, Y.B.; Ma, S.G.; Li, Y.; Lv, H.N.; Yu, S.S. Diterpenoids and sesquiterpenoids from the roots of *Illicium majus*. *J. Nat. Prod.* **2013**, *76*, 1976–1983. [CrossRef] [PubMed]

10. Pferschy-Wenzig, E.M.; Kunert, O.; Presser, A.; Bauer, R. In vitro anti-inflammatory activity of larch (*Larix* decidua L.) sawdust. *J. Agric. Food Chem.* **2008**, *56*, 11688–11693. [CrossRef] [PubMed]

11. Efferth, T.; Kahl, S.; Paulus, K.; Adams, M.; Rauh, R.; Boechzelt, H.; Hao, X.; Kaina, B.; Bauer, R. Phytochemistry and pharmacogenomics of natural products derived from traditional chinese medicine and chinese materia medica with activity against tumor cells. *Mol. Cancer Ther.* **2008**, *7*, 152–161. [CrossRef] [PubMed]

12. Johnson, J.J. Carnosol: A promising anti-cancer and anti-inflammatory agent. *Cancer Lett.* **2011**, *305*, 1–7. [CrossRef] [PubMed]

13. Zhou, Z.L.; Yang, Y.X.; Ding, J.; Li, Y.C.; Miao, Z.H. Triptolide: Structural modifications, structure-activity relationships, bioactivities, clinical development and mechanisms. *Nat. Prod. Rep.* **2012**, *29*, 457–475. [CrossRef]

14. Zhou, X.; Wang, Y.; Lee, W.Y.; Or, P.M.; Wan, D.C.; Kwan, Y.W.; Yeung, J.H. Miltirone is a dual inhibitor of P-glycoprotein and cell growth in doxorubicin-resistant hepG2 cells. *J. Nat. Prod.* **2015**, *78*, 2266–2275. [CrossRef] [PubMed]

15. Fan, D.; Zhu, G.Y.; Chen, M.; Xie, L.M.; Jiang, Z.H.; Xu, L.; Bai, L.P. Dihydro-β-agarofuran sesquiterpene polyesters isolated from the stems of *Tripterygium regelii*. *Fitoterapia* **2016**, *112*, 1–8. [CrossRef] [PubMed]

16. Fan, D.; Parhira, S.; Zhu, G.Y.; Jiang, Z.H.; Bai, L.P. Triterpenoids from the stems of *Tripterygium regelii*. *Fitoterapia* **2016**, *113*, 69–73. [CrossRef] [PubMed]

17. Fan, D.; Zhu, G.Y.; Li, T.; Jiang, Z.H.; Bai, L.P. Dimacrolide sesquiterpene pyridine alkaloids from the stems of *Tripterygium regelii*. *Molecules* **2016**, *21*, 1146. [CrossRef] [PubMed]

18. Shishido, K.; Nakano, K.; Wariishi, N.; Tateishi, H.; Omodani, T.; Shibuya, M.; Goto, K.; Ono, Y.; Takaishi, Y. *Tripterygium wilfordii* var. Regelii which are interleukin-1 inhibitors. *Phytochemistry* **1994**, *35*, 731–737. [CrossRef]

19. Xu, Y.; Ma, Y.; Zhou, L.; Sun, H. Abietane quinones from Rabdosia lophanthoides. *Phytochemistry* **1988**, *27*, 3681–3682. [CrossRef]

20. Becke, A.D. A new mixing of hartree—Fock and local density-functional theories. *J. Chem. Phys.* **1993**, *98*, 1372–1377. [CrossRef]

21. Godbout, N.; Salahub, D.R.; Andzelm, J.; Wimmer, E. Optimization of gaussian-type basis sets for local spin density functional calculations. Part I. Boron through neon, optimization technique and validation. *Can. J. Chem.* **1992**, *70*, 560–571. [CrossRef]

22. Shen, Q.; Takaishi, Y.; Zhang, Y.W.; Duan, H.Q. Immunosuppressive terpenoids from *Tripterygium wilfordii*. *Chin. Chem. Lett.* **2008**, *19*, 453–456. [CrossRef]

23. Takaishi, Y.; Wariishi, N.; Tateishi, H.; Kawazoe, K.; Miyagi, K.; Li, K.; Duan, H. Phenolic diterpenes from *Tripterygium wilfordii* var. Regelii. *Phytochemistry* **1997**, *45*, 979–984. [CrossRef]

24. Zhou, B.; Zhu, D.; Deng, F.; Huang, C.; Kutney, J.P.; Roberts, M. Studies on new components and stereochemistry of diterpenoids from Trypterygium wilfordii. *Planta Med.* **1988**, *54*, 330–332. [CrossRef] [PubMed]

25. Pacheco, A.G.; Machado de Oliveira, P.; Piló-Veloso, D.; Flávio de Carvalho Alcântara, A. ^{13}C-NMR data of diterpenes isolated from Aristolochia species. *Molecules* **2009**, *14*, 1245–1262. [CrossRef] [PubMed]

26. Gonzalez, A.G.; Fraga, B.M.; Hernandez, M.G.; Hanson, J.R. The ^{13}C NMR spectra of some ent-18-hydroxykaur-16-enes. *Phytochemistry* **1981**, *20*, 846–847. [CrossRef]

27. Bohlmann, F.; Rao, N. Natürlich vorkommende Terpen-Derivate, XXI. Über die Inhaltsstoffe von *Anona squamosa* L. *Chem. Ber.* **1973**, *106*, 841–844. [CrossRef]

28. Duan, H.; Takaishi, Y.; Momota, H.; Ohmoto, Y.; Taki, T.; Jia, Y.; Li, D. Immunosuppressive diterpenoids from *Tripterygium* wilfordii. *J. Nat. Prod.* **1999**, *62*, 1522–1525. [CrossRef] [PubMed]

29. Morota, T.; Qin, W.Z.; Takagi, K.; Xu, L.H.; Maruno, M.; Yang, B.H. Diterpenoids from *Tripterigium wilfordii*. *Phytochemistry* **1995**, *40*, 865–870. [CrossRef]

30. Duan, H.; Kawazoe, K.; Bando, M.; Kido, M.; Takaishi, Y. Di-and triterpenoids from *Tripterygium hypoglaucum*. *Phytochemistry* **1997**, *46*, 535–543. [CrossRef]

31. Zhou, W.; Xie, H.; Wu, P.; Wei, X. Abietane diterpenoids from Isodon lophanthoides var. Graciliflorus and their cytotoxicity. *Food Chem.* **2013**, *136*, 1110–1116. [CrossRef] [PubMed]

32. *Gaussian 09*; Revision A. 1; Gaussian Inc.: Wallingford, CT, USA, 2009.

33. Berova, N.; di Bari, L.; Pescitelli, G. Application of electronic circular dichroism in configurational and conformational analysis of organic compounds. *Chem. Soc. Rev.* **2007**, *36*, 914–931. [CrossRef] [PubMed]

International Journal of
Molecular Sciences

MDPI

Article

Affinin (Spilanthol), Isolated from *Heliopsis longipes*, Induces Vasodilation via Activation of Gasotransmitters and Prostacyclin Signaling Pathways

Jesús Eduardo Castro-Ruiz [1,2], Alejandra Rojas-Molina [2], Francisco J. Luna-Vázquez [2], Fausto Rivero-Cruz [3], Teresa García-Gasca [1,*] and César Ibarra-Alvarado [2,*]

[1] Laboratorio de Biología Celular y Molecular, Facultad de Ciencias Naturales, Universidad Autónoma de Querétaro, Campus Juriquilla, 76230 Querétaro, Qro., Mexico; dentaqro@live.com.mx

[2] Laboratorio de Investigación Química y Farmacológica de Productos Naturales, Facultad de Ciencias Químicas, Universidad Autónoma de Querétaro, Centro Universitario, 76010 Querétaro, Qro., Mexico; rojasa@uaq.mx (A.R.-M.); fjlunavz@yahoo.com.mx (F.J.L.-V.)

[3] Departamento de Farmacia, Facultad de Química, Universidad Nacional Autónoma de México, Ciudad Universitaria, 04510 México, D.F., Mexico; joserc@unam.mx

* Correspondence: tggasca@gmail.com (T.G.-G.); cibarra@uaq.mx (C.I.-A.);
Tel.: +52-442-1921-200 (ext. 5301) (T.G.-G.); +52-442-1921-200 (ext. 5527) (C.I.-A.)

Academic Editor: Toshio Morikawa
Received: 28 November 2016; Accepted: 13 January 2017; Published: 22 January 2017

Abstract: *Heliopsis longipes* roots have been widely used in Mexican traditional medicine to relieve pain, mainly, toothaches. Previous studies have shown that affinin, the major alkamide of these roots, induces potent antinociceptive and anti-inflammatory activities. However, the effect of *H. longipes* root extracts and affinin on the cardiovascular system have not been investigated so far. In the present study, we demonstrated that the dichloromethane and ethanolic extracts of *H. longipes* roots, and affinin, isolated from these roots, produce a concentration-dependent vasodilation of rat aorta. Affinin-induced vasorelaxation was partly dependent on the presence of endothelium and was significantly blocked in the presence of inhibitors of NO, H_2S, and CO synthesis (N^G-nitro-L-arginine methyl ester (L-NAME), DL-propargylglycine (PAG), and chromium mesoporphyrin (CrMP), respectively); K^+ channel blockers (glibenclamide (Gli) and tetraethyl ammonium (TEA)), and guanylate cyclase and cyclooxygenase inhibitors ($1H$-[1,2,4]oxadiazolo[4,3-a]quinoxalin-1-one (ODQ) and indomethacin (INDO), respectively). Our results demonstrate, for the first time, that affinin induces vasodilation by mechanisms that involve gasotransmitters, and prostacyclin signaling pathways. These findings indicate that this natural alkamide has therapeutic potential in the treatment of cardiovascular diseases.

Keywords: *Heliopsis longipes*; affinin; vasodilation; rat aorta; gasotransmitters; prostacyclin

1. Introduction

Heliopsis longipes (A. Gray) S. F. Blake (Asteraceae) (*H. longipes*) is an herbaceous plant native to Mexico, that grows particularly in the states of Querétaro, Guanajuato, and San Luis Potosí, where it is known by common names including "Chilcuague", "Chilcuán", "Chilmecatl", "Aztec root", "Golden root", among others [1–4]. In Central Mexico, the roots of this species are widely used as a spice, home insecticide, and for the treatment of some illnesses, which include toothaches, gingival disease, and muscular pain [5–8]. When *H. longipes* roots come into contact with oral cavity tissues, they produce numbness and a tingling sensation of the tongue, associated with a significant increase in salivary flow [9,10]. The predominant bioactive molecules found in *H. longipes* roots are *N*-alkylamides or alkamides, mainly *N*-isobutyl-2*E*,6*Z*,8*E*-decatrienamide, also known as affinin or spilanthol [7,11–16]. This alkamide is not only found in *H. longipes* roots, it has also been identified in other plants, including *Spilanthes* species (Synonym: *Acmella* species) [17–24]. A variety of biological activities such as larvicidal (10–14 µg/mL) [25], antimicrobial (25–300 µg/mL) [4], fungistatic, and bacteriostatic (5–150 µg/mL) [8] effects have been attributed to this compound. In addition, several pharmacological studies have demonstrated that affinin displays analgesic (ED_{50} = 1 mg/kg intraperitoneal (i.p.) in mice) [5,16], antinociceptive (ED_{50} = 6.98 mg/kg *per os* (p.o.); ED_{50} = 36 ± 5 mg/kg i.p. in mice) [6,26], anti-inflammatory (90–180 µM in macrophage cell line) [18], anxiolytic (3–30 mg/kg i.p. in mice) [6], and diuretic (800 mg/kg p.o. in mice) [27] properties. Some of these pharmacological activities have been also reported for crude organic extracts of *H. longipes* roots [5,6,26,28–31].

Affinin has an adequate lipophilicity. An in vitro permeability test showed that this alkamide (10 µg/mL) permeates through CaCo-2 cell monolayer cultures via passive diffusion. Whereas in vivo assays demonstrated that it is able to permeate skin and oral mucosa, and subsequently reach blood circulation, and cross the blood-brain barrier in high amounts (~98%) [23,32]. Therefore, this compound might be considered a valuable potential drug candidate [13,18,23,33].

With respect to safety assessment studies, the acute toxicity of affinin was evaluated on ICR mice and the determined median lethal dose (LD_{50} = 113 mg/kg) was significantly higher than the doses required to elicit antinociception [6,26]. No mutagenic effects were observed by using the Ames test [6] and antimutagenic effects of affinin were observed at 25 and 50 µg/mL [10]. The cytotoxic effect of affinin was determined on human HEK293 kidney cells and the calculated mean inhibitory concentration (IC_{50}) was 260 µg/mL, while the concentration used to observe biological effects was 100 µg/mL [27]. No cytotoxic effects of affinin, which elicits a stimulatory effect on nitric oxide (NO) production in RAW 264.7 murine macrophages, were observed at concentrations up to 40 µg/mL [18].

Regarding the mechanism of action underlying the antinociceptive effect of affinin, Déciga-Campos et al. [26] showed that this effect might be due to activation of opiodergic, serotoninergic, and GABAergic systems, and also involves participation of the NO/cGMP/potassium channel pathway. It has been well documented that this signaling pathway plays an important role in vascular tone regulation [34–39]. This physiological process is also regulated by other gasotransmitters, such as hydrogen sulfide (H_2S) and carbon monoxide (CO) [40–54]. Together with gasotransmitters, vascular endothelium releases prostacyclin, which also represents a key piece in the vasodilation process [55–57].

Considering involvement of the NO/cGMP/KATP pathway in the antinociceptive effect of affinin, we hypothesized that this compound might exert a vasodilator effect via activation of gasotransmitters and prostacyclin signaling pathways. Therefore, the aim of this study was to investigate whether affinin, isolated from *H. longipes* roots, was capable of inducing vasodilation and to explore its mechanism of action.

2. Results

2.1. Phytochemical Study of the Dichloromethane Extract Obtained from H. longipes Roots and Isolation of Affinin

Dichloromethane provided a higher yield of extract (19 g/kg roots dry weight) compared to ethanol (17 g/kg roots dry weight). Considering vasodilator potency, the dichloromethane extract was chosen to isolate the bioactive compounds. This extract (100 g) was fractionated by open column chromatography to obtain 21 fractions. Subsequent chromatography of fractions 8–17 resulted in the isolation of 28.5 g of pure affinin (Figure 1).

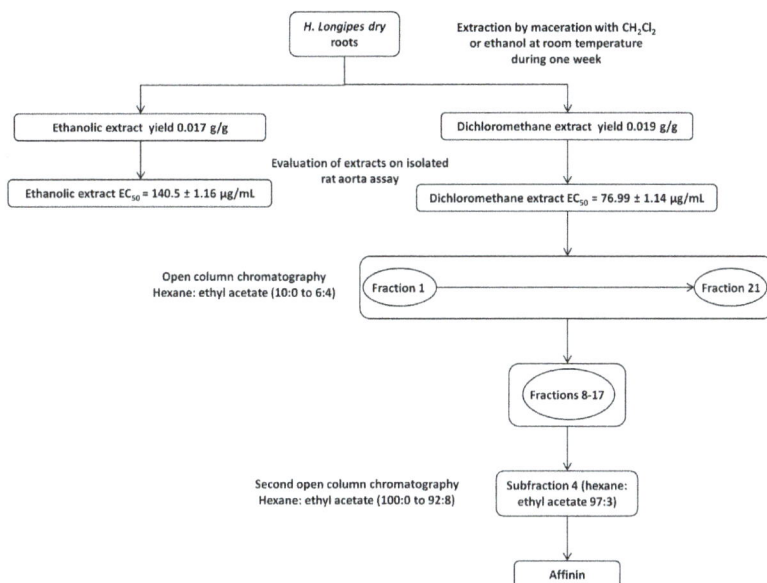

Figure 1. Diagram of the isolation of affinin from the dichloromethane extract of *H. longipes* roots.

Affinin (Figure 2) was identified by comparison with an authentic sample and by comparing its spectroscopic data (^{1}H-NMR and ^{13}C-NMR) with those previously reported in the literature (Table 1). High performance liquid chromatography/photodiode array detector (HPLC-PDA) analysis of affinin revealed a purity >94%.

Table 1. ^{13}C-NMR (400 MHz) and ^{1}H-NMR (400 MHz) spectral data of affinin.

H	δ_{ppm}	C
1	-	166.15
2	5.80 (1H, br d, J = 16.0, 8.0 Hz)	124.30
3	6.80 (1H, dt, J = 16.0, 8.0 Hz)	143.51
4	2.28 (4H, m)	32.20
5	2.28 (4H, m)	26.49
6	5.25 (1H, dt, J = 10.7, 7.1 Hz)	127.73
7	5.94 (1H, dd, J = 12.0 Hz)	129.52
8	6.25 (1H, br dd, J = 16.0, 4.0 Hz)	126.79
9	5.67 (1H, dq, J = 16.0, 6.0 Hz)	130.00
10	1.76 (3H, d, J = 6.0 Hz)	18.39
NH	5.47 (br s)	-
1'	3.13 (2H, dd, J = 6.0, 6.0 Hz)	46.97
2'	1.80 (1H, m)	28.68
3'	0.93 (6H, d, J = 6.7 Hz)	20.23
4'	0.93 (6H, d, J = 6.7 Hz)	18.40

Affinin was recorded in CDCl$_3$. Integrations, multiplicity, and coupling constants of protons are shown in parentheses.

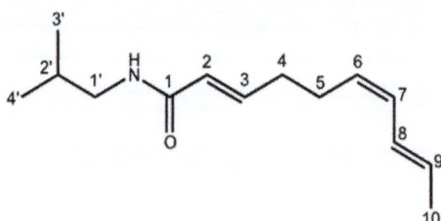

Figure 2. Chemical structure of affinin, the major alkamide in *H. longipes* roots.

2.2. Determination of the Vasodilator Effect of H. longipes Extracts and Affinin, and Elucidation of the Mechanism of Action of Affinin

2.2.1. Vasodilator Effect of *H. longipes* Roots Extracts and Affinin

The dichloromethane and ethanolic extracts of *H. longipes* roots, and affinin, induced a concentration- dependent relaxation of aortic rings with functional endothelium. Figure 3A shows the concentration-response curves for both extracts, affinin, and acetylcholine (ACh), which was used as a positive control. The dichloromethane extract (E_{max} = 100% ± 3.11% and EC_{50} = 76.99 ± 1.14 µg/mL) was approximately two fold more potent than the ethanolic extract (E_{max} = 100% ± 4.54% and EC_{50} = 140.5 ± 1.16 µg/mL), whereas affinin was significantly more potent than both extracts (E_{max} = 100% ± 3.10% and EC_{50} = 27.38 ± 1.20 µg/mL). Affinin turned out to be approximately twenty-five fold less potent than acetylcholine (E_{max} = 70.02% ± 1.43% and EC_{50} = 1.094 ± 1.14 µg/mL), however, this alkamide elicited a maximum vasodilator effect greater than that of the positive control (Table 2). Carboxymethylcellulose 1% (CMC), employed as a vehicle, did not show any significant vasodilator effect.

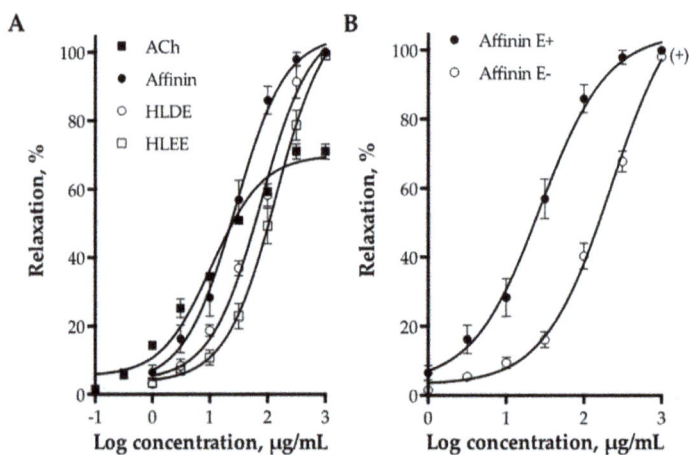

Figure 3. (**A**) Vasodilator effect of the dichloromethane extract (HLDE), the ethanolic extract (HLEE), and affinin from *Heliopsis longipes* roots on intact aortic rings. Acetylcholine (ACh) was used as positive control; (**B**) Concentration-response curves of the vasodilator effect of affinin in the presence (E+) and absence (E−) of endothelium. Values are expressed as mean ± standard error of the mean (SEM) (*n* = 6); + *p* < 0.01.

Table 2. Vasodilator effect of *Heliopsis longipes* roots extracts and affinin on rat aorta.

Compound	E_{max} (%)	EC_{50} (μg/mL)
Dichloromethane extract	100 ± 3.11	76.99 ± 1.14
Ethanolic extract	100 ± 4.54	140.5 ± 1.16
Affinin	100 ± 3.10	27.38 ± 1.20
ACh	70.02 ± 1.43	1.094 ± 1.14

Data are expressed as mean $\pm$ SEM ($n = 6$). Acetylcholine (ACh) is presented as positive control.

2.2.2. Role of Vascular Endothelium in the Vasodilation Induced by Affinin

Endothelial denudation caused a significant rightward shift in the concentration-response curve of affinin, without affecting the maximal response (E_{max} = 100% $\pm$ 4.5% and EC_{50} = 231.2 $\pm$ 1.13 μg/mL, $p < 0.01$) (Figure 3B).

2.2.3. Involvement of Gasotransmitters in the Vasodilation Produced by Affinin

The vasorelaxant effect of affinin was significantly reduced by inhibiting endothelial NO synthase (eNOS) with N^G-nitro-L-arginine methyl ester (L-NAME, 100 μM), heme-oxygenase (HO) with chromium mesoporphyrin IX (CrMP, 15 μM), and cystathionine-γ-lyase (CSE) with DL-propargylglycine (PAG, 1 mM), which indicated that the NO/cGMP, the CO/cGMP, and the H_2S/K_{ATP} pathways are involved in this effect (Figure 4A). The vasodilator effect of affinin was also significantly reduced by $1H$-[1,2,4]oxadiazolo[4,3-*a*]quinoxalin-1-one (ODQ, 10 μM), an inhibitor of soluble guanylate cyclase (sGC).

2.2.4. Involvement of K⁺ Channels in Affinin-Evoked Vasodilation

To determine whether activation of K⁺ channels participated in the vasodilatory effect of affinin, the effects of glibenclamide (Gli, a specific blocker of the K_{ATP} channels) and tetraethyl ammonium (TEA, a non-selective K⁺ channel inhibitor) were assessed. Both blockers significantly shifted to the right the concentration-response curve of the vasodilator effect of affinin (Figure 4B), indicating that these channels are involved in its effect.

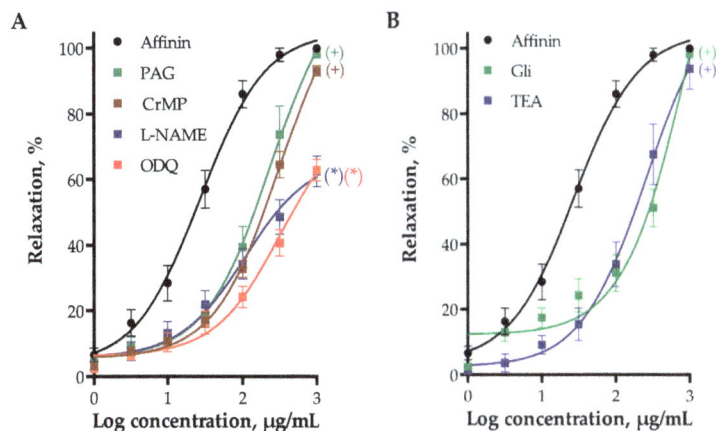

Figure 4. (**A**) Vasodilator effect of affinin in the absence (control) and presence of PAG (1 mM), chromium mesoporphyrin (CrMP, 15 μM), N^G-nitro-L-arginine methyl ester (L-NAME, 100 μM), and $1H$-[1,2,4]oxadiazolo[4,3-*a*]quinoxalin-1-one (ODQ, 10 μM) in rat aortic rings; (**B**) Vasodilator effect of affinin in the absence (control) and presence of glibenclamide (Gli, 10 μM) and tetraethyl ammonium (TEA, 1 mM) in rat aortic rings. Values are expressed as mean $\pm$ SEM ($n = 6$); + $p < 0.01$; * $p < 0.001$.

2.2.5. Effect of PGI$_2$/cAMP Pathway on Affinin-Induced Dilation of Rat Aorta

To test whether the PGI$_2$/cAMP pathway was implicated in affinin-induced relaxation, indomethacin (INDO, 10 µM) was used to inhibit cyclooxygenase (COX). INDO pre-treatment significantly reduced the affinin-vasorelaxant effect (Figure 5).

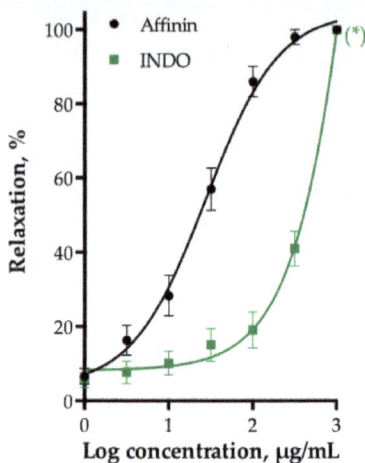

Figure 5. Vasodilatory effect of affinin in the absence (control) and presence of indomethacin (INDO, 10 µM) in rat aortic rings. Values are expressed as mean ± SEM ($n = 6$); * $p < 0.001$.

3. Discussion

H. longipes roots have a long tradition of culinary and medicinal use in Mexico. A number of studies have evidenced that organic extracts obtained from *H. longipes* roots and affinin, their major component, possess interesting biological and pharmacological activities [4–6,8,9,16,25,26,28–31,58]. However, currently, no investigation has been directed toward examining the effect of *H. longipes* root extracts and affinin on the vascular tone.

In the present study, both the dichloromethane and ethanolic extracts from *H. longipes* roots were found to significantly relax the isolated rat aorta. To our knowledge, this has not been previously reported. In 2008, Wongsawatkul et al. [59] described the vasorelaxant effect of four organic extracts (hexane, chloroform, ethyl acetate, and methanol extracts) prepared from aerial parts of *Spilanthes acmella* (Synonym: *Acmella oleracea*) on rat aorta rings. In that study, the ethyl acetate extract exhibited the most potent vasorelaxant effect and, according to the authors, such an effect could be attributed to the presence of polar phenolic and triterpenoid ester compounds. Additionally, the chloroform extract showed the highest maximum vasodilator response. The authors ascribed such effect to triterpenoids and fatty alcohols or esters present in the chloroformic extract. Furthermore, vasodilation induced by the *S. acmella* extracts was completely abolished in the absence of endothelium and significantly reduced in the presence of L-NAME (1 µM) and indomethacin (1 µM), which strongly suggested the participation of the NO and the PGI$_2$ pathways [59].

Other studies were carried out to test the effect of oral administration of the *S. acmella* ethanolic flower extract at doses ranging from 50 to 150 mg/kg on sexual performance in male rats. The extract was administered during 28 days and no toxic effects were observed. One of the main findings was the dose-dependent erectile function improvement induced by the extract and its capacity to produce long-term effects, even by day 15 after cessation of the treatment. In the same study, a good correlation was found between these results and a raise on NO levels determined in DS-1 cells (a human corpus cavernosum cell line) cultures stimulated with the *S. acmella* ethanolic extract (100 µg/mL). The authors suggested a possible contribution of affinin and other alkamides present in the extract on the observed effect of improved sexual function [22]. It is a well-known fact that erectile function is mediated by a complex integration of signals, where NO, synthesized by endothelial, inducible, and neuronal NOS, is the most important factor that contributes to vasodilation of the erectile vasculature of the penis [60–62].

Regarding our research, based on the vasodilator potency, we selected the dichloromethane extract of *H. longipes* roots to carry out a phytochemical study in order to isolate the bioactive constituents. Chromatographic analysis of the extract led to the isolation of affinin as the major component. This result is consistent with previous studies that have demonstrated that affinin is the main alkamide found in *H. longipes* roots [4,6,8,11,16,25,26]. Of great interest was the finding that affinin elicited a significant vasodilator effect, which was approximately three fold more potent than that of the crude extract. This finding represents the first demonstration that affinin is capable of relaxing the arterial smooth muscle. Considering that affinin was the most abundant constituent of the *H. longipes* dichloromethane extract, it can be inferred that this compound is responsible for the vasodilation induced by the crude extract.

Removal of endothelium significantly decreased, but did not completely block, the vasorelaxation induced by affinin, indicating that both, endothelial-dependent and independent vasodilation pathways are involved in its mechanism of action. The vasorelaxing effect was significantly reduced in the presence of NOS, CSE, and HO inhibitors, which evidenced that activation of the NO/cGMP, H$_2$S/KATP, and CO/cGMP pathways contribute to affinin-induced vasodilation. The most relevant inhibition was observed when aortas were preincubated with L-NAME ($p < 0.001$), suggesting that activation of the NO/cGMP pathway plays a more prominent role in the effect of this alkamide than that played by the other two gasotransmitters pathways. Moreover, inhibition of sGC by ODQ ($p < 0.001$) significantly reduced the vasodilatory effect of affinin, revealing that it might directly activate sGC, the main receptor of NO [36]. It is important to point out that CO-sensitive sGC isoforms exist in the vascular smooth muscle [46], therefore CO is also considered to be an important activator of this class of enzymes [37]. Activation of sGC on the smooth muscle cells might underlie, at least in part, endothelium-independent vasodilation caused by affinin.

One of the key mechanisms by which NO, CO, and H$_2$S, synthesized in endothelial cells, induce vasodilation is activation of potassium channels located in vascular smooth muscle cells. Regarding nitric oxide-cGMP induced vasodilation, it has been shown that cGMP-dependent protein kinase (PKG) phosphorylates calcium-activated potassium channels (K$_{Ca}$) on the smooth muscle cell membrane leading to a decrease in intracellular calcium concentration [38,39]. This same mechanism is involved in the vasorelaxation produced by CO. Moreover, this gasotransmitter is able to directly activate potassium channels, in particular K$_{Ca}$ [51,52]. On the other hand, H$_2$S mediates vasorelaxation via direct opening of K$_{ATP}$ channels [47–50]. In this study, we assessed whether potassium channel blockers impaired vasodilation provoked by affinin. Glibenclamide and TEA significantly decreased affinin-evoked vasodilation, which confirmed the activation of signaling pathways for NO, CO, and H$_2$S.

Indomethacin ($p < 0.001$) caused a significant reduction in affinin-induced vasodilation, suggesting that activation of the PGI$_2$/cAMP pathway is also involved in the mechanism of vasorelaxation caused by this alkamide. Along with gasotransmitters, PGI$_2$ is secreted by endothelial cells and elicits smooth muscle relaxation by stimulating adenylate cyclase, which subsequently increases cAMP levels. This second messenger activates calcium-activated potassium channels (K$_{Ca}$) via PKA-dependent phosphorylation [55,56]. Evidence from some previous studies suggest that cAMP also enhances K$_{Ca}$ activity by "cross-activation" of PKG [38,54].

According to our results, it is evident that affinin does not only act on a specific type of cell receptor in the arteries. Since its vasodilator effect is not completely blocked in the absence of endothelium, it is clear that this compound activates both endothelium dependent and independent pathways. Our results indicated that affinin is able to activate the NO/cGMP, CO/cGMP, H$_2$S/KATP, and PGI2/cAMP signaling pathways, and considering that the triggering of these four signaling pathways depends on the activation of molecular targets located on the endothelium layer, it is very likely that this alkamide might be acting on molecular targets, whose activation leads to an increase in Ca^{2+} levels in the endothelial cells. The chemical structure of N-alkylamides or alkamides [63], like affinin [13,64,65] resembles that of fatty acid amides [66–69], and the endogenous cannabinoid N-arachidonylethanolamine or anandamide [70]. This molecule produces a potent vasodilator effect through several proposed mechanisms that include activation of TRPV1 channels and G-coupled receptors, such as CB_1, CB_2, and endothelial non-CB_1/non-CB_2 [71–73]. Herradón et al. [72] showed that the vasodilator effect of anandamide in rat aorta is mainly produced by activation of the endothelial non-CB_1/non-CB_2 cannabinoid receptor, which in turn activates the NO/cGMP pathway. Therefore, considering the similarity between the chemical structures of anandamide and affinin, it is quite possible that endothelial non-CB_1/non-CB_2 or/and TRP channels may be the putative molecular targets for affinin in the endothelial cells.

Concerning endothelium-independent relaxation induced by affinin, our results indicate that this molecule might directly activate sGC. We can speculate that affinin might also directly activate R_{PGI}, although this possibility needs to be confirmed.

Figure 6 shows the proposed signaling pathways involved in the vasodilatory effect of affinin.

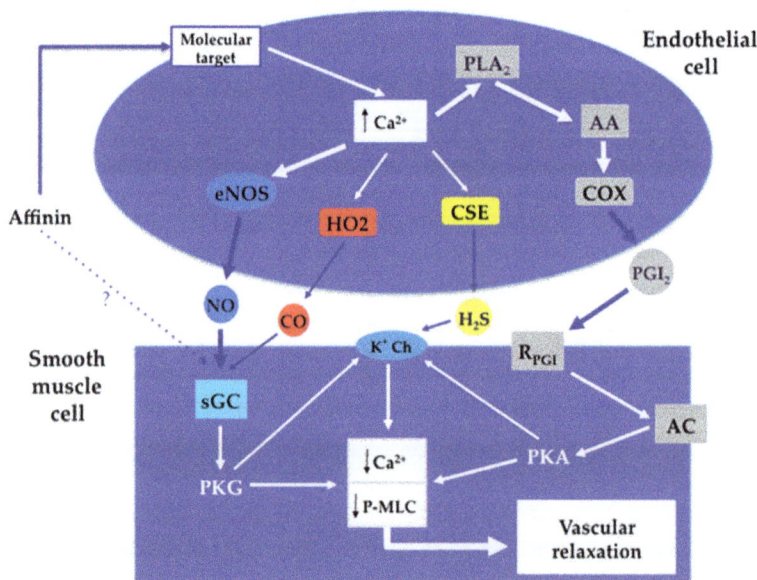

Figure 6. Pathways involved in the vasodilator effect of affinin. PLA_2, phospholipase A_2; AA, arachidonic acid; COX, cyclooxygenase; eNOS, endothelial NO synthase; HO2, heme-oxygenase 2; CSE, cystathionine-γ-lyase; sGC, soluble guanylate cyclase; PKG, protein kinase G; AC, adenylate cyclase; PKA, protein kinase A; K^+ Ch, K^+ channel; P-MLC, phosphorylated myosin light chain. Black upwards arrow, increased levels; Black downwards arrows, decreased levels. ?, pathway that remains to be confirmed.

The results of the present study have clearly shown that affinin, obtained from *H. longipes* roots, produces vasodilation of rat aorta by activating the NO/cGMP, CO/cGMP, H_2S/KATP, and PGI2/cAMP signaling pathways. This is the first report describing the vasodilator effect of this alkamide and some of the processes involved in its mechanism of action. The median effective concentration to produce vasodilation (EC_{50} = 27.38 µg/mL) falls within the concentration ranges at which this compound elicits other biological and pharmacological activities. Moreover, the EC_{50} obtained for the vasodilator effect of affinin is within the non-cytotoxic concentration range for mammalian cells, however, more cytotoxic studies must be performed in order to establish its possible adverse effects. Besides the other pharmacological properties that have been attributed to affinin, the vasodilator effect is a new interesting activity that might be ascribed to this alkamide. Undoubtedly, these results contribute to support the great therapeutic potential of *Heliopsis longipes* roots and affinin, their main constituent.

4. Materials and Methods

4.1. Reagents and Chemicals

Reagents and solvents used in the chemical study of *H. longipes* roots were purchased from JT Baker (Phillisburg, NJ, USA). Standards and solvents for the pharmacological assays were obtained from Sigma-Aldrich (St. Louis, MO, USA). CrMP was purchased from Porphyrin Products, Inc. (Logan, UT, USA).

4.2. Animals

All experimental protocols were performed in accordance with guidelines of the Mexican Official Standard NOM-062-ZOO-1999 [74], and approved by the Bioethics Committee of the Faculty of Natural Sciences, Autonomous University of Querétaro, México. Wistar male rats (250–300 g) were used for the pharmacological study; they were provided by the Institute of Neurobiology of the National Autonomous University of Mexico, Campus Juriquilla, Querétaro, Qro., Mexico. Animals were housed in standard cages under controlled temperature conditions with a 12:12 h light-dark cycle. Water and food were provided ad libitum.

4.3. Plant Material

H. longipes (Asteraceae) roots were collected in Peñamiller, Querétaro, Qro., Mexico. The specimens were identified (*H. longipes* vouchers J.E. Castro R.1. and R.2.) and deposited in the Herbario Jerzy Rzedowski (QMEX), Facultad de Ciencias Naturales, Universidad Autónoma de Querétaro, Querétaro, Qro., Mexico.

4.4. Preparation of the Extracts Employed for the Pharmacological Evaluation

Air dried *H. longipes* roots were ground to a fine powder. For the preparation of *H. longipes* root extracts, ground plant material (10 g) was subjected to maceration with either dichloromethane or absolute ethanol for one week in a 1:10 ratio (w/v). This process was repeated three times with fresh solvent. Thereafter, the plant material was filtered and the solvents were removed by rotary evaporation. The extraction yields were: 0.019 g extract/g dried roots for the dichloromethane extract and 0.017 g extract/g dried roots for the ethanolic extract.

4.5. Chemical Study of the Dichloromethane Extract Obtained from H. longipes Roots

4.5.1. Fractionation of the Dichloromethane Extract Obtained from *H. longipes* Roots and Purification of Affinin

Dried and ground plant material (7 kg) was extracted with dichloromethane as described above. One hundred grams of the dichloromethane extract were fractionated by column chromatography on normal phase using an open silica gel column (1 kg, Kiesegel 60 Merck 100–230 mesh, 8 × 110 cm). Hexane and ethyl acetate were used as eluents in ratios from 100:0 to 40:60. From this procedure, 472 fractions (250 mL) were collected, monitored by thin layer chromatography (TLC), and grouped into 21 pools according to their chromatographic similarity. TLC analysis of pools 8–17 revealed the presence of a main dark gray spot (*Rf* = 0.3, hexane: ethyl acetate 3:2), visualized with an ultraviolet lamp at 254 nm. Spraying TLC plates with a spray solution of anisaldehyde/sulfuric acid (0.5 mL anisaldehyde in 50 mL glacial acetic acid and 1 mL 97% sulfuric acid) developed a bright purple spot, as reported for other olefinic isobutyl-amides [75].

Pools 8–17 were combined (45 g) and further analyzed by open column chromatography (450 g, Kiesegel 60 Merck 100–230 mesh, 4.5 × 120 cm) using a step gradient of hexane and ethyl acetate 100:0 to 90:10. Based on their chromatographic similarity, determined by TLC, fractions eluted with hexane: ethyl acetate 97:3 were combined and evaporated to dryness in vacuo leaving a residue of 28.5 g of an apparently pure compound. The purity of the isolated compound was confirmed by HPLC-PDA, using an HPLC chromatograph (Waters 600 Associates, Milford, MA, USA) coupled to a photodiode array detector (Waters 2998). This analysis was carried out on a XBridge C18 (4.6 × 100 mm 3.5 µm) column. The flow rate of the mobile phase (acetonitrile/water 44:56 v/v) was 0.5 mL/min with column temperature of 30 °C and detection wavelength of 229 nm.

4.5.2. Determination of the Chemical Structure of Affinin

Chemical structure of the purified compound was elucidated by analysis of its proton nuclear magnetic resonance (^{1}H-NMR) and carbon-13 (^{13}C-NMR) spectra (Table 1). Nuclear magnetic resonance (NMR) spectra were taken on a Varian VNMRS 400 spectrometer with tetramethylsilane (TMS) as internal standard. Affinin was identified by comparing its spectroscopic constants with those reported in the literature [20,21].

4.6. Determination of the Vasodilator Effect and Elucidation of the Mechanism of Action of Affinin

4.6.1. Isolated Rat Aorta Assay

The rats were killed by decapitation. The thoracic aorta was surgically removed and placed in a Petri dish containing ice-cold (4 °C) Krebs-Henseleit solution with the following composition (mM): 126.8 NaCl; 5.9 KCl; 1.2 KH$_2$PO$_4$; 1.2 MgSO$_4$; 5.0 D-glucose; 30 NaHCO$_3$; 2.5 CaCl$_2$ (pH 7.4), bubbled with a mixture of carbogen (95% O$_2$ and 5% CO$_2$). Then, the intraluminal space of aorta was rinsed with fresh solution to prevent clot formation, cleaned from surrounding connective tissue, and sliced into rings (3–4 mm in length). Aortic rings were mounted between two metallic hooks, with one being fixed and the other attached to an isometric transducer, and placed into organ baths chambers containing pre-warmed Krebs-Henseleit solution (37 °C) gassed with carbogen. The aortic segments were allowed to equilibrate for 60 min under a resting tension of 1.5 g. During the resting period, the organ bath solution was exchanged every 10 min. In order to stimulate the vascular smooth muscle, the tissues were contracted with KCl solution (100 mM). Once a stable contractile tone was reached, the bathing medium was replaced every 10 min to restore the initial resting tension of 1.5 g. Afterwards, the aortic rings were contracted with 1 µM L-phenylephrine (Phe); the contractile force induced was defined as 100%, and once the plateau was reached, the test substances were cumulatively added. Acetylcholine (ACh), dissolved in distilled water, was evaluated in a concentration range of 0.2 ng/mL–2 mg/mL; while affinin and the extracts, dissolved in vehicle (carboxymethylcellulose 1% in distilled water), were tested in a concentration range of 1 µg/mL–1 mg/mL. When used, pharmacological inhibitors were added to the organ bath chambers 20 min before the addition of Phe. The changes in tension caused by the tested concentrations were detected by Grass FT03 force transducers coupled to a Grass 7D Polygraph; they were expressed as percentages of relaxation based on the contraction generated by adding Phe [76].

4.6.2. Participation of the Endothelium in the Vasodilator Response of Affinin

To determine whether the vasodilator response of affinin was dependent on the vascular endothelium, assays on aorta segments without endothelium were performed. In these experiments the endothelial layer was removed by flushing the lumen of aorta with 0.2% desoxycholate in saline solution 0.9%, as reported previously [76]. The absence of endothelium was confirmed at the start of the experiment, showing that the addition of 1 μM of acetylcholine (ACh) did not induce more than 5% relaxation. Once the cumulative concentrations of affinin were added to the bath chambers, as described above, sodium nitroprusside (100 μM) was added to the chambers to demonstrate that the artery was still capable of relaxation.

4.6.3. Evaluation of the Participation of the Gasotransmitters and Prostacyclin Signaling Pathways in the Vasodilator Response of Affinin

Involvement of the main gasotransmitters pathways in the vasodilator effect evoked by affinin was assessed by incubating intact endothelium aortic rings for 20 min in the presence of inhibitors of specific enzymes of each of these pathways: (1) NO/cGMP pathway: 100 μM N^G-nitro-L-arginine methyl ester (L-NAME, inhibitor of eNOS) or 10 μM 1*H*-[1,2,4]oxadiazolo[4,3-*a*]quinoxalin-1-one (ODQ, inhibitor of sGC); (2) H_2S/K_{ATP} channel pathway: 1 mM DL-propargylglycine (PAG, inhibitor of CSE); and (3) HO/CO pathway: 15 μM chromium mesoporphyrin IX (CrMP, inhibitor of HO) [40–44,49,52,77].

To determine the involvement of the prostacyclin pathway in the vasodilator effect of affinin, aortic segments were pre-incubated for 20 min in the presence of 1 μM indomethacin (INDO, inhibitor of COX) [78,79]. In addition, to assess whether activation of K^+ channels was involved in the vasodilation produced by affinin, the effect of pretreatment with the non-selective potassium channel blocker, 1 mM tetraethyl ammonium (TEA) and 10 μM glibenclamide (a specific blocker of the K_{ATP} channels) was evaluated [80,81].

4.7. Statistical Analysis

Evaluations of each concentration of the tested substances were performed on aortas obtained from at least three different rats ($n = 6$). All values are expressed as the mean ± standard error of the mean (SEM). The resulting data obtained from each evaluation were fitted to a sigmoidal equation, plotted, and analyzed to calculate EC_{50} (GraphPad Prism 7.02, San Diego, CA, USA). These results were subjected to one-way analysis of variance (ANOVA) using the statistical program GraphPad Prism 7.02, followed by the Tukey test to evaluate any significant differences between the means. Values of + $p < 0.01$ or * $p < 0.001$ were considered to be significant.

5. Conclusions

Our study provides a heretofore unknown evidence that affinin, isolated from *H. longipes* roots, is capable of inducing vasodilation via mechanisms that involve activation of gasotransmitters and prostacyclin signaling pathways. The NO/cGMP and PGI_2/cAMP pathways appear to play a more prominent role than either the H_2S/KATP pathway or the CO/cGMP pathway in affinin-evoked vasorelaxation. Undoubtedly, this molecule deserves further investigation in order to completely understand its mechanism of action. The results derived from this study suggest that affinin is a promising molecule for the development of drugs useful in the prevention and/or treatment of cardiovascular diseases, particularly when considering that it has an adequate lipophilicity that allows it to permeate skin and oral mucosa, and reach blood circulation.

Acknowledgments: Jesús E. Castro-Ruiz acknowledges Consejo Nacional de Ciencia y Tecnología (CONACYT) for his doctoral grant. The authors would like to thank Josué López Martínez and Yolanda Rodríguez Asa for their technical assistance.

Author Contributions: Jesús Eduardo Castro-Ruiz carried out the phytochemical study of *H. longipes* roots, conducted the pharmacological assays, and wrote the manuscript. Alejandra Rojas-Molina supervised the phytochemical study of *H. longipes* roots and contributed with the preparation of the manuscript. Francisco J. Luna-Vázquez supervised and helped in carrying out the pharmacological assays. Fausto Rivero-Cruz conducted the final identification of affinin. César Ibarra-Alvarado and Teresa García-Gasca designed this project, coordinated all the activities, and contributed with the preparation of the manuscript.

Conflicts of Interest: The authors declare no conflict of interest.

References

1. Little, E.J. *Heliopsis longipes*, a Mexican insecticidal plant species. *J. Wash. Acad. Sci.* **1948**, *38*, 269–274. [PubMed]

2. Correa, J.; Roquet, S.; Díaz, E. Multiple NMR analysis of the affinin. *Org. Magn. Reson.* **1971**, *3*, 1–5. [CrossRef]

3. Martínez, M. Chilmecatl. In *Las Plantas Medicinales de México*; Ediciones Botas: México, D.F., Mexico, 1989.

4. Molina-Torres, J.; García-Chávez, A.; Ramírez-Chávez, E. Antimicrobial properties of alkamides present in flavouring plants traditionally used in Mesoamerica: Affinin and capsaicin. *J. Ethnopharmacol.* **1999**, *64*, 241–248. [CrossRef]

5. Cilia-López, V.G.; Juárez-Flores, B.I.; Aguirre-Rivera, J.R.; Reyes-Agüero, J.A. Analgesic activity of *Heliopsis longipes* and its effect on the nervous system. *Pharm. Biol.* **2010**, *48*, 195–200. [CrossRef] [PubMed]

6. Déciga-Campos, M.; Arriaga-Alba, M.; Ventura-Martínez, R.; Aguilar-Guadarrama, B.; Rios, M.Y. Pharmacological and Toxicological Profile of Extract from *Heliopsis longipes* and Affinin. *Drug Dev. Res.* **2012**, *73*, 130–137. [CrossRef]

7. Acree, F.; Jacobson, M.J.; Haller, H.L. An amide posessing insecticidial properties from the roots of *Erigeron affinins* DC. *J. Org. Chem.* **1945**, 236–242. [CrossRef]

8. Molina-Torres, J.; Salazar-Cabrera, C.; Armenta-Salinas, C.; Ramírez-Chávez, E. Fungistatic and bacteriostatic activities of alkamides from *Heliopsis longipes* roots: Affinin and reduced amides. *J. Agric. Food Chem.* **2004**, *52*, 4700–4704. [CrossRef] [PubMed]

9. Ogura, M.; Cordell, G.A.; Quinn, M.L.; Leon, C.; Benoit, P.S.; Soejarto, D.D.; Farnsworth, N.R. Ethnopharmacologic studies. I. Rapid solution to a problem—Oral use of *Heliopsis longipes*—By means of a multidisciplinary approach. *J. Ethnopharmacol.* **1982**, *5*, 215–219. [CrossRef]

10. Arriaga-Alba, M.; Rios, M.Y.; Déciga-Campos, M. Antimutagenic properties of affinin isolated from *Heliopsis longipes* extract. *Pharm. Biol.* **2013**, *51*, 1035–1039. [CrossRef] [PubMed]

11. Molina-Torres, J.; Salgado-Garciglia, R.; Ramírez-Chávez, E.; del Río, R.E. Purely Olefinic Alkamides in *Heliopsis longipes* and *Acmella* (*Spilanthes*) *oppositifolia*. *Biochem. Syst. Ecol.* **1996**, *24*, 43–47. [CrossRef]

12. López-Martínez, S.; Aguilar-Guadarrama, B.; Ríos, M.Y. Minor alkamides from *Heliopsis longipes* S.F. Blake (Asteraceae) fresh roots. *Phytochem. Lett.* **2011**, *4*, 275–279. [CrossRef]

13. Ríos, M.Y.; Olivo, H.F. Natural and Synthetic Alkamides: Applications in Pain Therapy. *Stud. Nat. Prod. Chem.* **2014**, *43*, 79–121.

14. Greger, H. Alkamides: Structural relationships, distribution and biological activity. *Planta Med.* **1984**, *50*, 366–375. [CrossRef] [PubMed]

15. Greger, H. Alkamides: A critical reconsideration of a multifunctional class of unsaturated fatty acid amides. *Phytochem. Rev.* **2016**, *15*, 729–770. [CrossRef]

16. Ríos, M.Y.; Aguilar-Guadarrama, A.B.; Gutiérrez, M.D.C. Analgesic activity of affinin, an alkamide from *Heliopsis longipes* (*Compositae*). *J. Ethnopharmacol.* **2007**, *110*, 364–367. [CrossRef] [PubMed]

17. Johns, T.; Graham, K.; Towers, G.H.N. Molluscicidal activity of affinin and other isobutylamides from the asteraceae. *Phytochemistry* **1982**, *21*, 2737–2738. [CrossRef]

18. Wu, L.C.; Fan, N.C.; Lin, M.H.; Chu, I.R.; Huang, S.J.; Hu, C.Y.; Han, S.Y. Anti-inflammatory effect of spilanthol from *Spilanthes acmella* on murine macrophage by down-regulating LPS-induced inflammatory mediators. *J. Agric. Food Chem.* **2008**, *56*, 2341–2349. [CrossRef] [PubMed]

19. Boonen, J.; Baert, B.; Burvenich, C.; Blondeel, P.; de Saeger, S.; de Spiegeleer, B. LC-MS profiling of N-alkylamides in *Spilanthes acmella* extract and the transmucosal behaviour of its main bio-active spilanthol. *J. Pharm. Biomed. Anal.* **2010**, *53*, 243–249. [CrossRef] [PubMed]

20. Yasuda, I.; Takeya, K.; Itokawa, H. The geometric structure of spilanthol. *Chem. Pharm. Bull.* **1980**, *28*, 2251–2253. [CrossRef]

21. Nakatani, N.; Nagashima, M. Pungent Alkamides from *Spilanthes acmella* L. var. *oleracea* Clarke. *Biosci. Biotechnol. Biochem.* **1992**, *56*, 759–762. [CrossRef] [PubMed]

22. Sharma, V.; Boonen, J.; Chauhan, N.S.; Thakur, M.; de Spiegeleer, B.; Dixit, V.K. Spilanthes acmella ethanolic flower extract: LC–MS alkylamide profiling and its effects on sexual behavior in male rats. *Phytomedicine* **2011**, *18*, 1161–1169. [CrossRef] [PubMed]

23. Veryser, L.; Taevernier, L.; Joshi, T.; Tatke, P.; Wynendaele, E.; Bracke, N.; Stalmans, S.; Peremans, K.; Burvenich, C.; Risseeuw, M.; et al. Mucosal and blood-brain barrier transport kinetics of the plant *N*-alkylamide spilanthol using in vitro and in vivo models. *BMC Complement. Altern. Med.* **2016**, *16*, 177. [CrossRef] [PubMed]
24. Bae, S.S.; Ehrmann, B.M.; Ettefagh, K.A.; Cech, N.B. A validated liquid chromatography-electrospray ionization-mass spectrometry method for quantification of spilanthol in *Spilanthes acmella* (L.) Murr. *Phytochem. Anal.* **2010**, *21*, 438–443. [CrossRef] [PubMed]
25. Hernández-Morales, A.; Arvizu-Gómez, J.L.; Carranza-Álvarez, C.; Gómez-Luna, B.E.; Alvarado-Sánchez, B.; Ramírez-Chávez, E.; Molina-Torres, J. Larvicidal activity of affinin and its derived amides from *Heliopsis longipes* A. Gray Blake against *Anopheles albimanus* and *Aedes aegypti*. *J. Asia. Pac. Entomol.* **2015**, *18*, 227–231. [CrossRef]
26. Déciga-Campos, M.; Rios, M.Y.; Aguilar-Guadarrama, A.B. Antinociceptive effect of *Heliopsis longipes* extract and affinin in mice. *Planta Med.* **2010**, *76*, 665–670. [CrossRef] [PubMed]
27. Gerbino, A.; Schena, G.; Milano, S.; Milella, L.; Franco Barbosa, A.; Armentano, F.; Procino, G.; Svelto, M.; Carmosino, M. Spilanthol from *Acmella oleracea* lowers the intracellular levels of cAMP impairing NKCC2 phosphorylation and water channel AQP2 membrane expression in mouse kidney. *PLoS ONE* **2016**, *11*, e0156021. [CrossRef] [PubMed]
28. Cariño-Cortés, R.; Gayosso-De-Lucio, J.A.; Ortiz, M.I.; Sánchez-Gutiérrez, M.; García-Reyna, P.B.; Cilia-López, V.G.; Pérez-Hernández, N.; Moreno, E.; Ponce-Monter, H. Antinociceptive, genotoxic and histopathological study of *Heliopsis longipes* S.F. Blake in mice. *J. Ethnopharmacol.* **2010**, *130*, 216–221. [CrossRef] [PubMed]
29. Acosta-Madrid, I.I.; Castañeda-Hernández, G.; Cilia-López, V.G.; Cariño-Cortés, R.; Pérez-Hernández, N.; Fernández-Martínez, E.; Ortiz, M.I. Interaction between *Heliopsis longipes* extract and diclofenac on the thermal hyperalgesia test. *Phytomedicine* **2009**, *16*, 336–341. [CrossRef] [PubMed]
30. Hernández, I.; Márquez, L.; Martínez, I.; Dieguez, R.; Delporte, C.; Prieto, S.; Molina-Torres, J.; Garrido, G. Anti-inflammatory effects of ethanolic extract and alkamides-derived from *Heliopsis longipes* roots. *J. Ethnopharmacol.* **2009**, *124*, 649–652. [CrossRef] [PubMed]
31. Hernández, I.; Lemus, Y.; Prieto, S.; Molina-Torres, J.; Garrido, G. Anti-inflammatory effect of an ethanolic root extract of *Heliopsis longipes* in vitro. *Boletín Latinoam. Caribe Plantas Med. Aromáticas* **2009**, *8*, 160–164.
32. Veryser, L.; Wynendaele, E.; Taevernier, L.; Verbeke, F.; Joshi, T.; Tatke, P.; de Spiegeleer, B. *N*-alkylamides: From plant to brain. *Funct. Foods Heal. Dis.* **2014**, *4*, 264–275.
33. Boonen, J.; Baert, B.; Roche, N.; Burvenich, C.; de Spiegeleer, B. Transdermal behaviour of the *N*-alkylamide spilanthol (affinin) from *Spilanthes acmella* (*Compositae*) extracts. *J. Ethnopharmacol.* **2010**, *127*, 77–84. [CrossRef] [PubMed]
34. Coletta, C.; Papapetropoulos, A.; Erdelyi, K.; Olah, G.; Modis, K.; Panopoulos, P.; Asimakopoulou, A.; Gero, D.; Sharina, I.; Martin, E.; et al. Hydrogen sulfide and nitric oxide are mutually dependent in the regulation of angiogenesis and endothelium-dependent vasorelaxation. *Proc. Natl. Acad. Sci. USA* **2012**, *109*, 9161–9166. [CrossRef] [PubMed]
35. Bohlen, H.G. Nitric oxide and the cardiovascular system. *Compr. Physiol.* **2015**, *5*, 808–823. [PubMed]
36. Zhao, Y.; Vanhoutte, P.M.; Leung, S.W.S. Vascular nitric oxide: Beyond eNOS. *J. Pharmacol. Sci.* **2015**, *129*, 83–94. [CrossRef] [PubMed]
37. Derbyshire, E.R.; Marletta, M.A. Structure and Regulation of Soluble Guanylate Cyclase. *Annu. Rev. Biochem.* **2012**, *81*, 533–559. [CrossRef] [PubMed]
38. White, R.E.; Kryman, J.P.; El-Mowafy, A.M.; Han, G.; Carrier, G.O. cAMP-dependent vasodilators cross-activate the cGMP-dependent protein kinase to stimulate BK(Ca) channel activity in coronary artery smooth muscle cells. *Circ. Res.* **2000**, *86*, 897–905. [CrossRef] [PubMed]
39. Boerth, N.J.; Dey, N.B.; Cornwell, T.L.; Lincoln, T.M. Cyclic GMP-dependent protein kinase regulates vascular smooth muscle cell phenotype. *J. Vasc. Res.* **1997**, *34*, 245–259. [CrossRef] [PubMed]
40. Ahmad, A.; Sattar, M.A.; Rathore, H.A.; Khan, S.A.; Lazhari, M.I.; Afzal, S.; Hashmi, F.; Abdullah, N.A.; Johns, E.J. A critical review of pharmacological significance of Hydrogen Sulfide in hypertension. *Indian J. Pharmacol.* **2015**, *47*, 243–247. [PubMed]
41. Holwerda, K.M.; Karumanchi, S.A.; Lely, A.T. Hydrogen sulfide: Role in vascular physiology and pathology. *Curr. Opin. Nephrol. Hypertens.* **2015**, *24*, 170–176. [CrossRef] [PubMed]

42. Bełtowski, J.; Jamroz-Wiśniewska, A. Hydrogen sulfide and endothelium-dependent vasorelaxation. *Molecules* **2014**, *19*, 21506–21528. [CrossRef] [PubMed]
43. Wang, R.; Szabo, C.; Ichinose, F.; Ahmed, A.; Whiteman, M.; Papapetropoulos, A. The role of H_2S bioavailability in endothelial dysfunction. *Trends Pharmacol. Sci.* **2015**, *36*, 568–578. [CrossRef] [PubMed]
44. Durante, W.; Johnson, F.K.; Johnson, R.A. Role of carbon monoxide in cardiovascular function. *J. Cell. Mol. Med.* **2006**, *10*, 672–686. [CrossRef] [PubMed]
45. Durante, W. Carbon monoxide and bile pigments: Surprising mediators of vascular function. *Vasc. Med.* **2002**, *7*, 195–202. [CrossRef] [PubMed]
46. Purohit, R.; Fritz, B.G.; The, J.; Issaian, A.; Weichsel, A.; David, C.L.; Campbell, E.; Hausrath, A.C.; Rassouli-Taylor, L.; Garcin, E.D.; et al. YC-1 binding to the β subunit of soluble guanylyl cyclase overcomes allosteric inhibition by the α subunit. *Biochemistry* **2014**, *53*, 101–114. [CrossRef] [PubMed]
47. Zhang, Z.; Huang, H.; Liu, P.; Tang, C.; Wang, J. Hydrogen sulfide contributes to cardioprotection during ischemia-reperfusion injury by opening K ATP channels. *Can. J. Physiol. Pharmacol.* **2007**, *85*, 1248–1253. [CrossRef] [PubMed]
48. Zhao, W.; Zhang, J.; Lu, Y.; Wang, R. The vasorelaxant effect of H_2S as a novel endogenous gaseous KATP channel opener. *EMBO J.* **2001**, *20*, 6008–6016. [CrossRef] [PubMed]
49. Yoo, D.; Jupiter, R.C.; Pankey, E.A.; Reddy, V.G.; Edward, J.A.; Swan, K.W.; Peak, T.C.; Mostany, R.; Kadowitz, P.J. Analysis of Cardiovascular Responses to the H_2S donors Na_2S and NaHS in the Rat. *Am. J. Physiol. Heart Circ. Physiol.* **2015**, *309*, H605–H614. [CrossRef] [PubMed]
50. Mustafa, A.K.; Gadalla, M.M.; Snyder, S.H. Signaling by gasotransmitters. *Sci. Signal.* **2009**, *2*, re2. [CrossRef] [PubMed]
51. Wang, R.; Wu, L.; Wang, Z. The direct effect of carbon monoxide on K_{Ca} channels in vascular smooth muscle cells. *Pflug. Arch.* **1997**, *434*, 285–291. [CrossRef]
52. Decaluwé, K.; Pauwels, B.; Verpoest, S.; van de Voorde, J. Divergent mechanisms involved in CO and CORM-2 induced vasorelaxation. *Eur. J. Pharmacol.* **2012**, *674*, 370–377. [CrossRef] [PubMed]
53. Wang, R. Gasotransmitters: Growing pains and joys. *Trends Biochem. Sci.* **2014**, *39*, 227–232. [CrossRef] [PubMed]
54. Ibarra-Alvarado, C.; Galle, J.; Melichar, V.O.; Mameghani, A.; Schmidt, H.H. Phosphorylation of blood vessel vasodilator-stimulated phosphoprotein at serine 239 as a functional biochemical marker of endothelial nitric oxide/cyclic GMP signaling. *Mol. Pharmacol.* **2002**, *61*, 312–319. [CrossRef] [PubMed]
55. Stoner, L.; Erickson, M.L.; Young, J.M.; Fryer, S.; Sabatier, M.J.; Faulkner, J.; Lambrick, D.M.; McCully, K.K. There's more to flow-mediated dilation than nitric oxide. *J. Atheroscler. Thromb.* **2012**, *19*, 589–600. [CrossRef] [PubMed]
56. Giles, T.D.; Sander, G.E.; Nossaman, B.D.; Kadowitz, P.J. Impaired Vasodilation in the Pathogenesis of Hypertension: Focus on Nitric Oxide, Endothelial-Derived Hyperpolarizing Factors, and Prostaglandins. *J. Clin. Hypertens.* **2012**, *14*, 198–205. [CrossRef] [PubMed]
57. Kawabe, J.; Ushikubi, F.; Hasebe, N. Prostacyclin in Vascular Diseases. *Circ. J.* **2010**, *74*, 836–843. [CrossRef] [PubMed]
58. Ortiz, M.I.; Cariño-Cortés, R.; Pérez-Hernández, N.; Ponce-Monter, H.; Fernández-Martínez, E.; Castañeda-Hernández, G.; Acosta-Madrid, I.I.; Cilia-López, V.G. Antihyperalgesia induced by *Heliopsis longipes* extract. *Proc. West. Pharmacol. Soc.* **2009**, *52*, 75–77. [PubMed]
59. Wongsawatkul, O.; Prachayasittikul, S.; Isarankura-Na-Ayudhya, C.; Satayavivad, J.; Ruchirawat, S.; Prachayasittikul, V. Vasorelaxant and antioxidant activities of *Spilanthes acmella* Murr. *Int. J. Mol. Sci.* **2008**, *9*, 2724–2744. [CrossRef] [PubMed]
60. Yetik-Anacak, G.; Sorrentino, R.; Linder, A.E.; Murat, N. Gas what: NO is not the only answer to sexual function. *Br. J. Pharmacol.* **2015**, *172*, 1434–1454. [CrossRef] [PubMed]
61. Nangle, M.R.; Cotter, M.A.; Cameron, N.E. An in vitro study of corpus cavernosum and aorta from mice lacking the inducible nitric oxide synthase gene. *Nitric Oxide* **2003**, *9*, 194–200. [CrossRef] [PubMed]
62. Dalaklioglu, S.; Ozbey, G. The potent relaxant effect of resveratrol in rat corpus cavernosum and its underlying mechanisms. *Int. J. Impot. Res.* **2013**, *25*, 188–193. [CrossRef] [PubMed]
63. Boonen, J.; Bronselaer, A.; Nielandt, J.; Veryser, L.; de Tré, G.; de Spiegeleer, B. Alkamid database: Chemistry, occurrence and functionality of plant *N*-alkylamides. *J. Ethnopharmacol.* **2012**, *142*, 563–590. [CrossRef] [PubMed]

64. Martínez-Loredo, E.; Izquierdo-Vega, J.A.; Cariño-Cortés, R.; Cilia-López, V.G.; Madrigal-Santillán, E.O.; Zuñiga-Pérez, C.; Valadez-Vega, C.; Moreno, E.; Sánchez-Gutiérrez, M. Effects of *Heliopsis longipes* ethanolic extract on mouse spermatozoa in vitro. *Pharm. Biol.* **2016**, *54*, 266–271. [CrossRef] [PubMed]
65. Chicca, A.; Raduner, S.; Pellati, F.; Strompen, T.; Altmann, K.-H.; Schoop, R.; Gertsch, J. Synergistic immunomopharmacological effects of N-alkylamides in *Echinacea purpurea* herbal extracts. *Int. Immunopharmacol.* **2009**, *9*, 850–858. [CrossRef] [PubMed]
66. Sudhahar, V.; Shaw, S.; Imig, J.D. Mechanisms involved in oleamide-induced vasorelaxation in rat mesenteric resistance arteries. *Eur. J. Pharmacol.* **2009**, *607*, 143–150. [CrossRef] [PubMed]
67. Raboune, S.; Stuart, J.M.; Leishman, E.; Takacs, S.M.; Rhodes, B.; Basnet, A.; Jameyfield, E.; McHugh, D.; Widlanski, T.; Bradshaw, H.B. Novel endogenous N-acyl amides activate TRPV1–4 receptors, BV-2 microglia, and are regulated in brain in an acute model of inflammation. *Front. Cell. Neurosci.* **2014**, *8*, 195. [CrossRef] [PubMed]
68. Raduner, S.; Majewska, A.; Chen, J.; Xie, X.; Faller, B.; Altmann, K.; Hamon, J. Alkylamides from *Echinacea* Are a New Class of Cannabinomimetics. *J. Biol. Chem.* **2006**, *281*, 14192–14206. [CrossRef] [PubMed]
69. Rios, M. Natural Alkamides: Pharmacology, Chemistry and Distribution. In *Drug Discovery Research in Pharmacognosy*; InTech: Vienna, Austria, 2013; pp. 107–144.
70. Lu, H.C.; MacKie, K. An introduction to the endogenous cannabinoid system. *Biol. Psychiatry* **2016**, *79*, 516–525. [CrossRef] [PubMed]
71. Zygmunt, P.M.; Petersson, J.; Andersson, D.A.; Chuang, H.; Sørgård, M.; Di Marzo, V.; Julius, D.; Högestätt, E.D. Vanilloid receptors on sensory nerves mediate the vasodilator action of anandamide. *Nature* **1999**, *400*, 452–457. [PubMed]
72. Herradón, E.; Martín, M.I.; López-Miranda, V. Characterization of the vasorelaxant mechanisms of the endocannabinoid anandamide in rat aorta. *Br. J. Pharmacol.* **2007**, *152*, 699–708. [CrossRef] [PubMed]
73. O'Sullivan, S.E.; Kendall, D.A.; Randall, M.D. Vascular effects of Δ9-tetrahydrocannabinol (THC), anandamide and N-arachidonoyldopamine (NADA) in the rat isolated aorta. *Eur. J. Pharmacol.* **2005**, *507*, 211–221. [CrossRef] [PubMed]
74. Norma Oficial Mexicana, NOM-062-ZOO-1999, Especificaciones Técnicas Para la Producción, Cuidado y uso de los Animales de Laboratorio. Available online: http://www.fmvz.unam.mx/fmvz/principal/archivos/062ZOO.PDF (accessed on 19 January 2017).
75. Bauer, R.; Remiger, P. TLC and HPLC Analysis of Alkamides in *Echinacea* Drugs1,2. *Planta Med.* **1989**, *55*, 367–371. [CrossRef] [PubMed]
76. Ibarra-Alvarado, C.; Rojas, A.; Mendoza, S.; Bah, M.; Gutiérrez, D.M.; Hernández-Sandoval, L.; Martínez, M. Vasoactive and antioxidant activities of plants used in Mexican traditional medicine for the treatment of cardiovascular diseases. *Pharm. Biol.* **2010**, *48*, 732–739. [CrossRef] [PubMed]
77. Andresen, J.J.; Shafi, N.I.; Durante, W.; Bryan, R.M. Effects of carbon monoxide and heme oxygenase inhibitors in cerebral vessels of rats and mice. *Am. J. Physiol. Heart Circ. Physiol.* **2006**, *291*, H223–H230. [CrossRef] [PubMed]
78. Gonzalez, C.; Rosas-Hernandez, H.; Jurado-manzano, B.; Ramirez-Lee, M.A.; Salazar-Garcia, S.; Martinez-Cuevas, P.P.; Velarde-salcedo, A.J.; Morales-Loredo, H.; Espinosa-Tanguma, R.; Ali, S.F.; et al. The prolactin family hormones regulate vascular tone through NO and prostacyclin production in isolated rat aortic rings. *Acta Pharmacol. Sin.* **2015**, *36*, 572–586. [CrossRef] [PubMed]
79. Majed, B.H.; Khalil, R.A. Molecular mechanisms regulating the vascular prostacyclin pathways and their adaptation during pregnancy and in the newborn. *Pharmacol. Rev.* **2012**, *64*, 540–582. [CrossRef] [PubMed]
80. Shen, M.; Zhao, L.; Wu, R.; Yue, S.; Pei, J. The vasorelaxing effect of resveratrol on abdominal aorta from rats and its underlying mechanisms. *Vasc. Pharmacol.* **2013**, *58*, 64–70. [CrossRef] [PubMed]
81. Stott, J.B.; Jepps, T.A.; Greenwood, I.A. KV7 potassium channels: A new therapeutic target in smooth muscle disorders. *Drug Discov. Today* **2014**, *19*, 413–424. [CrossRef] [PubMed]

International Journal of
Molecular Sciences

MDPI

Article

Polyphenolic Extract of *Euphorbia supina* Attenuates Manganese-Induced Neurotoxicity by Enhancing Antioxidant Activity through Regulation of ER Stress and ER Stress-Mediated Apoptosis

Entaz Bahar [1], Geum-Hwa Lee [2], Kashi Raj Bhattarai [2], Hwa-Young Lee [2], Min-Kyung Choi [2], Harun-Or Rashid [2], Ji-Ye Kim [3], Han-Jung Chae [2,*] and Hyonok Yoon [1,*]

[1] College of Pharmacy, Research Institute of Pharmaceutical Sciences, Gyeongsang National University, Jinju 52828, Gyeongnam, Korea; entaz_bahar@yahoo.com
[2] Department of Pharmacology, Medical School, Chonbuk National University, Jeonju 54896, Jeonbuk, Korea; heloin@jbnu.ac.kr (G.-H.L.); meekasik@jbnu.ac.kr (K.R.B.); youngat84@gmail.com (H.-Y.L.); mkelf78@nate.com (M.-K.C.); rashid@jbnu.ac.kr (H.-O.R.)
[3] Department of Pathology, Severance Hospital and Yonsei University College of Medicine, Seoul 03722, Korea; alucion@gmail.com
*   Correspondence: hjchae@jbnu.ac.kr (H.-J.C.); hoyoon@gnu.ac.kr (H.Y.);
    Tel.: +82-63-270-3092 (H.-J.C.); +82-55-772-2422 (H.Y.); Fax: +82-63-275-2855 (H.-J.C.); +82-55-772-2409 (H.Y.)

Academic Editor: Toshio Morikawa
Received: 10 November 2016; Accepted: 24 January 2017; Published: 30 January 2017

Abstract: Manganese (Mn) is an important trace element present in human body, which acts as an enzyme co-factor or activator in various metabolic reactions. While essential in trace amounts, excess levels of Mn in human brain can produce neurotoxicity, including idiopathic Parkinson's disease (PD)-like extrapyramidal manganism symptoms. This study aimed to investigate the protective role of polyphenolic extract of *Euphorbia supina* (PPEES) on Mn-induced neurotoxicity and the underlying mechanism in human neuroblastoma SKNMC cells and Sprague-Dawley (SD) male rat brain. PPEES possessed significant amount of total phenolic and flavonoid contents. PPEES also showed significant antioxidant activity in 1,1-diphenyl-2-picrylhydrazyl (DPPH) radical scavenging and reducing power capacity (RPC) assays. Our results showed that Mn treatment significantly reduced cell viability and increased lactate dehydrogenase (LDH) level, which was attenuated by PPEES pretreatment at 100 and 200 μg/mL. Additionally, PPEES pretreatment markedly attenuated Mn-induced antioxidant status alteration by resolving the ROS, MDA and GSH levels and SOD and CAT activities. PPEES pretreatment also significantly attenuated Mn-induced mitochondrial membrane potential ($\Delta\Psi$m) and apoptosis. Meanwhile, PPEES pretreatment significantly reversed the Mn-induced alteration in the GRP78, GADD34, XBP-1, CHOP, Bcl-2, Bax and caspase-3 activities. Furthermore, administration of PPEES (100 and 200 mg/kg) to Mn exposed rats showed improvement of histopathological alteration in comparison to Mn-treated rats. Moreover, administration of PPEES to Mn exposed rats showed significant reduction of 8-OHdG and Bax immunoreactivity. The results suggest that PPEES treatment reduces Mn-induced oxidative stress and neuronal cell loss in SKNMC cells and in the rat brain. Therefore, PPEES may be considered as potential treat-ment in Mn-intoxicated patients.

Keywords: manganese; *Euphorbia supina*; neurotoxicity; antioxidant; neuroprotection

---

1. Introduction

Manganese (Mn) is a vital trace element for normal function and development of human body [1]. Mn binds to and/or regulates several important body enzymes such as Mn-superoxide dismutase

(Mn-SOD) and pyruvate carboxylase in the growth and development of central nervous system (CNS) [2]. In micronutrient studies, Mn deficiency has been found in parenteral nutrition patients [3,4]. Exposure to excess levels of Mn produces cognitive, psychiatric, and motor abnormalities [3]. It has been reported that overexposure to the Mn could produce neurodegenerative damage, resulting in development of manganism symptoms such as cognitive, psychiatric, and motor abnormalities, similar to idiopathic Parkinson's disease (PD) [4–8]. Chronic exposure of Mn causes toxic Mn accumulation in brain regions, especially in striatum [9–11]. Mn toxicity has been identified through overexposure of Mn in occupational (e.g., welders and smelters), environmental, medical and dietary routes [6,8,12]. It has been noted that Mn causes toxic effect mainly in the CNS and lungs, as well as in heart, liver, and reproductive organs and during embryonic stage [4,7,13–21]. Some countries use anti-knock agent methylcyclopentadienyl manganese tricarbonyl as a fuel additive, which could increase Mn overexposure to human [22,23]. A number of studies have identified the possible underlying mechanisms of Mn-induced neurotoxicity with some different aspects but it remains unclear. Mn has the ability to induce reactive oxygen species (ROS) generation, lead to mitochondrial dysfunction, impairs endoplasmic reticulum (ER) homeostasis and promotes apoptosis [24–27]. Mn can also induce excitotoxic cell death through alteration of neurotransmitters levels [28–30]. Mn can induce protease activation and apoptotic cell death [31,32]. Recently, involvement of ER stress and ER stress-mediated apoptosis has been found in Mn-induced neurotoxicity in the rat striatum in vivo [33].

The Korean prostrate spurge *Euphorbia supina (E. supina)*, in the family Euphorbiaceae, is characterized as a broadleaf weed, with pinkish stems, dense hair and spotted spurge exude a milky sap when injured. It has been used as folk medicine against various diseases such as bronchitis, hepatitis, hemorrhage, etc. It was reported that the plant contained a variety of biologically active components, such as terpenoids, tannins, and polyphenols [34]. Polyphenols have a great interest to researcher as they possess many biological benefits to human health, especially in neurodegenerative diseases including PD and Alzheimer's disease (AD) [34–36]. *E. supine* is abundant in polyphenols and, by using high-performance liquid chromatography-tandem mass spectrometry (HPLC-MS/MS), nine biologically interesting polyphenols were isolated and identified from this plant: gallic acid, protocatechuic acid, nodakenin, quercetin 3-*O*-hexoside, quercetin 3-*O*-pentoside, kaempferol 3-*O*-hexoside, kaempferol 3-*O*-pentoside, quercetin and kaempferol [37]. Polyphenols such as quercetin and kaempferol derivatives from *E. supina* have strong antioxidant properties [37]. Recently, it has been found that polyphenolic compounds of *E. supina* markedly inhibit metastatic cancer in MDA-MB-231 breast cancer cells [38]. A number of studies identified strong antioxidant activity of *E. supina* in presence of several key polyphenols [37]. Moreover, no systematic studies have been conducted to validate the pharmacological efficacy of polyphenols of *E. supina*. The present study aimed to investigate the protective effect of polyphenols of *E. supina* on Mn-induced oxidative stress and the underlying mechanism in human neuroblastoma SKNMC cells and Sprague-Dawley (SD) male rats.

2. Results

2.1. Total Phenol and Flavonoid Content

The PPEES possessed significant amount of total phenol and flavonoid content expressed as gallic acid and quercetin equivalents, respectively (Table 1). The phenolic content of PPEES was 175.53 ± 5.94 mg GAE/g. Flavonoid content of the PPEES was 98.48 ± 17.73 mg QE/g.

Table 1. Total phenolic content and flavonoid content of PPEES.

TPC in PPEES (mg·GAE/g)	TFC in PPEES (mg·QE/g)
175.53 ± 5.94	98.48 ± 7.73

2.2. DPPH Scavenging and RPC of PPEES

The DPPH activity of PPEES was found to increase in dose dependent manner. The IC_{50} value of the PPEES was 145.04 ± 6.2 µg/mL, while the IC_{50} of ascorbic acid was 14.27 ± 1.06 µg/mL. The RPC of PPEES was found to increase in dose dependent manner. The IC_{50} value of the PPEES was 86.052 ± 3.94 µg/mL, while the IC_{50} of ascorbic acid was 10.05 ± 0.64 µg/mL In comparison to ascorbic acid, PPEES showed strong antioxidant activity, as determined using DPPH and RPC (Table 2).

Table 2. Antioxidant capacity of PPEES.

DPPH Radical Scavenging Activity; IC50 (µg/mL)		Reducing Capacity of PPEES; IC50 (µg/mL)	
PPEES	Ascorbic acid	PPEES	Ascorbic acid
145.044 ± 6.2	14.27 ± 1.06	86.0517 ± 3.94	10.05 ± 0.64

2.3. Effect of PPEES on SKNMC Cell Lines

The cytotoxic effect of the PPEES on human neuroblastoma cell line SKNMC was evaluated by incubating it with various concentrations of extract (1–1000 µg/mL). The toxicity results revealed a decrease in percentage of viability at higher concentrations of the extract and the IC_{50} value was found to be 1181.281 ± 8.1 µg/mL.

2.4. Protective Effect of PPEES on Mn-Induced Cytotoxicity

The effect of PPEES on the viability of SKNMC cells under Mn-induced toxicity conditions was measured by crystal violet assay. Pretreatment of SKNMC cells with PPEES at concentrations of 50–200 µg/mL significantly ($p < 0.05$ or $p < 0.01$) protected SKNMC cells from Mn toxicity. An increase in cell viability was observed in treated cells compared to Mn alone group (Figure 1A). The result displayed that PPEES doses of 100 µg/mL and 200 µg/mL possessed the best protective effects. Correspondingly, PPEES pretreatment significantly decreased ($p < 0.05$ or $p < 0.01$) the Mn-caused LDH release (Figure 1B). No change of the cell viability and LDH activity was observed in control and PPEES groups (Figure 1).

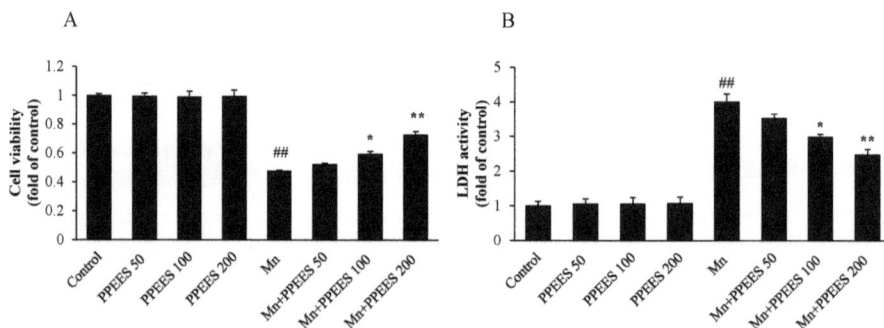

Figure 1. Protective effect of PPEES on Mn -induced cytotoxicity in SKNMC cell lines: (**A**) cell viability; and (**B**) LDH activity. Values were represented as mean $\pm$ SD ($n = 3$). ## $p < 0.01$ as compared with the control group; * $p < 0.05$; ** $p < 0.01$ as compared with the Mn alone group.

2.5. PPEES Attenuated Mn-Induced Oxidative Stress in SKNMC Cells

As shown in Figure 2A, the intracellular ROS level was markedly increased to 2.88-fold ($p < 0.01$) in SKNMC cells with the treatment Mn compared to the control. PPEES pretreatment with different concentrations (50, 100 and 200 µg/mL) significantly reduced the ROS level to 2.51, 2.31 ($p < 0.05$), and

1.75 fold ($p < 0.01$) of the control value, respectively. Similarly, the cells were pretreated with different concentrations of PPEES (50, 100 and 200 µg/mL) in the presence of Mn (500 µM) for 24 h significantly reduced ($p < 0.01$) the MDA levels from 309.08% to 254.81%, 227.71% ($p < 0.05$) and 174.15% ($p < 0.01$) (Figure 2B), respectively. Correspondingly, pretreatment of PPEES at the concentration of 100 and 200 µg/mL significantly increased the activities of SOD and CAT and the GSH level ($p < 0.05$ or $p < 0.01$) (Figure 2C–E). PPEES treatment alone at 50, 100 and 200 µg/mL had no effect on cellular oxidative stress.

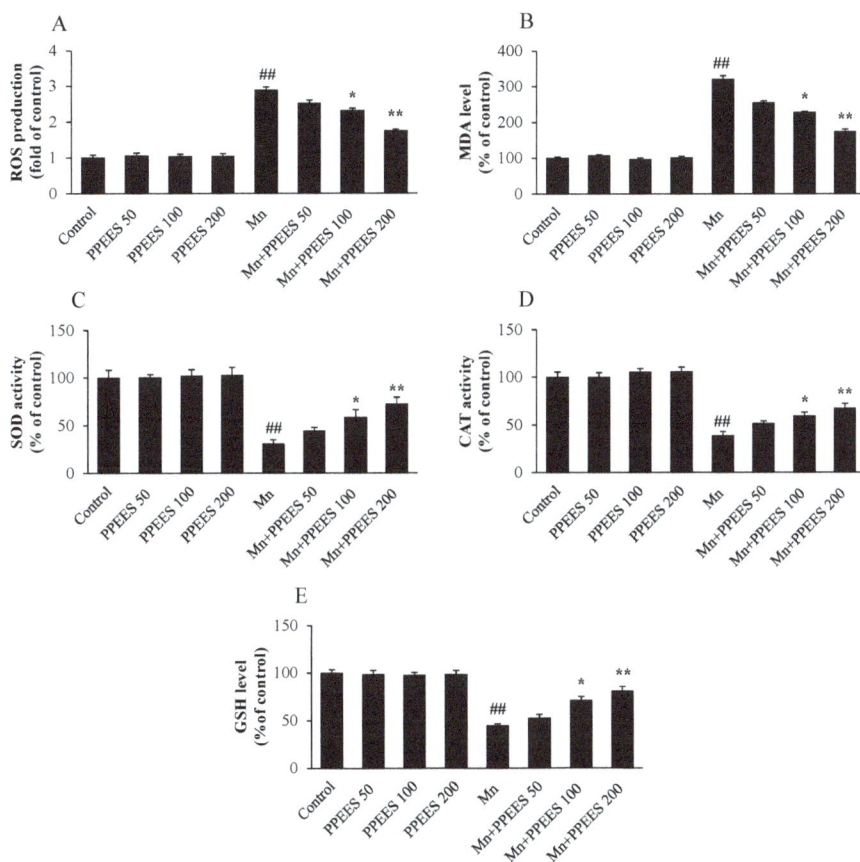

Figure 2. Protective Effect of PPEES on Mn-induced oxidative stress in SKNMC cell lines: (**A**) ROS; (**B**) malondialdehyde (MDA) levels; (**C**) superoxide dismutase (SOD) activity; (**D**) catalase (CAT) activity; and (**E**) glutathione (GSH) levels. Values were represented as mean ± SD ($n = 3$). ## $p < 0.01$ as compared with the control group; * $p < 0.05$ and ** $p < 0.01$ as compared with the Mn alone group.

2.6. PPEES Attenuates Mn-Induced Mitochondrial Dysfuction

The loss of mitochondrial membrane potential ($\Delta\Psi_m$) was observed using JC-1, a sensitive fluorescent dye. Mn exposure significantly reduced ($p < 0.01$) the $\Delta\Psi_m$ value in SKNMC cells (Figure 3). In comparison with the control group, the Mn group showed a reduced $\Delta\Psi_m$ at 45.5%, which could be rescued to 51.36%, 61.34% ($p < 0.05$) and 70.94% ($p < 0.01$) with the pretreatment of PPEES at the concentrations of 50, 100 and 200 µg/mL, respectively. No change of $\Delta\Psi_m$ was observed in control and PPEES alone groups (Figure 3).

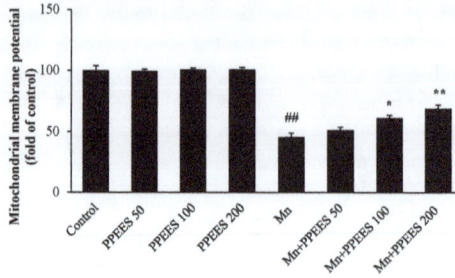

Figure 3. The protective effect of PPEES against Mn-induced mitochondrial dysfunction in SKNMC cells. JC-1 fluorescent dye was used to measure the loss of mitochondrial membrane potential ($\Delta\Psi_m$). Values were represented as mean ± SD ($n = 3$). ## $p < 0.01$, compared to the control group; * $p < 0.05$; ** $p < 0.01$, compared to the Mn alone group.

2.7. PPEES Reduced Apoptosis on Manganese-Induced Apoptosis in SKNMC Cells

SKNMC cells treated with Mn (500 µM) for 24 h showed typical properties of apoptosis, including chromatin condensation, fragmentation and nuclei shrinkage using Hoechst 33342 staining (Figure 4A). The amount and rate of apoptotic cells were significantly increased ($p < 0.01$) compared to the control and PPEES alone groups. However, the number of apoptotic cell was significantly reduced ($p < 0.05$ or $p < 0.01$) with PPEES pretreatment at 100 and 200 µg/mL in the presence of Mn (Figure 4B).

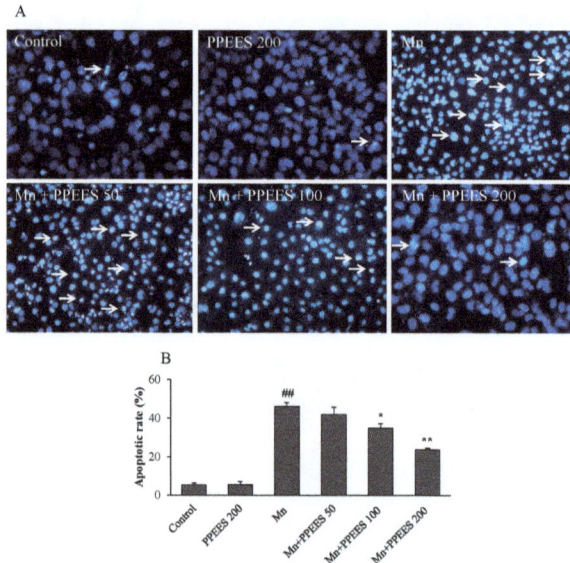

Figure 4. Protective effects of PPEES against Mn-induced apoptosis in SKNMC cells: (**A**) representative pictures showing the apoptotic cells (Hoechst-positive cells) in arrowheads; and (**B**) representative percentage of the apoptotic rate, measured by calculating the percent of Hoechst positive cells over the total number of cells. Values were represented as mean ± SD ($n = 3$). ## $p < 0.01$, compared to the control group; * $p < 0.05$; ** $p < 0.01$, compared to the Mn alone group.

2.8. PPEES Decreased Mn-Induced ER Stress and ER Stress-Mediated Apoptosis

Western blot and RT-PCR analyses were performed to investigate the effects of PPEES on Mn-induced ER stress and ER stress-mediated apoptosis. Western blot analyses were performed to investigate the effects of Mn and PPEES on the expression of GRP78, GADD34 and cleaved caspase-3 proteins in the SKNMC cell line. The ER stress biomarkers GRP78 and GADD34 were markedly increased ($p < 0.01$) in Mn-treated group, while PPEES pretreatment significantly reduced ($p < 0.05$ or $p < 0.01$) the Mn-induced changes in GRP78 and GADD34 to levels similar to that of both untreated control and PPEES treated only groups (Figure 5). The results showed that the Mn administration significantly increased ($p < 0.01$) the levels of apoptotic hallmark protein cleaved caspase-3 and that the treatment with PPEES significantly reduced ($p < 0.05$ or $p < 0.01$) the cleaved caspase-3 to levels similar to that of both untreated control and PPEES-treated only groups (Figure 5). RT-PCR analyses were performed to investigate the effects of Mn and PPEES on the expression of mRNA levels of XBP-1, CHOP, Bcl-2 and Bax in the SKNMC cell line. Our results showed that compared with the control treatment, the Mn treatment significantly increased ($p < 0.01$) the mRNA expression of XBP-1, CHOP and Bax, while significantly decreased ($p < 0.01$) the mRNA expression of the anti-apoptotic protein Bcl-2 (Figure 6). Interestingly, the treatment with PPEES significantly reversed ($p < 0.05$ or $p < 0.01$) the Mn-induced changes in the XBP-1, CHOP Bax and Bcl-2 mRNA expression levels (Figure 6).

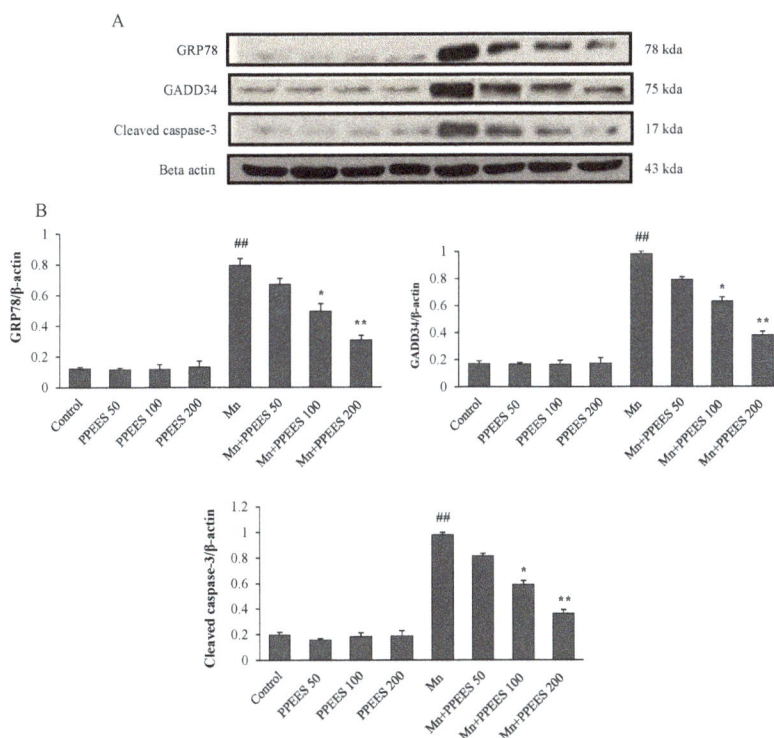

Figure 5. (**A**,**B**) Western blotting was performed to examine the effect of PPEES on the increased protein expression of GRP78, GADD34 and cleaved caspase-3, induced by administration of MnCl$_2$ (500 µM). Protein expression was normalized against β-actin. Values were represented as mean ± SD (n = 3). ## $p < 0.01$, compared to the control group; * $p < 0.05$; ** $p < 0.01$, compared to the Mn group.

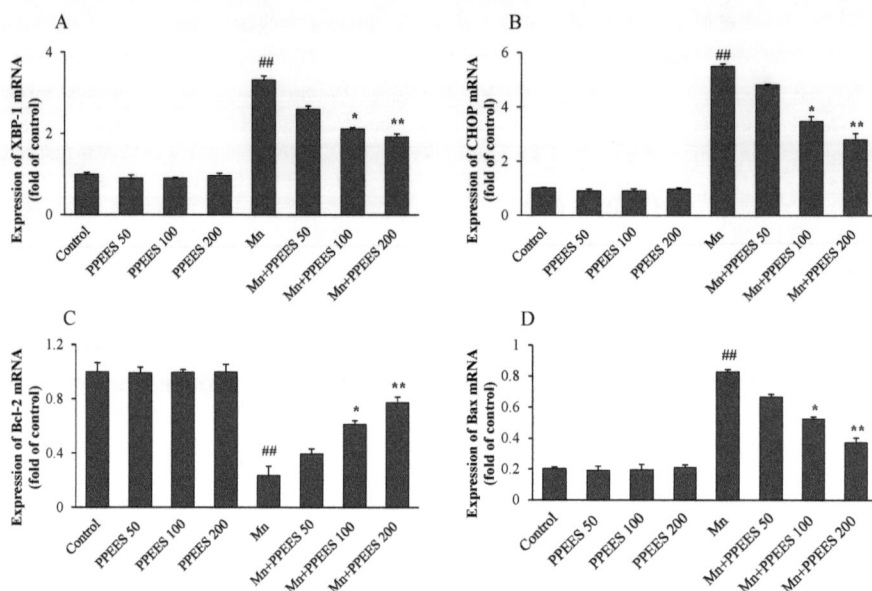

Figure 6. (A–D) RT-PCR was performed to examine the effect of PPEES on the mRNA expression of XBP-1, CHOP, Bcl-2 and Bax in SKNMC cell that result from Mn treatment. GAPDH served as an internal control. The transcriptive levels of XBP1-1, CHOP, Bcl-2 and Bax were normalized against GAPDH. Values were represented as mean $\pm$ SD (n = 3). ## $p < 0.01$, compared to the control group; * $p < 0.05$; ** $p < 0.01$, compared to the Mn group.

2.9. The OF Test

The OF test showed that Mn-treated rats spent significantly less ($p < 0.01$) time in the center of arena (Figure 7A) and displayed significantly less ($p < 0.01$) locomotors activity (Figure 7B) compared to normal control and Mn + PPEES groups. PPEES treatment significantly improved ($p < 0.05$ or $p < 0.01$) the time spent in the center area ($p < 0.05$ or $p < 0.01$) and the locomotors activity compared to the Mn-treated animals. To avoid possible unwanted olfactory influences on tested animals, the OF was thoroughly cleaned with a 10% ethanol solution.

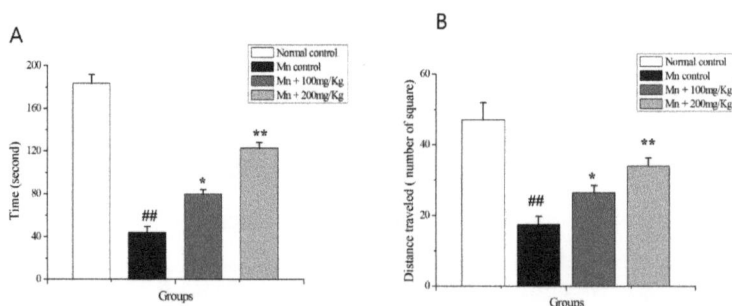

Figure 7. The locomotors activity of normal control, Mn control and PPEES-treated rats was evaluated by open field OF test: **(A)** representation of time spent in the center of the arena; and **(B)** representation of number of squares traveled. Values were represented as mean $\pm$ SD (n = 5). ## $p < 0.01$ (normal control versus Mn-exposed rats), * $p < 0.05$, ** $p < 0.01$ (Mn-exposed versus Mn + PPEES rats).

2.10. The Beneficial Effect of PPEES Treatment on Mn-Induced Histopathological and Immunohistochemically Altered Rats Brain

The histopathology examination was conducted using striatum part of brain under microscope (Figure 8A–D). Exposure of Mn led to marked histopathological alterations in the striatum characterized by neuronal damaged and present of ghost cells, hemorrhage and vacuolated cytoplasm. PPEES treatment showed beneficial effect compared to Mn-treated group. There was no histopathological alteration in striatum of normal control (Figure 8A).

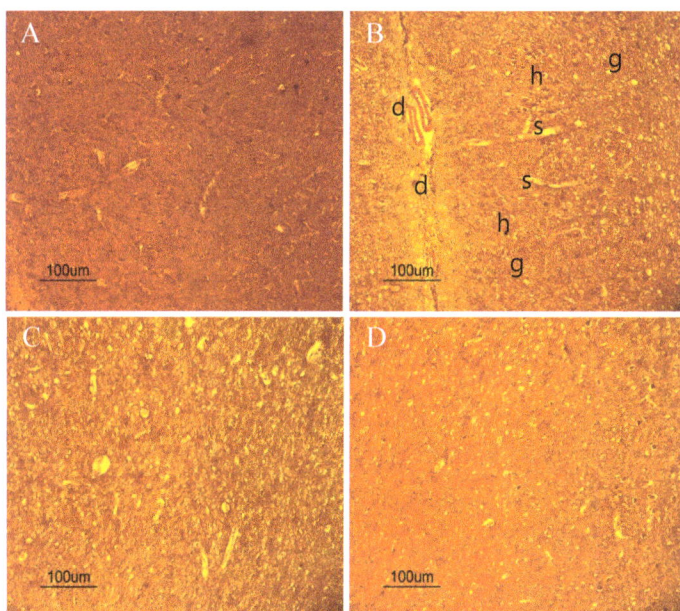

Figure 8. Histopathological images showing the beneficial effects of PPEES on Mn-induced changes in rat striatum: (**A**) control group (Normal saline); (**B**) manganese chloride (15 mg/kg) treated group; (**C**) manganese chloride (15 mg/kg) + PPEES (100 mg/kg); and (**D**) manganese chloride (15 mg/kg) + PPEES (200 mg/kg) treated group (magnification at 10×). Damage (d); Ghost cells (g); hemorrhage (h); and vacuolated cytoplasm (s) (magnification at 10×).

8-OHdG and Bax immunoreactivity was significantly increased in Mn-treated rats brain with respect to control group (Figures 9B and 10B). Moreover, PPEES treatment significantly reduced Mn-induced immunoreactivity in rat brain (Figure 9C,D and Figure 10C,D). In normal control rats, there was no immunoreactivity (Figures 9D and 10D).

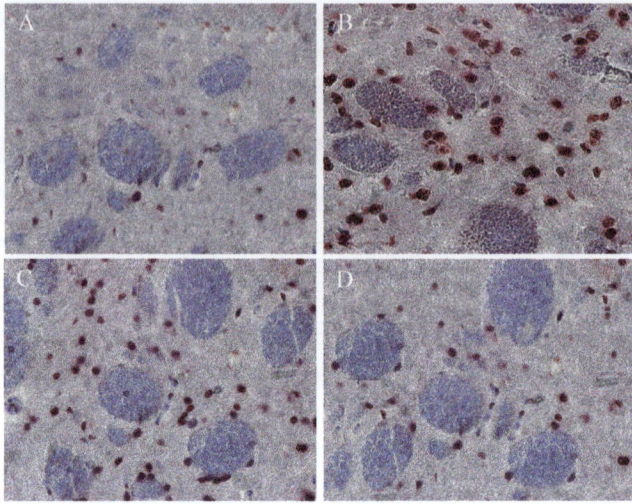

Figure 9. IHC staining showing the protective effect of PPEES against Mn-induced neurodegenetive disease by reducing oxidized RNA in neurons. PPEES treatment significantly reduced 8-hydroxy-2′-deoxyguanosine (8-OHdG) expression that result from Mn exposure in striatum: (**A**) control group (Normal saline); (**B**) manganese chloride (15 mg/kg) treated group; (**C**) manganese chloride (15 mg/kg) + PPEES (100 mg/kg); and (**D**) manganese chloride (15 mg/kg) + PPEES (200 mg/kg) treated group (magnification at 40×).

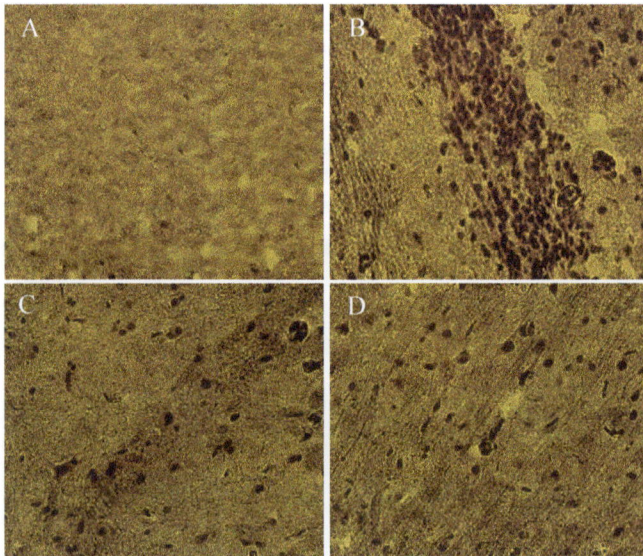

Figure 10. Photographs showing protective effect of PPEES against Mn-induced neurotoxicity by reducing Bax expression in cortex of rats: (**A**) control group (Normal saline); (**B**) manganese chloride (15 mg/kg) treated group; (**C**) manganese chloride (15 mg/kg) + PPEES (100 mg/kg); and (**D**) manganese chloride (15 mg/kg) + PPEES (200 mg/kg) treated group (magnification at 40×).

Int. J. Mol. Sci. **2017**, *18*, 300

3. Discussion

To investigate the protective role of PPEES on Mn-induced neurotoxicity, we used human neuroblastoma SKNMC cells and SD male rats. Our results revealed that Mn treatment could induce ROS generation, oxidative stress, mitochondrial dysfunction, apoptosis and neurotoxicity, while PPEES treatment could effectively resolved these undesired neurotoxicity.

We determined the beneficial effects of PPEES against the Mn-induced toxicity in SKNMC cells. The SKNMC cells were treated with Mn (500 μM) and three different concentrations (50, 100 and 200 μg/mL) of PPEES for 24 h. The results showed that PPEES at all three concentrations was nontoxic to SKNMC cells and that there was no significant difference between the cell viability and LDH activity of the control cells and the PPEES alone cells (Figure 1). Mn exposure at 500 μM significantly decreased ($p < 0.01$) cell viability and increased ($p < 0.01$) LDH activity in SKNMC cells after 24 h. However, pretreatment with PPEES at 100 mg/mL and 200 mg/mL concentrations significantly increased ($p < 0.05$ or $p < 0.01$) the cell viability and decreased ($p < 0.05$ or $p < 0.01$) LDH activity of Mn-exposed cells, which could protect the SKNMC cell from cytotoxicity [39].

To understand the antioxidant effects of PPEES against the Mn-induced oxidative stress, we quantified the intracellular ROS, MDA and GSH level and SOD and CAT activities in vitro. Mn treatment resulted in a significant increase ($p < 0.01$) in the ROS and MDA levels, decreased ($p < 0.01$) GSH levels, and decreased SOD and CAT activities compared with the control treatment and PPEES alone treatment. All three different concentrations (50, 100 and 200 μg/mL) was nontoxic to SKNMC cells and the concentrations of PPEES (100 mg/mL and 200 mg/mL) significantly reduced ($p < 0.05$ or $p < 0.01$) the ROS and MDA levels, increased ($p < 0.05$ or $p < 0.01$) GHS levels, and increased SOD and CAT activities, which suggested PPEES acted as a potent antioxidant (Figure 2) [39,40].

The main target of ROS-induced oxidative damage is mitochondria [41,42]. In the present study, Mn treatment significantly decreased ($p < 0.01$) $\Delta\Psi_m$ of SKNMC cells (Figure 3), which led to mitochondrial dysfunction. PPEES pretreatment significantly attenuated ($p < 0.05$ or $p < 0.01$) the disruption of $\Delta\Psi_m$, which led to initiating mitochondria mediated apoptosis through intrinsic and extrinsic apoptotic [43].

There are three main apoptosis pathways: mitochondrial pathway, death receptor pathway, and ER pathway [44]. It has been demonstrated that Mn treatment could induce apoptosis in SKNMC cell via the involvement of ER stress and mitochondria dysfunction [24]. In the present study, we found that Mn treatment significantly increased ($p < 0.01$) the apoptotic rates [44]. Interestingly, pretreatment of PPEES attenuated ($p < 0.05$ or $p < 0.01$) Mn-induced cytotoxicity involving the inhibition of cell apoptosis (Figure 4).

To investigate the protective role of PPEES on Mn-induced ER stress and ER stress-mediated apoptosis markers (GRP78, GADD34, XBP-1, CHOP, Bcl-2, Bax and cleaved caspase-3), both western blot and RT-PCR analyses were performed. ER stress-signaling pathways regulated by GRP78, leading to UPR survival, cell fate and apoptosis responses [45]. ER stress induces XBP-1 to produce a more active transcription factor [46]. Induction of CHOP expression is most sensitive to ER stress condition and led to DNA damage [45]. Overexpression of GADD34 can initiate or enhance apoptosis through various signals [47,48]. It has been reported that the pro-apoptotic Bax and the anti-apoptotic Bcl-2 proteins regulate the mitochondrial apoptotic pathway [49,50]. Both intrinsic and extrinsic apoptotic pathways can activate caspase-3, a key biomarker of apoptosis and consequently lead to DNA breakdown [43,51]. Our results showed that the Mn administration significantly ($p < 0.01$) altered the GRP78, XBP-1, CHOP, Bcl-2, Bax and caspase-3 activities. Importantly PPEES pretreatment significantly reversed ($p < 0.05$ or $p < 0.01$) the Mn-induced alteration in the GRP78, GADD34, XBP-1, CHOP, Bcl-2, Bax and caspase-3 activities (Figures 5 and 6).

To further understand this, we investigated the protective effects of PPEES on Mn-induced neuronal toxicity in the striatum and cerebral cortex of SD male rats treated with Mn at a dose regimen of 15 mg/kg. The striatum and cerebral cortex parts were chosen because Mn affects more severely the striatum and cortex regions than any other region of the CNS [33]. Exposure of Mn caused histopathological alterations in the striatum. Administration of PPEES (100 and 200 mg/kg) to Mn-exposed rats showed improvement of histopathological alteration in comparison to Mn-treated rats (Figure 8).

Oxidative stress is one of the most important factors in the pathogenesis of neurological disorders (e.g., AD and PD), and 8-OHdG is a common oxidative stress marker produced by oxidation of DNA bases [52]. We found that 8-OHdG expression in striatum region was significantly increased in Mn-exposed rats compared to normal control rats. 8-OHdG expression was significantly reduced in PPEES-treated rats (Figure 9). It has been established that several observed alterations of neuron in the vulnerable brain regions of AD patients are due to upregulation of Bax immunoreactivity [53,54]. Administration of PPEES to Mn-exposed rats showed significant reduction of Bax immunoreactivity (Figure 10).

4. Materials and Methods

4.1. Plant Material

E. supina plants were purchased from the Jirisan Medicinal Herbs (Jirisan, Republic of Korea). The identities of the plants were authenticated by Professor Chang Young-Nam, Ph.D., Department of Chinese Medicine Resource, College of Environmental & Bioresource sciences, Chonbuk National University, Jeonju, Republic of Korea. A voucher of herbarium sheet was deposited in the above-mentioned entity. The aerial parts of *E. supina* were separated and air dried without exposure to sunlight at room temperature for one week to make coarse powder. The coarse powder (500 g) was macerated with 2000 mL of hydro-alcoholic mixture (water:methanol 30%:70% vol/vol) for 10 days at room temperature. The resulted solution was filtered through Whatman filter paper (No. 4) and concentrated using a rotary evaporator under reduced pressure at <35 °C. The total dried plant extract was storage for further stage.

4.2. Extraction of Polyphenol Enriched Extracts of E. supina (PPEES)

To obtain polyphenol extracts of *E. supina* (PPEES), the total plant extract residue was first dissolved in water and 200 mL of petroleum ether (four times) was added to obtain a clear upper layer (petroleum ether). Then lower layer (aqueous water layer) was washed with 200 mL of ethyl acetate containing glacial acetic acid (10 mL/L) (four times). Finally, the resulted solutions were combined, and ethyl acetate was evaporated to obtain PPEES, which was stored at −20 °C [55]. Details of the Identification of polyphenolic compounds are described in the Supplementary Materials (Table S1, Figure S1).

4.3. Total Phenolic Content (TPC)

The total phenolic content of the PPEES was measured using the Folin–Ciocalteu reagent [56,57]. Briefly, PPEES was oxidized by adding Folin–Ciocalteu reagent and then sodium carbonate was added to neutralize the reaction. After 30 min, the absorbance of resulting solution was read at 760 nm using gallic acid as standard. Total phenolic content was expressed as mg gallic acid equivalent (GAE)/gm of extract.

4.4. Total Flavonoid Content (TFC)

The flavonoid content was determined using quercetin (Q) as reference standard [58,59]. The PPEES was mixed with aluminum trichloride and one drop of acetic acid was added. Then, the mixture was diluted with ethanol. After 40 min, the absorbance of resulting solution was measured at 415 nm. The absorbances of blank sample and standard quercetin solution were measured under the same condition. Total flavonoid content was expressed as mg quercetin equivalent (QE)/gm of extract.

4.5. 1,1-Diphenyl-2-picrylhydrazyl (DPPH) Assay

The 1,1-diphenyl-2-picrylhydrazyl (DPPH) radical scavenging activity assay was conducted following previously described methods with minor adjustments [60]. The PPEES was mixed with the 0.004% methanol solution of DPPH and incubated for 30 min. After incubation, the absorbances of all the samples were determined at 517 nm. Inhibition of DPPH (D) was measured using Equation (1).

$$D(\%) = \{(Ac - As)/Ac\} * 100 \qquad (1)$$

where Ac and As are the absorbance of control and test samples, respectively.

4.6. Reducing Power Capacity (RPC)

The assay was conducted following previously described method based on the measurement of absorbance of Pearl's Prussian blue [61]. The PPEES was added to a mixture of PBS (pH 6.6) and potassium ferricyanide [$K_3Fe(CN)_6$]. The resultant mixture was incubated and centrifuged after adding trichloroacetic. The upper layer was mixed with distilled water and $FeCl_3$. The absorbance was measured at 700 nm.

4.7. Cell Culture

SKNMC, a human neuroblastoma cell line, was obtained from the American type culture collection (Manassas, VA, USA). The cells were grown in Dulbecco's modified eagle medium (DMEM) supplemented with 10% fetal bovine serum (FBS), 4.5 g/L D-glucose, 2 mmol/L L-glutamine, 110 mg/L sodium pyruvate, 100 U/mL penicillin, and 100 µg/mL streptomycin at 37 °C in a humidified atmosphere containing 95% air and 5% CO_2.

4.8. Cytotoxicity of PPEES

PPEES was dissolved in dimethyl sulphoxide (DMSO) (D2650, Sigma, Saint Louis, MO, USA) to obtain a stock solution of 20 mg/mL, and 0.2 mg/mL of sub stock solution was prepared by diluting 10 µL of the stock solution into 990 µL serum-free DMEM medium, and prepared at different concentrations (the percentage of DMSO in the experiment should not exceed 0.5). Stock and sub stock solutions were both stored at 4 °C. Cell viability was determined by crystal violet assay. Briefly, SKNMC cells were seeded onto 24-well plate (5×10^4 cells/well), incubated overnight and pretreated with various concentrations of PPEES (0–1000 µg/mL) for 24 h. Then, the medium was removed and cells were washed with phosphate buffer solution (PBS). Two hundred microliters of 0.2% crystal violet solution was added to each well and incubated for 10 min at room temperature then washed with water and 100 µL 1% SDS was added to solubilize the stain solution until color was uniform and there were no areas of dense coloration in bottom of wells. The samples were read at 590 nm in microplate reader (Spectra MAX, Gemini EM, Molecular Device, Sunnyvale, CA, USA).

4.9. Lactate Dehydrogenase (LDH) Activity

The LDH activity assay was conducted based on reduction of nicotinamide adenine dinucleotide (NAD) by LDH. LDH release into the media was taken as an indicator of cell damage and the assay is based on the principle of reduction of nicotinamide adenine dinucleotide (NAD) by LDH. The stoichiometric conversion of a tetrazolium dye utilized the reduced NAD (NADH) that was measured spectrophotometrically using an assay kit Tox-7 (Sigma, Saint Louis, MO, USA). Briefly, SKNMC cells were seeded (5×10^4 cells/well)) and cultured in 24-well culture plates. The cells were then preincubated with or without different concentrations of PPEES (50, 100, and 200 µg/mL) at 37 °C for 6 h followed by incubation with 500 µM $MnCl_2$ (CAS: 7773-01-5, Sigma, Saint Louis, MO, USA) for 24 h. After treatment, cells were centrifuged at $240 \times g$ for 4 min and supernatant solution was transferred to assay plate. The plate was wrapped in foil and incubated at room temperature for 30 min.

After incubation, by adding stop solution the reaction was terminated and the plate read at 490 nm and at a reference wavelength of 690 nm in microplate reader (Spectra MAX, Gemini EM, Molecular Device, Sunnyvale, CA, USA). The amount of LDH release is expressed as the fold of absorbance of control.

4.10. Cell Viability

Mn-induced cell survival was determined by crystal violet assay. Briefly, SKNMC cells were seeded (5×10^4 cells/well)) and cultured in 24-well culture plates. The cells were then preincubated with or without different concentrations of extract (50, 100, and 200 µg/mL) at 37 °C in a humidified atmosphere of 5% CO_2/95% air for 6 h followed by incubation with 500 µM Mn for 24 h. Afterwards, the medium was removed and cells were washed with phosphate buffer solution (PBS) and 200 µL of 0.2% crystal violet solution was added to each well and incubated for 10 min at room temperature, and then wash with water and 100 µL 1% SDS was added to solubilize the stain solution until the color was uniform and there were no areas of dense coloration in bottom of wells. The samples were read at 590 nm in a microplate reader (Spectra MAX, Gemini EM, Molecular Device, Sunnyvale, CA, USA).

4.11. Measurement of Intracellular Reactive Oxygen Species (ROS) Level

The intracellular ROS generation was measured based on enzymatic conversion of a non-fluorescent compound dichloro-dihydro-fluorescein diacetate (DCFH-DA) to highly fluorescent compound DCF, following the previously described method [62]. Briefly, the cells were harvested and seeded onto 6-well plate with 2×10^5 cells per well in culture media and allowed to attach overnight. The cells were pretreated with the doses of PPEES at 50, 100 and 200 µg/mL at 37 °C for 6 h and washed with PBS. Then, the cells were treated with Mn (500 µM) for additional 24 h. Finally, after washing, the cells were seeded on the 6-well plate with PBS once and incubated with DCFH-DA (10 µmol/L) for 30 min at 37 °C in the dark. The fluorescence intensity was measured in the microplate reader (Spectra MAX, Gemini EM, Molecular Device, Sunnyvale, CA, USA) at an excitation wave length of 485 nm and an emission wave length of 538 nm after the cells were washed three times with PBS to remove the extracellular DCFH-DA. The level of intracellular ROS is shown as a fold of control.

4.12. Antioxidant Status

Antioxidant status of PPEES was examined by measurement of intracellular malondialdehyde (MDA) and glutathione (GSH) levels, and superoxide dismutase (SOD) and catalase (CAT) activities, using specific assay kits (Nanjing Jiancheng Co., Ltd., Nanjing, China) according to the manufacturer's instructions. In brief, SKNMC cells were seeded (2×10^5 cells/well) into 12-well plates and pre-treated with PPEES (50, 100 or 200 µg/mL) at 37 °C for 6 h. The cells were incubated with or without Mn (500 µM) for 24 h after removing PPEES containing medium. Then cells were washed with cold PBS and lysed using the cell lysis buffer. The cell lysates were centrifuged at $14,000 \times g$ for 10 min at 4 °C and supernatant solutions were used for measuring the levels of MDA and GSH, and the activities of SOD and CAT. Protein concentrations were measured using the BCA protein assay kit (Intron Biotechnology, Inc., Gyeonggi, Korea).

4.13. Measurement of Mitochondrial Membrane Potential ($\Delta\Psi_m$)

Harvested SKNMC cells the day before the experiment and seeded onto 6-well plate with 2×10^5 cells per well in culture media and allowed to attach overnight. The cells were pretreated with the doses of PPEES at 50, 100 and 200 µg/mL at 37 °C for 6 h and washed with PBS. Then, the cells were treated with Mn (500 µM) for an additional 24 h. Finally, after washing, the cells were seeded on a 6-well plate with PBS once and incubated with JC-1 (10 mM final concentration) for 30 min at 37 °C in the dark. The JC-1 green fluorescence intensity was measured in the microplate reader (Spectra MAX, Gemini EM, Molecular Device) at an excitation wave length of 488 nm and an emission wave length of 530 nm after the cells were washed two times with PBS to remove the extracellular JC-1. Monomeric JC-1 green fluorescence emission and aggregate were measured at excitation wavelength

488 nm, emission wavelength 530 nm on a microplate reader (Spectra MAX, Gemini EM, Molecular Device, Sunnyvale, CA, USA).

4.14. Apoptosis Assay

Hoechst33342 staining was conducted based on qualitative and quantitative measurements of the apoptotic cells by distinguishing apoptotic cells from normal cells. SKNMC cells were cultured in 6-well plates for 24 h. After treatment, the cells were incubated with 5 μg/mL Hoechst 33342 for 15 min, then washed twice with PBS and finally visualized by inverted fluorescence microscopy (Axioskop 2 plus microscope, Carl Zeiss, Oberkochen, Germany). The apoptotic cells were counted by observation of minimum 200 cells from five non-overlapping fields in all groups, and expressed as a percentage (%) of the total number of cells counted.

4.15. Real Time Polymerase Chain Reaction (RT-PCR)

To examine the protective mechanism of PPEES, the expression of X-box binding protein-1 (XBP-1), C/EBP homologous protein (CHOP), Bcl-2 and Bax was measured by real time quantitative polymerase chain reaction (RT-qPCR). The total RNA was extracted from SKNMC cells using trizol reagent (sigma-Aldrich, Saint Louis, MO, USA). The integrity of mRNA was measured spectrophotometrically examined according to its A260/A280 absorption. Subsequently, reverse transcription was used to obtain cDNA. RT-qPCR was conducted on Mastercycler ep realplex (Eppendorf, Hamburg, Germany) using housekeeping gene GAPDH as an internal control. Briefly, the amplification of primer was carried out with 40 cycles at a melting temperature of 94 °C for 15 s, an annealing temperature of 60 °C for 1 min, and an extension temperature of 72 °C for 50 s. The primers used in the amplification were as follow: XBP-1, forward primer: 5′-AAACAGAGTAGCAGCGCAGACTGC-3′, reverse primer: 5′-GGATCTCTAAAACTAGAGGCTTGGTG-3′; CHOP, forward primer: 5′-GAA AGCAGAAACCGGTCCAAT-3′, reverse primer: 5′- GGATGAGATATAGGTGCCCCC-3′; Bcl-2, forward primer: 5′-CCAGGTCTCCGATGAACTTTT-3′, reverse primer: 5′-CAGTGGTTCCATCTC CTTGTTG-3′; Bax, forward primer: 5′-TTTGCTTCAGGGTTTCATCC-3′, reverse primer: 5′-GCCAC TCGGAAAAAGACCTC-3′; GAPDH, forward primer: 5′-TGGAGTCTACTGGCGTCTT-3′, reverse primer: 5′-TGTCATATTTCTCGTGGTTCA-3′. The fold or percentage of change in the relative expression of the mRNA of target gene l was measured by the $2^{-\Delta\Delta Ct}$ method.

4.16. Western Blotting

The total proteins were extracted from SKNMC cells by using radioimmunoprecipitation assay (RIPA) lysis buffer (Intron Biotechnology, Inc., Gyeonggi, Korea), and the protein concentration was measured using bicinchoninic acid (BCA) kit (Intron Biotechnology, Inc., Gyeonggi, Korea). The separation of proteins was carried out on 8% and 12% polyacrylamide gels, and nitrocellulose (Bio-Rad, Hercules, CA, USA) membranes were used for electro-transferred in a semi-dried environment. Blots were blocked by 5% skim milk (tris-buffer and 0.1% Tween-20) and then incubated with primary anti-GRP78 (1:1000; SC-13539, Santa Cruz Biotechnology, Inc., Dallas, TX, USA), anit-GADD34 (1:1000, ab9869, Abcam, Cambridge, UK) and anti-cleaved caspase (1:1000, Asp175, 9661, cell signaling, Danvers, MA, USA) antibodies at 4 °C overnight. Subsequently, the blots were incubated with anti-mouse (#115-035-003; Jackson ImmunoResearch laboratories, Inc., West Grove, PA, USA), anti-goat (SC-2020, Santa Cruz Biotechnology, Inc.), and anti-rabbit (SC-2004, Santa Cruz Biotechnology, Inc.) secondary antibodies at room temperature for 1 h. Then, the blots were developed with EZ-Western Lumi Plus solution (ATTO Corporation, Tokyo, Japan) (Millipore Corporation, Billerica, MA, USA) and analyzed with Ez-Capture ST (ATTO Corporation, Tokyo, Japan).

4.17. Experimental Animal and Treatments

Seven-week-old Sprague-Dawley (SD) male rats, weighing 220–250 g each were purchased from DBL (Eumseong, South Korea). They were kept in clean and dry polypropylene cages with 12-h light–dark cycle at 25 ± 2 °C and 45%–55% relative humidity in the animal house, Pharmacology Department, Chonbuk National University. The rats were fed with a standard laboratory diet and water ad libitum. After a week of adaptation, the rats were randomly divided into four groups (each group, $n = 5$). The protocol used for this study in the rat as an animal model was carried out with the guidelines of the Institutional Animal Care and Usage Committee (IACUC) with approval from ethical committee of Chonbuk National University, Korea for using animals by describing the protocols of the study (Approved number: CBNU 2015-099).

The rats were divided into four groups, each group with 5 rats. Group I for normal control, other groups for Mn which were treated by 15 mg $MnCl_2$/kg body weight of rats through intraperitoneal (i.p.) injection five days/week for three weeks. Then, the rats designated for PPEES groups (Group III and IV) followed a daily oral dose of 100 and 200 mg/kg for another four weeks, while the rats in Mn-exposed (group II) and normal control groups received normal saline orally. Details of the treatment pattern and groups are described in the Supplementary Materials. Body weight and food consumption were measured daily (Figures S2 and S3).

4.18. The Open Field (OF) Test

The OF test was performed based on observation of locomotor activity of rats in an ideal environment. Briefly, rats were placed in an ideal environment in an OF box for a 30 min session. We recorded the locomotors activity throughout the experiment, spent time and number of squares traveled in the central area of the OF as indices of animal anxiety level.

4.19. Collection of Brain

The rats were deeply anesthetized with ketamine and normal saline (0.9%) was used for transcardial perfusion. The brain tissues were fixed using 4% paraformaldehyde (pH 7.4) solution for 12 h; incubated overnight at 4 °C in 100 mM sodium phosphate buffer (pH 7.4) containing 15% sucrose followed by 30% sucrose; and embedded in optimal cutting temperature (OCT, Leica Biosystems Melbourne Pty Ltd., DB Maarn, the Netherland) medium. Coronal sections (20 μm) from cryofixed tissue were collected on silane-coated slides (Muto Pure Chemical Co., Ltd., Tokyo, Japan) and stored at −70 °C.

4.20. Histopathology and Immunohistochemistry

Histopathological examination was conducted on the striatum part of brain from SD rats after embedded in OCT medium, following the previously described method [33]. The histopathological alterations were observed using hematoxylin and eosin (HE) staining under light microscope.

To examine the protective effect of the PPEES treatments on Mn-induced immunoreativity of oxidative protein 8-hydroxy-2′-deoxyguanosine (8-OHdG), pro-apoptotic protein Bax and apoptotic protein caspase-3 immunohistochemistry (IHC) was performed in the striatum and cortex of all treatment groups. Sections (14 μm) were prepared from OCT embedded samples described above. The sections were treated with mouse polyclonal anti-8-OHdG (1:500, N45.1, ab48508, Abcam) and rabbit polyclonal anti-Bax (1:500; P-19, sc-526, Santa Cruz Biotechnology, Inc.) antibodies at 4 °C overnight. Subsequently, these were incubated with biotinylated goat anti-mouse (1:30, code: D0314, Dako, Burlington, ON, Canada) and goat anti-rabbit (1:80, code: D0487, Dako) immunoglobulins and latter visualized with substrate chromogen (code: K3464, Dako), followed by hematoxylin and mounted with aqueous mount medium. The sections were cover slipped after drying and observed under a microscope, and images were taken by Nikon Differential Interference Contrast Inverted Microscope (Nikon, Kanagawa, Japan) equipped with Narishige micromanipulators (Narishige, Tokyo, Japan).

4.21. Statistical Data Analysis

All data were expressed as mean $\pm$ SD and one-way ANOVA (Analysis of variance) followed by Dunnett's test was used for the statistical analysis using SPSS software (version 16). * $p < 0.05$ and ** $p < 0.01$ were considered significant.

5. Conclusions

In conclusion, the present study reveals that PPEES could effectively inhibit Mn-induced neurotoxicity through antioxidant properties via regulation of ER stress and ER stress-mediated apoptosis (Figure 11). Further studies are also needed to elucidate the precise mechanism of action of PPEES and to evaluate its neuroprotective effects in various neurological disorders.

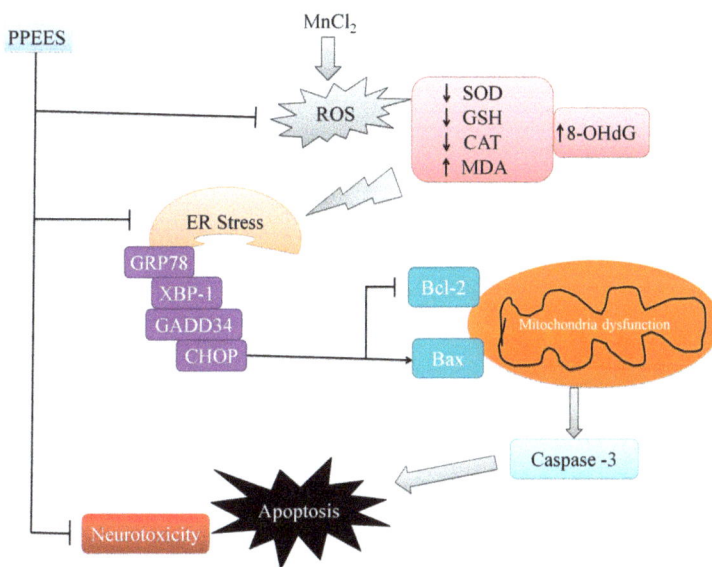

Figure 11. The proposed mechanism of PPEES against Mn-induced toxicity. The schematic diagram shows Mn could exceed ROS; subsequently, altering activity of SOD and CAT. Changing GSH, MDA and 8-OHdG levels led to ER stress, followed by apoptosis through mitochondrial dysfunction. This diagram shows that PPEES prevents the Mn-induced neurotoxicity through regulation of ER stress and ER stress-mediated apoptosis.

Supplementary Materials: The following are available online at www.mdpi.com/1422-0067/18/2/300/s1.

Acknowledgments: This work was supported by Development Fund Foundation, Gyeongsang National University, 2015, Jinju, Gyeongnam, Korea.

Author Contributions: This study was designed, directed and coordinated by Entaz Bahar and Hyonok Yoon. Han-Jung Chae acted as the principal investigator, and provided conceptual and technical guidance for all aspects of the project. Entaz Bahar planned and performed in vitro experiments with Harun-Or Rashid. Performed and analyzed in vivo rat experiments with Geum-Hwa Lee, Kashi Raj Bhattarai, Hwa-Young Lee, and Min Kyung Choi. Histopathological examination was done with Ji-Ye Kim. The manuscript was written by Entaz Bahar and Hyonok Yoon, and commented on by all authors.

Conflicts of Interest: The authors declare no conflict of interest.

References

1. Underwood, E.J. Trace metals in human and animal health. *J. Hum. Nutr.* **1981**, *35*, 37–48. [CrossRef] [PubMed]
2. Stephenson, A.P.; Schneider, J.A.; Nelson, B.C.; Atha, D.H.; Jain, A.; Soliman, K.F.; Aschner, M.; Mazzio, E.; Renee Reams, R. Manganese-induced oxidative DNA damage in neuronal SH-SY5Y cells: Attenuation of thymine base lesions by glutathione and N-acetylcysteine. *Toxicol. Lett.* **2013**, *218*, 299–307. [CrossRef] [PubMed]
3. Burton, N.C.; Guilarte, T.R. Manganese neurotoxicity: Lessons learned from longitudinal studies in nonhuman primates. *Environ. Health Perspect.* **2009**, *117*, 325–332. [CrossRef] [PubMed]
4. Hudnell, H.K. Effects from environmental Mn exposures: A review of the evidence from non-occupational exposure studies. *Neurotoxicology* **1999**, *20*, 379–397. [PubMed]
5. Barbeau, A. Manganese and extrapyramidal disorders (a critical review and tribute to Dr. George C. Cotzias). *Neurotoxicology* **1984**, *5*, 13–35. [PubMed]
6. Mena, I.; Marin, O.; Fuenzalida, S.; Cotzias, G.C. Chronic manganese poisoning. Clinical picture and manganese turnover. *Neurology* **1967**, *17*, 128–136. [CrossRef] [PubMed]
7. Iregren, A. Manganese neurotoxicity in industrial exposures: Proof of effects, critical exposure level, and sensitive tests. *Neurotoxicology* **1999**, *20*, 315–323. [PubMed]
8. Calne, D.B.; Chu, N.S.; Huang, C.C.; Lu, C.S.; Olanow, W. Manganism and idiopathic parkinsonism: Similarities and differences. *Neurology* **1994**, *44*, 1583–1586. [CrossRef] [PubMed]
9. Olanow, C.W.; Good, P.F.; Shinotoh, H.; Hewitt, K.A.; Vingerhoets, F.; Snow, B.J.; Beal, M.F.; Calne, D.B.; Perl, D.P. Manganese intoxication in the rhesus monkey: A clinical, imaging, pathologic, and biochemical study. *Neurology* **1996**, *46*, 492–498. [CrossRef] [PubMed]
10. Walter, U.; Niehaus, L.; Probst, T.; Benecke, R.; Meyer, B.U.; Dressler, D. Brain parenchyma sonography discriminates Parkinson's disease and atypical parkinsonian syndromes. *Neurology* **2003**, *60*, 74–77. [CrossRef] [PubMed]
11. Yamada, M.; Ohno, S.; Okayasu, I.; Okeda, R.; Hatakeyama, S.; Watanabe, H.; Ushio, K.; Tsukagoshi, H. Chronic manganese poisoning: A neuropathological study with determination of manganese distribution in the brain. *Acta Neuropathol.* **1986**, *70*, 273–278. [CrossRef] [PubMed]
12. Ferraz, H.B.; Bertolucci, P.H.; Pereira, J.S.; Lima, J.G.; Andrade, L.A. Chronic exposure to the fungicide maneb may produce symptoms and signs of CNS manganese intoxication. *Neurology* **1988**, *38*, 550–553. [CrossRef] [PubMed]
13. Brurok, H.; Berg, K.; Sneen, L.; Grant, D.; Karlsson, J.O.; Jynge, P. Cardiac metal contents after infusions of manganese. An experimental evaluation in the isolated rat heart. *Investig. Radiol.* **1999**, *34*, 470–476. [CrossRef]
14. Hunter, D.R.; Haworth, R.A.; Berkoff, H.A. Cellular manganese uptake by the isolated perfused rat heart: A probe for the sarcolemma calcium channel. *J. Mol. Cell. Cardiol.* **1981**, *13*, 823–832. [CrossRef]
15. Brurok, H.; Schjott, J.; Berg, K.; Karlsson, J.O.; Jynge, P. Manganese and the heart: Acute cardiodepression and myocardial accumulation of manganese. *Acta Physiol. Scand.* **1997**, *159*, 33–40. [CrossRef] [PubMed]
16. Brurok, H.; Skoglund, T.; Berg, K.; Skarra, S.; Karlsson, J.O.; Jynge, P. Myocardial manganese elevation and proton relaxivity enhancement with manganese dipyridoxyl diphosphate. Ex vivo assessments in normally perfused and ischemic guinea pig hearts. *NMR Biomed.* **1999**, *12*, 364–372. [CrossRef]
17. Jynge, P.; Brurok, H.; Asplund, A.; Towart, R.; Refsum, H.; Karlsson, J.O. Cardiovascular safety of MnDPDP and MnCl2. *Acta Radiol.* **1997**, *38*, 740–749. [CrossRef] [PubMed]
18. Goering, P.L. The road to elucidating the mechanism of manganese-bilirubin-induced cholestasis. *Toxicol. Sci.* **2003**, *73*, 216–219. [CrossRef] [PubMed]
19. Symonds, H.W.; Hall, E.D. Acute manganese toxicity and the absorption and biliary excretion of manganese in cattle. *Res. Vet. Sci.* **1983**, *35*, 5–13. [PubMed]
20. Treinen, K.A.; Gray, T.J.; Blazak, W.F. Developmental toxicity of mangafodipir trisodium and manganese chloride in Sprague-Dawley rats. *Teratology* **1995**, *52*, 109–115. [CrossRef] [PubMed]
21. Sanchez, D.J.; Domingo, J.L.; Llobet, J.M.; Keen, C.L. Maternal and developmental toxicity of manganese in the mouse. *Toxicol. Lett.* **1993**, *69*, 45–52. [CrossRef]

22. Gerber, G.B.; Leonard, A.; Hantson, P. Carcinogenicity, mutagenicity and teratogenicity of manganese compounds. *Crit. Rev. Oncol. Hematol.* **2002**, *42*, 25–34. [CrossRef]

23. Li, Y.; Sun, L.; Cai, T.; Zhang, Y.; Lv, S.; Wang, Y.; Ye, L. alpha-Synuclein overexpression during manganese-induced apoptosis in SH-SY5Y neuroblastoma cells. *Brain Res. Bull.* **2010**, *81*, 428–433. [CrossRef] [PubMed]

24. Yoon, H.; Kim, D.S.; Lee, G.H.; Kim, K.W.; Kim, H.R.; Chae, H.J. Apoptosis Induced by Manganese on Neuronal SK-N-MC Cell Line: Endoplasmic Reticulum (ER) Stress and Mitochondria Dysfunction. *Environ. Health Toxicol.* **2011**, *26*. [CrossRef] [PubMed]

25. Yoon, H.; Lee, G.H.; Kim, D.S.; Kim, K.W.; Kim, H.R.; Chae, H.J. The effects of 3, 4 or 5 amino salicylic acids on manganese-induced neuronal death: ER stress and mitochondrial complexes. *Toxicol. In Vitro* **2011**, *25*, 1259–1268. [CrossRef] [PubMed]

26. Zhang, S.; Zhou, Z.; Fu, J. Effect of manganese chloride exposure on liver and brain mitochondria function in rats. *Environ. Res.* **2003**, *93*, 149–157. [CrossRef]

27. Zhang, S.; Fu, J.; Zhou, Z. Changes in the brain mitochondrial proteome of male Sprague-Dawley rats treated with manganese chloride. *Toxicol. Appl. Pharmacol.* **2005**, *202*, 13–17. [CrossRef] [PubMed]

28. O'Neal, S.L.; Lee, J.W.; Zheng, W.; Cannon, J.R. Subacute manganese exposure in rats is a neurochemical model of early manganese toxicity. *Neurotoxicology* **2014**, *44*, 303–313. [CrossRef] [PubMed]

29. Vorhees, C.V.; Graham, D.L.; Amos-Kroohs, R.M.; Braun, A.A.; Grace, C.E.; Schaefer, T.L.; Skelton, M.R.; Erikson, K.M.; Aschner, M.; Williams, M.T. Effects of developmental manganese, stress, and the combination of both on monoamines, growth, and corticosterone. *Toxicol. Rep.* **2014**, *1*, 1046–1061. [CrossRef] [PubMed]

30. Moberly, A.H.; Czarnecki, L.A.; Pottackal, J.; Rubinstein, T.; Turkel, D.J.; Kass, M.D.; McGann, J.P. Intranasal exposure to manganese disrupts neurotransmitter release from glutamatergic synapses in the central nervous system in vivo. *Neurotoxicology* **2012**, *33*, 996–1004. [CrossRef] [PubMed]

31. Shi, S.; Zhao, J.; Yang, L.; Nie, X.; Han, J.; Ma, X.; Wan, C.; Jiang, J. KHSRP participates in manganese-induced neurotoxicity in rat striatum and PC12 cells. *J. Mol. Neurosci.* **2015**, *55*, 454–465. [CrossRef] [PubMed]

32. Alaimo, A.; Gorojod, R.M.; Beauquis, J.; Munoz, M.J.; Saravia, F.; Kotler, M.L. Deregulation of mitochondria-shaping proteins Opa-1 and Drp-1 in manganese-induced apoptosis. *PLoS ONE* **2014**, *9*, e91848. [CrossRef] [PubMed]

33. Wang, T.; Li, X.; Yang, D.; Zhang, H.; Zhao, P.; Fu, J.; Yao, B.; Zhou, Z. ER stress and ER stress-mediated apoptosis are involved in manganese-induced neurotoxicity in the rat striatum in vivo. *Neurotoxicology* **2015**, *48*, 109–119. [CrossRef] [PubMed]

34. Dhanalakshmi, C.; Manivasagam, T.; Nataraj, J.; Justin Thenmozhi, A.; Essa, M.M. Neurosupportive Role of Vanillin, a Natural Phenolic Compound, on Rotenone Induced Neurotoxicity in SH-SY5Y Neuroblastoma Cells. *Evid.-Based Complement. Altern. Med.* **2015**, *2015*. [CrossRef]

35. Okada, Y.; Okada, M. Protective effects of plant seed extracts against amyloid beta-induced neurotoxicity in cultured hippocampal neurons. *J. Pharm. Bioallied Sci.* **2013**, *5*, 141–147. [CrossRef] [PubMed]

36. Obrenovich, M.E.; Nair, N.G.; Beyaz, A.; Aliev, G.; Reddy, V.P. The role of polyphenolic antioxidants in health, disease, and aging. *Rejuvenation Res.* **2010**, *13*, 631–643. [CrossRef] [PubMed]

37. Song, Y.; Jeong, S.W.; Lee, W.S.; Park, S.; Kim, Y.H.; Kim, G.S.; Lee, S.J.; Jin, J.S.; Kim, C.Y.; Lee, J.E.; et al. Determination of Polyphenol Components of Korean Prostrate Spurge (Euphorbia supina) by Using Liquid Chromatography-Tandem Mass Spectrometry: Overall Contribution to Antioxidant Activity. *J. Anal. Methods Chem.* **2014**, *2014*. [CrossRef] [PubMed]

38. Ko, Y.S.; Lee, W.S.; Joo, Y.N.; Choi, Y.H.; Kim, G.S.; Jung, J.M.; Ryu, C.H.; Shin, S.C.; Kim, H.J. Polyphenol mixtures of Euphorbia supina the inhibit invasion and metastasis of highly metastatic breast cancer MDA-MB-231 cells. *Oncol. Rep.* **2015**, *34*, 3035–3042. [CrossRef] [PubMed]

39. Kumarappan, C.T.; Thilagam, E.; Vijayakumar, M.; Mandal, S.C. Modulatory effect of polyphenolic extracts of Ichnocarpus frutescens on oxidative stress in rats with experimentally induced diabetes. *Indian J. Med. Res.* **2012**, *136*, 815–821. [PubMed]

40. Chun, O.K.; Kim, D.O.; Lee, C.Y. Superoxide radical scavenging activity of the major polyphenols in fresh plums. *J. Agric. Food Chem.* **2003**, *51*, 8067–8072. [CrossRef] [PubMed]

41. Fattahi, S.; Zabihi, E.; Abedian, Z.; Pourbagher, R.; Motevalizadeh Ardekani, A.; Mostafazadeh, A.; Akhavan-Niaki, H. Total Phenolic and Flavonoid Contents of Aqueous Extract of Stinging Nettle and In Vitro Antiproliferative Effect on Hela and BT-474 Cell Lines. *Int. J. Mol. Cell. Med.* **2014**, *3*, 102–107. [PubMed]

42. Sen, S.; De, B.; Devanna, N.; Chakraborty, R. Total phenolic, total flavonoid content, and antioxidant capacity of the leaves of Meyna spinosa Roxb., an Indian medicinal plant. *Chin. J. Nat. Med.* **2013**, *11*, 149–157. [CrossRef]

43. Bahar, E.; Siddika, M.; Nath, B.; Yoon, H. Evaluation of In vitro Antioxidant and In vivo Antihyperlipidemic Activities of Methanol Extract of Aerial Part of Crassocephalum crepidioides (Asteraceae) Benth S Moore. *Trop. J. Pharm. Res.* **2016**, *15*, 481–488. [CrossRef]

44. Ganie, S.A.; Zargar, B.A.; Masood, A.; Zargar, M.A. Hepatoprotective and antioxidant activity of rhizome of Podophyllum hexandrum against carbon tetra chloride induced hepatotoxicity in rats. *Biomed. Environ. Sci.* **2013**, *26*, 209–221. [CrossRef] [PubMed]

45. Zargar, B.A.; Masoodi, M.H.; Ahmed, B.; Ganie, S.A. Antihyperlipidemic and Antioxidant Potential of Paeonia emodi Royle against High-Fat Diet Induced Oxidative Stress. *ISRN Pharmacol.* **2014**, *2014*. [CrossRef] [PubMed]

46. Yokozawa, T.; Kim, Y.A.; Kim, H.Y.; Okamoto, T.; Sei, Y. Protective effect of the Chinese prescription Kangen-karyu against high glucose-induced oxidative stress in LLC-PK1 cells. *J. Ethnopharmacol.* **2007**, *109*, 113–120. [CrossRef]

47. Dai, C.; Li, B.; Zhou, Y.; Li, D.; Zhang, S.; Li, H.; Xiao, X.; Tang, S. Curcumin attenuates quinocetone induced apoptosis and inflammation via the opposite modulation of Nrf2/HO-1 and NF-kB pathway in human hepatocyte L02 cells. *Food Chem. Toxicol.* **2016**, *95*, 52–63. [CrossRef] [PubMed]

48. Dai, C.; Tang, S.; Deng, S.; Zhang, S.; Zhou, Y.; Velkov, T.; Li, J.; Xiao, X. Lycopene attenuates colistin-induced nephrotoxicity in mice via activation of the Nrf2/HO-1 pathway. *Antimicrob. Agents Chemother.* **2015**, *59*, 579–585. [CrossRef] [PubMed]

49. Dai, C.; Tang, S.; Li, D.; Zhao, K.; Xiao, X. Curcumin attenuates quinocetone-induced oxidative stress and genotoxicity in human hepatocyte L02 cells. *Toxicol. Mech. Methods* **2015**, *25*, 340–346. [CrossRef] [PubMed]

50. Tait, S.W.; Green, D.R. Mitochondria and cell death: Outer membrane permeabilization and beyond. *Nat. Rev. Mol. Cell Biol.* **2010**, *11*, 621–632. [CrossRef] [PubMed]

51. Khosravi-Far, R.; Esposti, M.D. Death receptor signals to mitochondria. *Cancer Biol. Ther.* **2004**, *3*, 1051–1057. [CrossRef] [PubMed]

52. Wang, H.; Liu, H.; Zheng, Z.M.; Zhang, K.B.; Wang, T.P.; Sribastav, S.S.; Liu, W.S.; Liu, T. Role of death receptor, mitochondrial and endoplasmic reticulum pathways in different stages of degenerative human lumbar disc. *Apoptosis* **2011**, *16*, 990–1003. [CrossRef] [PubMed]

53. Wang, M.; Wey, S.; Zhang, Y.; Ye, R.; Lee, A.S. Role of the unfolded protein response regulator GRP78/BiP in development, cancer, and neurological disorders. *Antioxid. Redox Signal.* **2009**, *11*, 2307–2316. [CrossRef] [PubMed]

54. Yoshida, H.; Matsui, T.; Yamamoto, A.; Okada, T.; Mori, K. XBP1 mRNA is induced by ATF6 and spliced by IRE1 in response to ER stress to produce a highly active transcription factor. *Cell* **2001**, *107*, 881–891. [CrossRef]

55. Brush, M.H.; Weiser, D.C.; Shenolikar, S. Growth arrest and DNA damage-inducible protein GADD34 targets protein phosphatase 1 alpha to the endoplasmic reticulum and promotes dephosphorylation of the alpha subunit of eukaryotic translation initiation factor 2. *Mol. Cell. Biol.* **2003**, *23*, 1292–1303. [CrossRef] [PubMed]

56. Adler, H.T.; Chinery, R.; Wu, D.Y.; Kussick, S.J.; Payne, J.M.; Fornace, A.J., Jr.; Tkachuk, D.C. Leukemic HRX fusion proteins inhibit GADD34-induced apoptosis and associate with the GADD34 and hSNF5/INI1 proteins. *Mol. Cell. Biol.* **1999**, *19*, 7050–7060. [CrossRef] [PubMed]

57. Korsmeyer, S.J. BCL-2 gene family and the regulation of programmed cell death. *Cancer Res.* **1999**, *59* (Suppl. 7), 1693s–1700s. [CrossRef]

58. Chao, D.T.; Korsmeyer, S.J. BCL-2 family: Regulators of cell death. *Annu. Rev. Immunol.* **1998**, *16*, 395–419. [CrossRef] [PubMed]

59. Bahar, E.; Kim, H.; Yoon, H. ER Stress-Mediated Signaling: Action Potential and Ca(2+) as Key Players. *Int. J. Mol. Sci.* **2016**, *17*. [CrossRef] [PubMed]

60. Persson, T.; Popescu, B.O.; Cedazo-Minguez, A. Oxidative stress in Alzheimer's disease: Why did antioxidant therapy fail? *Oxidative Med. Cell. Longev.* **2014**, *2014*. [CrossRef]

61. Amos-Kroohs, R.M.; Bloor, C.P.; Qureshi, M.A.; Vorhees, C.V.; Williams, M.T. Effects of developmental exposure to manganese and/or low iron diet: Changes to metal transporters, sucrose preference, elevated zero-maze, open-field, and locomotion in response to fenfluramine, amphetamine, and MK-801. *Toxicol. Rep.* **2015**, *2*, 1046–1056. [CrossRef] [PubMed]

62. Lu, L.; Zhang, L.L.; Li, G.J.; Guo, W.; Liang, W.; Zheng, W. Alteration of serum concentrations of manganese, iron, ferritin, and transferrin receptor following exposure to welding fumes among career welders. *Neurotoxicology* **2005**, *26*, 257–265. [CrossRef] [PubMed]

International Journal of
Molecular Sciences

MDPI

Article

Phytochemical Analysis of *Agrimonia pilosa* Ledeb, Its Antioxidant Activity and Aldose Reductase Inhibitory Potential

Set Byeol Kim [1,†], Seung Hwan Hwang [1,†], Hong-Won Suh [2] and Soon Sung Lim [1,2,*]

1   Department of Food Science and Nutrition, Hallym University, 1 Hallymdaehak-gil, Chuncheon, Gangwon-do 24252, Korea; jwsbcb0187@naver.com (S.B.K.); isohsh@gmail.com (S.H.H.)
2   Institute of Natural Medicine, College of Medicine, Hallym University, 1 Hallymdaehak-gil, Chuncheon, Gangwon-do 24252, Korea; hwsuh@hallym.ac.kr
*   Correspondence: limss@hallym.ac.kr; Tel.: +82-33-248-2133; Fax: +82-33-251-0663
†   These authors contributed equally to this work.

Academic Editor: Toshio Morikawa
Received: 23 December 2016; Accepted: 6 February 2017; Published: 10 February 2017

Abstract: The aim of this study was to determine aldose reductase (AR) inhibitory activity and 1,1-diphenyl-2-picrylhydrazyl (DPPH) free radical scavenging activity of compounds from *Agrimonia pilosa* Ledeb (AP). We isolated agrimoniin (AM), four flavonoid glucosides and two flavonoid glucuronides from the *n*-butanol fraction of AP 50% methanol extract. In addition to isolated compounds, the AR-inhibitory activity and the DPPH free radical scavenging activity of catechin, 5-flavonoids, and 4-flavonoid glucosides (known components of AP) against rat lens AR (RLAR) and DPPH assay were measured. AM showed IC_{50} values of 1.6 and 13.0 µM against RLAR and DPPH scavenging activity, respectively. Additionally, AM, luteolin-7-*O*-glucuronide (LGN), quercitrin (QU), luteolin (LT) and afzelin (AZ) showed high inhibitory activity against AR and were first observed to decrease sorbitol accumulation in the rat lens under high-sorbitol conditions ex vivo with inhibitory values of 47.6%, 91.8%, 76.9%, 91.8% and 93.2%, respectively. Inhibition of recombinant human AR by AM, LGN and AZ exhibited a noncompetitive inhibition pattern. Based on our results, AP and its constituents may play partial roles in RLAR and oxidative radical inhibition. Our results suggest that AM, LGN, QU, LT and AZ may potentially be used as natural drugs for treating diabetic complications.

Keywords: *Agrimonia pilosa* Ledeb; aldose reductase; flavonoids; 1,1-diphenyl-2-picrylhydrazyl (DPPH); diabetic complication

---

1. Introduction

Aldose reductase (AR, EC.1.1.1.21) is a key enzyme in the polyol pathway that catalyzes the conversion of glucose to sorbitol in the hyperglycemic state and oxidoreductase-induced nicotinamide adenine dinucleotide phosphate (NADPH) to $NADP^+$ [1]. Accumulation of sorbitol leads to the generation of osmotic stress, an influx of water and causes of diabetic complications such as cataracts and retinopathy. In addition, the conversion of sorbitol to fructose is catalyzed by nicotinamide adenine dinucleotide (NADH)-dependent sorbitol dehydrogenase. Increased fructose formation leads to the formation of reactive dicarbonyl species such as glucosones, glyoxal, and methylglyoxal, which are important factors in advanced glycation end products [2]. Oxidative stress causes an imbalance between the formation of free radicals and the body's antioxidant potential. Free radicals are defined as atoms or molecules that contain one or more unpaired electrons [3]. Diabetes mellitus and its complications, such as retinopathy, nephropathy, neuropathy, and atherosclerosis, are caused by an imbalance in cells and free radicals, and this imbalance is mainly responsible for the auto-oxidation of glucose and glycosylated proteins [4,5] Therefore, the development of diabetic complications could be controlled by inhibiting AR activity and also by increasing antioxidant activity in the body.

Agrimonia pilosa Ledeb (*A. pilosa*, AP), belonging to the Rosaceae family, is famous in traditional Chinese medicine. According to pharmacological studies, it has anti-nociceptive, anti-inflammatory, antioxidant, anticancer and α-glucosidase inhibitory activity [6,7]. The known constituents of AP are 3-methoxy quercetin, quercitrin (QU), quercetin (QC), tiliroside, ursolic acid, tormentic acid and corosolic acid [8,9]. The major known flavonoids of AP are catechin (CT), luteolin (LT), QC, isoquercetin (IQC), hyperin (HP), apigenin (AG), vitexin (VT), kaempferol (KP), astragalin (AS), and afzelin (AZ) [10–12]. Generally, these flavonoids are involved in plant metabolism and possess antioxidant, antidiabetic, anticancer, and various inhibitory activities [13,14].

The isolation and purification of active compounds from complex plant extracts takes a long time. In the past few years, several online high performance liquid chromatography (HPLC) methods that use post-column derivative method, which are based on online detection by 1,1-diphenyl-2-picrylhydrazyl (DPPH) or 2,2′-azino-bis(3-ethylbenzothiazoline-6-sulphonic acid) (ABTS) radicals, have been utilized to screen antioxidants in some complex plant extracts to avoid the aforementioned problem. These methods required an online instrumental system and technical skills that were complex and available. Recently, the more convenient offline DPPH-HPLC method was successfully developed by spiking the complex plant extracts [15,16].

To date, no data have been published on the inhibitory effects of AP on rat lens AR (RLAR), DPPH radical scavenging capacity and offline DPPH-HPLC assay. Therefore, the inhibitory effects of ten known flavonoids from the literature as well as seven compounds isolated from AP 50% methanol (MeOH) extract (APE) were investigated to evaluate their use in treatment of RLAR-related diabetic complications. The active compounds of APE that showed antioxidant properties were investigated by offline DPPH-HPLC assay. Additionally, the ability of the major active compounds to decrease sorbitol accumulation in rat lens in ex vivo high-sorbitol conditions as well as the recombinant human AR (RHAR) inhibition type of the compounds were assessed.

2. Results

2.1. Structural Determination of Isolated Compounds

The effects of APE on RLAR and DPPH free radical scavenging activity were further investigated. APE exhibited inhibitory activity against RLAR, with 51.4% inhibition at a concentration of 10.0 μg/mL. Moreover, APE showed 53.4% inhibition of DPPH at a concentration of 7.1 μg/mL. Consequently, APE was further partitioned by systematic fractionation. Among the resulting fractions, the ethyl acetate (EtOAc) and *n*-butanol (*n*-BuOH)-soluble fractions exhibited potent inhibitory activity against RLAR with 84.4% and 92.4% inhibition, respectively, compared with the positive control tetramethylene glutaric acid (TMG; 99.7% inhibition) at a concentration of 1.0 μg/mL. The EtOAc and *n*-BuOH fractions also showed inhibitory activity against DPPH, with 62.3% and 61.0% inhibition, respectively, compared with the positive control L-ascorbic acid (81.0% inhibition) (Table 1).

Therefore, this study focused on the isolation of the AR inhibitor from the *n*-BuOH fraction. The seven compounds isolated from *n*-BuOH were identified as compound 1 (agrimoniin, AM, 69.3 mg) [17], compound 2 (rutin, RT, 30.3 mg) [18], compound 3 (luteolin-7-*O*-glucoside, LGC, 96.1 mg) [19], compound 4 (luteolin-7-*O*-glucuronide, LGN, 150.5 mg) [20], compound 5 (quercitrin, QU, 11.6 mg) [21], compound 6 (apigenin-7-*O*-glucoside, AGC, 21.3 mg) [22] and compound 7 (apigenin-7-*O*-glucuronide, AGN, 251.6 mg) [23] on the basis of the 1D and 2D NMR spectral data (Table S1), as well as by comparison with published spectral data (Figures 1 and 2).

Table 1. Inhibitory effect of 50% MeOH extract of *Agrimonia pilosa* Ledeb. on rat lens aldose reductase (RLAR) and 1,1-diphenyl-2-picrylhydrazyl (DPPH) free radical scavenging activity. TMG: tetramethylene glutaric acid.

Extract and Fractions		Inhibition (%)	
		RLAR	DPPH
Methylene chloride extract		6.8 ± 0.20	4.19 ± 0.14
50% MeOH ext.	Crude extract	51.4 ± 0.10	53.4 ± 0.14
	EtOAc fraction	84.4 ± 0.27	62.3 ± 0.04
	n-BuOH fraction	92.4 ± 0.14	61.0 ± 0.42
	Water fraction	37.9 ± 0.47	33.0 ± 0.10
RLAR	TMG	99.7 ± 0.11	-
DPPH	Ascorbic acid	-	81.0 ± 0.01

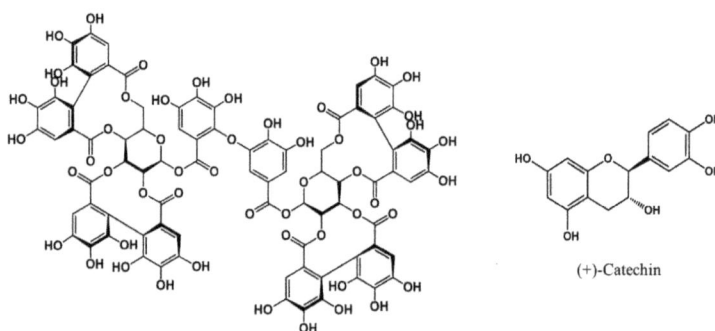

Agrimoniin (1)

(+)-Catechin

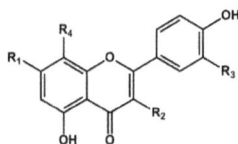

Flavonoid derivatives

No.	Compound name	R₁	R₂	R₃	R₄
	Rutin	OH	-O-α-L-rhamnopyranosyl-(1→6)-β-D-glucopyranose	OH	H
IC[a]	Luteolin-7-O-glucoside	-O-β-D-glucopyranose	H	OH	H
	Luteolin-7-O-glucuronide	-O-β-D-glucopyranosiduronic acid	H	OH	H
	Quercitrin	OH	-O-α-L-rhamnopyranose	OH	H
	Apigenin-7-O-glucoside	-O-β-D-glucopyranose	H	H	H
	Apigenin-7-O-glucuronide	-O-β-D-glucopyranosiduronic acid	H	H	H
	Luteolin	OH	H	OH	H
	Quercetin	OH	OH	OH	H
	Isoquercitrin	OH	-O-β-D-glucopyranose	OH	H
	Hyperin	OH	-O-β-D-galactopyranose	OH	H
KNC[b]	Apigenin	OH	H	H	H
	Vitexin	OH	H	H	-C-β-D-glucopyranose
	Kaempferol	OH	OH	H	H
	Astragalin	OH	-O-β-D-glucopyranose	H	H
	Afzelin	OH	-O-α-L-rhamnopyranose	H	H

Figure 1. The structure of the compounds known and isolated from the *n*-BuOH fraction of *A. pilosa* Ledeb; [a] IC is the compounds isolated from *A. pilosa* Ledeb; [b] KNC is the known compounds isolated from *A. pilosa* Ledeb.

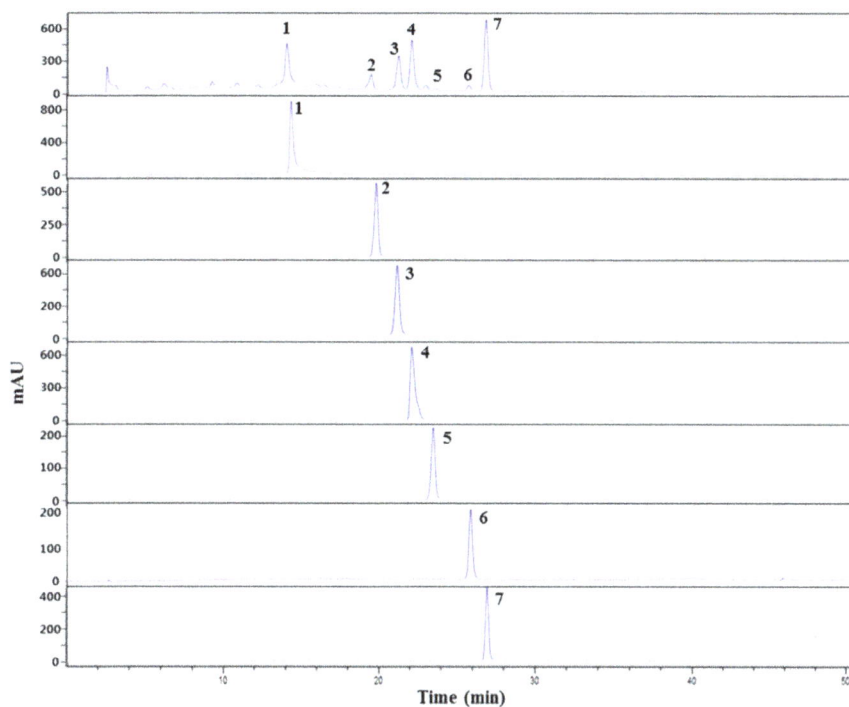

Figure 2. HPLC chromatogram of the compounds isolated from the *n*-BuOH fraction of *A. pilosa* Ledeb. at 254 nm; Peak 1: agrimoniin; Peak 2: rutin; Peak 3: luteolin-7-*O*-glucoside; Peak 4: luteolin-7-*O*-glucuronide; Peak 5: quercitrin; Peak 6: apigenin-7-*O*-glucoside; Peak 7: apigenin-7-*O*-glucuronide.

2.2. Inhibitory Effect of Isolated Compounds on RLAR

We compared the ability of the isolated compounds and TMG (a positive control) to inhibit RLAR activity (Table 2). Among the isolated constituents, RT, LGC and AGC exhibited RLAR inhibitory activity, with 50% inhibition concentration (IC$_{50}$) values of 9.5, 8.1 and 4.3 µM, respectively. QU had the highest IC$_{50}$ value of 0.2 µM, which was 2.5 times higher than the positive control (IC$_{50}$ of TMG = 0.5 µM). AM and LGN had high IC$_{50}$ values at 1.6 and 0.7 µM, respectively, while AGN was inactive. In addition, previous investigations of flavonoids isolated from AP by Jiang et al., Kato et al., and Liu et al. reported that CT, LT, QC, IQC, HP, AG, VT, KP, AS, and AZ were isolated from the leaves of AP [10–12]. The RLAR inhibitory effects of ten known compounds were evaluated using TMC as a positive control. LT and AZ had the strongest RLAR inhibitory activity, with IC$_{50}$ values of 0.6 and 1.0 µM, respectively. QC, IOC, HP, AG and AS also exhibited potent inhibitory activity, with IC$_{50}$ values ranging from 3.2 to 15.2 µM. CT and VT showed lower inhibitory activity, with 7.2% and 12.2% inhibition, respectively, against RLAR at a concentration of 10.0 µg/mL. The RLAR inhibitory effect of ten known compounds was similar to previous data from the literature [24,25].

Table 2. Inhibitory effect of compounds referenced and isolated from *A. pilosa* Ledeb. on rat lens aldose reductase (RLAR) and DPPH free radical scavenging activity.

Entry	Compounds		DPPH		RLAR	
			Experiments			References
		IC$_{50}$ (µM)	Inhibition (%)	IC$_{50}$ (µM)	IC$_{50}$ (µM)	
	Agrimoniin (AM)	13.0 ± 0.06	35.0 ± 0.41	1.6 ± 0.12	-	
	Rutin (RT)	66.8 ± 0.34	31.7 ± 0.65	9.5 ± 0.75	9.0 [24] [b]	
	Luteolin-7-*O*-glucoside (LGC)	71.5 ± 0.29	46.9 ± 0.95	8.1 ± 0.72	7.5 [26]	
IC [a]	Luteolin-7-*O*-glucuronide (LGN)	80.6 ± 0.38	83.3 ± 0.88	0.7 ± 0.54	3.1 [27]	
	Apigenin-7-*O*-glucoside (AGC)	>250	40.2 ± 0.56	4.3 ± 0.14	23.0 [28]	
	Quercitrin (QU)	77.9 ± 0.27	97.4 ± 1.38	0.2 ± 0.02	0.2 [29]	
	Apigenin-7-*O*-glucuronide (AGN)	>250	<0	>30	-	
	Catechin (CT)	106.7 ± 0.43	7.2 ± 1.02	>30	>30 [24]	
	Kaempferol (KP)	91.6 ± 0.68	11.8 ± 0.81	15.2 ± 1.32	10 [24]	
	Quercetin (QC)	70.4 ± 0.15	74.1 ± 0.85	3.2 ± 0.13	2.2 [24]	
	Isoquercitrin (IQC)	65.9 ± 0.46	41.0 ± 1.07	5.1 ± 0.88	4.5 [24]	
KNC [b]	Hyperin (HP)	73.3 ± 0.23	90.8 ± 0.96	4.1 ± 0.32	3.0 [24]	
	Apigenin (AG)	156.3 ± 1.21	81.8 ± 1.20	3.2 ± 0.21	2.2 [24]	
	Vitexin (VT)	>250	12.2 ± 0.95	>30	>30 [25]	
	Astragalin (AS)	>250	53.3 ± 1.14	5.1 ± 0.89	>30 [25]	
	Luteolin (LT)	88.2 ± 0.52	80.2 ± 0.90	0.6 ± 0.03	0.5 [27]	
	Afzelin (AZ)	>250	86.2 ± 0.38	1.0 ± 0.27	0.3 [30]	
Positive control	DPPH	L-Ascorbic acid	147.3 ± 0.43	-	-	-
	RLAR	TMG	-	119.7 ± 0.22	0.5 ± 0.05	1.0 [30]

[a] ICs are the compounds isolated from *A. pilosa* Ledeb; [b] KNCs are the known compounds isolated from *A. pilosa* Ledeb; [b] [Number] is reference number.

2.3. DPPH and Off-Line DPPH HPLC Assay

The *n*-BuOH fraction showed high DPPH radical scavenging activity. The chromatogram of the *n*-BuOH fraction without DPPH (blue line) and with DPPH (red line) is shown in Figure 3A, which presents the peak areas of seven compounds isolated reduced obviously. As shown in Figure 3B, seven compounds showed peak area reduction (PAR) between 13.6%–37.4%. In addition, Zeng et al. reported that the rear eluting peak of the 34-min retention time is of DPPH [15]. Among these seven compounds, AM and QU (PAR 23%–37%) would be more potent antioxidants than RT, LGC, LGN, AGC and AGN, which showed PAR lower than 20.0%. The results of seven compounds in *n*-BuOH fraction suggested antioxidant activity. The scavenging activity of the seven compounds isolated from the *n*-BuOH fraction of APE was evaluated by measuring DPPH free radical scavenging activity (Table 3). Of the tested compounds, AM had the highest IC$_{50}$ value at 13.0 µM. RT, LGC, LGN and QU also showed strong scavenging activity with IC$_{50}$ values of 66.8–80.6 µM, compared to the positive control, L-ascorbic acid (IC$_{50}$ = 147.3 µM). However, AGC and AGN had almost no DPPH free radical scavenging activity. Scavenging activity of the known compounds from AP was evaluated using L-ascorbic acid. Of the tested known compounds, LT, QC, IOC, HP, and KP showed strong scavenging activity, with IC$_{50}$ values of 88.2, 70.4, 65.9, 73.3, and 91.6 µM, respectively, which were higher than those of the positive control (L-ascorbic acid = 147.3 µM). On the other hand, VT, AS, and AZ showed no DPPH free radical scavenging activity.

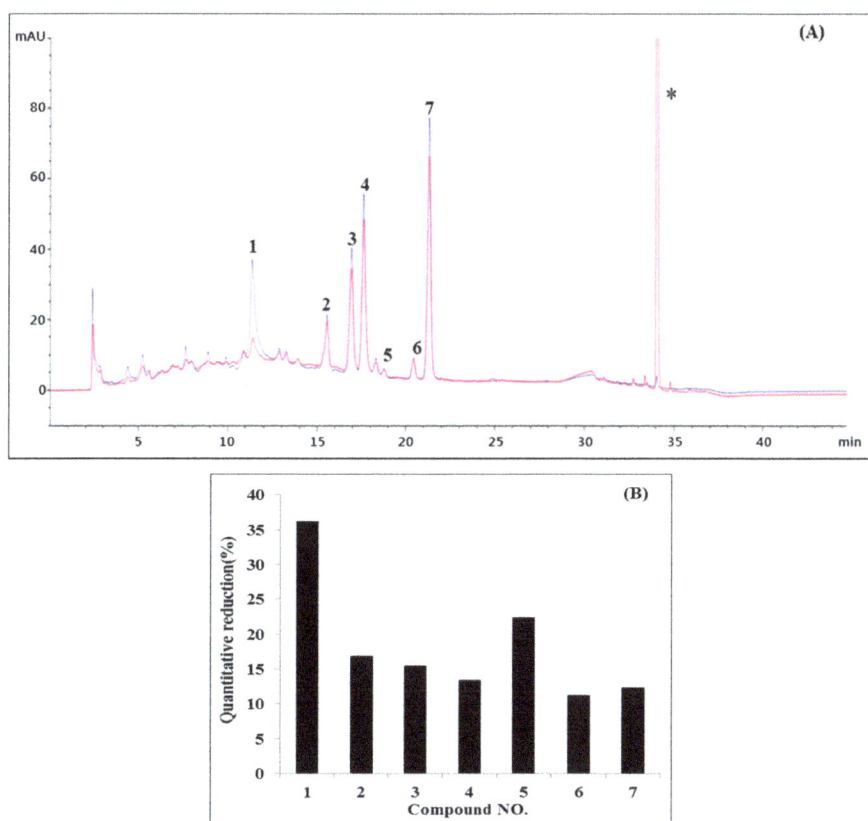

Figure 3. HPLC-ultraviolet (UV) (before reaction: blue) and DPPH-HPLC-UV (after reaction: red) of the *n*-BuOH fraction from *A. pilosa* Ledeb. at 254 nm (**A**) and quantitative reduction (%) in the peak areas of compounds designated as follows (**B**); Peak 1: Agrimoniin; Peak 2: Rutin; Peak 3: Luteolin-7-*O*-glucoside; Peak 4: Luteolin-7-*O*-glucuronide; Peak 5: Quercitrin; Peak 6: Apigenin-7-*O*-glucoside; Peak 7: Apigenin-7-*O*-glucuronide; * is DPPH.

Table 3. Inhibitory effect of the constituents on the sorbitol accumulation in rat lenses and inhibition type by active compound.

Compounds	Sorbitol Content (mg)/ Lens Wet Weight (g) [a]	Inhibition (%)	Inhibition Types (References)
Sorbitol free	No detection	-	-
Control	1.47 ± 0.04		
Quercetin [a]	0.21 ± 0.02	85.7 ± 8.32	Noncompetitive [31]
Agrimoniin (AM)	0.77 ± 0.02	47.6 ± 1.34	Noncompetitive
Luteolin-7-*O*-glucuronide (LGN)	0.12 ± 0.01	91.8 ± 9.01	Noncompetitive
Quercitrin (QU)	0.34 ± 0.02	76.9 ± 5.32	Uncompetitive [29]
Luteolin (LT)	0.12 ± 0.01	91.8 ± 7.91	Mixed type [30]
Afzelin (AZ)	0.10 ± 0.01	93.2 ± 8.67	Noncompetitive

[a] Quercetin was used as positive control; [b] [Number] is reference number.

2.4. Kinetic-Type RHAR Inhibition by the Active Compounds

A kinetic study using DL-glyceraldehyde as a substrate at a concentration range of 1.0 to 25.0 mM was performed to determine the type of inhibition that AM, LGN and AZ exhibited, which showed the highest activity. The kinetic analysis of RHAR inhibition shown in Figure 4 was conducted with AM, LGN and AZ using Lineweaver–Burk plots of 1/velocity and 1/concentration of substrate. With the change of the concentration of the substrate DL-glyceraldehyde, the slopes obtained with the uninhibited enzyme and the three different concentrations of each compound were found to be parallel. The results showed that the inhibition of RHAR by AM, LGN and AZ were competitive and mixed-type. In addition, Lee et al., Ha et al., and Chethan et al. reported that QU, LT, and QC showed uncompetitive, mixed-type, and noncompetitive inhibition patterns, respectively, against RHAR (Table 3) [31–33].

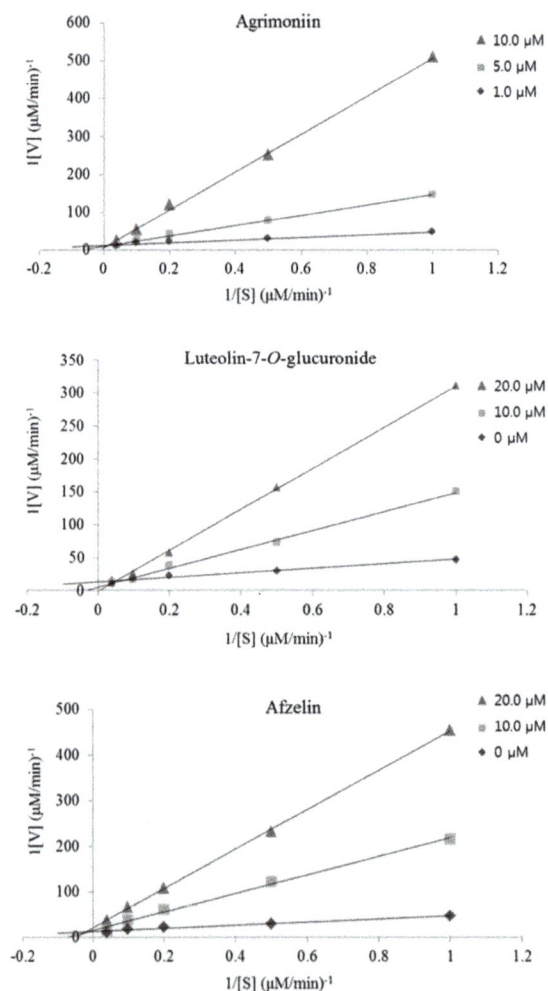

Figure 4. Lineweaver–Burk plots showing the reciprocal of the velocity (1/V) of recombinant rat lens aldose reductase versus the reciprocal of the substrate concentration (1/S) with DL-glyceraldehyde as the substrate at concentrations of 1.0 to 25.0 mM.

2.5. Lens Culture and Intracellular Sorbitol Measurement

We also investigated the effect of RLAR inhibitory compounds (including the known inhibitory compounds LT and AZ) on the sorbitol accumulation in isolated rat lens; the results are shown in Table 3. AM, LGN, QU, LT and AZ inhibited sorbitol accumulation by 47.6%, 91.8%, 76.9%, 91.8%, and 93.2%, respectively, at a concentration of 5.0 μg/mL. In addition, QC as a positive control, which inhibits sorbitol accumulation in isolated rat lens by 85.7%, reduced the sorbitol level in culture medium containing a high glucose concentration. These results indicated that RLAR inhibitors isolated from APE are effective in either preventing or slowing sugar cataract formation associated with diabetes.

3. Discussion

The results of the RLAR and DPPH revealed that all tested APEs have a potent inhibitory effect on RLAR and protect against oxidative stress (Figure 5); these results are shown in Table 1. In addition, the inhibitory effect of the *n*-BuOH fraction of APE on RLAR inhibition was comparable to that of the positive control TMS. In previous studies, VT, RT, HP, LT-7-*O*-β-D-glucopyranoside, QC, tiliroside, LT, AP, and KP isolated from AP were analyzed by HPLC-UV and showed α-glucosidase inhibitory activity, ABTS$^+$ radical scavenging activity, and hydroxyl radical scavenging activity [12]. QC-3-*O*-β-D-glucopyranoside, QC-3-*O*-α-L-rhamnopyranoside, (2S,3S)-(−)-taxifolin-3-*O*-β-D-glucopyranoside, KP-3-*O*-α-L-rhamnopyranoside, 1-butanoyl-3,5-dimethyl-phloroglucinyl-6-*O*-D-glucopyranoside, CT, tiliroside, AG, and agrimonolide in AP were established for characterization and simultaneous quantification by the HPLC-diode array detector-electrospray ionization mass spectrometry (MS)/MS method [10]. Kato et al successfully separated three new compounds and nine known compounds, including (−)-aromadendrin-3-*O*-β-D-glucopyranoside, desmethylagrimonolide-6-*O*-β-D-glucopyranoside, and 5,7-dihydroxy-2-propylchromone-7-*O*-β-D-glucopyranoside, agrimonolide-6-*O*-glucoside, takanechromone C, AT, AZ, tiliroside, LT, QC, IQC, and AGC from AP's aerial parts MeOH extract [11].

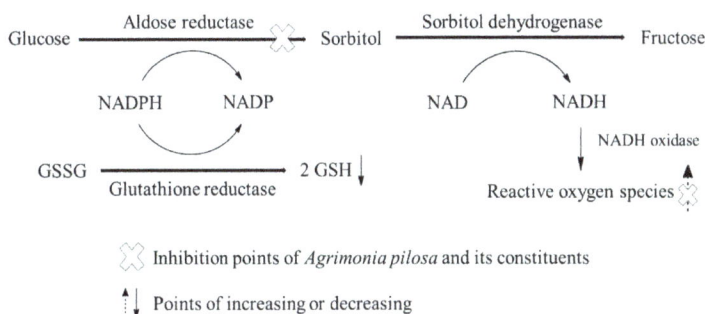

Figure 5. Inhibition points of *A. pilosa* Ledeb and its constituents on polyol pathway. GSH: glutathione, GSSG: glutathione disulfide, NAD: nicotinamide adenine dinucleotide, NADH: oxidoreductase-induced nicotinamide adenine dinucleotide, NADP: nicotinamide adenine dinucleotide phosphate, NADPH: oxidoreductase-induced nicotinamide adenine dinucleotide phosphate.

Various flavonoid constituents were isolated as active compounds from AP. Based on the literature, we evaluated the effect of ten known flavonoids and isolated compounds from the *n*-BuOH fraction of APE on RLAR. Among the compounds isolated, compound 4 was isolated for the first time from this plant and AM (IC$_{50}$ = 1.6 μM) was evaluated in RLAR for the first time. Except for AGN, all compounds showed a potent inhibitory effect, with IC$_{50}$ values of 0.2–9.5 μM. Among active compounds, LGN and QU had similar or higher activity than the positive control TMS. Previous flavonoid RLAR studies reported that LGC (7.5 μM) [26], LGN (3.1 μM), LT (0.5 μM) [27], and AGC (23.0 μM) [28] were isolated from plant sources. Matsuda et al. reported potent IC_{50} values as follows: QC (2.2 µM), IQC (4.5 µM), HP (3.0 µM), AG (2.2 µM), KP (10.0 µM), and RT (9.0 µM) [24]. QU had an IC_{50} value of 0.2 µM [29]. These reported data were similar to our experimental data (shown in Table 2). The sorbitol accumulation was not significantly correlated with RLAR activity. AM, LGN, QU, LT, AZ, and QC showed different RLAR inhibitory effects (IC_{50} values) as structures (QU, 0.2 µM > LT, 0.6 µM > LGN, 0.7 µM > AZ, 1.0 µM > AM, 1.6 µM > QC, 3.2 µM). On the other hand, high inhibition of sorbitol accumulation was observed in the following order: AZ (93.2%), LGN (91.8%) and LT (91.8%), QC (85.7%), QU (76.9%), and AM (47.6%). According to structures of flavonoids, different inhibitory effects were seen in vitro and ex vivo. Therefore, this result suggests that bioavailability may be affected by structures of flavonoids.

Based on these results, the RLAR inhibitory effect of flavonoid derivatives and the structure activity relationship (SAR) were investigated using the RLAR assay. RLAR inhibitory effects of flavonoid derivatives depend on the position and sugar type of the aromatic A and C ring at a catechol moiety. RT, QC, IQC, HP, and QU were isomers of flavonol and had rhamnoside, no sugar, galactoside, glucoside, and rutinoside, respectively, in the same position. However, these compounds showed different RLAR inhibitory effects and different IC_{50} values. RT showed IC_{50} values 15.5, 20.0, 25.5, and 47.5 times higher than those of QC, HP, IQC, and QU, respectively. In addition, LT derivatives showed different RLAR inhibitory effects according to sugar types (LGN > LT > LGC). Flavone derivatives showed different patterns on SAR. AZ, AG, AGN, AG, KP, and AGN have no hydroxyl at the 4′ position at a catechol moiety B ring, and showed lower activity than the flavonol structure. However, rhamnoside at the 3-position in the A ring of flavonol/flavone structure showed stronger activity than other sugar types, and glucuronide and glucoside at the 7′ position showed higher activity than glucoside at the 3′ and 7′ positions. A previous SAR study demonstrated that the inhibitory activity of flavonol/flavone was different according to 3′,4′-hydroxyl moiety in a catechol moiety at the B ring, and suggested that sugar type and hydroxyl moieties at the 3′ and 7′ position increased the activity of RLAR [26].

Offline DPPH-HPLC method is able to rapidly screen antioxidants from complex mixtures, especially for natural products with minimum sample preparation. Reduction or disappearance of the peak areas in the HPLC chromatogram certify potential antioxidant activity of the compounds, while there was no change of peak areas for compounds with no antioxidant activity after their reaction with DPPH. Zhang et al. reported that eighteen antioxidants were screened and identified from *Pueraria lobata* flowers by the offline DPPH-HPLC-MS/MS method [34] Moreover, seven antioxidant compounds in *Eucommia ulmoides* Olive were analyzed by offline DPPH-HPLC [35]. As shown in Figure 3, our offline DPPH-HPLC method results suggested that this method is a good strategy for selecting antioxidant compounds from crude plant extracts. Many studies were done for evaluating the antioxidant activities of flavonoids, which showed the ability to quench free radicals through several mechanisms, including the donation of electrons and hydrogen atoms, and chelate transition metals [36]. Thus, we evaluated the antioxidant activity of seven isolated compounds with offline DPPH-HPLC, as well as the DPPH radical scavenging activity of ten known flavonoids. The *n*-BuOH fraction of AP showed the capacity to scavenge DPPH radicals. In addition, AM, RT, LGC, LGN and QU showed potent DPPH inhibitory activity, with IC_{50} values of 13.0, 66.8, 71.5, 80.6, and 77.9 µM, respectively. Among ten known flavonoids, seven compounds (except for VT, AG, and AZ) exhibited potent DPPH inhibitory activity, with IC_{50} values of 65.9–156.3 µM, compared to L-ascorbic acid (147.3 µM, Table 2). Although activity results of offline DPPH-HPLC and DPPH assay showed different activity patterns, we believe that the offline DPPH method can be very efficient and fast for screening antioxidant compounds from complex mixtures (natural products, food, or materials).

4. Materials and Methods

4.1. General

^{1}H and ^{13}C NMR spectra and correlation 2D NMR spectra were obtained from a Bruker Avance DPX 400 (or 600) spectrometer. These spectra were obtained at operating frequencies of 400 MHz (^{1}H) and 100 (or 150) MHz (^{13}C) with CD_3OD, $(CD_3)_2SO$, $(CD_3)_2CO$, or D_2O and TMS used as an internal standard; chemical shifts were reported in δ values. Isolated compounds were analyzed by electron ionization-MS in a low resolution-MS equipped with JMS-700. A semi prep-HPLC system for separation identification (recycling preparative LC908-C60, JAI, Tokyo, Japan) was used.

4.2. Chemicals and Reagents

L-Ascorbic acid, DPPH, NADPH, DL-glyceraldehyde dimer, TMG, glucose, and the reference compounds used in this study (CT, LT, QC, IQC, HP, AG, VT, KP, AS, and AZ) were purchased from Sigma-Aldrich (St. Louis, MO, USA). RHAR was purchased from Wako Pure Chemical Industries (Osaka, Japan). All other chemicals and reagents used were of analytical grade.

4.3. Plant Materials

Dried bark of AP was purchased at a local market in Yeongcheon, Gyeongsangbuk-do Province, Korea (June 2015). The AP was identified by Hyung Joon Chi at Seoul National University, and a voucher specimen (No. RIC-2015-0615) was deposited at the Regional Innovation Center, Hallym University, Korea.

4.4. Extraction, Fractionation, and Isolation

The dried bark of AP (10 kg) was extracted twice with methylene chloride (50.0 L × 2 times) for 48 h at room temperature. The suspension was filtered and evaporated under reduced pressure at 40 °C to give methylene chloride extract (yield: 1.9%, 194.0 g). The residue was extracted with APE. The suspension was filtered and evaporated under reduced pressure at 40 °C to give the APE (yield: 7.6%, 762.9 g). This extract was suspended in distilled water and then successively partitioned with EtOAc, *n*-BuOH and water to yield EtOAc (17.2%, 110.0 g), *n*-BuOH (20.6%, 132.0 g) and water fractions (59.5%, 381.3 g), respectively. These fractions were concentrated to dryness by rotary evaporation at 40 °C, while the water fraction was freeze-dried. The *n*-BuOH fraction showed strong inhibitory activity against RLAR. Thus, the *n*-BuOH fraction was applied to an open glass column packed with Diaion HP-20 with MeOH-H_2O in a gradient of 30%–100% MeOH, thereby yielding 15 sub-fractions (HP-S1 to HP-S15). It was then eluted with water to wash any sugars or impure components. Fraction HP-S8 (200.0 mg) was purified to yield compound 1 (69.3 mg) and compound 2 (30.3 mg) by recycle HPLC with a gradient system from 30%–35% MeOH. Other fractions obtained from Diaion HP-20 were applied to a Sephadex LH-20 column (90 cm × 3 cm, internal diameter). Fraction HP-S9 (888.6 mg) was separated with 70% MeOH to obtain compound 3 (96.1 mg) and compound 4 (150.5 mg). Fraction HP-S11 (797.1 mg) was separated with a 60% MeOH system to obtain compound 5 (11.6 mg) and compound 6 (21.3 mg). Fraction HP-S12 (465.9 mg) was isolated with a 100% MeOH system to yield compound 7 (251.6 mg).

4.5. Preparation of Aldose Reductase

Crude RLAR was prepared as follows: lenses weighing 250–280 g were removed from Sprague–Dawley rats and frozen at −70 °C until use. This was approved by the University of Hallym Animal Care and Use Committee (registration number: Hallym 2015-06-08). Non-cataractous transparent lenses were pooled and a homogenate was prepared in 0.1 M phosphate-buffered saline (pH 6.2). RLAR homogenate was then centrifuged at 10,000× *g* for 20 min at 4 °C in a refrigerated centrifuge. The supernatant was collected and used as the RLAR. All procedures were carried out at 4 °C [37].

4.6. Determination of RLAR Inhibition In Vitro

RLAR activity was assayed spectrophotometrically by measuring the decrease in the absorption of NADPH at 340 nm over a 3-min period using $_{DL}$-glyceraldehyde as the substrate. Each 1.0 mL cuvette contained equal units of the enzyme, 0.10 M potassium phosphate buffer (pH 6.2), 1.6 mM NADPH, 25 mM $_{DL}$-glyceraldehyde (the substrate), and an inhibitor or dimethyl sulfoxide (DMSO). The inhibition of RLAR (%) was calculated with the following equation: $[1 - (\Delta A \text{ sample/min}) - (\Delta A \text{ blank/min})/(\Delta A \text{ control/min}) - (\Delta A \text{ blank/min})] \times 100\%$, where ΔA sample/min is the reduction of absorbance for 3 min with reaction solution, the test sample, and substrate, and ΔA control/min is the same but with DMSO instead of the test sample [38].

4.7. HPLC Analysis

The sample was analyzed using an Agilent Technologies modular model 1200 system with a vacuum degasser (G1322A), a quaternary pump (G1311A), an auto-sampler (G1329A), a thermo-statted column compartment (G1316A), and a variable wavelength detector (VWD, G1314D) system. The separation was achieved on an Eclipse XDB-phenyl column (150 mm × 4.6 mm, 3.5 μm) maintained at 30 °C. The elution solvents were 0.1% trifluoroacetic acid (A) and MeOH (B) with the following gradient: 20%–30% B (0–3 min), 30%–40% B (3–10 min), 40%–50% B (10–20 min), 50%–60% B (25–35 min), 60%–100% B (25–35 min), 100%–100% B (35–38 min), 100%–20% B (38–40 min), and 20%–20% B (40–45 min). Injection volume was 10 μL (sample concentration: 1 mg/mL) and UV wavelength was 254 nm.

4.8. Evaluation of DPPH Radical Scavenging Capacity

The stable free radical was used to determine the free radical-scavenging activity of the extracts [39]. Briefly, a 0.32 mM solution of DPPH in MeOH was prepared, and 180 μL of this solution was mixed with 30 μL of each sample at concentrations of 0.05–1.0 mg/mL in DMSO. After 20 min of incubation in the darkroom, the decrease in the absorbance of the solution was measured at 570 nm on a microplate reader (EL800 Universal Microplate reader, Bio-Tek instruments, Winooski, VT, USA). DPPH inhibitory activity was expressed as the percentage inhibition (%) of DPPH in the aforementioned assay system, and was calculated as $[1 - (A_{sample} - A_{blank}/A_{control} - A_{blank})] \times 100\%$, where $A_{control}$ is the absorbance of DPPH solution (180 μL) with methanol (30 μL); A_{blank} is the absorbance of distilled water (180 μL) with methanol (30 μL); A_{sample} is the absorbance of DPPH solution (180 μL) with sample solution (30 μL).

4.9. OffLine DPPH HPLC Assay

The offline DPPH HPLC assay was performed by modifying a previously-described protocol [40]. Thirty microliters (20 mg/mL in MeOH) of the *n*-BuOH fraction from APE were mixed with 180 μL prepared DPPH solution (0.32 mM). The mixture was incubated in the dark for 20 min, then filtered through a 0.45-μm filter for HPLC analysis. The *n*-BuOH (20 mg/mL in MeOH) was used as a control. The extent of peak decrease is expressed as a quantitative reduction.

4.10. Determination of Inhibition-Type of RHAR by Active Compound

Reaction mixtures consisted of 0.1 M potassium phosphate, 1.6 mM NADPH, and 2 mM of RHAR with varied concentrations of substrate $_{DL}$-glyceraldehyde and AR inhibitor in a total volume of 600 μL. Concentrations ranged from 0 to 25 mM for DL-glyceraldehyde and from 0 to 20 μM for the active compound. RHAR activity was assayed by measuring the decrease in absorption of NADPH after substrate addition at 340 nm using a Bio Tek Power Wave XS spectrophotometer (Bio Tek Instruments, Winooski, VT, USA) [41].

4.11. Lens Culture and Intracellular Sorbitol Measurement

Lenses isolated from 10-week-old Sprague–Dawley rats using the registration number mentioned in the section of 4.5 were cultured for 6 d in TC-199 medium that contained 15% fetal bovine serum, 100 units/mL penicillin, and 0.1 mg/mL streptomycin, under sterile conditions and an atmosphere of 5% CO_2 and 95% air at 37 °C. Samples were dissolved in DMSO. The lenses were divided into three groups and cultured in medium containing 30 mM glucose and RLAR-active compounds. Each lens was placed in a well containing 2.0 mL of medium. Sorbitol was determined by HPLC using the methods mentioned in the section of 4.7 after its derivatization by reaction with benzoic acid to a fluorescent compound [42].

4.12. Statistical Analysis

Inhibition rates were calculated as percentages (%) with respect to the control value, and the IC_{50} value was defined as the concentration at which 50% inhibition occurred. Data are expressed as mean values ± standard deviation of triplicate experiments.

5. Conclusions

In summary, seven compounds isolated from the *n*-BuOH fraction of the APE and ten flavonoids known as ingredients of AP were evaluated for RLAR inhibitory activity and DPPH radical scavenging activity. Additionally, antioxidant compounds in the *n*-BuOH fraction of APE were investigated with a DPPH offline HPLC assay. Of the compounds tested, AM, LGN, QU, LT, and AZ showed strong inhibitory activity against RLAR and sorbitol accumulation. Consequently, we conclude that APE and its constituents may play partial roles in RLAR and oxidative radical inhibition. Our results suggest that AP may potentially be used as a herbal drug to treat diabetic complications.

Supplementary Materials: Supplementary materials can be found at www.mdpi.com/1422-0067/18/2/379/s1.

Acknowledgments: This work was supported by Korea Institute of Planning and Evaluation for Technology in Food, Agriculture, Forestry and Fisheries (IPET) through High Value-Added Food Technology Development Program, funded by the Ministry of Agriculture, Food and Rural Affairs (MAFRA) (115001-3).

Author Contributions: Soon Sung Lim and Hong-Won Suh designed the experiments; Set Byeol Kim prepared the extract sample, isolated its compounds, and prepared aldose reductase and DPPH assays; Seung Hwan Hwang conducted the structure identification and data analysis; Seung Hwan Hwang wrote the first draft, and Soon Sung Lim revised the manuscript; All the authors read and approved the final manuscript and all authors name added in manuscript.

Conflicts of Interest: The authors declare no conflict of interest.

Abbreviations

AR	aldose reductase
NADPH	nicotinamide adenine dinucleotide phosphate
NADH	nicotinamide adenine dinucleotide
CT	catechin
AP	*Agrimonia pilosa* Ledeb
LT	luteolin
QC	quercetin
IQC	isoquercetin
HP	hyperin
AG	apigenin
VT	vitexin
KP	kaempferol
AS	astragalin
AZ	afzelin
RLAR	rat lens aldose reductase
APE	*Agrimonia pilosa* 50% methanol (MeOH) extract
RHAR	recombinant human aldose reductase
ICs	isolated compounds isolated from *Agrimonia pilosa*
KNCs	known compounds isolated from *Agrimonia pilosa*
TMG	tetramethylene glutaric acid
PAR	peak area reduction
CH_2Cl_2	methylene chloride
EtOAc	ethyl acetate
n-BuOH	*n*-butanol
DPPH	1,1-diphenyl-2-picrylhydrazyl

References

1. Fatmawati, S.; Kurashiki, K.; Takeno, S.; Kim, Y.U.; Shimizu, K.; Sato, M.; Imaizumi, K.; Takahashi, K.; Kamiya, S.; Kaneko, S.; et al. The inhibitory effect on aldose reductase by an extract of *Ganoderma lucidum*. *Phytother. Res.* **2009**, *23*, 28–32. [CrossRef] [PubMed]
2. Peyroux, J.; Sternberg, M. Advanced glycation endproducts (AGEs): Pharmacological inhibition in diabetes. *Pathol. Biol.* **2006**, *54*, 405–419. [CrossRef] [PubMed]
3. Abdollahi, M.; Ranjbar, A.; Shadnia, S.; Nikfar, S.; Rezaie, A. Pesticides and oxidative stress: A review. *Med. Sci. Monit.* **2004**, *10*, 141–147.
4. Sundaram, R.K.; Bhaskar, A.; Vijayalingam, S.; Viswanathan, M.; Mohan, R.; Shanmugasundaram, K.R. Antioxidant status and lipid peroxidation in type II diabetes mellitus with and without complications. *Clin. Sci.* **1996**, *90*, 255–260. [CrossRef] [PubMed]
5. Matough, F.A.; Budin, S.B.; Hamid, Z.A.; Alwahaibi, N.; Mohamed, J. The role of oxidative stress and antioxidants in diabetic complications. *Sultan Qaboos Univ. Med. J.* **2012**, *12*, 5–18. [CrossRef] [PubMed]
6. Zhu, L.; Tana, J.; Wang, B.; Hea, R.; Liuc, Y.; Zhenga, C. Antioxidant activities of aqueous extract from *Agrimonia pilosa* Ledeb and its fractions. *Chem. Biodivers.* **2009**, *6*, 1716–1726. [CrossRef] [PubMed]
7. Kim, J.J.; Jiang, J.; Shim, D.W.; Kwon, S.C.; Kim, T.J.; Ye, S.K.; Kim, M.K.; Shin, Y.K.; Koppula, S.; Kang, T.B.; et al. Anti-inflammatory and anti-allergic effects of *Agrimonia pilosa* Ledeb extract on murine cell lines and OVA-induced airway inflammation. *J. Ethnopharmacol.* **2012**, *140*, 213–221. [CrossRef] [PubMed]
8. Jung, M.K.; Park, M.S. Acetylcholinesterase inhibition by flavonoids from *Agrimonia pilosa*. *Molecules* **2007**, *12*, 2130–2139. [CrossRef] [PubMed]
9. An, R.B.; Kim, H.C.; Jeong, G.S.; Oh, S.H.; Oh, H.; Kim, Y.C. Constituents of the aerial parts of *Agrimonia pilosa*. *Nat. Prod. Sci.* **2005**, *11*, 196–198.
10. Jiang, Q.; Ma, J.; Wang, Y.; Ding, L.; Chen, L.; Qiu, F. Simultaneous determination of nine major constituents in *Agrimonia pilosa* Ledeb. by HPLC-DAD-ESI-MS/MS. *Anal. Methods* **2014**, *6*, 4373–4379. [CrossRef]

11. Kato, H.; Li, W.; Koike, M.; Wang, Y.; Koike, K. Phenolic glycosides from *Agrimonia pilosa*. *Phytochemistry* **2010**, *71*, 1925–1929. [CrossRef] [PubMed]
12. Liu, X.; Zhu, L.; Tan, J.; Zhou, X.; Xiao, L.; Yang, X.; Wang, B. Glucosidase inhibitory activity and antioxidant activity of flavonoid compound and triterpenoid compound from *Agrimonia pilosa* Ledeb. *BMC Complement. Altern. Med.* **2014**, *14*, 12. [CrossRef] [PubMed]
13. Veeresham, C.; Rama Rao, A.; Asres, K. Aldose reductase inhibitors of plant origin. *Phytother. Res.* **2013**, *28*, 317–333. [CrossRef] [PubMed]
14. Jung, H.A.; Isslam, M.D.N.; Kwon, Y.S.; Jin, S.E.; Son, Y.K.; Park, J.J.; Sohn, H.S.; Choi, J.S. Extraction and identification of three major aldose reductase inhibitors from *Artemisia montana*. *Food Chem. Toxicol.* **2011**, *49*, 376–384. [CrossRef] [PubMed]
15. Zeng, H.; Liu, Q.; Wang, M.; Jiang, M.; Jiang, S.; Zhang, L.; He, L.; Wang, J.; Chen, X. Target-guided separation of antioxidants from Semen cassia via off-line two-dimensional high-speed counter-current chromatography combined with complexation and extrusion elution mode. *J. Chromatogr. B* **2015**, *1001*, 58–65. [CrossRef] [PubMed]
16. Tang, D.; Li, H.J.; Chen, J.; Guo, C.W.; Li, P. Rapid and simple method for screening of natural antioxidants from Chinese herb Flos Lonicerae japonicae by DPPH-HPLC-DAD-TOF/MS. *J. Sep. Sci.* **2008**, *31*, 3519–3526. [CrossRef] [PubMed]
17. Olennikov, D.N.; Kashchenko, N.I.; Chirikova, N.K.; Kuzmina, S.S. Phenolic profile of *Potentilla anserina* L. (Rosaceae) herb of Siberian origin and development of a rapid method for simultaneous determination of major phenolics in *P. anserina* pharmaceutical products by microcolumn RP-HPLC-UV. *Molecules* **2015**, *20*, 224–248. [CrossRef] [PubMed]
18. Fathiazad, F.; Delazar, A.; Amiri, R.; Sarker, S.D. Extraction of flavonoids and quantification of rutin from waste tobacco leaves. *Iran. J. Pharm. Res.* **2006**, *3*, 222–227.
19. Lee, S.H.; Choi, M.J.; Choi, J.M.; Lee, S.; Kim, H.Y.; Cho, E.J. Flavonoids from *Taraxacum coreanum* protect from radical-induced oxidative damage. *J. Med. Plants Res.* **2012**, *6*, 5377–5384.
20. Lu, Y.; Foo, L.Y. Flavonoid and phenolic glycosides from *Salvia officinalis*. *Phytochemistry* **2000**, *55*, 263–267. [CrossRef]
21. Lu, Y.; Foo, L.Y. Identiücation and quantiücation of major polyphenols in apple pomace. *Food Chem.* **1997**, *59*, 187–194. [CrossRef]
22. Pieroni, A.; Heimler, D.; Pieters, L.; van Poel, B.; Vlietnick, A.J. In vitro anti-complementary activity of flavonoids from olive (*Olea europaea* L.) leaves. *Pharmazie* **1996**, *51*, 765–768. [PubMed]
23. Baris, O.; Karadayi, M.; Yanmis, D.; Guvenalp, Z.; Bal, T.; Gulluce, M. Isolation of 3 flavonoids from *Mentha longifolia* (L.) Hudson sucsp. *longifolia* and determination of their genotoxic potentials by using the *E. coli* WP2 test system. *J. Food Sci.* **2011**, *76*, 212–217.
24. Matsuda, H.; Morikawa, T.; Toguchida, I.; Yoshikawa, M. Structural requirements of flavonoids and related compounds for aldose reductase inhibitory activity. *Chem. Pharm. Bull.* **2002**, *50*, 788–795. [CrossRef] [PubMed]
25. Naeem, S.; Hylands, P.; Barlow, D. Construction of an Indonesian herbal constituents database and its use in Random Forest modelling in a search for inhibitors of aldose reductase. *Bioorg. Med. Chem.* **2012**, *20*, 1251–1258. [CrossRef] [PubMed]
26. Li, H.M.; Hwang, S.H.; Kang, B.G.; Hong, J.S.; Lim, S.S. Inhibitory effects of *Colocasia esculenta* (L.) Schott constituents on aldose reductase. *Molecules* **2014**, *19*, 13212–13224. [CrossRef] [PubMed]
27. Yoshikawa, M.; Morikawa, T.; Murakami, T.; Toguchida, I.; Harima, S.; Matsuda, H. Medical flowers, I. Aldose reductase inhibitors and three new eudesmane-type sesquiterpenes, kikkanols A, B, and C, from the flowers of *Chrysanthemum indicum* L. *Chem. Pharm. Bull.* **1999**, *47*, 340–345. [CrossRef] [PubMed]
28. Matsuda, H.; Morikawa, T.; Toguchida, I.; Harima, S.; Yoshikawa, M. Medicinal flowers. VI. Absolute stereostructures of two new flavanone glycosides and a phenylbutanoid glycoside from the Flowers of *Chrysanthemum indicum* L.: Their inhibitory activities for rat lens aldose reductase. *Chem. Pharm. Bull.* **2002**, *50*, 972–975. [CrossRef] [PubMed]
29. Yoshikawa, M.; Shimada, H.; Nishida, N.; Li, Y.; Toguchida, I.; Yamahara, J.; Matsuda, H. Antidiabetic principles of natural medicines. II. aldose reductase and α-glucosidase inhibitors from Brazilian natural medicine, the leaves of *Myrcia multiflora* DC. (Myrtacae): Structures of myrciacitrins I and II and myrciaphenones A and B. *Chem. Pharm. Bull.* **1998**, *46*, 113–119. [CrossRef] [PubMed]

30. Mok, S.Y.; Lee, S.H. Identification of flavonoids and flavonoid rhamnosides from *Rhododendron mucronulatum* for *albiflorum* and their inhibitory activities against aldose reductase. *Food Chem.* **2013**, *136*, 969–974. [CrossRef] [PubMed]

31. Lee, E.H.; Song, D.G.; Lee, J.Y.; Pan, C.H.; Um, B.H.; Jung, S.H. Flavonoids from the leaves of *Thuja orientalis* inhibit the aldose reductase and the formation of advanced glycation endproducts. *J. Korean Soc. Appl. Biol. Chem.* **2009**, *52*, 448–455. [CrossRef]

32. Ha, T.J.; Lee, J.H.; Lee, M.H.; Lee, B.W.; Kwon, H.S.; Park, C.H.; Shim, K.B.; Kim, H.T.; Baek, I.Y.; Jang, D.S. Isolation and identification of phenolic compounds from the seeds of *Perilla frutescens* (L.) and their inhibitory activities against α-glucosidase and aldose reductase. *Food Chem.* **2012**, *135*, 1397–1403. [CrossRef] [PubMed]

33. Chethan, S.; Dharmesh, S.M.; Malleshi, N.G. Inhibition of aldose reductase from cataracted eye lenses by finger millet (*Eleusine coracana*) polyphenols. *Bioorg. Med. Chem.* **2008**, *16*, 10085–10090. [CrossRef] [PubMed]

34. Zhang, Y.P.; Shi, S.Y.; Xiong, X.; Chen, X.Q.; Peng, M.J. Comparative evaluation of three methods based on high-performance liquid chromatography analysis combined with a 2,2′-diphenyl-1-picrylhydrazyl assay for the rapid screening of antioxidants from *Pueraria lobate* flowers. *Anal. Bioanal. Chem.* **2012**, *402*, 2965–2976. [CrossRef] [PubMed]

35. Dia, X.; Huang, Q.H.; Zhou, B.; Gong, Z.; Liu, Z.; Shi, S. Preparative isolation and purification of seven main antioxidants from *Eucommia ulmoides* Oliv. (Du-zhong) leaves using HSCCC guided by DPPH-HPLC experiment. *Food Chem.* **2013**, *139*, 563–570.

36. Agati, G.; Azzarello, E.; Pollastri, S.; Tattini, M. Flavonoids as antioxidants in plants: Location and functional significance. *Plant Sci.* **2012**, *196*, 67–76. [CrossRef] [PubMed]

37. Yoon, H.N.; Lee, M.Y.; Kim, J.K.; Suh, H.W.; Lim, S.S. Aldose reductase inhibitory compounds from *Xanthium strumarium*. *Arch. Pharm. Res.* **2013**, *36*, 1090–1095. [CrossRef] [PubMed]

38. Peak, J.H.; Lim, S.S. Preparative isolation of aldose reductase inhibitory compounds from *Nardostachys chinensis* by elution-extrusion counter-current chromatography. *Arch. Pharm. Res.* **2014**, *37*, 1271–1279. [CrossRef] [PubMed]

39. Choi, C.W.; Kim, S.C.; Hwang, S.S.; Choi, B.K.; Ahn, H.J.; Lee, M.Y.; Park, S.H.; Kim, S.K. Antioxidant activity and free radical scavenging capacity between Korean medicinal plants and flavonoids by assay-guided comparison. *Plant Sci.* **2002**, *163*, 1161–1168. [CrossRef]

40. Shi, S.Y.; Ma, Y.J.; Zhang, Y.P.; Liu, L.L.; Liu, Q.; Peng, M.J.; Xiong, X. Systematic separation and purification of 18 antioxidants from *Pueraria lobata* flower using HSCCC target-guided by DPPH-HPLC experiment. *Sep. Purif. Technol.* **2012**, *89*, 225–233. [CrossRef]

41. Kim, T.H.; Kim, J.K.; Kang, Y.H.; Lee, J.Y.; Kang, I.J.; Lim, S.S. Aldose reductase inhibitory activity of compounds from *Zea mays* L. *Biomed. Res. Int.* **2013**, *2013*, 8. [CrossRef] [PubMed]

42. Lee, Y.S.; Kim, S.H.; Jung, S.H.; Kim, J.K.; Pan, C.H.; Lim, S.S. Aldose reductase inhibitory compounds from *Glycyrrhiza uralensis*. *Biol. Pharm. Bull.* **2010**, *33*, 917–921. [CrossRef] [PubMed]

International Journal of
Molecular Sciences

MDPI

Article

Hazelnut (*Corylus avellana* L.) Shells Extract: Phenolic Composition, Antioxidant Effect and Cytotoxic Activity on Human Cancer Cell Lines

Tiziana Esposito [1,2], Francesca Sansone [1], Silvia Franceschelli [1], Pasquale Del Gaudio [1], Patrizia Picerno [1], Rita Patrizia Aquino [1] and Teresa Mencherini [1,*]

[1] Department of Pharmacy, University of Salerno, Via Giovanni Paolo II, 132, I-84084 Fisciano (SA), Italy; tesposito@unisa.it (T.E.); fsansone@unisa.it (F.S.); sfranceschelli@unisa.it (S.F.); pdelgaudio@unisa.it (P.D.G.); ppicerno@unisa.it (P.P.); aquinorp@unisa.it (R.P.A.)

[2] Ph.D. Program in Drug Discovery and Development, University of Salerno, Via Giovanni Paolo II 132, I-84084 Fisciano (SA), Italy

* Correspondence: tmencherini@unisa.it; Tel.: +39-089-968-294

Academic Editor: Toshio Morikawa
Received: 29 December 2016; Accepted: 7 February 2017; Published: 13 February 2017

Abstract: Hazelnut shells, a by-product of the kernel industry processing, are reported to contain high amount of polyphenols. However, studies on the chemical composition and potential effects on human health are lacking. A methanol hazelnut shells extract was prepared and dried. Our investigation allowed the isolation and characterization of different classes of phenolic compounds, including neolignans, and a diarylheptanoid, which contribute to a high total polyphenol content (193.8 ± 3.6 mg of gallic acid equivalents (GAE)/g of extract). Neolignans, lawsonicin and cedrusin, a cyclic diarylheptanoid, carpinontriol B, and two phenol derivatives, C-veratroylglycol, and β-hydroxypropiovanillone, were the main components of the extract (0.71%–2.93%, w/w). The biological assays suggested that the extract could be useful as a functional ingredient in food technology and pharmaceutical industry showing an in vitro scavenging activity against the radical 1,1-diphenyl-2-picrylhydrazyl radical (DPPH) ($EC_{50} = 31.7$ µg/mL with respect to α-tocopherol $EC_{50} = 10.1$ µg/mL), and an inhibitory effect on the growth of human cancer cell lines A375, SK-Mel-28 and HeLa ($IC_{50} = 584, 459$, and 526 µg/mL, respectively). The expression of cleaved forms of caspase-3 and poly(ADP-ribose) polymerase-1 (PARP-1) suggested that the extract induced apoptosis through caspase-3 activation in both human malignant melanoma (SK-Mel-28) and human cervical cancer (HeLa) cell lines. The cytotoxic activity relies on the presence of the neolignans (balanophonin), and phenol derivatives (gallic acid), showing a pro-apoptotic effect on the tested cell lines, and the neolignan, cedrusin, with a cytotoxic effect on A375 and HeLa cells.

Keywords: hazelnut by-product; neolignans; diaryleptanoid; DPPH radical; caspase-3; PARP-1

1. Introduction

Hazelnut (*Corylus avellana* L., Betulaceae family) is one of the most cultivated and marketed nuts in the world. Italy is the second largest hazelnut-producing area (about 105,000 t/year), behind Turkey (about 600,000 t/year) [1]. About 10% of the world crop production is sold as in-shell product consumed fresh or roasted, and the remaining 90% as shelled hazelnuts and used as an ingredient in food (bakery, confectionary industry, and chocolate) processing industries [2,3]. During the kernel harvesting and industrial processing, a large amount of by-products, including green leafy cover, shell and skin, is obtained. Their disposal represents both an economic problem for the producers and a serious environmental problem due to the combustion of the crop residues [4–6]. The ligno-cellulose shells, obtained after cracking the kernel, account for the majority of this waste, and they are used as a heat source, for mulching, and furfural production in dye manufacturing [7]. The reported antioxidant potential of both hazelnut kernel and shell extracts might be related to the presence of phenolic acids and tannins [8–11]. Polyphenols have received great attention for their human health benefits due to antioxidant properties [8,11–13]. Intake of foods or vegetable products rich in polyphenols is generally recognized as useful for the prevention and treatment of cancer, and cardiovascular, inflammatory, microbial, and age-related diseases [14]. In particular, the chemopreventive efficacy of these natural antioxidants has been demonstrated against several human cancer cell lines [15]. Therefore, recovery and upgrading of hazelnut shells seems to be consistent with the growing demand for ingredients that have beneficial effects on human health. Nevertheless, the information about the chemical profile of hazelnut shells is limited to the identification of free and bound phenolic compounds, such as flavonoid glycosides and aromatic acids, in hazelnuts cultivated in Poland [16]. Therefore, the aim of the present study was to define the chemical composition and biological activities of the methanol extract from hazelnut shells (HSE). The research led to the isolation and characterization by Nuclear Magnetic Resonance (NMR) and Elettrspray Mass Spectrometry (ESI-MS) of four neolignans with a dihydro[*b*]benzofuran skeleton, seven phenolic derivatives, and a cyclic diarylheptanoid. The in vitro free radical scavenging activity of HSE and isolated compounds was determined by DPPH test. The antiproliferative activity of HSE and its major components against human melanoma (primary and metastatic, A375, and SK-Mel-28, respectively) and cervical cancer (HeLa) cell lines was evaluated by MTT bioassay. The potential pro-apoptotic mechanism of action, as well as the involvement of caspase-3 and its major substrate PARP-1 in the apoptotic process, was investigated.

2. Results and Discussion

2.1. Extract Preparation, Chemical Composition, and Quantitative Analysis

In order to investigate the chemical profile and biological activities of hazelnut shells, a methanol extract (HSE) from powdered and defatted shells was prepared. The extraction yield, after maceration (3 times × 24 h) at room temperature of shells, was about 2.08%. This result is comparable to that reported by Shahidi et al. (2007) [9] and Contini et al. (2008) [8] using aqueous ethanol, methanol, or acetone as solvent systems and hot-reflux extractor (80 °C) or a long maceration at room temperature as extraction procedures. A portion of HSE (1.5 g) was subjected to chromatography by Sephadex LH-20 and RP-HPLC to obtain twelve major constituents belonging to different phenolic subclasses. The structures of the isolated compounds (Figure 1) were established by their NMR and MS data in comparison to those found in the literature. They include four dihydro[*b*]benzofuran-type neolignans (**1–4**), lawsonicin (**1**) [17], cedrusin (**2**) [18], balanophonin (**3**) [19], and ficusal (**4**) [20]; seven phenolic derivatives, dihydroconiferyl alcohol (**5**) [21], veratric acid (**6**) [22], vanillic acid (**7**) [17], gallic acid (**8**), methyl gallate (**9**) [23], C-veratroylglycol (**11**) [24], and β-hydroxypropiovanillone (**12**) [25]; and a cyclic diarylheptanoid, carpinontriol B (**10**) [26]. Vanillic and gallic acids (**7–8**) have been previously identified in hazelnut kernel and shells [16], while the presence in hazelnut of compounds **1–6** and **9–12** was revealed for the first time.

The major components of HSE, neolignans (**1**) and (**2**), cyclic diarylheptanoid (**10**), and phenols (**11**) and (**12**), were selected as markers of the extract and their quantitative analysis was performed by High-Performance Liquid Chromatography with Diode-Array Detection (HPLC-DAD) using the isolated compounds as the standards for calibration curves. The HPLC fingerprint is reported in Figure 2. Lawsonicin (**1**), cedrusin (**2**), carpinontriol B (**10**), C-veratroylglycol (**11**), and β-hydroxypropiovanillone (**12**) were found to be 1.98%, 1.79%, 1.41%, 2.93%, and 0.71%, *w/w* of the extract, respectively. Other isolated compounds (**3–9**) were not quantified.

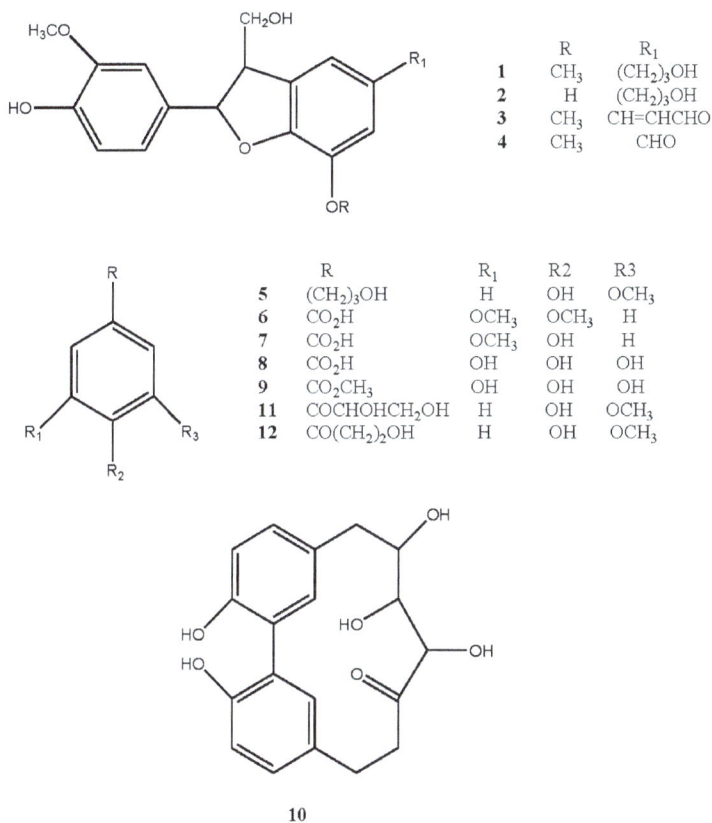

	R	R_1		
1	CH_3	$(CH_2)_3OH$		
2	H	$(CH_2)_3OH$		
3	CH_3	CH=CHCHO		
4	CH_3	CHO		

	R	R_1	R2	R3
5	$(CH_2)_3OH$	H	OH	OCH_3
6	CO_2H	OCH_3	OCH_3	H
7	CO_2H	OCH_3	OH	H
8	CO_2H	OH	OH	OH
9	CO_2CH_3	OH	OH	OH
11	$COCHOHCH_2OH$	H	OH	OCH_3
12	$CO(CH_2)_2OH$	H	OH	OCH_3

10

Figure 1. Structures of compounds (**1**–**12**) isolated from hazelnut shells extract (HSE).

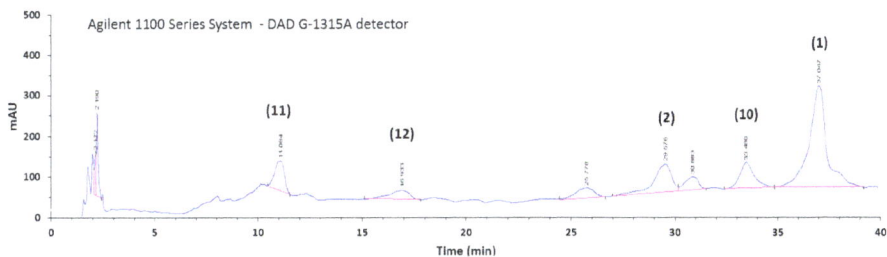

Figure 2. HPLC-DAD fingerprint (230 nm) of hazelnut shells extract (HSE). The peak numbers in this figure correspond to compounds in Figure 1.

2.2. Free Radical Scavenging Activity

The well known antioxidant activity of phenolic compounds is generally thought to be due to redox properties, which can play an important role in neutralizing free radicals, quenching singlet and triplet oxygen, or decomposing peroxides [27]. Considering their occurrence in HSE, the free-radical scavenging activity of the extract was verified by DPPH test. This method evaluates the ability of a sample to scavenge the chromogen long-lived DPPH free radical [28]. Results (Table 1) showed that the extract possessed a significant and concentration-dependent free radical scavenging (EC_{50} = 31.7 µg/mL) which may be correlated to its high polyphenol content, evaluated by Folin–Ciocalteu method, and expressed as gallic acid equivalent (193.8 mg GAE/g of the extract). Moreover, the free-radical scavenging activity of all isolated compounds was also evaluated, with the aim to identify the compounds responsible for HSE activity. As previously reported [23], gallic acid (**8**) and methyl gallate (**9**), water-soluble polyphenols, were very effective in quenching free-radicals, exhibiting an EC_{50} of 1.2 and 1.4 µg/mL, respectively, 10-fold higher than α-tocopherol (EC_{50} = 10.1 µg/mL) used as positive control (Table 1). Neolignans, lawsonicin (**1**), cedrusin (**2**), and balanophonin (**3**), phenolic acid derivatives, vanillic (**6**) and veratric (**7**) acids, and cyclic diarylheptanoid, carpinotriol B (**10**) had EC_{50} values ranging from 42.7 to 89.2 µg/mL (Table 1). Only ficusal (**4**) and dihydroconiferyl alcohol (**5**) were about 10-fold less active than α-tocopherol (Table 1). Results were in agreement with the observation that the structure of polyphenols is the key determinant of their antioxidant activity [29]. The strong effect of phenolic acids such as gallic acid (**8**) and methyl gallate (**9**) is due to three free hydroxyl groups at position 3, 4 and 5 on the aromatic ring [30]. The loss of a hydroxyl group and/or the presence of one or more methoxy groups on the aromatic ring reduced drastically the activity as observed for veratric (**7**) and vanillic (**6**) acids (Table 1), respectively. Moreover, in the series of di-ortho phenolic derivatives, β-hydroxypropiovanillone (**12**) was more active than C-veratroylglycol (**11**) and dihydroconiferyl alcohol (**5**) (Table 1), probably due to modification in the side chain. Considering the structures of neolignans **1–4**, the free-radical scavenging activity was as follows **2** > **3** > **1** > **4**, suggesting that the effect could be related to the 3-phenylpropan-1-ol unit and free hydroxyl group at position C-3' [31]. The presence of a methoxy group at C-3' (lawsonicin, **1** and balanophonin, **3**) decreases the efficacy; and the activity disappeared in ficusal (**4**) which shows the loss of the side chain and presence of an aldheide function at R_1. In conclusion, the significant free radical scavenging activity of the hazelnut shells extract could be ascribed to the additive and synergistic effect of its phenols, which may exert, in combination, a better antiradical effect than individual compound [32].

Table 1. Total Phenolic Content and free-radical scavenging activity of hazelnut shells extract (HSE) and compounds **1–12**.

Extract and Compounds	Phenol Content (mg/g Extract) [a]	EC_{50} [b] (µg/mL)
HSE	193.8 ± 3.6 [c]	31.7 ± 1.4 [c]
1		74.3 ± 3.8
2		42.7 ± 2.5
3		59.2 ± 2.9
4		160.0 ± 4.5
5		118.7 ± 3.5
6		55.4 ± 1.2
7		58.6 ± 3.5
8		1.2 ± 0.2
9		1.9 ± 0.8
10		78.2 ± 2.1
11		89.2 ± 3.2
12		54.6 ± 2.8
α-Tocopherol [d]		10.1 ± 1.3

[a] Gallic acid equivalent; [b] EC_{50} ± standard deviation (data from three experiments in triplicate); [c] Mean ± SD of three determination by the Folin–Ciocalteu method; [d] Positive control of the DPPH assay.

2.3. Cytotoxic Activity of Hazelnut Shells Extract (HSE) and Isolated Compounds

The treatment of melanoma and cervical cancer with conventional chemotherapy, surgery, and radiation, alone or in combination, is rather unsatisfactory [15,33]. Therefore, the research on functional foods fortified and enriched with natural potential chemopreventive additives, dietary supplements, and nutraceuticals able to decrease the incidence of these cancers, is raising a great interest. The cytotoxic activities of gallic acid and neolignans with a dihydro[*b*]benzofuran against several cancer cell lines have been reported [14,16,31,34]. In the present study, the activity of hazelnut extract and its constituents in inhibiting cell proliferation was evaluated by MTT assay against human melanoma (primary A375, metastatic SK-Mel-28), and cervical cancer (HeLa) cell lines. The total extract (HSE) exhibited a significant ($p < 0.05$) and concentration-dependent inhibitory effect on the tumor cell lines growth (IC_{50} 459–584 µg/mL, Table 2). Balanophonin (**3**), and gallic acid (**8**) were cytotoxic on all cell lines with IC_{50} values ranging from 142 to 200 µM (Table 2). The neolignan cedrusin (**2**) was found active in A375 and HeLa cells (IC_{50} = 130 and 141 µM, respectively) for the first time. On the contrary, other neolignans, lawsonicin (**1**) and ficusal (**4**), phenol derivatives, dihydroconyferyl alcohol (**5**), veratric acid (**6**), vanillic acid (**7**), C-veratroylglycol (**11**), and β-hydroxypropiovanillone (**12**), and cyclic diarylheptanoid, carpinontriol B (**10**) were not cytotoxic up to 1000 µM (Table 2). Results indicated that the effect of HSE on cancer cell growth might be due to a synergy of action of the neolignans, cedrusin (**2**) and balanophonin (**3**), and gallic acid (**8**). However, it cannot be excluded that not isolated or interfering constituents may contribute to the extract activity.

Table 2. Effect of hazelnut shells extract (HSE) and its compounds on human cancer cell lines.

Extract or Compound	Cell Line		
	A375 [a] (IC_{50}) [b]	SK-Mel-28 (IC_{50})	HeLa (IC_{50})
HSE	584.0 ± 9.0 [c]	459.0 ± 8.3	526.0 ± 8.9
1	NA [d]	NA	NA
2	130.0 ± 4.2	NA	141.0 ± 3.8
3	142.0 ± 3.6	150.0 ± 4.1	143.0 ± 4.4
4	NA	NA	NA
5	NA	NA	NA
6	NA	NA	NA
7	NA	NA	NA
8	170.0 ± 3.2	150.0 ± 4.0	200.0 ± 3.3
10	NA	NA	NA
11	NA	NA	NA
12	NA	NA	NA

[a] A375 and SK-Mel-28, melanoma cells; HeLa, cervical cancer cells; [b] IC_{50}, required concentration of hazelnut shells extract or pure compound to inhibit cell proliferation by 50% expressed as µg/mL for extract and µM for compounds; [c] IC_{50} ± standard deviation (data from three experiments in triplicate); and [d] Not active (IC_{50} > 1000 µM).

Gallic acid has been shown to induce apoptosis in cancer cells and it has been recognized as a chemopreventive agent [33,35,36]. However, there is no study in the literature supporting the possible mechanism of action of neolignans, cedrusin (**2**), and balanophonin (**3**), in the induction of human cancer cell death. Therefore, the potential apoptotic effect of the extract, HSE, and the most cytotoxic neolignans (**2**) and (**3**), and gallic acid (**8**) was investigated evaluating the presence of hypodiploid nuclei in the cells by flow-cytometric analysis, after incubating with the extract (100–500 µg/mL) or compounds (each 100–500 µM) for 24 h [17].

Figure 3 shows that the extract induced apoptosis in all treated cancer cells increasing in a dose-dependent manner the percentage of hypodiploid nuclei. Notably, this effect was significant ($p < 0.05$) from 250 µg/mL and was more evident in A375 cells compared to SK-Mel-28 and HeLa cells. Moreover, compounds (**2**), (**3**) and (**8**) exhibited a pro-apoptotic effect (data not shown).

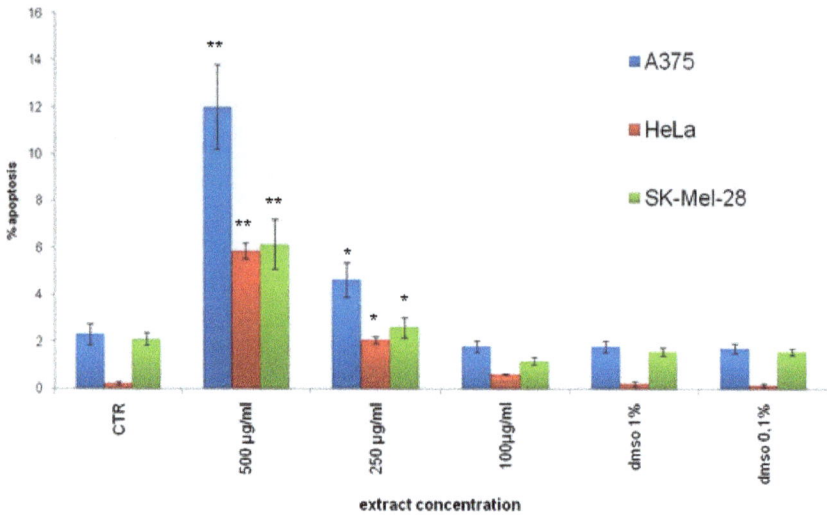

Figure 3. Effects of hazelnut shells extract (HSE) on apoptosis of A375, SK-Mel-28 and HeLa cells. Analysis of percentage of nuclei in apoptosis was performed with propidium iodide staining. Cancer cells were incubated with different concentrations of hazelnut shells extract (HSE) (100–500 μg/mL) for 24 h. Cells were then collected, and the percentage of hypodiploid nuclei was analyzed by flow cytometry (* $p < 0.05$, ** $p < 0.01$ vs. control cells). All results are shown as mean ± standard deviation of three experiments performed in triplicate. Statistical comparison between groups were made using ANOVA followed by the Bonferroni parametric test. Differences were considered significant if $p < 0.05$.

One of the most common signaling cascades involved in apoptosis is the activation of caspases, a family of cysteinyl-aspartate proteases, usually present as inactive zymogen forms. Caspases cleave several proteins, during the execution phase of apoptosis, and among them, PARP-1 (poly(ADP-ribose) polymerase-1), a nuclear enzyme involved in DNA repair, DNA replication, and modulation of chromatin structure [37]. In response to genotoxic stress, PARP-1 is cleaved by caspase-3 and -7 into a ~25 kDa N-terminal fragment, containing the DNA binding domain (DBD), and a ~85 kDa C-terminal fragment that retains basal enzymatic activity PARP-1, recognizes DNA strand interruptions, and can complex with RNA inhibiting transcription. Through these processes, PARP-1 cleavage may help cells to commit to the apoptotic pathway [38,39].

In order to investigate the mechanism of apoptosis induction by both the extract (HSE) and compounds in A375, SK-Mel-28 and HeLa cancer cells, the expressed levels of caspase-3 and PARP-1 cleavage were further analyzed by Western blotting analysis. Results indicated that HSE, balanophonin (**3**), and gallic acid (**8**) induced PARP-1 cleavage after 24–48 h of treatment (Figure 4) in human cervical cancer (HeLa) and human malignant metastatic melanoma (SK-Mel-28) cell lines. Therefore, hazelnut extract and compounds-induced apoptosis is mediated by caspase-3 activation in the above cancer cells. Conversely, no activation of PARP-1 in human malignant melanoma (A375) cells (Figure 4) suggested that the pro-apoptotic mechanism of extract and compounds must be further investigated.

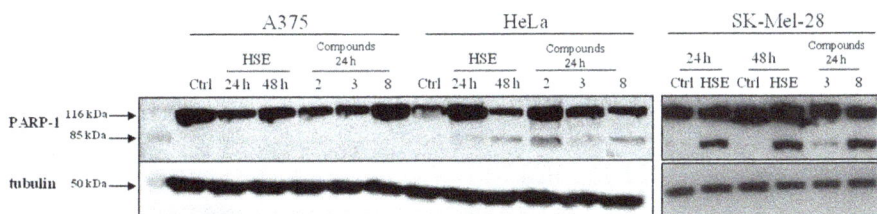

Figure 4. PARP-1 expression in A375, HeLa and SK-Mel-28 cell lines after 24 and 48 h of treatment. Hazelnut shells extract (HSE) and compounds (after 24 h) induce PARP-1 cleavage in HeLa and SK-Mel-28, but not in A375 cell lines.

3. Materials and Methods

3.1. Chemicals and Reagents

Analytical grade *n*-hexane, chloroform, *n*-butanol, and methanol employed for extraction and isolation procedures, methanol deuterated, Folin–Ciocalteu phenol reagent, 1,1-diphenyl-2-picrylhydrazyl radical (DPPH), α-tocopherol, and HPLC-grade methanol were purchased from Sigma-Aldrich (Milan, Lombardia, Italy). HPLC-grade water (18 mΩ) was prepared by a Milli-Q_{50} purification system (Millipore Corp., Bedford, MA, USA). Water and MeOH employed for the electrospray ionization ESI-MS analysis were of HPLC supergradient quality (Romil Ltd., Cambridge, UK). Human malignant melanoma (A375, and SK-Mel-28), human cervical cancer (HeLa), all reagents, and supplements for cell cultures were obtained from Gibco Life Technology Corp. (ThermoFischer Scientific, Milan, Italy). Sodium citrate, Triton X-100 and propidium iodide (PI) were purchased from (Sigma-Aldrich, St. Louis, MO, USA). PARP-1 (F-2) antibody was acquired from Santa Cruz Biotechnology, Inc. (Heidelberg, Germany).

3.2. General Experimental Procedures

A Bruker DRX-600 NMR spectrometer (Bruker Italia, Milano, Italia), operating at 599.19 MHz for ^{1}H and 150.858 MHz for ^{13}C, using the TopSpin 3.2 software package (Bruker Italia, Milano, Italy), was used for NMR experiments in CD_3OD. Chemical shifts are expressed in δ (parts per million) referring to the solvent peaks δ_H 3.31 and δ_C 49.05 for CD_3OD, with coupling constants, *J*, in Hertz. Conventional pulse sequences were used for ^{1}H-^{1}H DQF-COSY (Double Quantum Filter-Correlation Spectroscopy) ^{1}H-^{13}C HSQC (Heteronuclear Single Quantum Coherence), and HMBC (Heteronuclear Multiple Bond Correlation) experiments [27]. ESI-MS experiments were performed with a Finnigan LC-Q Deca spectrometer (Thermoquest, San Jose, CA, USA), equipped with Xcalibur 3.1 software (Thermoquest, San Jose, CA, USA). Chromatography was performed on Sephadex LH-20 (Pharmacia, Uppsala, Sweden). Thin-layer chromatography (TLC) analysis was performed with Macherey–Nagel precoated silica gel 60 F_{254} plates (Delchimica, Naples, Italy), and the spray reagent cerium sulfate (saturated solution in dilute H_2SO_4) and UV (254 and 366 nm) were used for the spot visualization. Preparative HPLC separations were conducted on a Waters 590 series pumping system, equipped with a Waters R401 refractive index detector and a Rheodyne injector (100 µL loop), using µ-Bondapak C_{18} (300 × 7.8 mm i.d., 10 µm, Waters) or Luna C_8 (250 × 10.0 mm i.d., 10 µm, Phenomenex, Torrance, CA, USA) as column. An Agilent 1100 series system (Agilent Technologies, Waldbronn, Germany), equipped with a Model G-1312 pump, a Rheodyne Model G-1322A loop (20 µL), and a DAD G-1315A detector was used for the HPLC quantitative analysis using a Nucleodur 100-5 C_{18} column (150 × 4.6 mm, 5 µm, Machery-Nagel). Peaks area were calculated with an Agilent Integrator (Agilent Technologies, Waldbronn, Germany).

3.3. Materials

Hazelnut shells were provided from a local company, Hazelnuts South Italy Manufacturing S.r.l. (Baiano, Avellino, Italy). They represented the waste of industrial processing carried out on two Italian varieties (90% Mortarella and 10% Lunga San Giovanni) at roasting temperature of 240 °C for 30 min. The shells were ground in a mortar grinder (RM 100, Retsch, Bergamo, Italy) for 5 min. The shells (1000 g) were sequentially defatted with *n*-hexane and chloroform, and extracted at room temperature (3 times × 1.6 L for 24 h) with methanol. The organic solvent was removed under vacuum at 40 °C in a rotary evaporator (Rotavapor R-200, Buchi Italia s.r.l, Cornaredo, Italy), to give 20.8 g of residue (HSE). The extraction yield, gravimetrically determined (balance Denver Instruments-PK-201; 15/30 °C), and expressed as the weight percentage of the dry matter compared to the total amount of the initial material, was 2.08%, *w/w*.

3.4. Isolation Procedure of Compounds **1**–**12**

A portion of the dried HSE (1.5 g) was fractionated using a Sephadex LH-20 column (1 m × 5 cm) with MeOH as eluent at flow rate 0.5 mL/min. Fractions of 8 mL each were collected, and combined into six major groups (**I**–**VI**) based on their TLC spots (Si-gel, *n*-BuOH–acetic acid–H_2O (60:15:25, *v/v/v*), $CHCl_3$–MeOH–H_2O (7:3:0.3, *v/v/v*). Fractions **I**, **III** and **V**–**VI** were purified by RP-HPLC on a C_8 column (flow rate 2.0 mL/min) with the elution solvent MeOH/H_2O 4:6 *v/v*. Fraction **I** (545.0 mg) yielded compounds **5** (6.3 mg, t_R = 15 min), and **2** (33.4 mg, t_R = 26 min), while fraction **III** (99.0 mg) afforded compounds **6** (9.2 mg, t_R = 20 min), and **7** (1.3 mg, t_R = 32 min). Fraction **V** (38.8 mg) consisted of compounds **8** (1.8 mg, t_R = 8 min), and **9** (0.2 mg, t_R = 14 min). Fraction **VI** (105.2 mg) gave compound **10** (2.8 mg, t_R = 42 min). Fraction **II** (114.8 mg) was separated by RP-HPLC on a C_8 column (flow rate 1.5 mL/min) using as solvent system MeOH/H_2O 4:6 *v/v* to afford compounds **11** (5.4 mg, t_R = 8 min), **12** (3.2 mg , t_R = 14 min), **2** (2.3 mg, t_R = 24 min), and **1** (2.5 mg, t_R = 54 min). Finally, fraction **IV** (53.8 mg) was purified by RP-HPLC using MeOH/H_2O 5:5 *v/v* on a C_{18} column (flow rate 2.0 mL/min) to obtain compounds **4** (1.7 mg, t_R = 14 min), and **3** (2.0 mg, t_R = 19 min).

3.5. Spectroscopic Data

Lawsonicin (**1**): NMR and optical rotation data were consistent with those previously reported [17]. ESI-MS (positive mode), *m/z* 361.4 [M + H]$^+$. Cedrusin (**2**): NMR and optical rotation data were consistent with previously reported [18]. ESI-MS (positive mode), *m/z* 347.3 [M + H]$^+$. Balanophonin (**3**): NMR data were consisted with previously reported [19]. ESI-MS (positive mode), *m/z* 357.3 [M + H]$^+$. Ficusal (**4**): NMR and optical rotation data were consistent with those previously reported [20]. ESI-MS (positive mode), *m/z* 331.1 [M + H]$^+$. Dihydroconyferyl alcohol (**5**): NMR data were consistent with previously reported [21]. ESI-MS (positive mode), *m/z* 183.2 [M + H]$^+$. Veratric acid (**6**): NMR data were consistent with previously reported [22]. ESI-MS (negative mode), *m/z* 181.1 [M − H]$^-$. Vanillic acid (**7**). NMR data were consisted with previously reported [17]. ESI-MS (negative mode), *m/z* 167.1 [M − H]$^-$. Gallic acid (**8**) and methyl gallate (**9**): NMR data were in agreement with those previously reported [23]. ESI-MS (negative mode), *m/z* 169.1 [M − H]$^-$ and 183.1 [M – H]$^-$, respectively. Carpinontriol B (**10**): NMR and optical rotation data were consistent with those previously reported [26]. ESI-MS (positive mode), *m/z* 344.1 [M + H]$^+$. C-veratroylglycol (**11**): NMR data were in agreement with those previously reported [24]. ESI-MS (positive mode), *m/z* 213.3 [M + H]$^+$. β-hydroxypropiovanillone (**12**): NMR data were in agreement with those previously reported [25]. ESI-MS (positive mode), *m/z* 197.0 [M + H]$^+$.

3.6. Quantitative Determination of Total Phenol Content

Total phenolic content (TPC) of hazelnut shells extract (HSE) was determined using the Folin–Ciocalteu colorimetric method [4]. TPC was expressed as gallic acid equivalents (GAE) mg/g of dried HSE (means ± standard deviation of three determinations).

3.7. Quantitative HPLC Analysis of HSE

Quantitative HPLC was carried out using as eluent system H_2O (solvent A) and MeOH (solvent B). The solvent gradient was as follows: $0{\rightarrow}3$ min, 5% B; $3{\rightarrow}7$ min, $5\%{\rightarrow}30\%$ B; $7{\rightarrow}17$ min, 30% B; $17{\rightarrow}35$ min, $30\%{\rightarrow}50\%$ B, $40{\rightarrow}50$ min 100% B. Elution was performed with a flow rate of 0.8 mL/min, injection volume of 20 μL, and DAD detector set at 230 nm. Analysis was carried out in triplicate. Lawsonicin (**1**), C-veratroylglycol (**11**), and cedrusin (**2**) (isolated from HSE and characterized by NMR, and MS data) were used to prepare three solutions at different concentration levels in the range 0.25–1.00 mg/mL for compounds **1** and **11**, and 0.25–2.00 mg/mL for **2**. The peak associated with each compound was identified by comparison of the retention times, and confirmed by co-injection of HSE with isolated compounds. Peak areas of isolated compounds lawsonicin (**1**), cedrusin (**2**), carpinontriol B (**10**), C-veratroylglycol (**11**), and β-hydroxypropiovanillone (**12**) (at each concentration) were plotted against the corresponding standard concentrations (mg/mL) using linear regression to generate standard curves (regression equation $y = 30885.7x - 1704.1$, $r = 0.9989$ for **1**; $y = 16723.0x - 1348.2$, $r = 0.9997$ for **2**; $y = 11109x + 213.7$, $r = 0.9980$ for **10**; $y = 8761.8x + 104.0$, $r = 1.0000$ for **11**, $y = 14914x - 460.22$, $r = 0.9984$ for **12**, where y is the peak area and x the concentration). HSE was dissolved in MeOH at 10 mg/mL, and analyzed under the same chromatographic conditions.

3.8. Antioxidant Activity

The radical scavenging activities of HSE and compounds **1–12** were assayed using stable 1,1-diphenyl-2-picrylhydrazyl radical (DPPH), according to our procedures previously reported [4]. Briefly, 1.5 mL of DPPH solution (25 mg/mL in methanol, prepared daily) was added to 0.375 mL of various concentrations, in MeOH solution, of each sample under investigation (ranged from 12 to 100 μg/mL). The mixtures were kept in the dark for 10 min at room temperature and the decrease in absorbance was measured at 517 nm against a blank consisting of an equal volume of methanol. α-Tocopherol was used as positive control. The DPPH concentration in the reaction medium was calculated from a calibration curve (range = 5–36 μg/mL) analyzed by linear regression ($y = 0.0228x - 0.0350$, $R^2 = 0.9999$), and EC_{50} (mean effective scavenging concentration) was determined as the concentration (in micrograms per milliliter) of sample necessary to decrease the initial DPPH concentration by 50%. All tests were performed in triplicate.

3.9. Cell Cultures

Human malignant melanoma (A375), and Human cervical cancer (HeLa) cell lines were grown at 37 °C in Dulbecco's modified Eagle's medium containing high glucose supplemented with 10% fetal calf serum, and 100 units/mL each of penicillin and streptomycin, and 2 mmol/L glutamine. Human melanoma (SK-Mel-28) cell line was grown at 37 °C in minimum essential medium (MEM) supplemented with 10% fetal calf serum and 100 units/mL each of penicillin and streptomycin. At the onset of each experiment, cells were placed in fresh medium and then cultured in the presence of different concentrations of HSE or its constituents. The experiments were repeated three times.

3.10. Cell Viability Assay

To perform the assay, the cells were grown in 96-well plates, in numbers of 7000 per well and after 24 h were treated with increasing concentrations of HSE from 10 μg/mL to 1 mg/mL and with isolated compounds from 10 nM to 500 μM, in triplicate for a given time (24 and 48 h). At the end of treatment, the plates were centrifuged at 1200 rpm for 5 minutes, the medium was aspirated and added 100 μL of 1 mg/mL MTT (3-[4,5-dimetiltiazol-2,5-diphenyl-2H-tetrazolium bromide]) to each well and the plates were kept at 37 °C for the time necessary to the formation of salt formazan (1–3 h depending on cell type). The solution was then removed from each well, and the formazan crystal within the cells were dissolved with 100 μL of DMSO. Absorption at 550 nm for each well was assessed by a Multiskan Spectrum Thermo Electron Corporation Reader. IC_{50} values were calculated from cell viability dose–response curves and defined as the concentration resulting in 50% inhibition of cell survival compared to untreated cells.

3.11. Flow Cytometry Analysis

Apoptosis was analyzed by propidium iodide incorporation in permeabilized cells and flow cytometry [17,40]. After 24 h of culture in 24-wells plates, cancer cells (5×10^4) were treated with HSE or compounds at different doses, and re-cultured for 24 or 48 h. The apoptosis analysis was carried out in permeabilized cells labelled with propidium iodide (PI) by incubation at 4 °C for 30 min with a solution containing 0.1% sodium citrate, 0.1% Triton X-100 and 50 mg/mL PI. Subsequently, the cancer cells were analyzed by flow cytometry by a FACSCalibur flow cytometer (Becton Dickinson, North Ryde, NSW, Australia). Each experiment was repeated three times.

3.12. Western Blotting Analysis

Cells were lysed in modified RIPA buffer (Tris-HCl pH 7.4 10 mM, NaCl 150 mM, EDTA 1 mM, NP40 1% Na-deoxycholic 0.1%, PMSF 1 mM, protease inhibitor cocktail). Equal amounts of proteins were separated by 10%–12% SDS-PAGE and blotted on ECl Hybond nitro-cellulose membranes (GE Healthcare, Buckinghamshire, UK). Blots were blocked in PBS containing 10% non-fat dry milk and 0.1% Tween-20 and incubated overnight with optimal dilutions of PARP-1 (F-2) antibody for detection of full-lenght and the C-terminal cleavage product (95 kDa) of PARP-1. Anti-mouse IgG HRP conjugated were used as secondary antibody, bands were visualized by autoradiography of ECL reaction (Pierce, Thermo Scientific, Rockford, IL, USA), and anti α-tubulin antibody were used as control for equal amounts of proteins loaded on the gel.

3.13. Statistical Analysis

All results are shown as mean ± standard deviation of three experiments performed in triplicate. Statistical comparison between groups were made using ANOVA followed by the Bonferroni parametric test. Differences were considered significant when $p < 0.05$.

4. Conclusions

Few chemical and biological studies on hazelnut shells, a waste product of industrial food processing, have been reported in the literature until now. The present research contributes to further understand the composition and bioactivity of hazelnut shells. Neolignans, dihydro[*b*]benzofuran-type (lawsonicin, cedrusin, balanophonin, and ficusal), phenolic derivatives (dihydroconyferil alcohol, veratric, vanillic and gallic acids, methyl gallate, C-veratroylglycol, and β-hydroxypropiovanillone), and a cyclic diarylheptanoid (carpinontriol B) are the main constituents of the hazelnut methanol extract and these phytochemicals, with the exception of vanillic and gallic acids, are found in hazelnut for the first time. The extract exhibited an in vitro significant free-radical scavenging activity that was mainly due to gallic acid and its methyl ester. Both compounds were proven to be potential free-radical scavengers in the methanol extracts. The hazelnut extract, some neolignans, cedrusin (**2**) and balanophonin (**3**) and gallic acid (**8**) are able to inhibit the growth of human cancer cells (primary melanoma, A375, metastatic melanoma, SK-Mel-28, and cervical cancer, HeLa) inducing apoptosis mediated by caspase-3 activation and PARP-1 cleavage. Thus, extracts from hazelnut shells might be useful as health-promoting ingredients potentially expandable in functional foods, nutraceuticals or dietary supplements.

Author Contributions: Tiziana Esposito performed the isolation and characterization of compounds; Francesca Sansone analyzed the data; Silvia Franceschelli performed the experiments in cancer cell lines; Pasquale Del Gaudio supported the literature studies; Patrizia Picerno assessed the radical scavenging activity. Teresa Mencherini and Rita Patrizia Aquino conceived and designed the experiments and wrote the paper.

Conflicts of Interest: The authors declare no conflict of interest.

References

1. FAOstat Agriculture Data. Available online: http://faostat3.fao.org (accessed on 12 May 2016).
2. Madesis, P.; Ganopoulos, I.; Bosmali, I.; Tsaftaris, A.; Barcode, H. Resolution Melting analysis for forensic uses in nuts: A case study on allergenic hazelnuts (*Corylus avellana*). *Food Res. Int.* **2013**, *50*, 351–360. [CrossRef]
3. Ciarmiello, L.F.; Mazzeo, M.F.; Minasi, P.; Peluso, A.; de Luca, A.; Piccirillo, P.; Siciliano, R.A.; Carbone, V. Analysis of Different European Hazelnut (*Corylus avellana* L.) Cultivars: Authentication, Phenotypic Features, and Phenolic Profiles. *J. Agric. Food Chem.* **2014**, *62*, 6236–6246. [CrossRef] [PubMed]
4. Piccinelli, A.L.; Pagano, I.; Esposito, T.; Mencherini, T.; Porta, A.; Petrone, A.M.; Gazzerro, P.; Picerno, P.; Snsone, F.; Rastrelli, L.; et al. HRMS profile of a hazelnut skin proanthocyanidin-rich fraction with antioxidant and anti-*Candida albicans* activities. *J. Agric. Food Chem.* **2016**, *64*, 585–595. [CrossRef] [PubMed]
5. Wijngaard, H.; Hossain, M.B.; Rai, D.K.; Brunton, N. Techniques to extract bioactive compounds from food by-products of plant origin. *Food Res. Int.* **2012**, *46*, 505–513. [CrossRef]
6. Kroyer, G. Impact of Food Processing on the Environment an Overview. *LWT Food Sci. Technol.* **1995**, *28*, 547–552. [CrossRef]
7. Stévigny, C.; Rolle, L.; Valentini, N.; Zeppa, G. Optimization of extraction of phenolic content from hazelnut shell using response surface methodology. *J. Sci. Food Agric.* **2007**, *87*, 2817–2822. [CrossRef]
8. Contini, M.; Baccelloni, M.; Massantini, R.; Anelli, G. Extraction of natural antioxidants from hazelnut (*Corylus avellana* L.) shell and skin wastes by long maceration at room temperature. *Food Chem.* **2008**, *110*, 659–669. [CrossRef]
9. Delgado, T.; Malheiro, R.; Pereira, J.A.; Ramalhosa, E. Hazelnut (*Corylus avellana* L.) kernels as a source of antioxidants and their potential in relation to other nuts. *Ind. Crops Prod.* **2010**, *32*, 621–626. [CrossRef]
10. Shahidi, F.; Alasalvar, C.; Liyana-Pathirana, C.M. Antioxidant phytochemicals in hazelnut kernel (*Corylus avellana* L.) and hazelnut byproducts. *J. Agric. Food Chem.* **2007**, *55*, 1212–1220. [CrossRef] [PubMed]
11. Kammerer, D.R.; Kammerer, J.; Valet, R.; Carle, R. Recovery of polyphenols from the by-products of plant food processing and application as valuable food ingredients. *Food Res. Int.* **2014**, *65*, 2–12. [CrossRef]
12. Alasalvar, C.; Karamać, M.; Kosińska, A.; Rybarczyk, A.; Shahidi, F.; Amarowicz, R. Antioxidant activity of hazelnut skin phenolics. *J. Agric. Food Chem.* **2009**, *57*, 4645–4650. [CrossRef] [PubMed]
13. Li, A.; Li, S.; Zhang, Y.; Xu, X.; Chen, Y.; Li, H. Resources and Biological Activities of Natural Polyphenols. *Nutrients* **2014**, *6*, 6020–6047. [CrossRef] [PubMed]
14. Fantini, M.; Benvenuto, M.; Masuelli, L.; Frajese, G.V.; Tresoldi, I.; Modesti, A.; Bei, R. In Vitro and in Vivo Antitumoral Effects of Combinations of Polyphenols, or Polyphenols and Anticancer Drugs: Perspectives on Cancer Treatment. *Int. J. Mol. Sci.* **2015**, *16*, 9236–9282. [CrossRef] [PubMed]
15. Di Domenico, F.; Foppoli, C.; Coccia, R.; Perluigi, M. Antioxidants in cervical cancer: Chemopreventive and chemotherapeutic effects of polyphenols. *Biochim. Biophys. Acta* **2012**, *1822*, 737–747. [CrossRef] [PubMed]
16. Ciemniewska-Żytkiewicz, H.; Verardo, V.; Pasini, F.; Bryś, J.; Koczoń, P.; Caboni, M.F. Determination of lipid and phenolic fraction in two hazelnut (*Corylus avellana* L.) cultivars grown in Poland. *Food Chem.* **2015**, *168*, 615–622. [CrossRef] [PubMed]
17. Mencherini, T.; Picerno, P.; Festa, M.; Russo, P.; Capasso, A.; Aquino, R. Triterpenoid constituents from the roots of *Paeonia rockii* ssp. *rockii*. *J. Nat. Prod.* **2011**, *74*, 2116–2121. [CrossRef] [PubMed]
18. Kim, T.H.; Ito, H.; Hayashi, K.; Hasegawa, T.; Machiguchi, T.; Yoshid, T. Aromatic Constituents from the Heartwood of *Santalum album* L. *Chem. Pharm. Bull.* **2005**, *53*, 641–644. [CrossRef] [PubMed]
19. Haruna, M.; Koube, T.; Ito, K.; Murata, H. Balanophonin, a new neo-lignan from *Balanophora japonica* Makino. *Chem. Pharm. Bull.* **1982**, *30*, 1525–1527. [CrossRef]
20. Li, Y.C.; Kuo, Y.H. Four new compounds, ficusal, ficusesquilignan A, B, and ficusolide diacetate from the heartwood of *Ficus microcarpa*. *Chem. Pharm. Bull.* **2000**, *48*, 1862–1865. [CrossRef] [PubMed]
21. Huang, Y.-H.; Zeng, W.M.; Li, G.Y.; Liu, G.Q.; Zhao, D.D.; Wang, J.; Zhang, Y.L. Characterization of a New Sesquiterpene and Antifungal Activities of Chemical Constituents from *Dryopteris fragrans* (L.) Schott. *Molecules* **2014**, *19*, 507–513. [CrossRef] [PubMed]
22. Crestini, C.; Caponi, M.C.; Argyropoulosc, D.S.; Saladino, R. Immobilized methyltrioxo rhenium (MTO)/H_2O_2 systems for the oxidation of lignin and lignin model compounds. *Bioorg. Med. Chem.* **2006**, *14*, 5292–5302. [CrossRef] [PubMed]

23. Picerno, P.; Mencherini, T.; Sansone, F.; del Gaudio, P.; Granata, I.; Porta, A.; Aquino, R.P. Screening of a polar extract of *Paeonia rockii*: Composition and antioxidant and antifungal activities. *J. Ethnopharm.* **2011**, *138*, 705–712. [CrossRef] [PubMed]
24. Li, L.; Seeram, N.P. Maple syrup phytochemicals include lignans, coumarins, a stilbene, and other previously unreported antioxidant phenolic compounds. *J. Nat. Prod.* **2010**, *58*, 11673–11679. [CrossRef] [PubMed]
25. Karonen, M.; Haemaelaeinen, M.; Nieminen, R.; Klika, K.D.; Loponen, J.; Ovcharenko, V.V.; Moilanen, E.; Pihlaja, K. Phenolic extracts from the bark of *Pinus sylvestris* L. and their effects on inflammatory mediators nitric oxide and prostaglandin E2. *J. Agric. Food Chem.* **2004**, *52*, 7532–7540. [CrossRef] [PubMed]
26. Lee, J.S.; Kim, H.J.; Park, H.; Lee, Y.S. New Diarylheptanoids from the Stems of *Carpinus cordata*. *J. Nat. Prod.* **2002**, *65*, 1367–1370. [CrossRef] [PubMed]
27. Kerbab, K.; Mekhelfi, T.; Zaiter, L.; Benayache, S.; Benayache, F.; Picerno, P.; Mencherini, T.; Sansone, F.; Aquino, R.P.; Rastrelli, L. Chemical composition and antioxidant activity of a polar extract of *Thymelaea microphylla* Coss. et Dur. *Nat. Prod. Res.* **2015**, *29*, 671–675. [CrossRef] [PubMed]
28. Slatnar, A.; Mikulic-Petkovsek, M.; Stampar, F.; Veberic, R.; Solar, A. Identification and quantification of phenolic compounds in kernels, oil and bagasse pellets of common walnut (*Juglans regia* L.). *Food Res. Int.* **2015**, *67*, 255–263. [CrossRef]
29. Balasundram, N.; Sundram, K.; Samman, S. Phenolic compounds in plants and agri-industrial by-products: Antioxidant activity, occurrence, and potential uses. *Food Chem.* **2006**, *99*, 191–203. [CrossRef]
30. Rice-Evans, C.A.; Miller, N.J.; Paganga, G. Structure-antioxidant activity relationships of flavonoids and phenolic acids. *Free Radic. Biol. Med.* **1996**, *20*, 933–956. [CrossRef]
31. Huang, X.X.; Zhou, C.C.; Li, L.Z.; Peng, Y.; Lou, L.L.; Liu, S.; Li, D.M.; Ikejima, T.; Song, S.J. Cytotoxic and antioxidant dihydrobenzofuran neolignans from the seeds of *Crataegus pinnatifida*. *Fitoterapia* **2013**, *91*, 217–223. [CrossRef] [PubMed]
32. Dai, J.; Mumper, R.J. Plant Phenolics: Extraction, Analysis and Their Antioxidant and Anticancer Properties. *Molecules* **2010**, *15*, 7313–7352. [CrossRef] [PubMed]
33. Lo, C.; Lai, T.Y.; Yang, J.H.; Yang, J.S.; Ma, Y.S.; Weng, S.W.; Chen, Y.Y.; Lin, J.G.; Chung, J.G. Gallic acid induces apoptosis in A375.S2 human melanoma cells through caspase-dependent and -independent pathways. *Int. J. Oncol.* **2010**, *37*, 377–385.
34. Chung, C.P.; Hsu, C.Y.; Lin, J.H.; Kuo, Y.H.; Chiang, W.; Lin, Y.L. Antiproliferative lactams and spiroenone from adlay bran in human breast cancer cell lines. *J. Agric. Food Chem.* **2011**, *59*, 1185–1194. [CrossRef] [PubMed]
35. You, B.R.; Park, W.H. The effects of mitogen-activated protein kinase inhibitors or small interfering RNAs on gallic acid-induced HeLa cell death in relation to reactive oxygen species and glutathione. *J. Agric. Food Chem.* **2011**, *59*, 763–771. [CrossRef] [PubMed]
36. Liu, C.; Lin, J.J.; Yang, Z.Y.; Tsai, C.C.; Hsu, J.L.; Wu, Y.J. Proteomic study reveals a co-occurrence of gallic acid-induced apoptosis and glycolysis in B16F10 melanoma cells. *J. Agric. Food Chem.* **2014**, *62*, 11672–11680. [CrossRef] [PubMed]
37. Talanian, R.V.; Brady, K.D.; Cryns, V.L. Caspases as Targets for Anti-Inflammatory and Anti-Apoptotic Drug Discovery. *J. Med. Chem.* **2000**, *43*, 3351–3371. [CrossRef] [PubMed]
38. Chaitanya, G.V.; Alexander, J.S.; Babu, P.P. PARP-1 cleavage fragments: Signatures of cell-death proteases in neurodegeneration. *Cell Commun. Signal.* **2010**, *8*, 1–31. [CrossRef] [PubMed]
39. Diamantopoulos, P.T.; Sofotasiou, M.; Papadopoulou, V.; Polonyfi, K.; Iliakis, T.; Viniou, N.A. PARP1-Driven Apoptosis in Chronic Lymphocytic Leukemia. *BioMed Res. Int.* **2014**, *2014*, 106713. [CrossRef] [PubMed]
40. Franceschelli, S.; Moltedo, O.; Amodio, G.; Tajana, G.; Remondelli, P. In the Huh7 Hepatoma Cells Diclofenac and Indomethacin Activate Differently the Unfolded Protein Response and Induce ER Stress Apoptosis. *Open Biochem. J.* **2011**, *5*, 45–51. [CrossRef] [PubMed]

International Journal of
Molecular Sciences

MDPI

Article

Quantitative Determination of Stilbenoids and Dihydroisocoumarins in *Shorea roxburghii* and Evaluation of Their Hepatoprotective Activity

Kiyofumi Ninomiya [1,2], Saowanee Chaipech [1,3], Yusuke Kunikata [1], Ryohei Yagi [1], Yutana Pongpiriyadacha [4], Osamu Muraoka [1,2] and Toshio Morikawa [1,2,*]

[1] Pharmaceutical Research and Technology Institute, Kindai University, 3-4-1 Kowakae, Higashi-osaka, Osaka 577-8502, Japan; ninomiya@phar.kindai.ac.jp (K.N.); chaipechann@hotmail.com (S.C.); yusuke.kunikata@takeda.com (Y.K.); poseidaon.acchi-kocchi-docchi@docomo.ne.jp (R.Y.); muraoka@phar.kindai.ac.jp (O.M.)
[2] Antiaging Center, Kindai University, 3-4-1 Kowakae, Higashi-osaka, Osaka 577-8502, Japan
[3] Faculty of Agro-Industry, Rajamangala University of Technology Srivijaya, Thungyai, Nakhon Si Thammarat 80240, Thailand
[4] Faculty of Science and Technology, Rajamangala University of Technology Srivijaya, Thungyai, Nakhon Si Thammarat 80240, Thailand; yutanap@hotmail.com
* Correspondence: morikawa@kindai.ac.jp; Tel.: +81-6-4307-4306; Fax: +81-6-6729-3577

Academic Editor: Maurizio Battino
Received: 28 December 2016; Accepted: 15 February 2017; Published: 20 February 2017

Abstract: A simultaneous quantitative analytical method for 13 stilbenoids including (−)-hopeaphenol (**1**), (+)-isohopeaphenol (**2**), hemsleyanol D (**3**), (−)-ampelopsin H (**4**), vaticanols A (**5**), E (**6**), and G (**7**), (+)-α-viniferin (**8**), pauciflorol A (**9**), hopeafuran (**10**), (−)-balanocarpol (**11**), (−)-ampelopsin A (**12**), and *trans*-resveratrol 10-C-β-ᴅ-glucopyranoside (**13**), and two dihydroisocoumarins, phayomphenols A₁ (**14**) and A₂ (**15**) in the extract of *Shorea roxburghii* (dipterocarpaceae) was developed. According to the established protocol, distributions of these 15 polyphenols (**1–15**) in the bark and wood parts of *S. roxburghii* and a related plant *Cotylelobium melanoxylon* were evaluated. In addition, the principal polyphenols (**1, 2, 8, 13–15**) exhibited hepatoprotective effects against ᴅ-galactosamine (ᴅ-galN)/lipopolysaccharide (LPS)-induced liver injury in mice at a dose of 100 or 200 mg/kg, p.o. To characterize the mechanisms of action, the isolates were examined in in vitro studies assessing their effects on (i) ᴅ-GalN-induced cytotoxicity in primary cultured mouse hepatocytes; (ii) LPS-induced nitric oxide (NO) production in mouse peritoneal macrophages; and (iii) tumor necrosis factor-α (TNF-α)-induced cytotoxicity in L929 cells. The mechanisms of action of these polyphenols (**1, 2,** and **8**) were suggested to be dependent on the inhibition of LPS-induced macrophage activation and reduction of sensitivity of hepatocytes to TNF-α. However, none of the isolates reduced the cytotoxicity caused by ᴅ-GalN.

Keywords: *Shorea roxburghii*; stilbenoid; dihydroisocoumarin; quantitative analysis; hepatoprotective effect; mechanism of action

1. Introduction

Stilbenoids, a family of polyphenols, are known for the complexity of their structure and diverse biological activities [1]. They occur with a limited but heterogeneous distribution in the plant kingdom. Some phylogenetically distant botanical families (e.g., Gnetaceae, Pinaceae, Cyperaceae, Fabaceae, Dipterocarpaceae, and Vitaceae) are well recognized as rich sources of stilbenoids and their oligomers (oligostilbenoids) [1]. *Trans*-resveratrol (3,5,4′-trihydroxy-*trans*-stilbene), one of the most popular naturally occurring stilbenoids, has been reported to have anti-aging properties as well as beneficial health effects in patients with cancer, cardiovascular, inflammatory, and central nervous system diseases [2–7]. The majority of the oligostilbenoids are regioselectively synthesized by phenoxy radical coupling of resveratrol [8–12]. A Dipterocarpaceae plant *Shorea roxburghii* G. Don is widely distributed in Thailand and its neighboring countries such as Cambodia, India, Laos, Malaysia, Myanmar, and Vietnam. In Thailand, this plant is locally called "Phayom", and its bark has been used as an astringent and preservative for traditional beverages. In the course of our studies on bioactive constituents from *S. roxburghii*, we have isolated the following 13 stilbenoids, (−)-hopeaphenol (1), (+)-isohopeaphenol (2), hemsleyanol D (3), (−)-ampelopsin H (4), vaticanols A (5), E (6), and G (7), (+)-α-viniferin (8), pauciflorol A (9), hopeafuran (10), (−)-balanocarpol (11), (−)-ampelopsin A (12), and *trans*-resveratrol 10-C-β-D-glucopyranoside (13), and two dihydroisocoumarins, phayomphenols A_1 (14) and A_2 (15), from methanol extract of the bark part as they were present in relative abundance (Figure 1) [13,14]. We also revealed that the methanol extract and its constituents displayed anti-hyperlipidemic and anti-diabetogenic effects in olive-oil and sucrose-loaded mice, respectively. These effects were noted in their corresponding target enzymatic inhibitory activities, such as pancreatic lipase, small intestinal α-glucosidase, and lens aldose reductase [13,14]. Furthermore, we have recently reported that several oligostilbenoids (1, 3, and 8) possess more potent anti-proliferative properties than that of a corresponding monomer, *trans*-resveratrol, against SK-MEL-28 human malignant melanoma cells [15]. Thus, the plant *S. roxburghii* is considered a promising abundant resource for these bioactive oligostilbenoids. In this paper, we propose a simple, rapid, and precise analytical method for high performance liguid chromatography (HPLC) simultaneous quantitative determination of 13 stilbenoids (1–13) and two dihydroisocoumarins (14 and 15) in *S. roxburghii* using a one-step sample preparation procedure. In addition, we also describe the hepatoprotective effects of the principal isolates from the bark of *S. roxburghii* as well as their possible mechanisms of action.

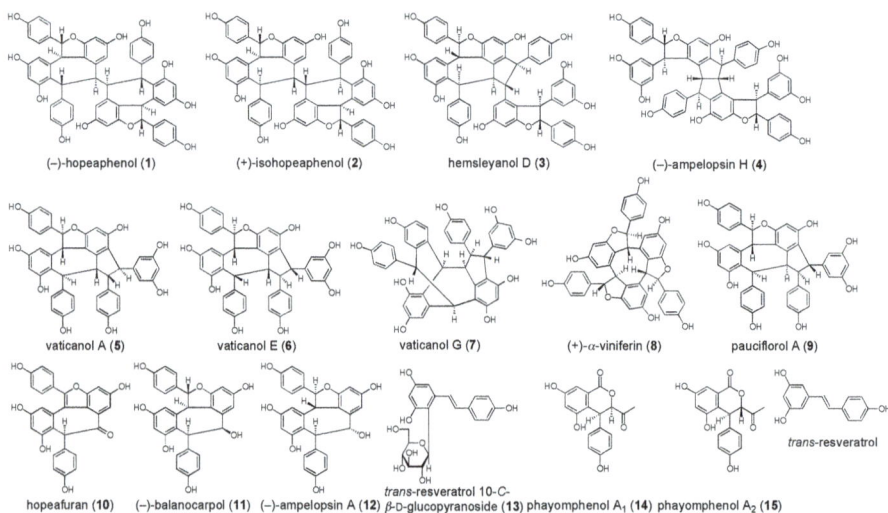

Figure 1. Stilbenoids and dihydroisocoumarins (1–15) from the bark of *Shorea roxburghii*.

2. Results and Discussion

2.1. Isolation

Studies have shown that the principal compounds (1–15) as well as following compounds—vaticanols B and C, malibatols A and B, (+)-parviflorol, *cis*-resveratrol 10-C-β-D-glucopyranoside, *trans*-piceid, and 1'S-dihydrophayomphenol A_2 (Figure S1)—were obtained from the bark of *S. roxburghii* [13,14].

2.2. Simultaneous Quantitative Analysis

As shown in Figure 2, a typical HPLC chromatogram for a standard solution mixture (**1**–**15**) under UV detection (284 nm) demonstrated good baseline separation for all peaks. Each peak was observed at the following retention time (t_R): **1**: 25.5 min; **2**: 26.5 min; **3**: 28.0 min; **4**: 28.1 min; **5**: 23.2 min; **6**: 26.2 min; **7**: 17.7 min; **8**: 30.2 min; **9**: 22.5 min; **10**: 27.0 min; **11**: 19.5 min; **12**: 16.8 min; **13**: 10.3 min; **14**: 14.8 min; **15**: 16.4 min. These peaks were unambiguously assigned by comparing of their retention times with those of authentic specimens.

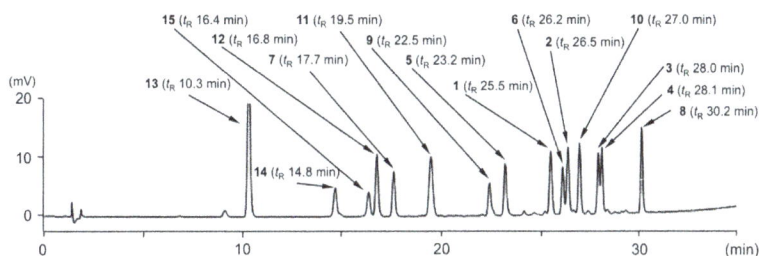

Figure 2. Typical HPLC chromatogram (UV, 284 nm) of standard solution mixture (each 25 μg/mL).

In order to optimize the extraction condition, the quality of the extracts in association with the contents of the stilbenoids (**1**–**13**) and dihydroisocoumarins (**14** and **15**) were examined. The extraction efficacies were compared for three solvent systems (methanol, 50% aqueous methanol, and water) under two different conditions (reflux for 120 min or sonication for 30 min, each twice). As shown in Table 1, "reflux in methanol" afforded the highest total contents of these polyphenols (**1**–**15**). Therefore, all of the analytical samples were prepared using the method "reflux in methanol for 120 min, twice". As shown in Table 2, analytical parameters such as linearity, limits of detection and quantitation, and precision of the developed method were evaluated. The calibration curves were linear in the range studied (2.5–50 μg/mL), showing the correlation coefficients (R^2) of more than 0.9994 for each analyte. Linear regression equations of their calibration curves for each analyte were described in Table 2. The detection and quantitation limits were estimated to be 0.05–0.26 and 0.14–0.80 ng, respectively, indicating sufficient sensitivity of this method. The relative standard deviation (RSD) values were 0.13%–1.54% for intra-day assays and 0.13%–1.58% for inter-day assays. Accuracy was determined in recovery experiments using the methanol under reflux extract from the bark of *S. roxburghii*. As shown in Table 3, overall recovery rate was observed in the range of 95.1%–104.9% with RSD values lower than 1.3%. According to the established protocol, contents of the stilbenoids (**1**–**13**) and dihydroisocoumarins (**14** and **15**) in both bark and wood of *S. roxburghii*, as well as a related plant classified the same Dipterocarpaceae family *Cotylelobium melanoxylon*, which have all been reported to possess the common oligostilbenoids, were evaluated [16] (Figure S2). The assays proved to be reproducible, precise, and readily applicable to the quality evaluation of these extracts. As shown in Table 4, total polyphenol content in the bark of *S. roxburghii* (72.60 mg/g in dry material) was found to be three-fold higher than the wood part (21.20 mg/g). Among them, two resveratrol tetramers, (−)-hopeaphenol (**1**, 13.31 mg/g in dry material) and (+)-isohopeaphenol (**2**, 10.21 mg/g), a resveratrol trimer, vaticanol E (**6**, 11.57 mg/g), and a dihydroisocoumarin, phayomphenol A$_1$ (**13**, 13.81 mg/g) were present relatively in abundance in the bark of *S. roxburghii*. As for *C. melanoxylon*, both of the total oligostilbenoid contents in the bark (286.73 mg/g) and wood (197.50 mg/g) were higher than those of *S. roxburghii*, and their distributions were biased towards vaticanols A (**5**, 76.45 mg/g in the bark), E (**6**, 120.75 mg/g in the bark), and G (**7**, 63.81 mg/g in the bark; 181.69 mg/g in the wood), which is supported by our previous report [16]. On the basis of this evidence, these Dipterocarpaceaeous plants, *S. roxburghii* and *C. melanoxylon*, have been shown to be useful as abundant resources for obtaining the bioactive oligostilbenoids.

Table 1. Extraction efficiency of stilbenoids (**1–13**) and dihydroisocoumarins (**14** and **15**) from the bark of *Shorea roxburghii*.

Extraction Method	Extraction Yield (%)	Contents (mg/g in Dry Material)															Total [a]
		1	2	3	4	5	6	7	8	9	10	11	12	13	14	15	
Methanol, reflux	17.74	13.31	10.21	7.89	2.41	4.84	11.57	1.25	2.35	0.91	0.52	1.71	0.34	13.81	1.03	0.45	72.60 (100)
50% Methanol, reflux	20.21	13.31	10.17	7.90	2.39	4.85	11.56	1.26	2.23	0.92	0.52	1.71	0.31	14.01	1.01	0.39	72.54 (100)
H$_2$O, reflux	12.28	4.72	4.64	1.62	0.57	1.65	3.51	0.65	0.15	0.37	0.17	0.69	0.16	9.66	0.67	0.36	29.59 (41)
Methanol, sonication	13.47	11.46	8.49	6.81	2.00	4.08	9.93	1.04	1.68	0.80	0.43	1.35	0.25	11.00	0.71	0.36	60.39 (83)
50% Methanol, sonication	11.67	8.42	6.47	4.83	1.49	3.06	6.18	0.85	1.12	0.61	0.36	1.00	0.20	10.95	0.65	0.31	46.50 (64)
H$_2$O, sonication	12.71	6.51	6.11	2.07	0.74	2.23	4.93	0.89	n.d. [b]	0.70	0.18	0.80	0.16	10.45	0.63	0.32	36.72 (51)

Extraction efficiency was tested using the bark of *Shorea roxburghii* (loss of drying 7.44%); [a] value (%) relative to the content obtained by methanol under reflux is given in parenthesis; and [b] less than the quantitation limit.

Table 2. Linearities, detection and quantitation limits, and precisions for stilbenoids (**1–13**) and dihydroisocoumarins (**14** and **15**) from the bark of *Shorea roxburghii*.

Analyte	Regression Equation [a]	Correlation Coefficient	Detection Limit [b] (ng)	Quantitation Limit [b] (ng)	Precision [c] (RSD, %)	
					Intra-Day	Inter-Day
(−)-Hopeaphenol (**1**)	$y = 4404 x - 822$	0.9997	0.11	0.34	0.30	0.28
(+)-Isohopeaphenol (**2**)	$y = 4465 x - 850$	0.9998	0.11	0.32	0.90	1.08
Hemsleyanol D (**3**)	$y = 3506 x - 923$	0.9999	0.11	0.34	1.09	0.81
(−)-Ampelopsin H (**4**)	$y = 3743 x - 885$	0.9998	0.11	0.34	0.80	0.78
Vaticanol A (**5**)	$y = 3747 x - 507$	1.000	0.14	0.42	1.32	0.13
Vaticanol E (**6**)	$y = 3009 x - 765$	0.9994	0.14	0.44	0.73	0.91
Vaticanol G (**7**)	$y = 3195 x - 487$	1.000	0.14	0.44	0.49	1.25
(+)-α-Viniferin (**8**)	$y = 4585 x - 166$	0.9999	0.08	0.26	1.54	1.23
Pauciflorol A (**9**)	$y = 2592 x - 441$	1.000	0.22	0.66	0.46	0.78
Hopeafuran (**10**)	$y = 4518 x - 150$	1.000	0.10	0.30	0.84	1.29
(−)-Balanocarpol (**11**)	$y = 5666 x + 325$	0.9999	0.12	0.36	0.46	1.58
(−)-Ampelopsin (**12**)	$y = 4516 x - 427$	1.000	0.10	0.32	0.35	0.32
Trans-resveratrol 10-C-Glc (**13**)	$y = 11461 x - 872$	1.000	0.05	0.14	0.13	0.96
Phayomphenol A$_1$ (**14**)	$y = 2700 x - 564$	0.9999	0.24	0.72	0.93	0.60
Phayomphenol A$_2$ (**15**)	$y = 2059 x - 189$	0.9999	0.26	0.80	0.36	0.93

[a] In the regression equation, x is the concentration of the analyte solution (µg/mL), and y is the peak area of the analyte; [b] values are the amount of the analyte injected on-column; and [c] precision of the analytical method were tested using the methanol under reflux extract of the bark of *Shorea roxburghii* ($n = 5$).

Table 3. Recoveries for stilbenoids (**1–13**) and dihydroisocoumarins (**14** and **15**) from the bark of *Shorea roxburghii*.

Add (µg/mL)	Recovery [a] (%)														
	1	2	3	4	5	6	7	8	9	10	11	12	13	14	15
50	96.1 ± 1.0	97.1 ± 1.1	95.1 ± 1.1	101.1 ± 1.0	98.0 ± 0.9	95.8 ± 0.7	99.8 ± 0.9	100.6 ± 1.1	104.9 ± 1.2	102.7 ± 1.0	96.2 ± 1.3	98.8 ± 0.8	95.7 ± 0.9	98.1 ± 0.5	98.3 ± 1.0
125	104.9 ± 0.9	103.1 ± 1.2	101.6 ± 0.6	100.7 ± 1.0	99.0 ± 1.1	100.7 ± 1.0	96.3 ± 1.1	95.9 ± 1.0	104.3 ± 1.0	103.7 ± 1.2	101.5 ± 0.9	96.5 ± 0.5	101.2 ± 0.5	102.4 ± 0.6	99.4 ± 0.3

[a] The recovery rates were determined by adding analytes of two different concentrations (50 and 125 µg/mL) to the sample solution; recoveries spiked with the methanol under reflux extract of the bark of *S. roxburghii* (*n* = 3).

Table 4. Contents of stilbenoids (**1–13**) and dihydroisocoumarins (**14** and **15**) in the extracts from the bark or wood of *Shorea roxburghii* and *Cotylelobium melanoxylon*.

Extraction Method	Loss of Drying [a] (%)	Extraction Yield [b] (%)	Contents (mg/g in Dry Material)															Total
			1	2	3	4	5	6	7	8	9	10	11	12	13	14	15	
S. roxburghii, bark	7.44	17.74	13.31	10.21	7.89	2.41	4.84	11.57	1.25	2.35	0.91	0.52	1.71	0.34	13.81	1.03	0.45	72.60
S. roxburghii, wood	6.76	8.32	3.39	3.03	1.83	0.51	2.04	3.75	0.40	n.d. [c]	n.d. [c]	n.d. [c]	2.03	0.28	3.94	n.d. [c]	n.d. [c]	21.20
C. melanoxylon, bark	8.96	31.40	n.d. [c]	n.d. [c]	24.59	n.d. [c]	76.45	120.75	63.81	n.d. [c]	1.13	n.d. [c]	n.d. [c]	n.d. [c]	n.d. [c]	n.d. [c]	n.d. [c]	286.73
C. melanoxylon, wood	7.77	27.89	n.d. [c]	n.d. [c]	4.91	n.d. [c]	7.45	3.45	181.69	n.d. [c]	n.d. [c]	n.d. [c]	n.d. [c]	n.d. [c]	n.d. [c]	n.d. [c]	n.d. [c]	197.50

[a] Each powdered sample was dried 105 °C for 8 h; [b] each powdered sample was extracted two times with methanol under reflux for 120 min; and [c] less than the quantiation limit.

*2.3. Protective Effects of Principal Polyphenols (**1**, **2**, **8**, and **13–15**) on Liver Injury Induced by*
D-*GalN/LPS in Mice*

Infection with hepatitis C virus and chronic consumption of alcohol are major causes of liver injury, cirrhosis, and hepatocellular carcinoma worldwide. Tumor necrosis factor-α (TNF-α) is known to mediate organ injuries through its induction of cellular inflammatory responses. In the liver, the biological effects of TNF-α have been implicated in hepatic injuries associated with hepatic toxins, ischemia/reperfusion, viral hepatitis, and alcoholic liver disease or alcohol-related disorders [17–19]. Therefore, TNF-α is considered as an important target in the attempt to discover anti-inflammatory and hepatoprotective agents. The D-GalN/lipopolysaccharide (LPS)-induced liver injury model is recognized to develop via immunological responses [20]. This model causes liver injury in two steps. First, expression of inhibitors of apoptosis proteins (IAPs) is inhibited by administration of D-GalN through depletion of uridine triphosphate in hepatocytes. Second, pro-inflammatory mediators such as nitric oxide (NO), reactive oxygen species (ROS), and TNF-α are released from LPS-activated macrophages (Kupffer's cells). Apoptosis of hepatocytes induced by TNF-α plays an important role in D-GalN/LPS-induced liver injury [21]. In our previous studies on hepatoprotective properties of compounds obtained from natural resources, we have already reported that sesquiterpenes [22–25], triterpenes [26], limonoids [27], coumarins [28], acid amides [29–31], phenylethanoids [32], and saponins [33] exhibited significant protective effects against liver injuries induced by D-GalN/LPS in mice. In the present study, we investigated the protective effects of principal polyphenols, (−)-hopeaphenol (**1**), (+)-isohopeaphenol (**2**), (+)-α-viniferin (**8**), *trans*-resveratrol 10-C-β-D-glucopyranoside (**13**), and phayomphenols A$_1$ (**14**) and A$_2$ (**15**), on the D-GalN/LPS-induced liver injury. As shown in Table 5, all of the tested compounds (**1**, **2**, **8**, and **13–15**) at a dose of 100 or 200 mg/kg, p.o. significantly inhibited the increase in serum levels of aspartate amino transaminase (sAST) and alanine amino transaminase (sALT), which served as markers of acute liver injury [34–36]. A corresponding stilbene monomer resveratrol has been reported to ameliorate hepatotoxicity in several in vivo liver injury models, such as streptozotocin-induced diabetic [37], acetoaminiphen-induced [38], and ethanol-induced oxidative stress models [39]. The hepatoprotective activities of compounds **1**, **2**, **8**, and **13–15** in this model were equivalent to or more potent than *trans*-resveratrol.

Table 5. Inhibitory effects of principal polyphenol constituents (**1**, **2**, **8**, and **13—15**) on D-GalN/lipopolysaccharide (LPS)-induced liver injury in mice.

Treatment	Dose (mg/kg, p.o.)	n	Inhibition (%)	
			sAST	sALT
(−)-Hopeaphenol (**1**)	100	6	92.2 ± 5.4 **	90.6 ± 6.5 **
(+)-Isohopeaphenol (**2**)	100	6	80.6 ± 4.1 **	79.8 ± 3.6 **
(+)-α-Viniferin (**8**)	100	6	70.1 ± 4.0 **	69.5 ± 6.2 **
–	200	6	75.3 ± 5.5 **	71.9 ± 2.6 **
Trans-resveratrol 10-C-Glc (**13**)	100	7	47.8 ± 4.8	43.9 ± 11.8
–	200	7	65.7 ± 3.4 **	67.5 ± 12.3 **
Phayomphenol A$_1$ (**14**)	100	7	48.0 ± 6.0 *	47.0 ± 6.5 *
Phayomphenol A$_2$ (**15**)	100	7	61.0 ± 8.1 **	64.7 ± 6.9 **
Trans-resveratrol	100	7	52.0 ± 9.0 *	51.2 ± 7.9 *
–	200	5	55.7 ± 9.9 *	63.1 ± 5.0 **
Curcumin [26,27]	12.5	10	21.1 ± 20.0	24.0 ± 2.6
–	25	10	47.8 ± 16.0	50.9 ± 14.6
–	50	9	63.8 ± 9.1 *	71.2 ± 7.1 *
Silybin [27]	500	8	71.1 ± 6.8 **	71.9 ± 3.1 **

Each value represents the mean ± SEM; asterisks denote significant differences from the control group, * $p < 0.05$, ** $p < 0.01$; commercial resveratrol was purchased from Wako Pure Chemical Industries, Ltd. (Osaka, Japan), whereas silybin was from Funakoshi Co., Ltd. (Tokyo, Japan).

2.4. Effects on D-GalN-Induced Cytotoxicity in Primary Cultured Mouse Hepatocytes

To characterize the mechanisms responsible for the hepatoprotective activity, the inhibitory effect of 23 polyphenol constituents, including principal polyphenols (**1–15**) isolated from the bark of *S. roxburghii* [13,14], on D-GalN-induced cytotoxicity in primary cultured mouse hepatocytes were examined. As shown in Table 6, none of the isolates led to a reduction in the cytotoxicity at concentrations of up to 100 µM. These results are similar to what was seen with curcumin [22,23,25,27], a naturally occurring hepatoprotective product obtained from turmeric [40]. Thus, the principal polyphenols (**1–15**) did not affect D-GalN-induced cytotoxicity. In contrast, *trans*-resveratrol inhibited the cytotoxicity (IC$_{50}$ = 40.8 µM), which was equivalent to that of silybin (IC$_{50}$ = 38.8 µM) [26,27,30,32], a naturally occurring hepatoprotective product obtained from milk thistle [41,42]. This evidence led us to confirm that *trans*-resveratrol and its oligomers do not have similar effects on D-GalN-induced cytotoxicity in hepatocytes.

Table 6. Inhibitory effects of constituents from the bark of *Shorea roxburghii* on D-GalN-induced cytotoxicity in mouse primary hepatocytes.

Treatment	Inhibition (%)				
	0 µM	3 µM	10 µM	30 µM	100 µM
(−)-Hopeaphenol (**1**)	0.0 ± 0.6	−5.8 ± 1.6	−4.5 ± 1.3	−5.0 ± 0.3	−5.8 ± 1.5
(+)-Isohopeaphenol (**2**)	0.0 ± 1.4	−4.9 ± 0.4	−7.9 ± 1.2	−10.2 ± 0.4	−12.5 ± 0.9
Hemsleyanol D (**3**)	0.0 ± 1.3	−7.2 ± 0.5	−12.1 ± 0.5	−7.8 ± 0.9	−23.2 ± 0.5
(−)-Ampelopsin H (**4**)	0.0 ± 0.2	−7.1 ± 0.7	−7.1 ± 1.2	−10.5 ± 1.3	−17.3 ± 0.3
Vaticanol A (**5**)	0.0 ± 0.9	1.6 ± 0.5	−1.3 ± 0.5	2.2 ± 1.1	−2.0 ± 1.1
Vaticanol E (**6**)	0.0 ± 0.4	−3.1 ± 0.7	−3.0 ± 1.0	−2.4 ± 0.6	−0.3 ± 1.0
Vaticanol G (**7**)	0.0 ± 0.6	−7.5 ± 0.7	−7.0 ± 1.7	−6.6 ± 1.7	−12.7 ± 1.0
(+)-α-Viniferin (**8**)	0.0 ± 2.1	−3.3 ± 1.9	10.1 ± 2.1	32.4 ± 4.0 **	−29.9 ± 0.6
Pauciflorol A (**9**)	0.0 ± 1.4	−1.5 ± 0.9	−5.0 ± 0.7	−6.1 ± 0.4	−10.2 ± 0.4
Hopeafuran (**10**)	0.0 ± 1.4	−6.7 ± 0.4	0.8 ± 1.4	13.0 ± 0.9 **	−22.0 ± 0.6
(−)-Balanocarpol (**11**)	0.0 ± 1.5	−6.4 ± 0.3	−3.1 ± 2.0	11.8 ± 2.7 **	13.5 ± 0.8 **
(−)-Ampelopsin A (**12**)	0.0 ± 1.8	−2.9 ± 0.8	−1.4 ± 1.2	22.5 ± 1.5 **	29.4 ± 1.0 **
Trans-resveratrol 10-C-Glc (**13**)	0.0 ± 2.0	7.1 ± 1.9	15.4 ± 2.5 **	12.8 ± 2.3 **	2.5 ± 0.8
Phayomphenol A$_1$ (**14**)	0.0 ± 1.3	5.1 ± 6.8	12.2 ± 4.8	26.9 ± 3.2 **	42.7 ± 4.3 **
Phayomphenol A$_2$ (**15**)	0.0 ± 2.2	1.7 ± 2.8	13.0 ± 1.9 *	16.1 ± 3.0 **	33.9 ± 3.0 **
Vaticanol B	0.0 ± 1.4	−1.5 ± 0.9	−5.0 ± 0.7	−6.1 ± 0.4	−10.2 ± 0.4
Vaticanol C	0.0 ± 2.1	−5.7 ± 2.0	−4.3 ± 1.5	−5.6 ± 2.2	−12.0 ± 1.3
Malibatol A	0.0 ± 0.7	−1.4 ± 0.9	4.2 ± 0.4	8.9 ± 1.3 *	30.8 ± 2.9 **
Malibatol B	0.0 ± 0.9	−0.8 ± 1.3	−6.1 ± 0.6	−10.9 ± 0.7	−17.0 ± 0.4
(+)-Parviflorol	0.0 ± 1.1	9.1 ± 2.5	21.2 ± 0.6 **	23.8 ± 1.0 **	−20.8 ± 0.2
Cis-resveratrol 10-C-Glc	0.0 ± 0.5	9.1 ± 2.0	20.9 ± 1.3 **	29.6 ± 2.6 **	33.6 ± 1.8 **
Trans-piceid	0.0 ± 3.3	1.3 ± 2.1	13.0 ± 2.6	19.9 ± 6.1 *	33.9 ± 3.6 **
1′S-Dihydrophayomphenol A$_2$	0.0 ± 1.9	−0.8 ± 1.4	−1.2 ± 1.1	14.4 ± 0.5 **	38.0 ± 4.8 **
Trans-resveratrol	0.0 ± 1.7	8.5 ± 0.4	14.1 ± 0.9 **	37.5 ± 3.7 **	57.3 ± 2.5 **
Curcumin [22,23,25,27]	0.0 ± 3.7	0.1 ± 3.8	1.1 ± 2.2	−17.7 ± 1.3	−44.3 ± 0.3
Silybin [26,27,30,32]	0.0 ± 0.3	4.8 ± 1.1	7.7 ± 0.7	45.2 ± 8.8 **	77.0 ± 5.5 **

Each value represents the mean ± SEM (*n* = 4); asterisks denote significant differences from the control group, * *p* < 0.05, ** *p* < 0.01; commercial *trans*-resveratrol was purchased from Wako Pure Chemical Industries, Ltd. (Osaka, Japan), whereas silybin was from Funakoshi Co., Ltd. (Tokyo, Japan).

2.5. Effects on LPS-Induced NO Production in Mouse Peritoneal Macrophages

The effects of the polyphenol constituents from the bark of *S. roxburghii* on NO production were examined to provide an estimation of macrophage activation levels in LPS-treated mouse peritoneal macrophages. As shown in Table 7, resveratrol tetramers, (−)-hopeaphenol (**1**, IC$_{50}$ = 4.6 µM), (+)-isohopeaphenol (**2**, 38.5 µM), (−)-ampelopsin H (**4**, 18.6 µM), and vaticanols B (26.8 µM) and C (14.5 µM), the trimers, (+)-α-viniferin (**8**, 9.7 µM) and pauciflorol A (**9**, 17.8 µM), the dimers, hopeafuran (**10**, 45.9 µM), and malibatols A (23.0 µM) and B (18.5 µM) significantly inhibited NO production without notable cytotoxic effects at the effective concentration. The potencies of the aforementioned oligostilbenoids were higher than those of the NO synthase inhibitor, N^G-monomethyl-L-arginine (L-NMMA, IC_{50} = 36.0 μM). The inhibitory activity of *trans*-resveratrol (17.8 μM) was equivalent to that of caffeic acid phenethyl ester (CAPE, 11.0 μM), an inhibitor of nuclear factor-κB activation.

Table 7. Inhibitory effects of the constituents from bark of *Shorea roxburghiii* on LPS-activated NO production in mouse peritoneal macrophages.

Treatment	Inhibition (%)					IC_{50} (μM)
	0 μM	3 μM	10 μM	30 μM	100 μM	
(−)-Hopeaphenol (1)	0.0 ± 4.0 (100.0 ± 1.6)	42.3 ± 2.5 ** (118.9 ± 3.1)	64.9 ± 3.0 ** (119.1 ± 4.9)	73.9 ± 1.5 ** (130.6 ± 4.5)	80.5 ± 2.3 ** (111.5 ± 6.2)	4.6
(+)-Isohopeaphenol (2)	0.0 ± 2.9 (100.0 ± 3.5)	38.5 ± 4.6 ** (132.6 ± 1.1)	30.7 ± 4.0 ** (129.6 ± 5.6)	41.7 ± 6.3 ** (139.3 ± 4.4)	95.7 ± 1.0 ** (35.2 ± 1.5 #)	38.5
Hemsleyanol D (3)	0.0 ± 3.8 (100.0 ± 2.5)	−4.3 ± 3.7 (120.1 ± 2.5)	8.4 ± 7.8 (135.3 ± 1.8)	34.8 ± 4.4 ** (93.3 ± 1.3)	-34.0 ± 1.8 (32.2 ± 2.9 #)	
(−)-Ampelopsin H (4)	0.0 ± 5.4 (100.0 ± 2.4)	29.2 ± 1.8 ** (102.1 ± 1.0)	39.0 ± 2.8 ** (110.0 ± 4.0)	52.0 ± 3.9 ** (139.0 ± 0.3)	99.7 ± 0.5 ** (85.4 ± 0.9)	18.6
Vaticanol A (5)	0.0 ± 5.0 (100.0 ± 1.3)	30.3 ± 4.5 ** (122.1 ± 4.3)	32.2 ± 5.5 ** (132.2 ± 2.7)	14.5 ± 3.8 (138.4 ± 5.6)	0.5 ± 4.7 (136.2 ± 10.9)	
Vaticanol E (6)	0.0 ± 5.7 (100.0 ± 1.3)	-4.0 ± 8.9 (121.1 ± 2.9)	14.7 ± 5.9 (111.7 ± 7.2)	29.5 ± 5.3 ** (136.2 ± 4.0)	-21.7 ± 5.6 (122.1 ± 10.2)	
Vaticanol G (7)	0.0 ± 3.0 (100.0 ± 2.9)	8.2 ± 5.6 (130.2 ± 5.0)	25.1 ± 2.5 ** (120.3 ± 3.0)	26.7 ± 3.0 ** (123.3 ± 5.4)	44.1 ± 2.3 ** (131.6 ± 9.5)	
(+)-α-Viniferin (8)	0.0 ± 2.9 (100.0 ± 4.3)	27.8 ± 5.7 ** (123.4 ± 8.8)	46.3 ± 3.5 ** (98.8 ± 10.5)	75.4 ± 1.3 ** (87.7 ± 2.9)	97.9 ± 0.6 ** (36.5 ± 3.2 #)	9.7
Pauciflorol A (9)	0.0 ± 4.1 (100.0 ± 4.5)	38.3 ± 3.8 ** (120.7 ± 5.6)	43.5 ± 3.2 ** (129.4 ± 3.8)	53.2 ± 2.2 ** (119.6 ± 2.2)	80.1 ± 1.4 ** (119.0 ± 2.4)	17.8
Hopeafuran (10)	0.0 ± 5.5 (100.0 ± 1.0)	30.6 ± 3.6 ** (125.7 ± 3.8)	33.3 ± 1.8 ** (128.7 ± 3.9)	33.6 ± 5.7 ** (128.9 ± 2.6)	67.7 ± 3.7 ** (117.5 ± 14.1)	45.9
(−)-Balanocarpol (11)	0.0 ± 7.2 (100.0 ± 6.5)	26.1 ± 7.6 ** (129.8 ± 7.4)	41.0 ± 5.2 ** (127.0 ± 4.8)	41.9 ± 5.2 ** (89.6 ± 4.8)	42.5 ± 2.7 ** (101.3 ± 2.9)	
(−)-Ampelopsin A (12)	0.0 ± 6.8 (100.0 ± 4.4)	16.5 ± 2.9 (127.3 ± 4.4)	23.1 ± 10.3 (139.0 ± 6.4)	32.0 ± 2.3 ** (128.3 ± 5.8)	32.9 ± 7.5 ** (116.3 ± 6.1)	
Trans-resveratrol 10-C-Glc (13)	0.0 ± 4.3 (100.0 ± 2.4)	39.0 ± 5.1 ** (126.6 ± 4.4)	43.6 ± 4.4 ** (125.2 ± 9.1)	38.2 ± 4.3 ** (119.8 ± 4.1)	45.4 ± 3.8 ** (113.0 ± 11.0)	
Phayomphenol A_1 (14)	0.0 ± 4.9 (100.0 ± 2.0)	33.6 ± 6.0 ** (111.3 ± 1.1)	40.7 ± 6.5 ** (108.0 ± 1.9)	43.8 ± 4.2 ** (113.0 ± 2.1)	40.3 ± 3.8 ** (107.5 ± 5.3)	
Phayomphenol A_2 (15)	0.0 ± 2.0 (100.0 ± 6.4)	35.2 ± 4.2 ** (102.0 ± 3.8)	38.1 ± 1.4 ** (102.9 ± 7.3)	42.4 ± 1.3 ** (120.9 ± 7.7)	45.4 ± 4.8 ** (104.7 ± 5.6)	
Vaticanol B	0.0 ± 5.4 (100.0 ± 2.0)	28.6 ± 2.6 ** (133.8 ± 2.7)	31.0 ± 4.8 ** (111.5 ± 7.1)	51.9 ± 1.7 ** (99.6 ± 5.1)	75.7 ± 1.0 ** (56.5 ± 2.8 #)	26.8
Vaticanol C	0.0 ± 3.1 (100.0 ± 2.7)	35.4 ± 5.6 ** (124.4 ± 9.7)	45.3 ± 4.2 ** (105.4 ± 8.0)	59.2 ± 2.2 ** (94.2 ± 4.0)	96.5 ± 2.3 ** (11.8 ± 0.8 #)	14.5
Malibatol A	0.0 ± 4.8 (100.0 ± 2.6)	28.5 ± 7.9 ** (133.8 ± 3.7)	42.5 ± 8.6 ** (131.6 ± 3.8)	47.8 ± 4.1 ** (106.3 ± 4.6)	77.3 ± 2.1 ** (96.0 ± 5.2)	23.0
Malibatol B	0.0 ± 5.7 (100.0 ± 41)	39.0 ± 5.7 ** (119.1 ± 3.4)	44.4 ± 4.8 ** (122.3 ± 2.3)	49.7 ± 2.3 ** (105.8 ± 7.0)	80.5 ± 3.3 ** (91.9 ± 4.6)	18.5
(+)-Parviflorol	0.0 ± 8.0 (100.0 ± 6.4)	34.8 ± 2.2 ** (122.8 ± 7.4)	48.6 ± 1.7 ** (130.1 ± 7.4)	49.7 ± 1.6 ** (92.7 ± 8.6)	50.9 ± 1.7 ** (95.7 ± 10.5)	40.9
Cis-resveratrol 10-C-Glc	0.0 ± 2.0 (100.0 ± 3.6)	26.5 ± 2.0 ** (102.9 ± 4.4)	35.5 ± 2.4 ** (109.6 ± 2.9)	47.0 ± 2.8 ** (99.6 ± 5.6)	46.5 ± 2.4 ** (91.8 ± 0.7)	
Trans-piceid	0.0 ± 3.5 (100.0 ± 1.8)	38.6 ± 3.1 ** (91.9 ± 7.7)	41.0 ± 2.7 ** (87.8 ± 5.5)	48.8 ± 2.1 ** (81.5 ± 3.3)	51.4 ± 5.3 ** (66.1 ± 4.8 #)	59.8
1'S-Dihydrophayomphenol A_2	0.0 ± 2.7 (100.0 ± 3.0)	40.1 ± 2.3 ** (117.3 ± 4.0)	40.6 ± 1.0 ** (115.6 ± 5.5)	48.6 ± 1.8 ** (109.9 ± 3.5)	49.4 ± 3.6 ** (113.5 ± 1.8)	
Trans-resveratrol	0.0 ± 2.6 (100.0 ± 1.4)	38.2 ± 6.3 ** (120.4 ± 1.7)	45.6 ± 1.3 ** (128.7 ± 3.6)	78.8 ± 1.1 ** (96.7 ± 10.1)	88.8 ± 1.0 ** (84.4 ± 3.0)	17.8
L-NMMA [26,27,30]	0.0 ± 3.1 (100.0 ± 0.9)	1.4 ± 2.8 (101.1 ± 5.7)	19.9 ± 2.8 ** (100.7 ± 6.2)	43.0 ± 2.1 ** (102.6 ± 4.2)	70.9 ± 1.6 ** (106.4 ± 4.6)	36.0
CAPE [26,27,30]	0.0 ± 2.1 (100.0 ± 1.5)	5.9 ± 5.2 (95.4 ± 0.7)	44.4 ± 3.2 ** (70.0 ± 4.0 #)	86.2 ± 1.1 ** (71.4 ± 6.0 #)	99.6 ± 0.1 ** (53.0 ± 1.4 #)	11.0

Each value represents the mean ± SEM (*n* = 4); asterisks denote significant differences from the control group, * *p* < 0.05, ** *p* < 0.01; # cytotoxic effects were observed, and values in parentheses indicate cell viability (%) in MTT assay; commercial *trans*-resveratrol was purchased from Wako Pure Chemical Industries, Ltd. (Osaka, Japan), whereas L-NMMA and CAPE were from Sigma-Aldrich Chemical Co., LLC. (St. Louis, MO, USA).

2.6. Effects on TNF-α-Induced Cytotoxicity in L929 Cells

The effects of the isolates on the sensitivity of hepatocytes to TNF-α were assessed by measuring TNF-α-induced decreases in the viability of L929 cells, a TNF-α-sensitive cell line, using the MTT assay [27]. As shown in Table 8, oligostilbenoids such as vaticanol G (**7**, IC_{50} = 86.6 μM), (+)-α-viniferin (**8**, 15.0 μM), pauciflorol A (**9**, 26.7 μM), hopeafuran (**10**, 22.0 μM), and malibatols A (**12.3** μM) and B (10.2 μM) inhibited the decrease in cell viability to a greater degree than that of silybin (IC_{50} = 60.4 μM).

Table 8. Inhibitory effects of the constituents from bark of *Shorea roxburghii* on TNF-α-induced cytotoxicity in L929 cells.

Treatment	Inhibition (%)					IC_{50} (μM)
	0 μM	3 μM	10 μM	30 μM	100 μM	
(−)-Hopeaphenol (**1**)	0.0 ± 0.2	3.8 ± 1.5	28.9 ± 0.2 **	−1.8 ± 0.7	−6.7 ± 0.8	
(+)-Isohopeaphenol (**2**)	0.0 ± 0.5	−0.6 ± 1.1	0.3 ± 1.0	69.8 ± 1.4 **	−6.2 ± 0.5	
Hemsleyanol D (**3**)	0.0 ± 1.1	6.7 ± 1.8	17.6 ± 1.7 **	91.5 ± 2.1 **	10.0 ± 1.8	
(−)-Ampelopsin H (**4**)	0.0 ± 0.8	1.0 ± 2.1	−10.7 ± 0.3	−16.4 ± 0.4	−18.9 ± 0.3	
Vaticanol A (**5**)	0.0 ± 0.3	3.6 ± 0.4	4.0 ± 1.2	7.8 ± 1.5	28.9 ± 1.2 **	
Vaticanol E (**6**)	0.0 ± 0.4	6.7 ± 1.1	9.4 ± 0.6	3.7 ± 0.8	4.9 ± 0.5	
Vaticanol G (**7**)	0.0 ± 0.4	7.9 ± 0.8	9.7 ± 0.7	12.7 ± 0.6 *	57.6 ± 4.2 **	86.6
(+)-α-Viniferin (**8**)	0.0 ± 1.2	12.2 ± 1.2 *	38.3 ± 0.6 **	84.4 ± 2.5 **	78.5 ± 1.9 **	15.0
Pauciflorol A (**9**)	0.0 ± 1.0	8.2 ± 0.9	16.1 ± 0.7 **	54.6 ± 5.0 **	94.7 ± 4.6 **	26.7
Hopeafuran (**10**)	0.0 ± 0.7	8.3 ± 1.3	17.2 ± 1.3 **	66.4 ± 1.5 **	85.1 ± 1.2 **	22.0
(−)-Balanocarpol (**11**)	0.0 ± 1.1	4.5 ± 0.8	7.5 ± 1.7	10.4 ± 0.9 *	17.7 ± 3.2 **	
(−)-Ampelopsin A (**12**)	0.0 ± 1.1	4.1 ± 1.5	8.4 ± 1.1	14.6 ± 2.1 *	35.5 ± 1.5 **	
Trans-resveratrol 10-C-Glc (**13**)	0.0 ± 1.4	4.5 ± 2.2	4.4 ± 1.9	6.6 ± 1.6	6.6 ± 2.3	
Phayomphenol A_1 (**14**)	0.0 ± 0.7	0.0 ± 0.6	1.8 ± 0.6	1.9 ± 0.6	1.5 ± 0.4	
Phayomphenol A_2 (**15**)	0.0 ± 0.8	0.7 ± 0.7	1.4 ± 0.9	0.6 ± 0.5	−3.1 ± 0.7	
Vaticanol B	0.0 ± 0.7	0.0 ± 0.7	12.1 ± 1.1 *	7.9 ± 0.4	−9.8 ± 0.5	
Vaticanol C	0.0 ± 0.6	9.3 ± 0.3	14.4 ± 1.0 **	94.3 ± 2.6 **	−8.2 ± 0.2	18.2
Malibatol A	0.0 ± 0.5	8.1 ± 0.9	30.7 ± 3.8 **	86.9 ± 2.9 **	81.8 ± 4.3 **	12.3
Malibatol B	0.0 ± 1.2	16.8 ± 0.6 **	37.7 ± 3.3 **	90.7 ± 0.3 **	77.2 ± 2.9 **	10.2
(+)-Parviflorol	0.0 ± 0.8	4.5 ± 0.9	9.4 ± 2.2	13.0 ± 0.4 **	24.3 ± 0.9 **	
Cis-resveratrol 10-C-Glc	0.0 ± 0.7	4.8 ± 0.2	9.7 ± 1.2	14.5 ± 1.7 *	31.6 ± 1.5 **	
Trans-piceid	0.0 ± 0.9	3.4 ± 1.4	6.7 ± 2.6	3.6 ± 1.8	1.6 ± 1.7	
1′S-Dihydrophayomphenol A_2	0.0 ± 0.3	2.0 ± 0.4	3.8 ± 0.5	2.5 ± 1.2	4.7 ± 0.5	
Trans-resveratrol	0.0 ± 1.4	1.4 ± 0.5	1.8 ± 0.8	2.1 ± 1.2	5.3 ± 0.6	
Silybin [27,36]	0.0 ± 2.6	5.3 ± 2.8	22.0 ± 3.8 **	48.0 ± 4.1 **	50.8 ± 3.9 **	60.4

Each value represents the mean ± SEM (*n* = 4); asterisks denote significant differences from the control group, * $p < 0.05$, ** $p < 0.01$.; commercial *trans*-resveratrol was purchased from Wako Pure Chemical Industries, Ltd. (Osaka, Japan), whereas silybin was from Funakoshi Co., Ltd. (Tokyo, Japan).

3. Materials and Methods

3.1. Chemicals and Reagents

Methanol, acetic acid, and distilled water for HPLC were purchased from Nacalai Tesque Inc., Kyoto, Japan. All other chemicals were reagent grade, and were purchased from Wako Pure Chemical Industries, Ltd., Tokyo, Japan or Nacalai Tesque Inc., Kyoto, Japan.

3.2. Plant Materials

The bark and wood parts of *Shorea roxburghii* and *Cotylelobium melanoxylon* were collected from Phatthalung Province, Thailand, in September 2006 or September 2007 as described previously [13,16]. All of the plant materials were identified by one of the authors, Yutana Pongpiriyadacha. Voucher specimens (2006.09. Raj-02, 2006.09. Raj-02#, 2007.09. Raj-04, and 2006.09. Raj-04#, respectively) were deposited in the Garden of Medicinal Plants, Kindai University. The materials were air-dried in a shaded room for more than a month.

3.3. Standard Solution Preparation

An accurately weighed 20.0 mg of each compound (**1–15**) was introduced into a 20 mL volumetric flask, and methanol was added to make up the volume of the stock standard solution (1000 μg/mL). Aliquots of 50, 100, 250, and 500 μL of the stock standard solution were transferred into 10 mL volumetric flasks and the volume was made up with 50% aqueous methanol for use as working solutions (5.0, 10, 25, and 50 μg/mL, respectively) for constructing calibration curves. For calibration, an aliquot of 2 μL of each solution was injected into the HPLC system. Each peak was observed at the following retention times: **1** (t_R 25.5 min), **2** (t_R 26.5 min), **3** (t_R 28.0 min), **4** (t_R 28.1 min), **5** (t_R 23.2 min), **6** (t_R 26.2 min), **7** (t_R 17.7 min), **8** (t_R 30.2 min), **9** (t_R 22.5 min), **10** (t_R 27.0 min), **11** (t_R 19.5 min), **12** (t_R 16.8 min), **13** (t_R 10.3 min), **14** (t_R 14.8 min), and **15** (t_R 16.4 min).

3.4. Sample Solution Preparation

An accurately weighed pulverized sample powder (ca. 2 g, conversion with loss on drying) was extracted with 20 mL of three solvent systems (methanol, 50% aqueous methanol, or water) under two conditions (reflux for 120 min or sonication for 30 min, each twice). The extracts were combined and centrifuged at 3000 rpm for 5 min, then the supernatants were diluted to 100 mL with the extraction solvent. An aliquot (1 mL) of the extract solution was transferred into a 10 mL volumetric flask and 50% aqueous methanol was added to make up the volume. The solution was filtered through a syringe filter (0.45 μm), and an aliquot of 2 μL was subjected to the HPLC analysis. To calculate the extraction yields, the remaining extraction solution (90 mL) was evaporated in vacuo (Table 1).

3.5. HPLC Instruments and Conditions

All analytical experiments in this study were performed with an LC-20A series Prominence HPLC system (Shimadzu Co., Kyoto, Japan), which consists of a UV-VIS detector, a binary pump, a degasser, an autosampler, a thermostated column compartment, and a control module. The chromatographic separation was performed on an Inertsil ODS-3 column (3 μm particle size, 2.1 mm i.d. × 100 mm, GL Sciences Inc., Tokyo, Japan) operated at 40 °C with mobile phase A (acetonitrile) and B (H_2O containing 0.1% formic acid). The gradient program was as follows: 0 min (A:B = 10:90, v/v) → 20 min (30:70, v/v) → 30–40 min (50:50, v/v, hold). The flow rate was 0.2 mL/min with UV detection at 284 nm and the injection volume was 2 μL.

3.6. Calibration and Validation

The standard curves were prepared with four concentration levels in the range of 0.5–50 μg/mL. Standard curves were made on each day of analysis. Linearity for each compound was plotted using linear regression of the peak area versus concentration. The coefficient of correlation (R^2) was used to judge the linearity. The detection and quantitation limits for each analyte were determined by the signal-to-noise (S/N) ratio for each analyte by analyzing a series of diluted standard solutions until the S/N ratios were about 3 and 10, respectively, based on a 2 μL injection. Precision and accuracy of the analytical method were tested using a methanol under reflux extract of the bark of *S. roxburghii*. The intra- and inter-day precisions were determined by estimating the corresponding responses five times on the same day and on five different days, respectively (Table 2). The recovery rates were determined by adding analytes of two different concentrations (50 and 125 μg/mL) to the sample solution of the homogeneous extract (Table 3).

3.7. Reagents for Bioassays

LPS (from *Salmonella enteritidis*), minimum essential medium (MEM), and William's E medium were purchased from Sigma-Aldrich Chemical (St. Louis, MO, USA); fetal bovine serum (FBS) was from Life Technologies (Rockville, MD, USA); other chemicals were from Nakalai Tesque Inc. (Kyoto,

Japan) or Wako Pure Chemical Industries, Co., Ltd. (Osaka, Japan). The 96-well microplates were purchased from Sumitomo Bakelite Co., Ltd. (Tokyo, Japan).

3.8. Animals

Male ddY mice were purchased from Kiwa Laboratory Animal Co., Ltd. (Wakayama, Japan). The animals were housed at a constant temperature of $23 \pm 2\,°C$ and fed a standard laboratory chow (MF, Oriental Yeast Co., Ltd., Tokyo, Japan). All experiments were performed with conscious mice unless otherwise noted. The experimental protocol was approved by Kindai University's Committee for the Care and Use of Laboratory Animals (KAPR-26-001, 1 April 2014).

3.9. Effects on D-GalN/LPS-Induced Liver Injury in Mice

Protective effects on D-GalN/LPS-induced liver injury in mice were determined according to the previously described protocol [27]. *Trans*-resveratrol, curcumin, and silybin were used as reference compounds.

3.10. Effects on Cytotoxicity Induced by D-GalN in Primary Cultured Mouse Hepatocytes

Hepatoprotective effects on D-GalN-induced cytotoxicity in primary cultured mouse hepatocytes were determined according to the protocol described previously [27]. *Trans*-resveratrol, curcumin, and silybin were used as reference compounds.

3.11. Effects on Production of NO in LPS-Induced Mouse Peritoneal Macrophages

Assays for NO production in TGC-induced mouse peritoneal macrophages were performed as described previously [27]. *Trans*-resveratrol, N^G-Monomethyl-L-arginine (L-NMMA), and caffeic acid phenethyl ester (CAPE) were used as reference compounds.

3.12. Effects on Cytotoxicity Induced by TNF-α in L929 Cells

Inhibitory effects on TNF-α-induced cytotoxicity in L929 cells were determined according to the protocol described previously [27]. *Trans*-resveratrol and silybin were used as reference compounds.

3.13. Statistics

Values are expressed as means $\pm$ SEM. One-way analysis of variance (ANOVA) followed by Dunnett's test was used for statistical analysis. Probability (*p*) values less than 0.05 were considered significant.

4. Conclusions

We have developed a practical method for the simultaneous quantitative determination of 13 stilbenoids, (−)-hopeaphenol (**1**), (+)-isohopeaphenol (**2**), hemsleyanol D (**3**), (−)-ampelopsin H (**4**), vaticanols A (**5**), E (**6**), and G (**7**), (+)-α-viniferin (**8**), pauciflorol A (**9**), hopeafuran (**10**), (−)-balanocarpol (**11**), (−)-ampelopsin A (**12**), and *trans*-resveratrol 10-C-β-D-glucopyranoside (**13**), and two dihydroisocoumarins, phayomphenols A_1 (**14**) and A_2 (**15**), in the bark and wood parts of *Shorea roxburghii* and *Cotylelobium melanoxylon*. The method was validated with respect to linearity, detection limit, precision, and accuracy. The assay was reproducible and precise, and could be useful to obtain abundant resources of the bioactive oligostilbenoids. Among the isolates from the bark of *S. roxburghii*, the principal polyphenols (**1**, **2**, **8**, and **13**–**15**) exhibited protective effects against liver injury induced by D-GalN/LPS in mice at a dose of 100 or 200 mg/kg, p.o. The mechanisms of action are likely dependent on inhibition of LPS-induced macrophage activation and a reduction in sensitivity of hepatocytes to TNF-α. They did not affect the reduction of cytotoxicity caused by D-GalN, as summarized in Figure 3. It is well known that activation of NF-κB is a key factor in both activation of macrophages and TNF-α-induced cell death. The detailed mechanisms of action for the hepatoprotective effects, including the influence on NF-κB activation, require further study.

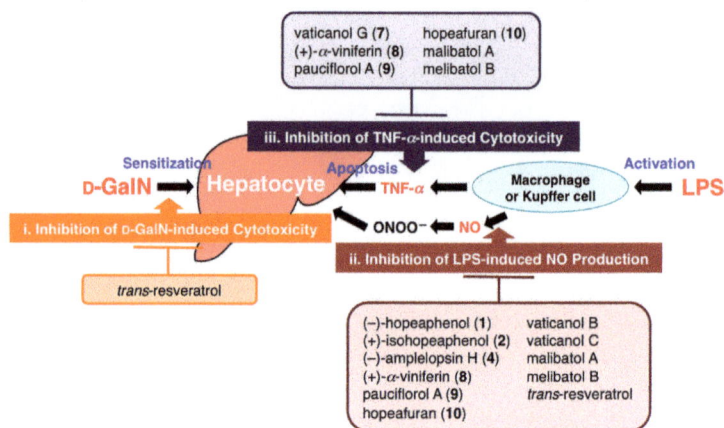

Figure 3. Plausible mechanisms of action of stilbenoids on D-GalN/LPS-induced liver injury.

Supplementary Materials: Supplementary materials can be found at www.mdpi.com/1422-0067/18/2/451/s1.

Acknowledgments: This work was supported by the MEXT-Supported Program for the Strategic Research Foundation at Private Universities, 2014–2018, Japan (S1411037, Toshio Morikawa), as well as JSPS KAKENHI, Japan, Grant Numbers 15K08008 (Toshio Morikawa), 15K08009 (Kiyofumi Ninomiya), and 16K08313 (Osamu Muraoka). Thanks are also due to the Kobayashi International Scholarship Foundation, Japan for the financial support (Toshio Morikawa).

Author Contributions: Kiyofumi Ninomiya, Saowanee Chaipech, Yutana Pongpiriyadacha, Osamu Muraoka, and Toshio Morikawa conceived and designed the experiments; Kiyofumi Ninomiya, Saowanee Chaipech, Yusuke Kunikata, Ryohei Yagi, and Toshio Morikawa performed the experiments; Kiyofumi Ninomiya and Toshio Morikawa analyzed the data; Saowanee Chaipech, Yutana Pongpiriyadacha, and Toshio Morikawa contributed the materials; Kiyofumi Ninomiya and Toshio Morikawa wrote the paper.

Conflicts of Interest: The authors declare no conflict of interest.

References

1. Riviére, C.; Pawlus, A.D.; Mérillon, J.-M. Natural stilbenoids: Distribution in the plant kingdom and chemotaxonomic interest in Vitaceae. *Nat. Prod. Rep.* **2012**, *29*, 1317–1333. [CrossRef] [PubMed]
2. Kovacic, P.; Somanathan, R. Multifaceted approach to resveratrol bioactivity—Focus on antioxidant action, cell signaling and safety. *Oxid. Med. Cell. Longev.* **2010**, *3*, 86–100. [CrossRef] [PubMed]
3. Vang, O.; Ahmad, N.; Baile, C.A.; Baur, J.A.; Brown, K.; Csiszar, A.; Das, D.K.; Delmas, D.; Gottfried, C.; Lin, H.-Y.; et al. What is new for an old molecule? Systematic review and recommendations on the use of resveratrol. *PLoS ONE* **2011**, *6*, e19881. [CrossRef] [PubMed]
4. Tomé-Carneiro, J.; Gonzálvez, M.; Larrosa, M.; Yáñez-Gascón, M.J.; García-Almagro, F.J.; Ruiz-Ros, J.A.; Tomás-Barberán, F.A.; García-Conesa, M.T.; Espín, J.C. Resveratrol in primary and secondary prevention of cardiovascular disease: A dietary and clinical perspective. *Ann. N. Y. Acad. Sci.* **2013**, *1290*, 37–51. [CrossRef] [PubMed]
5. Tomé-Carneiro, J.; Larrosa, M.; González-Sarrías, A.; Tomás-Barberán, F.A.; García-Conesa, M.T.; Espín, J.C. Resveratrol and clinical trials: The crossroad from in vitro studies to human evidence. *Curr. Pharm. Design* **2013**, *19*, 6064–6093. [CrossRef]
6. Lançon, A.; Frazzi, R.; Latruffe, N. Anti-oxidant, anti-inflammatory and anti-angiogenic properties of resveratrol in ocular diseases. *Molecules* **2016**, *21*, 304. [CrossRef] [PubMed]

7. Li, C.; Xu, X.; Tao, Z.; Sun, C.; Pan, Y. Resveratrol derivatives: An updated patent review (2012–2015). *Expert Opin. Ther. Pat.* **2016**, *26*, 1189–1200. [CrossRef] [PubMed]
8. Ito, T. Structures of oligostilbenoids in Dipterocarpaceaeous plants and their biological activities. *Yakugaku Zasshi* **2011**, *131*, 93–100. [CrossRef] [PubMed]
9. Ito, T.; Ito, H.; Nehira, T.; Sawa, R.; Iinuma, M. Structure elucidation of highly condensed stilbenoids: Chiroptical properties and absolute configuration. *Tetrahedron* **2014**, *70*, 5640–5649. [CrossRef]
10. Ito, T.; Hoshino, R.; Iinuma, M. Absolute configuration of resveratrol oligomers isolated from *Hopea utilis*. *Helv. Chim. Acta* **2015**, *98*, 32–46. [CrossRef]
11. Ito, T.; Iinuma, M. Occurrence of non-heterocyclic resveratrol tetramer in *Vatica chinensis*. *Phytochem. Lett.* **2016**, *15*, 37–41. [CrossRef]
12. Ito, T.; Hara, Y.; Kubota, Y.; Sawa, R.; Iinuma, M. Absolute structure of resveratrol hexamers in Dipterocarpaceaeous plants. *Tetrahedron* **2016**, *72*, 891–899. [CrossRef]
13. Morikawa, T.; Chaipech, S.; Matsuda, H.; Hamao, M.; Umeda, Y.; Sato, H.; Tamura, H.; Ninomiya, K.; Yoshikawa, M.; Pongpiriyadacha, Y.; et al. Anti-hyperlipidemic constituents from the bark of *Shorea roxburghii*. *J. Nat. Med.* **2012**, *66*, 516–524. [CrossRef] [PubMed]
14. Morikawa, T.; Chaipech, S.; Matsuda, H.; Hamao, M.; Umeda, Y.; Sato, H.; Tamura, H.; Kon'i, H.; Ninomiya, K.; Yoshikawa, M.; et al. Antidiabetogenic oligostilbenoids and 3-ethyl-4-phenyl-3,4-dihydroisocoumarins from the bark of *Shorea roxburghii*. *Bioorg. Med. Chem.* **2012**, *20*, 823–840. [CrossRef] [PubMed]
15. Moriyama, H.; Moriyama, M.; Ninomiya, K.; Morikawa, T.; Hayakawa, T. Inhibitory effects of oligostilbenoids from the bark of *Shorea roxburghii* on malignant melanoma cell growth: Implications for novel topical anticancer candidates. *Biol. Pharm. Bull.* **2016**, *39*, 1675–1682. [CrossRef] [PubMed]
16. Matsuda, H.; Asao, Y.; Nakamura, S.; Hamao, M.; Sugimoto, S.; Hongo, M.; Pongpiriyadacha, Y.; Yoshikawa, M. Antidiabetogenic constituents from the Thai traditional medicine *Cotylelobium melanoxylon*. *Chem. Pharm. Bull.* **2009**, *57*, 487–494. [CrossRef] [PubMed]
17. Wang, Y.; Singh, R.; Lefkowitch, J.H.; Rigoli, R.M.; Czaja, M.J. Tumor necrosis factor-induced toxic liver injury results from JNK2-dependent activation of caspase-8 and the mitochondrial death pathway. *J. Biol. Chem.* **2006**, *281*, 15258–15267. [CrossRef] [PubMed]
18. Tilg, H.; Day, C.P. Management strategies in alcoholic liver disease. *Nat. Rev. Gastroenterol. Hepatol.* **2007**, *4*, 24–34. [CrossRef] [PubMed]
19. Seronello, S.; Sheikh, M.Y.; Choi, J. Redox regulation of hepatitis C in nonalcoholic and alcoholic liver. *Free Rad. Biol. Med.* **2007**, *43*, 869–882. [CrossRef] [PubMed]
20. Freudenberg, M.A.; Galanos, C. Tumor necrosis factor α mediates lethal activity of killed Gram-negative and Gram-positive bacteria in D-galactosamine-treated mice. *Infect. Immun.* **1991**, *59*, 2110–2115. [PubMed]
21. Josephs, M.D.; Bahjat, F.R.; Fukuzuka, K.; Ksontini, R.; Solorzano, C.C.; Edwards, C.K., 3rd; Tannahill, C.L.; MacKay, S.L.; Copeland, E.M., 3rd; Moldawer, L.L. Lipopolysaccharide and D-galactosamine-induced hepatic injury is mediated by TNF-α and not by Fas ligand. *Am. J. Physiol. Regul. Integr. Comp. Physiol.* **2000**, *278*, R1196–R1201. [PubMed]
22. Matsuda, H.; Ninomiya, K.; Morikawa, T.; Yoshikawa, M. Inhibitory effect and action mechanism of sesquiterpenes from Zedoariae Rhizoma on D-galactosamine/lipopolysaccharide-induced liver injury. *Bioorg. Med. Chem. Lett.* **1998**, *8*, 339–344. [CrossRef]
23. Matsuda, H.; Morikawa, T.; Ninomiya, K.; Yoshikawa, M. Hepatoprotective constituents from Zedoariae Rhizoma: Absolute stereostructures of three new carabrane-type sesquiterpenes, curcumenolactonea A, B, and C. *Bioorg. Med. Chem.* **2001**, *9*, 909–916. [CrossRef]
24. Morikawa, T.; Matsuda, H.; Ninomiya, K.; Yoshikawa, M. Medicinal foodstuffs. XXIX. Potent protective effects of sesquiterpenes and curcumin from Zedoariae Rhizoma on liver injury induced by D-galactosamine/lipopolysaccharide or tumor necrosis factor-α. *Biol. Pharm. Bull.* **2002**, *25*, 627–631. [CrossRef] [PubMed]
25. Morikawa, T. Search for bioactive constituents from several medicinal food: Hepatoprotective, antidiabetic, and antiallergic activities. *J. Nat. Med.* **2007**, *61*, 112–126. [CrossRef]
26. Morikawa, T.; Ninomiya, K.; Imura, K.; Yamaguchi, T.; Aakgi, Y.; Yoshikawa, M.; Hayakawa, T.; Muraoka, O. Hepatoprotective triterpene from traditional Tibetan medicine *Potentilla anserina*. *Phytochemistry* **2014**, *102*, 169–181. [CrossRef] [PubMed]

27. Ninomiya, K.; Miyazawa, S.; Ozeki, K.; Matsuo, N.; Muraoka, O.; Kikuchi, T.; Yamada, T.; Tanaka, R.; Morikawa, T. Hepatoprotective limonoids from Andiroba (*Carapa guianensis*). *Int. J. Mol. Sci.* **2016**, *17*, 591. [CrossRef] [PubMed]

28. Yoshikawa, M.; Nishida, N.; Ninomiya, K.; Ohgushi, T.; Kubo, M.; Morikawa, T.; Matsuda, H. Inhibitory effects of coumarin and acetylene constituents from the roots of *Angellica furcijuga* on D-galactosamine/lipopolysaccharide-induced liver injury in mice and on nitric oxide production in lipopolysaccharide-activated mouse peritoneal macrophages. *Bioorg. Med. Chem.* **2006**, *14*, 456–463. [CrossRef] [PubMed]

29. Matsuda, H.; Ninomiya, K.; Morikawa, T.; Yasuda, D.; Yamaguchi, I.; Yoshikawa, M. Protective effects of amide constituents from the fruit of *Piper chaba* on D-galactosamine/TNF-α-induced cell death in mouse hepatocytes. *Bioorg. Med. Chem. Lett.* **2008**, *18*, 2038–2042. [CrossRef] [PubMed]

30. Matsuda, H.; Ninomiya, K.; Morikawa, T.; Yasuda, D.; Yamaguchi, I.; Yoshikawa, M. Hepatoprotective amide constituents from the fruit of *Piper chaba*: Structural requirements, mode of action, and new amides. *Bioorg. Med. Chem.* **2009**, *17*, 7313–7323. [CrossRef] [PubMed]

31. Morikawa, T. Search for TNF-α sensitivety degradation principles from medicinal foods—Hepatoprotective amide constituents from Thai natural medicine *Piper chaba*. *Yakugaku Zasshi* **2010**, *130*, 785–791. [CrossRef] [PubMed]

32. Morikawa, T.; Pan, Y.; Ninomiya, K.; Imura, K.; Matsuda, H.; Yoshikawa, M.; Yuan, D.; Muraoka, O. Acylated phenylethanoid oligoglycosides with hepatoprotective activity from the desert plant *Cistanche tubulosa*. *Bioorg. Med. Chem.* **2010**, *18*, 1882–1890. [CrossRef] [PubMed]

33. Yoshikawa, M.; Morikawa, T.; Kashima, Y.; Ninomiya, K.; Matsuda, H. Structures of new dammarane-type triterpene saponins from the flower buds of *Panax notoginseng* and hepatoprotective effects of principal ginseng saponins. *J. Nat. Prod.* **2003**, *66*, 922–927. [CrossRef] [PubMed]

34. Osumi, W.; Jin, D.; Imai, Y.; Tashiro, K.; Li, Z.-L.; Otsuki, Y.; Maemura, K.; Komeda, K.; Hirokawa, F.; Hayashi, M.; et al. Recombinant human soluble thrombomodulin improved lipopolysaccharide/ D-galactosamine-induced acute liver failure in mice. *J. Pharmacol. Sci.* **2015**, *129*, 233–239. [CrossRef] [PubMed]

35. Cho, H.I.; Hong, J.M.; Choi, J.W.; Choi, H.S.; Kwak, J.H.; Lee, D.U.; Lee, S.K.; Lee, S.M. β-Caryophyllene alleviates D-galactosamine and lipopolysaccharide-induced hepatic injury through suppression of the TLR4 and RAGE signaling pathways. *Eur. J. Pharmacol.* **2015**, *764*, 613–621. [CrossRef] [PubMed]

36. Sass, G.; Heinlein, S.; Agli, A.; Bang, R.; Schümann, J.; Tiegs, G. Cytokine expression in three mouse models of experimental hepatitis. *Cytokine* **2002**, *19*, 115–120. [CrossRef] [PubMed]

37. Hamadi, N.; Mansour, A.; Hassan, M.H.; Khalifi-Touhami, F.; Badary, O. Ameliorative effects of resveratrol on liver injury in streptozotocin-induced diabetic rats. *J. Biochem. Mol. Toxicol.* **2012**, *26*, 384–392. [CrossRef] [PubMed]

38. Wang, Y.; Jiang, Y.; Fan, X.; Tan, H.; Zeng, H.; Wang, Y.; Chen, P.; Huang, M.; Bi, H. Hepato-protective effect of resveratrol against acetaminophen-induced liver injury is associated with inhibition of CYP-mediated bioactivation and regulation of SIRT1-p53 signaling pathways. *Toxicol. Lett.* **2015**, *236*, 82–89. [CrossRef] [PubMed]

39. Chen, W.-M.; Shaw, L.-H.; Chang, P.-J.; Tung, S.-Y.; Chang, T.-S.; Shen, C.-H.; Hsieh, Y.-Y.; Wei, K.-L. Hepatoprotective effect of resveratrol against ethanol-induced oxidative stress through induction of superoxide dismutase in vivo and in vitro. *Exp. Ther. Med.* **2016**, *11*, 1231–1238. [CrossRef] [PubMed]

40. Nabavi, S.F.; Daglia, M.; Moghaddam, A.H.; Habtemariam, S.; Nabavi, S.M. Curcumin and liver disease: From chemistry to medicine. *Compr. Rev. Food Sci. Food Saf.* **2013**, *13*, 62–77. [CrossRef]

41. Fraschini, F.; Demartini, G.; Esposti, D. Pharmacology of silymarin. *Clin. Drug Investig.* **2002**, *22*, 51–65. [CrossRef]

42. Kren, V.; Walteravá, D. Silybin and silymarin—New effects and applications. *Biomed. Pap.* **2005**, *149*, 29–41. [CrossRef]

International Journal of
Molecular Sciences

MDPI

Article

Vanillin Suppresses Cell Motility by Inhibiting STAT3-Mediated HIF-1α mRNA Expression in Malignant Melanoma Cells

Eun-Ji Park [1,†], Yoon-Mi Lee [2,†], Taek-In Oh [1], Byeong Mo Kim [3], Beong-Ou Lim [1] and Ji-Hong Lim [1,2,*]

[1]   Department of Biomedical Chemistry, College of Biomedical & Health Science, Konkuk University, Chungju 380-701, Korea; peunji0503@kku.ac.kr (E.-J.P.); dk1050@kku.ac.kr (T.-I.O.); beongou@kku.ac.kr (B.-O.L.)
[2]   Interdisciplinary Research Center for Health, Konkuk University, Chungju 380-701, Korea; yoonmilee@kku.ac.kr
[3]   Severance Integrative Research Institute for Cerebral and Cardiovascular Diseases (SIRIC), Yonsei University College of Medicine, Seodamun-gu, Seoul 03722, Korea; BKIM2@yuhs.ac
*   Correspondence: jhlim@kku.ac.kr; Tel.: +82-43-840-3567; Fax: +82-43-840-3929
†   These authors contributed equally to this work.

Academic Editor: Toshio Morikawa
Received: 6 January 2017; Accepted: 24 February 2017; Published: 1 March 2017

Abstract: Recent studies have shown that vanillin has anti-cancer, anti-mutagenic, and anti-metastatic activity; however, the precise molecular mechanism whereby vanillin inhibits metastasis and cancer progression is not fully elucidated. In this study, we examined whether vanillin has anti-cancer and anti-metastatic activities via inhibition of hypoxia-inducible factor-1α (HIF-1α) in A2058 and A375 human malignant melanoma cells. Immunoblotting and quantitative real time (RT)-PCR analysis revealed that vanillin down-regulates HIF-1α protein accumulation and the transcripts of HIF-1α target genes related to cancer metastasis including fibronectin 1 (*FN1*), lysyl oxidase-like 2 (*LOXL2*), and urokinase plasminogen activator receptor (*uPAR*). It was also found that vanillin significantly suppresses HIF-1α mRNA expression and de novo HIF-1α protein synthesis. To understand the suppressive mechanism of vanillin on HIF-1α expression, chromatin immunoprecipitation was performed. Consequently, it was found that vanillin causes inhibition of promoter occupancy by signal transducer and activator of transcription 3 (STAT3), but not nuclear factor-κB (NF-κB), on *HIF1A*. Furthermore, an in vitro migration assay revealed that the motility of melanoma cells stimulated by hypoxia was attenuated by vanillin treatment. In conclusion, we demonstrate that vanillin might be a potential anti-metastatic agent that suppresses metastatic gene expression and migration activity under hypoxia via the STAT3-HIF-1α signaling pathway.

Keywords: vanillin; HIF-1α; STAT3; migration; melanoma

1. Introduction

Malignant melanoma is a skin cancer that develops from the abnormal growth and differentiation of melanocytes with hyperpigmentation; the incidence of melanoma cases has been increasing, and this particular skin cancer is associated with a high rate of mortality caused by early and rapid metastasis [1]. Significant therapeutic advances have been made using small molecule inhibitors that target melanoma, but challenges to eradicate these solid tumors still persist.

Given the fact that a hypoxic microenvironment is a major feature in multiple types of solid cancers including melanoma, hypoxia-inducible factor-1 (HIF-1) composed of α and β subunits is a pivotal transcription factor in the adaptation of cells to low oxygen conditions. HIF-1α protein is tightly regulated by oxygen levels despite the constitutive expression of HIF-1α mRNA. Under normoxic conditions, HIF-1α gets degraded before it can be translocated to the nucleus, associate with HIF-1β, and begin the hypoxic response that is conducive to tumor formation. For this degradation to occur, HIF-1α is hydroxylated by the prolyl 4-hydroxylase (P4H) enzyme into proline 402 and 564 residues, which cause HIF-1α to bind to the von Hippel-Lindau protein (VHL) E3-ubiquitin ligase complex, leading to ubiquitination of this complex and subsequent signaling for proteasomal degradation. Conversely, hypoxic conditions arrest this oxygen-dependent reaction, resulting in the dimerization of α and β subunits in the nucleus. Additionally, oncogenic signaling pathways such as the PI3K (phosphoinositide 3-kinase)-AKT-mTOR (mammalian target of rapamycin)axis and mitogen-activated protein kinase (MAPK) pathway are involved in de novo HIF-1α protein synthesis via 5′-cap-dependent translation initiation [2,3], and various transcription factors including signal transducer and activator of transcription 3 (STAT3) and nuclear factor-κB (NF-κB) are critical for regulating HIF-1α mRNA expression with promoter occupancy at the proximal region of *HIF1A* [4–7]. The HIF-1α and HIF-1β complex binds to the hypoxia-response element (HRE), thereby affecting various downstream genes associated with cancer cell angiogenesis, migration, and metastasis. During cancer metastasis, HIF-1α in hypoxic microenvironments transcriptionally increases various transcripts related to stimulation of cell migration and invasion, including fibronectin 1 (*FN1*), urokinase plasminogen activator receptor (*uPAR*), lysyl oxidase-like 2 (*LOXL2*), and matrix metalloproteinases (*MMPs*) [8].

Vanillin is a major component of vanilla bean extract and is widely used as a flavoring in foods. Because of its antioxidant properties, many biological activities of vanillin have been studied. Vanillin inhibited mutagen induced-DNA damage or spontaneous mutation in bacteria and human cells by eliciting DNA repair [9–11]. Recent reports have shown that vanillin has anti-cancer effects through increased apoptosis and cell cycle arrest in melanoma, colon, and cervical cancer cells [12–14]. In addition, vanillin has also been reported to exhibit anti-invasive and anti-metastatic activities by suppressing the phosphoinositide 3-kinase (PI3K) and NF-κB signaling pathways in lung, breast, and liver cancer cells [15–17]. Nevertheless, the precise molecular mechanism by which vanillin suppresses cancer growth and metastatic potential has not yet been elucidated. Taking into consideration all of the above facts, this study focused on the role of vanillin in the suppression of cancer cell motility and the mechanism of HIF-1α inhibition under hypoxic environments in A2058 and A375 malignant melanoma cells.

In the present study, we evaluated the inhibitory effects of vanillin on hypoxia-inducible factor (HIF)-1α accumulation and cancer cell motility under hypoxia by abrogating STAT3-mediated HIF1A mRNA expression. Our results suggest that vanillin is a potential therapeutic compound that can be used to develop anti-metastatic agents or preventive functional foods for malignant melanoma.

2. Results

2.1. Anti-Cancer Effects of Vanillin in Human A2058 and A375 Malignant Melanoma Cells

Because the anti-cancer effect of vanillin under hypoxic conditions has not been previously reported, we measured cell viability under both normoxia and hypoxia in the absence or presence of vanillin. Vanillin did not have any significant effect on the viability of A2058 and A375 melanoma cells under hypoxia (Figure 1).

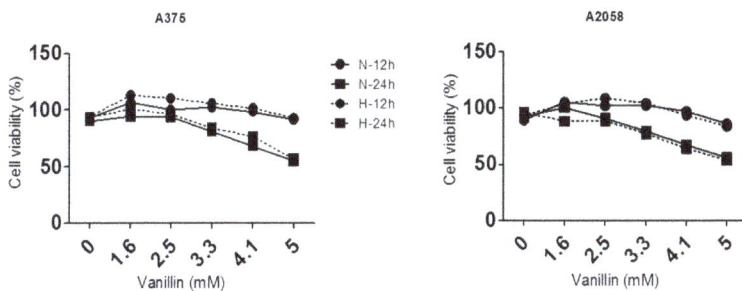

Figure 1. Anti-cancer effects of vanillin upon normoxia or hypoxia. Anti-cancer effects of vanillin in A375 and A2058 malignant melanoma cells. Cells were incubated with 1.6, 2.5, 3.3, 4.1, and 5.0 mM of vanillin for 12 h or 24 h under normoxic (N) or hypoxic (H) condition. Cell viability was measured by crystal violet assay as described in the materials and methods section.

2.2. Vanillin Decreases HIF-1α Protein Levels under Hypoxia in A2058 and A375 Malignant Melanoma Cells

We investigated the inhibitory effect of vanillin on hypoxia-induced HIF-1α accumulation. Based on our results, vanillin strongly suppressed hypoxia-induced HIF-1α accumulation, and this effect was independent of any toxicity to A2058 and A375 cells (Figure 2A). To determine whether vanillin also decreases nuclear HIF-1α protein, we measured nuclear HIF-1α protein levels under both normoxia and hypoxia in the absence or presence of vanillin. We found that vanillin inhibits HIF-1α accumulation in the nucleus (Figure 2B), suggesting that vanillin could suppress both HIF-1α transcriptional activity and HIF-1α protein levels. In addition, HIF-1β protein levels were slightly increased under vanillin treatment in the cytoplasm, but decreased in the nucleus. Because HIF-1β is translocated into the nucleus through its interaction with HIF-1α, the suppression of HIF-1α protein levels by vanillin treatment may change HIF-1β protein levels in the cytoplasm and nucleus (Figure 2B).

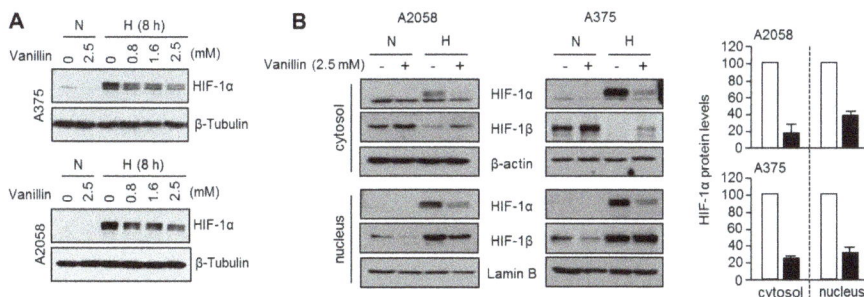

Figure 2. Vanillin decreases hypoxia-inducible factor (HIF)-1α protein levels upon hypoxia. (**A**) Vanillin suppresses hypoxia-induced HIF-1α protein levels. Cells were incubated in the absence or presence of vanillin for 8 h under normoxic or hypoxic condition. HIF-1α protein levels were measured by immunoblotting using anti-HIF-1α antibody; (**B**) Nuclear HIF-1α protein was decreased by vanillin treatment under hypoxia. Protein levels were quantified using Image J 1.49v software (National Institutes of Health, Bethesda, MD, USA). Values represent the mean ± standard deviations of three independent experiments performed.

2.3. Vanillin Attenuates HIF-1α Protein Synthesis

To understand the suppression mechanism of vanillin on HIF-1α accumulation, we investigated whether the presence of vanillin could decrease HIF-1α whose increased levels were induced by using an iron chelator and 26S proteasome inhibition. Consequently, increased HIF-1α expression by both iron chelator (Figure 3A) and proteasome inhibitor (Figure 3B), as observed in the control cells, was also dramatically decreased by vanillin treatment, suggesting that neither hydroxylation nor proteasomal degradation of HIF-1α is associated with the suppressive effect of vanillin on HIF-1α accumulation. Therefore, we next investigated whether vanillin could attenuate de novo protein synthesis of HIF-1α. The results, shown in Figure 3C, demonstrate that vanillin strongly attenuates HIF-1α accumulation increased by MG132 (which blocks all proteolytic activity of the 26S proteasome complex) after blocking the de novo synthesis of HIF-1α by using CHX, as protein translation inhibitor, in both A2058 and A375 melanoma cells, suggesting that vanillin causes inhibition of de novo HIF-1α protein synthesis. Because the mammalian target of rapamycin (mTOR)-mediated phosphorylation of eukaryotic translation initiation factor 4E-binding protein 1 (4E-BP1), eukaryotic translation initiation factor 4E (eIF4E), and S6 kinase ribosomal protein stimulate de novo HIF-1α protein synthesis [2,3], we further elucidated the phosphorylation status of 4E-BP1, eIF4E, and S6 kinase after vanillin treatment. From Figure 3D, it was observed that vanillin does not alter phosphorylation of 4E-BP1, eIF4E, and S6 kinase, suggesting that mTOR-mediated de novo HIF-1α protein synthesis is not responsible for the reduction of HIF-1α by vanillin.

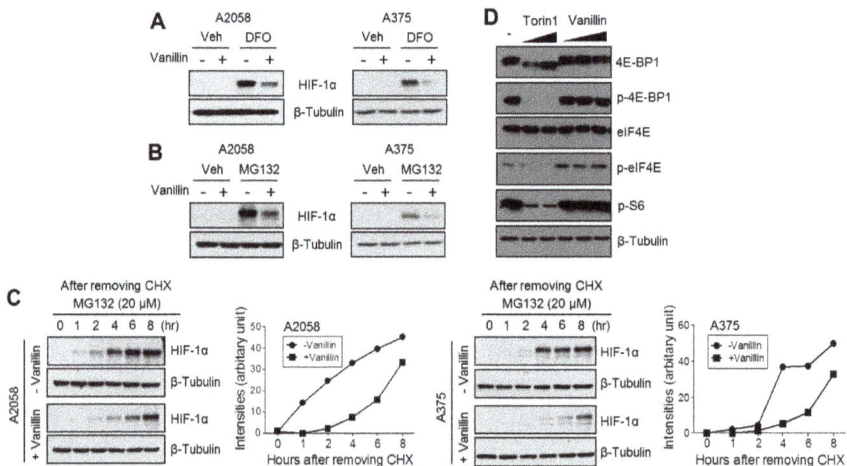

Figure 3. Vanillin attenuates HIF-1α protein synthesis. (**A**) Vanillin suppresses HIF-1α increased by treatment with the iron chelator, deferoxamine (DFO), under normoxic condition. A2058 and A375 cells were pre-incubated with 2.5 mM of vanillin for 1 h, and then further incubated in the absence or presence of 50 μM of DFO for 6 h; (**B**) Vanillin suppresses HIF-1α increased by proteasome inhibitor (MG132) treatment. Cells were pre-incubated with 2.5 mM of vanillin for 1 h, and then further incubated in the absence or presence of 20 μM of MG132 for 6 h; (**C**) Vanillin attenuates de novo HIF-1α protein synthesis. Cells were pre-incubated with 100 μM cycloheximide (CHX) for 12 h and then washed with phosphate-buffered saline (PBS). Cells were incubated with fresh culture medium containing 20 μM of MG132 and 2.5 mM of vanillin (or dimethyl sulfoxide (DMSO)) for indicated time. HIF-1α protein levels were detected by immunoblotting, and then protein levels were quantified using Image J software; (**D**) Vanillin does not regulate the signal transduction pathway related to 5′-cap-dependent translation.

2.4. Vanillin Decreases HIF-1α mRNA Levels by Inhibiting Promoter Occupancy of STAT3 at HIF1A

Because decreased mRNA levels also cause retardation of de novo protein synthesis, we next measured the HIF-1α mRNA levels after vanillin treatment of A2058 and A375 melanoma cells. We found that the HIF-1α mRNA levels were significantly decreased by vanillin treatment (Figure 4A). It is becoming clear that STAT3 and NF-κB are critical transcription factors for HIF-1α mRNA expression through direct interaction with the proximal promoter of *HIF1A* [5–7]. Therefore, we investigated the suppressive effect of vanillin on STAT3 activation and its promoter occupancy on *HIF1A*. Vanillin significantly decreases STAT3 phosphorylation in both A2058 and A375 melanoma cells (Figure 4B). In addition, it was also found that the proximal promoter of *HIF1A* is dissociated from STAT3, but not from NF-κB (Figure 4C). These results suggest that the inactivation of STAT3 is responsible for the decreased HIF-1α levels in response to treatment with vanillin.

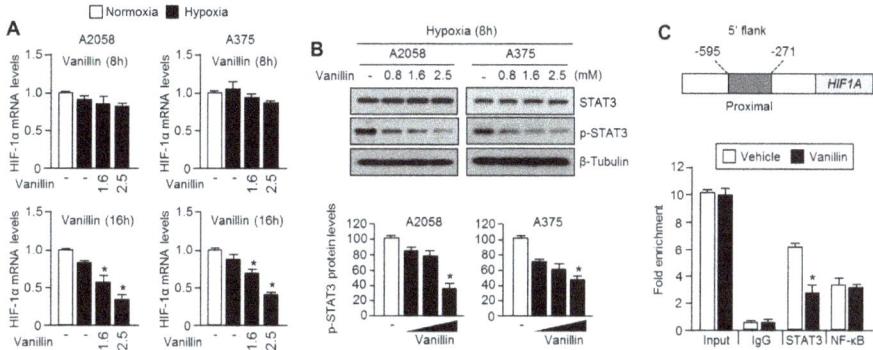

Figure 4. Vanillin decreases HIF-1α mRNA expression by inhibiting signal transducer and activator of transcription 3 (STAT3). (**A**) Vanillin decreases HIF-1α mRNA levels in A2058 and A375 human melanoma cells. Cells were incubated in the absence or presence of vanillin for 8 h or 16 h under hypoxic condition. HIF-1α mRNA levels were measured using quantitative real time (RT)-PCR. Values represent the mean ± SD of three independent experiments performed in duplicate; * $p < 0.05$; (**B**) Vanillin decreases STAT3 phosphorylation under hypoxia. A2058 and A375 melanoma cells were incubated with vanillin for 8 h under hypoxia. STAT3 protein levels were detected by immunoblotting, and then protein levels were quantified using Image J software. Values represent the mean ± SD of three independent experiments performed; * $p < 0.05$; (**C**) Vanillin causes dissociation of STAT3 from the HIF-1α promoter region. A2058 cells were incubated with 2.5 mM of vanillin for 8 h and then fixed with formalin. Chromatin was immunoprecipitated with non-immunized serum (IgG) or the antisera as indicated. The proximal region of HIF-1α promoter was amplified using quantitative PCR. Values represent the mean ± SD of two independent experiments performed in triplicate; * $p < 0.05$.

2.5. Vanillin Down-Regulates HIF-1α Target Gene Expression and Causes Suppression of Cell Motility

To determine whether vanillin functionally suppresses HIF-1α transcriptional activity as well as protein levels, HIF-1α responsive promoter activity (HRE- or vascular endothelial growth factor (VEGF)-luciferase) was measured. In this case, vanillin dramatically down-regulates HIF-1α promoter activity (Figure 5A) and its target genes involved in glycolytic metabolism (*CA-IX*, *PDK1*, *GLUT1*, and *LDHA*: carbonic anhydrase 9, pyruvate dehydrogenase kinase 1, glucose transporter 1, and lactate dehydrogenase A) and cancer metastasis (*FN1*, *LOXL2*, and *uPAR*) under hypoxia in A2058 and A375 cells (Figure 5B,C). Because HIF-1α stimulates cancer cell motility and invasiveness under hypoxia [8], the inhibitory effect of vanillin on cell migration increased by hypoxia was tested. Consequently, it was found that vanillin strongly attenuates cell migration under hypoxia in A2058 and A375 cells (Figure 5D). These results demonstrate that vanillin suppresses cancer cell motility by inhibiting HIF-1α target gene expression associated with cancer metastasis.

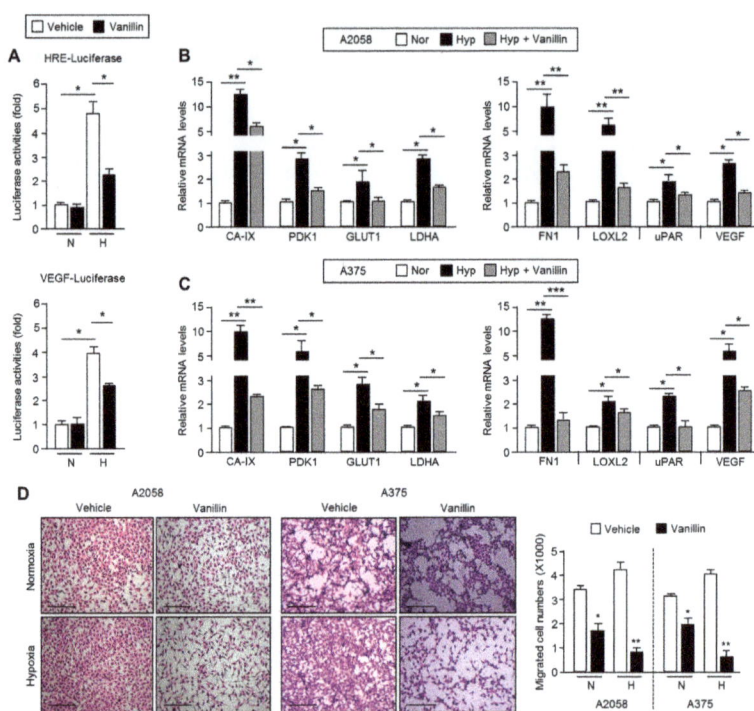

Figure 5. Vanillin inhibits HIF-1α transcriptional activity and cell motility in malignant melanoma cells. (**A**) Inhibitory effect of vanillin on HIF-1α transcriptional activity. A2058 cells were transiently transfected with the hypoxia-responsive element (HRE) or vascular endothelial growth factor (VEGF)–luciferase vector and then incubated for 24 h. Transfected cells were incubated under normoxia or hypoxia in the absence or presence of vanillin (2.5 mM) for 24 h. Values represent the mean ± standard deviation of three independent experiments performed in duplicate; * $p < 0.05$; (**B,C**) Vanillin suppresses hypoxia-induced HIF-1α target gene expression. A2058 and A375 cells were incubated under normoxia or hypoxia in the absence or presence of vanillin (2.5 mM) for 24 h. HIF-1α target gene expression was measured using quantitative RT-PCR. Values represent the mean ± standard deviation of two independent experiments performed in triplicate; * $p < 0.05$, ** $p < 0.01$, and *** $p < 0.001$; (**D**) Inhibitory effect of vanillin on cell migration. A2058 and A375 cells were seeded into transwell chambers and incubated under normoxia or hypoxia for 16 h in the absence or presence of vanillin (2.5 mM). Scale bar representing 200 μm. Migrated cell numbers were counted and values represent the mean ± standard deviation of two independent experiments performed in triplicate; * $p < 0.05$ and ** $p < 0.01$.

3. Discussion

Because intratumoral hypoxia causes HIF-1α overexpression, genetic alterations of HIF-1α are commonly observed in malignant solid cancers and closely associated with treatment failure and increased mortality; therefore, it is important to identify HIF-1α inhibitors and test their efficacy as anticancer therapeutics [8]. A growing number of HIF-1α inhibitors derived from natural products, low molecular weight secondary metabolites produced by plants and microbes, have recently been identified as HIF-1α inhibitors [18]. For example, it has been reported that apigenin (4′,5,7-trihydroxyflavone) and resveratrol (*trans*-3,4,5′-trihydroxystilbene) promote HIF-1α protein degradation in a manner that is independent of the microenvironment oxygen levels [19,20]. Pleurotin and genistein (4′,5,7-Trihydroxyisoflavone) inhibit the accumulation of HIF-1α protein by suppressing protein synthesis under both normoxic and hypoxic conditions [21,22]. Many of the currently identified HIF-1α inhibitors derived from natural products affect protein accumulation or degradation. Interestingly, we propose that vanillin may be a promising HIF-1α inhibitor that acts to reduce HIF-1α levels by suppressing HIF-1α mRNA expression.

Vanillin, a widely used flavoring agent from vanilla, has been shown to exhibit multiple biological effects, including anti-cancer, anti-mutagenic, and anti-bacterial activity in mammalian cells [9,14,23]. In this study, we demonstrate that vanillin effectively decreases HIF-1α protein levels and the expression of its target genes related to cell motility, angiogenesis, and glycolytic metabolism in A2058 and A375 malignant melanoma cells.

To understand the precise molecular mechanism by which vanillin decreases HIF-1α protein levels, we investigated whether vanillin attenuates HIF-1α protein synthesis or promotes proteasomal degradation. We found that vanillin dramatically attenuates de novo HIF-1α protein synthesis. In addition, MG132, a 26S proteasome inhibitor, did not rescue vanillin-mediated HIF-1α reduction in melanoma cells, suggesting that vanillin reduces HIF-1α using a mechanism that is independent of proteasomal degradation. Because the growth factor-mediated PI3K-mTOR signaling pathway activates HIF-1α protein synthesis via 4E-BP1, eIF4E, and S6 kinase linked 5′-cap-dependent translation initiation [2,3], we tested whether vanillin regulates the phosphorylation status of 4E-BP1, eIF4E, and S6 kinase using torin1, a selective mTOR inhibitor, as a positive control. Unlike torin1, vanillin did not alter the phosphorylation status of 4E-BP1, eIF4E, and S6 kinase, suggesting that vanillin does not participate in PI3K-mTOR-mediated protein synthesis. Therefore, we further investigated the inhibitory effect of vanillin on HIF-1α mRNA expression. Interestingly, HIF-1α mRNA was significantly decreased by vanillin treatment in A2058 melanoma cells.

How does vanillin decrease HIF-1α mRNA expression? To answer this question, we investigated the inhibitory effect of vanillin on STAT3-mediated HIF-1α mRNA expression, because STAT3 is one of the transcription factors that is associated with the proximal promoter of *HIF1A* [5,6]. Vanillin reduces STAT3 phosphorylation and promoter occupancy on the 5′-flank of *HIF1A*. These results suggest that vanillin decreases HIF-1α by suppressing STAT3-mediated transcription. Nevertheless, we did not provide a precise molecular mechanism to explain how vanillin inhibits STAT3 phosphorylation and proximal promoter occupancy on the 5′ flanking region of *HIF1A*. Therefore, how vanillin suppresses STAT3 phosphorylation and transcriptional activity should be further investigated.

Although the anti-metastatic effect of vanillin by inhibiting MMP-9 expression in breast and hepatocellular carcinoma cells has recently been reported, the molecular mechanism by which vanillin attenuates migration and invasion in cancer cells is not fully demonstrated [15–17]. In the present study, we provide insight into some part of the mechanism that involves the STAT3-HIF-1α axis on the vanillin-mediated suppression of cancer cell migration and invasion. Indeed, cell migration was significantly decreased by vanillin by approximately 50% under normoxic condition. Under hypoxia, vanillin suppressed cell migration by approximately 75%, suggesting that vanillin could sensitively block cell motility in malignant tumors with hypoxic microenvironments.

On the basis of previous studies of the anti-metastatic effects of millimolar-range vanillin used to treat cells in vivo, mice were administered 100 mg/kg/day vanillin. Although 100 mg/kg/day can be regarded as a high concentration, no side effects were observed [17]. Variable concentrations below 100 mg/kg/day can be considered for animal studies. Several methods can be used to determine the effects low-concentration vanillin. Vanillin derivatives can be developed to improve delivery efficacy at low concentrations, which are more effective for preventing malignant melanoma metastasis. Indeed, a recent report showed that 60 mg/kg/day of *o*-vanillin, a vanillin isomer, strongly suppressed tumor growth in mice bearing A375 human malignant melanoma xenografts [14]. For clinical application, it should be further evaluated whether *o*-vanillin has anti-metastatic effects via the inhibition of HIF-1α accumulation. In addition, vanillin may be useful as a functional food and not limited to chemotherapy. Our results provide a foundation for further analysis of vanillin for the prevention and treatment of malignant melanoma.

4. Materials and Methods

4.1. Reagents and Antibodies

Vanillin (V1104), deferoxamine (D9533), MG132 (M7449), dimethyl sulfoxide (DMSO), protease inhibitor cocktail, and cycloheximide (01810) were purchased from Sigma-Aldrich (St. Louis, MO, USA). Antibodies against 4E-BP1 (9452), phospho-4E-BP1 (2855), eIF4E (9742), phospho-eIF4E (9741), phospho-S6 (4857), STAT3 (12640), and phospho-STAT3 (9131) were obtained from Cell Signaling Technology (Danvers, MA, USA). Antibodies against β-tubulin (sc-9104) and β-actin (sc-47778) were purchased from Santa Cruz Biotechnology (Santa Cruz, CA, USA). Anti-HIF-1α and anti-HIF-1β antibodies were kindly provided by Jong-Wan Park of Seoul National University, Seoul, Korea.

4.2. Cell Culture and Treatment

A2058 and A375 melanoma cell lines were purchased from American Type Culture Collection (ATCC, Manassas, VA, USA). Cells were cultured in Dulbecco's modified Eagle's medium (DMEM) containing 10% fetal bovine serum (FBS) (Gibco, Carlsbad, CA, USA) and 25 mM glucose in a humidified atmosphere of 5% CO_2 at 37 °C. The oxygen level in the hypoxia incubator chamber was maintained at 1% by continuously injecting N_2 gas. Vanillin at various stock concentrations (0.8, 1.6, 2.5 M) was dissolved in dimethyl sulfoxide (DMSO) and diluted into culture medium prior to treatment.

4.3. Cell Viability Assay

To determine the cell viability, we used crystal violet staining in this study. Cells were seeded in 24-well plates, and incubated for 24 h, followed by treatment with vanillin in increasing concentrations for 12 and 24 h at 20% (normoxia) or 1% (hypoxia) oxygen conditions. The cells were fixed with 4% paraformaldehyde for 15 min and stained with 0.05% crystal violet staining solution (HT90132, Sigma-Aldrich) for 15 min. To measure optical density, 1% sodium dodecyl sulfate (SDS) solution was added to the cells, and these were further incubated for 15 min at room temperature. The dissolved solutions were transferred to a 96-well plate and measured at 595 nm using a microplate reader (BioTek, Winooski, VT, USA).

4.4. Cytosolic and Nuclear Extract Preparation

Cells were washed using cold phosphate-buffered saline and harvested by centrifugation at 1000 rpm for 5 min at 4 °C. The harvested cell pellets were resuspended and incubated with ice-cold buffer A (20 mM Tris at pH 7.8, 1.5 mM $MgCl_2$, 10 mM KCl, 0.2 mM ethylenediaminetetraacetic acid (EDTA), 0.5 mM dithiothreitol (DTT), and protease inhibitor cocktail) for 5 min on ice, and then cells were collected by centrifugation at 1000 rpm for 5 min at 4 °C. Next, the cell pellets were lysed using 0.06% NP-40 containing buffer A for 10 min on ice. The cell lysates were centrifuged at 3000 rpm for 5 min, and then the supernatants containing cytosolic proteins were frozen. After obtaining the cytosolic fraction, the pellets were incubated and lysed using buffer B (20 mM Tris-Cl at pH 7.8, 1.5 mM $MgCl_2$, 0.2 mM EDTA, 0.5 mM DTT, and 20% glycerol) containing 400 mM NaCl for 30 min on ice. During incubation, the cells were homogenized with a glass homogenizer. The incubated samples were centrifuged at 14,000 rpm for 30 min at 4 °C, and then supernatants containing the nuclear proteins were transferred into fresh tubes.

4.5. Immunoblotting

Total proteins were extracted using cell lysis buffer (1% NP-40, 150 mM NaCl, 50 mM Tris pH 7.4, 2 mM EDTA, and protease inhibitor cocktail). Cell lysates were separated by 7.5% or 10% SDS-polyacrylamide gel electrophoresis (PAGE). Separated proteins were transferred onto an Immobilon-P membrane (Millipore, Billerica, MA, USA). The transferred membranes were blocked with 5% skim milk in Tris-buffered saline containing 0.05% Tween-20 (TBS-T) for 1 h at room temperature, and then incubated overnight with primary antibodies diluted at 1:1000 or 1:5000 in 5% skim milk in TBS-T at 4 °C. Horseradish peroxidase-conjugated secondary antibodies were incubated for 1 h at room temperature, and then protein levels were visualized using an Enhanced Chemiluminescence Prime kit (GE Healthcare, Little Chalfont, UK).

4.6. Quantitative Real-Time PCR

Total RNA was isolated with TRIzol and 2 μg of this RNA was used to synthesize cDNA using a high capacity cDNA reverse transcription kit (Applied Biosystems). The cDNA was amplified over 40 cycles (95 °C for 15 s, 60 °C for 1 min). Experimental C_q values were normalized to *H36B4* and relative mRNA levels were calculated on the basis of H36B4 mRNA levels. The sequences of the PCR primers (5′–3′) were: ATGGAGCCCAGCAGCAA and GGCATTGATGACTCCAGTGTT for *GLUT1*; CCACTCCAGCAGGGAAGG and GCGACGCAGCCTTTGAAT for *CA-IX*; TGAACATTCTGGCT GGTGACAGGA and ATGATGTCATTCCCACAATGGCCC for *PDK1*; CTACCTCCACCATGCCAAGT and AGCTGCGCTGATAGACATCC for *VEGF*; CCATAAAGGGCAACCAAGAG and ACCTCGG TGTTGTAAGGTGG for *FN1*; CACTGCGGATCCCTGAAAC and CCTGTCTTCGGGCTGATG for *LOXL2*; AGCCTTACCGAGGTTGTGTG and AAATGCATTCGAGGTAACGG for *uPAR*.

4.7. In Vitro Migration Assay

In vitro cell migration assays were performed using a Transwell chamber from Sigma-Aldrich (St. Louis, MO, USA). The underside of the Transwell insert membrane was coated with collagen and incubated at room temperature until it was dry. Cells in 0.1 mL of fetal bovine serum-free medium were seeded into the upper chamber, and the lower chamber was filled with 6% fetal bovine serum-containing medium as a chemotactic source, and then cells were incubated for 16 h at 37 °C. After incubation, the Transwell chambers were quickly washed using phosphate-buffered saline and stained with hematoxylin and eosin. The Transwell insert membranes, containing the migrated cells, were placed on a slide glass and analyzed using a microscope (Olympus, Tokyo, Japan). To quantify the migrated cell numbers, two random fields under 40× magnification were quantified by counting the cell numbers.

4.8. Chromatin Immunoprecipitation

Cultured cells were fixed with 1% formaldehyde to cross-link chromatin and proteins, and soluble chromatin and protein complex samples were incubated overnight at 4 °C with antibodies against STAT3 and NF-κB p65. STAT3 or NF-κB p65 interacting DNA was eluted, and then occupancy of STAT3 on the proximal region of the HIF-1α promoter was measured using quantitative PCR. The sequences of PCR primers for the quantitative ChIP assay (5′–3′) are ATCTGAGCAACGAGACCAAA and CACGTGCTCGTCTGTGTTTA.

4.9. Luciferase Activity Assay

Hypoxia-responsive element (HRE) or VEGF (vascular endothelial growth factor) promoter-luciferase reporter plasmids were a gift from Navdeep Chandel (Addgene plasmid # 26731) and Jong-Wan Park (Seoul National University, Korea). Cells were transfected with reporter plasmids using Polyfect (QIAGEN, Valencia, CA, USA) and then incubated for 48 h for stabilization. After incubation, luciferase activity was measured using a luminometer (Berthold Technologies, Bad Wildbad, Germany) and normalized against β-gal activities to account for transfection efficiency.

4.10. Statistical Analysis

All data were analyzed using an unpaired Student's *t*-test for two experimental comparisons and a one-way ANOVA (analysis of variance) followed by Tukey's post hoc test for multiple comparisons using GraphPad Prism 5.01 (GraphPad Software Inc., La Jolla, CA, USA). Data are represented as means ± standard deviations (SDs). Differences between mean values were considered statistically significant when the associated *p*-value was less than 0.05.

5. Conclusions

In the present study, the major findings were that vanillin: (i) decreases HIF-1α protein levels and that this effect is independent of proteasomal degradation; (ii) suppresses STAT3 phosphorylation and its promoter occupancy on the proximal region of HIF-1α; and (iii) attenuates cell migration by down-regulating HIF-1α target genes associated with cancer metastasis in A375 and A2058 human malignant melanoma cells. Taken together, these results may provide useful information for the development of vanillin derivatives as anti-metastatic agents or functional foods for preventing malignant melanoma metastasis.

Acknowledgments: This study was supported by the National Research Foundation of Korea (NRF) grant funded by the Korean Government (2014R1A2A2A04007791 and 2015R1A2A2A01002483).

Author Contributions: Ji-Hong Lim conceived and designed the experiments; Eun-Ji Park, Yoon-Mi Lee, Taek-In Oh, and Byeong Mo Kim performed the experiments; Beong-Ou Lim supported the materials; Ji-Hong Lim, and Yoon-Mi Lee analyzed data; Ji-Hong Lim and Yoon-Mi Lee wrote the manuscript.

Conflicts of Interest: The authors declare no conflict of interest.

Abbreviations

HIF-1α	Hypoxia-inducible factor-1α
STAT3	Signal transducer and activator of transcription 3
NF-κB	Nuclear factor-κB
FN1	Fibronectin 1
LOXL2	Lysyl oxidase-like 2
uPAR	Urokinase plasminogen activator receptor
VHL	Von Hippel-Landau protein
MAPK	Mitogen activated protein kinase
HRE	Hypoxia-response element
MMPs	Matrix metalloproteinases
PI3K	Phosphoinositide 3-kinase
mTOR	Mammalian target of rapamycin
4E-BP1	Eukaryotic translation initiation factor 4E-binding protein 1
eIF4E	Eukaryotic translation initiation factor 4E

References

1. Haq, R.; Fisher, D.E. Targeting melanoma by small molecules: Challenges ahead. *Pigment Cell. Melanoma Res.* **2013**, *26*, 464–469. [CrossRef] [PubMed]
2. Laplante, M.; Sabatini, D.M. mTOR signaling at a glance. *J. Cell. Sci.* **2009**, *122*, 3589–3594. [CrossRef] [PubMed]
3. Lim, J.H.; Lee, Y.M.; Lee, G.; Choi, Y.J.; Lim, B.O.; Kim, Y.J.; Choi, D.K.; Park, J.W. PRMT5 is essential for the eIF4E-mediated 5'-cap dependent translation. *Biochem. Biophys. Res. Commun.* **2014**, *452*, 1016–1021. [CrossRef] [PubMed]
4. Dodd, K.M.; Tee, A.R. STAT3 and mTOR: Co-operating to drive HIF and angiogenesis. *Oncoscience* **2015**, *2*, 913–914. [PubMed]
5. Lee, Y.M.; Lim, J.H.; Yoon, H.; Chun, Y.S.; Park, J.W. Antihepatoma activity of chaetocin due to deregulated splicing of hypoxia-inducible factor 1α pre-mRNA in mice and in vitro. *Hepatology* **2011**, *53*, 171–180. [CrossRef] [PubMed]
6. Niu, G.; Briggs, J.; Deng, J.; Ma, Y.; Lee, H.; Kortylewski, M.; Kujawski, M.; Kay, H.; Cress, W.D.; Jove, R.; et al. Signal transducer and activator of transcription 3 is required for hypoxia-inducible factor-1α RNA expression in both tumor cells and tumor-associated myeloid cells. *Mol. Cancer Res.* **2008**, *6*, 1099–1105. [CrossRef] [PubMed]

7. Van Uden, P.; Kenneth, N.S.; Rocha, S. Regulation of hypoxia-inducible factor-1α by NF-κB. *Biochem. J.* **2008**, *412*, 477–484. [CrossRef] [PubMed]
8. Semenza, G.L. Targeting HIF-1 for cancer therapy. *Nat. Rev.* **2003**, *3*, 721–732. [CrossRef] [PubMed]
9. Akagi, K.; Hirose, M.; Hoshiya, T.; Mizoguchi, Y.; Ito, N.; Shirai, T. Modulating effects of ellagic acid, vanillin and quercetin in a rat medium term multi-organ carcinogenesis model. *Cancer Lett.* **1995**, *94*, 113–121. [CrossRef]
10. King, A.A.; Shaughnessy, D.T.; Mure, K.; Leszczynska, J.; Ward, W.O.; Umbach, D.M.; Xu, Z.; Ducharme, D.; Taylor, J.A.; Demarini, D.M.; et al. Antimutagenicity of cinnamaldehyde and vanillin in human cells: Global gene expression and possible role of DNA damage and repair. *Mutat. Res.* **2007**, *616*, 60–69. [CrossRef] [PubMed]
11. Shaughnessy, D.T.; Schaaper, R.M.; Umbach, D.M.; DeMarini, D.M. Inhibition of spontaneous mutagenesis by vanillin and cinnamaldehyde in *Escherichia coli*: Dependence on recombinational repair. *Mutat. Res.* **2006**, *602*, 54–64. [CrossRef] [PubMed]
12. Ho, K.; Yazan, L.S.; Ismail, N.; Ismail, M. Apoptosis and cell cycle arrest of human colorectal cancer cell line HT-29 induced by vanillin. *Cancer Epidemiol.* **2009**, *33*, 155–160. [CrossRef] [PubMed]
13. Lirdprapamongkol, K.; Sakurai, H.; Suzuki, S.; Koizumi, K.; Prangsaengtong, O.; Viriyaroj, A.; Ruchirawat, S.; Svasti, J.; Saiki, I. Vanillin enhances TRAIL-induced apoptosis in cancer cells through inhibition of NF-κB activation. *In Vivo* **2010**, *24*, 501–506. [PubMed]
14. Marton, A.; Kusz, E.; Kolozsi, C.; Tubak, V.; Zagotto, G.; Buzas, K.; Quintieri, L.; Vizler, C. Vanillin Analogues O-vanillin and 2,4,6-Trihydroxybenzaldehyde Inhibit NF-κB Activation and Suppress Growth of A375 Human Melanoma. *Anticancer Res.* **2016**, *36*, 5743–5750. [CrossRef] [PubMed]
15. Liang, J.A.; Wu, S.L.; Lo, H.Y.; Hsiang, C.Y.; Ho, T.Y. Vanillin inhibits matrix metalloproteinase-9 expression through down-regulation of nuclear factor-κB signaling pathway in human hepatocellular carcinoma cells. *Mol. Pharmacol.* **2009**, *75*, 151–157. [CrossRef] [PubMed]
16. Lirdprapamongkol, K.; Kramb, J.P.; Suthiphongchai, T.; Surarit, R.; Srisomsap, C.; Dannhardt, G.; Svasti, J. Vanillin suppresses metastatic potential of human cancer cells through PI3K inhibition and decreases angiogenesis in vivo. *J. Agric. Food Chem.* **2009**, *57*, 3055–3063. [CrossRef] [PubMed]
17. Lirdprapamongkol, K.; Sakurai, H.; Kawasaki, N.; Choo, M.K.; Saitoh, Y.; Aozuka, Y.; Singhirunnusorn, P.; Ruchirawat, S.; Svasti, J.; Saiki, I. Vanillin suppresses in vitro invasion and in vivo metastasis of mouse breast cancer cells. *Eur. J. Pharm. Sci.* **2005**, *25*, 57–65. [CrossRef] [PubMed]
18. Nagle, D.G.; Zhou, Y.D. Natural product-based inhibitors of hypoxia-inducible factor-1 (HIF-1). *Curr. Drug Targets* **2006**, *7*, 355–369. [CrossRef] [PubMed]
19. Cao, Z.; Fang, J.; Xia, C.; Shi, X.; Jiang, B.H. Trans-3,4,5′-Trihydroxystibene inhibits hypoxia-inducible factor 1α and vascular endothelial growth factor expression in human ovarian cancer cells. *Clin. Cancer Res.* **2004**, *10*, 5253–5263. [CrossRef] [PubMed]
20. Osada, M.; Imaoka, S.; Funae, Y. Apigenin suppresses the expression of VEGF, an important factor for angiogenesis, in endothelial cells via degradation of HIF-1α protein. *FEBS Lett.* **2004**, *575*, 59–63. [CrossRef] [PubMed]
21. Wang, G.L.; Jiang, B.H.; Semenza, G.L. Effect of protein kinase and phosphatase inhibitors on expression of hypoxia-inducible factor 1. *Biochem. Biophys. Res. Commun.* **1995**, *216*, 669–675. [CrossRef] [PubMed]
22. Welsh, S.J.; Williams, R.R.; Birmingham, A.; Newman, D.J.; Kirkpatrick, D.L.; Powis, G. The thioredoxin redox inhibitors 1-methylpropyl 2-imidazolyl disulfide and pleurotin inhibit hypoxia-induced factor 1α and vascular endothelial growth factor formation. *Mol. Cancer Ther.* **2003**, *2*, 235–243. [PubMed]
23. Gupta, S.C.; Kim, J.H.; Prasad, S.; Aggarwal, B.B. Regulation of survival, proliferation, invasion, angiogenesis, and metastasis of tumor cells through modulation of inflammatory pathways by nutraceuticals. *Cancer Metastasis Rev.* **2010**, *29*, 405–434. [CrossRef] [PubMed]

International Journal of
Molecular Sciences

MDPI

Article

Hepatoprotective Effects of Nicotiflorin from *Nymphaea candida* against Concanavalin A-Induced and D-Galactosamine-Induced Liver Injury in Mice

Jun Zhao [1], Shilei Zhang [2], Shuping You [2], Tao Liu [2,*], Fang Xu [1], Tengfei Ji [3,*] and Zhengyi Gu [1,*]

[1]   Key Laboratory for Uighur Medicine, Institute of Materia Medica of Xinjiang, Urumqi 830004, China;
     zhaojun21cn@163.com (J.Z.); xufangxj@163.com (F.X.)
[2]   School of Public Health, Xinjiang Medical University, Urumqi 830011, China; zhangsl6191@163.com (S.Z.);
     youshuping@163.com (S.Y.)
[3]   State Key Laboratory of Bioactive Substance and Function of Natural Medicines, Institute of Materia Medica,
     Chinese Academy of Medical Sciences and Peking Union Medical College, Beijing 100050, China
*   Correspondences: xjmult@163.com (T.L.); jitf@imm.ac.cn (T.J.); zhengyi087@126.com (Z.G.);
     Tel.: +86-991-436-5561 (T.L.); +86-10-6021-2117 (T.J.); +86-991-232-8537 (Z.G.)

Academic Editor: Toshio Morikawa
Received: 22 January 2017; Accepted: 2 March 2017; Published: 8 March 2017

Abstract: *Nymphaea candida* was used to treat hepatitis in Ugyhur medicine, and nicotiflorin (kaempferol 3-*O*-β-rutinoside) is the main characteristic component in this plant. In this study, The the hepatoprotective activities of nicotiflorin from *N. candida* were investigated by Concanavalin A (Con A, 20 mg/kg bw)- and D-Galactosamine (D-GalN, 800 mg/kg bw)-induced acute liver injury in mice. Pretreatment with nicotiflorin (25, 50, 100 mg/kg bw/day, p.o.) for ten days significantly reduced the impact of Con A toxicity (20 mg/kg bw) on the serum markers of liver injury, aspartate aminotransferase (AST), and alanine aminotransferase (ALT). The hepatic anti-oxidant parameters (malondialdehyde, MDA; superoxide dismutase, SOD; glutathione, GSH; and nitric oxide, NO) in mice with nicotiflorin treatment were significantly antagonized for the pro-oxidant effects of Con A. Moreover, pretreatment with nicotiflorin (100 mg/kg bw) significantly decreased Con A-induced elevation in the serum levels of pro-inflammatory cytokines interleukin-1β (IL-1β), interleukin-6 (IL-6), tumor necrosis factor-α (TNF-α), and interferon-γ (IFN-γ) ($p < 0.05$). A protective effect was reconfirmed against D-GalN-induced chemical liver injury, elevated serum enzymatic and cytokines levels were significantly decreased by nicotiflorin, and liver homogenate antioxidant indicators were significantly restored toward normal levels. Both histopathological studies also supported the protective effects of nicotiflorin. Therefore, the presented results suggest that nicotiflorin is the potent hepatoprotective agent that could protect the liver against acute immunological and chemical injury; this ability might be attributed to its antioxidant and immunoregulation potential.

Keywords: nicotiflorin; *Nymphaea candida*; hepatoprotective; Concanavalin A; D-galactosamine

1. Introduction

Liver, the major organ for the detoxification and metabolism of xenobiotics, is susceptibly injured by various factors such as toxic chemicals, excess consumption of alcohol, infections, and autoimmune disorders [1]. Moreover, liver injury is also a commonly pathological state of various liver diseases, and its long-term existence often leads to liver fibrosis, liver cirrhosis, and liver cancer [2]. Therefore, the prevention of liver injury is an important means of liver disease treatment [3]. It has always been one of focuses of pharmaceutical research to find significant hepatoprotective compounds from natural plants and traditional folk medicine. In recent years, more and more natural products with hepatoprotection have been isolated from various medicinal plants, such as silymarin, oleanolic acid, and curcumin [4–6].

In Xinjiang, China, *Nymphaea candida* has been used as a folk medicine for head pains, coughs, hepatitis, and hypertension [7]. The previous study showed that extracts from the flowers of *N. candida* Presl have better free radical scavenging and hepatoprotective activities, and nicotiflorin was one of the main characteristic compounds in this plant [8]. Nicotiflorin (Figure 1), namely kaempferol 3-*O*-β-rutinoside, a flavonol glycoside isolated from a variety of plants (*Edgeworthia chrysantha*, *Carthamus tinctorius*, *N. candida*, etc.) [9–11], has been reported to have various pharmaceutical effects, such as antioxidant, anti-inflammatory, and neuroprotective effects [12–15]. In the previous study, nicotiflorin at the doses of 200 and 400 mg/kg bw showed preventive effects on CCl_4-induced liver injury in mice [16]. This study aimed to investigate further the hepatoprotective effects of nicotiflorin (Doses as 25, 50 and 100 mg/kg bw) and its mechanisms by Concanavalin A (Con A)-induced and D-galactosamine (D-GalN)-induced liver injury in mice for the development and application of this compound as well as *N. candida*.

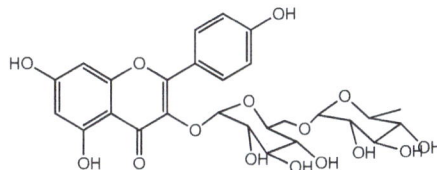

Figure 1. Chemical structure of nicotiflorin.

2. Results and Discussion

2.1. Protective Effect of Nicotiflorin on Con A Induced Hepatotoxicity in Mice

Concanavalin A (Con A), a lectin derived from jack bean seeds, has been widely used to establish an experimental murine model of hepatitis, and this model can mimic many pathological features of viral and autoimmune hepatitis in humans [17,18]. This reproducible liver injury is easily induced by a one-shot intravenous injection of Con A, and this damage could significantly increase the serum levels of transaminases as well as the filtration of neutrophils, macrophages, and T cells [19]. As a T cell mitogen, Con A can activate T cells to proliferate and produce pro-inflammatory cytokines including tumor necrosis factor-α (TNF-α), interferon-γ (IFN-γ), interleukin-1 (IL-1), and interleukin-6 (IL-6) [20].

The serum alanine aminotransferase (ALT) and aspartate aminotransferase (AST) activities are biochemical markers of liver damage [21]. Figure 2 shows that, after Con A injection, a statistically significant increasement in the serum ALT and AST levels was observed at 8 h, compared to the control group, as well as the liver index and spleen index ($p < 0.01$). After treatment with the drug for ten consecutive days, nicotiflorin at three different doses (25, 50 and 100 mg/kg bw) could remarkably prevent the Con A-induced increases of the serum activities of ALT and AST ($p < 0.01$, $p < 0.05$, Figure 2). Organ indexes of the liver and spleen were evaluated in mice. Compared with the Con A group, the elevated liver index and spleen index were also significantly reduced by nicotiflorin at different doses (25, 50, and 100 mg/kg bw) ($p < 0.01$, Table 1). Moreover, there were no significant differences in the changes of body weight before and after the experiment between each group ($p > 0.05$, Table 1).

Table 1. Effects of nicotiflorin on the body weight and liver, speen index in Concanavalin A (Con A)-intoxicated mice.

Group	Inital BW (g)	Final BW (g)	Liver Index	Speen Index
Control	19.83 ± 0.37	26.49 ± 0.39	48.12 ± 0.80	3.65 ± 0.18
Con A	20.41 ± 0.49	27.44 ± 0.65	79.59 ± 2.20 [##]	7.19 ± 1.42 [##]
DDB (150 mg/kg bw) + Con A	20.01 ± 0.53	26.53 ± 0.57	55.82 ± 0.63 **	5.01 ± 0.22 **
nicotiflorin (25 mg/kg bw) + Con A	19.84 ± 0.47	26.64 ± 0.40	61.82 ± 1.22 **	5.48 ± 0.27 **
nicotiflorin (50 mg/kg bw) + Con A	20.11 ± 0.70	27.09 ± 0.41	57.39 ± 2.15 **	4.89 ± 0.36 **
nicotiflorin (100 mg/kg bw) + Con A	21.69 ± 0.31	28.18 ± 0.37	53.98 ± 1.13 **	4.24 ± 0.20 **

Values are mean ± S.E.M., $n = 10$; [##] $p < 0.01$ compared with control group; ** $p < 0.01$ compared with Con A group; BW, body weight.

Histological changes in liver tissues shown by hematoxylin-eosin (HE) staining confirmed the preventive effect of nicotiflorin against Con A-induced immunological liver injury (Figure 3). For livers in the control group, the extent of liver injury was grade 0; the hepatic lobule structure integrity, and a well-preserved cytoplasm, prominent nucleus, and nucleolus were shown (Figure 2A). After Con A injection, liver sections in the model group revealed extensive liver damage such as liver cells with severe edema, condensed nuclei, increased vacuole formation, acidophilic degeneration, inflammatory cells infiltration, centrilobular fatty changes, and widespread hepatocellular necrosis (grades III, Figure 3B). In contrast, mice pretreated with nicotiflorin (25, 50, and 100 mg/kg bw, Figure 3D–F), showed protective effects, and the injury scores of vacuole formation and hepatocellular necrosis were significantly decreased, with main liver damage grades of 0 and I (Table 2). The positive control drug, biphenyl dicarboxylate (DDB, 150 mg/kg bw), also significantly ameliorated liver damage induced by Con A ($p < 0.05$).

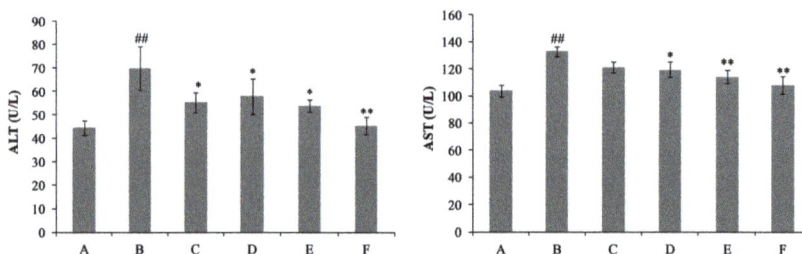

Figure 2. Effects of nicotiflorin on serum alanine aminotransferase (ALT) and aspartate aminotransferase (AST) in Con A- intoxicated mice. (A) Control group; (B) Con A-treated group; (C) Con A and biphenyl dicarboxylate (DDB, 150 mg/kg bw)-treated group; (D) Con A and nicotiflorin (25 mg/kg bw)-treated group; (E) Con A and nicotiflorin (50 mg/kg bw)-treated group; and (F) Con A and nicotiflorin (100 mg/kg bw)-treated group. Values are mean ± S.E.M., $n = 10$. [##] $p < 0.01$ compared with control group; * $p < 0.05$, ** $p < 0.01$ compared with Con A group.

Table 2. Effects of nicotiflorin on the pathological grading of Con A-intoxicated mice.

Group	0	I	II	III	*p*-Value
Control	10	0	0	0	-
Con A	0	0	2	8	#
DDB (150 mg/kg bw) + Con A	1	5	2	2	*
nicotiflorin (25 mg/kg bw) + Con A	2	6	1	1	*
nicotiflorin (50 mg/kg bw) + Con A	2	5	2	1	*
nicotiflorin (100 mg/kg bw) + Con A	2	6	1	1	*

$n = 10$; # $p < 0.01$ compared with control group; * $p < 0.05$, compared with Con A group.

Malondialdehyde (MDA), a major degradation product of lipid hydroperoxides, has attracted much attention as a indicator for assessing the extent of lipid peroxidation in oxidative liver damage. In this study, the hepatic level of malondialdehyde (MDA) was analyzed by the thiobarbituric acid (TBA) method [22]. Con A treatment markedly increased the hepatic MDA level compared with the control group, whereas the pre-administration of nicotiflorin (25, 50, and 100 mg/kg bw) significantly decreased the MDA levels ($p < 0.05$, Figure 4). The antioxidant (glutathione, GSH) content and antioxidant enzyme (superoxide dismutase, SOD) activity in the liver was also measured. The hepatic levels of GSH and SOD were conspicuously decreased in Con A-treated mice compared with those in the control group, whereas the pre-administration of nicotiflorin significantly reversed the decreased activities of GSH and SOD ($p < 0.05$, $p < 0.01$, Figure 4). Compared with control group, Con A injection significantly increased hepatic homogenate nitric oxide (NO) content ($p < 0.01$, Figure 4). Pretreatment with nicotiflorin (50 and 100 mg/kg) significantly decreased serum NO content ($p < 0.05$, $p < 0.01$, Figure 4). The positive control drug, DDB (150 mg/kg), also significantly decreased serum NO content ($p < 0.05$).

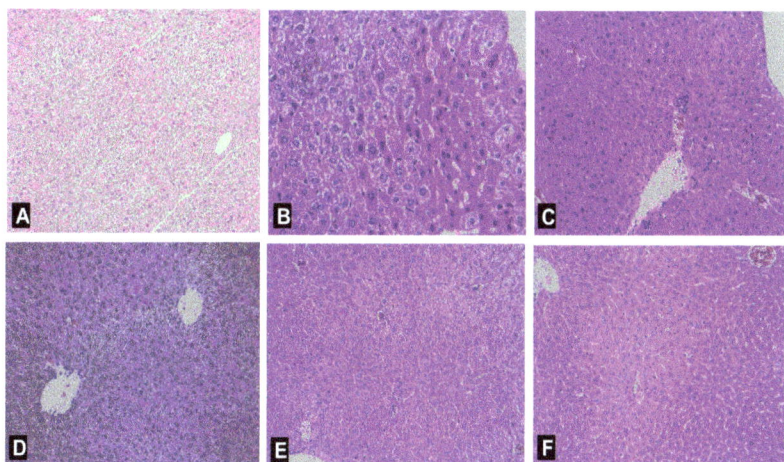

Figure 3. Histological analysis of the livers after Con A administration. Typical images were chosen from the different experimental groups (original magnification 10×20). (**A**) Control group; (**B**) Con A-treated group; (**C**) Con A and DDB (150 mg/kg bw)-treated group; (**D**) Con A and nicotiflorin (25 mg/kg bw)-treated group; (**E**) Con A and nicotiflorin (50 mg/kg bw)-treated group; and (**F**) Con A and nicotiflorin (100 mg/kg bw)-treated group.

The hepatic natural killer T cells play important roles in Con A-induced liver injury by releasing a variety of cytokines, such as IFN-γ, TNF-α, IL-1β, and IL-6 [23]. Among the various cytokines released by Con A-activated T-cells, TNF-α and IFN-γ are considered to play critical roles in the development of massive hepatocellular apoptosis and necrosis [24]. In this study, compared with the Con A group, pretreatment with nicotiflorin at the middle and high doses (50 and 100 mg/kg bw) significantly decreased serum IL-1β ($p < 0.01$) and TNF-α ($p < 0.01$) levels, of which nicotiflorin at the low dose (25 mg/kg bw) also significantly decreased serum IL-1β levels ($p < 0.05$) (Figure 5). Moreover, nicotiflorin (100 mg/kg bw) could significantly decrease the elevated serum IFN-γ level by Con A ($p < 0.01$). The positive control drug, DDB, also significantly decreased serum IL-1β, TNF-α and IFN-γ levels compared to the Con A group ($p < 0.05$). Therefore, nicotiflorin might alleviate the uncontrolled immune response through immunomodulation to play a role of hepatoprotection.

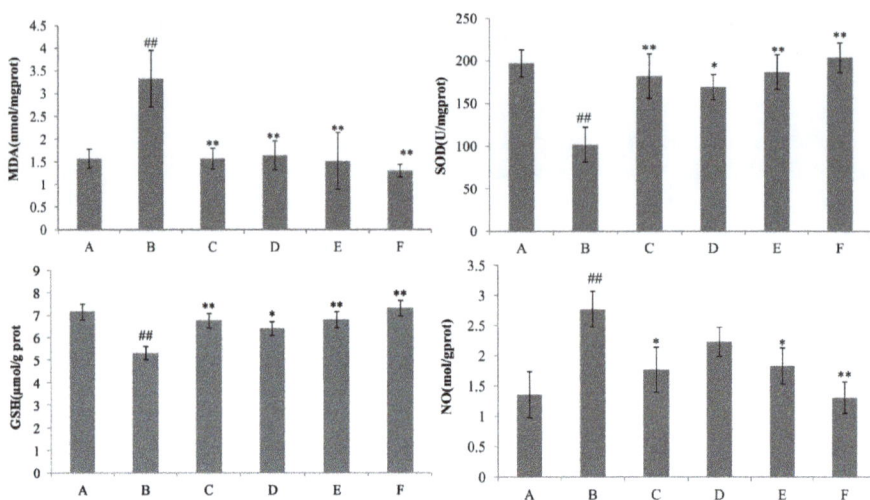

Figure 4. Effects of nicotiflorin on hepatic homogenate superoxide dismutase (SOD), malondialdehyde (MDA), glutathione (GSH), and nitric oxide (NO) in Con A-intoxicated mice. (A) Control group; (B) Con A-treated group; (C) Con A and DDB (150 mg/kg bw)-treated group; (D) Con A and nicotiflorin (25 mg/kg bw)-treated group; (E) Con A and nicotiflorin (50 mg/kg bw)-treated group; and (F) Con A and nicotiflorin (100 mg/kg bw)-treated group. Values are mean ± S.E.M., n = 10. ## $p < 0.01$ compared with the control group; * $p < 0.05$, ** $p < 0.01$ compared with the Con A group.

2.2. Protective Effect of Nicotiflorin on D-GalN-Induced Hepatotoxicity in Mice

To further confirm the hepatoprotective activity of nicotiflorin, we investigated whether nicotiflorin protects against D-GalN-induced acute chemical liver injury. As a well-established hepatotoxicant, D-GalN can induce a liver injury similar to human viral hepatitis in its morphologic and functional features. Therefore, it is very useful for the evaluation of hepatoprotection to construct a liver injury model by D-GalN [25–27]. In this study, mice intoxicated with D-GalN developed severe hepatocellular injuries with a significant elevation in serum AST and ALT activities when compared to the control group ($p < 0.01$). Treatment with nicotiflorin at all doses (25, 50, 100 mg/kg bw) significantly prevented the elevation of serum AST compared to the D-GalN group ($p < 0.01$); and nicotiflorin at a high dose (100 mg/kg bw) significantly prevented the elevation of serum ALT compared to the D-GalN group ($p < 0.05$) (Figure 6). Moreover, nicotiflorin at all doses (25, 50, 100 mg/kg bw) significantly decreased the elevation of the liver index and speen index compared to the D-GalN group. The changes of body weight before and after the experiment did not show a significant difference between the groups ($p > 0.05$, Table 3).

Table 3. Effects of nicotiflorin on the body weight and the liver and speen indexes in D-GalN-intoxicated mice.

Group	Inital BW (g)	Final BW (g)	Liver Index	Speen Index
Control	20.72 ± 0.59	25.02 ± 1.11	45.12 ± 1.13	4.25 ± 0.22
D-GalN	20.28 ± 0.34	26.35 ± 0.72	84.44 ± 1.47 ##	8.26 ± 0.20 ##
DDB (150 mg/kg bw) + D-GalN	21.36 ± 0.42	26.38 ± 0.58	53.75 ± 1.09 **	5.54 ± 0.18 **
nicotiflorin (25 mg/kg bw) + D-GalN	21.25 ± 0.48	25.64 ± 0.46	65.44 ± 0.89 **	6.43 ± 0.57 **
nicotiflorin (50 mg/kg bw) + D-GalN	21.26 ± 0.59	26.13 ± 0.80	57.48 ± 2.46 **	5.55 ± 0.30 **
nicotiflorin (100 mg/kg bw) + D-GalN	20.71 ± 0.59	26.75 ± 0.56	49.35 ± 1.50 **	4.88 ± 0.23 **

Values are mean ± S.E.M., n = 10. ## $p < 0.01$ compared with the control group. ** $p < 0.01$ compared with the Con A group. BW, body weight.

In this study, we further examined liver histopathological characters to explore the protective effects of nicotiflorin on D-GalN-intoxicated mice. A photomicrograph of control mice liver showed the hepatic lobule structure integrity and the liver cells in mice to radiate out from central vein at the center (Figure 7A). A photomicrograph of D-GalN-intoxicated mice liver section showed swelling, loose cytoplasm, acidophilic degeneration, visible extensive hepatocytesteatosis, and lymphocytic infiltration (grade III, Figure 7B). As demonstrated in Table 4 and Figure 7D–F, nicotiflorin at different doses (25, 50, and 100 mg/kg bw) showed liver structure damage prevention effects at various levels against a D-GalN challenge. The histological observations basically supported the results obtained from biochemical index.

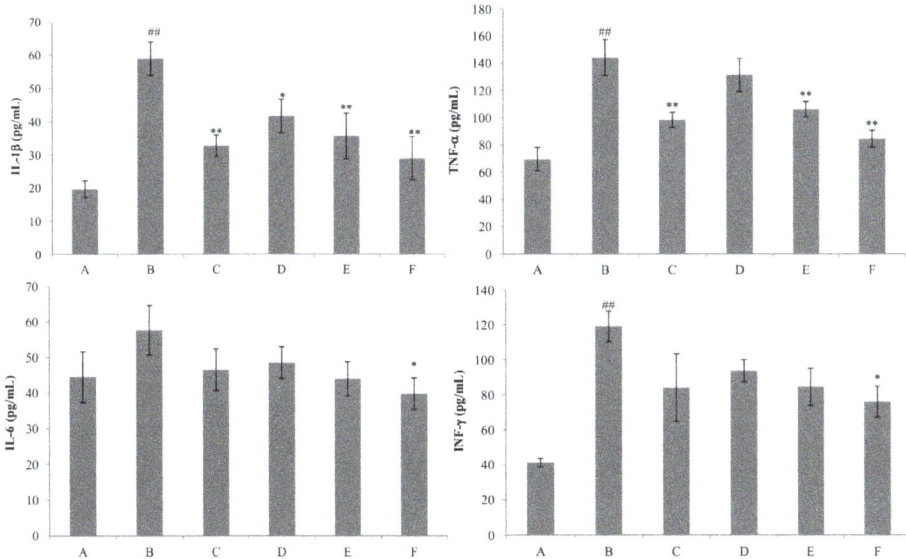

Figure 5. Effects of nicotiflorin on the serum interleukin-1β (IL-1β), tumor necrosis factor-α (TNF-α), interleukin-6 (IL-6), and interferon-γ (IFN-γ) in Con A-intoxicated mice. (A) Control group; (B) Con A-treated group; (C) Con A and DDB (150 mg/kg bw)-treated group. (D) Con A and nicotiflorin (25 mg/kg bw)-treated group; (E) Con A and nicotiflorin (50 mg/kg bw)-treated groupl and (F) Con A and nicotiflorin (100 mg/kg bw)-treated group. Values are mean ± S.E.M., n = 10. ## $p < 0.01$ compared with the control group. * $p < 0.05$, ** $p < 0.01$ compared with the Con A group.

Table 4. Effects of nicotiflorin on the pathological grading of D-GalN-intoxicated mice.

Group	0	I	II	III	p-Value
Control	10	0	0	0	-
D-GalN	0	0	3	7	#
DDB (150 mg/kg bw) + D-GalN	1	6	2	1	*
nicotiflorin (25 mg/kg bw) + D-GalN	1	5	3	1	*
nicotiflorin (50 mg/kg bw) + D-GalN	2	4	3	1	*
nicotiflorin (100 mg/kg bw) + D-GalN	3	5	1	1	*

n = 10; # $p < 0.01$ compared with control group; * $p < 0.05$, compared with D-GalN group.

Liver injury induced by D-GalN provoked a significant reduction of SOD and GSH activities ($p < 0.01$) and a significant increment of MDA and NO content ($p < 0.01$) in the liver homogenate of the D-GalN group as compared to the control group (Figure 8). The results showed that the content of GSH was significantly increased by nicotiflorin at the doses of 25, 50, and 100 mg/kg bw ($p < 0.05$,

$p < 0.01$). Treatment with nicotiflorin (50 and 100 mg/kg bw) significantly prevented the reduction of SOD activity ($p < 0.05$, $p < 0.01$) and the increase of MDA content ($p < 0.05$, $p < 0.01$) induced by D-GalN intoxication. Compared with control group, Con A significantly increased hepatic homogenate NO content ($p < 0.01$). Nicotiflorin (25, 50, 100 mg/kg bw) could significantly decrease serum NO content compared to the D-GalN group ($p < 0.01$), and the positive control drug, DDB (150 mg/kg bw), also significantly decreased serum NO levels ($p < 0.01$). In the D-GalN group, the serum IL-1β, IL-6, TNF-α, and IFN-γ levels were significantly higher than that of the control group ($p < 0.01$). Treatment with nicotiflorin (50, 100 mg/kg bw) significantly reduced the increased serum IL-1β by D-GalN ($p < 0.01$). Moreover, compared with the D-GalN group, nicotiflorin (100 mg/kg bw) could significantly decrease serum TNF-α, IL-6 and IFN-γ levels ($p < 0.05$) (Figure 9).

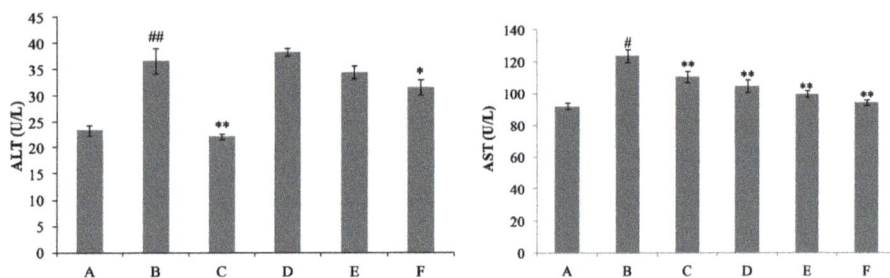

Figure 6. Effects of nicotiflorin on hepatic homogenate ALT and AST in D-GalN-intoxicated mice. (A) Control group; (B) D-GalN-treated group; (C) D-GalN and DDB (150 mg/kg bw)-treated group; (D) D-GalN and nicotiflorin (25 mg/kg bw)-treated group; (E) D-GalN and nicotiflorin (50 mg/kg bw)-treated group; and (F) D-GalN and nicotiflorin (100 mg/kg bw)-treated group. Values are mean ± S.E.M., $n = 10$. # $p < 0.05$, ## $p < 0.01$ compared with the control group. * $p < 0.01$, ** $p < 0.01$ compared with the D-GalN group.

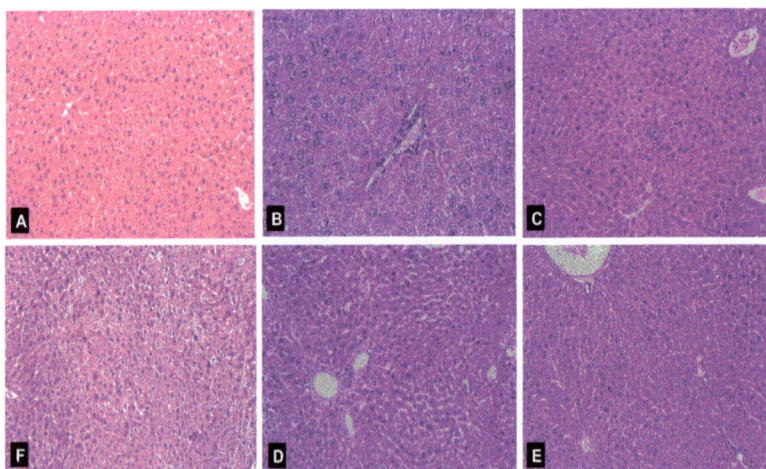

Figure 7. Histological analysis of the livers after D-GalN administration. Typical images were chosen from the different experimental groups (original magnification 10×20). (**A**) Control group; (**B**) D-GalN-treated group; (**C**) D-GalN and DDB (150 mg/kg bw)-treated group; (**D**) D-GalN and nicotiflorin (25 mg/kg bw)-treated group; (**E**) D-GalN and nicotiflorin (50 mg/kg bw)-treated group; and (**F**) D-GalN and nicotiflorin (100 mg/kg bw)-treated group.

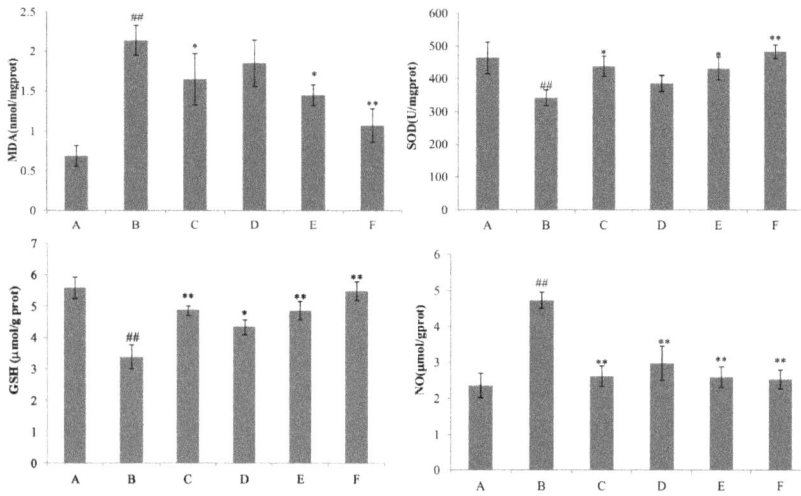

Figure 8. Effects of nicotiflorin on hepatic homogenate SOD, MDA, GSH, and NO in D-GalN-intoxicated mice. (A) Control group; (B) D-GalN-treated group; (C) D-GalN and DDB (150 mg/kg bw)-treated group; (D) D-GalN and nicotiflorin (25 mg/kg bw)-treated group; (E) D-GalN and nicotiflorin (50 mg/kg bw)-treated group; and (F) D-GalN and nicotiflorin (100 mg/kg bw)-treated group. Values are mean ± S.E.M., n = 10. $^{\#\#}$ $p < 0.01$ compared with the control group. * $p < 0.05$, ** $p < 0.01$ compared with the D-GalN group.

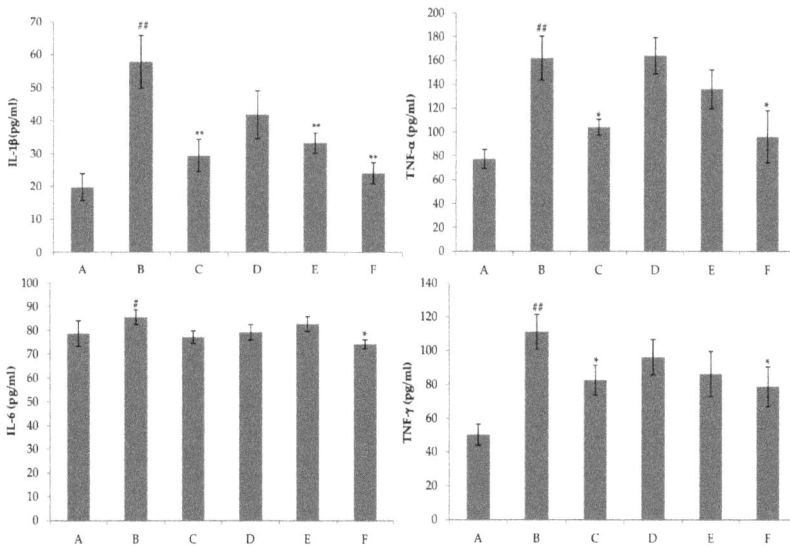

Figure 9. Effects of nicotiflorin on the serum IL-1β, TNF-α, IL-6, and IFN-γ in D-GalN-intoxicated mice. (A) Control group; (B) D-GalN-treated group; (C) D-GalN and DDB (150 mg/kg)-treated group; (D) D-GalN and nicotiflorin (25 mg/kg bw)-treated group; (E) D-GalN and nicotiflorin (50 mg/kg bw)-treated group; and (F) D-GalN and nicotiflorin (100 mg/kg bw)-treated group. Values are mean ± S.E.M., n = 10. $^{\#}$ $p < 0.05$, $^{\#\#}$ $p < 0.01$ compared with the control group. * $p < 0.05$, ** $p < 0.01$ compared with the D-GalN group.

3. Materials and Methods

3.1. Chemicals and Reagents

Concanavalin A (Con A) and D-galactosamine (D-GalN) were purchased from Sigma Chemical Co. (St. Louis, MO, USA). Biphenyl dicarboxylate (DDB) was obtained from Dezhou Deyao Pharmaceutical Co. Assay kits for aspartate aminotransferase (AST) and alanine aminotransferase (ALT) were provided by Zhongsheng Tech. (Beijing, China). Commercial kits used for determining (MDA), superoxide dismutase (SOD), glutathione (GSH), and nitric oxide (NO) activities were obtained from the Jiancheng Institute of Biotechnology (Nanjing, China). Elisa kits for interleukin-1β (IL-1β), interleukin-6 (IL-6), tumor necrosis factor-α (TNF-α), and interferon-γ (INF-γ) were supplied by Biosource Co. (St. Louis, MO, USA). All other chemicals were of analytical grade and were purchased from a local reagent retailer.

3.2. Plant Material and Preparation of Nicotiflorin

The dried flower buds of *N. candida* were purchased from the Traditional Uighur Medicine Hospital in Urumqi and identified by associate researcher Jiang He, Institute of Materia Medica of Xingjiang in China. Ten kilograms of this plant were extracted with 70% ethanol under reflux for 1 h three times, and 70% ethanol extracts was evaporated under vacuum. The 70% ethanol extracts were purified by D101 resin to obtain the extracts as follows: water and 30%, 50%, and 95% ethanol eluates, of which 50% of the ethanol eluates were applied to an ODS RP-18 column and eluted with mixtures of MeOH/H_2O (0%–100%) successively. Nicotiflorin was obtained from 40% methanol eluates by Sephadex LH-20 chromatography repeatedly. The chromatographic analysis of nicotiflorin was performed using an high performance liquid chromatography (HPLC) system consisting of a Shimadzu LC-10ATvp and Phenomenex Gemini column (250 mm × 4.6 mm, 5 μm, with precolumn). The mobile phase was composed of A: 0.2% phosphoric acid aqueous solution and B: acetonitrile. The gradient program was set as follows; 0–5 min, 5% B in A; 5–10 min, 5%–11% B in A; 10–30 min, 11%–14% B in A; 30–60 min, and 14%–20% B in A. The detection wavelength was 266 nm, and the flow rate was 1.0 mL/min. The nicotiflorin purity content was quantified as 98.12% by peak area normalization method.

3.3. Animals

Kunming mice, weighing 20.0 ± 2.0 g, supplied by the Experimental Animal Centre of Xinjiang Medical University in China (No. SYXK(xin) 2011-0004). The mice were housed in plastic cages with a room temperature of 25 ± 1 °C under a 12 h light–dark cycle and were provided with rodent chow and water ad libitum. All procedures were approved by the Animal Care and Used Committee (No. 20150812-1; 12 August 2015) of Institute of Materia Medica of Xinjiang (Urumqi, China).

3.4. Concanavalin A (Con A)-Induced Hepatotoxicity

The mice were randomly divided into six groups; the control group, the Con A-induced liver injury model group, the positive control group (DDB, 150 mg/kg bw), and the nicotiflorin groups (25, 50, and 100 mg/kg bw). Mice in the control and model groups were given distilled water by intragastric administration (ig) (0.2 mL/10 g bw, once daily). Mice in the DDB and nicotiflorin groups received DDB (150 mg/kg bw, ig, once daily) and nicotiflorin (25, 50, and 100 mg/kg bw, ig, once daily), respectively. All administrations were conducted for ten consecutive days. One hour after the last administration on the seventh day, mice in the control group received saline (0.1 mL/10 g bw, iv) while mice in the other groups were injected with Con A (20 mg/kg bw) [28]. Mice were sacrificed after fasting for 8 h, blood samples were collected, and serum was isolated for further tests; the livers were removed for biochemical studies and histopathological analysis.

3.5. D-*Galactosamine (D-GalN)-Induced Hepatotoxicity*

To study the effect of nicotiflorin on D-GalN-induced liver injury, mice were randomly divided into six groups with 10 mice per group as follows; control group, model group, DDB group (150 mg/kg bw), and the nicotiflorin groups (25, 50 or 100 mg/kg bw). The mice in the pre-treatment groups were administered by intragastric gavage (0.1 mL/10 g bw) with different doses of nicotiflorin, respectively, once a day for 7 days, while the control and model groups were given distilled water only. On the seventh day, 1 h after of the last administration, mice in various groups were given an intraperitoneal injection of D-GalN in normal saline (800 mg/kg, 0.2 mL/10 g b.w.), while the control group was injected intraperitoneally with an equal amount of normal saline solution. Mice were sacrificed after fasting for 8 h. Blood samples were collected and serum was isolated for further tests. The livers were removed for biochemical studies and histopathological analysis [29].

3.6. *Measurement of Liver Index, Spleen Index and Body Weight in Mice*

The body weight of the animals was weighed before and after the experiment. The liver index and spleen index were calculated as liver weight (mg) and spleen weight (mg) divided by the body weight of the mice (10 g), respectively.

3.7. *Measurement of Aminotransferase and Cytokine Levels in the Serum*

The blood samples were collected by retroorbital bleeding, the collected blood was centrifuged at 3000 r/min for 10 min and 4 °C, and serum was obtained. The activities of serum enzymes alanine aminotransferase (ALT) and aspartate aminotransferase (AST) were determined using the commercial assay kits. Enzyme activities were expressed as an international unit (U/L). The levels of IL-1β, IL-4, TNF-α, and IFN-γ in plasma were determined using enzyme-linked immunosorbent assay (ELISA) kits according to the kit introduction.

3.8. *Measurement of Liver Homogenate Contents of MDA, SOD, GSH and NO*

Liver samples were homogenized in normal saline to give a 10% (w/v) liver homogenate and then centrifuged at 3000 rpm for 10 min at 4 °C. Supernatant was used to determine the MDA, GSH, SOD, and total protein concentrations by using the detection kits according to the manufacturer's protocols. The levels of NO in liver homogenate were measured using nitrate reductase assay according to the kit introduction.

3.9. *Histopathological Examination*

For the histological investigations, liver tissues were removed from a portion of the left lobe, fixed in 10% formalin, embedded in paraffin, sliced in 5 μm sections, and stained with hematoxylin and eosin (H and E) according to standard protocols. The slides were observed for conventional morphological evaluation under a light microscope (Olympus BX43, Olympus, Tokyo, Japan) and photographed at 10 × 20 magnification. The degree of liver histological damage was scored as follows on a scale of 0–III: grade 0, no necrosis with normal liver tissue structure; I, part of the liver tissue swelling accompanied with sporadic dotted necrosis and necrosis <1/4 of the hepatic lobule area; grade II, liver cell swelling, visible in the spotty necrosis and minimal necrosis, inflammatory cells infiltration in the portal area, and necrosis <1/2 of the hepatic lobule area; III, liver cell swelling, massive necrosis, inflammatory cell infiltration, and necrosis >1/2 of the hepatic lobule area [30].

3.10. *Statistical Analysis*

All data were expressed as the mean ± standard error (S.E.M.). The differences between different groups were analyzed using one-way analysis of variance (ANOVA) (SPSS software package for windows, version 13.0, Chicago, IL, USA). * $p < 0.05$ and ** $p < 0.01$ were taken as statistically significant.

4. Conclusions

The present study clearly demonstrates that nicotiflorin possess significant *in vivo* anti-hepatotoxic activities, and this is confirmed in two experimental animal models. Therefore, nicotiflorin could be used as a protective agent against acute liver injury, and its mechanism might be attributed to its antioxidant and immunoregulatory capacities.

Acknowledgments: This work is supported by grants from the science and technology support program of Xinjiang Uygur Autonomous Region in China (No. 201333118).

Author Contributions: Jun Zhao, Tao Liu, Tengfei Ji, and Zhengyi Gu conceived and designed the experiments; Shilei Zhang, Jun Zhao, Shuping You, and Fang Xu performed the experiments; Jun Zhao, Tao Liu, and Shilei Zhang analyzed the data; Tao Liu, Tengfei Ji, and Zhengyi Gu contributed reagents, materials, and analysis tools; and Jun Zhao wrote the paper.

Conflicts of Interest: The authors declare no conflict of interest.

References

1. Ghabril, M.; Chalasani, N.; Björnsson, E. Drug-induced liver injury: A clinical update. *Curr. Opin. Gastroenterol.* **2010**, *26*, 222–226. [CrossRef] [PubMed]

2. Giannelli, G.; Quaranta, V.; Antonaci, S. Tissue remodelling in liver diseases. *Histol. Histopathol.* **2003**, *18*, 1267–1274. [PubMed]

3. Upur, H.; Amat, N.; Blazekovic, B.; Talip, A. Protective effect of *Cichorium glandulosum* root extract on carbon tetrachloride-induced and galactosamine-induced hepatotoxicity in mice. *Food Chem. Toxicol.* **2009**, *47*, 2022–2030. [CrossRef] [PubMed]

4. Hsiang, C.Y.; Lin, L.J.; Kao, S.T.; Lo, H.Y.; Chou, S.T.; Ho, T.Y. Glycyrrhizin, silymarin, and ursodeoxycholic acid regulate a common hepatoprotective pathway in HepG2 cells. *Phytomedicine* **2015**, *22*, 768–777. [CrossRef] [PubMed]

5. Liu, J.; Wu, Q.; Lu, Y.F.; Pi, J.B. New insights into generalized hepatoprotective effects of oleanolic acid: Key roles of metallothionein and Nrf2 induction. *Biochem. Pharm.* **2008**, *76*, 922–928. [CrossRef] [PubMed]

6. Černý, D.; Lekić, N.; Váňová, K.; Muchová, L.; Hořínek, A.; Kmoníčková, E.; Zídek, Z.; Kameníková, L.; Farghali, H. Hepatoprotective effect of curcumin in lipopolysaccharide/galactosamine model of liver injury in rats: Relationship to HO-1/CO antioxidant system. *Fitoterapia* **2011**, *82*, 786–791.

7. Tiegs, G.; Hentschel, J.; Wendel, A. A T cell-dependent experimental liver injury in mice inducible by concanavalin A. *J. Clin. Investig.* **1992**, *90*, 196–203. [CrossRef] [PubMed]

8. Zhao, J.; Liu, T.; Ma, L.; Yan, M.; Gu, Z.Y.; Huang, Y.; Xu, F.; Zhao, Y. Antioxidant and preventive effects of extract from *Nymphaea candida* flower on in vitro immunological liver injury of rat primary hepatocyte cultures. *Evid. Complement. Altern. Med.* **2009**. [CrossRef]

9. Tong, S.; Yan, J.; Chen, G.; Lou, J. Purification of rutin and nicotiflorin from the flowers of *Edgeworthia chrysantha* Lindl. by high speed counter-current chromatography. *J. Chromatogr. Sci.* **2009**, *47*, 341–344. [PubMed]

10. Olajide, O.; Li, S.S.; Liu, H.T.; Chai, X.; Wang, Y.F.; Gao, X.M. Studies on chemical constituents and DPPH Free radical scavenging activity of *Carthamus tinctorius* L. *Nat. Prod. Res. Dev.* **2014**, *26*, 60–63.

11. Zhao, J.; Yan, M.; He, J.H.; Huang, Y.; Zhao, Y. Flavonol glycosides from the flowers of *Nymphaea Candida*. *Chin. JMAP* **2008**, *25*, 115–117.

12. Huang, J.L.; Fu, S.T.; Jiang, Y.Y.; Cao, Y.B.; Guo, M.L.; Wang, Y.; Xu, Z. Protective effects of nicotiflorin on reducing memory dysfunction, energy metabolism failure and oxidative stress in multi-infarct dementia model rats. *Pharmcol. Biochem. Behav.* **2007**, *86*, 741–748. [CrossRef] [PubMed]

13. Li, R.P.; Guo, M.L.; Zhang, G.; Xu, X.F.; Li, Q. Nicotiflorin reduces cerebral ischemic damage and upregulates endothelial nitric oxide synthase in primarily cultured ratcerebral blood vessel endothelial cells. *J. Ethnopharmacol.* **2006**, *107*, 143–150. [CrossRef] [PubMed]

14. Li, R.P.; Guo, M.L.; Zhang, G.; Xu, X.F.; Li, Q. Neuroprotection of nicotiflorin inpermanent focal cerebral ischemiaand in neuronal cultures. *Biol. Pharm. Bull.* **2006**, *29*, 1868–1872. [CrossRef] [PubMed]

15. Habtemariam, S. A-glucosidase inhibitory activity of kaempferol-3-*O*-rutinoside. *Nat. Prod. Commun.* **2011**, *6*, 201–203. [PubMed]

16. Wang, Y.; Tang, C.Y.; Zhang, H. Hepatoprotective effects of kaempferol 3-*O*-rutinoside and kaempferol 3-*O*-glucoside from *Carthamus tinctorius* L. on CCl$_4$-induced oxidative liver injury in mice. *J. Food Drug Anal.* **2015**, *23*, 310–317. [CrossRef]

17. Imose, M.; Nagaki, M.; Kimura, K.; Takai, S.; Imao, M.; Naiki, T. Leflunomide protects from T-cell-mediated liver injury in mice through inhibition of nuclear factor κB. *Hepatology* **2004**, *40*, 1160–1169. [CrossRef] [PubMed]

18. Chen, F.; Zhu, H.H.; Zhou, L.F.; Li, J.; Zhao, L.Y.; Wu, S.S.; Wang, J.; Liu, W.; Chen, Z. Genes related to the very early stage of Con A-induced fulminant hepatitis: A gene-chip-based study in a mouse model. *BMC Genom.* **2010**, *11*, 240. [CrossRef] [PubMed]

19. Zhou, Y.; Chen, K.; He, L.; Xia, Y.; Dai, W.; Wang, F.; Li, J.; Li, S.; Liu, T.; Zheng, Y.; et al. The protective effect of resveratrol on concanavalin-A induced acute hepatic injury in mice. *Gastroenterol. Res. Pract.* **2015**, *2015*, 506390. [CrossRef] [PubMed]

20. Sass, G.; Heinlein, S.; Agli, A.; Bang, R.; Schumann, J.; Tiegs, G. Cytokine expression in three mouse models of experimental hepatitis. *Cytokine* **2002**, *19*, 115–120. [CrossRef] [PubMed]

21. Yang, Y.; Qin, X.Y.; Guo, Z.; Liu, T.Y.; Pan, R.Y. Protective effect of ginsenoside Rg1 on immue-mediated liver injury in mice. *Chin. J. Public Health* **2015**, *31*, 309–311.

22. Pan, C.W.; Zhou, G.Y.; Chen, W.L.; Lu, Z.G.; Jin, L.X.; Zheng, Y.; Lin, W.; Pan, Z.Z. Protective effect of forsythiaside A on lipopolysaccharide/D-galactosamine-induced liver injury. *Int. Immunopharmacol.* **2015**, *26*, 80–85. [CrossRef] [PubMed]

23. Ksontini, R.; Colagiovanni, D.B.; Josephs, M.D.; Edwards, C.K.; Tannahill, C.L.; Solorzano, U. Disparate roles for TNF-α and fasligand in concanavalin A-induced hepatitis. *J. Immunol.* **1998**, *160*, 4082–4089. [PubMed]

24. Kimikide, N.; Mitsuyoshi, O.; Masashi, Y.; Shujiro, T.; Yukiomi, N.; Keisuke, T.; Kazunobu, A.; Isao, M. Macrophage inflammatory protein-2 induced by TNF-α plays a pivotal role in concanavalin A-induced liver injury in mice. *J. Hepatol.* **2001**, *35*, 217–224.

25. Sinha, M.; Manna, P.; Sil, P.C. Amelioration of galactosamine-induced nephrotoxicity by a protein isolated from the leaves of the herb, *Cajanus indicus* L. *BMC Complement. Altern. Med.* **2007**, *7*, 11. [CrossRef] [PubMed]

26. Abe, K.; Ijiri, M.; Suzuki, T.; Taguchi, K.; Koyama, Y.; Isemura, M. Green tea with a high catechin content suppresses inflammatory cytokine expression in the galactosamine-injured rat liver. *Biomed. Res.* **2005**, *26*, 187–192. [CrossRef] [PubMed]

27. Choi, J.H.; Kang, J.W.; Kim, D.W.; Sung, Y.K.; Lee, S.M. Protective effects of Mg-CUD against D-galactosamine-induced hepatotoxicity in rats. *Eur. J. Pharm.* **2011**, *657*, 138–143. [CrossRef] [PubMed]

28. Wang, J.H.; Huang, Z.M.; Yang, X.B.; Chen, H.Y. Protective effect of hyperin on immunological liver injury in mice. *Chin. J. Exp. Tradit. Med. Fomulae* **2015**, *21*, 137–141.

29. Ai, G.; Huang, Z.M.; Liu, Q.C.; Han, Y.Q.; Chen, X. The protective effect of total phenolics from *Oenanthe Javanica* on acute liver failure induced by D-galactosamine. *J. Ethnopharmacol.* **2016**, *186*, 53–60. [CrossRef] [PubMed]

30. Zhang, L.; Li, J.; Wang, J.Q.; Xia, L.J.; Jiang, H.; Jing, D. The protective effects of total flavonoids Chrysanthemum indicum on CCl$_4$-induced acute liver injury in mice. *Acta Univ. Med. Anhui* **2007**, *42*, 412–414.

International Journal of
Molecular Sciences

MDPI

Article

Profile of Polyphenol Compounds of Five Muscadine Grapes Cultivated in the United States and in Newly Adapted Locations in China

Zheng Wei [1,2,†], Jianming Luo [1,†], Yu Huang [3], Wenfeng Guo [2], Yali Zhang [1], Huan Guan [2], Changmou Xu [4] and Jiang Lu [1,2,5,*]

1   College of Food Science and Nutritional Engineering, China Agricultural University, Beijing 100083, China; weizheng76096@sina.com (Z.W.); baiding86@126.com (J.L.); zhangyali@cau.edu.cn (Y.Z.)
2   Guangxi Crop Genetic Improvement and Biotechnology Laboratory, Guangxi Academy of Agricultural Sciences, Nanning 530007, China; gwenfeng@163.com (W.G.); guanhuan2010@163.com (H.G.)
3   Grape and Wine Research Institute, Guangxi Academy of Agricultural Sciences, Nanning 530007, China; hy0611@163.com
4   Department of Food Science and Technology, University of Nebraska-Lincoln, Lincoln, NE 68508, USA; cxu13@unl.edu
5   Center for Viticulture and Enology, School of Agriculture and Biology, Shanghai Jiao Tong University, Shanghai 200240, China
*   Correspondence: jiang.lu@sjtu.edu.cn; Tel.: +86-21-3420-6005
†   These authors contributed equally to this work.

Academic Editor: Toshio Morikawa
Received: 27 January 2017; Accepted: 10 March 2017; Published: 14 March 2017

Abstract: Polyphenol compositions and concentrations in skins and seeds of five muscadine grapes (cv. "Noble", "Alachua", "Carlos", "Fry", and "Granny Val") cultivated in the United States (Tallahassee-Florida, TA-FL) and South China (Nanning-Guangxi, NN-GX and Pu'er-Yunnan, PE-YN) were investigated, using ultra performance liquid chromatography tandem triple quadrupole time-of-flight mass spectrometry (UPLC Triple TOF MS/MS). Fourteen ellagitannins were newly identified in these muscadine grapes. The grapes grown in NN-GX accumulated higher levels of ellagic acid, methyl brevifolin carboxylate, and ellagic acid glucoside in skins, and penta-O-galloyl-glucose in seeds. In PE-YN, more flavonols were detected in skins, and higher contents of flavan-3-ols, ellagic acid, and methyl gallate were identified in seeds. Abundant seed gallic acid and flavonols were found among the grapes grown in TA-FL. Based on principal component analysis (PCA) of 54 evaluation parameters, various cultivars grown in different locations could be grouped together and vice versa for the same cultivar cultivated in different regions. This is the result of the interaction between genotype and environmental conditions, which apparently influences the polyphenol synthesis and accumulation.

Keywords: polyphenols; ellagic acids; muscadine grapes; geographical location; UPLC-Triple TOF-MS/MS

1. Introduction

Muscadinia rotundifolia Michx., commonly called as muscadine grape, is a genus from the Vitaceae family. Muscadine grapes are indigenous to the southeast United States and are well-adapted to the warm, humid environment [1,2]. They have important economic value, mainly due to their resistance to Pierce's disease (*Xylella fastidiosa*), *Plasmopara viticola*, and *Elsinoe ampelina* [3–6]. They also have multiple health benefits, including antioxidant, antiviral agents, anticancer, antibacterial, and anti-inflammatory properties [7–9]. The muscadine grape currently comprises over 100 cultivars such as "Noble", "Alachua", "Carlos", "Fry", and "Granny Val", among many others [2]. All commercial muscadine grape production occurs in the United States [7,10]. Recently, some of these muscadine grape cultivars have been introduced to South China for the first time. These cultivars provide better possibilities, adapting to local climates for table and wine grape production in areas unsuitable for the growth of *Vitis vinifera* [11,12].

Polyphenol compounds are secondary metabolites and represent the third most abundant constituent in grapes after carbohydrates and fruit acids. They are mainly distributed in seeds (60%–70%) and skins (28%–35%), with less than 10% in pulp [5,13]. Major polyphenols present in muscadine grapes have been previously reported. For example, the skins and seeds of these grapes contained large amounts of ellagic acids (ellagic acid derivatives and ellagitannins) [13,14]. The anthocyanin levels in red skins of "Noble" and "Alachua" ranged from 292.0 to 554.9 mg as malvidin-3-glucoside per 100 g dry weight (DM) [12]. In addition, their skins also had flavonols such as glycosides of quercetin, kaempferol, and myricetin [3,5]. Meanwhile, abundant flavan-3-ols like procyanidins (epicatechin and epicatechin gallate) and condensed tannins were identified in muscadine seeds [13,15,16]. However, these identifications have not been fully elucidated, and the contents varied greatly in different reports [1,10,13,14,17]. These differences were mainly because of the complexity, diversity, and polymeric and isomeric forms of large size of these compounds, as well as the lack of complete fragmentation data and suitable quantitative standards [18,19]. In addition, materials from different growing conditions may also contribute to the variance.

The synthesis and accumulation of polyphenol compounds in fruits are regulated by genetic and climatic conditions, as well as agricultural practices [20]. Marshall et al. [5] examined the polyphenol concentrations in muscadine fruits of 21 cultivars, and found that the stilbene, ellagic acid, and flavonol differed significantly among cultivars. Chen [7] investigated skin extracts among 17 muscadine cultivars (6 bronze and 11 dark) and discovered that dark-skinned cultivars had higher contents than the bronze ones with respect to polyphenol and ellagic acids. Meanwhile, the impact of climatic factors on polyphenol compositions and concentrations of grape berries has been widely studied. It was reported that these compounds of grape berries varied in different vineyards, and variations were also presented among cultivars being grown in the same vineyard for two consecutive years [21]. Additionally, agronomic strategies such as alteration of environmental conditions (light, temperature, mineral nutrition, and water management), application of elicitors, stimulating agents and plant activators have been employed to enhance the biosynthesis of polyphenols [18,22–25].

To our knowledge, the muscadine grapes have never been reported as growing commercially in countries outside the United States. Therefore, the aim of this study is to characterize and distinguish the polyphenol compositions of five muscadine grape cultivars grown in China (Nanning-Guangxi and Pu'er-Yunnan) and the native United States (Tallahassee-Florida), respectively, and to link the differences to the growing conditions in different geographical regions in order to elucidate factors related to the polyphenol accumulation. The overall goal of this study is to develop a strategy to grow muscadine grapes with richer polyphenol compositions in China.

2. Results

2.1. Total Phenolic Content (TPC)

Muscadine grapes cv. "Noble", "Alachua", "Carlos", and "Fry" harvested in 2012 season from Nanning-Guangxi (NN-GX) and Pu'er-Yunnan (PE-YN), China, had higher total phenolic content (TPC) in skins and lower in seeds than those grown in their native origin of Tallahassee-Florida (TA-FL), United States (Table 1). "Granny Val", as an exception, had less TPC in skins but was richer in TPC in its seeds in China than in the USA. Nevertheless, the combined TPCs of the muscadine grape skins and seeds grown in China were a little bit lower than those of grapes grown in the USA. For all cultivars grown in China and USA, TPCs of grape seeds were about 3-fold higher than those of skins. For example, the TPC in "Noble" seeds was 96.65 mg gallic acid equivalent (GAE)/g dry weight (DW), while in skins it was 38.96 mg GAE/g DW. This is because the precursors of ellagic acid biosynthesis and condensed tannins were mainly detected in seeds, which resulting a higher TPC in seeds than in skins.

Table 1. Total phenolic content (TPC) in skins and seeds of five muscadine grapes cultivated in three different regions.

Cultivar	Region	Year	TPC (mg GAE/g DW)		
			Skins	Seeds	Skins + Seeds
Red cultivars					
Noble (N)	TA-FL	2012	38.96 ± 2.57 [d]	96.65 ± 3.30 [a]	135.61 ± 5.17 [c]
	NN-GX	2012	42.06 ± 2.00 [c]	82.10 ± 5.15 [b]	124.16 ± 6.55 [d]
		2013	48.08 ± 0.67 [b]	95.07 ± 1.07 [a]	143.14 ± 0.95 [b]
	PE-YN	2012	41.40 ± 1.93 [c]	85.44 ± 3.40 [b]	126.84 ± 4.32 [d]
		2013	52.68 ± 1.91 [a]	97.55 ± 3.04 [a]	150.23 ± 3.19 [a]
	Sig. [a]		**	***	***
Alachua (A)	TA-FL	2012	36.21 ± 1.70 [b]	118.59 ± 2.51 [a]	154.80 ± 3.33 [b]
	NN-GX	2012	45.03 ± 1.35 [a]	107.05 ± 3.48 [b]	152.08 ± 4.28 [b]
		2013	46.12 ± 2.82 [a]	119.41 ± 4.14 [a]	165.53 ± 1.49 [a]
	Sig. [a]		***	***	ns
	Sig. [b]		**, **	***, ***	***, ***
Bronze cultivars					
Carlos (C)	TA-FL	2012	30.86 ± 4.88 [c]	103.74 ± 4.50 [b]	134.60 ± 7.92 [c]
	NN-GX	2012	35.98 ± 1.58 [b]	90.32 ± 5.06 [c]	126.30 ± 5.38 [d]
		2013	56.44 ± 1.37 [a]	122.55 ± 1.61 [a]	178.99 ± 1.96 [a]
	PE-YN	2012	30.22 ± 1.50 [c]	83.83 ± 4.80 [d]	114.05 ± 4.94 [e]
		2013	57.33 ± 2.07 [a]	102.87 ± 3.01 [b]	160.20 ± 3.58 [b]
	Sig. [a]		***	***	***
Fry (F)	TA-FL	2012	27.43 ± 4.25 [b]	118.74 ± 3.47 [a]	146.17 ± 6.69 [b]
	NN-GX	2012	38.68 ± 1.83 [a]	77.88 ± 3.81 [c]	116.56 ± 3.54 [c]
		2013	39.80 ± 2.02 [a]	113.49 ± 2.78 [b]	153.29 ± 3.70 [a]
	Sig. [a]		***	***	***
Granny Val (G)	TA-FL	2012	34.44 ± 2.02 [b]	76.56 ± 2.28 [d]	111.00 ± 3.83 [c]
	NN-GX	2012	29.34 ± 2.41 [c]	81.14 ± 1.73 [c]	110.48 ± 3.21 [c]
		2013	44.92 ± 1.08 [a]	116.36 ± 1.21 [a]	161.28 ± 0.94 [a]
	PE-YN	2012	20.53 ± 0.78 [d]	85.78 ± 2.94 [b]	106.31 ± 3.24 [d]
		2013	35.38 ± 1.08 [b]	114.13 ± 3.46 [a]	149.51 ± 3.85 [b]
	Sig. [a]		***	***	*
	Sig. [b]		***, ***	***, ***	***, ***

Values are expressed as means of triplicate determinations ± standard deviation S.D. For each cultivar, different small letters within column indicate significant differences (Duncan's test, $p = 0.05$). Significance among different regions for each cultivar in the 2012 season (Sig. [a]), and among different color cultivars grown in TA-FL and NN-GX in 2012 (Sig. [b]) was tested for $p < 0.001$ (***), $p < 0.01$ (**), $p < 0.05$ (*), and not significant (ns), respectively. TA-FL: Tallahassee-Florida; NN-GX: Nanning, Guangxi; PE-YN: Pu'er-Yunnan; GAE: gallic acid equivalent.

2.2. Polyphenol Composition and Accumulation

2.2.1. Ellagic Acids and Precursors Profiles

Ellagic acids (ellagic acid derivatives and ellagitannins) and their precursors (gallic acid derivatives) were the chief polyphenol compounds identified in muscadine skins and seeds (Table 2). In 2012 season, the grapes grown in China exhibited significantly higher contents of ellagic acids and precursors than those grown in the USA, except "Noble", which possessed significantly lower levels of ellagic acids and precursors in seeds. The highest values of skin ellagic acids and precursors were observed in cultivars "Granny Val" and "Alachua" from NN-GX (248.73 and 220.99 mg GAE/100 g DW, respectively), whereas the lowest were found in cultivar "Granny Val" from TA-FL (85.42 mg GAE/100 g DW). The seeds of red cultivars from TA-FL and NN-GX exhibited higher contents of ellagic acids and precursors than those in PE-YN. For instance, "Noble" from TA-FL had 636.75 mg GAE/100 g DW, while this value was 380.61 mg GAE/100 g DW in PE-YN.

Int. J. Mol. Sci. **2017**, *18*, 631

Table 2. Different categories of polyphenol contents in skins and seeds of five muscadine grapes cultivated in three different regions.

Cultivar	Region	Year	Ellagic Acids and Precursors (mg GAE/100 g DW)		Flavonols (mg RE/100 g DW)		Benzoic Acids (mg GAE/100 g DW)		Flavan-3-ols (mg EE/100 g DW)		Stilbenes (mg REE/100 g DW)		Cinnamic Acids (mg CAE/100 g DW)	
			Skins	Seeds	Skins	Seeds	Skins	Seeds	Skins	Seeds	Skins	Seeds	Skins	Seeds
Red cultivars														
Noble (N)	TA-FL	2012	174.30 b	636.75 a	35.42 c	18.96 a	19.49 b	4.20 d	nd	89.72 c	0.06 b	nd	nd	nd
	NN-GX	2012	197.25 b	557.76 b	26.28 d	7.03 c	36.60 a	19.10 b	0.08 b	128.43 b	nd	nd	nd	nd
		2013	255.66 a	670.73 a	57.52 b	13.03 b	39.00 a	34.90 a	nd	216.25 a	nd	nd	nd	nd
	PE-YN	2012	196.85 b	380.61 d	38.89 c	10.48 b,c	18.00 b	7.18 c,d	0.05 b	136.88 b	0.06 b	nd	nd	nd
		2013	299.12 a	468.22 c	66.00 a	14.76 a,b	40.36 a	10.96 c	nd	230.45 a	0.11 a	n/a	nd	n/a
	Sig. a		ns	***	**	***	***	**	n/a	*	n/a	n/a	n/a	n/a
Alachua (A)	TA-FL	2012	174.31 c	510.76 b	37.74 b	21.26 a	17.04 a	6.75 b	0.02 b	197.61 b	nd	nd	nd	nd
	NN-GX	2012	220.99 b	577.89 a	24.54 c	9.50 b	7.51 b	19.11 a	0.05 a	250.96 b	0.06 a	nd	nd	nd
		2013	294.57 a	512.46 b	55.20 a	28.85 a	9.72 b	0.36 c	0.03 a,b	369.42 a	0.06 a	nd	nd	nd
	Sig. a		*	ns	ns	ns	ns	**	**	ns	n/a	n/a	n/a	n/a
	Sig. b		ns, **	*, ns	ns, ns	ns, *	ns, ***	*, ns	***, **	**, ***	n/a, n/a	n/a, n/a	n/a, n/a	n/a, n/a
Bronze cultivars														
Carlos (C)	TA-FL	2012	144.94 d	257.21 c	21.52 c	20.69 c	12.28 b,c	4.93 b,c	0.05 b	186.56 b	nd	nd	nd	nd
	NN-GX	2012	169.89 c	406.42 b	20.79 c	10.52 d	21.42 a	7.22 a	0.08 b	250.65 a	0.18 a	nd	0.14 b	nd
		2013	212.85 b	457.55 a,b	31.66 b	41.60 a	16.12 b	4.34 c	0.24 a	266.21 a	0.24 a	nd	nd	nd
	PE-YN	2012	155.07 c,d	381.49 d	33.10 b	14.67 c,d	6.81 d	6.14 a,b	0.06 b	177.49 b	nd	nd	0.22 a	nd
		2013	264.76 a	514.50 a	54.75 a	29.06 b	7.77 c,d	1.71 d	0.02 b	258.52 a	0.03 b	n/a	nd	n/a
	Sig. a		*	ns	*	***	**	*	*	***	n/a	n/a	n/a	n/a
Fry (F)	TA-FL	2012	132.30 c	482.50 a	30.06 c	27.89 b	13.31 b	9.14 a	0.02 b	246.94 a	nd	nd	0.11	nd
	NN-GX	2012	201.26 b	486.00 a	39.95 b	7.98 c	22.37 a	6.62 b	0.07 a	167.81 b	nd	nd	nd	nd
		2013	273.25 a	475.86 a	60.60 a	38.97 a	6.59 c	5.50 b	nd	240.49 a	0.12	nd	nd	nd
	Sig. a		*	ns	*	**	**	*	***	*	n/a	n/a	n/a	n/a
Granny Val (G)	TA-FL	2012	85.42 d	153.14 d	16.07 c	10.17 b	16.41 b,c	4.93 b	0.02 c,d	212.01 b,c	0.06 a	nd	0.19 b	nd
	NN-GX	2012	248.73 b	398.08 b	33.90 b	9.82 b	33.97 a	19.65 a	0.04 b,c	172.55 c	0.06 a	nd	nd	nd
		2013	295.44 a	582.68 a	44.83 a,b	31.56 a	14.80 b,c	11.66 a,b	0.06 a,b	250.59 b	0.09 a	nd	nd	nd
	PE-YN	2012	201.41 c	224.19 c	35.69 a,b	12.13 b	11.90 a	18.83 a	0.09 a	312.45 a	0.06 a	nd	0.59 a	nd
		2013	268.67 a,b	281.51 c	47.86 a	23.60 a	23.53 a,b	10.83 a,b	nd	319.56 a	nd	n/a	nd	n/a
	Sig. a		***	***	ns	ns	*	ns	**	**	ns	n/a	n/a	n/a
	Sig. b		*, **	***, ns	ns, **	***, ns	ns, ns	***, ns	***, **	ns, *	n/a, n/a	n/a, n/a	n/a, n/a	n/a, n/a

Values are expressed as means of triplicate determinations. For each cultivar, different small letters within column indicate significant differences (Duncan's test, $p = 0.05$). Significance among different regions for each cultivar in the 2012 season (Sig. [a]), and among different color cultivars grown in TA-FL and NN-GX in 2012 (Sig. [b]) was tested for $p < 0.001$ (***), $p < 0.01$ (**), $p < 0.05$ (*), not significant (ns), and not applicable (n/a), respectively. nd: not detected; CAE: caffeic acid equivalent; EE: (−)-epicatechin equivalent; RE: rutin equivalent; REE: resveratrol equivalent. See Table 1 for the abbreviation of regions.

A total of 46 different ellagic acids and precursors were identified in 2012's skin samples, whereas only 33 different kinds were detected in seeds (Tables S1–S3). These compounds varied among muscadine grape cultivars. The "Noble" had more ellagic acids and precursors in both skins and seeds than the other cultivars. For example, there were 40 different compounds in skins and 26 in seeds of "Noble" grape grown in PE-YN. The constituents of ellagic acids and precursors differed between the skins and seeds. For example, the ellagic acid derivatives were the primary form found in skins (Figure 1a), while the precursors were the main type detected in seeds (Figure 1b). Significant content variations of precursors were found among cultivars growing in different regions. For example, "Alachua" from NN-GX contained the highest level of seed precursors (514.99 mg GAE/100 g DW), whereas "Granny Val" from TA-FL had the lowest (129.29 mg GAE/100 g DW).

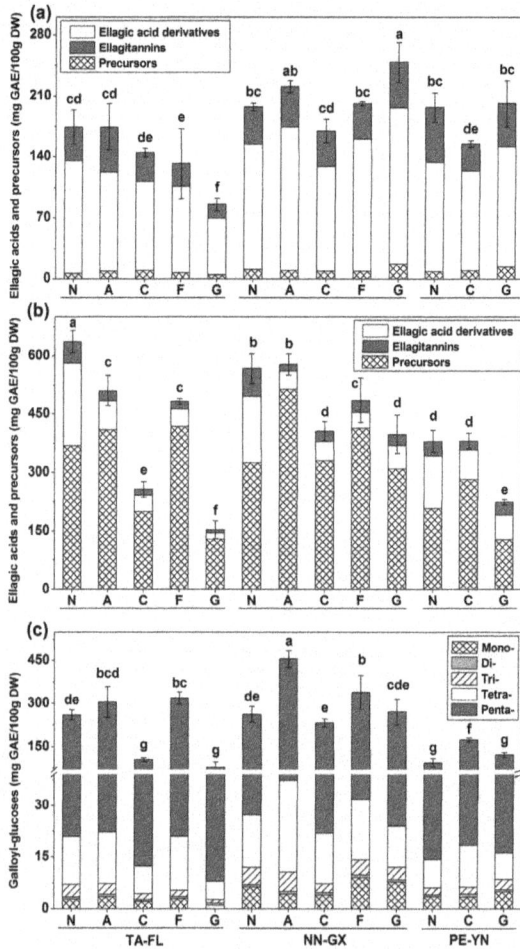

Figure 1. Ellagic acids and precursors distribution in five muscadine skins (**a**) and seeds (**b**), and galloyl-glucoses distribution in muscadine seeds (**c**), among three different regions in the 2012 season. On the top of each column, standard deviation is show for the total content, and different small letters indicate significant differences (Duncan's test, $p = 0.05$). See Table 1 for the abbreviation of cultivars and regions.

Skin ellagic acid accounted for over 55% of total ellagic acids and precursors in all of the muscadine grapes grown in China and USA (Tables S1–S3). Methyl brevifolin carboxylate and tri-*O*-methyl ellagic acid were common in the grapes from TA-FL and NN-GX, whereas ducheside A and B were identified mostly from PE-YN. Ellagic acid glucoside and diglucoside were both detected in skins, and the glucoside/diglucoside ratio was higher in NN-GX.

Thirteen unknown ellagitannins with m/z 443 to 957 were first reported in muscadine skins from this study. Among them, ellagitannin m/z 643 was the most common one, particularly in "Noble" grapes grown in PE-YN. In addition, mono-*O*-methy ellagic acid, ellagitannin m/z 681 and 689 were almost exclusively found in red cultivars, yet tri-*O*-galloyl-glucose was only detected in bronze ones among samples from all the growing regions.

Interestingly, no tetra- and penta-*O*-galloyl-glucoses were detected in the muscadine skins, while five galloyl-glucoses (from mono- to penta-*O*-) were identified in the seeds, of which the overall sum of the five galloyl-glucoses accounted for above 50% of the total seed ellagic acids and precursors in the three regions (Figure 1c). Additionally, muscadine grapes in NN-GX demonstrated the highest content of penta-*O*-galloyl-glucose, while the grapes in TA-FL exhibited higher levels of gallic acid, and methyl gallate was dominant among the muscadine grapes produced in PE-YN. Moreover, an unkown ellagitannin m/z 967 was quantified in muscadine seeds for the first time.

2.2.2. Flavonols Profiles

Flavonols were the second-highest polyphenol in terms of content detected in muscadine skins in our study, whereas their contents in seeds were relatively low (Table 2). The cultivar "Noble" from PE-YN and "Fry" from NN-GX were characterized with significantly higher skin flavonols (38.92 and 38.89 mg rutin equivalent (RE)/100 g DW, respectively) than others. However, the seed flavonols significantly differed among the growing regions. In general, higher flavonols were found among grape cultivars grown in the USA. (10.17–27.89 mg RE/100 g DW in TA-FL) than in China (10.48–14.47 mg RE/100 g DW in PE-YN, and 7.03–10.52 mg RE/100 g DW in NN-GX).

The contents of skin flavonol glycosides ranked in order as following: aglycones > monoglucosides > acylates > diglucosides, according to the concentrations (Tables S1–S3). The region of PE-YN exhibited significantly higher contents of aglycones in comparison to other regions. Additionally, the monoglucosides (glucoside and glucuronide) were richer in red cultivars than in bronze ones, while the diglucosides (quercetin-3,4′-*O*-diglucoside) were only present in red cultivars, consistently in China and USA. Interestingly, the ratio of mono-/di- glucoside flavonols varied among the muscadine grapes growing in different regions, for example, the ratio ranged from 2.3 in TA-FL growing "Noble" to 7.9 in NN-GX growing "Alachua". Furthermore, a higher content of acylated form was identified in "Noble" grape, which was likely benefited from a richer content of dihydroquercetin caffeoyl glucoside.

Myricetin derivative was the dominant type among the skin flavonol derivatives, followed by quercetin, syringetin and kaempferol derivatives, which differed among cultivars and growing regions. Myricetin derivative was significantly higher in cv. "Fry" and "Granny Val" from NN-GX (Figure 2a), while the quercetin derivative was higher in "Noble" grape. Interestingly, kaempferol derivative was mainly observed in red cultivars from TA-FL, whereas the syringetin derivative was significantly accumulated in bronze cultivars from PE-YN. Quercetin, isorhamnetin and kaempferol derivatives were the main flavonols in the muscadine seeds.

The 3′,4′,5′-substituted flavonol (myricetin, laricitrin, and syringetin derivatives) content was the highest in skins (Figure 2b). This substituted flavonol was significantly higher in bronze cultivars than the red ones. However, the 3′,4′-substituted (quercetin and isorhamnetin derivatives) was the second most common flavonol, particularly in red cultivars. In addition, 3′,4′-substituted flavonol was the main one in seeds, followed by 4′-substituted flavonol (kaempferol derivatives).

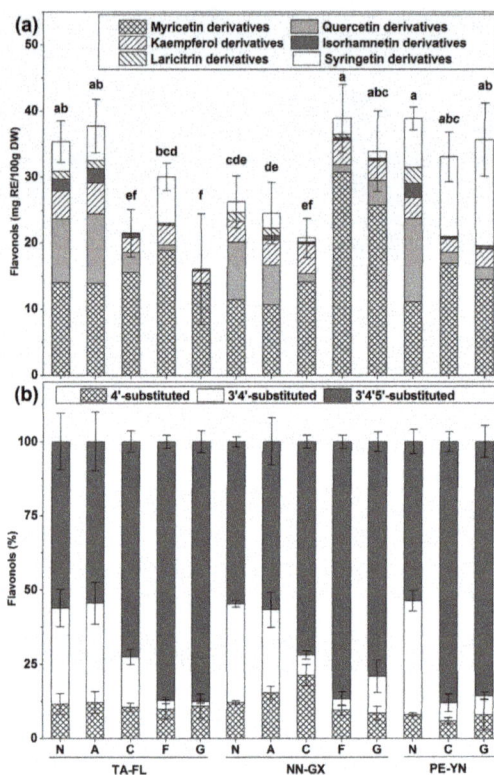

Figure 2. Flavonols derivatives (**a**) and substituted flavonols (**b**) distribution in five muscadine grape skins among three different regions in the 2012 season. On the top of each column, standard deviation is shown for the total content, and different small letters indicate significant differences (Duncan's test, $p = 0.05$). See Table 1 for the abbreviation of cultivars and regions.

2.2.3. Benzoic Acid Profiles

There were great differences of benzoic acids contents among the muscadine grapes growing in the three regions (Table 2). In the 2012 season, the grapes from NN-GX showed the highest levels of skin benzoic acids, especially for "Noble" cultivar that had 36.60 mg GAE/100 g DW. This result may be due to the higher brevifolin carboxylic acid content in the muscadine skins from NN-GX (Table S2). The seeds possessed significantly lower content of benzoic acids in comparison to the skins. In addition, both di- and mono- hydroxy benzoic acid were detected among the muscadine grapes investigated. There was higher ratio of di-/mono- hydroxy benzoic acid in the skins than in the seeds.

2.2.4. Flavan-3-ols Profiles

A small amount of flavan-3-ols was detected in muscadine skins (<0.1 mg (−)-epicatechin equivalent (EE)/100 g DW), while flavan-3-ols was the second highest polyphenol found in the seeds (> 89 mg EE/100 g DW) (Table 2). The seed flavan-3-ols levels also varied significantly among cultivars grown in China and the USA, with the exception of "Alachua", which were consistent among all the regions studied in the 2012 season. Overall, with a few exceptions, the bronze cultivars possessed higher contents of flavan-3-ols than the red ones. For instance, the "Granny Val" in PE-YN had 312.45 mg EE/100 g DW of seed flavan-3-ols.

For flavan-3-ols constitutes, seven flavan-3-ols, mostly in their monomer forms (Tables S1–S3), were found in the skin samples. On the other hand, there were eleven flavan-3-ols detected in muscadine seeds, including two monomers, three gallates, one hexoside, three dimers, and two trimers. Monomers and gallates were the major types in all of the tested samples, accounting for 52.44%–71.27% and 18.66%–41.15% of total flavan-3-ols, respectively (Figure 3). Furthermore, the highest and lowest monomers were observed in cultivar "Granny Val" from PE-YN and "Noble" from TA-FL, respectively, whereas cultivars "Noble" and "Fry" from TA-FL showed extremely high and low levels of gallates, respectively.

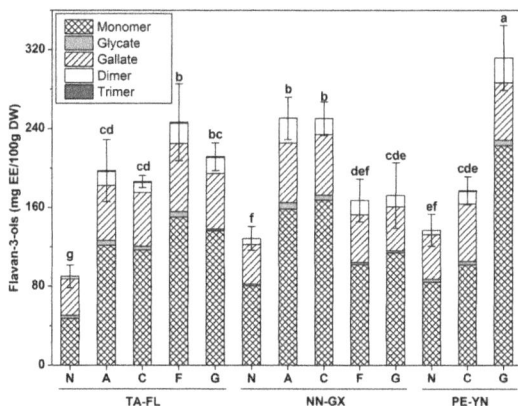

Figure 3. Flavan-3-ols distribution in five muscadine seeds among three different regions in the 2012 season. On the top of each column, standard deviation is show for the total content, and different small letters indicate significant differences (Duncan's test, *p* = 0.05). See Table 1 for the abbreviation of cultivars and regions.

2.2.5. Stilbenes and Cinnamon Acids Profiles

The stilbene contents in the five muscadine skins ranging from 0 to 0.24 mg resveratrol equivalent (REE)/100 g DW among different samples, were considerably low in this study (Table 2). The 'Carlos' from NN-GX had the highest levels of skin stilbenes (0.18 and 0.24 mg REE/100 g DW in 2012 and 2013 season, respectively) than others. With the two individual stilbenes, resveratrol was commonly found in all the cultivars, whereas resveratrol-3-*O*-glucoside was only detected in "Carlos" from NN-GX (Table S2). A small amount of cinnamon acids was only discovered for bronze cultivars in 2012 season. Nevertheless, the stilbenes and cinnamic acids were not identified in seeds in our study.

2.3. Principal Component Analysis (PCA)

Fifty-four evaluation parameters (Table 3) in all the samples studied were subjected to principal component analysis (PCA), in order to separate grapes according to their polyphenol characteristics. In the skins, the first three principal components (PCs) explained 83.7% of total variance (56.4%, 14.4%, and 12.9% for PC1, PC2, and PC3, respectively), which indicated that these factors were sufficient for further discussion. Figure 4a shows the skin distribution of the PCA biplot (loadings plot combined with scores plot) in two-dimensional space for PC1 and PC2. In this figure it is possible to note that the skin samples revealed three distinct groups. Group A, located in quadrants II and III, included three bronze cultivars from TA-FL and cultivar 'Carlos' from PE-YN. This group was matched through low values of total ellagic acids and precursors, total ellagic acid derivatives, total ellagitannins, ellagic acid, and ellagic acid glucoside/diglucoside (Table 3). Group B, in quadrant IV, was comprised of cultivar "Fry" from NN-GX, and "Granny Val" from NN-GX and PE-YN. This group was correlated with high total ellagic acids and precursors, total ellagic acid derivatives, ellagic acid, ellagic acid glucoside/diglucoside, ducheside B, total flavonols, myricetin derivatives, and 3',4',5'-substituted flavonols, and low TPC. Red cultivars from all the three regions and "Carlos" form NN-GX comprised group C, located in quadrants I and II. This group was characterized by the abundant TPC, total ellagitannins, and quercetin derivatives, as well as the lower ducheside B and 3′,4′,5′-substituted flavonols.

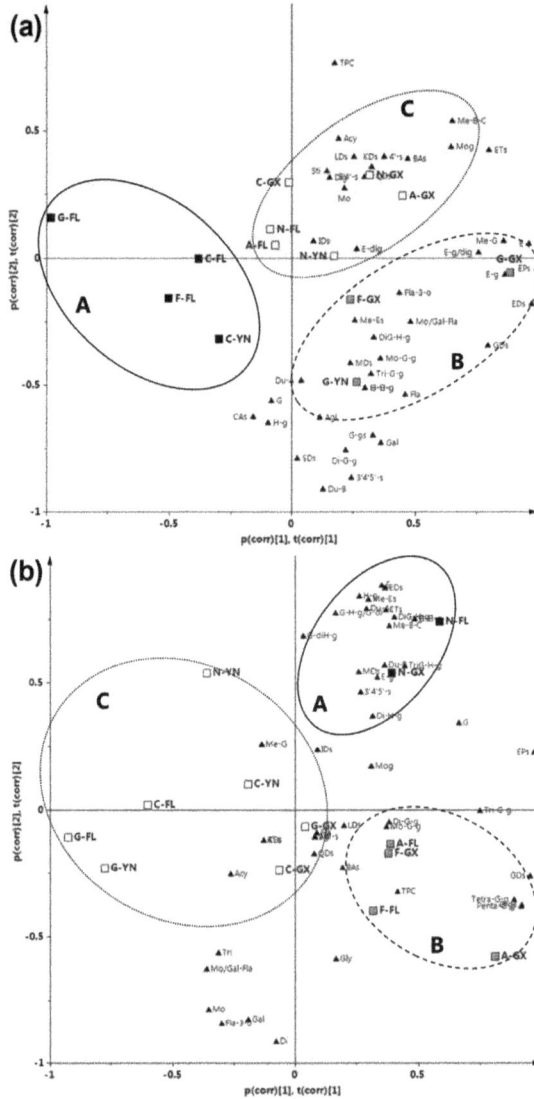

Figure 4. PCA biplot of loadings plot (triangle, polyphenol compounds, see Table 3) and scores plot (box, cultivars and regions) of five muscadine grape skins (**a**) and seeds (**b**). The dates were pareto scaled and the groups were classed by HCA through Ward's method. N (A, C, F, G)-FL (GX, YN): cv. "Noble" ("Alachua", "Carlos", "Fry", "Granny Val") from Tallahassee- Florida (Nanning-Guangxi, Pu'er-Yunnan).

Table 3. The contribution scores of 54 evaluation parameters from different groups of muscadine skins and seeds based on principal components analysis (PCA) and hierarchical cluster analysis (HCA) in the 2012 season.

No.	Compounds	Abbreviation	Skins			Seeds		
			Group A	Group B	Group C	Group A	Group B	Group C
1	Total phenolic content	TPC	-0.4043	-1.2307	1.4201	-0.1302	0.4540	-0.2132
2	Ellagic acids and precursors	EPs	-4.0271	2.4873	0.3189	2.5816	1.8111	-2.6843
3	Ellagic acid derivatives	EDs	-3.0158	2.5967	0.1427	5.5202	-1.1154	-0.2678
4	Ellagic acid	E	-2.7306	1.4526	0.2713	4.8190	-1.0040	-0.2151
5	Methyl-ellagic acids (1–3)	Me-Es	-0.1053	0.5372	-0.2334	1.3958	-0.2875	-0.0571
6	Ducheside A	Du-A	-0.0512	0.1191	0.0386	0.9210	-0.2086	-0.0345
7	Ducheside B	Du-B	0.1316	1.5802	-1.0522	1.0886	-0.1932	-0.0910
8	Ellagic acid glucoside	E-g	-0.8667	0.8782	-0.0100	0.3727	-0.0648	-0.0317
9	Ellagic acid-dihexoside	E-dig	-0.0232	0.0385	-0.0194	—	—	—
10	Ellagic acid glu/diglu	E-g/dig	-1.6530	1.4763	0.0737	—	—	—
11	Ellagitannins	ETs	-1.9865	0.2552	1.1640	1.6039	-0.2092	-0.1179
12	Methyl brevifolin carboxylate	Me-B-C	-0.9633	0.4421	0.6616	0.8561	-0.0601	-0.0858
13	HHDP-glucose	H-g	0.0546	0.0342	-0.1208	0.7131	-0.1203	-0.0299
14	Di-HHDP-glucose	Di-H-g	—	—	—	0.1157	-0.0191	-0.0138
15	Mono-/di-HHDP-glucose	Mo/di-H-g	-0.0325	0.1229	-0.0254	0.4223	-0.0227	-0.0541
16	Pedunculagin α/β isomer (Di-HHDP-glucose)	G-H-g	-0.0307	0.1229	-0.0254	0.2189	-0.0866	0.0002
17	Galloyl-HHDP-glucose (Corlagin, Strictinin)	G-diH-g	—	—	—	0.4264	-0.1078	-0.0077
18	Galloyl-bis-HHDP-glucose (Casuarinin)	G-H-g/G-diH-g	—	—	—	0.4223	-0.0227	-0.0541
19	Galloyl-HHDP-glucose/Galloyl-bis-HHDP-glucose	H-G-g	—	—	—	0.8165	-0.0951	-0.0365
20	HHDP-galloyl-glucose (Isostrictinin)	DiG-H-g	—	0.0334	0.0139	0.1536	0.0236	-0.0365
21	Tellimagrain I (Digalloyl-HHDP-glucose)	TriG-H-g	—	—	—	0.5890	2.7441	-2.3172
22	Tellimagrain II (Trigalloyl-HHDP-glucose)	CbS	-0.4620	0.9006	-0.1423	0.5468	0.3028	-0.3943
23	Precursors Gallic acid derivatives)	G	0.0934	0.0532	-0.1932	-0.1558	-0.0158	0.0232
24	Gallic acid	Me-G	-0.6455	0.6317	0.0035	0.0113	0.0393	-0.0362
25	Methyl gallate	Mo-G-g	-0.0028	0.0315	-0.0143	-0.0016	0.0059	-0.0037
26	Mono-O-galloyl-glucose	Di-G-g	0.0170	0.2667	-0.1653	0.1246	0.0980	-0.1513
27	Di-O-galloyl-glucose	Tri-G-g	0.0035	0.0996	-0.0766	0.1164	0.5409	-0.4237
28	Tri-O-galloyl-glucose	Tetra-G-g	—	—	—	0.5180	2.7015	-2.1336
29	Tetra-O-galloyl-glucose	Penta-G-g	—	—	—	0.5533	0.2601	-0.0142
30	Penta-O-galloyl-glucose	G-gs	0.0137	0.2838	-0.1641	-0.0143	0.0376	-0.0153
31	Galloyl-glucoses (1–5)	Fla	-0.5623	1.0174	0.0858	0.2699	-0.0538	-0.0118
32	Flavonols	MDs	0.0022	1.0660	-0.7312	-0.0641	0.0734	0.0107
33	Myricetin derivatives	QDs	-0.3407	-0.3765	1.1126	-0.0013	-0.0213	-0.0067
34	Quercetin derivatives	KDs	-0.2456	-0.0330	0.2954	0.0296	0.0184	-0.0028
35	Kaempferol derivatives	IDs	-0.0266	-0.0245	0.3417	0.0073	-0.0007	0.0107
36	Isorhamnetin derivatives	LDs	-0.1156	-0.1750	0.4492	-0.0013	-0.0213	-0.0140
37	Laricitrin derivatives	SDs	0.0174	0.5389	-0.3337	-0.0317	0.0497	-0.0211
38	Syringetin derivatives	4'-s	-0.2456	-0.0330	0.2954	0.2033	-0.0183	-0.0141
39	4'-substituted flavonol	3'4'-s	-0.3238	-0.3563	1.1856	-0.0351	0.0521	-0.0228
40	3'4'-substituted flavonol	3'4'5'-s	-0.0401	2.3605	-1.1655	0.1012	-0.0196	0.0072
41	3'4'5'-substituted flavonol	Agl	-0.0720	0.2056	0.0156	-0.0109	-0.0100	—
42	Flavonol aglycone	Mog	-0.6273	-0.1335	0.7024	—	—	—
43	Flavonol monoglucoside	Dig	-0.1058	-0.2083	0.6461	—	—	—
44	Flavonol diglucoside	Acy	-0.2102	-0.4368	0.9267	—	—	—

Table 3. Cont.

No.	Compounds	Abbreviation	Skins			Seeds		
			Group A	Group B	Group C	Group A	Group B	Group C
45	Benzoic acids	BAs	-0.5820	0.5540	0.2280	0.0510	0.0038	-0.0110
46	Flavan-3-ols	Fla-3-o	-0.0175	0.0228	-0.0020	-3.6329	0.7254	0.1494
47	Flavan-3-ol monomer	Mo	-0.0021	-0.0014	0.0037	-2.7396	0.3694	0.1878
48	Flavan-3-ol glycate	Gly	–	–	–	-0.2402	0.1468	-0.0084
49	Flavan-3-ol gallate	Gal	-0.0077	0.1030	-0.0425	-1.5488	0.4488	0.0222
50	Flavan-3-ol dimer	Di	–	–	–	-0.4692	0.1855	-0.00004
51	Flavan-3-ol trimer	Tri	–	–	–	-0.2596	0.0910	0.0051
52	Flavan-3-ol monomer/gallate	Mo/Gal	-0.0103	0.1002	-0.0349	-0.2456	0.0016	0.0313
53	Stilbenes	Sti	-0.0164	-0.0003	0.0271	–	–	–
54	Cinnamic acids	CAs	0.0181	0.1248	-0.0987	–	–	–

The score contribution is described as group-average weight of first three principal components (PCs) for skins and first two PCs for seeds. –: no contribution. In bold, compounds with absolute value of contribution score >1.

For the seeds, the first two PCs described 83.8% of total variance (55.2% for PC1, and 28.6% for PC2). The PCA biplots of seed samples also were classified into three distinct groups (Figure 4b). Group A, consisting of cultivar "Noble" from TA-FL and NN-GX, located in quadrant I, profited from high values of total ellagic acids and precursors, total ellagic acid derivatives, total ellagitannins, ellagic acid, sum of three methyl-ellagic acids, ducheside B, and low levels of total flavan-3-ols, monomer and gallate of flavan-3-ols (Table 3). Group B, in quadrant IV, was comprised of cultivars "Alachua" and "Fry" from TA-FL and NN-GX. This group was contributed mostly not only by high contents of total ellagic acids and precursors, total gallic acid derivatives, sum of five galloyl-glucoses, and penta-*O*-galloyl-glucose, and also by low total ellagic acid derivatives, and ellagic acid. Others comprised group C, located in quadrants II, III, and IV, and were linked by low levels of total ellagic acids and precursors, total gallic acid derivatives, sum of five galloyl-glucoses, and penta-*O*-galloyl-glucose.

3. Discussion

The chemical diversity of grapes is mostly affected by secondary metabolites represented by different phytochemical groups such as polyphenols, terpenoids, and tannins, among others, which have been attracted research interest owing to their biological activity [26]. As expected, polyphenol synthesis and accumulation in grapes is influenced by multiple factors, for instance, genotype/cultivar, geographic origin, and environmental conditions [27]. Among them, genotype plays a pivotal role in the polyphenol contents of grapes [5,7,12,28]. In this study, "Alachua" possessed the highest TPC among the muscadine cultivars investigated, which is in agreement with the findings of Sandhu and Gu [13].

Grape polyphenol compounds are greatly affected by environmental conditions such as temperature, solar radiation, sunshine duration, and rainfall [11,21]. However, a quick search of the literature demonstrates that only a few reports have addressed the effects of these parameters on TPC, ellagic acids and precursors concentration in muscadine grapes. Seasonal changes in temperature and day length are considered to be the main factors influencing the content of ellagic acids [29]. For example, there were negative correlations of overall average temperature with TPC, total tannin, and punicalagin concentration in pomegranate, which means the lower average temperature during maturing and harvest periods could promote primary polyphenol accumulation [30]. The day and night temperatures as well as the difference between them were found to play an important role in polyphenol accumulation of *Vitis* grapes [11]. Wang and Camp [29] investigated the influence of day/night temperature combinations after blooming on strawberry growth and fruit quality, and found the fruit grown at 18/12 °C contained greater amounts of ellagic acid. As the day/night temperature increased, fruit ellagic acid content decreased, while the anthocyanin content increased. This finding showed that the decrease in ellagic acid at high temperature may be due to the inhibition of ellagic acid biosynthesis or enhancement of degradation. In this sense, the average temperature and the difference between day and night in TA-FL of the USA (33.4/22.2 °C, range of temperature 11.2 °C) was larger than that in PE-YN (26.3/19.4 °C, range of temperature 6.9 °C) and NN-GX (32.9/25.4 °C, range of temperature 7.5 °C) of China before 30 days of harvest, which might lead to lower total and individual ellagic acids contents in muscadine grape skins from TA-FL.

Solar radiation and sunshine duration influence grape polyphenol synthesis and accumulation [11]. The relatively long illumination time could accumulate higher polyphenol in grapes [23,31]. For example, grapes exposed to increased daylight are capable of increased flavan-3-ol biosynthesis [32]. However, the stronger sunlight intensity results in higher berry temperature, which could cause a decrease of polyphenol in the berry [11,33]. Based on this concern, in the three regions, PE-YN (22.82° N) and NN-GX (22.47° N) was in the south of TA-FL (30.43° N), in addition, the sunshine duration for the Hengduan Mountains Region of PE-YN in the southwest of China was longer (average annual sunshine 2038.4 h), contrasting with other two plain regions of NN-GX and TA-FL (1585 and 1883 h, respectively), therefore, the former two had abundant solar radiation and sunshine duration. These differentiations should be considered as an important factor leading to increase photosynthesis and result in higher seed flavan-3-ol contents in PE-YN.

Water availability is another climatic factor affecting the synthesis and accumulation of polyphenol in grape berries. Appropriate vine water stress could lead to high anthocyanin and tannin contents [34]. One possible explanation for these results is the enhancement of TPC in water-stressed berries triggered by a reduced berry size and weight, followed by a higher proportion of fruit achene to flesh [35]. Another reason is that under these conditions, carbon should be preferentially allocated to the synthesis of primary metabolites, the level of which are not detrimental but promote the synthesis of carbon-based secondary metabolites, primarily polyphenols [30]. In addition, water deficits also increased the expression of many genes responsible for the biosynthesis of trihydroxylated anthocyanins in grape berries [36]. In our study, the average monthly rainfall before 30 days of harvest in the USA (TA-FL) was 184.9 mm (average of 1981–2010), which was lower than in China (218.8 mm in NN-GX, and 324.3 mm in PE-YN), as was the content of accumulated abundant upstream trihydroxylated anthocyanin compounds such as 3′,4′,5′-droxylated flavonols in TA-FL seeds. It is noteworthy that less than 100 mm of rainfall between veraison and maturing is an important indicator for the selection of superior wine-producing areas in *V. vinifera* [11], whereas the muscadine grapes could exhibit dominant polyphenol with the rainfall over 180 mm. This appeared to reflect better potentialities for adapting to humid climates for muscadine grapes.

In general, less is known about the impact of altitude factor on phytochemical composition. Doumet et al. [37] considered no significant differences were detected with respect to polyphenol contents and radical scavenging activity, in *Fragaria vesca* grown under the same environmental conditions but different altitude. Nevertheless, Guerrero–Chavez et al. [38] found anthocyanin concentration correlated negatively with the increase of altitude, and the strawberry antioxidant potential, measured by flow injection analysis with amperometric detection, was lower in fruits grown at higher altitude (900 vs. 1500 m). In our study, PE-YN had the highest flavonols of syringetin derivatives, which might due to the higher altitude (>1300 m) in comparison to NN-GX and TA-FL (<80 m).

However, the variation in metabolite content of ellagic acids is often correlated with data derived from genetics, transcriptomics, and environmental factors. Based on PCA and hierarchical cluster analysis (HCA), various cultivars grown in different locations could be grouped together and vice versa. The same cultivars could fall in different groups when they were cultivated in different regions. This is the result of interaction between genotype and environmental conditions which apparently influence the polyphenol synthesis and accumulation. More research is needed and multiple factors must be coordinately considered at the same time, in order to better understand the role played by the independent and/or mutual influence factors of ellagic acids biosynthesis in muscadine grapes, especially when global warming is intensifying, which affects many viticulture areas in the world.

4. Materials and Methods

4.1. Chemicals

All chemicals and standards mentioned below were of HPLC grade. Acetonitrile and methanol were purchased from Thermo Fisher Scientific (Waltham, MA, USA). Formic acid was supplied from VWR (Helsinki, Finland). Standard caffeic acid (≥98%), ellagic acid (≥98%), (−)-epicatechin (≥90%), gallic acid (≥98%), kaempferol (≥97%), myricetin (≥96%), quercetin (≥98%), and rutin (≥95%) were obtained from Sigma-Aldrich (St. Louis, MO, USA). Penta-O-galloyl-glucose (≥99%) and resveratrol (≥98%) were purchased from Solarbio (Beijing, China). Ultra-pure water from a Millipore Synergy water purification system (Merck KGaA, Darmstadt, Germany) with conductivity of 18 MΩ was used throughout. Other reagents were analytical pure and obtained from Beijing Chemical Works (Beijing, China).

4.2. Grape Materials

Fruits of five fully ripened muscadine grape cultivars (red cv. "Noble" and "Alachua", and bronze cv. "Carlos", "Fry", and "Granny Val") were collected from Tallahassee-Florida, United States (TA-FL) and Nanning-Guangxi, China (NN-GX), likewise, grapes of three cultivars (cv. "Noble", "Carlos", and "Granny Val") were obtained from Pu'er-Yunnan, China (PE-YN), for two consecutive years (2012 and 2013). Grapes from three to four clusters per vine, and ten vines per target cultivar were picked randomly throughout the vineyard, taking into account the balance between shadow and sun exposure, and following a z-shaped pattern to avoid edge and center effects [26]. Samples were transported immediately under refrigeration (ca. 2–5 °C) to the laboratory. Skin and seed fractions were separated manually, freeze-dried (LGJ-18, Songyuan Huaxing Biotechnology Co., LTD, Beijing, China) 48 h, and stored at −80 °C.

To illustrate the regional difference of the three sampling locations, consider the geographical distribution presented in Figure 5. Based on the data from the America's National Oceanic and Atmospheric Administration and China Meteorological National Administration (1981–2010), all these regions have a warm and humid subtropical climate, with long summers and short, mild winters. TA-FL is located at 30.43° N latitude, 84.26° W longitude, and has an altitude of 62 m. The annual average temperature is 19.8 °C (27.8 °C in July, and 10.7 °C in January), the annual rainfall is 1506 mm, and the sunshine lasts 1883 h. NN-GX is located at 22.47° N latitude, 108.21° E longitude, with an altitude of 80 m. The annual average temperature is 21.8 °C (28.4 °C in July, and 12.9 °C in January), the annual rainfall is 1310 mm, and the sunshine lasts 1585 h. PE-YN is located at 22.82° N latitude, 100.97° E longitude, and has an altitude above 1306 m. The annual average temperature is 19.5 °C (23.2 °C in July, and 13.7 °C in January), the annual rainfall is 1497 mm, and the city receives 2038 hours of bright sunshine annually.

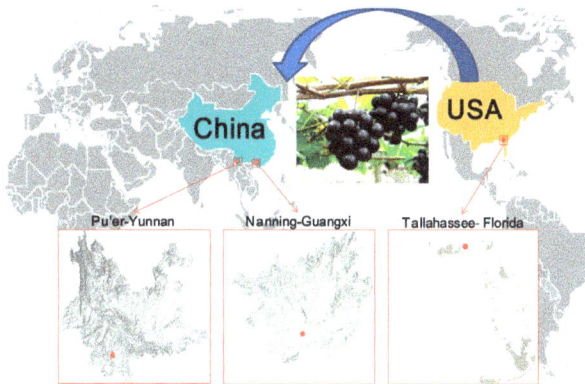

Figure 5. The three region distribution of muscadine grapes used in this study.

4.3. Extraction and Determination of Polyphenol in Muscadine Grapes

4.3.1. Preparation of Berries Extraction

Freeze-dried grape seeds were crushed and then defatted as previously reported [11,39]. Briefly, freeze dried grape seeds were moderately crushed by a stainless-steel grinder (FW-135, Tester Co., LTD, Tianjin, China), then defatted with petroleum ether at a ratio of 1:20 (w/v). After 12 h extraction at room temperature and in the dark, the liquid was separated from the solid by vacuum filtration (T-50, Jinteng Experiment Equipment Co., LTD, Tianjin, China) through a sintered glass filter (Pyrex, porosity 10–15 μm). The defatted procedure was carried out twice and the solid residue was evenly distributed over a culture dish, then kept in dark 6 h for evaporation of petroleum ether. The ultimate defatted grape seeds were put into a mortar containing liquid nitrogen and ground into powder as fine as possible. Freeze-dried grape skins were ground with the stainless-steel grinder to pass 60 sieve sizes (0.25 mm), then both were stored at −80 °C.

4.3.2. Extraction of Polyphenols in Muscadine Grapes

Polyphenol compounds were extracted from skins and seeds as previously reported in our laboratory [11,39]. Briefly, 0.5 g freeze-dried skins or defatted seeds were placed into a 50-mL centrifuge tube with methanol/water/hydrochloric acid (70:29:1, $v/v/v$) at a ratio of 1:30 (w/v), vortexed for 15 s (HMQL-VORTEX-5, Midwest Group, Beijing, China), then extracted at 616 W for 28 min in an ultrasonic cleaning machine (SB-5200, Ningbo Scientz Biotechnology Co., LTD, Ningbo, China) at 25 °C. After centrifuging at 7600 rpm for 20 min (Allegra X-30R, Beckman Coulter Inc., Brea, CA, USA), the supernatant was collected and the precipitate was re-extracted two more times. The supernatant was combined and the solvent was removed by vacuum evaporation (RE-52, Shanghai YaRong Biochemical Instrument Factory, Shanghai, China) at 35 °C. The solids obtained after evaporation were re-dissolved in 5 mL of methanol (1% formic acid) and stored at −80 °C. Extractions were performed in three replicates for all individual samples. Samples were filtered by 0.22 μm of cellulose membrane (Jinteng Experiment Equipment Co., LTD, Tianjin, China) and then detected by UPLC Triple TOF-MS/MS.

4.3.3. Determination of TPC

The TPC was determined by Folin–Ciocalteu colourimetric method as previously reported in our laboratory [11,39]. Briefly, all samples were diluted. Folin–Ciocalteu reagent and sodium carbonate were successively added. The solution was reacted at 40 °C for 30 min, and then the absorbance was read at 760 nm by a UV-2800 spectrometer (UNICO, Suite E Dayton, NJ, USA). Gallic acid was used as standard and values were expressed as gallic acid equivalent dry weight (mg GAE/g DW), with the linearity range 50–1000 μg/mL ($R^2 > 0.999$).

4.3.4. UPLC-Triple TOF-MS/MS Analysis

Sample analysis was carried out using an Eksigent ultraLC 110 and a Triple TOF 4600-MS/MS (AB SCIEX, Framingham, MA, USA) coupled with a Duospray Ion Source interface and automatic Calibrant Delivery System (CDS). The ultraLC 110 includes an online degasser, a double pump, an autosampler, and a thermostatic column control system, all of which were controlled by Analyst® TF 1.6 software. The UPLC separation was performed on a reversed-phase Zorbax SB-C18 column (2.1 mm × 150 mm × 5 μm, Agilent Technologies, Santa Clara, CA, USA) at 35 °C. The mobile phase was water with 0.5% formic acid (A) and acetonitrile (B) at the following gradient: 0–1 min, 5% B; 1–20 min, 5%–60% B; 20–21 min, 60%–95% B; 21–30 min, column wash and stabilization. The flow rate was 0.3 mL/min and the injection volume was 10 μL. MS conditions: ion source gas 1 and 2 (air), 55 psi; curtain gas (N_2), 30 psi; source temperature, 550 °C; ionspray voltage, −4.5 kv; collision energy, −40 ± 10 eV, scan from m/z 100 to 2000. Tuning Solution in Installation Kit and APCI Negative Calibration Solution (AB SCIEX) were used to monitor the stability of the ionization efficiency of the mass spectrometer and the m/z values of Triple TOF systems.

Data acquisition and processing was performed using Peak View 2.0 and Marker View 1.2.1 software (AB SCIEX). The polyphenol compounds were identified based on total ion chromatogram, retention time, exact molecular weight, and Triple TOF MS/MS fragmentation characteristics, such as the representative description in Figures S1–S3 and Table S4.

Multi Quant 3.0 software (AB SCIEX) was used for quantitative analysis. Caffeic acid, ellagic acid, (−)-epicatechin, gallic acid, kaempferol, myricetin, penta-*O*-galloyl-glucose, quercetin, resveratrol, and rutin were quantified by their standards, respectively, and expressed as mg/100 g DW. Other benzoic acids, cinnamic acids, ellagic acids, flavan-3-ols, flavonols, and stilbens were respectively expressed as micrograms of gallic acid equivalent (GAE), caffeic acid equivalent (CAE), GAE, (−)-epicatechin equivalent (EE), rutin equivalent (RE), and resveratrol equivalent (REE)/100 g DW. The CDS was adjusted every five hours when sample running, the standard curves were produced every day with three parallel measurements. The linear ranges of different standards were 0–5, 0–10, and 10–50 mg/L, corresponding to the concentrations ($R^2 > 0.999$).

4.4. Statistical Analysis

Results were expressed as means of three parallel measurements $\pm$ S.D. Microsoft Excel 2010, SPSS 20.0 (IBM, Armonk, NY, USA), and Origin 8.5 (Origin Lab, Northampton, MA, USA) software were used for data processing and graphing. Significance difference was tested by ANOVA (Duncan's test, $p = 0.05$). Principal component analysis (PCA) and hierarchical cluster analysis (HCA) were performed by SIMCA-P 13.0 (Umetrics, Malmö, Sweden).

5. Conclusions

Ellagic acids and precursors were the characteristic polyphenols detected in muscadine grapes. Our research was the first study to analyze the ellagic acids and precursor composition and accumulation in muscadine grapes grown in South China (NN-GX, and PE-YN). Fourteen new ellagitannins (thirteen in skins, and one in seeds) were identified in muscadine grapes for the first time. Multiple factors influenced the polyphenol synthesis and accumulation. Differences were observed varied within and between grape genotype/cultivars (white and red) and grape fractions (skins and seeds), as well as in different regions under different environmental conditions. Based on PCA, the cultivars from different regions were classified into three distinct groups, in both skins and seeds, presenting characteristic and discriminative variances. These results indicated that muscadine grapes could be grown well in countries besides the USA.

Supplementary Materials: Supplementary materials can be found at www.mdpi.com/1422-0067/18/3/631/s1.

Acknowledgments: The authors gratefully acknowledge the financial support provided by the earmarked fund for the China Agriculture Research System (grant number CARS-30-yz-2), China Agricultural University Scientific Fund (grant number 2012RC019), Construction Project of Guangxi Crop Genetic Improvement and Biotechnology Laboratory (grant number 15-140-33-3), and Guangxi Bagui Scholar Special Foundation.

Author Contributions: Zheng Wei conceived and designed the experiments, and carried out the experimental analyses, data interpretation and manuscript writing together with Jianming Luo; Wenfeng Guo contributed to the statistical analysis and focused on the date interpretation of PCA and HCA; Yu Huang and Huan Guan helped in sample collection, pretreatment and performed the qualitative analysis of muscadine grape polyphenol; and Yali Zhang, Changmou Xu and Jiang Lu gave their valuable support in the critical revision of the manuscript and conceptual content.

Conflicts of Interest: The authors declare no conflict of interest.

References

1. Pastrana-Bonilla, E.; Akoh, C.C.; Sellappan, S.; Krewer, G. Phenolic content and antioxidant capacity of muscadine grapes. *J. Agric. Food Chem.* **2003**, *51*, 5497–5503. [CrossRef] [PubMed]
2. Conner, P.J. Characteristics of promising muscadine grape (*Vitis rotundifolia* Michx.) selections from the University of Georgia (USA) Breeding Program. In Proceedings of the X International Conference on Grapevine Breeding and Genetics, 2014; Volume 1046, pp. 303–307.
3. Talcott, S.T.; Lee, J. Ellagic acid and flavonoid antioxidant content of muscadine wine and juice. *J. Agric. Food Chem.* **2002**, *50*, 3186–3192. [CrossRef] [PubMed]
4. Louime, C.; Lu, J.; Onokpise, O.; Vasanthaiah, H.K.N.; Kambiranda, D.; Basha, S.M.; Yun, H.K. Resistance to *Elsinoe Ampelina* and expression of related resistant genes in *Vitis Rotundifolia* Michx. grapes. *Int. J. Mol. Sci.* **2011**, *12*, 3473–3488. [CrossRef] [PubMed]
5. Marshall, D.A.; Stringer, S.J.; Spiers, J.D. Stilbene, ellagic acid, flavonol, and phenolic content of muscadine grape (*Vitis rotundifolia* Michx.) cultivars. *Pharm. Crops* **2012**, *3*, 69–77. [CrossRef]
6. Yu, Y.; Wu, J.; Fu, S.; Yin, L.; Zhang, Y.; Lu, J. Callose synthase family genes involved in the grapevine defense response to downy mildew disease. *Phytopathology* **2016**, *106*, 56–64. [CrossRef] [PubMed]
7. Chen, W.W. Antimicrobial and Antioxidant Activity of Muscadine (*Vitis rotundifolia* Michx.) Extracts as Influenced by Solvent Extraction Methods and Cultivars. Master's Thesis, Mississippi State University, Starkville, MI, USA, 2011.
8. Xu, C.; Yagiz, Y.; Hsu, W.; Simonne, A.; Lu, J.; Marshall, M.R. Antioxidant, antibacterial, and antibiofilm properties of polyphenols from muscadine grape (*Vitis rotundifolia* Michx.) pomace against selected foodborne pathogens. *J. Agric. Food Chem.* **2014**, *62*, 6640–6649. [CrossRef] [PubMed]

9.   Xu, C.; Yavuz, Y.; Marshall, S.; Li, Z.; Simonne, A.H.; Lu, J.; Marshall, M.R. Application of muscadine grape (*Vitis rotundifolia*) pomace extract to reduce carcinogenic acrylamide. *Food Chem.* **2015**, *182*, 200–208. [CrossRef] [PubMed]

10. You, Q. Biological Properties Evaluation and Chemical Profiles of Phenolic Compounds in Muscadine Grapes (*Vitis rotundifolia*). Ph.D. Thesis, Clemson University, Clemson, SC, USA, 2012.

11. Xu, C.; Zhang, Y.; Zhu, L.; Huang, Y.; Lu, J. Influence of growing season on phenolic compounds and antioxidant properties of grape berries from vines grown in subtropical climate. *J. Agric. Food Chem.* **2011**, *59*, 1078–1086. [CrossRef] [PubMed]

12. Zhu, L.; Zhang, Y.; Lu, J. Phenolic contents and compositions in skins of red wine grape cultivars among various genetic backgrounds and originations. *Int. J. Mol. Sci.* **2012**, *13*, 3492–3510. [CrossRef] [PubMed]

13. Sandhu, A.K.; Gu, L. Antioxidant capacity, phenolic content, and profiling of phenolic compounds in the seeds, skin, and pulp of *Vitis rotundifolia* (Muscadine Grapes) as determined by HPLC-DAD-ESI-MS(n). *J. Agric. Food Chem.* **2010**, *58*, 4681–4692. [CrossRef] [PubMed]

14. Lee, J.; Johnson, J.V.; Talcott, S.T. Identification of ellagic acid conjugates and other polyphenolics in muscadine grapes by HPLC-ESI-MS. *J. Agric. Food Chem.* **2005**, *53*, 6003–6010. [CrossRef] [PubMed]

15. Lorrain, B.; Chira, K.; Teissedre, P. Phenolic composition of Merlot and Cabernet-Sauvignon grapes from Bordeaux vineyard for the 2009-vintage: Comparison to 2006, 2007 and 2008 vintages. *Food Chem.* **2011**, *126*, 1991–1999. [CrossRef] [PubMed]

16. Narduzzi, L.; Stanstrup, J.; Mattivi, F. Comparing wild American grapes with *Vitis vinifera*: A metabolomics study of grape composition. *J. Agric. Food Chem.* **2011**, *63*, 6823–6834. [CrossRef] [PubMed]

17. Marshall-Shaw, D.A.; Stringer, S.J.; Sampson, B.J.; Spiers, J.D. Storage retention of stilbene, ellagic acid, flavonol, and phenolic content of muscadine grape (*Vitis rotundifolia* Michx.) cultivars. *J. Food Chem. Nutr.* **2014**, *2*, 81–92.

18. Hager, T.J.; Howard, L.R.; Liyanage, R.; Lay, J.O.; Prior, R.L. Ellagitannin composition of blackberry as determined by HPLC-ESI-MS and MALDI-TOF-MS. *J. Agric. Food Chem.* **2008**, *56*, 661–669. [CrossRef] [PubMed]

19. Quideau, S.; Deffieux, D.; Douat-Casassus, C.; Pouységu, L. Plant polyphenols: Chemical properties, biological activities, and synthesis. *Angew. Chem. Int. Edit.* **2011**, *50*, 586–621. [CrossRef] [PubMed]

20. García-ESstévez, I.; Andrés-García, P.; Alcalde-Eon, C.; Giacosa, S.; Rolle, L.; Rivas-Gonzalo, J.C.; Quijada-Morín, N.; Escribano-Bailón, M.T. Relationship between agronomic parameters, phenolic composition of grape skin, and texture properties of *Vitis vinifera* L. cv. Tempranillo. *J. Agric. Food Chem.* **2015**, *63*, 7663–7669. [CrossRef] [PubMed]

21. Downey, M.O.; Dokoozlian, N.K.; Krstic, M.P. Cultural practice and environmental impacts on the flavonoid composition of grapes and wine: A review of recent research. *Am. J. Enol. Viticult.* **2006**, *57*, 257–268.

22. Sandhu, A.K.; Gray, D.J.; Lu, J.; Gu, L. Effects of exogenous abscisic acid on antioxidant capacities, anthocyanins, and flavonol contents of muscadine grape (*Vitis rotundifolia*) skin. *Food Chem.* **2011**, *126*, 982–988. [CrossRef]

23. Lu, Z.; Liu, Y.; Zhao, L.; Jiang, X.; Li, M.; Wang, Y.; Xu, Y.; Gao, L.; Xia, T. Effect of low-intensity white light mediated de-etiolation on the biosynthesis of polyphenols in tea seedslings. *Plant Physiol. Biochem.* **2014**, *80*, 328–336. [CrossRef] [PubMed]

24. Artem, V.; Antoce, A.O.; Namolosanu, I.; Ranca, A.; Petrescu, A. The influence of the vine cultivation technology on the phenolic composition of red grapes. *Horticulture* **2015**, *59*, 117–122.

25. Zhu, L.; Zhang, Y.; Zhang, W.; Lu, J. Effects of exogenous abscisic acid on phenolic characteristics of red *Vitis vinifera* grapes and wines. *Food Sci. Biotechnol.* **2016**, *25*, 361–370. [CrossRef]

26. Perestrelo, R.; Barros, A.S.; Rocha, S.M.; Câmara, J.S. Establishment of the varietal profile of *Vitis vinifera* L. grape varieties from different geographical regions based on HS-SPME/GC-qMS combined with chemometric tools. *Microchem. J.* **2014**, *116*, 107–117. [CrossRef]

27. Silva, J.K.; Cazarin, C.B.B.; Correa, L.C.; Batista, A.G.; Furlan, C.P.B.; Biasoto, A.C.T.; Pereira, G.E.; Camargo, A.C.; Maróstica Junior, M.R. Bioactive compounds of juices from two Brazilian grape cultivars. *J. Sci. Food Agric.* **2016**, *96*, 1990–1996. [CrossRef] [PubMed]

28. Heras-Roger, J.; Díaz-Romero, C.; Darias-Martín, J. A comprehensive study of red wine properties according to variety. *Food Chem.* **2016**, *196*, 1224–1231. [CrossRef] [PubMed]

29. Wang, S.Y.; Camp, M.J. Temperatures after bloom affect plant growth and fruit quality of strawberry. *Sci. Hortic.* **2000**, *85*, 183–199. [CrossRef]

30. Li, X.; Wasila, H.; Liu, L.; Yuan, T.; Gao, Z.; Zhao, B.; Ahmad, I. Physicochemical characteristics, polyphenol compositions and antioxidant potential of pomegranate juices from 10 Chinese cultivars and the environmental factors analysis. *Food Chem.* **2015**, *175*, 575–584. [CrossRef] [PubMed]

31. Spayd, S.E.; Tarara, J.M.; Mee, D.L.; Ferguson, J.C. Separation of sunlight and temperature effects on the composition of *Vitis vinifera* cv. Merlot berries. *Am. J. Enol. Viticult.* **2012**, *53*, 171–182.

32. Sun, X.; Li, L.; Ma, T.; Liu, X.; Huang, W.; Zhan, J. Profiles of phenolic acids and flavan-3-ols for select Chinese red wines: A comparison and differentiation according to geographic origin and grape variety. *J. Food Sci.* **2015**, *80*, 2170–2179. [CrossRef] [PubMed]

33. Yamane, T.; Shibayama, K. Effects of trunk girdling and crop load levels on fruit quality and root elongation in 'Aki Queen' grapevines. *J. Jap. Soc. Hortic. Sci.* **2006**, *75*, 439–444. [CrossRef]

34. Roby, G.; Harbertson, J.F.; Adams, D.A.; Matthews, M.A. Berry size and vine water deficits as factor in winegrape composition: Anthocyanins and tannins. *Aust. J. Grape Wine R.* **2004**, *10*, 100–107. [CrossRef]

35. Terry, L.A.; Chope, G.A.; Giné Bordonaba, J. Effect of water deficit irrigation and inoculation with: Botrytis cinerea on strawberry (*Fragaria × ananassa*) fruit quality. *J. Agric. Food Chem.* **2007**, *55*, 10812–10819. [CrossRef] [PubMed]

36. Castellarin, S.D.; Matthews, M.A.; Gaspero, G.D.; Gambetta, G.A. Water deficits accelerate ripening and induce changes in gene expression regulating flavonoid biosynthesis in grape berries. *Planta* **2007**, *227*, 101–112. [CrossRef] [PubMed]

37. Doumett, S.; Fibbi, D.; Cincinelli, A.; Giordani, E.; Nin, S.; Del Bubba, M. Comparison of nutritional and nutraceutical properties in cultivated fruits of *Fragaria vesca* L. produced in Italy. *Food Res. Int.* **2011**, *44*, 1209–1216. [CrossRef]

38. Guerrero-Chavez, G.; Scampicchio, M.; Andreotti, C. Influence of the site altitude on strawberry phenolic composition and quality. *Sci. Hortic.* **2015**, *192*, 21–28. [CrossRef]

39. Wei, Z.; Zhao, Y.; Huang, Y.; Zhang, Y.; Lu, J. Optimization of ultrasound-assisted extraction of ellagic acid and total phenols from muscadine (*Vitis rotundifolia*) by response surface methodology. *Food Sci.* **2015**, *36*, 29–35. (In Chinese)

International Journal of
Molecular Sciences

MDPI

Review

Understanding the Effectiveness of Natural Compound Mixtures in Cancer through Their Molecular Mode of Action

Thazin Nwe Aung [1,2], Zhipeng Qu [1,2], R. Daniel Kortschak [1,2] and David L. Adelson [1,2,*]

[1] Department of Genetics and Evolution, School of Biological Sciences, The University of Adelaide, Adelaide, South Australia 5005, Australia; thazin.nweaung@adelaide.edu.au (T.N.A.); zhipeng.qu@adelaide.edu.au (Z.Q.); dan.kortschak@adelaide.edu.au (R.D.K.)
[2] Zhendong Australia China Centre for Molecular Chinese Medicine, The University of Adelaide, Adelaide, South Australia 5005, Australia
* Correspondence: david.adelson@adelaide.edu.au; Tel.: +61-8-8313-7555

Academic Editor: Toshio Morikawa
Received: 15 February 2017; Accepted: 15 March 2017; Published: 17 March 2017

Abstract: Many approaches to cancer management are often ineffective due to adverse reactions, drug resistance, or inadequate target specificity of single anti-cancer agents. In contrast, a combinatorial approach with the application of two or more anti-cancer agents at their respective effective dosages can achieve a synergistic effect that boosts cytotoxicity to cancer cells. In cancer, aberrant apoptotic pathways allow cells that should be killed to survive with genetic abnormalities, leading to cancer progression. Mutations in apoptotic mechanism arising during the treatment of cancer through cancer progression can consequently lead to chemoresistance. Natural compound mixtures that are believed to have multiple specific targets with minimal acceptable side-effects are now of interest to many researchers due to their cytotoxic and chemosensitizing activities. Synergistic interactions within a drug mixture enhance the search for potential molecular targets in cancer cells. Nonetheless, biased/flawed scientific evidence from natural products can suggest false positive therapeutic benefits during drug screening. In this review, we have taken these factors into consideration when discussing the evidence for these compounds and their synergistic therapeutic benefits in cancer. While there is limited evidence for clinical efficacy for these mixtures, in vitro data suggest that these preparations merit further investigation, both in vitro and in vivo.

Keywords: cancer; apoptosis; chemosensitization; microRNA; natural compound mixtures; metal derivatized natural compounds

1. Introduction

Cancer remains one of the highest causes of death globally. Various types of chemotherapies fail due to adverse reactions, drug resistance, and target specificity of some types of drugs. There is now emerging interest in developing drugs that overcome the problems stated above by using natural compounds, which may affect multiple targets with reduced side effects and which are effective against several cancer types. Natural compounds from various sources including plants, animals, and microorganisms offer a great opportunity for discovery of novel therapeutic candidates for the treatment of cancer [1]. Apoptosis is a self-destructive programmed sequence of signal transduction events that destroys cells that become a threat, or are no longer necessary to the organism [2]. When there is aberrant apoptosis, cells that should be killed instead become immortal, leading to the pathogenesis of many diseases including cancer.

Apoptosis falls into two categories: extrinsic and intrinsic apoptosis. Extrinsic apoptosis occurs when cell death is triggered by binding of extracellular stress ligands to transmembrane receptors such as death receptors CD95 (APO-1/Fas) and tumor necrosis factor (receptor1) [3] as well as dependence receptors such as netrin-1 receptor UNC5H4 [4,5]. In contrast, intrinsic apoptosis occurs in the mitochondria through heterogeneous signaling cascades dependent or independent of caspases [4]. Failure to trigger complete apoptosis in the unhealthy cell population is a cause for cells to grow out of control, leading to cancer [6]. When apoptosis is defective in one of the main apoptotic pathways, it increases the likelihood of the cell becoming cancerous. Various well-established treatments have been designed to destroy cancer cells through apoptosis. Another important mechanism of cell death in cancer cells in response to chemotherapy is autophagy, which takes places in the lysosome by self-degrading intracellular proteins and organelles. It triggers cell death in the absence of apoptotic regulators, but in the presence of important autophagy-regulated genes such as *BECN1* [7] and *ATG5* [8]. In this review, we primarily confine our discussion to apoptotic cell death and autophagic cell death caused by natural chemotherapeutic agents in the context of cancer.

Resistance to treatments that target apoptotic cell death is indicative of treatment failure. Anti-apoptotic mutations during cancer progression reduce chemotherapy-induced apoptosis in spontaneous murine tumors [9] and produce multi-drug resistance [10]. Therefore, understanding how to induce cell cytotoxicity via chemosensitization is as important as how to trigger apoptosis in cancer cells with chemotherapies. It has been reported that natural compounds such as quercetin [11] and tetrandrine [12], known to have anti-tumor activities, are able to not only kill cancer cells but also restore drug sensitivity [13,14]. Moreover, there is evidence that natural compounds including rhamnetin and cirsiliol can radiosensitize in non-small cell lung cancer (NSCLC) [15]. This suggests that natural compounds can have therapeutic effects in cancer chemo-radiotherapy.

Effective development of an anti-cancer drug needs to consider different sets of upregulated, downregulated, and mutated genes and their regulatory pathways in cancer cells. Computational genomics is a powerful tool to identify differential gene expression based on cancer treatment, as it improves our understanding of challenging mechanistic changes in cancer cells and facilitates treatment with a wide range of molecular targets. Whole transcriptome sequencing comprehensively investigates messenger RNA (mRNA)-Seq and small/non-coding RNA-sequencing (RNA-Seq), analyzing tens of thousands of RNA transcripts to uncover their genetic functions. Transcriptomic results subjected to Gene Ontology (GO) clustering and annotation identify differentially expressed genes and can further identify candidate target pathways [16]. Here we highlight the efficacy of complex natural compound mixtures by using molecular approaches with specific emphasis on cancer apoptosis and chemosensitization.

2. Treatment of Cancer through Targeting Apoptosis

There are many therapies for treating cancer, including surgery, radiation therapy, hormone therapy, chemotherapy, and targeted therapies such as immunotherapy and monoclonal antibody therapy. Depending on the type of cancer and underlying biological conditions in the patient, therapy consists of either a single or combination of classical treatments such as surgery, chemotherapy, and/or radiotherapy.

Chemotherapy is a treatment that uses anti-cancer drugs to damage DNA in unhealthy and rapidly dividing cancer cells. Chemotherapy with a defined dosage is usually used to trigger cancer cell cytotoxicity at desirable apoptotic rates. The effectiveness of chemotherapeutic agents depends on their type, dosage, and any adverse reactions in patients. There are several anti-cancer drugs used alone or in combination with other agents to kill cancerous cells. Chemotherapeutic drugs that include synthetic, semi-synthetic, and naturally occurring compounds are cytotoxic, and can destroy both cancerous cells and rapidly dividing normal cells. These agents signal through both death receptors and mitochondrial pathways to induce one or more of the apoptotic pathways [17]. They are characterized based on their structure, derivation, and mechanism of action. Some affect parts of the cell cycle, while others are not phase specific. Depending on the mechanism of action, they are categorized into different groups including alkylating antineoplastic agents, kinase inhibitors, vinca alkaloids, anthracyclines, antimetabolites, aromatase inhibitors, and topoisomerase inhibitors [18]. Nonetheless, the pharmacokinetic variability of synthetic drugs in patients often limits optimal effectiveness with minimal toxic side effects. On the other hand, treatment of cancer by natural compounds and their semi-synthetic analogues both in vitro and in vivo shows promising results against different malignancies [19,20]. Natural compounds such as sesquiterpenes, flavonoids, alkaloids, diterpenoids, saponins, and polyphenolic compounds [11,21] can be substituted for, or applied in combination with, existing drugs.

3. Natural Compounds as Anti-Cancer Agents

Natural compounds with potent anti-cancer activities are widely available from different plant tissues. Eighty percent of the population worldwide traditionally use natural compounds contained in medicinal plants [22] and are largely dependent on them. Naturally occurring compounds target tumor cells by regulating cell death pathways such as extrinsic and intrinsic apoptosis pathways and autophagic pathways. Evidence from in vitro and in vivo studies in prostate cancer treatment with isoflavones and phytoestrogens from soy showed NF-κB deactivation, apoptosis induction, and angiogenesis inhibition [23,24]. A collection of plant-derived natural anti-cancer compounds can be found at Naturally Occurring Plant-based Anti-cancer Compound-Activity-Target Database (NPACT, http://crdd.osdd.net/raghava/npact/) where approximately 1980 experimentally validated compound-target interactions are documented [25]. Millimouno et al. also reviewed promising natural compounds and their related natural sources, pharmacological actions, and molecular targets in details [21].

4. Traditional Chinese Medicines (TCMs) as Anti-Cancer Agents

Due to the complex etiology and pathophysiology of cancer, it is relatively difficult to treat the disease with just single target drugs. Moreover, regardless of the specificity and efficiency of single target therapy, it is difficult to achieve optimal cytotoxic effects on cancer cells because of their rapid molecular adaptations. In contrast, synergistic interactions within multi-component drug preparations allow us to broaden the search for potential molecular targets in cancer cells. Traditional Chinese Medicines (TCM) is formulated based on the compatibility and interrelationships between herbal ingredients that render synergistic therapeutic benefits [26]. TCM uses a combinatorial approach where the application of two or more agents at their respective effective concentrations achieves a synergistic effect that boosts cytotoxicity to cancer cells and can have additional effects on the tumor environment and the immune response to tumors. Therefore, TCM has been used as an alternative or complementary medicine worldwide, and has long been used to treat cancer in China. Chinese herbal medicinal products have been used for cancer prevention and treatment for many years [27], and there is evidence to suggest that TCMs are effective against cancer recurrence and metastasis and can enhance quality of life (QoL), and prolong survival time [27]. For instance, Bioactive polysaccharides with β-1,3, β-1,4, and β-1,6 side branches in TCM stimulate the immune system, thereby indirectly suppressing tumors [28]. TCM is also used to reduce the side effects of conventional chemotherapy for advanced pancreatic cancer, advanced colorectal cancer, and breast cancer [29–31]. There are a range of TCM extracts from *Anemarrhena asphodeloides*, *Artemisia argyi*, *Commiphora myrrha*, *Duchesnea indica*, *Gleditsia sinensis*, *Ligustrum lucidum*, *Rheum palmatum*, *Rubia cordifolia*, *Salvia chinensis*, *Scutellaria barbata*, and *Uncaria rhychophylla* that specifically inhibit cancer cell proliferation from breast, lung, pancreas, and prostate tissues of human and mouse, but show limited inhibition against normal human mammary epithelial cell growth [32]. Artemisinin derivatives artesunate (ART) and dihydroartemisinin have been shown to inhibit cancer cell proliferation and suppress angiogenesis in cervical, uterus chorion, embryo transversal cancer, and ovarian cancer [33]. Because naturally occurring compounds such as plant extracts in TCM are highly chemically diverse, they have become highly significant in the discovery and development of effective therapeutic anti-cancer drugs. TCM preparations can contain alkaloids, flavonoids, saponins, terpenes, polyphenols, fatty acids, and essential oils as bioactive ingredients [34,35].

4.1. Natural Compounds from TCM as Cancer Therapeutics

The main components of TCM such as alkaloids, flavonoids, and saponins are used either individually or as mixtures to treat different types of cancer. Below we list compounds contained in various TCMs that have been found to have anti-cancer activities, including triggering apoptosis.

4.2. Alkaloids

Alkaloids are more abundantly found in broad ranges of the plant kingdom than other classes of natural plant products [36] and are active against various cancers. Alkaloids commonly consist of a nitrogen atom within a heterocyclic ring [37] and are of relatively low toxicity. Several alkaloids have a wide range of significant biological functions including anti-inflammatory, anti-bacterial, anti-diabetic, and anti-cancer activities [38–40]. Some well-developed semi-synthetic anti-cancer drugs are alkaloid derivatives including vinblastine, vinorelbine, vincristine, and vindesine. They are the most important active ingredients in traditional medicine and have been approved for cancer treatment in the United States and Europe [41].

Matrine is a major quinolizindine alkaloid found in the *Sophora flavescens* Aiton plant [42]. Matrine stimulates major apoptotic cascades by upregulating Fas/FasL and Bax, and downregulating Bcl-2 leading to the activation of caspase-3, -8, and -9 in MG-63, U-2OS, Saos-2, and MNNG/HOS human osteosarcoma cells [43]. It also represses cancer metastasis via vascular endothelial growth factor (VEGF)-Protein Kinase B (Akt)-nuclear factor kappa-light-chain-enhancer of activated B cells (NF-κB) signaling in MDA-MB-231 breast cancer cells. The reduction of Bcl-2/Bax protein and mRNA levels by matrine leads to an increase of cell cycle arrest in cancer cells [44]. In human medulloblastoma D341 cells, increased expression of Bcl-2 and decreased expression of Bax is triggered by matrine through caspase-3 and -9 mediated apoptotic pathways [45]. In HepG2 cells, matrine induces tumor suppressor transcription factor p53 through the adenosine monophosphate-activated protein kinase (AMPK) signal transduction pathway, resulting in autophagic cell death through the p53/AMPK signaling pathway [46]. Interestingly, this research reported that downregulation of AMPK leads to a switch to apoptotic cell death from autophagic cell death [46]. The sequential signal transduction leading from autophagy to apoptosis via the activation of p53 was discussed by Guillermo Mariño et al. [47]. Furthermore, metabolomics analysis of matrine treated HepG2 cells identified lipid droplet metabolites, which are substrates for macro autophagy that may partly drive immunity and apoptosis [48]. Li et al. also reported that matrine treatment reduced the level of glutathione (GSH), and the elevated level of GSH is related to chemoresistance in cancer [49]. The above results provide evidence that matrine alone can induce cell death and can be effective against various tumor types.

Oxymatrine is another major quinolizindine alkaloid found in the *Sophora flavescens* Aiton plant. It is cytotoxic to SW1116 human colon cancer cells by downregulating human telomerase reverse transcriptase (hTERT) and upregulating *p53* as well as *mad1* in a concentration dependent manner [50]. It also inhibits the growth of GBC-SD and SGC-996 gallbladder cancer cells via the activation of caspase-3 together with Bax and the suppression of Bcl-2 and NF-κB [51]. It is also known that a mixture of oxymatrine and micellar nanoparticles is an effective proliferation inhibitor of SMM7721 cells [52]. Oxymatrine treatment significantly induces apoptosis by increasing Bax protein expression and reducing Bcl-2 in human lung cancer A549 cells [53]. Proteomic analysis has shown that oxymatrine induces apoptosis in HeLa cells by inhibiting inosine monophosphate dehydrogenase type II (IMPDH2), mitochondrial related apoptotic protein [54]. These studies suggest that oxymatrine may be a useful drug candidate for cancer therapy.

Another type of natural alkaloid is tetrandrine, contained in the Chinese medicinal plant Hang-Fang-Chi, *Radix Stephania tetrandra* S. Moore. This compound has anti-inflammatory, immunosuppressive, and anti-cancer activities [55]. It was reported that a derivative (H1) of tetrandrine has the ability to reverse multi-drug resistance [56]. It inhibits cancer cell proliferation and induces apoptosis in human esophageal cancer cell lines ECa109, Eca109-C3, and human monoblastic leukemic U937 cells. It is also effective in reversing multi-drug resistance in Adriamycin-resistant human breast cancer MCF-7/Adr and human nasopharyngeal cancer KB_{v200} [12]. The molecular mechanisms of action of tetrandrine in cancer cells include upregulation of Bax, Bak, Bad, and apaf-1, downregulation of Bcl-2 and Bcl-xl, releasing cytochrome *c*, and activation of caspase-3 and -9 in the apoptotic mitochondrial pathway [56]. The efficacy of tetrandrine with respect to activation of the intrinsic apoptosis pathway highlights its potential importance as a therapeutic agent.

The semi-synthetic alkaloid analogue vinblastine is an anti-mitotic drug that was originally isolated from the periwinkle plant *Catharanthus roseus* (L.) G. Don. It kills cancer cells by shortening microtubules, disrupting microtubule function resulting in the disappearance of the mitotic spindle, thereby inhibiting cell proliferation [57]. Low concentrations of vinblastine have been shown to slow down or block mitosis in HeLa and BSC cells [57]. Vinblastine is highly potent in relapsed/refractory anaplastic large-cell lymphoma (ALCL) with 65% five-year overall survival [58].

4.3. Flavonoids

Flavonoids are plant secondary metabolites, widely present in fruits and vegetables that are consumed daily. They generally have a sixteen-carbon skeleton and the structures vary around the heterocyclic oxygen ring [59]. Research has shown that flavonoids inhibit cell proliferation and angiogenesis, cause cell cycle arrest, induce cell apoptosis, and reverse multi-drug resistance and/or a combination of the aforementioned mechanisms [60].

Trifolirhizin, a pterocarpan flavonoid, is present in the *Sophora flavescens* Aiton plant. It was shown that trifolirhizin reduces the expression of pro-inflammatory cytokines such as TNF-α, cyclooxygenase-2 (COX-2) and IL6 in experimentally lipopolysaccharide (LPS)-stimulated mouse J774A.1 macrophage [61]. Zhou et al. also showed that it was able to inhibit the growth of human ovarian A2780 and lung H23 cancer cells in vitro. The compound also has anti-proliferative activity in oral carcinoma SCC2095 cells [62]. A combination of trifolirhizin together with maackiain (a constituent of *Trifolium pratense*) has been shown to induce apoptosis in human leukemia HL-60cells. This mixture of compounds resulted in the degradation of DNA into oligonucleosome-size fragments in a time- and dose- dependent manner [63]. These results suggest that trifolirhizin may possibly be developed as an anti-inflammatory nutraceutical for cancer prevention as well as for the mitigation of DNA damage through apoptosis.

Curcumin is a traditional medicine and main curcuminoid of *Curcuma longa* and has been implicated in the perturbation of several genetic pathways [64,65]. It was reported that it selectively targets tumor cells rather than normal cells in vitro and activates different apoptotic pathways including caspase, induction of death receptors and DNA fragmentation, mitochondrial activation, autophagy pathways, inhibition of NF-κB, and inhibition of COX-2 and 5 LOX [65]. In proteomic identification of curcumin treated MCF-7 cells, 3-PGDH and ERP29 were found to be upregulated and TDP-43, SF2/ASF, and eIF3i were downregulated, suggesting curcumin induced apoptosis in breast cancer [64]. However, due to its lack of aqueous solubility, high concentrations are required to demonstrate potential chemotherapeutic efficacy [66]. While many researchers are optimistic regarding curcumin's potential effectiveness against cancer, there is evidence to show that curcumin has no therapeutic benefits despite many published articles and clinical trials [67]. Skepticism about curcumin is based on both poor characterization of curcumin and pan-assay interference in many experiments that indicated that curcumin was a promising compound for cancer treatment. Yet, in spite of no significant effects in trials, researchers still think that because curcumin can interact with many proteins and because of suggestive trends in trial results, there is still justification for further study [68]. Until better experiments are carried out, the anti-cancer activity of curcumin remains unconfirmed.

A flavonoid, quercetin, is abundant in daily-consumed foods such as onions (*Allium cepa*) with a wide range of anecdotally reported health benefits that include anti-oxidant, anti-inflammatory, and anti-cancer activities in vitro and is effective against various cancer cells [69]. Quercetin mainly targets the cell cycle at G1/S and G2/M check points by inducing the p21 CDK inhibitor while decreasing pRb phosphorylation, thereby blocking E2F1, which is an important transcription factor of DNA synthesis proteins [70]. Deng et al. reported that apoptosis-mediated cell death from quercetin treatment resulted from arresting the cell cycle at G0/G1 phase in MCF-7 breast cancer cells. They also showed that increasing concentrations of quercetin were directly proportional to the decreasing concentrations of survivin, a member of a protein family that negatively regulates apoptosis [71]. Proteomic analysis revealed that quercetin treatment suppressed cell proliferation while arresting mitosis leading to apoptosis by downregulating IQGAP1 and β-tubulin and their interactions with other proteins in HepG2 cells [72]. Despite the abundance of quercetin, it has not been investigated in cancer clinical trials.

4.4. Saponins

Saponins are found not only in a wide range of plants but also in animals, and have different carbon backbones that classify them as either steroids or triterpenes. They are secondary metabolites with potent biological functions. These compounds are active against several tumors not only as single compounds but also in combination with conventional therapies by causing cell cycle arrest and triggering apoptosis [73].

Chikusetsusaponin IVa butyl ester (CS-IVa-Be) is an apoptotic triterpenoid saponin extracted from *Acanthopanas gracilistylus* herb. The extract from this Chinese medicinal herb has been found to cause cell cycle arrest at G0/G1 stage in a variety of cancer cell lines including MT-2, Raji, HL-60, TMK-1, and HSC-2 [74]. The compound induces apoptosis in MDA-MB-231 cells by inhibiting IL-6 family induced STAT3 activity through the IL-6/JAK/STAT3 signaling pathway. It also sensitizes the Tumor necrosis factor (TNF)-related apoptosis-inducing ligand (TRAIL), a specific inducer of cancer cell apoptosis, in TRAIL resistant MDA-MB-231 cells by upregulating death receptor 5 (DR5) [75]. Because of this, CS-IVa-Be induces apoptosis upon treatment with TRAIL in TRAIL resistant MDA-MB-231 cells.

Polyphyllin D is a promising anti-proliferative steroidal saponin extracted from the traditional Chinese medicinal plant *Paris polyphylla*. The cytotoxic activity of polyphyllin D was observed via induction of DNA fragmentation and dissipation of mitochondrial membrane potential Δψm, resulting in mitochondrial dysfunction and loss of membrane integrity in MCF-7 and MDA-MB-231 cells [76]. A 50% reduction in tumor growth from 10 consecutive days of polyphyllin D administration in mice was also documented.

Diosgenin is effective against HCT-116 human colon cancer cells by reducing both mRNA and protein expression of 3-hydroxy-3-methylglutaryl CoA reductase, resulting in apoptosis [77]. It truncated the poly (ADP-ribose) polymerase protein from 116-kDa to an 85 kDA fragment, which leads to the induction of apoptosis. This indicates that Diosgenin is a potent apoptosis inducer in HCT-116 cancer cells. Diosgenin arrested the cell cycle at sub-G1 phase, suppressed FAS expression, and inhibited mammalian target of rapamycin (mTOR) phosphorylation in HER2 overexpressing human AU565 breast cancer cells, inhibiting cell proliferation [78].

Another apoptotic triterpenoid saponin is Macranthoside B (MB), extracted from *Lonicera macranthoids*. It is strongly effective in various tumors via mitochondrially mediated apoptosis resulting from an increased Bax/Bcl-2 ratio [79]. Furthermore, MB induced apoptosis via autophagy through the ROS/AMPK/mTOR pathway, while elevating reactive oxygen species (ROS) together with 5′ AMPK, and reducing mTOR in human ovarian cancer A2780 cells [80].

With respect to the cytotoxic properties of saponins, a wide range of these compounds has been tested, with some shown to be potent apoptotic inducers. Yet the potential of this class of compounds remains to be fully explored, and they may also be effective in combination with other agents to synergistically enhance their therapeutic effect in cancer.

4.5. Drugs Based on Mixtures of Compounds

Compound Kushen Injection (CKI), approved by the State Food and Drug Administration of China, has been used to treat different types of cancer, including liver, gastric, and non-small cell lung carcinoma in combination with Western anti-cancer agents [81]. It contains alkaloids, flavonoids, saccharides, and organic acids [82] and is extracted from two medical herbs including *Radix Sophorae flavescentis* and *Rhizoma Smilacis Glabrae* [83]. It modulates immunity, decreases inflammation, relieves cancer pain, and, most importantly, has anti-neoplastic activity [83]. For example, CKI downregulates β-catenin through the Wnt signaling pathway, which in turn targets the oncogenes *c-MYC* and *CyclinD1* [84], leading to suppression of MCF-7 cancer stem cell-like side population (SP) cells [85]. A systematic review and meta-analysis reported that CKI could reduce adverse effects in cancer patients and improved total pain relief and QoL [86]. Transcriptome analysis of CKI treated MCF-7 cells by Qu et al. revealed that the mixture inhibited cell proliferation and induced apoptosis in a concentration-dependent manner by primarily targeting the cell cycle in MCF-7 cells [87]. Qu et al. also showed that long non-coding RNA (lncRNA) *H19* was dramatically downregulated in MCF-7 cells treated with CKI [87]. *H19* is overexpressed in several cancer types and is associated with tumor metastasis, for example, where lncRNA *H19* suppresses miR-630, perturbing the inhibition of EZH2 in nasopharyngeal carcinoma [88]. The primary effect of CKI on cancer cells is through the cell cycle, but it also affects many other pathways and it may be useful as both an anticancer and anti-inflammatory agent. It was claimed that CKI is effective in inhibiting metastasis and reversing multi-drug resistance (MDR) as well [83]. Yet, there is currently no research evidence supporting the effectiveness of CKI on reversing MDR in English-language journals. Therefore, in vivo and clinical relevance of the drug should be researched to establish the effective usage of CKI with respect to chemosensitizing activities.

Anti-tumor B (ATB), known as Zeng Sheng Ping, is also an herbal medicine which is formulated from six different medicinal plants, and its main constituents are flavones, alkaloids, phytosterols, sapogenins, triterpenes, and triterpenoids [89]. ATB decreased lung tumor load by approximately 60% in both wild-type and Ink4a/Arf tumor suppressor gene-deficient mice and 90% in p53 transgenic mice [89]. The drug markedly reduced cell proliferation by inhibiting the mitogen-activated protein kinase (MAPK) pathway while increasing apoptosis by reducing Bcl-2 in oral cancer in hamsters [90]. Lim et al. also showed that the Notch2 receptor, which is an important signaling regulator of brain tumors, and its downstream effector gene *Hes1* were downregulated by ATB [91]. ATB induced apoptosis in both U87 glioblastoma and DAOY medulloblastoma cells [91], suggesting that ATB might be effective against multiple tumors. Based on the results from animal models, ATB has shown chemo-preventive activities in hamsters and mice with carcinogen-induced oral cancers [92]. Microarrays, combined with GenMAPP analysis of mouse lung tumor models, showed that multiple genes affected by herbal medicine ATB are members of different genetic pathways such as ubiquitin-proteasome, Notch, Ras-MAPK, and G13 pathways [89], which are important in mitogenesis, neoplastic transformation [93] and apoptosis [94]. This gene expression microarray study showed that ATB is a potential tumor suppressor capable of targeting cell proliferation, differentiation, and apoptosis [89].

The proteomic profile of MCF-7 cells treated with Zilongjin, an herbal antitumor medicine, showed the downregulation of HSP27, a blocker of apoptosis [95]. Zilongjin also suppressed the expression of eIF3I and eIF1AY proteins, which are important regulators of translation initiation [95]. Collectively, the proteomic approach can identify translational perturbations in cancer cells and protein-wide changes from these perturbations in response to stimuli. Microarray based gene expression analysis of four different lung cancer cell lines treated by Zilongjin showed that 170 genes were upregulated and 313 were downregulated by the drug. Of these 483 genes, eleven genes including *HELLS*, *JUN*, *XIAP*, *MCM6*, *CDKN2C*, *CCNE2*, *HN1L*, *TFDP2*, *CCNG2*, *GADD45A*, and *CDKN1A* were found to be involved in cancer-related pathways such as apoptosis, cell cycle, and MAPK cascade [96].

Taken together, the findings based on the molecular and '–omics approaches' suggest that natural compound mixtures have multiple targets in cancer cells. Combined compounds from two or more sources are potential resources for the development of multi-targeted anti-cancer therapeutics and have a broader range of molecular targets in cancer cells. It is important to bear in mind that the clinical effectiveness of some of these combined drugs have only been reported in non-English language papers and these often lack compelling clinical data. Therefore, more work is needed to evaluate natural compound mixtures as cancer treatments. Figure 1 shows the molecular targets of two groups of single compounds, alkaloids and flavonoids, compared to a compound mixture, CKI, that contains both groups of natural compounds in an in vitro setting. Summaries of therapeutic effects from single or mixtures of natural compounds and their possible cellular mechanisms are shown in Table 1.

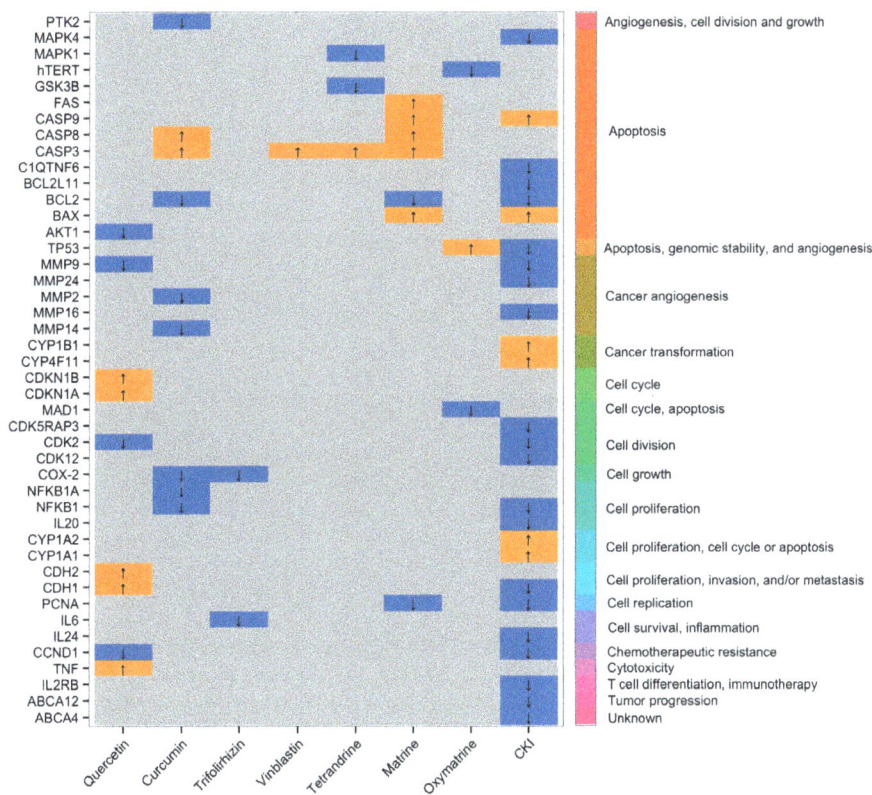

Figure 1. Differential gene expression in different cancer cell lines induced by flavonoids and alkaloids as well as Compound Kushen Injection (CKI). Left hand axis shows differentially expressed genes, bottom axis shows natural anti-cancer agent treatments (Quercetin, Curcumin, Trifolirhizin = flavonoids), (Vinblastine, Tetrandrine, Matrine, Oxymatrine = alkaloids), and (compound mixture = CKI) and the right hand axis shows Gene Ontology (GO) clustering and annotation of differentially expressed genes. Up and down-regulated genes (up = YELLOW and down = BLUE) known to be affected by natural compound anti-cancer agents were obtained from http://crdd.osdd.net/raghava/npact/browse.php and significantly differentially expressed genes from CKI treated MCF-7 cells were obtained from RNA-Seq experiments conducted by Qu et al. [87].

Table 1. Reported therapeutic effects by single or mixtures of natural compounds in different stages of cellular mechanisms.

Ref.	Herbal Medicines	Types of Cancer	Cell Lines/Model	Mechanisms of Actions
[97]	Curcumin	Colorectal	Colorectal cancer stem cells (CCSCs)	Apoptosis
[98]	Ginsenoside Rg3	Liver	Tumor bearing rats	Apoptosis, Immune responses
[99]	Curcumin	Breast	MCF-7	Anti-inflammation
[100]	Matrine	Lung	HepG2	Proliferation and metastasis chemosensitization
[72]	Quercetin	Lung	HepG2	Apoptosis
[95]	Zilongjin	Breast	MCF-7	Inhibits malignant proliferation, apoptosis
[101]	Triterpenes from *Ganoderma lucidum*	Cervical	HeLa	Cell death, oxidative stress, calcium signaling, and ER stress
[64]	Curcumin	Breast	MCF-7	Apoptosis
[102]	Triterpenes from *Patrinia heterophylla*	Leukemia	K562	Energy metabolism, oxidative stress, signal transduction, differential induction, protein biosynthesis, and apoptosis
[54]	Oxymatrine	Cervical	HeLa	Inhibits proliferation, apoptosis
[103]	Sanguinarine from Papaveraceae family	Pancreatic	BxPC-3, MIA PaCa-2	Decreases cellular hypoxia and cell proliferation, induces apoptosis leading to cancer cells inhibition
[89]	Zeng Sheng Ping (Antitumor B)	Lung	Mouse lung	Ubiquitin-proteasome, Notch, Ras-MAPK, G13 pathway, cell proliferation, differentiation, and apoptosis
[104]	Aidi injection	Breast	MCF-7	Inhibits proliferation, apoptosis
[96]	Zilongjin	Lung	A549, H446, H460, and H520	Cell cycle regulation, MAPK cascade, and apoptosis
[87]	Compound Kushen Injection	Breast	MCF-7	Cell cycle regulation, cell growth related pathway

5. Alternative Approach of Triggering Apoptosis Using Metal-Derivatized Natural Compounds

Structurally modified anti-cancer compounds can be highly effective in treating cancer due to their controlling chemo-, site- selectivity in cells. The heavy metal-based compound, cisplatin, has a broad spectrum of cytotoxicity that makes it among the most popular and effective chemotherapy drugs. Cisplatin and compounds of this type are used to treat approximately 50%–70% of all cancer patients [105]. Although metal-based anti-cancer drugs show significant effectiveness against cancer, they can also have severe adverse effects. Moreover, different mechanisms are used by cancer cells to resist cytotoxic drugs, complicating the development of novel potent anti-cancer drugs. Natural compounds have been shown to have low adverse reactions in normal cells of cancer patients, and natural compound-metal complexes have been confirmed as potential (pro) drugs [106]. The alkaloid liriodenine has anti-tumor activity and when combined with platinum (II) and ruthenium (II), the liriodenine-metal complexes covalently bind to DNA and enhance the cytotoxicity of liriodenine. When Gallium (III) and Tin (IV) are combined with matrine, these combinations further modulate cell cycle arrest at the G2/M phase, and Gold (III)-matrine complexes inhibit topoisomerase I [106], thereby causing DNA replication processes in cancer cells to malfunction. The actions of these different metal-based compounds vary depending on cell type. For instance, Gallium III-matrine and Gold III-matrine compounds have significant anti-proliferative activities in SW480 cells, HeLa cells, HepG2 cells, and MCF-7 cells, respectively [106]. It is worth noting that the efficacy of synthetic metal-natural compounds in vitro greatly exceeds those of cisplatin and matrine alone. These results indicate that metal-based cytotoxic natural compounds may hold promise as anti-cancer therapeutics, based on their multi-targeting effects on cancer cell regulatory networks. A summary of therapeutic effects by metal-derivatized natural compounds and their mechanisms of actions compared to their original forms and/or existing metal-based anticancer drugs are shown in Table 2.

Table 2. Mechanisms of action of metal-derivatized natural compounds compared to their original forms and/or existing metal-based anticancer drugs.

Groups of Natural Compounds	Metal Derivatized Natural Compounds	Source	Cell Lines/Model	Mechanisms of Actions	Remarks	Ref.
Alkaloids	GL331	**Compound:** Podophyllotoxin **Plant:** *Podophyllum* species	KB/VCR, MCF-7/ADR, and HL60/VCR	TOPO II inhibitor, cell cycle arrest at G2, cause DNA breakage and apoptosis via inhibiting protein tyrosine kinase	GL331 shows greater cytotoxicity in vitro and in vivo, and overcomes multi-drug resistance (MDR) compared to etoposide. GL331 is now in phase II clinical trial	[107]
	[H-MT][GaCl$_4$] [H-MT][AuCl$_4$] [Sn(H-MT)Cl$_5$]	**Compound:** MT (*Matrine*) **Plant:** *Sophora flavescens*	SW480, HeLa, HepG2, and MCF-7	Cell cycle arrest at the G$_2$/M phase	MT + Gallium (GaCl$_4$) and MT + Gold (AuCl$_4$) enhanced the cytotoxicity better than MT alone and cisplatin	[108]
	[Ru(N–N)$_2$ (Norharman)$_2$] (SO$_3$CF$_3$)$_2$	**Compound:** Norharman **Plant:** *Peganum harmala* L.	HepG2, HeLa, MCF-7, and MCF-10A	Cell cycle arrest at G0/G1, apoptosis via mitochondrial dysfunction and ROS accumulation	IC$_{50}$ value of the complex is much lower and the anti-proliferative activity is much higher than those of Norharman and cisplatin	[109]
	[L+H][AuCl$_4$] [AuCl$_3$ L]	**Compound:** Liriodenine (L) **Plant:** *Zanthoxylum nitidum*	MCF-7	TOPO I inhibitor, cell cycle arrest at S phase	Higher anti-proliferative activity than cisplatin, Adriamycin, liriodenine alone, and NaAuCl4	[110]
Flavonoids	hesperetin [CuL$_2$(H$_2$O)$_2$]nH$_2$O	**Compound:** hesperetin, **Plant:** *Stilbella fimetaria*	HepG2, and SGC-7901	Growth inhibition	DNA binding affinity of hesperetin-Cu(II) complex is stronger than that of free hesperetin	[111]
	Zn(morin)$_2$.3H$_2$O Cu(morin)$_2$.2H$_2$O	**Compound:** Morin **Plant:** *Maclura pomifera*	Hep-2, BBHK-2, BHK21, and HL-60	In vitro antitumor activity	Higher anti-proliferative activity than morin alone	[106]
	Cu(Que)$_2$(H$_2$O)$_2$	**Compound:** quercetin, **Plant:** various fruits and vegetables	A549	DNA breakage, apoptosis via generation of ROS and intercalation into DNA	Higher cytotoxic activity than that of quercetin alone	[112]

6. Chemoresistance in Cancer and Chemosensitization by Natural Compounds

The main reason for cancer treatment failure by chemotherapy is the emergence of drug-resistance during cancer progression. Collectively, MDR occurs when ABC transporters become overexpressed. Out of 48 human ABC transporters, the mechanisms of actions of P-gp, MDR1, ABCB1, MRP1/ABCC1, and ABCG2 have been widely reported clinically [113]. However, additional ABC transporters have been explored as targets for cancer therapeutics in drug discovery. For instance, four ABC transporter genes (*ABCA4*, *ABCC3*, *ABCC5*, and *ABCC8*) were upregulated in resistant MCF-7/AdVp3000 cells, whereas complete or partial downregulation of these genes was observed in the revertant MCF-7/AdVpRev cells [114]. Overexpression of drug resistance associated genes *ABCA4* [113,114] and *ABCA12* [115,116] were observed in human pancreatic cancers and MCF-7 breast cancer cell lines, respectively. Therefore, it might be productive to scrutinize the mechanisms of actions of less commonly studied ABC transporters from drug resistant cancer cell lines when treated with natural compounds.

To overcome drug resistance, natural compounds such as sesquiterpenes, flavonoids, alkaloids, diterpenoids, saponins, and polyphenolic compounds [11,21] are substituted or applied in combination with existing drugs. These compounds are known to have anti-tumor activities and can not only kill cancer cells but also restore drug sensitivity. For example, tetrandrine (bioactive alkaloid) modulates P-gp-mediated drug efflux in vitro and has anti-neoplastic activity when given together with doxorubicin to mice bearing resistant MCF-1/DOX cells in vivo [13,14]. Another natural product is Quercetin (a flavonoid [11]), which blocks *MDR1* transcription, thereby suppressing P-gp expression [117] and restoring daunorubicin chemosensitivity in HL-60/DOX and K562/DOX cell lines [118,119].

A different flavonoid, curcumin, inhibits the main ABC transporters such as P-gp, MRP1, and ABCG2, and increases vincristine chemosensitivity in SGC7901/VCR cell lines [120,121]. 20(*S*)-Ginsenoside Rg3 (saponins) chemosensitizes vincristine, doxorubicin, etoposide, and colchicine resistant KBV20 cell lines in a time- and dose-dependent manner [11]. Oxymatrine chemosensitizes cisplatin resistant HeLa/DDP cells by suppressing inosine monophosphate dehydrogenase type II (IMPDH2) levels and inducing apoptosis through the mitochondrial pathway [54]. CKI may also act as a stimulator of drug resistance reversal and chemosensitization in cancer. Transcriptome analysis from Qu et al. revealed that two drug resistant ABC transporter genes, *ABCA12* and *ABCA4*, were significantly downregulated in CKI treated MCF-7 cells [87]. These results support the assertion that natural compounds can reverse or alter chemoresistance.

7. Apoptosis in MDR Cells through Modulation of MicroRNA (miRNA) Networks by Natural Compounds

The compelling link between possible successful cancer chemotherapy and triggering apoptosis by chemosensitization is connected in this review by the involvement of microRNAs. miRNAs are highly conserved small non-coding RNA molecules that directly interact with their target mRNA, causing either mRNA transcriptional degradation or translational repression that results in the reduction of gene expression [122]. miRNAs have emerged as both potential therapeutic agents and biomarkers [123]. Some have oncogenic properties while others act as tumor suppressors. Miller et al. reported that eight upregulated miRNAs and seven downregulated miRNAs were found in tamoxifen resistant human breast cancer MCF-7 cells [124]. Moreover, increased expression of miR-415 is correlated with the decreased expression of *MDR1* genes and seems to elevate sensitivity to doxorubicin in DOX-resistant breast cancer cells [125]. In tumors, epithelial-to-mesenchymal transition (EMT) plays an important role of tumor invasion/migration and metastasis [126]. ZEB1 and ZEB2 repress E-cadherin expression [127,128] and downregulation of E-cadherin allows epithelial cells to undergo EMT, which contributes resistance to EGFR-directed therapy in cancer. Members of the miR-200 family, particularly miR-200c and miR-200b, target ZEB1 and ZEB2 and control EMT to sensitize EGFR therapy [129].

Growing evidence indicates that natural compounds are important regulators of miRNA mediated genetic pathways in cancer. For example, matrine significantly reduces the overexpression of miRNA-21 and stimulates apoptosis in HepG2 and Hep3B cells [130]. Matrine also downregulated 14 miRNAs and upregulated their target genes in the MAPK signaling pathway in SGC7901 human gastric cancer cells [131]. Another natural compound, oxymatrine, acts on human ovarian cancer OVCAR3 cells by upregulating miR-29b, which downregulates the expression level of matrix metalloproteinase-2 and induces apoptosis [132]. In the microarray analysis of MCF-7 cells, upregulation of 45 miRNAs was shown after treatment with TCM Aidi injection. Of these, miRNA-126 was found to be a suppressor of proliferation of MCF-7 cells [104]. Research showed that inhibition of miRNA-25 by a natural phenol, isoliquiritigenin, leads to autophagy in MCF-7/ADR cells via increased expression of autophagy regulator ULK1 [133]. In the transcriptome of CKI treated MCF-7 cells, the upregulation of hsa-miR-6879 and downregulation of its target drug resistant gene *ABCA12* were observed [87]. However, the prediction of the role of pre-miR-6879 in terms of the regulation of ABC transporters still needs to be experimentally confirmed. The link between natural compounds inducing miRNA mediated drug resistance reversal and chemosensitization, leading to cancer cells autophagy and apoptosis, supports the investigation of natural compounds in drug resistant cancer chemosensitization therapy. Whilst it is known that miRNAs are important in phytochemically mediated cancer cell death, little is known about full or partial reversal of MDR via the regulation of miRNAs by natural compounds. Hence, there is a gap between the use of natural compounds that regulate miRNA and MDR that is not well understood. A model of the regulatory interactions between drugs, miRNA, and target genes in the context of autophagy, drug resistance reversal, and apoptosis is shown in Figure 2.

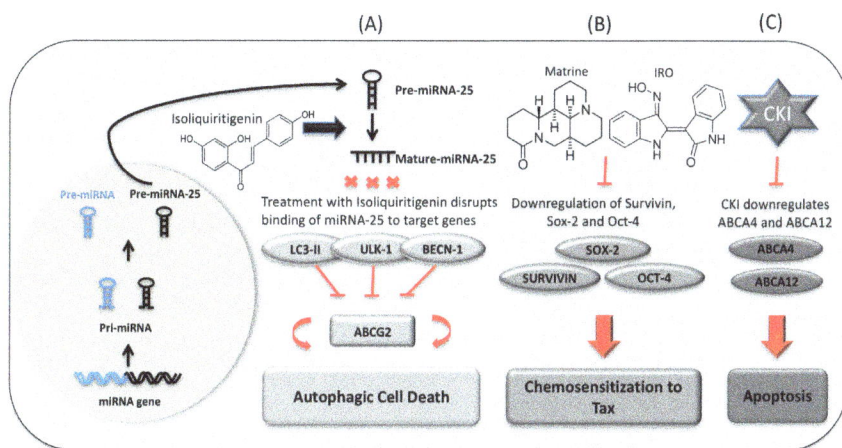

Figure 2. Three panels (**A–C**) show the influence of natural compounds on gene expression in terms of MDR regulation. (**A**) Treatment of Isoliquiritigenin blocks binding of autophagy-related miR-25 to the 3' UTR region of *ULK-1*, *LC3-II*, and *BECN-1* in killing drug-resistant breast cancer cells [133]; (**B**) Matrine and indirubin-3'-monoxime (IRO) reverse chemoresistance of paclitaxel (TAX) by downregulating the expression of Sox-2, Survivin, and Oct-4 proteins in human squamous cell carcinoma [134]; (**C**) CKI suppresses the expression of the drug resistant related genes *ABCA12* and *ABCA4* in cancer cell apoptosis. The involvement of miRNAs in the regulation of these genes with respect to drug resistance has not yet been confirmed [87].

8. Conclusions

Despite significant progress in the fields of cancer diagnosis and chemotherapy, cancer remains one of the greatest causes of death worldwide. Novel approaches to cancer management often fail due to frequent genetic alterations and mutations in cancer genomes. Because of the high frequency of side effects caused by chemotherapy, metastatic cancers still need new, more effective chemotherapeutics. There is emerging interest in developing drugs to tackle these problems by using natural compounds, which may affect multiple targets with lower side effects and be effective against several cancer types. Despite attempts in chemotherapy using natural compounds, many questions remain regarding their efficacy and potential modes of action. From the medicinal chemistry point of view, an important fact when applying the natural compounds to consider is the impurities of the extracted herbal medicine that have their own biological activity which may provide false positive signal of the molecules during drug screening. In addition, flawed scientific evidence from natural product mixtures can suggest false positive therapeutic benefits. Therefore, careful examination to approve the molecular target engagement while researching drug discovery is necessary. The best possible drug combinations are based on the understanding of the cancer-specific context of mutated oncogenes, tumor suppressor genes, and their regulatory pathways. To evaluate novel treatment approaches involving the use of mixtures of natural compounds, we must understand how coding and non-coding RNAs, oncogenes, downregulated tumor suppressor genes, and mutated genes, as well as their signal transduction pathways, respond to these drugs. Systems biology can help characterize the roles of functionally cryptic elements such as long and short ncRNAs in cancer. We have reviewed the effectiveness of natural compounds in cancer cells in terms of triggering apoptosis and chemosensitization through the application of molecular genetic and "-omics" approaches. There is evidence to suggest that natural compounds might be effective and less toxic in some circumstances. Furthermore, metal-derivatized natural compounds can also trigger apoptosis and natural compound-metal complexes have been confirmed as potential (pro) drugs [106]. This suggests that natural compounds and their derivatives merit further investigation as anti-cancer therapeutics.

The effect of multi-drug resistance during cancer progression is another important hurdle for chemotherapy. In resistance to chemotherapy-induced cancer cell death, genetic alterations in the drug-induced apoptotic program also cause MDR to biochemically unrelated drugs [135]. This is illustrated by the relationship between chemoresistance and targeted upregulation of apoptosis regulator *bcl-2* by multiple DNA damaging stimuli [135]. Therefore, it is important to study the role of natural compounds in chemosensitization, since this may create new avenues for novel drug development in cancer treatment. The investigation of natural compound mixtures as apoptosis-inducing and chemosensitizing cancer therapeutics will improve our understanding of the molecular changes in cancer cells and should provide clues about how the disease can be controlled. Our current knowledge of natural compounds' effects on cancer is mainly from cell-based experiments and partly from in vivo experiments. To date, the clinical effectiveness of some of these combined drugs has not been robustly demonstrated. As a result, additional research using in vivo systems and better clinical trials is needed to determine the safety and clinical usefulness of these drugs.

Acknowledgments: Thazin Nwe Aung acknowledges Saeed Nourmohammadi for valuable assistance and advice along with everyone in David L. Adelson's lab. Special thanks to Merry Wickes for her advice to improve the manuscript.

Author Contributions: Thazin Nwe Aung and David L. Adelson designed the study. Thazin Nwe Aung performed the literature review and wrote the manuscript. Zhipeng Qu prepared Figure 1. Zhipeng Qu, R. Daniel Kortschak, and David L. Adelson contributed to writing the paper. All authors read and approved the final manuscript.

Conflicts of Interest: The authors declare that the research was conducted in the absence of any commercial or financial relationships that could be construed as a potential conflict of interest. The Zhendong Centre is funded through a charitable donation of Zhendong Pharmaceutical Co., Ltd. Zhendong Pharmaceutical Co., Ltd. has had no control over the scientific research we carry out or the content of this article.

Abbreviations

CKI	Compound Kushen Injection
MDR	Multi-drug Resistance
ATB	Antitumor B
NPACT	Naturally Occurring Plant-based Anti-cancer Compound-Activity-Target Database
TCM	Traditional Chinese Medicine
QoL	Quality of Life
MT	Matrine
GO	Gene Ontology
IRO	Indirubin-3'-monoxime
CS-IVa-Be	Chikusetsusaponin IV a Butyl Ester
VEGF	Vascular Endothelial Growth Factor
Akt	Protein Kinase B
NF-κB	Nuclear Factor κ-light-chain-enhancer of activated B cells
ER	Endoplasmic Reticulum
MAPK	Mitogen Activated Protein Kinase

References

1. Newman, D.J.; Cragg, G.M.; Snader, K.M. Natural products as sources of new drugs over the period 1981–2002. *J. Nat. Prod.* **2003**, *66*, 1022–1037. [CrossRef] [PubMed]
2. Alberts, B.; Johnson, A.; Lewis, J.; Raff, M.; Roberts, K.; Walter, P. Programmed Cell Death (Apoptosis). In *Molecular Biology of the Cell*, 4th ed.; Garland Science: New York, NY, USA, 2002.
3. Wajant, H. The Fas signaling pathway: More than a paradigm. *Science* **2002**, *296*, 1635–1636. [CrossRef] [PubMed]

4. Galluzzi, L.; Vitale, I.; Abrams, J.; Alnemri, E.; Baehrecke, E.; Blagosklonny, M.; Dawson, T.M.; Dawson, V.; El-Deiry, W.; Fulda, S. Molecular definitions of cell death subroutines: Recommendations of the Nomenclature Committee on Cell Death 2012. *Cell. Death Differ.* **2012**, *19*, 107–120. [CrossRef] [PubMed]

5. Mehlen, P.; Bredesen, D.E. Dependence receptors: From basic research to drug development. *Sci. Signal.* **2011**, *4*, mr2. [CrossRef] [PubMed]

6. Wong, R. Apoptosis in cancer: From pathogenesis to treatment. *J. Exp. Clin. Cancer Res.* **2011**, *30*, 87. [CrossRef] [PubMed]

7. Chen, R.; Wang, H.; Liang, B.; Liu, G.; Tang, M.; Jia, R.; Fan, X.; Jing, W.; Zhou, X.; Wang, H. Downregulation of ASPP2 improves hepatocellular carcinoma cells survival via promoting BECN1-dependent autophagy initiation. *Cell Death Dis.* **2016**, *7*, e2512. [CrossRef] [PubMed]

8. Codogno, P.; Meijer, A.J. Atg5: More than an autophagy factor. *Nat. Cell Biol.* **2006**, *8*, 1045–1046. [CrossRef] [PubMed]

9. Schmitt, C.A.; McCurrach, M.E.; de Stanchina, E.; Wallace-Brodeur, R.R.; Lowe, S.W. INK4a/ARF mutations accelerate lymphomagenesis and promote chemoresistance by disabling p53. *Genes Dev.* **1999**, *13*, 2670–2677. [CrossRef] [PubMed]

10. Dive, C.; Hickman, J. Drug-target interactions: Only the first step in the commitment to a programmed cell death? *Br. J. Cancer* **1991**, *64*, 192. [PubMed]

11. Chai, S.; To, K.; Lin, G. Circumvention of multi-drug resistance of cancer cells by Chinese herbal medicines. *Chin. Med.* **2010**, *5*, 26. [CrossRef] [PubMed]

12. Yu-Jen, C. Potential role of tetrandrine in cancer therapy. *Acta Pharmacol. Sin.* **2002**, *23*, 1102–1106.

13. Tian, H.; Pan, Q. A comparative study on effect of two bisbenzylisoquinolines, tetrandrine and berbamine, on reversal of multidrug resistance. *Acta Pharm. Sin.* **1997**, *32*, 245–250.

14. Choi, S.-U.; Park, S.-H.; Kim, K.-H.; Choi, E.-J.; Kim, S.; Park, W.-K.; Zhang, Y.-H.; Kim, H.-S.; Jung, N.-P.; Lee, C.-O. The bis benzylisoquinoline alkaloids, tetrandine and fangchinoline, enhance the cytotoxicity of multidrug resistance-related drugs via modulation of P-glycoprotein. *Anti Cancer Drugs* **1998**, *9*, 255–262. [CrossRef] [PubMed]

15. Kang, J.; Kim, E.; Kim, W.; Seong, K.M.; Youn, H.; Kim, J.W.; Kim, J.; Youn, B. Rhamnetin and cirsiliol induce radiosensitization and inhibition of epithelial-mesenchymal transition (EMT) by miR-34a-mediated suppression of Notch-1 expression in non-small cell lung cancer cell lines. *J. Biol. Chem.* **2013**, *288*, 27343–27357. [CrossRef] [PubMed]

16. Graw, S.; Meier, R.; Minn, K.; Bloomer, C.; Godwin, A.K.; Fridley, B.; Vlad, A.; Beyerlein, P.; Chien, J. Robust gene expression and mutation analyses of RNA-sequencing of formalin-fixed diagnostic tumor samples. *Sci. Rep.* **2015**, *5*, 12335. [CrossRef] [PubMed]

17. Kaufmann, S.H.; Earnshaw, W.C. Induction of apoptosis by cancer chemotherapy. *Exp. Cell Res.* **2000**, *256*, 42–49. [CrossRef] [PubMed]

18. Espinosa, E.; Zamora, P.; Feliu, J.; Barón, M.G. Classification of anticancer drugs—A new system based on therapeutic targets. *Cancer Treat. Rev.* **2003**, *29*, 515–523. [CrossRef]

19. Manson, M.M. Cancer prevention–the potential for diet to modulate molecular signalling. *Trends Mol. Med.* **2003**, *9*, 11–18. [CrossRef]

20. Prakash, O.; Kumar, A.; Kumar, P. Anticancer potential of plants and natural products: A review. *Am. J. Pharmacol. Sci.* **2013**, *1*, 104–115. [CrossRef]

21. Millimouno, F.M.; Dong, J.; Yang, L.; Li, J.; Li, X. Targeting apoptosis pathways in cancer and perspectives with natural compounds from Mother Nature. *Cancer Prev. Res.* **2014**, *7*, 1081–1107. [CrossRef] [PubMed]

22. Fulda, S. Evasion of apoptosis as a cellular stress response in cancer. *Int. J. Cell Biol.* **2010**. [CrossRef] [PubMed]

23. Davis, J.N.; Kucuk, O.; Sarkar, F.H. Genistein inhibits NF-κB activation in prostate cancer cells. *Nutr. Cancer* **1999**, *35*, 167–174. [CrossRef] [PubMed]

24. Stephens, F.O. The rising incidence of breast cancer in women and prostate cancer in men. Dietary influences: A possible preventive role for nature's sex hormone modifiers-the phytoestrogens (review). *Oncol. Rep.* **1999**, *6*, 865–935. [CrossRef] [PubMed]

25. Mangal, M.; Sagar, P.; Singh, H.; Raghava, G.P.; Agarwal, S.M. NPACT: Naturally occurring plant-based anti-cancer compound-activity-target database. *Nucleic Acids Res.* **2013**, *41*, D1124–D1129. [CrossRef] [PubMed]

26. Zhou, X.; Seto, S.W.; Chang, D.; Kiat, H.; Razmovski-Naumovski, V.; Chan, K.; Bensoussan, A. Synergistic effects of Chinese herbal medicine: A comprehensive review of methodology and current research. *Front. Pharmacol.* **2016**, *7*, 201. [CrossRef] [PubMed]

27. Liu, J.; Wang, S.; Zhang, Y.; Fan, H.T.; Lin, H.S. Traditional Chinese medicine and cancer: History, present situation, and development. *Thorac. Cancer* **2015**, *6*, 561–569. [CrossRef] [PubMed]

28. Chang, R. Bioactive polysaccharides from traditional Chinese medicine herbs as anticancer adjuvants. *J. Altern. Complement. Med.* **2002**, *8*, 559–565. [CrossRef] [PubMed]

29. Ni, Y.H.; Li, X.; Xu, Y.; Liu, J.P. Chinese herbal medicine for advanced pancreatic cancer. *Cochrane Libr.* **2012**. [CrossRef]

30. Zhang, M.; Liu, X.; Li, J.; He, L.; Tripathy, D. Chinese medicinal herbs to treat the side-effects of chemotherapy in breast cancer patients. *Cochrane Libr.* **2007**. [CrossRef]

31. Guo, Z.; Jia, X.; Liu, J.P.; Liao, J.; Yang, Y. Herbal medicines for advanced colorectal cancer. *Cochrane Libr.* **2012**. [CrossRef]

32. Shoemaker, M.; Hamilton, B.; Dairkee, S.H.; Cohen, I.; Campbell, M.J. In vitro anticancer activity of twelve Chinese medicinal herbs. *Phytother. Res.* **2005**, *19*, 649–651. [CrossRef] [PubMed]

33. Chen, H.-H.; Zhou, H.-J.; Fang, X. Inhibition of human cancer cell line growth and human umbilical vein endothelial cell angiogenesis by artemisinin derivatives in vitro. *Pharmacol. Res.* **2003**, *48*, 231–236. [CrossRef]

34. Sun, M.; Cao, H.; Sun, L.; Dong, S.; Bian, Y.; Han, J.; Zhang, L.; Ren, S.; Hu, Y.; Liu, C. Antitumor activities of kushen: Literature review. *Evid. Based Complement. Altern. Med.* **2012**. [CrossRef] [PubMed]

35. Liu, J.; Henkel, T. Traditional Chinese medicine (TCM): Are polyphenols and saponins the key ingredients triggering biological activities? *Curr. Med. Chem.* **2002**, *9*, 1483–1485. [CrossRef] [PubMed]

36. Hartwell, J. Types of anticancer agents isolated from plants. *Cancer Treat. Rep.* **1976**, *60*, 1031–1067. [PubMed]

37. Lu, J.-J.; Bao, J.-L.; Chen, X.-P.; Huang, M.; Wang, Y.-T. Alkaloids isolated from natural herbs as the anticancer agents. *Evid. Based Complement. Altern. Med.* **2012**. [CrossRef] [PubMed]

38. Yu, H.-H.; Kim, K.-J.; Cha, J.-D.; Kim, H.-K.; Lee, Y.-E.; Choi, N.-Y.; You, Y.-O. Antimicrobial activity of berberine alone and in combination with ampicillin or oxacillin against methicillin-resistant Staphylococcus aureus. *J. Med. Food* **2005**, *8*, 454–461. [CrossRef] [PubMed]

39. Han, J.; Lin, H.; Huang, W. Modulating gut microbiota as an anti-diabetic mechanism of berberine. *Med. Sci. Monit. Basic Res.* **2011**, *17*, RA164–RA167. [CrossRef]

40. Ji, Y. *Active Ingredients of Traditional Chinese Medicine: Pharmacology and Application*; People's Medical Publishing House Cp., LTD: Shelton, CT, USA, 2011.

41. Moudi, M.; Go, R.; Yien, C.Y.S.; Nazre, M. Vinca alkaloids. *Int. J. Prev. Med.* **2013**, *4*, 1231–1235. [PubMed]

42. Liu, Y.; Xu, Y.; Ji, W.; Li, X.; Sun, B.; Gao, Q.; Su, C. Anti-tumor activities of matrine and oxymatrine: Literature review. *Tumor Biol.* **2014**, *35*, 5111–5119. [CrossRef] [PubMed]

43. Liang, C.Z.; Zhang, J.K.; Shi, Z.; Liu, B.; Shen, C.Q.; Tao, H.M. Matrine induces caspase-dependent apoptosis in human osteosarcoma cells in vitro and in vivo through the upregulation of Bax and Fas/FasL and downregulation of Bcl-2. *Cancer Chemother. Pharmacol.* **2012**, *69*, 317–331. [CrossRef] [PubMed]

44. Yu, P.; Liu, Q.; Liu, K.; Yagasaki, K.; Wu, E.; Zhang, G. Matrine suppresses breast cancer cell proliferation and invasion via VEGF-Akt-NF-κB signaling. *Cytotechnology* **2009**, *59*, 219–229. [CrossRef] [PubMed]

45. Zhou, K.; Ji, H.; Mao, T.; Bai, Z. Effects of matrine on the proliferation and apoptosis of human medulloblastoma cell line D341. *Int. J. Clin. Exp. Med.* **2014**, *7*, 911–918. [PubMed]

46. Xie, S.-B.; He, X.-X.; Yao, S.-K. Matrine-induced autophagy regulated by p53 through AMP-activated protein kinase in human hepatoma cells. *Int. J. Oncol.* **2015**, *47*, 517–526. [CrossRef] [PubMed]

47. Mariño, G.; Niso-Santano, M.; Baehrecke, E.H.; Kroemer, G. Self-consumption: The interplay of autophagy and apoptosis. *Nat. Rev. Mol. Cell Biol.* **2014**, *15*, 81–94. [CrossRef] [PubMed]

48. Rambold, A.S.; Cohen, S.; Lippincott-Schwartz, J. Fatty acid trafficking in starved cells: Regulation by lipid droplet lipolysis, autophagy, and mitochondrial fusion dynamics. *Dev. Cell* **2015**, *32*, 678–692. [CrossRef] [PubMed]

49. Estrela, J.M.; Ortega, A.; Obrador, E. Glutathione in cancer biology and therapy. *Crit. Rev. Clin. Lab. Sci.* **2006**, *43*, 143–181. [CrossRef] [PubMed]

50. Zou, J.; Ran, Z.H.; Xu, Q.; Xiao, S.D. Experimental study of the killing effects of oxymatrine on human colon cancer cell line SW1116. *Chin. J. Dig. Dis.* **2005**, *6*, 15–20. [CrossRef] [PubMed]

51. Wu, X.-S.; Yang, T.; Gu, J.; Li, M.-L.; Wu, W.-G.; Weng, H.; Ding, Q.; Mu, J.-S.; Bao, R.-F.; Shu, Y.-J. Effects of oxymatrine on the apoptosis and proliferation of gallbladder cancer cells. *Anti Cancer drugs* **2014**, *25*, 1007–1015. [CrossRef] [PubMed]

52. Jin, N.; Zhao, Y.-X.; Deng, S.-H.; Sun, Q. Preparation and in vitro anticancer activity of oxymatrine mixed micellar nanoparticles. *Die Pharm. Int. J. Pharm. Sci.* **2011**, *66*, 506–510.

53. Wang, B.; Han, Q.; Zhu, Y. Oxymatrine inhibited cell proliferation by inducing apoptosis in human lung cancer A549 cells. *Bio Med. Mater. Eng.* **2015**, *26*, 165–172. [CrossRef] [PubMed]

54. Li, M.; Su, B.-S.; Chang, L.-H.; Gao, Q.; Chen, K.-L.; An, P.; Huang, C.; Yang, J.; Li, Z.-F. Oxymatrine induces apoptosis in human cervical cancer cells through guanine nucleotide depletion. *Anti Cancer Drugs* **2014**, *25*, 161–173. [CrossRef] [PubMed]

55. Wu, J.-M.; Chen, Y.; Chen, J.-C.; Lin, T.-Y.; Tseng, S.-H. Tetrandrine induces apoptosis and growth suppression of colon cancer cells in mice. *Cancer Lett.* **2010**, *287*, 187–195. [CrossRef] [PubMed]

56. Qin, R.; Shen, H.; Cao, Y.; Fang, Y.; Li, H.; Chen, Q.; Xu, W. Tetrandrine induces mitochondria-mediated apoptosis in human gastric cancer BGC-823 cells. *PLoS ONE* **2013**, *8*, e76486. [CrossRef] [PubMed]

57. Panda, D.; Jordan, M.A.; Chu, K.C.; Wilson, L. Differential effects of vinblastine on polymerization and dynamics at opposite microtubule ends. *J. Biol. Chem.* **1996**, *271*, 29807–29812. [CrossRef] [PubMed]

58. Brugières, L.; Pacquement, H.; Le Deley, M.-C.; Leverger, G.; Lutz, P.; Paillard, C.; Baruchel, A.; Frappaz, D.; Nelken, B.; Lamant, L. Single-drug vinblastine as salvage treatment for refractory or relapsed anaplastic large-cell lymphoma: A report from the French Society of Pediatric Oncology. *J. Clin. Oncol.* **2009**, *27*, 5056–5061. [CrossRef] [PubMed]

59. Yao, L.H.; Jiang, Y.; Shi, J.; Tomas-Barberan, F.; Datta, N.; Singanusong, R.; Chen, S. Flavonoids in food and their health benefits. *Plant Foods Hum. Nutr.* **2004**, *59*, 113–122. [CrossRef] [PubMed]

60. Chahar, M.K.; Sharma, N.; Dobhal, M.P.; Joshi, Y.C. Flavonoids: A versatile source of anticancer drugs. *Pharmacogn. Rev.* **2011**, *5*, 1. [PubMed]

61. Zhou, H.; Lutterodt, H.; Cheng, Z.; Yu, L. Anti-inflammatory and antiproliferative activities of trifolirhizin, a flavonoid from Sophora flavescens roots. *J. Agric. Food Chem.* **2009**, *57*, 4580–4585. [CrossRef] [PubMed]

62. Yin, T.; Yang, G.; Ma, Y.; Xu, B.; Hu, M.; You, M.; Gao, S. Developing an activity and absorption-based quality control platform for Chinese traditional medicine: Application to Zeng-Sheng-Ping (Antitumor B). *J. Ethnopharmacol.* **2015**, *172*, 195–201. [CrossRef] [PubMed]

63. Aratanechemuge, Y.; Hibasami, H.; Katsuzaki, H.; Imai, K.; Komiya, T. Induction of apoptosis by maackiain and trifolirhizin (maackiain glycoside) isolated from sanzukon (Sophora Subprostrate Chen et T. Chen) in human promyelotic leukemia HL-60 cells. *Oncol. Rep.* **2004**, *12*, 1183–1188. [CrossRef] [PubMed]

64. Fang, H.; Chen, S.; Guo, D.; Pan, S.; Yu, Z. Proteomic identification of differentially expressed proteins in curcumin-treated MCF-7 cells. *Phytomedicine* **2011**, *18*, 697–703. [CrossRef] [PubMed]

65. Ravindran, J.; Prasad, S.; Aggarwal, B.B. Curcumin and cancer cells: How many ways can curry kill tumor cells selectively? *AAPS J.* **2009**, *11*, 495–510. [CrossRef] [PubMed]

66. Jordan, B.C.; Mock, C.D.; Thilagavathi, R.; Selvam, C. Molecular mechanisms of curcumin and its semisynthetic analogues in prostate cancer prevention and treatment. *Life Sci.* **2016**, *152*, 135–144. [CrossRef] [PubMed]

67. Nelson, K.M.; Dahlin, J.L.; Bisson, J.; Graham, J.; Pauli, G.F.; Walters, M.A. The Essential Medicinal Chemistry of Curcumin: Miniperspective. *J. Med. Chem.* **2017**, *60*, 1620–1637. [CrossRef] [PubMed]

68. Baker, M.M. Deceptive curcumin offers cautionary tale for chemists. *Nature* **2017**, *541*, 144–145. [CrossRef] [PubMed]

69. Sak, K. Site-specific anticancer effects of dietary flavonoid quercetin. *Nutr. Cancer* **2014**, *66*, 177–193. [CrossRef] [PubMed]

70. Jeong, J.H.; An, J.Y.; Kwon, Y.T.; Rhee, J.G.; Lee, Y.J. Effects of low dose quercetin: Cancer cell-specific inhibition of cell cycle progression. *J. Cell. Biochem.* **2009**, *106*, 73–82. [CrossRef] [PubMed]

71. Deng, X.-H.; Song, H.-Y.; Zhou, Y.-F.; Yuan, G.-Y.; Zheng, F.-J. Effects of quercetin on the proliferation of breast cancer cells and expression of survivin in vitro. *Exp. Ther. Med.* **2013**, *6*, 1155–1158. [PubMed]

72. Zhou, J.; Liang, S.; Fang, L.; Chen, L.; Tang, M.; Xu, Y.; Fu, A.; Yang, J.; Wei, Y. Quantitative proteomic analysis of HepG2 cells treated with quercetin suggests IQGAP1 involved in quercetin-induced regulation of cell proliferation and migration. *OMICS J. Integr. Biol.* **2009**, *13*, 93–103. [CrossRef] [PubMed]

73. Man, S.; Gao, W.; Zhang, Y.; Huang, L.; Liu, C. Chemical study and medical application of saponins as anti-cancer agents. *Fitoterapia* **2010**, *81*, 703–714. [CrossRef] [PubMed]

74. Shan, B.E.; Zeki, K.; Sugiura, T.; Yoshida, Y.; Yamashita, U. Chinese medicinal herb, Acanthopanax gracilistylus, extract induces cell cycle arrest of human tumor cells in vitro. *Jpn. J. Cancer Res.* **2000**, *91*, 383–389. [CrossRef] [PubMed]

75. Yang, J.; Qian, S.; Cai, X.; Lu, W.; Hu, C.; Sun, X.; Yang, Y.; Yu, Q.; Gao, S.P.; Cao, P. Chikusetsusaponin IVa butyl ester (CS-IVa-Be), a novel IL-6R antagonist, inhibits IL-6/STAT3 signaling pathway and induces cancer cell apoptosis. *Mol. Cancer Ther.* **2016**, *15*, 1190–1200. [CrossRef] [PubMed]

76. Lee, M.-S.; Chan, J.Y.-W.; Kong, S.-K.; Yu, B.; Eng-Choon, V.O.; Nai-Ching, H.W.; Mak Chung-Wai, T.; Fung, K.-P. Effects of polyphyllin D, a steroidal saponin in Paris polyphylla, in growth inhibition of human breast cancer cells and in xenograft. *Cancer Biol. Ther.* **2005**, *4*, 1248–1254. [CrossRef] [PubMed]

77. Raju, J.; Bird, R.P. Diosgenin, a naturally occurring furostanol saponin suppresses 3-hydroxy-3-methylglutaryl CoA reductase expression and induces apoptosis in HCT-116 human colon carcinoma cells. *Cancer Lett.* **2007**, *255*, 194–204. [CrossRef] [PubMed]

78. Chiang, C.-T.; Way, T.-D.; Tsai, S.-J.; Lin, J.-K. Diosgenin, a naturally occurring steroid, suppresses fatty acid synthase expression in HER2-overexpressing breast cancer cells through modulating Akt, mTOR and JNK phosphorylation. *FEBS Lett.* **2007**, *581*, 5735–5742. [CrossRef] [PubMed]

79. Wang, J.; Zhao, X.-Z.; Qi, Q.; Tao, L.; Zhao, Q.; Mu, R.; Gu, H.-Y.; Wang, M.; Feng, X.; Guo, Q.-L. Macranthoside B, a hederagenin saponin extracted from Lonicera macranthoides and its anti-tumor activities in vitro and in vivo. *Food Chem. Toxicol.* **2009**, *47*, 1716–1721. [CrossRef] [PubMed]

80. Shan, Y.; Guan, F.; Zhao, X.; Wang, M.; Chen, Y.; Wang, Q.; Feng, X. Macranthoside B Induces Apoptosis and Autophagy Via Reactive Oxygen Species Accumulation in Human Ovarian Cancer A2780 Cells. *Nutr. Cancer* **2016**, *68*, 280–289. [CrossRef] [PubMed]

81. Tan, C.-J.; Zhao, Y.; Goto, M.; Hsieh, K.-Y.; Yang, X.-M.; Morris-Natschke, S.L.; Liu, L.-N.; Zhao, B.-Y.; Lee, K.-H. Alkaloids from Oxytropis ochrocephala and Antiproliferative Activity of Sophoridine Derivatives Against Cancer Cell Lines. *Bioorganic Med. Chem. Lett.* **2015**, *26*, 1495–1497. [CrossRef] [PubMed]

82. Ma, Y.; Gao, H.; Liu, J.; Chen, L.; Zhang, Q.; Wang, Z. Identification and determination of the chemical constituents in a herbal preparation, Compound Kushen injection, by HPLC and LC-DAD-MS/MS. *J. Liq. Chromatogr. Relat. Technol.* **2014**, *37*, 207–220. [CrossRef]

83. Wang, W.; You, R.-L.; Qin, W.-J.; Hai, L.-N.; Fang, M.-J.; Huang, G.-H.; Kang, R.-X.; Li, M.-H.; Qiao, Y.-F.; Li, J.-W. Anti-tumor activities of active ingredients in Compound Kushen Injection. *Acta Pharmacol. Sin.* **2015**, *36*, 676–679. [CrossRef] [PubMed]

84. Polakis, P. Wnt signaling and cancer. *Genes Dev.* **2000**, *14*, 1837–1851. [CrossRef] [PubMed]

85. Xu, W.; Lin, H.; Zhang, Y.; Chen, X.; Hua, B.; Hou, W.; Qi, X.; Pei, Y.; Zhu, X.; Zhao, Z. Compound Kushen Injection suppresses human breast cancer stem-like cells by down-regulating the canonical Wnt/b-catenin pathway. *J. Exp. Clin. Cancer Res.* **2011**, *30*, 103. [CrossRef] [PubMed]

86. Guo, Y.-M.; Huang, Y.-X.; Shen, H.-H.; Sang, X.-X.; Ma, X.; Zhao, Y.-L.; Xiao, X.-H. Efficacy of Compound Kushen Injection in Relieving Cancer-Related Pain: A Systematic Review and Meta-Analysis. *Evid. Based Complement. Altern. Med.* **2015**. [CrossRef] [PubMed]

87. Qu, Z.; Cui, J.; Harata-Lee, Y.; Aung, T.N.; Feng, Q.; Raison, J.; Kortschak, R.D.; Adelson, D.L. Identification of Candidate Anti-Cancer Molecular Mechanisms Of Compound Kushen Injection Using Functional Genomics. *Oncotarget* **2016**, *7*, 66003–66019. [CrossRef] [PubMed]

88. Xudong, L.; Yan, L.; Xi, Y.; Xiaoguang, W.; Xiaoguang, H. Long noncoding RNA H19 regulates EZH2 expression by interacting with miR-630 and promotes cell invasion in nasopharyngeal carcinoma. *Biochem. Biophys. Res. Commun.* **2016**, *473*, 913–919.

89. Zhang, Z.; Wang, Y.; Yao, R.; Li, J.; Yan, Y.; La Regina, M.; Lemon, W.L.; Grubbs, C.J.; Lubet, R.A.; You, M. Cancer chemopreventive activity of a mixture of Chinese herbs (antitumor B) in mouse lung tumor models. *Oncogene* **2004**, *23*, 3841–3850. [CrossRef] [PubMed]

90. Guan, X.; Sun, Z.; Chen, X.; Wu, H.; Zhang, X. Inhibitory effects of Zengshengping fractions on DMBA-induced buccal pouch carcinogenesis in hamsters. *Chin. Med. J.* **2012**, *125*, 332–337. [CrossRef] [PubMed]

91. Lim, K.J.; Rajan, K.; Eberhart, C.G. Effects of Zeng Sheng Ping/ACAPHA on Malignant Brain Tumor Growth and Notch Signaling. *Anticancer Res.* **2012**, *32*, 2689–2696. [PubMed]

92. Sun, Z.; Guan, X.; Li, N.; Liu, X.; Chen, X. Chemoprevention of oral cancer in animal models, and effect on leukoplakias in human patients with ZengShengPing, a mixture of medicinal herbs. *Oral Oncol.* **2010**, *46*, 105–110. [CrossRef] [PubMed]

93. Voyno-Yasenetskaya, T.A.; Pace, A.M.; Bourne, H.R. Mutant alpha subunits of G12 and G13 proteins induce neoplastic transformation of Rat-1 fibroblasts. *Oncogene* **1994**, *9*, 2559–2565. [PubMed]

94. Berestetskaya, Y.V.; Faure, M.P.; Ichijo, H.; Voyno-Yasenetskaya, T.A. Regulation of apoptosis by α-subunits of G12 and G13 proteins via apoptosis signal-regulating kinase-1. *J. Biol. Chem.* **1998**, *273*, 27816–27823. [CrossRef] [PubMed]

95. Tian, Z.H.; Li, Z.F.; Zhou, S.B.; Liang, Y.Y.; He, D.C.; Wang, D.S. Differentially expressed proteins of MCF-7 human breast cancer cells affected by Zilongjin, a complementary Chinese herbal medicine. *Proteom. Clin. Appl.* **2010**, *4*, 550–559. [CrossRef] [PubMed]

96. Zhang, P.; Wang, X.; Xiong, S.; Wen, S.; Gao, S.; Wang, L.; Cao, B. Genome wide expression analysis of the effect of the Chinese patent medicine zilongjin tablet on four human lung carcinoma cell lines. *Phytother. Res.* **2011**, *25*, 1472–1479. [CrossRef] [PubMed]

97. Huang, L.-C.; Clarkin, K.C.; Wahl, G.M. Sensitivity and selectivity of the DNA damage sensor responsible for activating p53-dependent G1 arrest. *Proc. Natl. Acad. Sci. USA* **1996**, *93*, 4827–4832. [CrossRef] [PubMed]

98. Wang, Y.; Wang, J.; Yao, M.; Zhao, X.; Fritsche, J.; Schmitt-Kopplin, P.; Cai, Z.; Wan, D.; Lu, X.; Yang, S. Metabonomics study on the effects of the ginsenoside Rg3 in a β-cyclodextrin-based formulation on tumor-bearing rats by a fully automatic hydrophilic interaction/reversed-phase column-switching HPLC-ESI-MS approach. *Anal. Chem.* **2008**, *80*, 4680–4688. [CrossRef] [PubMed]

99. Bayet-Robert, M.; Morvan, D. Metabolomics reveals metabolic targets and biphasic responses in breast cancer cells treated by curcumin alone and in association with docetaxel. *PLoS ONE* **2013**, *8*, e57971. [CrossRef] [PubMed]

100. Li, Z.; Zheng, L.; Shi, J.; Zhang, G.; Lu, L.; Zhu, L.; Zhang, J.; Liu, Z. Toxic Markers of Matrine Determined Using ^{1}H-NMR-Based Metabolomics in Cultured Cells In Vitro and Rats In Vivo. *Evid. Based Complement. Altern. Med.* **2015**. [CrossRef]

101. Yue, Q.-X.; Song, X.-Y.; Ma, C.; Feng, L.-X.; Guan, S.-H.; Wu, W.-Y.; Yang, M.; Jiang, B.-H.; Liu, X.; Cui, Y.-J. Effects of triterpenes from Ganoderma lucidum on protein expression profile of HeLa cells. *Phytomedicine* **2010**, *17*, 606–613. [CrossRef] [PubMed]

102. Wei, D.-F.; Wei, Y.-X.; Cheng, W.-D.; Yan, M.-F.; Su, G.; Hu, Y.; Ma, Y.-Q.; Han, C.; Lu, Y.; Hui-Ming, C. Proteomic analysis of the effect of triterpenes from Patrinia heterophylla on leukemia K562 cells. *J. Ethnopharmacol.* **2012**, *144*, 576–583. [CrossRef] [PubMed]

103. Singh, C.K.; Kaur, S.; George, J.; Nihal, M.; Hahn, M.C.P.; Scarlett, C.O.; Ahmad, N. Molecular signatures of sanguinarine in human pancreatic cancer cells: A large scale label-free comparative proteomics approach. *Oncotarget* **2015**, *6*, 10335. [CrossRef] [PubMed]

104. Zhang, H.; Zhou, Q.-M.; Lu, Y.-Y.; Jia, D.; Su, S.-B. Aidi Injection () Alters the Expression Profiles of MicroRNAs in Human Breast Cancer Cells. *J. Tradit. Chin. Med.* **2011**, *31*, 10–16. [CrossRef]

105. Dyson, P.J.; Sava, G. Metal-based antitumour drugs in the post genomic era. *Dalton Trans.* **2006**, 1929–1933. [CrossRef] [PubMed]

106. Chen, Z.-F.; Liang, H.; Liu, Y.-C. *Traditional Chinese Medicine Active Ingredient-Metal Based Anticancer Agents*; INTECH Open Access Publisher: Rijeka, Croatia, 2012.

107. Dholwani, K.; Saluja, A.; Gupta, A.; Shah, D. A review on plant-derived natural products and their analogs with anti-tumor activity. *Indian J. Pharmacol.* **2008**, *40*, 49–58. [CrossRef]

108. Chen, Z.-F.; Mao, L.; Liu, L.-M.; Liu, Y.-C.; Peng, Y.; Hong, X.; Wang, H.-H.; Liu, H.-G.; Liang, H. Potential new inorganic antitumour agents from combining the anticancer traditional Chinese medicine (TCM) matrine with Ga (III), Au (III), Sn (IV) ions, and DNA binding studies. *J. Inorg. Biochem.* **2011**, *105*, 171–180. [CrossRef]

109. Tan, C.; Wu, S.; Lai, S.; Wang, M.; Chen, Y.; Zhou, L.; Zhu, Y.; Lian, W.; Peng, W.; Ji, L. Synthesis, structures, cellular uptake and apoptosis-inducing properties of highly cytotoxic ruthenium-Norharman complexes. *Dalton Trans.* **2011**, *40*, 8611–8621. [CrossRef] [PubMed]

110. Chen, Z.-F.; Liu, Y.-C.; Huang, K.-B.; Liang, H. Alkaloid-metal based anticancer agents. *Curr. Top. Med. Chem.* **2013**, *13*, 2104–2115. [CrossRef]

111. Tan, M.; Zhu, J.; Pan, Y.; Chen, Z.; Liang, H.; Liu, H.; Wang, H. Synthesis, cytotoxic activity, and DNA binding properties of copper (II) complexes with hesperetin, naringenin, and apigenin. *Bioinorg. Chem. Appl.* **2009**. [CrossRef] [PubMed]

112. Tan, J.; Wang, B.; Zhu, L. DNA binding and oxidative DNA damage induced by a quercetin copper (II) complex: Potential mechanism of its antitumor properties. *JBIC J. Biol. Inorg. Chem.* **2009**, *14*, 727–739. [CrossRef] [PubMed]

113. Glavinas, H.; Krajcsi, P.; Cserepes, J.; Sarkadi, B. The role of ABC transporters in drug resistance, metabolism and toxicity. *Curr. Drug Deliv.* **2004**, *1*, 27–42. [CrossRef] [PubMed]

114. Liu, Y.; Peng, H.; Zhang, J.-T. Expression profiling of ABC transporters in a drug-resistant breast cancer cell line using AmpArray. *Mol. Pharmacol.* **2005**, *68*, 430–438. [CrossRef] [PubMed]

115. Park, S.; Shimizu, C.; Shimoyama, T.; Takeda, M.; Ando, M.; Kohno, T.; Katsumata, N.; Kang, Y.-K.; Nishio, K.; Fujiwara, Y. Gene expression profiling of ATP-binding cassette (ABC) transporters as a predictor of the pathologic response to neoadjuvant chemotherapy in breast cancer patients. *Breast Cancer Res. Treat.* **2006**, *99*, 9–17. [CrossRef] [PubMed]

116. Sasaki, N.; Ishii, T.; Kamimura, R.; Kajiwara, M.; Machimoto, T.; Nakatsuji, N.; Suemori, H.; Ikai, I.; Yasuchika, K.; Uemoto, S. Alpha-fetoprotein-producing pancreatic cancer cells possess cancer stem cell characteristics. *Cancer Lett.* **2011**, *308*, 152–161. [CrossRef] [PubMed]

117. Kioka, N.; Hosokawa, N.; Komano, T.; Hirayoshi, K.; Nagate, K.; Ueda, K. Quercetin, a bioflavonoid, inhibits the increase of human multidrug resistance gene (MDR1) expression caused by arsenite. *FEBS Lett.* **1992**, *301*, 307–309. [CrossRef]

118. Cai, X.; Chen, F.; Han, J.; Gu, C.; Zhong, H.; Ouyang, R. Restorative effect of quercetin on subcellular distribution of daunorubicin in multidrug resistant leukemia cell lines K562/ADM and HL-60/ADM. *Chin. J. Cancer* **2004**, *23*, 1611–1615.

119. Cai, X.; Chen, F.; Han, J.; Gu, C.; Zhong, H.; Teng, Y.; Ouyang, R. Reversal of multidrug resistance of HL-60 adriamycin resistant leukemia cell line by quercetin and its mechanisms. *Chin. J. Oncol.* **2005**, *27*, 326–329.

120. Tang, X.-Q.; Bi, H.; Feng, J.-Q.; Cao, J.-G. Effect of curcumin on multidrug resistance in resistant human gastric carcinoma cell line SGC7901/VCR. *Acta Pharmacol. Sin.* **2005**, *26*, 1009–1016. [CrossRef] [PubMed]

121. Ganta, S.; Amiji, M. Coadministration of paclitaxel and curcumin in nanoemulsion formulations to overcome multidrug resistance in tumor cells. *Mol. Pharm.* **2009**, *6*, 928–939. [CrossRef] [PubMed]

122. Fix, L.N.; Shah, M.; Efferth, T.; Farwell, M.A.; Zhang, B. MicroRNA expression profile of MCF-7 human breast cancer cells and the effect of green tea polyphenon-60. *Cancer Genom. Proteom.* **2010**, *7*, 261–277.

123. Kasinski, A.L.; Slack, F.J. MicroRNAs en route to the clinic: Progress in validating and targeting microRNAs for cancer therapy. *Nat. Rev. Cancer* **2011**, *11*, 849–864. [CrossRef] [PubMed]

124. Miller, T.E.; Ghoshal, K.; Ramaswamy, B.; Roy, S.; Datta, J.; Shapiro, C.L.; Jacob, S.; Majumder, S. MicroRNA-221/222 confers tamoxifen resistance in breast cancer by targeting p27Kip1. *J. Biol. Chem.* **2008**, *283*, 29897–29903. [CrossRef] [PubMed]

125. Kovalchuk, O.; Filkowski, J.; Meservy, J.; Ilnytskyy, Y.; Tryndyak, V.P.; Vasyl'F, C.; Pogribny, I.P. Involvement of microRNA-451 in resistance of the MCF-7 breast cancer cells to chemotherapeutic drug doxorubicin. *Mol. Cancer Ther.* **2008**, *7*, 2152–2159. [CrossRef] [PubMed]

126. Mani, S.A.; Guo, W.; Liao, M.-J.; Eaton, E.N.; Ayyanan, A.; Zhou, A.Y.; Brooks, M.; Reinhard, F.; Zhang, C.C.; Shipitsin, M. The epithelial-mesenchymal transition generates cells with properties of stem cells. *Cell* **2008**, *133*, 704–715. [CrossRef] [PubMed]

127. Angst, B.D.; Marcozzi, C.; Magee, A.I. The cadherin superfamily. *J. Cell Sci.* **2001**, *114*, 625–626. [PubMed]

128. Hurteau, G.J.; Carlson, J.A.; Spivack, S.D.; Brock, G.J. Overexpression of the microRNA hsa-miR-200c leads to reduced expression of transcription factor 8 and increased expression of E-cadherin. *Cancer Res.* **2007**, *67*, 7972–7976. [CrossRef] [PubMed]

129. Adam, L.; Zhong, M.; Choi, W.; Qi, W.; Nicoloso, M.; Arora, A.; Calin, G.; Wang, H.; Siefker-Radtke, A.; McConkey, D. miR-200 expression regulates epithelial-to-mesenchymal transition in bladder cancer cells and reverses resistance to epidermal growth factor receptor therapy. *Clin. Cancer Res.* **2009**, *15*, 5060–5072. [CrossRef] [PubMed]

130. Lin, Y.; Lin, L.; Jin, Y.; Zhang, Y.; Wang, D.; Tan, Y.; Zheng, C. Combination of Matrine and Sorafenib Decreases the Aggressive Phenotypes of Hepatocellular Carcinoma Cells. *Chemotherapy* **2014**, *60*, 112–118. [CrossRef] [PubMed]

131. Li, H.; Xie, S.; Liu, X.; Wu, H.; Lin, X.; Gu, J.; Wang, H.; Duan, Y. Matrine alters microRNA expression profiles in SGC-7901 human gastric cancer cells. *Oncol. Rep.* **2014**, *32*, 2118–2126. [CrossRef] [PubMed]
132. Li, J.; Jiang, K.; Zhao, F. Oxymatrine suppresses proliferation and facilitates apoptosis of human ovarian cancer cells through upregulating microRNA-29b and downregulating matrix metalloproteinase-2 expression. *Mol. Med. Rep.* **2015**, *12*, 5369–5374. [CrossRef] [PubMed]
133. Wang, Z.; Wang, N.; Liu, P.; Chen, Q.; Situ, H.; Xie, T.; Zhang, J.; Peng, C.; Lin, Y.; Chen, J. MicroRNA-25 regulates chemoresistance-associated autophagy in breast cancer cells, a process modulated by the natural autophagy inducer isoliquiritigenin. *Oncotarget* **2014**, *5*, 7013–7026. [CrossRef] [PubMed]
134. Luo, S.; Deng, W.; Wang, X.; Lü, H.; Han, L.; Chen, B.; Chen, X.; Li, N. Molecular mechanism of indirubin-3′-monoxime and Matrine in the reversal of paclitaxel resistance in NCI-H520/TAX25 cell line. *Chin. Med. J.* **2013**, *126*, 925–929. [PubMed]
135. Schmitt, C.A.; Lowe, S.W. Apoptosis and chemoresistance in transgenic cancer models. *J. Mol. Med.* **2002**, *80*, 137–146. [CrossRef] [PubMed]

International Journal of
Molecular Sciences

MDPI

Article

Assessment of Antioxidant and Cytoprotective Potential of Jatropha (*Jatropha curcas*) Grown in Southern Italy

Teresa Papalia [1], Davide Barreca [2,*] and Maria Rosaria Panuccio [1]

[1] Department of Agricultural Science, "Mediterranea" University, Feo di Vito, 89124 Reggio Calabria, Italy; tpapalia@unime.it (T.P.); mpanuccio@unirc.it (M.R.P.)
[2] Department of Chemical, Biological, Pharmaceutical and Environmental Sciences, University of Messina, 98166 Messina, Italy
* Correspondence: dbarreca@unime.it; Tel.: +39-090-6765-187; Fax: +39-090-6765-186

Academic Editor: Toshio Morikawa
Received: 19 February 2017; Accepted: 15 March 2017; Published: 18 March 2017

Abstract: Jatropha (*Jatropha curcas* L.) is a plant native of Central and South America, but widely distributed in the wild or semi-cultivated areas in Africa, India, and South East Asia. Although studies are available in literature on the polyphenolic content and bioactivity of *Jatropha curcas* L., no information is currently available on plants grown in pedoclimatic and soil conditions different from the autochthon regions. The aim of the present work was to characterize the antioxidant system developed by the plant under a new growing condition and to evaluate the polyphenol amount in a methanolic extract of leaves. Along with these analyses we have also tested the antioxidant and cytoprotective activities on lymphocytes. RP-HPLC-DAD analysis of flavonoids revealed a chromatographic profile dominated by the presence of flavone C-glucosydes. Vitexin is the most abundant identified compound followed by vicenin-2, stellarin-2, rhoifolin, and traces of isovitexin and isorhoifolin. Methanolic extract had high scavenging activity in all antioxidant assays tested and cytoprotective activity on lymphocytes exposed to tertz-buthylhydroperoxide. The results highlighted a well-defined mechanism of adaptation of the plant and a significant content of secondary metabolites with antioxidant properties, which are of interest for their potential uses, especially as a rich source of biologically active products.

Keywords: *Jatropha curcas* L.; RP-HPLC-DAD analysis of flavonoids; cytoprotective activities; antioxidant; polyphenols

1. Introduction

Jatropha curcas L. also known as physic nut (family *Euphorbiaceae*) can be classified as a large shrub or a small perennial tree able to reach a height between three and ten meters [1]. This plant is widespread in tropical and subtropical regions of Southeast Africa, Central and Latin America, Asia and India. *Jatropha curcas* L. is a species that is able to grow in dry and hot conditions, as, for instance, in fringe areas of semi-arid regions, where many species do not survive [2,3].

The result of adaptations to living in relatively harsh environmental conditions is a crop that is useful for the study of key physiological mechanisms adopted by plant to overcome multiple stresses [3].

The main interest for this plant is in regards to its great potential for biodiesel production. In fact, the high content of oil in *Jatropha curcas* L. seeds (up to 60% dependent on geographical and climatic conditions) can be used directly or in transesterified form as a biodiesel [4,5]. In addition, this plant is gaining a lot of attention because of its multipurpose and noteworthy economic potential [6].

The coagulant capacity, for instance, of industrial effluent obtained by grounded seeds is well known for the control of environmental pollution [7]. For centuries preparations of all parts of the plant (such as seed, leaf, stem bark, fruit, and latex) have found wide utilization in traditional medicine and for veterinary purposes. Detoxified oil of *Jatropha curcas* L. represents a rich protein supplement in animal feed [8]. In the literature, several biological effects were reported for the plant such as wound-healing, anti-inflammatory, antimalaria, antiparasitic, antimicrobial, insecticidal, antioxidant, and anticancer activity [9–16]. Literature data are available on the composition and biomedical applications of *Jatropha curcas* L. leaves and the identified compounds include cyclic triterpenes, alkaloids, and flavonoids [17]. The leaves were used as remedy for malaria, rheumatic, and muscular pains [18,19].

In vivo studies on antihyperglycemic activity of methanolic extract of leaves of *Jatropha curcas*. L were also reported [20]. Knnappan et al. [21], demonstrated the in vivo antiulcer activity of alcoholic extract of leaves. Furthermore, methanolic and aqueous extracts of leaves of *Jatropha curcas* L. have been found to inhibit drug-resistant HIV strains and hemagglutinin protein of influenza virus [22,23].

The present study is part of a research project, funded by the Calabria Region, aimed to promote the cultivation of *Jatropha curcas* L. in Calabrian marginal areas, for agriculture and bioenergy purposes. The considerable potential of this plant, the low input requirements, and its lower CO_2 footprint in comparison with other oil-bearing crops, as well as the ability to prevent soil erosion problems, are the main advantages and the main reasons to promote *Jatropha curcas* L. cultivation in Calabrian marginal soils [24,25]. *Jatropha curcas* L. plants, originating from seeds of Kenyan trees were grown in hot and arid climatic conditions in Melito di Porto Salvo (Reggio Calabria, Italy) on a sandy-loam moderately alkaline soil. The objective was to evaluate phytochemical content and enzymatic mechanisms carried out by *Jatropha curcas* L. as strategies for its environmental adaptability. In order to improve the knowledge and to valorize this Calabrian population as a source of natural bioactive molecules, we have performed RP-HPLC-DAD analysis of a leaf methanol extract to evaluate polyphenol amount and, jointly, we have also tested antioxidant and cytoprotective activities on lymphocytes and erythrocyte membranes treated with tert-butylhydroperoxide (t-BOOH).

2. Results and Discussion

Jatropha curcas L. has a life expectancy of up to 50 years and is able to grow under a wide range of soil regimes (such as in deep, fertile, and loose soil), but it does not tolerate sticky, impermeable, and waterlogged soils. This plant requires sufficient sunshine, and cannot grow well under shade [2]. In this study we investigated how *Jatropha curcas* L. plants, originating from seeds of Kenyan trees, have adapted in Southern Italy, precisely in Melito Porto Salvo (Reggio Calabria). In this country the climate is warm, with an average temperature of about 18 °C and annual average rainfall of 767 mm. Chemical and physical characteristics of Melito soil evidenced a sandy-loam, moderately alkaline soil, with a low content of carbonates and a low salinity (Table 1). The amount and composition of soil organic matter (SOM) is strictly related to the performance of soil, in terms of quality and fertility, and a two percent SOM content (Table 1) is considered sufficient in these soils. The ratio of total organic carbon and total nitrogen (C/N ratio) is a traditional indicator to quantify the nature and the humification level of the organic matter present in soil. In general, in soils with a C/N ratio between 9 and 11, organic matter is well humified and quantitatively fairly stable over time. Results showed a C/N ratio lower than 9–10 indicating in Melito soil a prevalence of oxidation reactions leading to a decrease of the content of organic substance and in nitrogen release (Table 1).

Table 1. Chemical and physical characteristics of field for *Jatropha curcas* L. cultivation.

Texture	pH	E.C. (mS/cm)	Total Carbonates (%)	TOC (%)	SOM (%)	N (g/kg)	C/N
Loam-sandy	8.20	1.65	2.00	14.06	2.60	1.82	7.73

Electrical conductivity (E.C.); Soil organic matter (SOM); total organic carbon (TOC); Nitrogen (N); ratio of total carbon and total nitrogen (C/N).

2.1. Phytochemical Screening and Antioxidant Activity

In order to assess the degree of adaptation of *Jatropha curcas* L. plants located in Melito Porto Salvo, a phytochemical screening was performed. Since photosynthesis is one of the primary processes most affected by abiotic stresses [26,27], the evaluation of photosynthetic pigments and reactive oxygen species (ROS) content are considered traditional parameters to evaluate the performance and adaptation degree of a species. The high detected level of chlorophylls confirmed a good adaptation of plants in these soil and climatic conditions. ROS are generated as natural products of plant cellular photosynthetic and aerobic metabolism. Chloroplasts are a major site of ROS produced by energy transfer in photosynthetic electron transfer chains [28]. Peroxisomes and glyoxysomes also generate reactive oxygen species during metabolic pathways of photorespiration and fatty acid oxidation [29]. ROS have different roles in the organism and, at low concentration, for example, they behave as signal molecules for the activation/block of metabolic processes [30,31]. This mechanism of ROS homeostasis is maintained by enzymatic components such as superoxide dismutase (SOD), ascorbate peroxidase (APX), and catalase (CAT), and non-enzymatic compounds like ascorbic acid (ASA), reduced glutathione, a-tocopherol, carotenoids, phenolics, and flavonoids [32]. SODs are the only plant enzymes able to scavenge the superoxide anion. Moreover, in different cell compartments, Cat or APX (which utilize ascorbate as a reductant) eliminate H_2O_2 produced in the reaction catalyzed by SOD [33]. Catalase is unique among antioxidant enzymes in not requiring a reducing equivalent [34]. H_2O_2, being moderately reactive, does not cause extensive damage by itself; it can cross membranes and traverse considerable distance within the cell. At low concentration, H_2O_2 acts as regulatory signal for essential physiological processes, cell cycle, growth, and development [35]. Results on antioxidant enzymes showed significant modifications of dehydroascorbate reductase (DHA Rd), peroxidases (POX), and ascorbate peroxidase (APX) enzymes. Moreover, APX activity and ascorbate-glutathione cycle have a fundamental role in several cellular compartments such us peroxisomes, cytosol, chloroplasts, and mitochondria [33]. DHA Rd enzyme is responsible for the regeneration of ascorbic acid from an oxidized state in a reaction requiring glutathione. CAT activity in fresh leaves of *Jatropha curcas* L. was very low compared to other enzymes. However, CAT activity is generally low under normal growth conditions and it increases only at relatively high H_2O_2 concentrations or under stress conditions to support APX, SOD, and other peroxidases primarily involved in ROS homeostasis. The values obtained for APX and DHA Rd activities are in line with a high content of reduced glutathione and ascorbic acid detected in leaves of *Jatropha curcas* L. (Table 2). The amount of carotenoids, anthocyanins, and glutathione, and also a high ascorbic acid content with respect to the dehydroascorbic acid concentration (Table 2), indicate how the plant has developed its antioxidative defense system in the acclimation process for controlling ROS homeostasis. Plant phenolics constitute one of major groups of compounds acting as primary antioxidant or free radical terminators [32,36–46]. The leaf extracts of *Jatropha curcas* L. are also rich in phenolic compounds and tartaric acid ester derivatives (Table 2), which further contribute to the health promoting properties of this plant. The total amount of phenolic compounds are in line with the amounts detected in other water extracts of jatropha plants collected in different seasons [47], but are obviously inferior to the one obtained in organic solvent, where the total amount is notoriously higher than water extract [48–51]. The analyses of enzymatic and non-enzymatic antioxidants results show that in *Jatropha curcas* L. leaves there are remarkable amounts of these active components, allowing us to hypothesize a direct role in the ability of the plant to resist environmental stresses and improve survival potentiality in the new habitat.

Table 2. Analysis of phytochemical composition and enzymatic antioxidants of leaves of *Jatropha curcas* L. Value were expressed as mean ± standard error (*n* = 3).

Phytochemical Screening of *Jatropha curcas* L. Leaf	Value
Chlorophyll a (mg·g^{-1} Fresh Weight)	1.60 ± 0.10
Chlorophyll b (mg·g^{-1} Fresh Weight)	0.90 ± 0.03
Catalase (CAT) activity (nmol H_2O_2·g^{-1} Fresh Weight)	14.75 ± 1.20
Peroxidases (POX) activity (μmol guaiacol·g^{-1} Fresh Weight)	1.06 ± 0.04
Ascorbate peroxidase (APX) activity (μmol H_2O_2·g^{-1} Fresh Weight)	1.30 ± 0.04
Dehydroascorbate reductase (DHA-Rd) activity (μmol ASA·g^{-1} Fresh Weight)	0.77 ± 7.10
Ascorbic acid (ASA) (μmol ascorbic acid/g Dry Weight)	3.78 ± 0.19
Dehydroascorbic acid (μmol dehydroascorbic acid/g Dry Weight)	2.34 ± 0.20
Reduced glutathione (μmol GSH/g Dry Weight)	1.75 ± 0.14
Total phenols (mg tannic acid/g Dry Weight)	7.36 ± 0.60
Total carotenoids (mg/g Fresh Weight)	0.20 ± 0.03
Anthocyanins (μg anthocyanin·g^{-1} Fresh Weight)	9.42 ± 2.30
Tartaric acid esters derivatives (μg caffeic acid·g^{-1} Fresh Weight)	23.00 ± 0.10

2.2. Analysis of Anti-Peroxidative and Cytoprotective Activity

The health promoting properties of the compounds present in the methanol extract were also analyzed to check their anti-peroxidative and cytoprotective ability on erythrocyte membranes and lymphocytes treated with tert-butylhydroperoxide (t-BOOH). Erythrocyte membrane lipid peroxidation has been performed by TBARS assay, analyzing the amount of malondialdehyde formation. t-BOOH (100 μM) is able to induce a remarkable amount of damage corresponding to the formation of ~1.22 ± 0.1 μM of malondialdehyde. The compounds present in the methanol extract of jatropha are able to reduce ~40, 33, 10, and 1% the formation of this compound utilizing 1.0, 0.5, 025, and 0.1 μM gallic acid equivalents (GAE), respectively (Figure 1). This activity is most probably due to the flavanone structure identified by chromatographic separation and in particular to the presence of apigenin derivatives. These results have been further supported by the analysis of cytoprotective activity of the extract against lymphocyte- t-BOOH treatment. A preliminary evaluation, obtained via incubating the cells with the same final gallic acid equivalent of methanol extract utilized in our work, shows no effects on lymphocytes (data not shown). As can be seen in Figure 2, the incubation of lymphocytes for 24 h at 37 °C in the presence of this strong oxidant (100 μM) induced a decrease of cellular vitality by up to 62%. The presence of methanol extract (at the final concentration of 1.0 and 0.5 μM GAE) remarkably improved cell survival with an increase of viable cells by ~1.9 and 1.3-fold, respectively, following treatment with t-BOOH. It was observed that 0.25–0.1 μM GAE have few or no effects on the process, resulting in values almost completely superimposable to the one obtained with cells incubated in the presence of only t-BOOH in the case of 0.1 μM GAE. The cytoprotective effects of the compounds present in the extract have been further analyzed taking into account the release of lactate dehydrogenase (LDH) from lymphocytes and the inhibition of caspase 3 activation. As can be seen in Figure 2, we highlight a decrease in the amount of LDH released in the samples incubated with t-BOOH in the presence of 1.0 and 0.5 μM GAE of methanol extract, as well as in the activation of caspase 3 in the same samples. Lower concentrations (0.25 and 0.1 μM GAE) were not able to induce statistically significant changes in the two enzymes analyzed. LDH is a marker of cell survival and compound toxicity due to its release outside the cells upon membrane damage, while caspases are one of the main markers of apoptosis onset. The decrease in the LDH release supports the hypothesis that the compounds present in the extract can directly act on t-BOOH by decreasing its strong oxidant activity, well evident at level of fatty acids peroxidation, and scavenging the reactive species that originated at the membrane level. This action is further confirmed by the process of caspase 3 activation, where the elimination of reactive species cannot be the trigger for its activation.

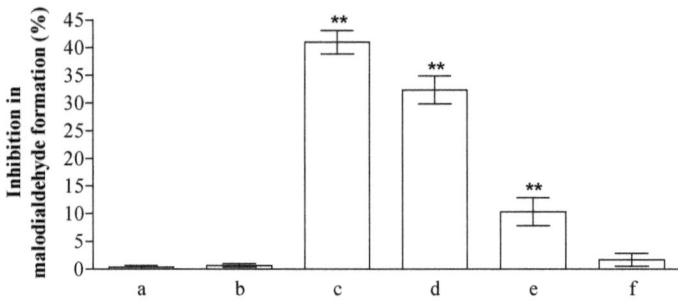

Figure 1. Inhibition (%) of erythrocyte membranes lipid peroxidation by *Jatropha curcas* L. methanol extract. Hemolysates plus 100 µM of tert-butylhydroperoxide (t-BOOH) were incubated for 30 min in the absence (a) or in the presence of 1.0, 0.5, 0.25, and 0.1 µM GAE (c–f). To check the possible influence of the solvent present in the extract, we incubated the hemolysates in the presence of the same amounts of methanol present in the samples (b). The values are expressed as mean ± SD (*n* = 3). The ** shows significant statistical differences (*p* < 0.05) with respect to erythrocyte membranes treated in the presence of only t-BOOH.

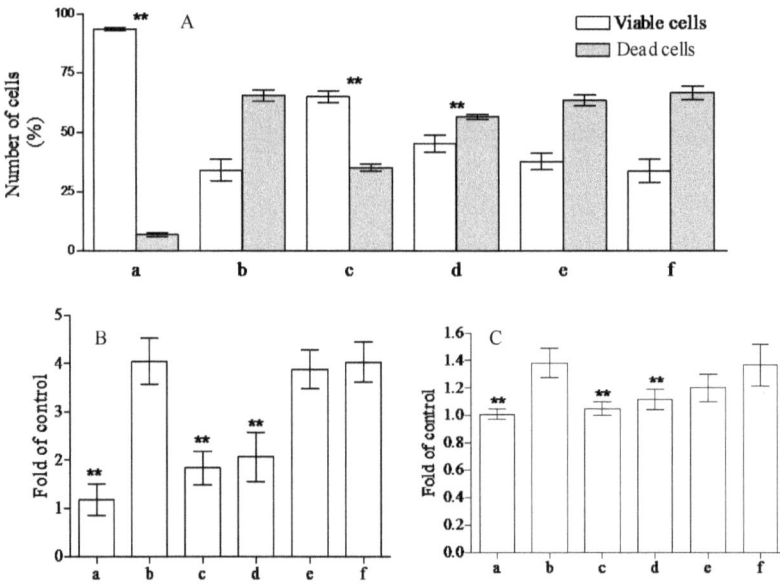

Figure 2. Cytoprotective effects of *Jatropha curcas* L. methanol extract on lymphocytes. Lymphocytes plus 100 µM of t-BOOH were incubated for 24 h in the absence (b) or in the presence of 1.0, 0.5, 0.25, and 0.1 µM GAE (c–f). To check the possible influence of the solvent present in the extract, we incubated the lymphocytes in the presence of the same amounts of methanol present in the samples (a). Cell vitality, integrity, and apoptotic events were analyzed by trypan blue staining (**A**); lactate dehydrogenase (LDH) release (**B**) and caspase 3 activation (**C**), respectively. The samples were analyzed by one-way ANOVA, followed by Tukey's test. Asterisks ** indicate significant differences (*p* < 0.05) with respect to lymphocytes treated in the presence of only t-BOOH. Each value represents mean ± SD (*n* = 3).

2.3. RP-DAD-HPLC Separation and Identification of Flavonoids Derivatives

In order to shed some light on the compounds that are present in the extract and responsible of such activities, we have performed a RP-DAD-HPLC separation to identify the presence of flavone and flavanone derivatives. The methanol extract was characterized by the presence of several well defined peaks belonging to flavonoids, as shown in the chromatograms recorded at 280 and 325 nm (Figure 3). This first approach let us to perform a preliminary screening based on the intense absorptions in the 270–280 nm region (Band II) of flavanone derivatives and the absorbance at the 320–330 nm region (Band I) where, principally, flavones and flavonols have remarkable absorption. The analyses of UV/visible spectrum of each peak show the presence of only flavone derivatives. Moreover, the identification of the compounds has been performed by means of acid hydrolysis and subsequent analysis of the aglycones and sugars. The chromatogram recorded for crude extracts after acidic hydrolysis (not shown) revealed that compounds 1–4 were resistant to HCl treatment, whereas 5–6 were hydrolyzed, providing evidence that the former flavonoids possessed C-linked saccharide moieties, whereas the latter bear O-linked glycosyl substituents. Moreover, the presence of a single aglycon molecule in the chromatograms revealed a pattern characterized by the presence of apigenin derivatives with the presence of glucose and rhamnose. The characteristic UV spectra, their retention time and co-elution with authentication standard let us to identify compounds as vicenin-2 (**1**); stellarin-2 (**2**); vitexin (**3**); isovitexin (**4**); isorhoifolin (**5**); and rhoifolin (**6**). Several of these compounds have already been reported in other leaves of *Jatropha curcas* L. (although grown in conditions different from the one tested in our experiment) and Jatropha genus suggesting a common pattern of flavonoids that are conserved in the species and may represent an indication of the endogenous adaptation of the plant to Calabrian marginal areas [17,52–58].

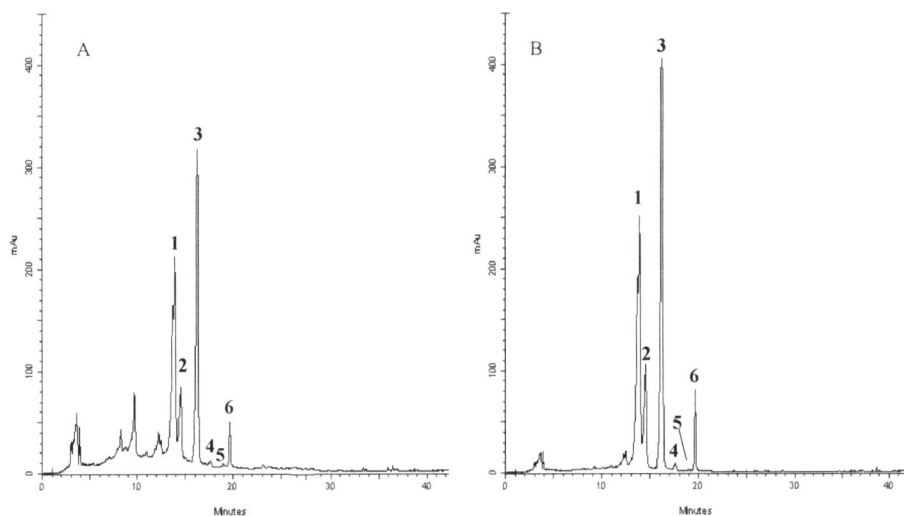

Figure 3. Representative HPLC chromatograms of flavonoids derivatives of *Jatropha curcas* L. methanol extract: absorbance at 278 nm (**A**) and 325 nm (**B**). Peak identification was performed by matching retention time and UV spectra against commercially available reference compounds. Peaks: vicenin-2 (**1**); stellarin-2 (**2**); vitexin (**3**); isovitexin (**4**); isorhoifolin (**5**); and rhoifolin (**6**).

The quantifications of the identified flavonoids are depicted in Table 3.

Table 3. Flavonoids content in methanol extract of *Jatropha curcas* L. leaves.

Compounds	mg/kg F.W.
Vicenin-2	3.7 ± 0.41
Stellarin-2	1.2 ± 0.23
Vitexin	6.0 ± 0.52
Isovitexin	0.13 ± 0.04
Isorhoifolin	Trace
Rhoifolin	2.2 ± 0.25

2.4. Antioxidant Capacity

On the basis of the remarkable content of flavonoids and the presence of substituted flavone structures in the methanolic extract obtained from the leaves, we performed an in vitro biological assay in order to evaluate the antioxidant (DPPH, ABTS, FRAP, Ferrozine assay) and the cytoprotective activity of this extract. The DPPH is a stable radical frequently used to examine radical scavenging activity of natural compounds, and it is one of the starting points to check propensity of compounds or extracts to react with radicals. It has a strong absorbance at 517 nm due to its unpaired electron, giving the radical a purple color. Upon reduction with an antioxidant, its absorption decreases due to the formation of its non-radical form, DPPH-H. The activities of crude methanolic extracts in the scavenging of DPPH radical were concentration dependent. For instance (Figure 4A), the samples with 10 µL of methanol extract were able to reach up to ~65% of inhibition, corresponding to 14.7 µM trolox equivalents (TE). These results (IC_{50} = 58.8 TE µg/mL) are in line with the one obtained from ethanolic extract from the leaves of plants grown in Java, inferior to the ones obtained from the methanolic extract of leaves collected from plants grown in Malaysia and Egypt, but higher than the one obtained with plants grown in Iraq [48–50,59]. According to the ability of compounds present in methanolic extract to scavenge DPPH, ABTS radical formation was also inhibited with an activity corresponding to 16.25 ± 0.68 µM TE. These results suggest that the methanolic extract of *Jatropha curcas* L. leaves contains compounds capable of donating hydrogen to a free radical to eliminate its reactivity. Iron has a pivotal role in the wellness of organisms and it is also one of the main elements involved in the formation of radical species, so we tested the capability of the compounds present in the methanol extract while maintaining it in ferrous state and chelating it. The Fe^{3+}–Fe^{2+} reducing power method is usually used in the determination of reducing power. The amount of Fe^{2+} can be determined by measuring the generation of Perl's Prussian blue at 593 nm. The reducing power of 2.5 µL of the extract corresponds to 15.48 ± 2.9 µM of ascorbic acid equivalent. The chelating power of the extract was also tested using the ferrozine assay. Free transition metals can give rise to the generation of several ROS, in living organisms, through the oxidation of lipids, proteins and genetic materials; The presence of chelating agents can help organisms to stabilize and decrease the reactivity of these elements. As can be seen in Figure 4B, the decrease of the ferrozine–Fe^{2+} complex is influenced by the presence of the extract, although its activity is clearly lower than that of ethylenediaminetetraacetic acid (EDTA) utilized as positive control able to chelate all the ferrous present in the solution. The calculation of Oxygen Radical Absorbance Capacity (ORAC), utilizing a calibration curve obtained with trolox, showed a value of 7.71 ± 0.68 µmol TE/mg. This value is comparable to the one obtained for acetate, ethanol, and water extracts of *Jatropha curcas* L. seed shell [60].

By four different methods of antioxidant activity determination, we can see that the extract of *J. curcas* leaves exhibited relatively strong antioxidant activities, which may be due, at least in part, to their high phenolic content. In particular, the flavone derivatives characterized by the presence of a double bond at the 2, 3 position of the C ring conjugated with the 4-oxo group in position 4 may have a pivotal role in the process [41–44]. Recent studies indicate that all parts of this plant are valuable for multiple purposes, improving its valorization for large-scale plantation.

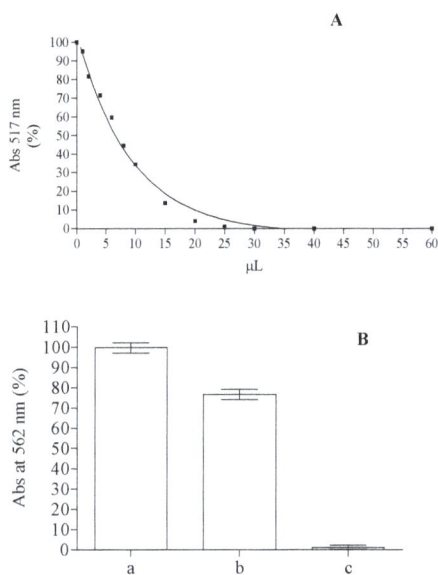

Figure 4. DPPH (**A**) and ferrozine assay (**B**) obtained with different amounts of *Jatropha curcas* L. methanol extract of *Jatropha curcas* L. leaves. Ferrozine assay without (a) or with 10 μL of methanol extract (b) or EDTA (c).

3. Materials and Methods

3.1. Reagents, Chemicals, and Instrumentation

HPLC-grade acetonitrile and methanol, as well as vicenin-2, vitexin, isovitexin, roipholin, isorhoifolin, and apigenin, were supplied by Sigma-Aldrich (St. Louis, MO, USA), while dimethylformamide (DMF) was supplied by Carlo Erba (Milano, Italy). All the other reagents and chemicals used in this study were of analytical grade and were purchased from Sigma (Sigma-Aldrich GmbH, Sternheim, Germany).

3.2. Chlorophyll and Carotenoid Pigments

Fresh leaves (0.050 g) were mixed with 2.5 mL of 100% ethanol in the dark for 24 h at 4 °C. Upon the conclusion of the incubation time the samples were centrifuged for 10 min at 7000 rpm. Lichtenthaler's equation was employed to analyze the concentration of chlorophyll and carotenoid, based on absorbance at 649, 665, and 470 nm.

3.3. Anthocyanins

Fresh leaves (0.02 g) were extracted with 0.5 mL of a methanol:HCl solution (99:1, *v:v*) and centrifuged at 4 °C for 10 min at 7000 rpm. The absorbance of the supernatant was recorded at 530 and 657 nm and anthocyanin concentration was calculated according to the Equation (1):

$$[A_{530\,nm} - (0.025 \times A_{657\,nm}) \times mL\ extract]/g\ fresh\ weight \qquad (1)$$

3.4. Tartaric Acid Esters and Total Phenols

Tartaric acid esters were tested by monitoring the absorbance change at 320 nm based on the procedure described by Romani et al. [61]. Fresh leaves (0.5 g) were extracted with 2 mL of methanol and centrifuged at 4 °C for 15 at 14,000 *g*. An aliquot of 25 μL of supernatant was diluted with

225 µL of 10% ethanol and 250 µL of 0.1% HCl in 95% ethanol, and 1 µL of 2% HCl was then added. The solution was mixed and tartaric acid ester were calculated at 320 nm as micrograms of caffeic acid/g fresh weight.

Total phenolic compounds have been analyzed by the Folin–Ciocalteu colorimetric method based on the procedure of Singleton et al. [62]. Dry leaves were extracted in water and the absorbance was recorded against blank at 765 nm and total phenols were expressed as mg tannic acid/g dry weight.

3.5. Reduced Glutathione

Reduced glutathione (GSH) level was determined by the method described by Jollow et al. [63]. Fresh leaf (0.5 g) homogenates in 3% of trichloroacetic acid were centrifugated at 3000 rpm at 4 °C. The supernatant was mixed with Ellman's reagent and the absorbance of supernatant recorded at 412 nm and related to a calibration curve of GSH solutions (0–500 µg/mL).

3.6. Ascorbic and Dehydroascorbic Acid

Fresh leaves (0.5 g) were extracted in a chilled mortar with 5% metaphosphoric acid at 4 °C. After centrifugation at 18,000 rpm at 4 °C the supernatant was used for the determination of dehydroascorbic acid (DHA) and ascorbic acid (ASC) according to Law et al. [64].

3.7. Enzyme Assays

Fresh leaves were ground using a chilled mortar and pestle and homogenized in 0.1 M phosphate buffer solution (pH 7.0) containing 100 mg soluble polyvinylpolypyrrolidone (PVPP) and 0.1 mM ethylenediaminetetraacetic acid (EDTA). The homogenate was filtered through two layers of muslin cloth and centrifuged at 10,000 rpm for 20 min at 4 °C. The resulting supernatant was used for all assays.

Catalase (CAT, EC 1.11.1.6). The disappearance of H_2O_2 at 240 nm was determined according to Beaumont et al. [65] by using extinction coefficient $(\varepsilon) = 0.036$ mM^{-1}·cm^{-1}. The reaction mixture contained 1 mL potassium phosphate buffer (50 mM, pH 7.0), 40 µL enzyme extract, and 5 µL H_2O_2.

Peroxidase (POX, EC 1.11.1.7). The reduction in guaiacol concentration was determined by reading the absorbance at 436 nm continuously for 90 s. POX activity was quantified by the amount of tetraguaiacol formed using its extinction coefficient $(\varepsilon) = 25.5$ mM^{-1}·cm^{-1} according to Panda et al. [66].

Dehydroascorbate reductase (DHA-Rd, EC 1.8.5.1). The reaction mixture contained 0.1 M K-phosphate buffer pH 6.5, 1 mM GSH, and 1 mM DHA. The activity was assayed following the increase in absorbance at 265 nm due to the production of ASC [67].

Ascorbate peroxidase (APX, EC 1.11.1.11). The decrease in absorbance at 290 nm, due to oxidation of ascorbate was determined according to Amako et al. [68]. The reaction mixture was 0.1 M K-phosphate buffer pH 6.5, 90 mM H_2O_2, and 50 mM ascorbate. Absorbance was recorded continuously for 90 s and APX activity was quantified by using the extinction coefficient, 14 mM^{-1}·cm^{-1}.

3.8. Preparation of Methanol Extract

The fresh leaves of *Jatropha curcas* L., harvested in summer 2013, were frozen at −20 °C. The frozen leaves were ground to a powder with a frozen mortar and ~10.0 g were extracted at room temperature under continuous stirring for 6 h with methanol (1:20 *w:v*). The samples were then centrifuged at 2500 rpm for 10 min and the supernatants were filtered with filter paper and evaporated to dryness in a rotavapor. This procedure was repeated three times and the powders obtained were resuspended in methanol to obtain a *w:v* ratio with the starting fresh leaves material of 1:1, with the end product utilized for RP-HPLC-DAD separation, antioxidant, and cytoprotective assays.

3.9. DPPH Radical Scavenging Assay

The antioxidant activity against 2,2-diphenyl-1-picrylhydrazyl DPPH radical was performed according to Molineux [69]. The assays were carried out by adding fixed amounts of extracts (0–60 µL) with DPPH solution (80 µM), resulting in the final volume of 1.0 mL. The reaction mixture was incubated for 30 min at 37 °C and, upon finishing the incubation time, the absorbance changes were recorded at 517 nm. The decrease in absorbance in percentage was analyzed utilizing the following equation:

$$\text{Absorbance decrease (\%)} = 100 \times (A_c - A_s)/A_c \tag{2}$$

where A_c is the absorbance of the control and A_s is the absorbance of the sample. Results have been expressed as Trolox equivalent (TE).

3.10. ABTS Radical Scavenging Assay

The 2,2'-azino-bis(3-ethylbenzothiazoline-6-sulphonic acid ABTS free radical-scavenging activity was carried out by a decolorization assay according to Re et al. [70]. Fixed amounts of the samples were added with the radical cation ABTS$^+$ and the absorbance changes at 734 nm were recorded in a spectrophotometer after 6 min. The activity was expressed as inhibition in percentage at 734 nm using Trolox (1.1, 1.7, 2.3, 2.9, 3.5 µg/mL) as the reference compound.

3.11. Ferric-Reducing Antioxidant Power (FRAP) Assay

The ferric reducing antioxidant power assay was performed according to the method described by Benzie and Strain [71]. The samples were repeated in triplicate and the absorbance recorded at 593 nm after 4 min incubation at 37 °C. The antioxidant abilities of the extracts were expressed as equivalents of ascorbic acid utilizing a calibration curve obtained with fresh solutions of known ascorbic acid concentrations (0.005–0.02 mM).

3.12. Ferrozine Assay

The potential chelating activity of the extracts toward ferrous ions was analyzed by the method of Dorman et al. [72] with little modifications. As a reference compound we utilized EDTA (0.1 mM final concentration). The activity of the extract was performed by adding 10 µL to a solution of 0.5 mM FeSO$_4$ (0.01 mL). After the addition of 5.0 mM ferrozine (0.4 mL) solution, the samples were shaken and left for 10 min at room temperature (RT). Finally, the absorbance at 562 nm was recorded with a spectrophotometer. The inhibition (%) of ferrozine Fe^{2+} complex formation was obtained using the following equation:

$$\text{\% Inhibition} = [(A_c - A_s)/A_c] \times 100 \tag{3}$$

where A_c is the absorbance of the control and A_s is the absorbance of the samples in the presence of the extracts.

3.13. Flavonoids Profile Identification

The identification of flavonoids present in the methanol extract was performed by utilizing a Shimazu Reverse Phase–Diode Array Detection–High Performance Liquid Chromatography (Shimadzu, Kyoto, Japan) with injection loop of 20 µL. The column was a BioDiscovery C18 (Supelco, Bellefonte, PA, USA) of 250 mm × 4.6 mm i.d., 5 µm and equipped with a 20 mm × 4.0 mm guard column. The temperature was set at 30 °C and flow-rate at 1.0 mL/min. The separation was performed utilizing a linear gradient of acetonitrile in H$_2$O as mobile phase. The gradient was: 0–15 min (5%–20% of acenitrile), 15–20 min (20%–30% of acetonitrile), 20–35 min (30%–100% of acetonitrile), 35–40 min (100% of acetonitrile), 40–45 min (100%–5% of acetonitrile), and 45–55 min (5% of acetonitrile). The chromatograms were recorded at 278 and 325 nm and UV/visible spectra of each peak were between 200 and 450 nm. The identification of the compounds were performed according to retention time, UV spectra, and co-elution with authentication standards. Quantitative analysis was carried out by integration of the areas of the peaks from the chromatogram at 325 nm and comparison with calibration curves obtained with the known concentration of a commercially available standard (0.1–10 mg/L).

3.14. Acid Hydrolysis

Acid hydrolysis of the samples was performed according to Hertog et al. [73].

3.15. Erythrocytes Lipid Peroxidation Assay

Hemolysates were prepared according to Barreca et al. [38] and lipid peroxidation was analyzed by thiobarbituric acid reactive substance (TBARS) assay [74].

3.16. Lymphocyte Isolation

Lymphocytes were isolated according to Barreca et al. [42,44,45] and utilized in the following tests.

3.17. Cytotoxicity Assays

To perform the cytotoxicity assay, we treated cells (1×10^6/mL) with 100 µM of t-BOOH in the absence or in the presence of 1.0, 0.5, 0.25, and 0.1 µM gallic acid equivalents (GAE) of the extracts for 24 h. Parallel controls were performed without t-BOOH, but in the presence of the same final gallic acid equivalents of methanol extract utilized during experimentation. Moreover, in all experiments blanks, without t-BOOH, were performed [38,75]. The cell viability, after finishing the incubation period were established with trypan blue staining. The cells were diluted 1:1 (*v:v*) with 0.4% trypan blue and counted with an haemocytometer. Results are expressed as the percentage of live or dead cells (ratio of unstained or stained cells to the total number of cells, respectively). To check cytotoxicity we also analyzed lactate dehydrogenase (LDH) release from damaged cells into culture medium with a commercially available kit from BioSystems (Barcelona, Spain). Extracts did not show interference with the determination of LDH at the concentration utilized in the experiments. For caspase activity determination, we followed the procedures described by Bellocco et al. [45].

3.18. Oxygen Radical Absorbance Capacity (ORAC) Assay

The ORAC assay was performed according to Dàvalos et al. [76] with few modifications. Twenty microliters of methanol extract were added to 120 µL of fresh fluorescein solution (117 nM). After a preincubation time of 15 min at 37 °C, we added 60 µL of freshly prepared 2,2'-Azobis(2-methylpropionamidine)dihydrochloride (AAPH) solution (40 mM). Fluorescence was recorded every 30 s for 90 min (λ_{ex}: 485; λ_{em}: 520). A blank using 20 µL of methanol instead of the sample was also analyzed, along with a reference calibration curve with Trolox (10–100 µM). The ORAC value was expressed as µmoles of Trolox Equivalent (TE)/mg of fresh weight (F.W.) sample. All assays were carried out in triplicate.

3.19. Statistical Analysis

The values of the data are expressed as means ± standard deviation. One-way analysis of variance (ANOVA) was performed on the obtained results. Tukey's test was run to check the significance of the difference between the samples and the respective controls. A $p < 0.05$ value indicates statistically significant difference.

4. Conclusions

In this study the obtained results concerning a phytochemical and enzymatic screening suggested that the *Jatropha curcas* L. plants, originated from Kenya and grown in Melito Porto Salvo, are well suited to the typical Mediterranean climate of Southern Italy. Moreover, the methanolic extract of the leaves shows very interesting antioxidant and cytoprotective activities, which can be attributed also to its flavonoids profile, which is dominated by the presence of flavone compounds, one of the most studied and promising forms of secondary metabolites for potential use as nutraceuticals. Therefore, methanolic extracts of *Jatropha curcas* L. leaves could represent a promising source of natural antioxidants compounds to employ in the pharmaceutical and cosmetic industries.

Acknowledgments: This work was supported by the regional project entitled "Si. Re. Ja." POR Calabria FESR 2007/2013.

Author Contributions: Teresa Papalia, Davide Barreca and Maria Rosaria Panuccio contributed in equal manner to the design and execution of experiments, data analysis, writing, and revision of the work.

Conflicts of Interest: The authors declare no conflict of interest.

References

1. Divakara, B.N.; Upadhyaya, H.D.; Wani, S.P.; Gowda, C.L.L. Biology and genetic improvement of *Jatropha curcas* L.: A review. *Appl. Energy* **2010**, *87*, 732–742. [CrossRef]
2. Francis, G.; Edinger, R.; Becker, K. A concept for simultaneous wasteland reclamation, fuel production, and socioeconomic development in degraded areas in India. Need, potential and perspectives of Jatropha plantations. *Nat. Res. Forum* **2005**, *29*, 12–24. [CrossRef]
3. Silva, E.N.; Ferreira-Silva, S.N.; Fontenelea, A.V.; Ribeiro, R.V.; Viegasc, R.A.; Silveira, J.A.G. Photosynthetic changes and protective mechanisms against oxidative damage subjected to isolated and combined drought and heat stresses in *Jatropha curcas* plants. *J. Plant Physiol.* **2010**, *167*, 1157–1164. [CrossRef] [PubMed]
4. Pramanik, K. Properties and use of *Jatropha curcas* oil and diesel fuel blends in compression ignition engine. *Renew. Energy* **2003**, *28*, 239–248. [CrossRef]
5. Kumar, A.; Sharma, S. An evaluation of multipurpose oil seed crop for industrial uses (*Jatropha curcas* L.): A review. *Ind. Crops Prod.* **2008**, *28*, 1–10. [CrossRef]
6. Ye, M.; Li, C.; Francis, G.; Makkar, H.P.S. Current situation and prospects of *Jatropha curcas* as a multipurpose tree in China. *Agrofor. Syst.* **2009**, *76*, 487–497. [CrossRef]
7. Pandey, V.C.; Singh, K.; Singh, J.S.; Kumar, A.; Singh, B.; Singh, R.P. *Jatropha curcas*: A potential biofuel plant for sustainable environmental development. *Renew. Sustain. Energy Rev.* **2012**, *16*, 2870–2883. [CrossRef]
8. Makkar, H.P.S.; Francis, G.; Becker, K. Preparation of protein concentrate from *Jatropha curcas* screw-pressed seed cake and toxic and antinutritional factors in protein concentrate. *J. Sci. Food Agric.* **2008**, *88*, 1542–1548. [CrossRef]
9. Shetty, S.; Udupa, S.L.; Udupa, A.L.; Vollala, V.R. Woundhealing activities of bark extract of *Jatropha curcas* Linn. in albino rats. *Saudi Med. J.* **2006**, *27*, 1473–1476. [PubMed]
10. Mujumdar, A.M.; Misar, A.V. Anti-inflammatory activity of *Jatropha curcas* roots in mice and rats. *J. Ethnopharmacol.* **2004**, *90*, 11–15. [CrossRef] [PubMed]
11. Ankrah, N.A.; Nyarko, A.K.; Addo, P.G.; Ofosuhene, M.; Dzokoto, C.; Marley, E.; Addae, M.M.; Ekuban, F.A. Evaluation of efficacy and safety of a herbal medicine used for the treatment of malaria. *Phytother. Res.* **2003**, *17*, 697–701. [CrossRef] [PubMed]
12. Fagbenro-Beyioku, A.F.; Oyibo, W.A.; Anuforom, B.C. Disinfectant/antiparasitic activities of *Jatropha curcas*. *East Afr. Med. J.* **1998**, *75*, 508–511. [PubMed]
13. Kalimuthu, K.; Vijayakumar, S.; Senthilkumar, R. Antimicrobial activity of the biodesel plant, *Jatropha curcas*. *Intern. J. Pharm. Biol. Sci.* **2010**, *1*, 1–5.
14. Adebowale, K.O.; Adedire, C.O. Chemical composition and insecticidal properties of the underutilized *Jatropha curcas* seed oil. *Afr. J. Biotechnol.* **2006**, *5*, 901–906.
15. Igbinosa, O.O.; Igbinosa, I.H.; Chigor, V.N.; Uzunuigbe, O.E.; Oyedemi, S.O.; Odjadiare, E.E.; Okoh, A.I.; Igbinosa, E.O. Polyphenolic contents and antioxidant potential of stem bark extracts of *Jatropha curcas* (Linn). *Int. J. Mol. Sci.* **2011**, *12*, 2958–2971. [CrossRef] [PubMed]
16. Muangman, S.; Thippornwong, M.; Tohtong, R. Anti-metastatic effects of curcusone B, a diterpene from *Jatropha curcas*. *In Vivo* **2005**, *19*, 265–268. [PubMed]
17. Sabandar, C.W.; Ahmat, N.; Mahf Jaafar, F.; Sahidin, I. Medicinal property, phytochemistry and pharmacology of several Jatropha species (Euphorbiaceae): A review. *Phytochemistry* **2013**, *85*, 7–29. [CrossRef] [PubMed]

18. Asase, A.; Oteng-Yeboah, A.A.; Odamtten, G.T.; Simmonds, M.S.J. Ethnobotanical study of some Ghananian anti-malarial plants. *J. Ethnopharmacol.* **2005**, *99*, 273–279. [CrossRef] [PubMed]
19. Thomas, R.; Sah, N.K.; Sharma, P.B. Therapeutic biology of *Jatropha curcas*: A mini review. *Curr. Pharm. Biotechnol.* **2008**, *9*, 315–324. [CrossRef] [PubMed]
20. El-Baz, F.K.; Aly, H.F.; Abd-Alla, H.; Saad, S.A. Bioactive flavonoid glycosides and antidiabetic activity of *Jatropha curcas* on streptozotocin-induced diabetic rats. *Int. J. Pharm. Sci. Rev. Res.* **2014**, *29*, 143–156.
21. Knnappan, N.; Jaikumar, S.; Manavalan, R.; Muthu, A.K. Antiulcer activity of methanolic extract of *Jatropha curcas* (Linn.) on aspirin-induced gastric lesions in wistar rats. *Pharmacologyonline* **2008**, *1*, 279–293.
22. Dahake, R.; Roy, S.; Patil, D.; Rajopadhye, S.; Chowdhary, A.; Deshmukh, R.A. Potential anti-HIV activity of *Jatropha curcas* Linn. leaf extracts. *J. Antivir. Antretrovir.* **2013**, *5*, 160–165. [CrossRef]
23. Patil, D.Y.; Roy, S.; Dahake, R.; Rajopadhye, S.; Kothari, S.; Deshmukh, R.; Chowdhary, A. Evaluation of *Jatropha curcas* Linn. leaf extracts for its cytotoxicity and potential to inhibit hemagglutinin protein of influenza virus. *Indian J. Virol.* **2013**, *24*, 220–226. [CrossRef] [PubMed]
24. Settineri, G.; Panuccio, M.R.; Muscolo, A. *Jatropha curcas* sludge valorization. *Procedia Soc. Behav. Sci.* **2016**, *223*, 865–887. [CrossRef]
25. De Rossi, A.; Vescio, R.; Russo, D.; Macrì, G. Potential use of *Jatropha curcas* L. on marginal lands of southern Italy. *Procedia Soc. Behav. Sci.* **2016**, *223*, 770–775. [CrossRef]
26. Liu, X.; Huang, B. Photosynthetic acclimation to high temperatures associated with heat tolerance in creeping bentgrass. *J. Plant Physiol.* **2008**, *165*, 1947–1953. [CrossRef] [PubMed]
27. Lawlor, D.W.; Cornic, G. Photosynthetic carbon assimilation and associated metabolism in relation to water deficits in higher plants. *Plant Cell. Environ.* **2002**, *25*, 275–294. [CrossRef] [PubMed]
28. Asada, K. Production and scavenging of reactive oxygen species in chloroplasts and their functions. *Plant Physiol.* **2006**, *141*, 391–396. [CrossRef] [PubMed]
29. Del Rio, L.A.; Sandalio, L.M.; Corpas, F.J.; Barroso, J.B. Reactive oxygen species and reactive nitrogen species in peroxisomes. Production, scavenging and role in cell signaling. *Plant Physiol.* **2006**, *141*, 330–335. [CrossRef] [PubMed]
30. Foyer, C.H.; Noctor, G. Redox regulation in photosynthetic organisms: Signaling, acclimation, and practical implications. *Antioxid. Redox Signal.* **2009**, *11*, 861–905. [CrossRef] [PubMed]
31. Jaleel, C.A.; Gopi, R.; Panneerselvam, R. Alterations in non-enzymatic antioxidant components of *Catharanthus roseus* exposed to paclobutrazol, gibberellic acid and *Pseudomonas fluorescens*. *Plant Omics J.* **2009**, *2*, 30–40.
32. Gechev, T.S.; van Breusegem, F.; Stone, J.M.; Denev, I.; Laloi, C. Reactive oxygen species as signals that modulate plant stress responses and programmed cell death. *Bioessays* **2006**, *28*, 1091–1101. [CrossRef] [PubMed]
33. Shigeoka, S.; Ishikawa, T.; Tamoi, M.; Miyagawa, Y.; Takeda, T.; Yabuta, Y.; Yoshimura, K. Regulation and function of ascorbate peroxidase isoenzymes. *J. Exp. Bot.* **2002**, *53*, 1305–1319. [CrossRef] [PubMed]
34. Mittler, R. Oxidative stress, antioxidants and stress tolerance. *Trends Plant Sci.* **2002**, *7*, 405–410. [CrossRef]
35. Das, M.K.; Roychoudhury, A. ROS and responses of antioxidant as ROS-scavengers during environmental stress in plants. *Front. Environ. Sci.* **2014**, *2*, 1–13. [CrossRef]
36. Smeriglio, A.; Barreca, D.; Bellocco, E.; Trombetta, D. Chemistry, pharmacology and health benefits of Anthocyanins. *Phytother. Res.* **2016**, *30*, 1265–1286. [CrossRef] [PubMed]
37. Panuccio, M.R.; Fazio, A.; Papalia, T.; Barreca, D. Antioxidant properties and flavonoid profile in leaves of Calabrian *Lavandula multifida* L., an autochthon plant of Mediterranean Southern regions. *Chem. Biodivers.* **2016**, *13*, 1–6. [CrossRef] [PubMed]
38. Barreca, D.; Laganà, G.; Leuzzi, U.; Smeriglio, A.; Trombetta, D.; Bellocco, E. Evaluation of the nutraceutical, antioxidant and cytoprotective properties of ripe pistachio (*Pistachia vera* L. variety Bronte) hulls. *Food Chem.* **2016**, *196*, 493–502. [CrossRef] [PubMed]
39. Barreca, D.; Gattuso, G.; Laganà, G.; Leuzzi, U.; Bellocco, E. *C*- and *O*-glycosyl flavonoids in Sanguinello and Tarocco blood orange (*Citrus sinensis* (L.) Osbeck) juice: Identification and influence on antioxidant properties and acetylcholinesterase activity. *Food Chem.* **2016**, *196*, 619–627. [CrossRef] [PubMed]
40. Smeriglio, A.; Barreca, D.; Bellocco, E.; Trombetta, D. Proanthocyanidins and hydrolysable tannins: Occurrence, dietary intake and pharmacological effects. *Br. J. Pharmacol.* **2016**. [CrossRef] [PubMed]

41. Bellocco, E.; Barreca, D.; Laganà, G.; Calderaro, A.; El Lekhlifi, Z.; Chebaibi, S.; Smeriglio, A.; Trombetta, D. Cyanidin-3-*O*-galactoside in ripe pistachio (*Pistachia vera* L. variety Bronte) hulls: Identification and evaluation of its antioxidant and cytoprotective activities. *J. Funct. Foods* **2016**, *27*, 376–385. [CrossRef]

42. Barreca, D.; Laganà, G.; Toscano, G.; Calandra, P.; Kiselev, M.A.; Lombardo, D.; Bellocco, E. The interaction and binding of flavonoids to human serum albumin modify its conformation, stability and resistance against aggregation and oxidative injuries. *BBA Gen. Subj.* **2016**. [CrossRef] [PubMed]

43. Barreca, D.; Laganà, G.; Tellone, E.; Ficarra, S.; Leuzzi, U.; Galtieri, A.; Bellocco, E. Influences of flavonoids on erythrocyte membrane and metabolic implication through anionic exchange modulation. *J. Membr. Biol.* **2009**, *230*, 163–171. [CrossRef] [PubMed]

44. Barreca, D.; Laganà, G.; Ficarra, S.; Tellone, E.; Leuzzi, U.; Galtieri, A.; Bellocco, E. Evaluation of the antioxidant and cytoprotective properties of the exotic fruit *Annona cherimola* Mill. (*Annonaceae*). *Food Res. Int.* **2011**, *44*, 2302–2310. [CrossRef]

45. Bellocco, E.; Barreca, D.; Laganà, G.; Leuzzi, U.; Tellone, E.; Ficarra, S.; Kotyk, A.; Galtieri, A. Influence of L-rhamnosyl-D-glucosyl derivatives on properties and biological interaction of flavonoids. *Mol. Cell Biochem.* **2009**, *321*, 165–171. [CrossRef] [PubMed]

46. Rice-Evans, C.A.; Miller, N.J.; Paganga, G. Structure-antioxidant activity relationships of flavonoids and phenolic acids. *Free Radic. Biol. Med.* **1996**, *20*, 933–966. [CrossRef]

47. Tomar, N.S.; Sharma, M.; Agarwal, R.M. Phytochemical analysis of *Jatropha curcas* L. during different seasons and developmental stages and seedling growth of wheat (*Triticum aestivum* L) as affected by extracts/leachates of *Jatropha curcas* L. *Physiol. Mol. Biol. Plants* **2015**, *21*, 83–92. [CrossRef] [PubMed]

48. Oskoueian, E.; Abdullah, N.; Saad, W.Z.; Omar, A.R.; Ahmad, S.; Kuan, W.B.; Zolkifli, N.A.; Hendra, R.; Ho, Y.W. Antioxidant, anti-inflammatory and anticancer activities of methanolic extracts from *Jatropha curcas* Linn. *J. Med. Plants Res.* **2011**, *5*, 49–57.

49. Najda, A.; Almehemdi, A.F.; Zabar, A.F. Chemical composition and nutritional value of *Jatropha curcas* L. leaves. *J. Genet. Environ. Res. Conserv.* **2013**, *1*, 221–226.

50. Diwani, G.E.; Rafie, S.E.; Hawash, S. Antioxidant activity of extracts obtained from residues of nodes leaves stem and root of Egyptian *Jatropha curcas*. *Afr. J. Pharm. Pharmacol.* **2009**, *3*, 521–530.

51. Rampadarath, S.; Puchooa, D.; Ranghoo-Sanmukhiya, V.M. A comparison of polyphenolic content, antioxidant activity and insecticidal properties of *Jatropha* species and wild *Ricinus communis* L. found in Mauritius. *Asian Pac. J. Trop. Med.* **2014**, *7*, S384–S390. [CrossRef]

52. Abd-Alla, H.I.; Moharram, F.A.; Gaara, A.H.; El-Safty, M.M. Phytoconstituents of *Jatropha curcas* L. leaves and their immunomodulatory activity on humoral and cell-mediated immune response in chicks. *Z. Naturforsch.* **2009**, *64*, 495–501. [CrossRef]

53. Masaoud, I.M.; Ripperger, H.; Porzel, A.; Adam, G. Flavonol glycosides from Jatropha variegate. *Z. Prakt. Chem.* **1995**, *337*, 43–45. [CrossRef]

54. Subramanian, S.S.; Nagarajan, S.; Sulochana, N. Flavonoids of the leaves of *Jatropha gossypiifolia*. *Phytochemistry* **1971**, *10*, 1690. [CrossRef]

55. Debnath, M.; Bisen, P.S. *Jatropha curcas* L., a multipurpose stress resistant plant with a potential for ethnomedicine and renewable energy. *Curr. Pharm. Biotechnol.* **2008**, *9*, 288–306. [CrossRef] [PubMed]

56. Huang, Q.; Guo, Y.; Fu, R.; Peng, T.; Zhang, Y.; Chen, F. Antioxidant activity of flavonoids from leaves of *Jatropha curcas*. *Sci. Asia* **2014**, *40*, 193–197. [CrossRef]

57. Félix-Silva, J.; Souza, T.; Menezes, Y.A.S.; Cabral, B.; Câmara, R.B.G.; Silva-Junior, A.A.; Rocha, H.A.O.; Rebecchi, I.M.M.; Zucolotto, S.M.; Fernandes-Pedrosa, M.F. Aqueous leaf extract of *Jatropha gossypiifolia* L. (*Euphorbiaceae*) inhibits enzymatic and biological actions of Bothrops jararaca snake venom. *PLoS ONE* **2014**, *9*, e104952. [CrossRef] [PubMed]

58. Pilon, A.C.; Carneiro, R.L.; Carnevale Neto, F.; Bolzani, V.S.; Castro-Gamboa, I. Interval multivariate curve resolution in the dereplication of HPLC-DAD data from *Jatropha gossypifolia*. *Phytochem. Anal.* **2013**, *24*, 401–406. [CrossRef] [PubMed]

59. Rofida, S. Antioxidant activity of *Jatropha curcas* and *Jatropha gossypifolia* by DPPH method. *Farmasains* **2015**, *2*, 281–284.

60. Fu, R.; Zhang, Y.; Guo, Y.; Liu, F.; Chen, F. Determination of phenolic contents and antioxidant activities of extracts of *Jatropha curcas* L. seed shell, a by-product, a new source of natural antioxidant. *Ind. Crops Prod.* **2014**, *58*, 265–270. [CrossRef]

61. Romani, A.; Mancini, P.; Tatti, S.; Vincieri, F. Polyphenols and polysaccharidies in Tuscan grapes and wines. *Ital. J. Food Sci.* **1996**, *8*, 13–24.

62. Singleton, V.R.; Orthofer, R.; Lamuela-Raventós, R.M. Analysis of total phenols and other oxidation substrates and antioxidants by means of Folin-Ciocalteu reagent. *Methods Enzymol.* **1999**, *299*, 152–178.

63. Jollow, D.J.; Mitchell, J.R.; Zampaglione, N.; Gillette, J.R. Bromobenzene-induced liver necrosis. Protective role of glutathione and evidence for 3,4-bromobenzene oxide as the hepatotoxic metabolite. *Pharmacology* **1974**, *11*, 151–169. [CrossRef] [PubMed]

64. Law, M.Y.; Charles, S.A.; Halliwell, B. Glutathione and ascorbic acid in spinach (*Spinacia oleracea*) chloroplasts. The effect of hydrogen peroxide and of Paraquat. *Biochem. J.* **1983**, *210*, 899–903. [CrossRef] [PubMed]

65. Beaumont, F.; Jouve, H.M.; Gagnon, J.; Gaillard, J.; Pelmont, J. Purification and properties of a catalase from potato tubers (*Solanum tuberosum*). *Plant Sci.* **1990**, *72*, 19–26. [CrossRef]

66. Panda, S.K.; Singha, L.B.; Khan, M.H. Does aluminium phytotoxicity induce oxidative stress in greengram (*Vigna radiata*). *Bulg. J. Plant Physiol.* **2003**, *29*, 77–86.

67. Doulis, A.G.; Debian, N.; Kingston-Smith, A.H.; Foyer, C.H. Differential localization of antioxidants in maize leaves. *Plant Physiol.* **1997**, *114*, 1031–1037. [CrossRef] [PubMed]

68. Amako, K.; Chen, G.X.; Asada, K. Separate assays specific for ascorbate peroxidase and guaiacol peroxidase and for the chloroplastic and cytosolic isozymes of ascorbate peroxidase in plants. *Plant Cell Physiol.* **1994**, *35*, 497–504.

69. Molineux, P. The use of the stable free radical diphenylpicrylhydrazyl (DPPH) for estimating antioxidant activity. *Songklanakarin J. Sci. Technol.* **2004**, *26*, 211–219.

70. Re, R.; Pellegrini, N.; Proteggente, A.; Pannala, A. Antioxidant activity applying an improved ABTS radical cation decolorization assay. *Free Radic. Biol. Med.* **1999**, *26*, 1231–1237. [CrossRef]

71. Benzie, F.F.; Strain, J.J. Ferric reducing antioxidant power assay: Direct measure of total antioxidant activity of biological fluids and modified version for simultaneous measurement of total antioxidant power and ascorbic acid concentration. *Methods Enzymol.* **1999**, *299*, 15–23. [PubMed]

72. Dorman, H.J.D.; Kosar, M.; Kahlos, K.; Holm, Y.; Hiltunen, R. Antioxidant properties and composition of aqueous extracts from *Mentha* species, hybrids, varieties, and cultivars. *J. Agric. Food Chem.* **2003**, *51*, 4563–4569. [CrossRef] [PubMed]

73. Hertog, M.G.L.; Hollman, P.C.H.; Venema, D.P. Optimization of quantitative HPLC determination of potentially anticarcinogenic flavonoids in fruits and vegetables. *J. Agric. Food Chem.* **1992**, *40*, 1591–1598. [CrossRef]

74. Yagi, K.; Rastogi, R. Assay for lipid peroxides in animal tissues by thiobarbituric acid reaction. *Annu. Rev. Biochem.* **1979**, *95*, 351–358.

75. Smeriglio, A.; Mandalari, G.; Bisignano, C.; Filocamo, A.; Barreca, D.; Bellocco, E.; Trombetta, D. Polyphenolic content and biological properties of Avola almond (*Prunus dulcis* Mill. D.A. Webb) skin and its industrial byproducts. *Ind. Crops Prod.* **2016**, *83*, 283–293. [CrossRef]

76. Dávalos, A.; Gómez-Cordovés, C.; Bartolomé, B. Extending applicability of the oxygen radical absorbance capacity (ORAC-fluorescein) assay. *J. Agric. Food Chem.* **2004**, *52*, 48–54. [CrossRef] [PubMed]

International Journal of
Molecular Sciences

MDPI

Article

Biosynthesis of α-Glucosidase Inhibitors by a Newly Isolated Bacterium, *Paenibacillus* sp. TKU042 and Its Effect on Reducing Plasma Glucose in a Mouse Model

Van Bon Nguyen [1,2], Anh Dzung Nguyen [3], Yao-Haur Kuo [4,5] and San-Lang Wang [1,*]

[1]   Department of Chemistry, Tamkang University, New Taipei City 25137, Taiwan; bondhtn@gmail.com
[2]   Department of Science and Technology, Tay Nguyen University, Buon Ma Thuot 630000, Vietnam
[3]   Institute of Biotechnology and Environment, Tay Nguyen University, Buon Ma Thuot 630000, Vietnam; nadzungtaynguyenuni@yahoo.com.vn
[4]   Division of Chinese Materia Medica Development, National Research Institute of Chinese Medicine, Taipei 11221, Taiwan; kuoyh@nricm.edu.tw
[5]   Graduate Institute of Integrated Medicine, College of Chinese Medicine, China Medical University, Taichung 40402, Taiwan
*   Correspondence: sabulo@mail.tku.edu.tw; Tel.: +886-2-2621-5656; Fax: +886-2-2620-9924

Academic Editor: Toshio Morikawa
Received: 14 February 2017; Accepted: 22 March 2017; Published: 25 March 2017

Abstract: *Paenibacillus* sp. TKU042, a bacterium isolated from Taiwanese soil, produced α-glucosidase inhibitors (aGIs) in the culture supernatant when commercial nutrient broth (NB) was used as the medium for fermentation. The supernatant of fermented NB (FNB) showed stronger inhibitory activities than acarbose, a commercial anti-diabetic drug. The IC_{50} and maximum α-glucosidase inhibitory activities (aGIA) of FNB and acarbose against α-glucosidase were 81 μg/mL, 92% and 1395 μg/mL, 63%, respectively. FNB was found to be strongly thermostable, retaining 95% of its relative activity, even after heating at 100 °C for 30 min. FNB was also stable at various pH values. Furthermore, FNB demonstrated antioxidant activity (IC_{50} = 2.23 mg/mL). In animal tests, FNB showed remarkable reductions in the plasma glucose of ICR (Institute of Cancer Research) mice at a concentration of 200 mg/kg. Combining FNB and acarbose enhanced the effect even more, with an added advantage of eliminating diarrhea. According to HPLC (High-performance liquid chromatography) fingerprinting, the *Paenibacillus* sp. TKU042 aGIs were not acarbose. All of the results suggest that *Paenibacillus* sp. TKU042 FNB could have potential use as a health food or to treat type 2 diabetes.

Keywords: α-glucosidase inhibitor; *Paenibacillus*; type 2 diabetes; acarbose; plasma glucose

1. Introduction

Diabetes mellitus (DM) is a chronic metabolic disorder and a serious worldwide health problem. DM is commonly classified as either type 1 (insulin-dependent diabetes mellitus, IDDM) or type 2 (non-insulin-dependent diabetes mellitus, NIDDM). Among diagnosed cases of DM, more than 90% are type 2 [1]. The estimated number of people worldwide suffering from type 2 diabetes will exceed 330 million by 2030 [2]. It has been reported that type 2 diabetes can be treated with α-glucosidase inhibitors (aGIs) [3]. Even now, some commercial aGIs, such as acarbose, miglitol, and voglibose, are used for the treatment of type 2 diabetes. However, these treatments have several side effects, including diarrhea, flatulence, and abdominal discomfort, prompting the exploration of original, sufficient, and natural sources of aGIs instead.

Various natural aGIs have been extracted from medicinal plants [4]. However, these aGIs are not easy to isolate or acquire on a large scale [5,6]. Though chemically synthesized aGIs can be obtained in great amounts, they may also result in hepatic disorders and other adverse effects [7]. Biological synthesis, on the other hand, may have potential as an alternative means of acquiring natural aGIs.

So far, aGIs have been described as being produced by microorganisms, including bacteria [8–16] and fungi [5,17–19]. Among these aGIs-producing microbes, only strains of *Streptomyces* [8,9], *bacillus* [8,10–14], *Stenotrophomonas maltrophilia* [15], and *Actinoplanes* spp. SE-50 [16] have been studied extensively. To the best of our knowledge, there are no reports in the literature on aGIs-producing strains belonging to the genus *Paenibacillus*.

According to our recent literature review, *Paenibacillus* species have rarely been reported to produce materials detrimental to humans [20,21]. Our preceding studies revealed that *Paenibacillus* strains isolated using squid pen powder (SPP) as the sole source of carbon/nitrogen (C/N) were able to transform SPP into bioactive materials, such as exopolysaccharides by *P. mucilaginosus* TKU032 [22], and *P. macerans* TKU029 [23], antioxidants by *Paenibacillus* sp. TKU036 [20], and biosurfactants by *P. macerans* TKU029 [23]. In this study, we isolated and identified an aGI-producing strain of *Paenibacillus* sp. TKU042 which secreted acarbose-comparable aGIs in the fermented nutrient broth (FNB). The optimization of culture conditions, pH and thermal stabilities, as well as the effects of FNB on mice, were subsequently explored.

2. Results and Discussion

2.1. Isolation, Screening, and Identification of Strain TKU042

More than 600 bacterial strains were isolated from the soils of Northern Taiwan using a medium that contained 1% squid pen powder (SPP) as the sole source of carbon/nitrogen. Of these, TKU042 demonstrated the strongest inhibitory activity (97%) with an IC_{50} value of 3.9 ± 0.12 mg/mL; it was, therefore, chosen for further investigation. This potent strain was initially identified as *Paenibacillus* sp. based on 16S rDNA sequences. The name of the species was identified using an analytical profile index (API); however no match was found. Therefore, the TKU042 strain was simply labeled as *Paenibacillus* sp. Many strains of *Paenibacillus* have been reported as possessing potential industrial, agricultural, medical, and health food applications, such as the production of enzymes from *P. curdlanolyticus* B-6 [24], *P. barengoltzii* [25], *P. barengoltzii* [26], and *P. ehimensis* MA2012 [27], exopolysaccharides from *P. mucilaginosus* TKU032 [22] and *P. macerans* TKU029 [23], antioxidants from *Paenibacillus* sp. TKU036 [20], biosurfactants from *P. macerans* TKU029 [23], biological control agents from *P. ehimensis* MA2012 [27] and *P. polymyxa* AC-1 [28] or biofertizlizers from *P. polymyxa* [29,30]. Based on our recent literature review, the biosynthesis of aGIs by the genus *Paenibacillus* has not yet been reported.

2.2. Effects of the C/N (Carbon/Nitrogen) Source on aGIs Production

During fermentation, the source of C/N was proposed as the significant factor in the synthesis of aGIs since it could influence the production of some aGI-related enzymes [31]. Similar phenomena were found in our previous reports, which showed SPP as the most suitable C/N source for the production of exopolysaccharides and antioxidants by isolated strains of *Paenibacillus* [20–23]. Three sources of C/N: 1% SPP, 1% shrimp head powder (SHP), and 0.8% nutrient broth (NB), were investigated for the production of aGIs by *Paenibacillus* sp. TKU042 (Figure 1). The inhibitory activities of the fermented NB, SPP, and SHP reached 100% (at day 1), 100% (at day 3), and 70% (at day 9), respectively (Figure 1A). To analyze aGI productivity, the culture supernatants were diluted appropriately to obtain aGI activity, and expressed as U/mL. As shown in Figure 1B, there were remarkable differences among the three culture supernatants. NB seemed to be the best C/N source for aGI production, showing productivity (1200 U/mL at day 4) approximately 5.3- and 10-fold higher than those of SPP (220 U/mL at day 3) and SHP (120 U/mL at day 6), respectively. Throughout fermentation, the cell growth of *Paenibacillus* sp. TKU 042 was also monitored by measuring the absorbance of the cell solution at 660 nm. As shown in Figure 1C, there was no relationship between cell growth and aGI productivity.

Soybean has been widely used in many studies as the sole source of C/N to synthesize aGIs [8,11,12,17,18]. To enhance aGI production, the soybean-containing medium has been supplemented with various carbohydrates. For example, *Streptomyces lavendulae* GC-148 produced an aGI in medium containing 2% glucose and 1.5% soybean [8], and the *Bacillus* species required 4% soybean peptone supplemented with 5% carbon source (glucose, galactose, lactose, or sorbitol) [11]. In this study, *Paenibacillus* sp. TKU 042 used NB for aGI production, diverging from the other reports.

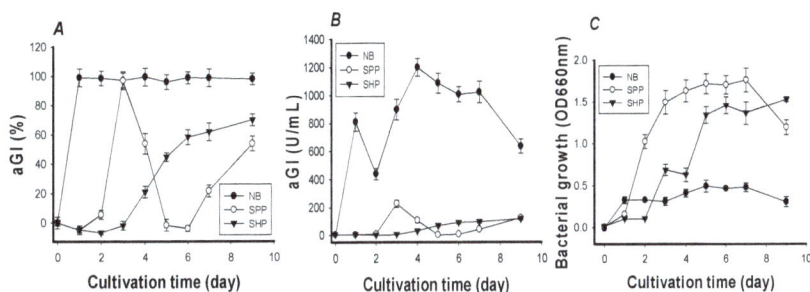

Figure 1. Screening C/N sources for fermentation. SPP, SHP (shrimp head powder), and NB were used as the sole sources of C/N with concentrations of 1%, 1% and 0.8%, respectively. Fermentation conditions were set at 30 °C, shaking speed of 150 rpm, 100/250 mL for the medium/flask volume ratio, and 1 mL of bacterial seed solution (OD$_{660nm}$ = 0.38). The inhibition against yeast α-glucosidase was then tested and calculated in % (**A**) or U/mL (**B**); the growth of *Paenibacillus* sp. TKU042 was detected at OD$_{660nm}$ (**C**). To eliminate the influence of SHP and SPP on OD$_{660nm}$ adsorption, any residual SHP and SPP present after fermentation were removed by centrifugation at 500 rpm for 10 min. *Saccharomyces cerevisiae* α-glucosidase was used to test aGI activity.

2.3. Optimization of Culture Conditions

NB was confirmed as the best source of C/N, and chosen to establish optimal parameters, including cultivation temperature, culture volume, concentration of NB, and the amount of seed culture inoculated (Figure 2).

The results showed that aGI productivity was highest on the sixth day of cultivation at 25 °C (1700 U/mL) and the fourth day of cultivation at 30 °C (1550 U/mL) (Figure 2A). Taking cultivation time into account, 30 °C was selected to study the effects of culture volume. As shown in Figure 2B, aGI productivity was highest on the sixth day of cultivation in 50 mL-medium (1800 U/mL) followed by the fourth day of cultivation in 100 mL-medium (1600 U/mL). After considering the recovered volume of culture supernatant, 100 mL-medium (at 30 °C) was selected to investigate the effects of NB concentration on aGI productivity. As seen in Figure 2C, the highest aGI productivity was found on the fourth day when either 0.4% or 0.8% NB was used. 0.8% NB was ultimately selected for further research since aGI productivity declined slightly after day 4 when 0.4% NB was used. The inoculated volume of the seed culture had no significant effect on aGI production (Figure 2D).

Overall, aGIs were effectively produced by *Paenibacillus* sp. TKU042 in the 0.8% NB-containing medium (100 mL of medium in a 250 mL-Erlenmeyer flask) at 30 °C and an initial pH of 6.85, using a reciprocal shaker at 150 rpm for 4 days. The culture conditions before and after the optimization study are summarized in Table 1; in short, the IC$_{50}$ and productivity of *Paenibacillus* sp. TKU 042 aGIs were increased approximately 23-fold after optimization.

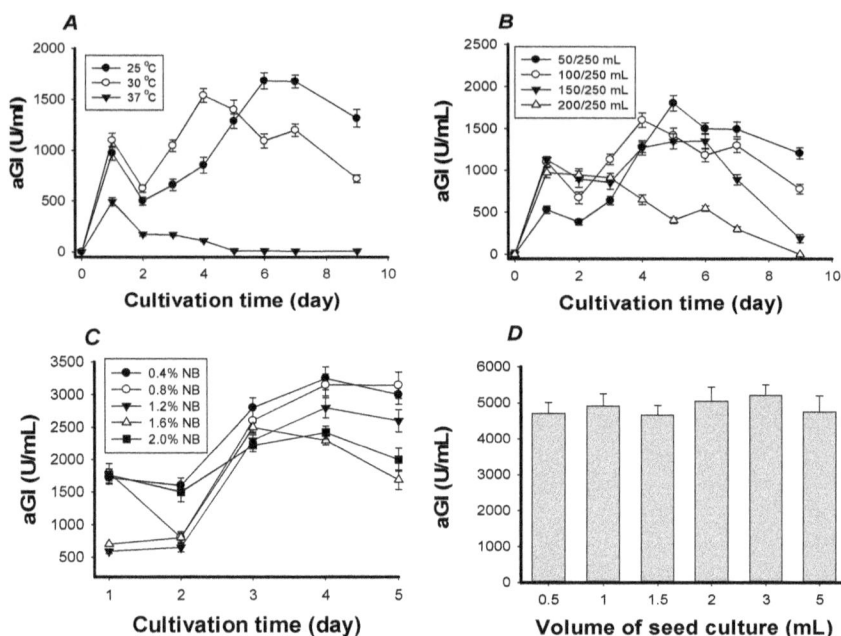

Figure 2. The effects of some parameters on aGIs production. (**A**) The effect of incubation temperature on aGI activity: Fermentation was conducted using 150 rpm shaking speed, a medium/flask volume ratio of 100/250 mL, 0.8% NB (C/N source), and 1 mL of seed culture (OD_{660nm} = 0.38); (**B**) the effect of medium/flask volume ratio on aGI productivity: fermentation was conducted at 30 °C, with a shaking speed of 150 rpm, and using 0.8% NB (C/N source), and 1 mL of seed culture (OD_{660nm} = 0.38); (**C**) the effect of NB concentration on aGI productivity: fermentation conducted at 30 °C, with a shaking speed of 150 rpm, a medium/flask volume ratio of 100/250 mL, and 1 mL of seed culture (OD_{660nm} = 0.38); and (**D**) the effect of seed culture volume on aGI productivity: fermentation was conducted at 30 °C, with a shaking speed of 150 rpm, a medium/flask volume ratio of 100/250 mL, and using 0.8% NB (C/N source). *Saccharomyces cerevisiae* α-glucosidase was then used to analyze the aGIA.

Table 1. Comparison of culture conditions before and after optimization.

Compared Factors	Before Optimization	After Optimization
C/N source	SPP	NB
Cultivation temperature (°C)	37	30
C/N Concentration (%)	1	0.8–1.2
Medium/flask volume ratio	2/5	2/5
Seed culture (%)	1	1
Inhibition (IC_{50} µg/mL)	3900 ± 120	81 ± 4.3
aGIs Productivity (U/mL)	220	5000

2.4. Specific aGI Activity and Antioxidant Activity of FNB

To evaluate the potential of *Paenibacillus* sp. TKU 042 aGIs as anti-diabetic drugs, the inhibitory specificity of FNB was tested against eight kinds of commercial enzymes. As shown in Table 2, FNB showed strong inhibitory activity against bacterial (*Bacillus stearothermophilus*) α-glucosidase (IC_{50} = 36.7 ± 1.7 µg/mL), yeast (*Saccharomyces cerevisiae*) α-glucosidase (IC_{50} = 81 ± 4.3 µg/mL), and rat α-glucosidase (IC_{50} = 101 ± 5.1 µg/mL), but had a weaker response against rice α-glucosidase (IC_{50} = 508 ± 17 µg/mL). On the other hand, the anti-diabetic drug acarbose demonstrated better inhibitory activity against α-glucosidases from bacteria (0.012 ± 0.001 µg/mL), rice (2.92 ± 0.67 µg/mL), and rat (107 ± 2.5 µg/mL), but showed weaker inhibitory activity against yeast α-glucosidase (1395 ± 5 µg/mL).

α-Glucosidase from *S. cerevisiae* has been commonly used to evaluate aGI activity in many studies [18,32–34]. Compared to yeast and other sources of α-glucosidases, rats may be a more valuable resource since the animal enzyme is closer to that of humans [32]. In this study, both FNB and acarbose inhibited α-glucosidases originating from rats and yeast. Among them, FNB showed much stronger aGI activity than acarbose against α-glucosidase from yeast but approximately the same amount of inhibition against rat α-glucosidase. Therefore, FNB could be a potential candidate for α-glucosidase inhibition.

α-Amylase inhibitors have been reported to be a useful treatment for type 2 diabetes [35]. Consequently, the inhibitory activities of FNB against α-amylases were also studied. As shown in Table 2, FNB showed no inhibitory activity against porcine pancreatic and *B. subtilis* α-amylases. Similarly, FNB demonstrated no inhibitory activity against proteases from bromelain or papain. It is well-known that there are several proteases in the gastrointestinal tract to help ingest proteins. Since FNB showed no inhibitory activity against these tested proteases, it may have the potential to avoid disturbing the ingestion of proteins by the use of aGIs.

Table 2. Specific inhibitory activity of FNB and acarbose against enzymes.

Enzyme	n	Inhibition of FNB		Inhibition of Acarbose	
		IC$_{50}$ (µg/mL)	Maximum Inhibitory Activity (%)	IC$_{50}$ (µg/mL)	Maximum Inhibitory Activity (%)
S. cerevisiae α-glucosidase	3	81 ± 4.3 [c,d]	92 ± 3.2	1395 ± 5 [a]	63 ± 2.5
Rat α-glucosidase	3	101 ± 5.1 [c]	95 ± 2.1	107 ± 2.5 [c]	89 ± 2.4
B. stearothermophilus α-glucosidase	3	36.7 ± 1.7 [c,d]	98.2 ± 1.3	0.012 ± 0.001 [d]	100 ± 1.2
Rice α-glucosidase	3	508 ± 17 [b]	94 ± 2.3	2.92 ± 0.67 [d]	100 ± 1.4
Porcine pancreatic α-amylase	3	-	-	ND	ND
B. subtilis α-amylase	3	-	-	ND	ND
Papain protease	3	-	-	ND	ND
Bromelain protease	3	-	-	ND	ND

(-) no inhibition; ND: not determined; CV = 12.74; LSD$_{0.01}$ = 86.385; n = 3: triplicates of each experiment. Mean of IC$_{50}$ values with the different letters are significantly different based on t test ranking.

Several previous studies [20,22] suggested that antioxidant compounds could be produced by the genus *Paenibacillus*. Therefore, the antioxidant property of FNB was also tested, using the 2,2-diphenyl-1-picrylhydrazyl (DPPH) radical scavenging activity assay mentioned in the experimental methods section, to provide more bioactive proof of FNB's usefulness. The antioxidant activity was calculated as both a percentage and IC$_{50}$ value. α-Tocopherol, a commercial antioxidant compound, showed stronger activity than FNB, generating IC$_{50}$ values of 0.0247 ± 0.0012 and 2.23 ± 0.14 mg/mL, respectively. However, FNB can achieve the same amount of activity (approximately 100%) as α-tocopherol. Overall, FNB could be used as a health food due to its valuable bio-functions, including its potential for α-glucosidase inhibition and acceptable antioxidant activity.

2.5. Confirmation that aGIs Contained in FNB Were Produced during NB Fermentation

The same concentrations (20 mg/mL) of unfermented NB (UNB) and fermented NB (FNB) solutions were analyzed by HPLC. The differences in the HPLC finger prints of NB before and after fermentation are clearly observed in Figure 3. After fermentation, some major peaks disappeared (the peaks at retention times of 6, 13.5, and 17.5 min) or reduced their area (the peak at 11 min), while some new peaks appeared at 6.5 and 13.2 min, and the peak at 17.3 min, enhanced its area. UNB was tested for α-glucosidase inhibition but no activity was observed. The differences in both the HPLC finger prints and the inhibitory activity of UNB and FNB suggest that the aGIs were produced by fermentation, since none existed in the NB beforehand. Each fraction of aGI activity in FNB was analyzed on the HPLC finger print; three fractions showed aGI activity ($\geq$85% at the concentration of 200 μg/mL). The isolation and identification of these aGIs will eventually be conducted. The HPLC finger print of acarbose was also analyzed; based on the difference in retention time, we confirmed that the active compound in FNB aGIs was not acarbose.

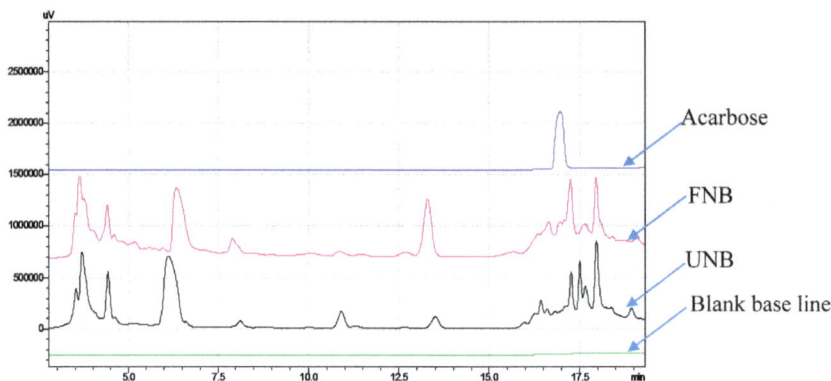

Figure 3. The HPLC finger prints of unfermented and fermented NB. Analysis conditions: mobile phase: 0–5′ (1% Acetonitrile (ACN), 5–20′ (1%–10% ACN), 20–25′ (10%–40% ACN), 25–40′ (40%–50% ACN); Ultraviolet (UV) detector: 240 nm; flow rate 0.8 mL/min, column temperature: 25 °C, using C18 column.

2.6. The Thermal and pH Stabilities of FNB aGIs

The thermal stability of FNB aGIs was determined by first treating FNB at 40–100 °C for 30 min, and then testing the residual aGI activity. FNB aGIs were strongly thermostable at all temperatures, with relative activities greater than 95% (Figure 4A). They maintained a relative activity of 92% even when heated at 100 °C for 30 min. The high thermal stability of FNB aGIs suggests they may have the potential to remain active for a long time.

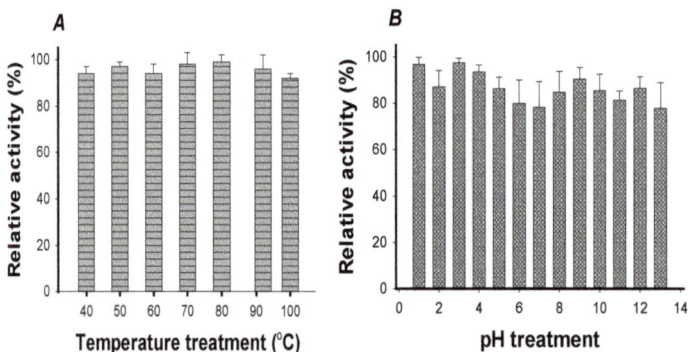

Figure 4. Thermal and pH stability of FNB. Thermal stability (**A**) was tested by treating FNB for 30 min at 40–100 °C and pH 6.8; the residual aGI activity was analyzed at 37 °C; The pH stability (**B**) was tested by treating FNB to a pH range of 1–13 at 37 °C for 30 min then evaluating the residual aGI activity at pH 6.8.

pH stability is also an important factor. After treating FNB with a large range of pH values at 37 °C for 30 min, the residual aGI activity was analyzed at pH 6.8. As shown in Figure 4B, aGI activity remained at 87%–96% after 30 min pre-incubation at pH ranges from 1–13. It is well-known that the stomach is very acidic; therefore, potential aGIs should be stable and able to work well in the acidic environments of the gastrointestinal tract. Bacteria-produced aGIs that show thermal and pH stability are rarely reported. In the previous reports [32,35], aGIs from ELC, a methanolic extract from a medicinal plant (*Euonymus laxiflorus* Champ), were reported to be heat-stable at 97 ± 3 °C (heating 30 min) with a relative activity of 90%, demonstrating the same thermal stability as FNB. However, the relative activity of this extract (50%) was lower than that of FNB (80%–93%) in the acidic environment (pH 4–6).

2.7. The Effects of FNB on Reducing Plasma Glucose in the Mouse Model

Three doses of FNB (100, 200, and 400 mg/kg) were used on ICR mice to evaluate their impact on plasma glucose. As shown in Figure 5A, significant reductions in plasma glucose were observed 0.5 and 1 h after sucrose loading for 100 and 400 mg/kg.

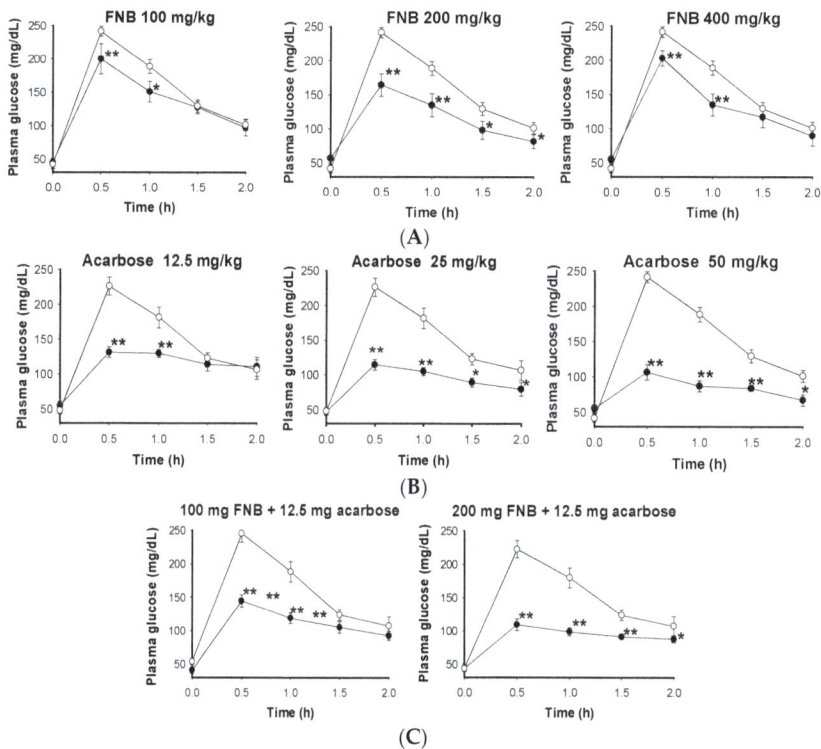

Figure 5. Effects of FNB and acarbose, alone or in combination, on the increase in plasma glucose levels following oral sucrose loading in ICR mice. (**A**) Oral FNB administration; (**B**) Oral acarbose administration; (**C**) Combined FNB and acarbose administration. FNB (100, 200, and 400 mg/kg bw), acarbose (12.5, 25 and 50 mg/kg bw), and a combination of FNB (100 and 200 mg/kg bw) and acarbose (12.5 mg/kg bw) were administered to mice (●, $n = 8$). In the control groups (○, $n = 8$), distilled water was administered. After loading the sample (or acarbose or distilled water) for 25 min, sucrose solution was orally administered at a concentration of 3 g/kg bw; blood was sampled and measured 0.5, 1, 1.5 and 2 h, thereafter. Statistical analyses were conducted using the Least Significant Difference (LSD) test. The significant differences for each experimental mean (in the comparison with its control) are illustrated with (*) at $p < 0.05$ and (**) at $p < 0.01$.

The FNB with 200 mg/kg body weight (bw) was ultimately the best, as it reduced blood sugar for 1.5 h. This result is comparable with a dose of acarbose at 25 mg/kg, but better than acarbose at 12.5 mg/kg (Figure 5B).

In the combined FNB and acarbose group, two doses of FNB (100 and 200 mg/kg) and one dose of acarbose (12.5 mg/kg) were used (Figure 5C). The combination of FNB at 100 mg and acarbose at 12.5 mg had no effect compared to acarbose (12.5 mg/kg) administration alone. On the other hand, the combination of FNB at 200 mg/kg and acarbose at 12.5 mg/kg was much better at reducing blood glucose than acarbose alone at 12.5 mg/kg, slightly higher than acarbose at 25 mg/kg, and comparable to arcabose at 50 mg/kg. The number of mice with diarrhea was also recorded. No ill effects were observed in the control, FNB at 100, 200, or 400 mg, acarbose at 12.5 mg or the combination of FNB (100, 200 mg) with acarbose at 12.5 mg. However, for the groups who received 25 and 50 mg of acarbose, diarrhea was observed in 3/8 mice (37.5%) and 5/8 mice (62.5%), respectively.

These results show that oral FNB loading can reduce plasma glucose levels in mice, and the combination of FNB (200 mg/kg) and acarbose (12.5 mg/kg) can produce approximately the same effect as acarbose alone at 25 and 50 mg. As such, the acarbose dosage can be reduced two- to four-fold, resulting in reduced side effects.

3. Materials and Methods

3.1. Materials

Nutrient broth was purchased from Creative Life Science Co., Taipei, Taiwan, squid pens were acquired from Shin-Ma Frozen Food Co. (Yilan, Taiwan), and shrimp head power (SHP) was obtained from Fwu-Sow Industry (Taichun, Taiwan). Rat α-glucosidase (intestinal acetone powders from rat) was purchased from Sigma Aldrich, Singapore. Acarbose, *Saccharomyces cerevisiae* α-glucosidase, *Bacillus stearothermophilus* α-glucosidase, and 2, 2-diphenyl-1-picrylhydrazyl (DPPH) were purchased from Sigma Chemical Co., St. Louis, MO, USA. Rice α-glucosidase (Type 4) and porcine pancreatic α-amylase (Type VI-B) were purchased from Sigma Aldrich, St. Louis, MO, USA. The proteases of bromelain and papain were obtained from Challenge Bioproducts Co., Ltd., Yunlin, Taiwan. When possible, reagents, solvents and other common chemicals were obtained at the highest grade.

3.2. Measurement of Rat α-Glucosidase Inhibition

To determine rat intestinal α-glucosidase inhibition, the techniques from the previous report were followed [36], with modifications. One hundred mg of rat-intestinal acetone powder was suspended in 4 mL of 0.1 mol/L potassium phosphate buffer (NPB), pH 7. The suspension was sonicated several times for 12 s at 4 °C and then centrifuged for 20 min at 10,000× *g*. The residues were twice suspended with 3 mL of 0.1 mol/L NPB at pH 7, as described above. The resulting supernatant was mixed together and dialyzed for 12 h at 4–6 °C, and then used for the assay. The α-glucosidase, sample solutions and buffer were pre-mixed at volumes of 50, 50 and 100 μL, respectively. After pre-incubation at 37 °C for 20 min, the reaction started when 50 μL of *p*-nitrophenyl glucopyranoside (10 mmol/L) was added to the mixture. After 20 min at 37 °C, 100 μL Na_2CO_3 solution (1 mol/L) was added to terminate the reaction. The absorbance of the final mixture was measured at 410 nm [37] and used to calculate inhibition (%) using the following equation:

$$\alpha - \text{glucosidase inhibitory activity } (\%) \ = \ (A \ - \ B)/A \ \times \ 100$$

where "*A*" is the absorbance of the control (buffer was used instead of sample with the same volume) and "*B*" is the absorbance of the final mixture containing the sample and α-glucosidase [38]. The IC_{50} value was defined and determined as per the previous study [32]. The α-glucosidase and the samples were prepared in 0.1 mol/L potassium phosphate buffer (pH 7).

General α-glucosidase and α-amylase inhibition assays were carried out as described in our previous papers ([32,35], respectively). Acarbose dissolved in the same buffer (pH 7) was also analyzed as a control.

3.3. DPPH Radical Scavenging Activity Assay

One hundred and twenty microliters of each sample (at different concentrations) was mixed with 30 μL of a methanolic solution containing 0.75 mM DPPH radicals in the well of a plate with 96 wells. The mixture was kept in the dark for 30 min before absorbance at 517 nm was measured [39]. Radical scavenging activity was calculated using the following formula:

$$\text{Antioxidant activity } (\%) \; = \; [(AB \, - \, AA)/AB] \, \times \, 100$$

where AB and AA stand for absorption of the blank sample ($t = 0$ min) and absorption of the tested sample solution ($t = 30$ min), respectively. α-Tocopherol dissolved in MeOH was also analyzed as a control.

3.4. Isolation and Screening of aGI-Producing Strains

About 60 soil samples were randomly collected from Northern Taiwan. For convenience, half of the soil samples were collected from the campus of Tamkang University (Tamsui, New Taipei City, Taiwan). Ten milliliters of sterile distilled water was added to 1 g of ground soil and stirred. Half a milliliter of the soil suspension was then added to medium containing 0.1% K_2HPO_4, 0.05% $MgSO_4 \cdot 7H_2O$, and 1% squid pen powder (SPP), supplemented with 1.5% agar in a plastic Petri dish containing 10 mL of medium (pH 7.2). After 1–3 days of incubation at 37 °C, the single bacterial colonies that appeared were sub-cultured several times on the same medium.

All of the isolated strains were incubated in a 250-mL Erlenmeyer flask with 50 mL of the same medium, but without agar. Fermentation was performed at 37 °C and a 150 rpm shaking speed. After 1–4 days of fermentation, the culture was centrifuged at 500 rpm for 10 min to remove the medium residue. The solution was then centrifuged again at 4000 rpm for 20 min to separate the cells from the culture supernatant. The supernatant was then used to test yeast α-glucosidase inhibitory activity. The cells were dissolved in the same volume of distilled water as the supernatant and measured at 660 nm to detect bacterial growth. The screening of active bacteria was based on the inhibitory activity of the supernatant.

3.5. Optimization of Culture Conditions for Synthesis of aGIs

Three sources of C/N (1% SPP, 1% SHP, and 0.8% NB) were used to cultivate *Paenibacillus* sp. TKU042; conditions were set at 30 °C, a shaking speed of 150 rpm and a 100/250 mL medium/flask volume ratio for nine days. NB appeared to be most suitable for synthesizing aGIs and was chosen to test optimization of specific parameters, including concentration of NB (0.4%, 0.8%, 1.2%, 1.6%, 2.0% w/v), medium volume (50, 100, 150, and 200 mL), culture temperature (25, 30, and 37 °C) and the amount of seed culture (0.5, 1, 1.5, 2, 3, and 5 mL of bacterial seed solution at $OD_{660nm} = 0.38$). The culture supernatant obtained was centrifuged again at 4000 rpm for 20 min, then used to test yeast α-glucosidase inhibitory activity.

3.6. Measurement of Inhibitor Stability

The pH stability of the sample was determined as per the methods described in our previous paper [35]. To determine thermal stability, the sample was exposed to a temperature range of 40–100 °C for 30 min before testing for inhibition against yeast α-glucosidase, using the general α-glucosidase inhibition assay [32].

3.7. Experimental Animal Protocol

Seventy-two seven week-old male ICR mice obtained from The National Laboratory Animal Center (No. 128, Sec. 2, Academia Rd., Nangang Dist., Taipei City 11529, Taiwan) were randomly divided into nine groups (8 mice per group). All had equivalent mean plasma glucose levels and body weights, including the control group (orally administered with distilled water only). Three experimental groups received oral FNB (100, 200, or 400/kg), three experimental groups received oral acarbose (12.5, 25 or 50 mg/kg), and two experimental groups received both FNB and acarbose (100 mg of FNB combined with 12.5 mg of acarbose, or 200 mg of FNB combined with 12.5 mg of acarbose). This animal study was conducted in accordance with the guidelines and approval of the Institutional Animal Care and Use Committee of the National Research Institute of Chinese Medicine, Ministry of Health and Welfare. (IACUC No.104-706-1, 29 December 2014).

The sucrose tolerance test was performed following the methods described in the previous report [40], with slight modifications. Mice were fasted for 16 h and were then given different concentrations of distilled water (control group), acarbose, or mixtures of the solutions described in detail above for 20 min. A sucrose solution was orally administered at 3 g/kg bw and blood was sampled and measured after 0.5, 1, 1.5 and 2 h.

Statistical Analysis Software (SAS-9.4, provided by SAS Institute Taiwan Ltd, Minsheng East Road, section 2, Taipei, Taiwan 149-8) was used to analyze the significant differences of the calculated IC_{50} values, the α-glucosidase inhibition (%) and the plasma glucose levels.

4. Conclusions

Of more than 600 bacterial strains isolated from Taiwanese soils, *Paenibacillus* sp. TKU042 showed the best potential as a source of aGIs. This bacterium produced aGIs and antioxidants when NB was used as the sole source of C/N. The aGIs of the fermented product (FNB) showed higher inhibitory activity than acarbose against rat and yeast α-glucosidases. The FNB aGIs also showed high thermal and pH stability, even when pre-incubated at 100 °C or exposed to acidic conditions (pH 1) for 30 min. FNB also had an acceptable effect on reducing plasma glucose in mice. All of these results suggest that NB is a potential carbon/nitrogen source for producing aGIs using *Paenibacillus* sp. TKU 042. Furthermore, the aGIs produced might be useful candidates for treating type 2 diabetes and obesity, as well as future use as a health food.

Acknowledgments: This work was supported in part by a grant from the Ministry of Science and Technology, Taiwan (MOST 105-2313-B-032-001).

Author Contributions: San-Lang Wang conceived and designed the experiments; Van Bon Nguyen performed the experiments; San-Lang Wang, Yao-Haur Kuo, and Anh Dzung Nguyen analyzed the data; San-Lang Wang and Yao-Haur Kuo contributed reagents/materials/analysis tools; Van Bon Nguyen and San-Lang Wang wrote the paper.

Conflicts of Interest: The authors declare no conflict of interest.

References

1. Cambell, R.K. Clarifying the role of incretin-based therapies in the treatment of type 2 diabetes mellitus. *Clin. Ther.* **2011**, *33*, 511–527. [CrossRef] [PubMed]
2. Lakshmanasenthil, S.; Vinothkumar, T.; Geetharamani, D.; Marudhupandi, T.; Suja, G.; Sindhu, N.S. Fucoidan—A novel α-amylase inhibitor from *Turbinaria ornata* with relevance to NIDDM therapy. *Biocatal. Agric. Biotechnol.* **2014**, *3*, 66–70. [CrossRef]
3. DeMelo, E.B.; Gomes, A.; Carvalha, I. α- and β-Glucosidase inhibitors: Chemical structure and biological activity. *J. Tetrahedron* **2006**, *62*, 10277–10302.
4. Yin, Z.; Zhang, W.; Feng, F.; Zhang, Y.; Kang, W. α-Glucosidase inhibitors isolated from medicinal plants. *Food Sci. Hum. Wellness* **2014**, *3*, 136–174. [CrossRef]
5. Chen, J.; Cheng, Y.Q.; Yamaki, K.; Li, L.T. Anti-α-glucosidase activity of Chinese traditionally fermented soybean (douchi). *Food Chem.* **2007**, *103*, 1091–1096. [CrossRef]

6. Fujita, H.; Yamagami, T.; Ohshima, K. Long-term ingestion of touchi-extract, a α-glucosidase inhibitor, by borderline and mild type-2 diabetic subjects is safe and significantly reduces blood glucose levels. *Nutr. Res.* **2003**, *23*, 713–722. [CrossRef]

7. Wang, G.; Peng, Z.; Wang, J.; Li, X.; Li, J. Synthesis, in vitro evaluation and molecular docking studies of novel triazine-triazole derivatives as potential α-glucosidase inhibitors. *Eur. J. Med. Chem.* **2017**, *125*, 423–429. [CrossRef] [PubMed]

8. Ezure, Y.; Maruo, S.; Miyazaki, K.; Kawamata, M. Moranoline (1-deoxynojirimycin) fermentation and its improvement. *Agric. Biol. Chem. Tokyo* **1985**, *49*, 1119–1125.

9. Kameda, Y.; Asano, N.; Yoshikawa, M.; Takeuchi, M.; Yamaguchi, T.; Matsui, K.; Horii, S.; Fukase, H. Valiolamine, a new α-glucosidase inhibiting amino-cyclitol produced by *Streptomyces hygroscopicus*. *J. Antibiot.* **1984**, *37*, 1301–1307. [CrossRef] [PubMed]

10. Kim, H.S.; Lee, J.Y.; Hwang, K.Y.; Cho, Y.S.; Park, Y.S.; Kang, K.D.; Seong, S.I. Isolation and identification of a Bacillus sp. producing α-glucosidase inhibitor 1-deoxynojirimycin. *Korean J. Microbiol. Biotechnol.* **2011**, *39*, 49–55.

11. Onose, S.; Ikeda, R.; Nakagawa, K.; Kimura, T.; Yamagishi, K.; Higuchi, O.; Miyazawa, T. Production of the α-glycosidase inhibitor 1-deoxynojirimycin from *Bacillus* species. *Food Chem.* **2013**, *138*, 516–523. [CrossRef] [PubMed]

12. Nam, H.; Jung, H.; Karuppasamy, S.; Park, Y.S.; Cho, Y.S.; Lee, J.Y.; Seong, S.; Suh, J.G. Anti-diabetic effect of the soybean extract fermented by *Bacillus subtilis* MORI in db/db mice. *Food Sci. Biotechnol.* **2012**, *21*, 1669–1676. [CrossRef]

13. Zhu, Y.P.; Yamaki, K.; Yoshihashi, T.; Ohnishi, K.M.; Li, X.T.; Cheng, Y.Q.; Mori, Y.; Li, L.T. Purification and identification of 1-deoxynojirimycin (DNJ) in okara fermented by *Bacillus subtilis* B2 from Chinese traditional food (meitaoza). *J. Agric. Food Chem.* **2010**, *58*, 4097–4103. [CrossRef] [PubMed]

14. Cho, Y.S.; Park, Y.S.; Lee, J.Y.; Kang, K.D.; Hwang, K.Y.; Seong, S.I. Hypoglycemic effect of culture broth of *Bacillus subtilis* S10 producing 1-deoxynojirimycin. *J. Korean Soc. Food Sci. Nutr.* **2008**, *37*, 1401–1407. [CrossRef]

15. Zheng, Y.G.; Xue, Y.P.; Shen, Y.C. Production of valienamine by a newly isolated strain: *Stenotrophomonas maltrophilia*. *Enzym. Microb. Technol.* **2006**, *39*, 1060–1065. [CrossRef]

16. Schmidt, D.D.; Frommer, W.; Junge, B.; Müller, L.; Wingender, W.; Truscheit, E.; Schäfer, D. α-Glucosidase inhibitors, new complex oligosaccharides of microbial origin. *Naturwissenschaften* **1977**, *64*, 535–536. [CrossRef] [PubMed]

17. Fujita, H.; Yamagami, T.; Ohshima, K. Efficacy and safety of Touchi extract, an a-glucosidase inhibitor derived from fermented soybeans, in non-insulin-dependent diabetic mellitus. *J. Nutr. Biochem.* **2001**, *12*, 351–356.

18. McCue, P.; Kwon, Y.I.; Shetty, K. Anti-diabetic and antihypertensive potential of sprouted and solid-state bioprocessed soybean. *Asian Pac. J. Clin. Nutr.* **2005**, *14*, 145–152.

19. Jing, L.; Zong, S.; Li, J.; Surhio, M.M.; Ye, M. Purification, structural features and inhibition activity on α-glucosidase of a novel polysaccharide from *Lachnum* YM406. *Process Biochem.* **2016**, *51*, 1706–1713. [CrossRef]

20. Wang, S.L.; Li, H.T.; Zhang, L.J.; Lin, Z.H.; Kuo, Y.H. Conversion of squid pen to homogentisic acid via *Paenibacillus* sp. TKU036 and the antioxidant and anti-inflammatory activities of homogentisic acid. *Mar. Drugs* **2016**, *14*, 183. [CrossRef] [PubMed]

21. Liang, T.W.; Wang, S.L. Recent advances in exopolysaccharides from *Paenibacillus* spp.: Production, isolation, structure, and bioactivities. *Mar. Drugs* **2015**, *13*, 1847–1863. [CrossRef] [PubMed]

22. Liang, T.W.; Tseng, S.C.; Wang, S.L. Production and characterization of antioxidant properties of exopolysaccharides from *Paenibacillus mucilaginosus* TKU032. *Mar. Drugs* **2016**, *14*, 40. [CrossRef] [PubMed]

23. Liang, T.W.; Wu, C.C.; Cheng, W.T.; Chen, Y.C.; Wang, C.L.; Wang, I.L.; Wang, S.L. Exopolysaccharides and antimicrobial biosurfactants produced by *Paenibacillus macerans* TKU029. *Appl. Biochem. Biotechnol.* **2014**, *172*, 933–950. [CrossRef] [PubMed]

24. Sermsathanaswadi, J.; Baramee, S.; Tachaapaikoon, C.; Pason, P.; Ratanakhanokchai, K.; Kosugi, A. The family 22 carbohydrate-binding module of bifunctional xylanase/β-glucanase Xyn10E from *Paenibacillus curdlanolyticus* B-6 has an important role in lignocellulose degradation. *Enzym. Microb. Technol.* **2017**, *96*, 75–84. [CrossRef] [PubMed]

25. Shi, R.; Liu, Y.; Mu, Q.; Jiang, Z.; Yang, S. Biochemical characterization of a novel L-asparaginase from *Paenibacillus barengoltzii* being suitable for acrylamide reduction in potato chips and mooncakes. *Int. J. Biol. Macromol.* **2017**, *96*, 93–99. [CrossRef] [PubMed]

26. Yang, S.; Fu, X.; Yan, Q.; Guo, Y.; Liu, Z.; Jiang, Z. Cloning, expression, purification and application of a novel chitinase from a thermophilic marine bacterium *Paenibacillus barengoltzii*. *Food Chem.* **2016**, *192*, 1041–1048. [CrossRef] [PubMed]

27. Seo, D.J.; Lee, Y.S.; Kim, K.Y.; Jung, W.J. Antifungal activity of chitinase obtained from *Paenibacillus ehimensis* MA2012 against conidial of *Collectotrichum gloeosporioides* in vitro. *Microb. Pathogenes.* **2016**, *96*, 10–14. [CrossRef] [PubMed]

28. Hong, C.E.; Kwon, S.Y.; Park, J.M. Biocontrol activity of *Paenibacillus polymyxa* AC-1 against *Pseudomonas syringae* and its interaction with *Arabidopsis thaliana*. *Microbiol. Res.* **2016**, *185*, 13–21. [CrossRef] [PubMed]

29. Lal, S.; Tabacchioni, S. Ecology and biotechnological potential of *Paenibacillus polymyxa*: A minirevie. *Indian J. Microbiol.* **2009**, *49*, 2–10. [CrossRef] [PubMed]

30. Puri, A.; Padda, K.P.; Chanway, C.P. Evidence of nitrogen fixation and growth promotion in canola (*Brassica napus* L.) by an endophytic diazotroph *Paenibacillus polymyxa* P2b-2R. *Biol. Fertil. Soils* **2016**, *52*, 119–125. [CrossRef]

31. Zhu, Y.P.; Yin, L.J.; Cheng, Y.Q.; Yamaki, K.; Mori, Y.; Su, Y.C.; Li, L.T. Effects of sources of carbon and nitrogen on production of α-glucosidase inhibitor by a newly isolated strain of *Bacillus subtilis* B2. *Food Chem.* **2008**, *109*, 737–742. [CrossRef]

32. Nguyen, V.B.; Nguyen, Q.V.; Nguyen, A.D.; Wang, S.L. Screening and evaluation of α-glucosidase inhibitors from indigenous medicinal plants in Dak Lak Province, Vietnam. *Res. Chem. Intermed.* **2015**. [CrossRef]

33. Zhang, L.; Hogan, S.; Li, J.; Sun, S.; Canning, C.; Zheng, S.J.; Zhou, K. Grape skin extract inhibits mammalian intestinal α-glucosidase activity and suppresses postprandial glycemic response in streptozocin-treated mice. *Food Chem.* **2011**, *126*, 466–471. [CrossRef]

34. Trinh, B.T.D.; Staerk, D.; Jäger, A.K. Screening for potential α-glucosidase and α-amylase inhibitory constituents from selected Vietnamese plants used to treat type 2 diabetes. *J. Ethnopharmacol.* **2016**, *186*, 189–195. [CrossRef] [PubMed]

35. Nguyen, V.B.; Nguyen, Q.V.; Nguyen, A.D.; Wang, S.L. Porcine pancreatic α-amylase inhibitors from *Euonymus laxiflorus* Champ. *Res. Chem. Intermed.* **2017**, *43*, 259–269. [CrossRef]

36. Kwon, Y.I.; Jang, H.D.; Shetty, K. Evaluation of *Rhodiola crenulata* and *Rhodiola rosea* for management of type II diabetes and hypertension. *Asia Pac. J. Clin. Nutr.* **2006**, *15*, 425–432. [PubMed]

37. Kim, Y.; Wang, M.; Rhee, H. A novel α-glucosidase inhibitor from pine bark. *J. Carbohydr. Res.* **2004**, *339*, 715–717. [CrossRef] [PubMed]

38. Yu, Z.; Yin, Y.; Zhao, W.; Yu, Y.; Liu, B.; Liu, J.; Chen, F. Novel peptides derived from egg white protein inhibiting α-glucosidase. *Food Chem.* **2011**, *129*, 1376–1382. [CrossRef]

39. Liang, T.W.; Chen, W.T.; Lin, Z.H.; Kuo, Y.H.; Nguyen, A.D.; Pan, P.S.; Wang, S.L. An amphiprotic novel chitosanase from *Bacillus mycoides* and its application in the production of chitooligomers with their antioxidant and anti-inflammatory evaluation. *Int. J. Mol. Sci.* **2016**, *17*, 1302. [CrossRef] [PubMed]

40. Arai, I.; Amagaya, S.; Komatsu, Y.; Okada, M.; Hayashi, T.; Kasai, M.; Arisawa, M.; Momose, Y. Improving effects of the extracts from *Eugenia uniflora* on hyperglycemia and hypertriglyceridemia in mice. *J. Ethnopharmacol.* **1999**, *68*, 307–314. [CrossRef]

International Journal of
Molecular Sciences

MDPI

Article

Wedelolactone Acts as Proteasome Inhibitor in Breast Cancer Cells

Tereza Nehybová [1,2], Jan Šmarda [1], Lukáš Daniel [2,3], Marek Stiborek [4], Viktor Kanický [4,5], Ivan Spasojevič [6], Jan Preisler [4,5], Jiří Damborský [2,3] and Petr Beneš [1,2,*]

1   Laboratory of Cell Differentiation, Department of Experimental Biology, Faculty of Science, Masaryk University, Kamenice 5, 625 00 Brno, Czech Republic; 322903@mail.muni.cz (T.N.); smarda@sci.muni.cz (J.Š.)
2   International Clinical Research Center, Center for Biological and Cellular Engineering, St. Anne's University Hospital, Pekarska 53, 656 91 Brno, Czech Republic; 211165@mail.muni.cz (L.D.); jiridamborsky0@gmail.com (J.D.)
3   Loschmidt Laboratories, Department of Experimental Biology and Research Centre for Toxic Compounds in the Environment RECETOX, Faculty of Science, Masaryk University, Kamenice 5, 625 00 Brno, Czech Republic
4   Department of Chemistry, Masaryk University, Kamenice 5, 625 00 Brno, Czech Republic; 408516@mail.muni.cz (M.S.); viktor.kanicky@ceitec.muni.cz (V.K.); preisler@chemi.muni.cz (J.P.)
5   CEITEC-Central European Institute of Technology, Masaryk University, Kamenice 5, 625 00 Brno, Czech Republic
6   Department of Life Sciences, Institute for Multidisciplinary Research, University of Belgrade, 11030 Belgrade, Serbia; redoxsci@gmail.com
*   Correspondence: pbenes@sci.muni.cz; Tel.: +420-54949-3125; Fax: +420-54949-5533

Academic Editor: Toshio Morikawa
Received: 1 February 2017; Accepted: 25 March 2017; Published: 29 March 2017

Abstract: Wedelolactone is a multi-target natural plant coumestan exhibiting cytotoxicity towards cancer cells. Although several molecular targets of wedelolactone have been recognized, the molecular mechanism of its cytotoxicity has not yet been elucidated. In this study, we show that wedelolactone acts as an inhibitor of chymotrypsin-like, trypsin-like, and caspase-like activities of proteasome in breast cancer cells. The proteasome inhibitory effect of wedelolactone was documented by (i) reduced cleavage of fluorogenic proteasome substrates; (ii) accumulation of polyubiquitinated proteins and proteins with rapid turnover in tumor cells; and (iii) molecular docking of wedelolactone into the active sites of proteasome catalytic subunits. Inhibition of proteasome by wedelolactone was independent on its ability to induce reactive oxygen species production by redox cycling with copper ions, suggesting that wedelolactone acts as copper-independent proteasome inhibitor. We conclude that the cytotoxicity of wedelolactone to breast cancer cells is partially mediated by targeting proteasomal protein degradation pathway. Understanding the structural basis for inhibitory mode of wedelolactone might help to open up new avenues for design of novel compounds efficiently inhibiting cancer cells.

Keywords: breast cancer; copper; proteasome; reactive oxygen species; wedelolactone

1. Introduction

The ubiquitin-proteasome system (UPS) controls a highly complex and tightly regulated process of cellular protein degradation. In contrast to rather non-specific extracellular/membrane protein degradation by lysosomes, the proteasomes destroy proteins labelled with polyubiquitin chains. UPS consists of numerous protein components (E1, E2, E3 enzymes, proteasome, deubiquitinases). The mammalian cytosolic 26S proteasome complex contains the core 20S proteasome capped with one or two 19S regulatory subunit(s). The proteolytic activities are located in β1, β2, and β5 subunits of the core 20S proteasome complex [1–3]. The eukaryotic proteasome possesses at least three distinct protease activities: chymotrypsin-like (cleavage after hydrophobic residues, located in β5 subunit), trypsin-like (cleavage after basic residues, located in β2 subunit), and caspase-like (cleavage after acidic residues, located in β1 subunit) [4]. The chymotrypsin-like activity is usually the strongest one [5].

Deregulation of UPS has severe effect on cellular function and homeostasis. Over 80% of cellular proteins are degraded via UPS including those involved in regulation of cell proliferation, differentiation, immune signaling, and cell response to stress [3,6]. Emerging evidence show that the targeting of UPS degradation pathway might be a viable anticancer strategy. Due to increased rate of genomic mutations, transformed cells accumulate large quantities of misfolded or overexpressed proteins. In response to such accumulation, malignant cells enhance the expression and activity of UPS [5,7]. Preclinical studies have confirmed a higher susceptibility of malignant cells to cytotoxic effects of UPS inhibitors when compared to normal cells [6,8]. Pertinent to this, several proteasome inhibitors have entered clinical trials, and some have already been approved for the treatment of aggressive hematopoietic tumors [9,10]. However, tumor cell resistance and high toxicity remain an issue in solid tumors leading to search for new UPS inhibitors [5].

Wedelolactone, a natural coumestan, is one of the bioactive compounds found in extracts of *Eclipta alba* and *Wedelia calendulacea* [11]. Recently, in vitro and in vivo anti-cancer properties of wedelolactone in solid tumors including breast, colon, prostate, hepatocellular, pituitary cancers, and neuroblastoma were described in a number of reports [12–19]. Wedelolactone is clearly a multi-target compound and its anti-cancer properties were primarily attributed to the inhibition of multiple kinases, androgen receptor, 5-lipoxygenase, and the c-Myc protein [13,15,17–21]. However, it was found recently that wedelolactone also inhibits topoisomerase IIα activity and blocks DNA synthesis in the breast cancer cells, and that these effects are promoted by copper ions, at least partially via redox interactions [12,22].

This study shows that wedelolactone acts as inhibitor of 20S/26S proteasome chymotrypsin-like and to lesser extent also trypsin-like and caspase-like activities. Treatment of breast cancer cells with wedelolactone resulted in accumulation of ubiquitinated proteins and proteins representing typical proteasomal targets, such as p21, p27, p53, and Bax. Molecular docking revealed a productive binding of wedelolactone to the active sites of β1, β2, and β5 proteasomal subunits with a stronger preference for β5 subunit. The proteasome inhibition by wedelolactone is not dependent on cellular copper level in breast cancer cells. This study concludes that wedelolactone acts as copper-independent inhibitor of proteasome.

2. Results

2.1. Wedelolactone Inhibits Proteolytic Activities of Proteasome in Breast Cancer Cell Lines

MDA-MB-231, MDA-MB-468, and T47D cells were exposed to increasing concentrations of wedelolactone to study its effect on proteasome in breast cancer cells. Chymotrypsin-like, trypsin-like and caspase-like activities of proteasome were evaluated in cell extracts using the activity-specific fluorogenic substrates. Wedelolactone inhibited all three proteolytic activities of proteasome with the highest potency for the chymotrypsin-like activity (IC_{50} values 27.8 μM for MDA-MB-231, 12.78 μM for MDA-MB-468 and 19.45 μM for T47D) (Figure 1).

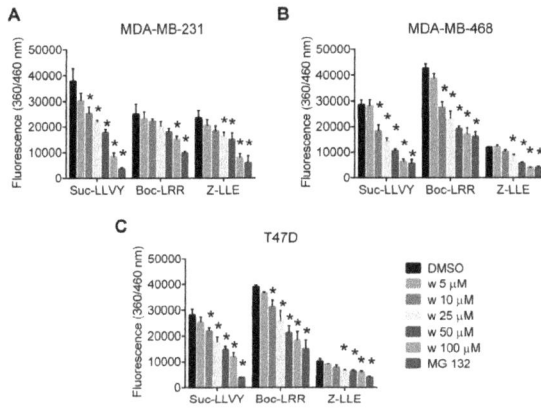

Figure 1. Wedelolactone inhibits chymotrypsin-like, trypsin-like and caspase-like activities in breast cancer cells. MDA-MB-231 (**A**); MDA-MB-468 (**B**); and T47D (**C**) cells were treated with various concentrations of wedelolactone (w) for 10 h. Proteasome activities were evaluated in cell extracts using the activity-specific fluorogenic substrates (Suc-LLVY-AMC for testing chymotrypsin-like, Z-LLE-AMC for caspase-like, and Boc-LRR-AMC for trypsin-like activities). Treatment with MG132 served as a positive control. The data represent the mean values from three independent experiments. Error bars indicate the SD. * indicates a significant ($p < 0.05$) difference between wedelolactone-/MG132- and DMSO-treated cells.

2.2. Wedelolactone Inhibits Proteolytic Activities of Purified 20S and 26S Proteasome Complexes In Vitro

The 26S proteasome purified from MDA-MB-231 cells and the commercially available 20S proteasome were incubated separately with the activity-specific fluorogenic substrates and wedelolactone in various concentrations to evaluate the ability of wedelolactone to inhibit their chymotrypsin-like, trypsin-like, and caspase-like activities. Wedelolactone inhibited all three proteasomal activities in vitro in a dose-dependent manner with the highest potency against the chymotrypsin-like activity (IC_{50} values 9.97 µM for 26S and 6.13 µM for 20S proteasome) (Figure 2).

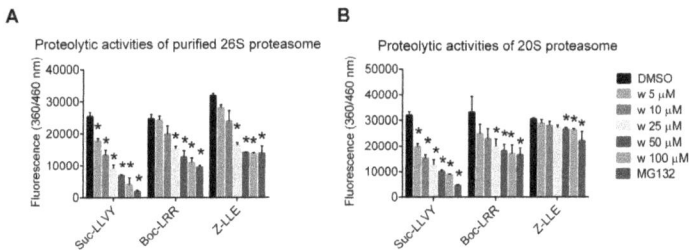

Figure 2. Wedelolactone inhibits chymotrypsin-like, trypsin-like, and caspase-like activities of purified 26S and 20S proteasome complexes in vitro. Wedelolactone (w) was added at various concentrations to reaction mixture containing either (**A**) 26S proteasome purified from MDA-MB-231 cells or (**B**) commercially available 20S proteasome, and fluorogenic substrate (Suc-LLVY-AMC for testing chymotrypsin-like, Z-LLE-AMC for caspase-like, and Boc-LRR-AMC for trypsin-like activities). Fluorescence was measured after 1 h incubation. MG132 was used as a positive control. The data represent the mean values from three independent experiments. Error bars indicate the SD. * indicates a significant ($p < 0.05$) difference in proteolytic activities between reaction mixtures containing wedelolactone/MG132 and DMSO.

2.3. Wedelolactone Causes Accumulation of Polyubiquitinated and Short-Lived Proteins in Breast Cancer Cells

The level of polyubiquitinated proteins and p21, p27, p53, and Bax proteins in wedelolactone-treated MDA-MB-231, MDA-MB-468, and T47D breast cancer cells were analyzed by immunoblotting to further confirm the inhibitory effect of wedelolactone on proteasome. We found the dose-dependent accumulation of p21, p27, p53, Bax as well as multiubiquitinated proteins occurring in all three cell lines tested (Figure 3A). To confirm that the wedelolactone-induced accumulation of p21, p27, p53, and Bax proteins is not caused by increased rate of their transcription/expression, the transcripts of corresponding genes were quantified using quantitative polymerase chain reaction (qPCR). We found that wedelolactone did not affect expression of any of these genes in all three breast cancer cell lines (Figure 3B).

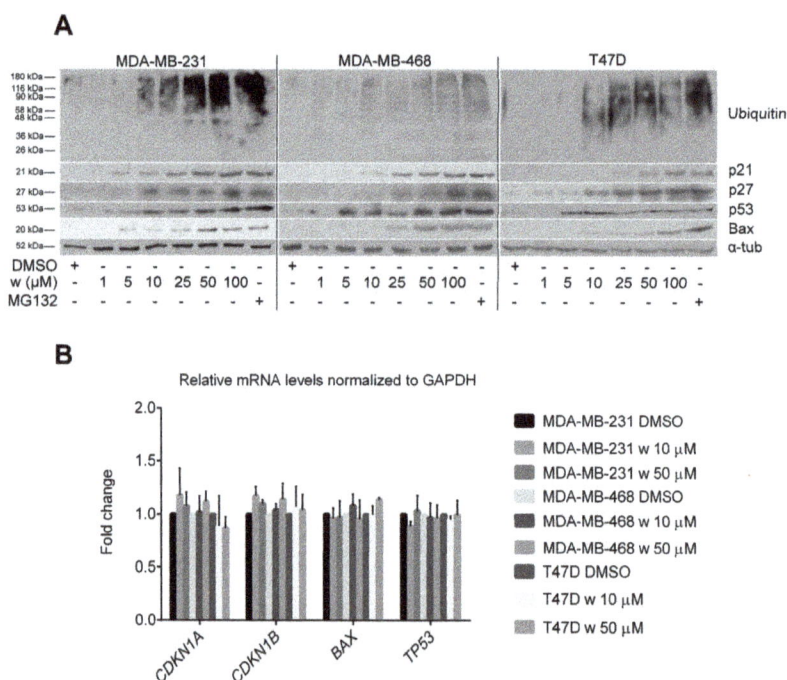

Figure 3. Multiubiquitinated and high turnover proteins accumulate in wedelolactone-treated breast cancer cells. MDA-MB-231, MDA-MB-468, and T47D cells were treated with various concentrations of wedelolactone or solvent for 10 h. (**A**) Protein extracts were subsequently analyzed by SDS-PAGE and immunoblotting using p21-, p27-, p53-, Bax-, and ubiquitin-specific antibodies. Treatment with MG132 served as a positive control; (**B**) Transcripts of *CDKN1A*, *CDKN1B*, *TP53* and *BAX* genes were quantified using qPCR.

2.4. Cytotoxicity of Wedelolactone Increases for Cells with High Content of Intracellular Copper

The authors of this study suggested previously that cytotoxicity of wedelolactone can be at least partly explained by redox-cycling with copper ions, reactive oxygen species (ROS) generation and promoted oxidative stress [22]. To confirm the role of copper ions in cytotoxicity of wedelolactone, breast cancer cells were transiently transfected with plasmid coding for human copper transporter *CTR1*. The transfection efficiency determined by flow-cytometry was $65.6\% \pm 3.7\%$, $43.6\% \pm 4.2\%$, and $46.6\% \pm 3.8\%$ in MDA-MB-231, MDA-MB-468, and T47D cells, respectively (Figure S1). Expression of exogenous CTR1 protein was confirmed by immunoblotting (Figure 4A). Transfected cells were exposed to copper sulfate for 24 h or left untreated. Analysis of relative copper concentrations in cell lysates revealed that only combination of *CTR1* overexpression/copper supplementation efficiently increases copper-loading of all three cell lines (Figure 4B). Therefore, for next set of experiments, copper-loaded cells were prepared by simultaneous *CTR1* overexpression/copper supplementation. Copper-loaded and control cells (both over-expressing exogenous *CTR1*) were exposed to wedelolactone or DMSO for 48 h and their mortality was assessed by PI exclusion assay using flow-cytometry. In agreement with our hypothesis, cytotoxicity induced by wedelolactone was enhanced by copper loading (Figure 4C, Figure S2). Furthermore, to analyze whether copper loading enhances the wedelolactone-induced ROS production, copper-loaded and control cells (both over-expressing exogenous *CTR1*) were treated with wedelolactone or solvent for 10 h and ROS production was analyzed after DHE staining by flow-cytometry. Copper-loaded cells produced more ROS in response to wedelolactone than controls (Figure 4D, Figure S3). No significant differences in cell mortality or ROS production was observed in mock-transfected cells that were either pre-incubated with copper or left untreated and subsequently exposed to wedelolactone (Figure S4). It is hypothesized that intracellular level of copper did not reach the required threshold in this case. These results support previous findings that cytotoxicity of wedelolactone is at least partially mediated via (redox) interactions with copper ions.

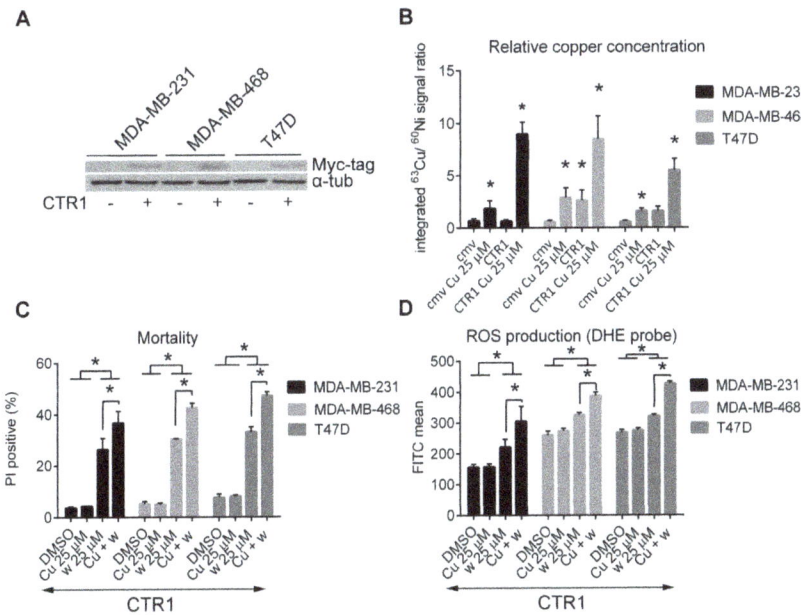

Figure 4. Copper loading enhances cytotoxicity and ROS production in wedelolactone-treated breast cancer cells. MDA-MB-231, MDA-MB-468, and T47D cells were transiently transfected with pCNDA3.1-hCTR1-N-Myc (CTR1) or control pCDNA3.1 plasmids (cmv), pretreated with copper sulfate (Cu) for 24 h and subsequently treated with wedelolactone (w) or solvent (DMSO) in fresh media. (**A**) Cells were harvested and expression of exogenous CTR1 protein was confirmed by SDS-PAGE followed by immunoblotting with the Myc-Tag antibody; (**B**) Relative copper concentration in cell lysates was analyzed by SALD ICP MS. Data for copper are presented as an integrated $^{63}Cu/^{60}Ni$ signal ratio. (**C**) Cell mortality and (**D**) ROS production were evaluated after PI/DHE staining using flow-cytometry. The data represent the mean values from three independent experiments. Error bars indicate the SD. * indicates a significant ($p < 0.05$) difference.

2.5. Copper Does Not Affect the Inhibition of Proteasome Activity by Wedelolactone in Breast Cancer Cell Lines

There are several copper-interacting compounds that have been shown to inhibit proteasome [23,24]. To analyze whether the inhibitory effect of wedelolactone on proteolytic activities of proteasome is also mediated by copper, we compared chymotrypsin-like, trypsin-like and caspase-like proteolytic activities of proteasome in copper-loaded and control breast cancer cells (both overexpressing exogenous *CTR1*). The inhibition of all three proteolytic activities by wedelolactone were found to be similar in copper-loaded and control cells (Figure 5), suggesting that the inhibition of proteasome activities by wedelolactone is a copper-independent process. No significant differences in proteolytic activities were observed in mock-transfected cells that were either pre-incubated with copper or left untreated and subsequently exposed to wedelolactone (Figure S5).

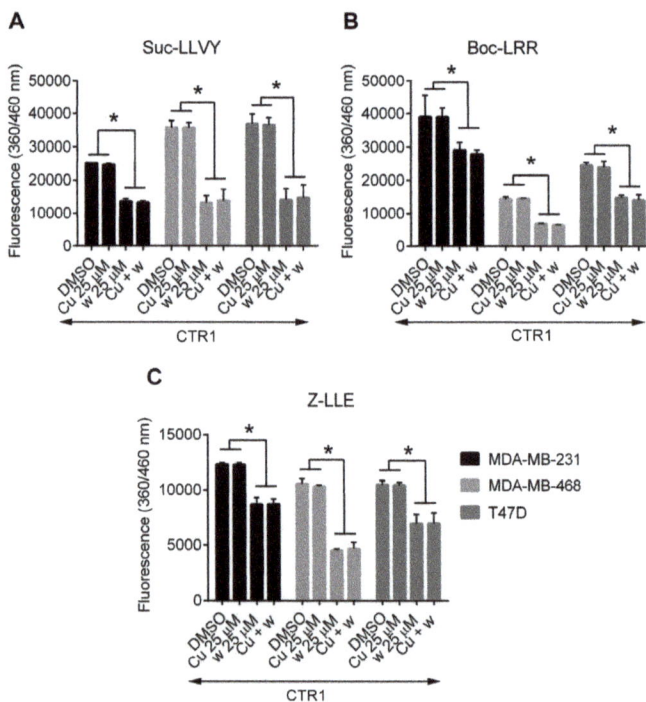

Figure 5. Copper loading does not affect inhibition of proteasome by wedelolactone. Cells were transfected with the pCNDA3.1-hCTR1-N-Myc plasmid, loaded with copper (Cu) or left untreated and exposed to various concentrations of wedelolactone (w) or solvent (DMSO) for 10 h. Proteasome activities were evaluated in cell extracts using the activity-specific fluorogenic substrates (Suc-LLVY-AMC for testing chymotrypsin-like (**A**), Z-LLE-AMC for caspase-like (**B**), and Boc-LRR-AMC for trypsin-like activities (**C**)). The data represent the mean values from three independent experiments. Error bars indicate the SD. * indicates a significant ($p < 0.05$) difference.

2.6. Molecular Docking of Wedelolactone to the Active Sites of Proteasome

To reveal the mechanism of proteasome inhibition by wedelolactone, in silico docking analysis was performed. Since the functional units of proteasome are located at each of the inner β rings, the blind docking was initially performed to assess the specificity of wedelolactone to the active sites. This analysis revealed that wedelolactone occupied the active site of the β5 subunit at least three times more often than the active sites in other units of the protein.

The focused docking identified similar binding mode of wedelolactone in the β1 and β2 subunits. Aside from the hydrophobic contacts, wedelolactone formed specific H-bond with the backbone of Thr21 and Gly47 in both subunits. Moreover, wedelolactone formed an additional H-bond with the side-chain of Thr20 in the β1 subunit. Wedelolactone sterically blocked the catalytic residue Thr1 in both β1 and β2 subunits, possibly modulating its proteolytic activity (Figure 6A,B). The focused docking identified a different binding mode of wedelolactone in the β5 subunit. Aside from the specific H-bond with the backbone of Gly47, two hydroxyl groups of wedelolactone were able to form H-bond with the catalytic residue Thr1 (Figure 6C). This specific interaction might be responsible for the preferred binding to the β5 active site over the β1 and β2 active sites, observed in the blind docking. Beside the steric hindrance, the interaction with the catalytic residue Thr1 might be responsible for the elevated inhibition of chymotrypsin-like activity observed experimentally.

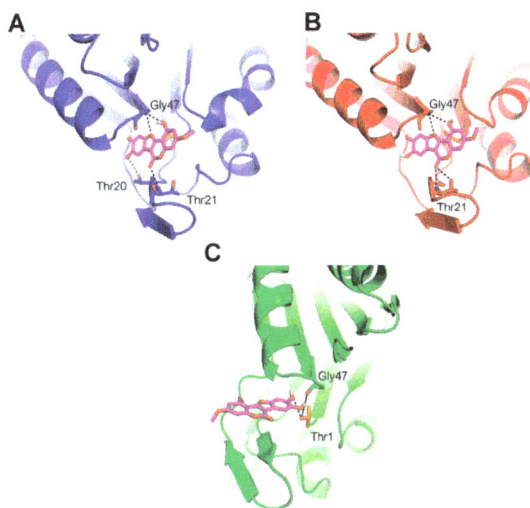

Figure 6. The binding modes of wedelolactone in the β1 (**A**), β2 (**B**), and β5 (**C**) subunits of the yeast 20S proteasome. Wedelolactone is represented as magenta sticks, β1, β2, and β5 subunits are represented as blue, red, and green, respectively. Dashed lines represent the specific H-bonds to the active sites residues. The catalytic residue Thr1 is represented as orange sticks.

3. Discussion

Wedelolactone is a natural polyphenolic catechol-type compound with anti-cancer effects that are exerted via multiple mechanisms/targets [12–21]. In our previous studies, we reported that cytotoxic effect of wedelolactone can be partially attributed to its pro-oxidative and DNA damage activity that is promoted by copper ions. Such activities most likely involve production of ROS and (semi)quinones [22] and were previously described for other polyphenolic compounds [25–27]. Very recently, copper-mediated cytotoxicity was confirmed also for coumestrol, another coumestan with structure similar to that of wedelolactone [28]. In accordance with this, cytotoxicity of wedelolactone was enhanced here by copper loading in three breast cancer cell lines. Previously, quinones and ROS, formed by oxidative metabolism of catechol-type polyphenol dopamine, were reported to act as proteasome inhibitors [29]. We found that copper overloading significantly enhanced wedelolactone-induced ROS production and cytotoxicity but it did not further enhance its proteasome inhibitory properties, suggesting the copper-independent mechanism of proteasome inhibition by wedelolactone.

The structural basis of proteasome inhibition by wedelolactone was subsequently revealed by molecular docking. Molecular structure of the core 20S proteasome is extremely conserved and is organized in four stacked rings, each formed by seven subunits in an $\alpha7\beta7\beta7\alpha7$ configuration. Seven distinct β subunits are carrying the enzyme active sites, specifically $\beta1$ carries caspase-like activity, $\beta2$ is responsible for trypsin-like activity and $\beta5$ encodes chymotrypsin-like activity [30]. Protein degradation is facilitated by nucleophilic N-terminal threonine (Thr1) residues of catalytic β subunits, in which the side chain hydroxyl group reacts with peptide bonds of substrates as well as functional groups of inhibitors [31]. Inhibitors of the 20S proteasome can be divided into two main groups based on whether or not they form a covalent bond with the active site Thr1 according to classification proposed by Kisselev et al. [10]. Molecular docking revealed that wedelolactone occupies the active sites of $\beta1$, $\beta2$, and $\beta5$ proteasomal subunits. While similar binding mode was predicted for $\beta1$ and $\beta2$ subunit, a specific interaction between both hydroxyl groups of wedelolactone with catalytic residue Thr1 was observed only in $\beta5$ subunit. These differences in binding modes are probably responsible also for predicted favored interaction with $\beta5$ subunit and might explain a stronger inhibition of proteasomal chymotrypsin-like activity compared to trypsin- and caspase-like activities.

The observed IC_{50} values for chymotrypsin-like inhibitory activity of wedelolactone were below 10 μM in vitro and within 10–25 μM range in cells. It is noteworthy that wedelolactone induced growth arrest and apoptosis in all three breast cancer cell line tested at concentrations corresponding to the above mentioned IC_{50} values [12,22]. This suggests that inhibition of proteasome may contribute significantly to cytotoxicity of this compound.

Inhibition of proteasome results in increased levels of polyubiquitinated proteins because most of the proteasome-mediated protein degradation pathways require ubiquitination [32]. Moreover proteins with high turnover, including p21, p27, p53, and Bax accumulates in cells in response to proteasomal inhibition [33–35]. Such accumulation was clearly documented here in wedelolactone-treated cells. It is important to note that some previous studies showing connection between the treatment with wedelolactone and altered expression of numerous proteins should be interpreted with caution as wedelolactone can affect not only protein expression but also protein degradation pathway.

This study concluded that natural coumestan wedelolactone acts as a copper-independent proteasome inhibitor with potency similar to other flavonoids. As cancer cells were reported to be more sensitive to proteasome inhibition, this novel function of wedelolactone might explain its preferred toxicity towards cancer cells observed previously [20]. Understanding a structural basis for inhibitory mode of wedelolactone might help to open up new avenues for design of novel compounds efficiently inhibiting cancer cells.

4. Material and Methods

4.1. Chemicals and Plasmids

Chemicals were obtained from commercial providers: wedelolactone, dimethyl sulfoxide (DMSO), propidium iodide (PI), and copper sulfate (Sigma-Aldrich, St. Louis, MO, USA), dihydroethidium (DHE; Cayman Pharma, Ann Arbor, MI, USA), Proteasome Activity Fluorometric Assay Kit II (UPBio, Aurora, CO, USA). The pCNDA3.1-hCTR1-N-Myc plasmid was kindly provided by Dennis J. Thiele [36].

4.2. Cell Culture

The human breast cancer cell lines MDA-MB-231, MDA-MB-468, and T47D were cultured in HEPES-modified RPMI 1640 medium (Sigma-Aldrich) supplemented with 10% fetal calf serum (FCS, Sigma-Aldrich), 2 mM L-glutamine, 100 U/mL penicillin, and 100 μg/mL streptomycin (Lonza, Verviers, Belgium) in a humidified atmosphere of 5% CO_2 at 37 °C. In all experiments, wedelolactone was applied at concentrations that have been shown previously to effectively induce cell death in breast cancer cell lines [12,22].

4.3. Proteasome Activity Assay

4.3.1. Purification of 26S Proteasome from MDA-MB-231 Cells

Human 26S proteasome was purified from 8×10^6 of MDA-MB-231 cells using The Rapid 26S Proteasome Purification Kit (J4310, UBPBio) according to manufacturer's instructions.

4.3.2. Proteasome Activity In Vitro

Chymotrypsin-like, trypsin-like and caspase-like proteasome activities were determined using Proteasome Activity Fluorometric Assay Kit II (J4120, UBPBio) according to manufacturer's instructions. Briefly, wedelolactone was added at various concentrations to 150 μL reaction mixture containing either purified 10 nM bovine 20S proteasome (A1400, UBPBio) or 5 μg of MDA-MB-231-purified 26S proteasome complex, and 50 μM fluorogenic substrate (Suc-LLVY-AMC to test chymotrypsin-like activity, Z-LLE-AMC to test caspase-like activity and Boc-LRR-AMC to test trypsin-like activity) in 1× Proteasome Assay Buffer (40 mM Tris, pH 7.1, 2 mM β-mercaptoethanol; UBPBio). MG132 at concentration of 10 μM and aliquots of DMSO were used as positive and negative controls, respectively. Fluorescence was measured by TECAN infinite 200 plate reader (TECAN, Mannedorf, Switzerland) for 1 h at 37 °C.

4.3.3. Proteasome Activity in Cancer Cell Lines

MDA-MB-231, MDA-MB-468 and T47D (6×10^5) cells were seeded in 5 mL of culture media, exposed to various concentrations of wedelolactone, DMSO or 10 μM MG132 for 10 h. Cells were harvested, resuspended in cell lysis buffer (40 mM Tris, pH 7.2, 50 mM NaCl, 2 mM β-mercaptoethanol, 2 mM ATP, 5 mM MgCl, 10% Glycerol) and briefly sonicated using an Ultrasonic Processor UP100H (Hielscher, Ringwood, NJ, USA). Cell lysates were cleared by centrifugation and protein concentration in supernatant was determined using DC protein assay (Biorad, Hercules, CA, USA). Protein extract (50 μg) was mixed with 50 μM of fluorogenic substrates (UBPBio) in 1× Proteasome Assay Buffer in a total volume of 100 μL. Fluorescence was measured by TECAN infinite 200 plate reader (TECAN) for 1 h at 37 °C.

4.3.4. Proteasome Activity after Copper-Overloading

MDA-MB-231, MDA-MB-468, and T47D cells (6×10^5) were seeded in 5 mL of culture media. Next day, transient transfection was performed, using 4 μL of Lipofectamine LTX reagent (Invitrogen, Carlsbad, CA, USA) with a mixture containing 2 μg of pCDNA3.1-hCTR1-N-Myc or control pCDNA3.1 plasmid and 2 μL of PLUS reagent (Invitrogen). Six hours later, the medium was replaced, cells were treated with 25 μM copper sulfate or left untreated for 24 h, and then were exposed to wedelolactone or DMSO for 10 h in fresh media. Cells were then harvested and proteasome activity was analyzed as described in 4.3.3.

4.4. Immunoblotting

5×10^5 cells were seeded in 6-well plates. The next day, the cells were exposed to various concentrations of wedelolactone, DMSO, or 10 μM MG132 for 10 h. Cells were harvested and lysed as described previously [37]. Cell lysates were subjected to SDS–PAGE and immunoblotted. Sample loading was normalized according to protein concentration determined by DC protein assay (Biorad). Blots were probed with anti-ubiquitin (3933S; Cell Signaling Technology, Inc., Beverly, MA, USA), anti-p21, anti-p27 (sc-817 and sc-528; Santa Cruz Biotechnology Inc., Santa Cruz, CA, USA), anti-p53, anti-Bax or anti-Myc-Tag (9282, 5023 and 2276S; Cell Signaling Technology), anti-α-tubulin antibodies (T9026; Sigma-Aldrich), and secondary antibodies conjugated with peroxidase (Sigma-Aldrich). Blots were developed with a standard ECL procedure with Immobilon Western Chemiluminiscent HRP Substrate (Millipore, Billerica, MA, USA).

4.5. RNA Isolation, cDNA Synthesis and qPCR

1×10^6 cells were seeded in 5 mL dishes. Next day, the cells were exposed to various concentrations of wedelolactone or DMSO for 10 h. Cells were harvested and total RNA was isolated using GenElute Mammalian Total RNA Miniprep kit (Sigma-Aldrich). For cDNA synthesis, 1 µg of total RNA was reverse-transcribed using the QuantiTect Reverse Transcription kit (Qiagen, Hilden, Germany) according to the manufacturer's instructions in a final reaction volume of 20 µL. Expression of *CDKN1A*, *CDKN1B*, *TP53* and *BAX* genes was determined using the target-specific primers (Table S1) and KAPA SYBR FAST qPCR MASTER MIX (KK460, Kapa Biosystems, Cambridge, MA, USA) on LightCycler 480 II (Roche, Basel, Switzerland). Expression of the reference *GAPDH* gene (probe 4326317E, ThermoFisher Scientific, Waltham, MA, USA) was used for data normalization.

4.6. Cell Mortality

3×10^5 cells were seeded in 6-well plates. After 24 h, transient transfection was performed, using 3 µL of Lipofectamine LTX reagent (Invitrogen) with a mixture containing 1.5 µg of pCNDA3.1-hCTR1-N-Myc or control pCDNA3.1 plasmid and 1.5 µL of PLUS reagent (Invitrogen). Six hours later, the medium was replaced, cells were either exposed to 25 µM copper sulfate for 24 h or left untreated and subsequently subjected to wedelolactone or DMSO for 48 h in fresh media. Cytotoxicity of wedelolactone was analyzed 48 h later by PI staining (1 µg/mL) using flow-cytometry as described previously [38].

4.7. Reactive Oxygen Species Production Analysis

3×10^5 cells were seeded in 6-well plates. After 24 h, transient transfection was performed as described in chapter 4.6. Medium was replaced after 6 h, cells were left untreated or pretreated with 25 µM copper sulfate for 24 h, and subsequently exposed to wedelolactone or DMSO for 10 h in fresh media. The cells were washed with PBS and stained with 10 µM DHE for 20 min at 37 °C in the dark. Reactive oxygen species (ROS) were measured using flow-cytometry (BD FACSVerse, BD Biosciences, Franklin Lakes, NJ, USA) at an excitation wavelength of 485 nm and an emission wavelength of 538 nm. Data were analyzed using BD FACSuite software (BD Biosciences).

4.8. Analysis of Copper Concentrations in Cells

For detection of intracellular copper concentrations, 6×10^5 cells were seeded in 5 mL of growth medium. The next day, the cells were transiently transfected with CTR1 or control pCDNA3.1 vector. Transfection was performed using 4 µL of the Lipofectamine LTX reagent (Invitrogen) with a mixture containing 2 µg of plasmid and 2 µL of PLUS reagent (Invitrogen). Six hours later, the medium was replaced with fresh one and cells were pretreated with 25 µM copper sulfate or left untreated for 24 h. Then, the cells were exposed to 25 µM wedelolactone or DMSO for 10 h in fresh media. Cells were harvested and cell lysates prepared as described previously [39]. Briefly, pelleted cells were washed twice with $1 \times$ PBS and 1.0×10^6 cell were lysed in a mixture of 3 M HCl/10% trichloroacetic acid at room temperature for 3 h followed by incubation at 70 °C for 5 h. The lysate was centrifuged (600 g/5 min) to remove cell debris and the total amount of copper in supernatant was determined by substrate-assisted laser desorption inductively-coupled plasma mass spectrometry (SALD ICP MS) [40].

Each sample was mixed with 400 µg/L aqueous solution containing nickel as an internal standard (ASTASOL-®Ni, CRM, ANALYTIKA®, Prague, Czech Republic) in the 1:1 ratio and spotted by a micropipette onto a polyethylene terephthalate plate (PET) as a 200 nL droplet in seven replicates. The sample plate was inserted into an ablation system (model UP 213, New Wave, Fremont, CA, USA) and spots were scanned by an Nd:YAG 213 nm laser beam in a zig-zag shaped raster with the raster spacing 190 µm; the laser beam waist was adjusted to the size ~250 µm. Size of the raster was selected according to the spot diameter to desorb the entire sample (typical spot diameter ~1.4 mm), and the analysis time of each sample was approximately 2 min. The ablation cell was flushed with a carrier gas (helium, flow rate 1.0 L/min), which transported the aerosol to an ICP mass spectrometer (model 7500CE, Agilent Technologies, Santa Clara, CA, USA). A sample gas flow of argon was admixed to the helium carrier gas flow subsequent to the laser ablation cell (0.6 L/min). Optimization of LA ICP MS conditions (gas flow rates, sampling depth, electrostatic lens voltages of the MS) was performed with the glass reference material NIST SRM 612 regarding the maximum signal-to-noise ratio and minimum oxide formation (ThO^+/Th^+ counts ratio 0.2%, U^+/Th^+ counts ratio 1.1%). Other ICP MS parameters were adjusted in compliance with the manufacturer's recommendations. The laser fluence was ~0.75 J/cm^2, the repetition rate 10 Hz, and the scan rate 160 µm/s. The ions were measured with an integration time 0.1 s. Both the flush time and the laser warm-up time were set to 10 s. The ion signal of two copper isotopes, ^{63}Cu and ^{65}Cu, was monitored to reveal possible polyatomic interferences. The signal ratio of the most abundant isotopes, ^{63}Cu and ^{60}Ni as the internal standard was used for data evaluation.

4.9. Molecular Docking

The three-dimensional structure of wedelolactone was downloaded from ZINC database [41], (ZINC ID: ZINC6483512). The output file in Sybyl mol2 format was converted into AutoDock Vina [42] compliant pdbqt format by MGLTools [43]. The crystal structure of yeast 20S proteasome (PDB ID: 5CZ4) was used as a target in molecular docking. All ligands and water molecules were removed from the target molecule. The hydrogen atoms were added to the target by PyMol [44]. The Gasteiger charges and AutoDock atom types were assigned to targets by MGLTools. The active site of β1 (caspase-like activity), β2 (trypsin-like activity) and β5 (chymotrypsin-like activity) subunits and both inner β rings were selected as target regions for molecular docking performed by AutoDock Vina. The region selected for focused docking was represented by a box of 22.5 Å × 22.5 Å × 22.5 Å centered at the catalytic residue Thr1. The entire protein surface was selected for a blind docking to assess the specificity of wedelolactone towards the enzyme active sites. The region selected for the blind docking was represented by a box with 87.5 Å × 87.5 Å × 87.5 Å dimension centered at the middle of the two inner β rings harboring the active sites. Ten and twenty conformations were produced by AutoDock Vina in the focused and blind docking, respectively. The docked conformations were re-scored by NNScore 2.0 [45], which predicts binding affinity of the conformation as an average over 20 distinct neural-networks.

4.10. Statistics

Values were expressed as means ± standard deviations (SD). To determine statistical significance, the values were compared using a two-tailed *t*-test for unpaired samples. Differences were considered to be statistically significant with the *p*-value < 0.05. IC$_{50}$ values were determined by nonlinear regression using GraphPad PRISM 6 software (GraphPad-San Diego, CA, USA). All results were reproduced at least in three independent experiments.

Supplementary Materials: Supplementary materials can be found at www.mdpi.com/1422-0067/18/4/729/s1.

Acknowledgments: This work was funded by the projects No. LQ1601, LQ1605 and LO1214 from the National Program of Sustainability II (MEYS CR), 15-05387S from the Czech Science Foundation, and MUNI/A/0967/2015 from Masaryk University. We thank Lucia Knopfova for help with manuscript preparation. The funders had no role in study design, data collection and analysis, decision to publish, or preparation of the manuscript.

Author Contributions: Tereza Nehybová, Petr Beneš, and Ivan Spasojevič conceived the project; Tereza Nehybová, Marek Stiborek, and Lukáš Daniel performed the experiments; Tereza Nehybová, Petr Beneš, Lukáš Daniel, Marek Stiborek, Jiří Damborský, and Jan Preisler analyzed the data; Viktor Kanický and Jan Šmarda supported the project with experimental techniques; Tereza Nehybová, Petr Beneš, Ivan Spasojevič, Jiří Damborský, Jan Šmarda, and Jan Preisler wrote and revised the manuscript; Petr Beneš, Jan Šmarda, Jan Preisler, and Jiří Damborský supervised the study. All authors read and approved the final manuscript.

Conflicts of Interest: The authors declare no conflict of interest.

Abbreviations

DHE	Dihydroethidium
DMSO	Dimethyl sulfoxide
ECL	Enhanced chemiluminescence
FCS	Fetal calf serum
CTR1	Human copper transporter 1
PDB	Protein data bank
PI	Propidium iodide
ROS	Reactive oxygen species
SALD ICP MS	Substrate-assisted laser desorption inductively-coupled plasma mass spectrometry
SDS-PAGE	Sodium dodecyl sulfate polyacrylamide gel electrophoresis
qPCR	Quantitative polymerase chain reaction
UPS	Ubiquitin-proteasome system

References

1. Groll, M.; Heinemeyer, W.; Jäger, S.; Ullrich, T.; Bochtler, M.; Wolf, D.H.; Huber, R. The catalytic sites of 20S proteasomes and their role in subunit maturation: A mutational and crystallographic study. *Proc. Natl. Acad. Sci. USA* **1999**, *96*, 10976–10983. [CrossRef] [PubMed]
2. Heinemeyer, W.; Fischer, M.; Krimmer, T.; Stachon, U.; Wolf, D.H. The active sites of the eukaryotic 20S proteasome and their involvement in subunit precursor processing. *J. Biol. Chem.* **1997**, *272*, 25200–25209. [CrossRef] [PubMed]
3. Shen, M.; Schmitt, S.; Buac, D.; Duo, Q.P. Targeting the ubiquitin–proteasome system for cancer therapy. *Expert Opin. Ther. Targets* **2013**, *17*, 1091–1108. [CrossRef] [PubMed]
4. Groll, M.; Huber, R. Purification, crystallization, and X-ray analysis of the yeast 20S proteasome. *Methods Enzymol.* **2005**, *398*, 329–336. [PubMed]
5. Grigoreva, T.A.; Tribulovich, V.G.; Garabadzhiu, A.V.; Melino, G.; Barlev, N.A. The 26S proteasome is a multifaceted target for anti-cancer therapies. *Oncotarget* **2015**, *6*, 24733–24749. [CrossRef] [PubMed]
6. Crawford, L.J.; Walker, B.; Irvine, A.E. Proteasome inhibitors in cancer therapy. *J. Cell Commun. Signal.* **2011**, *5*, 101–110. [PubMed]
7. Arlt, A.; Bauer, I.; Schafmayer, C.; Tepel, J.; Müerköster, S.S.; Brosch, M.; Röder, C.; Kalthoff, H.; Hampe, J.; Moyer, M.P.; et al. Increased proteasome subunit protein expression and proteasome activity in colon cancer relate to an enhanced activation of nuclear factor E2-related factor 2 (Nrf2). *Oncogene* **2009**, *28*, 3983–3996. [CrossRef] [PubMed]
8. Rajkumar, S.V.; Richardson, P.G.; Hideshima, T.; Anderson, K.C. Proteasome Inhibition As a Novel Therapeutic Target in Human Cancer. *J. Cell. Oncol.* **2005**, *23*, 630–639. [CrossRef] [PubMed]
9. Kubiczkova, L.; Pour, L.; Sedlarikova, L.; Hajek, R.; Sevcikova, S. Proteasome inhibitors—Molecular basis and current perspectives in multiple myeloma. *J. Cell. Mol. Med.* **2014**, *18*, 947–961. [CrossRef] [PubMed]
10. Kisselev, A.F.; van der Linden, W.A.; Overkleeft, H.S. Proteasome inhibitors: An expanding army attacking a unique target. *Chem. Biol.* **2012**, *19*, 99–115. [CrossRef] [PubMed]
11. Wagner, H.; Geyer, B.; Kiso, Y.; Hikino, H.; Rao, G.S. Coumestans as the main active principles of the liver drugs Eclipta alba and Wedelia calendulacea. *Planta Med.* **1986**, *5*, 370–374. [CrossRef]
12. Benes, P.; Knopfova, L.; Trcka, F.; Nemajerova, A.; Pinheiro, D.; Soucek, K.; Fojta, M.; Smarda, J. Inhibition of topoisomerase IIα: Novel function of wedelolactone. *Cancer Lett.* **2011**, *303*, 29–38. [CrossRef] [PubMed]
13. Chen, Z.; Sun, X.; Shen, S.; Zhang, H.; Ma, X.; Liu, J.; Kuang, S.; Yu, Q. Wedelolactone a naturally occurring coumestan, enhances interferon-G signaling through inhibiting STAT1 protein dephosphorylation. *J. Biol. Chem.* **2013**, *288*, 14417–14427. [CrossRef] [PubMed]
14. Idris, A.I.; Libouban, H.; Nyangoga, H.; Landao-Bassonga, E.; Chappard, D.; Ralston, S.H. Pharmacologic inhibitors of IkappaB kinase suppress growth and migration of mammary carcinosarcoma cells in vitro and prevent osteolytic bone metastasis in vivo. *Mol. Cancer Ther.* **2009**, *8*, 2339–2347. [CrossRef] [PubMed]
15. Lin, F.M.; Chen, L.R.; Lin, E.H.; Ke, F.C.; Chen, H.Y.; Tsai, M.J.; Hsiao, P.W. Compounds from Wedelia chinensis synergistically suppress androgen activity and growth in prostate cancer cells. *Carcinogenesis* **2007**, *28*, 2521–2529. [CrossRef] [PubMed]

16. Nehybova, T.; Smarda, J.; Benes, P. Plant coumestans: Recent advances and future perspectives in cancer therapy. *Anticancer Agents Med. Chem.* **2014**, *14*, 1351–1362. [CrossRef] [PubMed]

17. Sukumari-Ramesh, S.; Bentley, J.N.; Laird, M.D.; Singh, N.; Vender, J.R.; Dhandapani, K.M. Dietary phytochemicals induce p53- and caspase-independent cell death in human neuroblastoma cells. *Int. J. Dev. Neurosci.* **2011**, *29*, 701–710. [CrossRef] [PubMed]

18. Tsai, C.H.; Lin, F.M.; Yang, Y.C.; Lee, M.T.; Cha, T.L.; Wu, G.J.; Hsieh, S.C.; Hsiao, P.W. Herbal extract of Wedelia chinensis attenuates androgen receptor activity and orthotopic growth of prostate cancer in nude mice. *Clin. Cancer Res.* **2009**, *15*, 5435–5444. [CrossRef] [PubMed]

19. Vender, J.R.; Laird, M.D.; Dhandapani, K.M. Inhibition of NFκB reduces cellular viability in GH3 pituitary adenoma cells. *Neurosurgery* **2008**, *62*, 1122–1127. [CrossRef] [PubMed]

20. Sarveswaran, S.; Gautam, S.C.; Ghosh, J. Wedelolactone a medicinal plant-derived coumestan, induces caspase-dependent apoptosis in prostate cancer cells via downregulation of PKCe without inhibiting Akt. *Int. J. Oncol.* **2012**, *41*, 2191–2199. [PubMed]

21. Sarweswaran, S.; Ghosh, R.; Parikh, R.; Ghosh, J. Wedelolactone, an Anti-inflammatory Botanical, Interrupts c-Myc Oncogenic Signaling and Synergizes with Enzalutamide to Induce Apoptosis in Prostate Cancer Cells. *Mol. Cancer Ther.* **2016**, *15*, 2791–2801. [CrossRef] [PubMed]

22. Benes, P.; Alexova, P.; Knopfova, L.; Spanova, A.; Smarda, J. Redox state alters anti-cancer effects of wedelolactone. *Environ. Mol. Mutagen.* **2012**, *53*, 515–524. [CrossRef] [PubMed]

23. Daniel, K.G.; Gupta, P.; Harbach, R.H.; Guida, W.C.; Dou, Q.P. Organic copper complexes as a new class of proteasome inhibitors and apoptosis inducers in human cancer cells. *Biochem. Pharmacol.* **2004**, *67*, 1139–1151. [CrossRef] [PubMed]

24. Ding, W.Q.; Liu, B.; Vaught, J.L.; Yamauchi, H.; Lind, S.E. Anticancer activity of the antibiotic clioquinol. *Cancer Res.* **2005**, *65*, 3389–3395. [PubMed]

25. Arif, H.; Rehmani, N.; Farhan, M.; Ahmad, A.; Hadi, S.M. Mobilization of Copper ions by Flavonoids in Human Peripheral Lymphocytes Leads to Oxidative DNA Breakage: A Structure Activity Study. *Int. J. Mol. Sci.* **2015**, *16*, 26754–26769. [CrossRef] [PubMed]

26. Farhan, M.; Khan, H.Y.; Oves, M.; Al-Harrasi, A.; Rehmani, N.; Arif, H.; Hadi, S.M.; Ahmad, A. Cancer Therapy by Catechins Involves Redox Cycling of Copper Ions and Generation of Reactive Oxygen species. *Toxins* **2016**, *8*, 37. [CrossRef] [PubMed]

27. Khan, H.Y.; Zubair, H.; Faisal, M.; Ullah, M.F.; Farhan, M.; Sarkar, F.H.; Ahmad, A.; Hadi, S.M. Plant polyphenol induced cell death in human cancer cells involves mobilization of intracellular copper ions and reactive oxygen species generation: A mechanism for cancer chemopreventive action. *Mol. Nutr. Food Res.* **2014**, *58*, 437–446. [CrossRef] [PubMed]

28. Zafar, A.; Singh, S.; Naseem, I. Cytotoxic activity of soy phytoestrogen coumestrol against human breast cancer MCF-7 cells: Insights into the molecular mechanism. *Food Chem. Toxicol.* **2017**, *99*, 149–161. [CrossRef] [PubMed]

29. Zhou, Z.D.; Lim, T.M. Dopamine (DA) induced irreversible proteasome inhibition via DA derived quinones. *Free Radic. Res.* **2009**, *43*, 417–430. [CrossRef] [PubMed]

30. Adams, J. The proteasome: Structure, function, and role in the cell. *Cancer Treat. Rev.* **2003**, *29*, 3–9. [CrossRef]

31. Moore, B.S.; Eustáquio, A.S.; McGlinchey, R.P. Advances in and applications of proteasome inhibitors. *Curr. Opin. Chem. Biol.* **2008**, *12*, 434–440. [CrossRef] [PubMed]

32. Hochstrasser, M. Ubiquitin, proteasomes, and the regulation of intracellular protein degradation. *Curr. Opin. Cell Biol.* **1995**, *7*, 215–223. [CrossRef]

33. Bae, S.H.; Ryoo, H.M.; Kim, M.K.; Lee, K.H.; Sin, J.I.; Hyun, M.S. Effects of the proteasome inhibitor bortezomib alone and in combination with chemotherapeutic agents in gastric cancer cell lines. *Oncol. Rep.* **2008**, *19*, 1027–1032. [CrossRef] [PubMed]

34. Nam, S.; Smith, D.M.; Dou, P.D. Tannic Acid Potently Inhibits Tumor Cell Proteasome Activity, Increases p27 and Bax Expression, and Induces G_1 Arrest and Apoptosis. *Cancer Epidemiol. Biomark. Prev.* **2001**, *10*, 1083–1088.

35. Yang, H.; Landis-Piwowar, K.R.; Chen, D.; Milacic, V.; Dou, Q.P. Natural Compounds with Proteasome Inhibitory Activity for Cancer Prevention and Treatment. *Curr. Protein Pept. Sci.* **2008**, *9*, 227–239. [CrossRef] [PubMed]

36. Lee, J.; Peña, M.M.; Nose, Y.; Thiele, D.J. Biochemical characterization of the human copper transporter Ctr1. *J. Biol. Chem.* **2002**, *277*, 4380–4387. [CrossRef] [PubMed]

37. Nehybova, T.; Smarda, J.; Daniel, L.; Brezovsky, J.; Benes, P. Wedelolactone induces growth of breast cancer cells by stimulation of estrogen receptor signalling. *J. Steroid. Biochem. Mol. Biol.* **2015**, *152*, 76–83. [PubMed]

38. Jancekova, B.; Ondrouskova, E.; Knopfova, L.; Smarda, J.; Benes, P. Enzymatically active cathepsin D sensitizes breast carcinoma cells to TRAIL. *Tumour Biol.* **2016**, *37*, 10685–10696. [CrossRef] [PubMed]

39. Navratilova, J.; Hankeova, T.; Benes, P.; Smarda, J. Acidic pH of tumor microenvironment enhances cytotoxicity of the disulfiram/Cu^{2+} complex to breast and colon cancer cells. *Chemotherapy* **2013**, *59*, 112–120. [CrossRef] [PubMed]

40. Pes, O.; Jungova, P.; Vyhnanek, R.; Vaculovic, T.; Kanicky, V.; Preisler, J. Off-line coupling of capillary electrophoresis to substrate-assisted laser desorption inductively coupled plasma mass spectrometry. *Anal. Chem.* **2008**, *80*, 8725–8732. [CrossRef] [PubMed]

41. Irwin, J.J.; Shoichet, B.K. ZINC—A free database of commercially available compounds for virtual screening. *J. Chem. Inf. Model.* **2005**, *45*, 177–182. [CrossRef] [PubMed]

42. Trott, O.; Olson, A.J. AutoDock Vina: Improving the speed and accuracy of docking with a new scoring function, efficient optimization, and multithreading. *J. Comput. Chem.* **2010**, *31*, 455–461. [CrossRef] [PubMed]

43. Sanner, M.F. Python: A programming language for software integration and development. *J. Mol. Graph. Model.* **1999**, *17*, 57–61. [PubMed]

44. Delano, W.T. *The PyMol Molecular Graphics System, Version 1.5*; Schrödinger, LLC: New York, NY, USA, 2009.

45. Durrant, J.; McCammon, J. NNScore 2.0: A Neural-Network Receptor-ligand Scoring Function. *J. Chem. Inf. Model.* **2011**, *51*, 2897–2903. [CrossRef] [PubMed]

International Journal of
Molecular Sciences

MDPI

Article

Acteoside and Isoacteoside Protect Amyloid *β* Peptide Induced Cytotoxicity, Cognitive Deficit and Neurochemical Disturbances In Vitro and In Vivo

Young-Ji Shiao [1], Muh-Hwan Su [2,3], Hang-Ching Lin [2,3] and Chi-Rei Wu [4,*]

[1]   National Research Institute of Chinese Medicine, Ministry of Health and Welfare, Taipei 11490, Taiwan;
      yshiao@nricm.edu.tw
[2]   School of Pharmacy, National Defense Medical Center, Taipei 11490, Taiwan;
      smh1027@syncorebio.com (M.-H.S.); lhc@sinphar.com.tw (H.-C.L.)
[3]   Sinphar Pharmaceutical Co., Ltd., Sinphar Group (Taiwan), Research & Development Center,
      I-Lan 26944, Taiwan
[4]   Department of Chinese Pharmaceutical Sciences and Chinese Medicine Resources, College of Pharmacy,
      China Medical University, Taichung 40402, Taiwan
*   Correspondence: crw@mail.cmu.edu.tw; Tel.: +886-4-2205-3366 (ext. 5506)

Academic Editor: Toshio Morikawa
Received: 1 March 2017; Accepted: 20 April 2017; Published: 24 April 2017

Abstract: Acteoside and isoacteoside, two phenylethanoid glycosides, coexist in some plants. This study investigates the memory-improving and cytoprotective effects of acteoside and isoacteoside in amyloid β peptide 1-42 (Aβ 1-42)-infused rats and Aβ 1-42-treated SH-SY5Y cells. It further elucidates the role of amyloid cascade and central neuronal function in these effects. Acteoside and isoacteoside ameliorated cognitive deficits, decreased amyloid deposition, and reversed central cholinergic dysfunction that were caused by Aβ 1-42 in rats. Acteoside and isoacteoside further decreased extracellular Aβ 1-40 production and restored the cell viability that was decreased by Aβ 1-42 in SH-SY5Y cells. Acteoside and isoacteoside also promoted Aβ 1-40 degradation and inhibited Aβ 1-42 oligomerization in vitro. However, the memory-improving and cytoprotective effects of isoacteoside exceeded those of acteoside. Isoacteoside promoted exploratory behavior and restored cortical and hippocampal dopamine levels, but acteoside did not. We suggest that acteoside and isoacteoside ameliorated the cognitive dysfunction that was caused by Aβ 1-42 by blocking amyloid deposition via preventing amyloid oligomerization, and reversing central neuronal function via counteracting amyloid cytotoxicity.

Keywords: acteoside; isoacetoside; amyloid β peptide; Morris water maze; acetylcholine; amyloid cascade

1. Introduction

Alzheimer's disease (AD), the most epidemic progressive neurodegenerative disorder, is characterized by behavioral disturbances such as cognitive deficits and neuropathological symptoms such as neuronal loss, senile plaques and neurofibrillary tangles [1]. Senile plaque contains fibrils that are compounds of amyloid β peptide (Aβ), which is formed from amyloid precursor protein (APP) via the amyloidogenic pathway [1,2]. When Aβ is oligomerized to amyloid fibrils and deposited in the brain especial entorhinal cortex and hippocampus, it causes cerebral neuronal loss, and particularly the degeneration of cholinergic neuronal circuits in the basal forebrain (BF) (cholinergic dysfunction) [3]. Therefore, recent researchers have suggested potential therapeutic approaches against AD that involve several disease-modifying strategies, such as blocking the cellular production of Aβ, preventing Aβ oligomerization, promoting Aβ degradation, and counteracting Aβ cytotoxicity [2].

Acteoside and its isomeric phenylethanoid glycoside, isoacteoside, (Figure 1) co-exist in various plants, such as *Cistanches* spp., *Castilleja* spp. and *Plantago* spp. [4,5]. Acteoside has been found to have antioxidative, anti-inflammatory, anti-nociceptive, anti-metastatic, hepatoprotective and cytoprotective activities [6–12]. Reports have shown that acteoside can alleviate acquired learning disability in mice that is induced by scopolamine [13], and reduce cerebral injury in mice that is induced by D-galactose [14,15]. Acteoside also shortens the escape latency in the Morris water maze (MWM) and reduces the number of retention errors in the step-down test in D-galactose plus AlCl$_3$-induced mouse senescence model [16,17]. Acteoside protects neuronal damage caused by Aβ 25-35 in SH-SY5Y neuroblastoma cells [10,18] and inhibits the aggregation of Aβ 1-42 in vitro [19]. However, few studies have shown the effects of acteoside on Aβ 1-42-induced cognitive dysfunction in vivo and the pharmacological activities of isoacteoside. Therefore, this study investigates the effects of acteoside and isoacteoside on Aβ 1-42-induced behavioral changes following the osmotic intracisternal infusion of Aβ 1-42 into the lateral ventricle in rats. A meta-analysis of four behavioral tasks by Myhrer [20], acetylcholinergic and catecholaminergic activities strongly affect learning and memory. Aβ 1-42-infusion causes central acetylcholinergic and catecholaminergic dysfunction, which is closely related to memory deficits [21,22]. Therefore, this study further investigates the role of the central neurotransmitters in the acteoside- or isoacteoside-induced reversal of cognitive dysfunction that is caused by Aβ 1-42 infusion by measuring the levels of central neurotransmitters and the activities of related enzymes. Cognitive dysfunction and neurotransmitter disturbances in AD patients are closely associated with an amyloid cascade that involves amyloid generation, amyloid oligomerization and amyloid cytotoxicity [1,2,22]. The effects of acteoside and isoacteoside on amyloid generation, amyloid oligomerization and amyloid cytotoxicity are investigated in vitro to elucidate their memory-improving effects on Aβ 1-42-induced cognitive dysfunction.

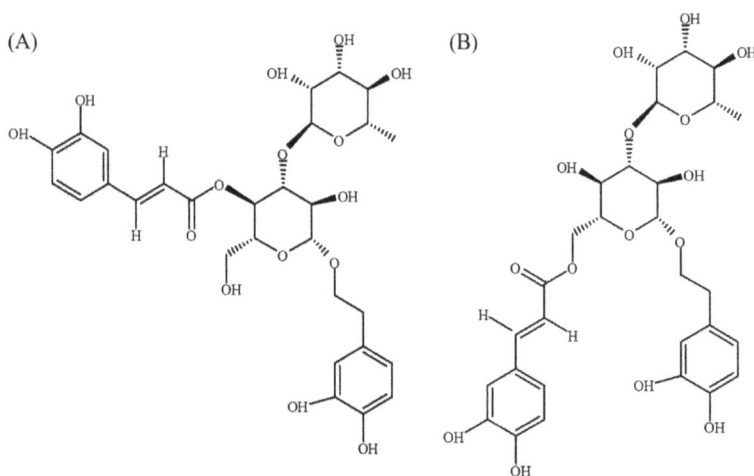

Figure 1. Structures of: (**A**) acteoside; and (**B**) isoacteoside.

2. Results

2.1. In Vivo Aβ 1-42-Infusion Model

2.1.1. Effects of Acteoside and Isoacteoside on Behavioral Dysfunction Induced by Aβ 1-42 in Rats

Aβ 1-42 infusion reduced the index of exploratory behavior, which incorporates time spent in the hole and the number of entries into the hole ($p < 0.01$, $p < 0.001$) (Figure 2A–C), but Aβ 1-42 infusion did not alter the movement time, distance or velocity of rats (Figure S1). Acteoside (2.5 mg/kg) or isoacteoside (2.5, 5.0 mg/kg) increased the time spent in the hole and the number of entry into the hole of Aβ 1-42-infused rats ($p < 0.01$) (Figure 2A,B), but only isoacteoside (2.5, 5.0 mg/kg) resorted the index of exploratory behavior ($p < 0.05$, $p < 0.01$) (Figure 2C). Neither acteoside nor isoacteoside at any dosage altered the motor activities of Aβ 1-42-infused rats (Figure S1).

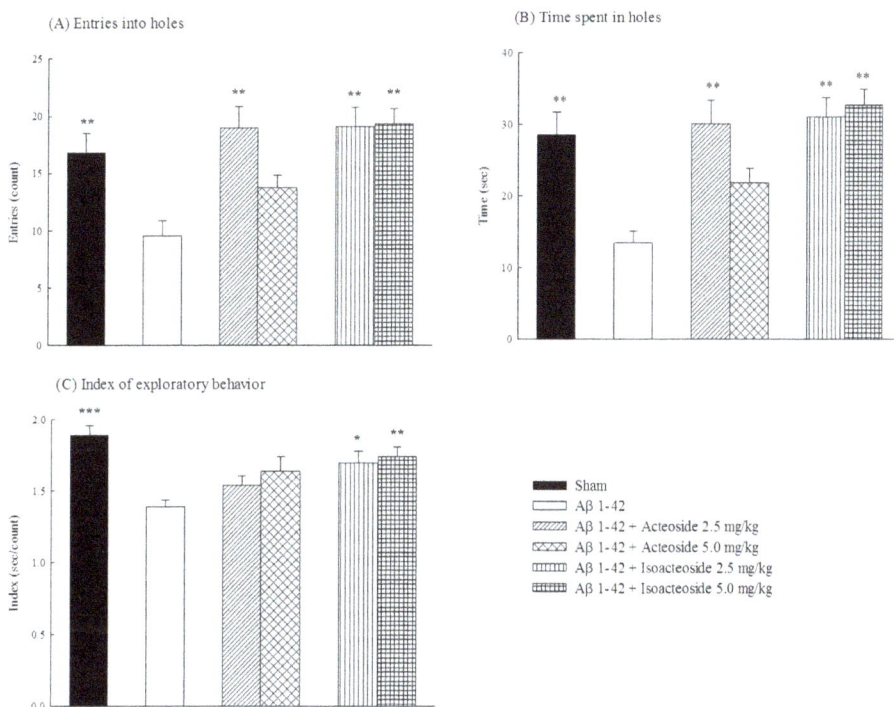

Figure 2. Effects of acteoside or isoacteoside (2.5, 5.0 mg/kg; po) on: (**A**) the number of entries into holes; and (**B**) time spent in holes; and (**C**) index of exploratory behavior in Aβ 1-42-infused rats. Exploratory test was performed on Day 7 following Aβ 1-42 infusion. Acteoside or isoacteoside was continuously administered after Aβ 1-42 infusion until all rats were sacrificed. Columns indicate mean ± SEM ($n = 12$). * $p < 0.05$, ** $p < 0.01$, *** $p < 0.001$ compared with Aβ 1-42-infused rats.

In a passive avoidance test, Aβ 1-42 shortened the latency of retention trial relative to the sham group ($p < 0.001$). Acteoside or isoacteoside (2.5, 5.0 mg/kg) prolonged the latency of retention trial in Aβ 1-42-infused rats ($p < 0.01$, $p < 0.001$) (Figure 3A). In MWM, the Aβ 1-42-infused group had a longer escape latency over eight trials on four training days (from Day 10 to Day 13 following Aβ 1-42 infusion) than the sham group ($p < 0.05$, $p < 0.01$). Aβ 1-42 infusion also shortened the time spent in the platform-quadrant from that of the sham group ($p < 0.001$) (Figure 3B,C). Both acteoside (5.0 mg/kg) and isoacteoside (2.5, 5.0 mg/kg) shortened the increase in escape latency that was caused by Aβ 1-42 infusion ($p < 0.05$, $p < 0.01$). Both acteoside and isoacteoside (2.5, 5.0 mg/kg) also prolonged the time spent in the platform-quadrant relative to Aβ 1-42-infused group ($p < 0.001$) (Figure 3B,C). However, the sham, Aβ 1-42-infused, acteoside- or isoacteoside-treated groups did not vary in swimming velocity (Figure 3D).

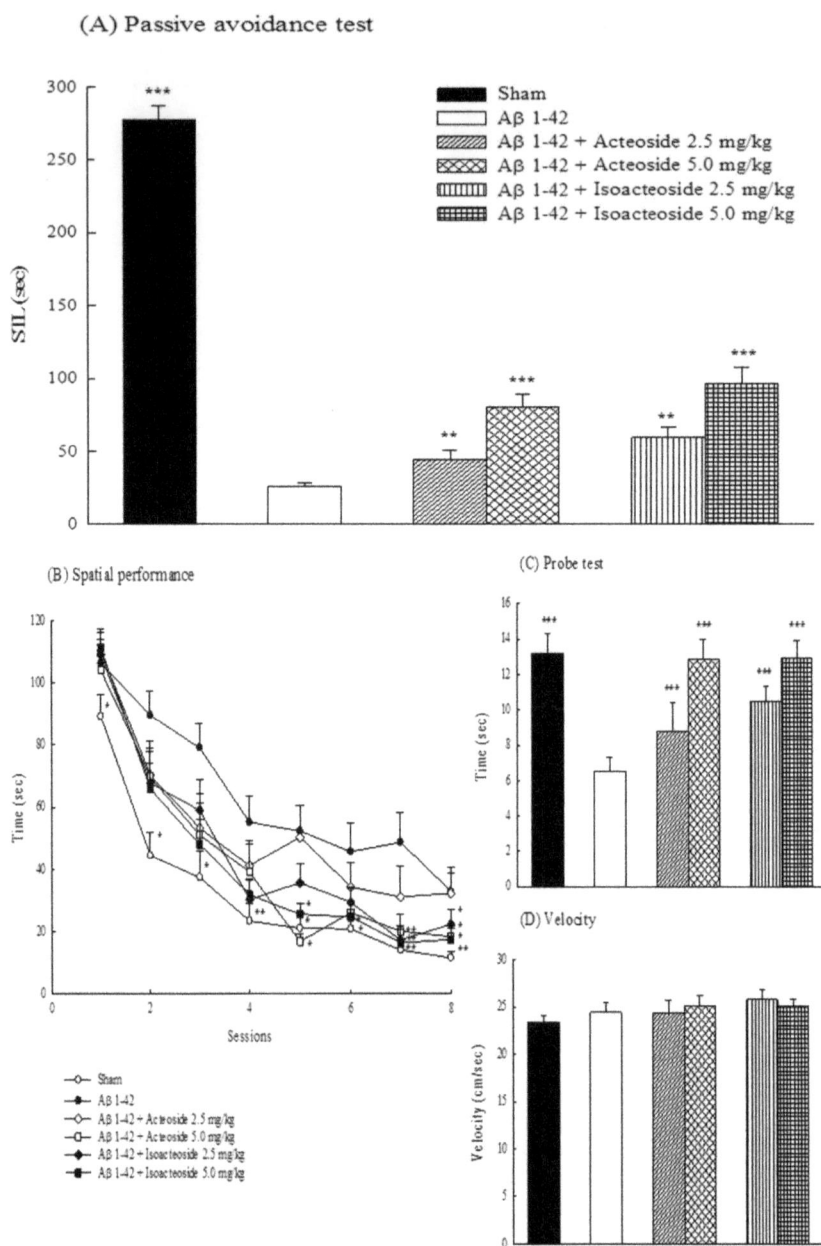

Figure 3. Effects of acteoside or isoacteoside (2.5, 5.0 mg/kg; po) on: (**A**) step-through latency (STL) of passive avoidance task; (**B**) spatial performance; (**C**) probe test; and (**D**) swimming velocity of MWM in Aβ 1-42-infused rats. Passive avoidance test was performed on Days 8–9 following Aβ 1-42 infusion. Spatial performance and probe test of MWM were performed on Days 10–14 following Aβ 1-42 infusion. Acteoside or isoacteoside was continuously administered after Aβ 1-42 infusion until all rats were sacrificed. Columns indicate mean ± SEM ($n = 12$). ** $p < 0.01$, *** $p < 0.001$ compared with Aβ 1-42-infused rats.

2.1.2. Effects of Acteoside and Isoacteoside on Amyloid Deposition and Neurochemical Disturbances Induced by Aβ 1-42 in Rats

Figure 4 displays photographs of immunological staining and ratio of Aβ 1-42 deposition in the brain. The Aβ 1-42-infused group exhibited a significantly greater ratio of Aβ 1-42 deposition in the brain than the sham group ($p < 0.01$). Acteoside (5.0 mg/kg) or isoacteoside (2.5, 5.0 mg/kg) reduced the ratio of Aβ 1-42 deposition in the brain ($p < 0.01$, $p < 0.001$) (Figure 4G).

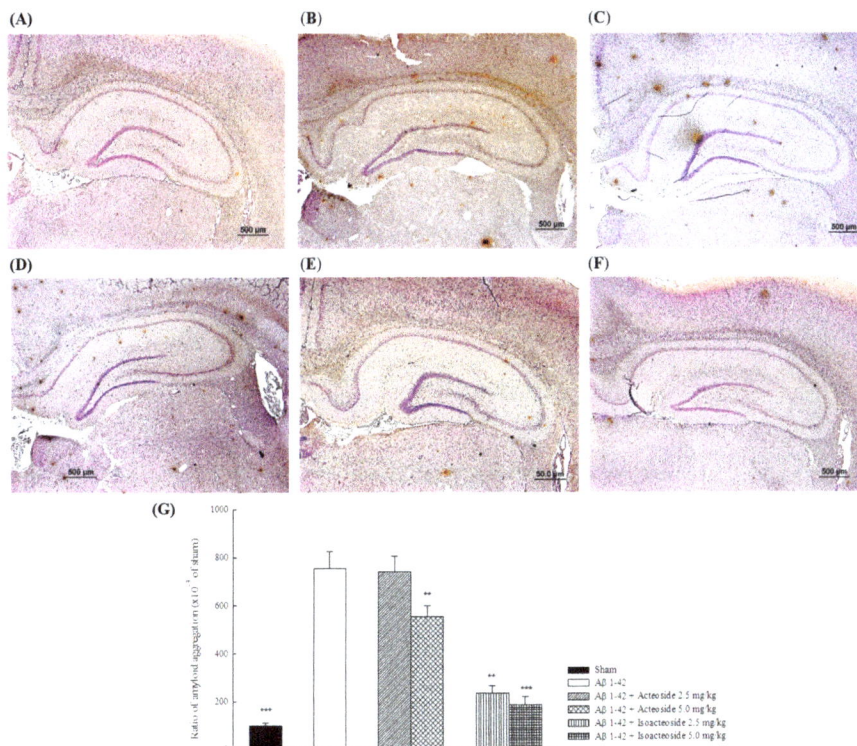

Figure 4. Effects of acteoside or isoacteoside (2.5, 5.0 mg/kg; po) on Aβ 1-42 deposition in Aβ 1-42-infused rats: (**A**) sham group; (**B**) Aβ 1-42-infused group; (**C**) acteoside (2.5 mg/kg)-treated group; (**D**) acteoside (5.0 mg/kg)-treated group; (**E**) isoacteoside (2.5 mg/kg)-treated group; and (**F**) isoacteoside (5.0 mg/kg)-treated group; and (**G**) ratio of amyloid deposition. Acteoside or isoacteoside was continuously administered after Aβ 1-42 infusion until all rats were sacrificed. Columns indicate mean ± SEM ($n = 6$). ** $p < 0.01$, *** $p < 0.001$ compared with Aβ 1-42-infused rats.

Aβ 1-42 infusion decreased the levels of cortical and hippocampal acetylcholine (Ach) ($p < 0.05$, $p < 0.001$) as well as hippocampal choline (Ch) ($p < 0.01$) (Figure 5A,B). Both acteoside and isoacteoside (2.5, 5.0 mg/kg) reversed the decline in hippocampal Ach levels that were caused by Aβ 1-42 infusion ($p < 0.05$, $p < 0.01$, $p < 0.001$), but only a dose of 5.0 mg/kg reversed the decrease in cortical Ach levels that was caused by Aβ 1-42 infusion ($p < 0.05$) (Figure 5B). Aβ 1-42 infusion reduced cortical and hippocampal dopamine (DA) levels ($p < 0.01$, $p < 0.001$), but only reduced hippocampal norepinephrine (NE) levels ($p < 0.01$) (Table 1). Only isoacteoside (2.5, 5.0 mg/kg) reversed the decline in hippocampal DA levels that was caused by Aβ 1-42 infusion ($p < 0.001$) (Table 1).

Aβ 1-42 infusion increased cortical and hippocampal acetylcholinesterase (AChE) activities ($p < 0.05$, $p < 0.01$). Both acteoside and isoacteoside (2.5, 5.0 mg/kg) prevented that increase in cortical and hippocampal AChE activity that would otherwise have been caused by Aβ 1-42 infusion ($p < 0.05$) (Figure 6A). Aβ 1-42 increased cortical monoamine oxidase-A (MAO-A) and MAO-B activities ($p < 0.05$, $p < 0.01$), but reduced hippocampal MAO-A and MAO-B activities in the rats ($p < 0.05$) (Figure 6B–C). Only isoacteoside at 5.0 mg/kg reversed the decrease in hippocampal MAO-A activity in Aβ 1-42-infused rats ($p < 0.01$) (Figure 6B,C).

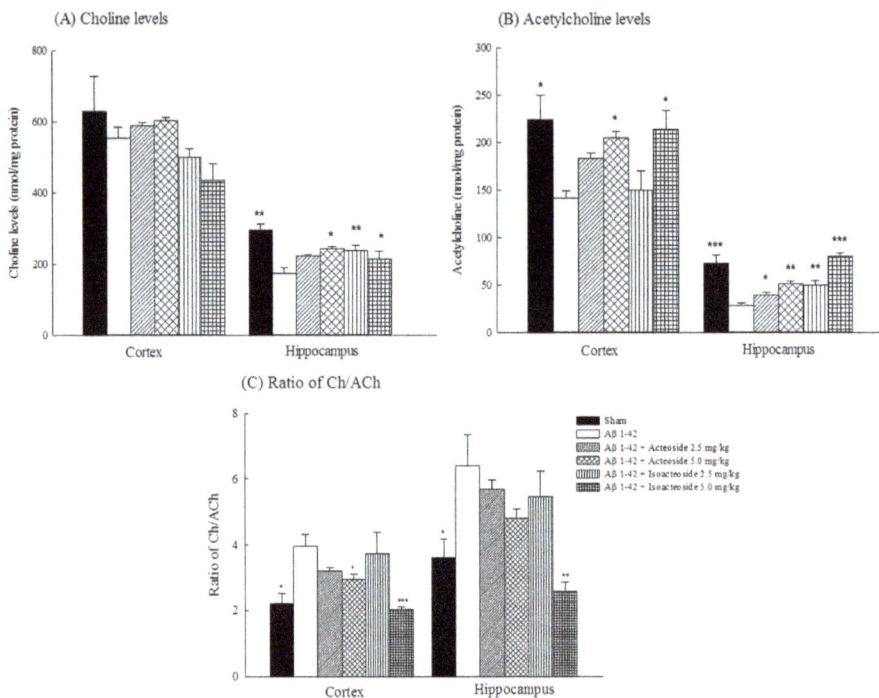

Figure 5. Effects of acteoside or isoacteoside (2.5, 5.0 mg/kg; po) on: (**A**) choline (Ch) levels; (**B**) acetylcholine (Ach) levels; and (**C**) ratio of Ch to Ach in cortex and hippocampus of Aβ 1-42-infused rats. Acteoside or isoacteoside was continuously administered after Aβ 1-42 infusion until all rats were sacrificed. Columns indicate mean ± SEM ($n = 6$). * $p < 0.05$, ** $p < 0.01$, *** $p < 0.001$ compared with Aβ 1-42-infused rats.

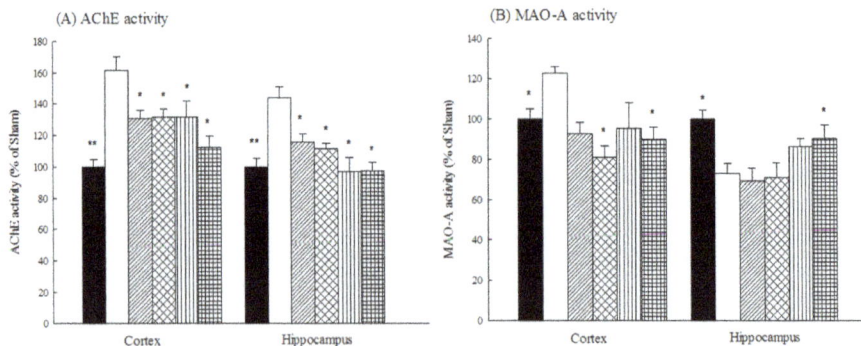

Figure 6. *Cont.*

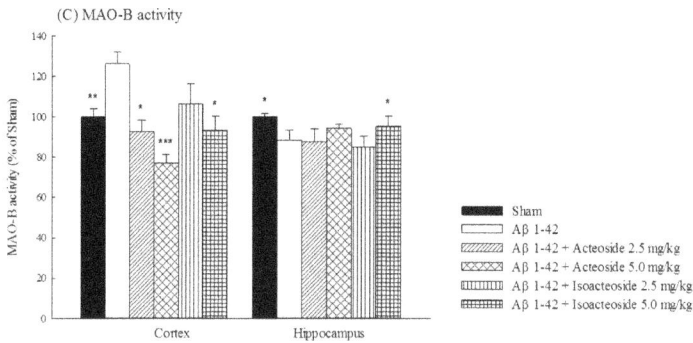

Figure 6. Effects of acteoside or isoacteoside (2.5, 5.0 mg/kg; po) on: (**A**) AChE; (**B**) MAO-A; and (**C**) MAO-B activities in cortex and hippocampus of Aβ 1-42-infused rats. Acteoside or isoacteoside was continuously administered after Aβ 1-42 infusion until all rats were sacrificed. Columns indicate mean ± SEM ($n = 6$). * $p < 0.05$, ** $p < 0.01$, *** $p < 0.001$ compared with Aβ 1-42-infused rats.

Table 1. Effects of acteoside and isoacteoside (2.5, 5.0 mg/kg; po) on the levels of cortical and hippocampal neurotransmitters and their metabolites in Aβ 1-42-infused rats.

The Levels of Cortical Neurotransmitters and Their Metabolites (ng/g Protein)					
-	MHPG	NE	DOPAC	HVA	DA
Vehicle	26.05 ± 1.11 *	15.63 ± 0.41	17.13 ± 0.85 *	3.76 ± 0.78 **	3.83 ± 0.16 **
Aβ 1-42	22.87 ± 0.81	13.61 ± 1.20	11.97 ± 0.66	2.90 ± 0.23	2.18 ± 0.11
Acteoside					
2.5 mg/kg	22.40 ± 1.24	13.08 ± 1.06	12.91 ± 0.66	2.87 ± 0.11	2.26 ± 0.06
5.0 mg/kg	23.84 ± 1.94	12.91 ± 0.67	13.25 ± 1.07	2.90 ± 0.15	2.23 ± 0.15
Isoacteoside					
2.5 mg/kg	31.71 ± 3.21 *	12.60 ± 1.56	12.53 ± 3.33	2.91 ± 0.58	2.67 ± 0.43
5.0 mg/kg	34.41 ± 4.81 *	14.18 ± 1.01	13.80 ± 3.03	2.74 ± 0.69	3.84 ± 0.47 *
The Levels of Hippocampal Neurotransmitters and Their Metabolites (ng/g Protein)					
	MHPG	NE	DOPAC	HVA	DA
Vehicle	622.30 ± 17.58	64.49 ± 1.65 **	5.90 ± 0.39	3.60 ± 0.20	5.41 ± 0.35 ***
Aβ 1-42	620.97 ± 25.79	51.96 ± 1.54	5.12 ± 0.36	3.24 ± 0.16	0.97 ± 0.07
Acteoside					
2.5 mg/kg	641.15 ± 23.10	53.16 ± 2.29	5.22 ± 0.27	3.26 ± 0.12	1.07 ± 0.15
5.0 mg/kg	631.13 ± 36.81	53.53 ± 1.83	5.32 ± 0.14	3.34 ± 0.10	1.17 ± 0.13
Isoacteoside					
2.5 mg/kg	689.99 ± 79.95	54.01 ± 6.20	4.98 ± 1.15	3.68 ± 0.46	5.28 ± 0.83 ***
5.0 mg/kg	574.02 ± 54.16	45.13 ± 3.36	4.13 ± 0.45	3.29 ± 0.25	5.66 ± 0.36 ***

Acteoside or isoacteoside was continuously administered after Aβ 1-42 infusion until all rats were sacrificed. Columns indicate mean ± SEM ($n = 6$). * $p < 0.05$, ** $p < 0.01$, *** $p < 0.001$ compared with Aβ 1-42-infused rats.

2.2. In Vitro Test on Amyloid Cacade

2.2.1. Effects of Acteoside and Isoacteoside on Neuronal Damage Induced by Aβ 1-42, and Intracellular and Extracellular Aβ 1-40 Levels in SH-SY5Y Cells

Incubation of SH-SY5Y cells with 20 µM Aβ 1-42 for 24 h reduced the cell viability to 52.73% of that of control cells ($p < 0.001$). Treatment with acteoside (50 µg/mL) or isoacteoside (50 µg/mL) recovered the cell viability that was reduced by Aβ 1-42 (20 µM) ($p < 0.001$) (Figure 7A). Both acteoside (50 µg/mL) and isoacteoside (50 µg/mL) reduced the extracellular Aβ 1-40 levels in SH-SY5Y cells ($p < 0.05$, $p < 0.001$), but did not alter the intracellular Aβ 1-40 levels (Figure 7B,C).

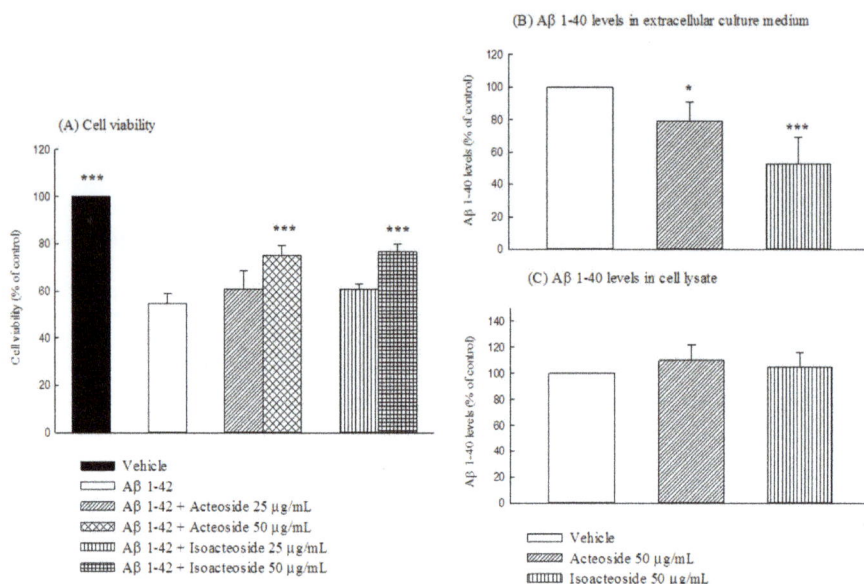

Figure 7. Effects of acteoside or isoacteoside (25, 50 µg/mL) on Aβ 1-42 toxicity, and extracellular and intracellular Aβ 1-40 levels in SH-SY5Y cells: (**A**) cell viability; (**B**) Aβ 1-40 levels in extracellular culture medium; and (**C**) Aβ 1-40 levels in cell lysate. Acteoside or isoacteoside was administered 1 h before treatment with Aβ 1-42. Columns indicate mean ± SD ($n = 4$). * $p < 0.05$, *** $p < 0.001$ compared with (**A**) Aβ 1-42-treated group or (**B,C**) vehicle group.

2.2.2. Effects of Acteoside and Isoacteoside on Aβ 1-40 Degradation and Aβ 1-42 Oligomerization In Vitro

Both acteoside (50 µg/mL) and isoacteoside (50 µg/mL) increased the degradation of added synthetic Aβ 1-40 (10 ng) in SH-SY5Y-conditioned cell free medium in vitro ($p < 0.05$) (Figure 8A). Both acteoside (50 µg/mL) and isoacteoside (50 µg/mL) also reduced Aβ 1-42 oligomerization, as determined from the thioflavin T (ThT) binding fluorescence intensity in vitro ($p < 0.001$) (Figure 8B).

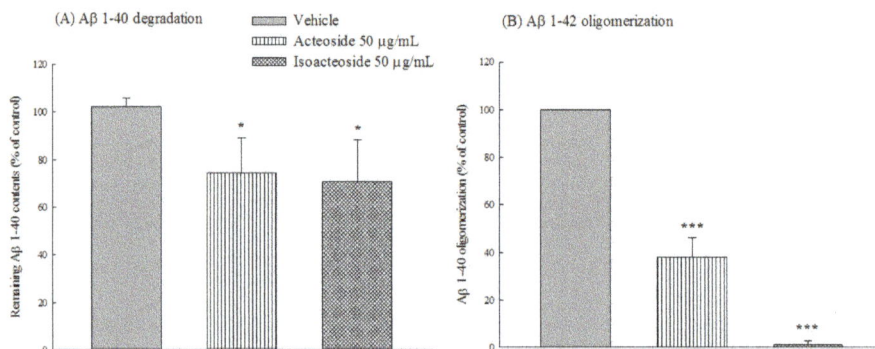

Figure 8. Effects of acteoside or isoacteoside (50 µg/mL) on: (**A**) Aβ 1-40 degradation; and (**B**) Aβ 1-42 oligomerization in vitro. Acteoside or isoacteoside was co-cultured with Aβ 1-40 or Aβ 1-42. Columns indicate mean ± SD ($n = 4$). * $p < 0.05$, *** $p < 0.001$ compared with vehicle group.

3. Discussion

Based on AD pathology, Aβ 1-42 is the critical protein in AD and intracisternal injection with Aβ 1-42 into rats produced memory impairment, morphological changes in the brain, and neuronal degeneration including cholinergic and monoaminergic systems [22,23]. The presented data reveal that intracisternal Aβ 1-42 infusion caused behavioral deficits including in the exploratory behavior, passive avoidance response, and spatial performance of MWM in rats. These results are consistent with our previous report and other reports [22–24]. Acteoside at a dose of 2.5–5.0 mg/kg ameliorated the deficits of passive avoidance learning and reference memory that were caused by Aβ 1-42, but only a dose of 5.0 mg/kg ameliorated the impairment of spatial performance. However, no dose of acteoside improved exploratory behavior. These memory-improving effects of acteoside are similar to those identified in other reports, which found that acteoside at 1.0–120 mg/kg reversed the memory impairment that was induced by scopolamine, D-galactose or D-galactose plus AlCl₃ [13,14,16,17,25]. This difference between the results obtained herein with those other reports may be related to the given route and duration, and various models. Isoacteoside at 2.5–5.0 mg/kg exhibited a similar therapeutic potential against Aβ 1-42-induced behavioral dysfunction, but this effect of isoacteoside may differ from that of acteoside because isoacteoside reversed memory impairment partially by promoting exploratory behavior. Thus, we suggest that acteoside and isoacteoside may be potential anti-dementia phenylethanoid glycosides, and that these two stereoisomeric compounds exhibit similar memory-improving potentials but different behavioral-improving patterns against Aβ 1-42-induced behavioral dysfunction.

AD patients have complex neurochemical disturbances including of the catecholaminergic, cholinergic and glutaminergic neuronal systems [26]. AD patients have higher MAO-B activity than healthy controls, and this increased MAO-B activity may reflect abnormalities in the dopaminergic system [27]. In an AD-like animal model, Aβ 1-42 infusion into the lateral ventricle also caused memory deficits which were closely related to Aβ deposition and a subsequent cascade that caused, for example central cholinergic dysfunction in BF, including a decline in Ach levels and an up-regulation of AChE activity [22,28]. Thus, we further investigated the effects of acteoside and isoacteoside on Aβ-induced pathological changes, including amyloid deposition and neurochemical disturbances in rats. Our present data revealed similar pathological and neurochemical symptoms. Additionally, Aβ 1-42 infusion herein reduced cortical and hippocampal DA levels and hippocampal NE levels. The neurochemical changes were similar to those observed elsewhere [3,21,24]. Aβ 1-42 infusion was also found to cause a differential alteration of cortical and hippocampal MAO activities, mainly by elevating cortical MAO activities and reducing hippocampal MAO activities. Most related investigations have indicated that MAO-B activity was elevated around Aβ plaques (especially plaque-associated astrocytes), and have suggested the existence of a close positive correlation between MAO-B activity and amyloid plaques in the frontal cortex [27,29,30]. However, immunohistochemical studies have demonstrated MAO-B activities reflect disease-specific cellular changes in AD brain, and reduced MAO-B activities in advanced AD patients [31]. Researchers hole differing opinions regarding the alteration of MAO-A activities in AD patient. Recent reports have indicated that the alteration of MAO-A activities in AD patients may be related to presenilin-1 variants. Based on our results and others, we suggest that the alteration of regional MAO-A/B activities following Aβ 1-42 infusion may involve the regional activation of astrocytes around plaques sites and the loss of astrocytes/neurons with the progress of AD [27,31]. Aβ deposition causes up-regulates AChE activity around senile plaques, which favors the assembly of Aβ into fibrils, which cause Aβ cytotoxicity and, in particular, cholinergic and dopaminergic dysfunction [3]. Acteoside at 2.5–5.0 mg/kg reversed hippocampal Ach levels and inhibited the up-regulation of hippocampal AChE activity, but only at 5.0 mg/kg did it reduce Aβ deposition and reverse the disturbances of cortical Ach levels in Aβ 1-42-infused rats. Isoacteoside at 2.5–5.0 mg/kg reduced Aβ deposition and restored hippocampal cholinergic and dopaminergic neuronal function, including by blocking AChE up-regulating activity, but only at 5.0 mg/kg did it reverse the alteration of cortical and hippocampal MAO activities in Aβ

1-42-infused rats. Furthermore, acteoside at 5.0 mg/kg only restored the turnover rate of cortical Ach in Aβ 1-42-infused rats, whereas isoacteoside at 5.0 mg/kg restored the turnover rate of both cortical and hippocampal Ach (Figure 5C). From these above results, we suggest that the effects of acteoside and isoacteoside against Aβ 1-42-induced cognitive deficit may be related to reducing Aβ deposition, and then leading to a reversal of cortical cholinergic function, including an increase in the cortical Ach levels and a decrease in the Ach utility, by inhibiting AChE activity. Isoacteoside reduces Aβ deposition and Ach utility more than does acteoside. Unlike acteoside, isoacteoside restored cortical and hippocampal DA levels that were decreased by Aβ 1-42 infusion. Some researchers have pointed out that frontal and striatal DA levels are related to exploratory behavior [32–34]. Therefore, the memory-improving effects of isoacteoside may be further related to an improvement in exploratory behavior by the restoration of the dopaminergic function.

According to the amyloid cascade hypothesis, Aβ monomers are generated from APP via amyloidogenic pathway and secreted into the extracellular medium. Aβ monomers aggregated to form progressively larger species such as Aβ oligomers or fibrils under various physiological conditions, and are then deposited into senile plaques, causing neuronal dysfunction, such as neuronal apoptosis, and a decrease in long-term potentiation [2,3]. The results in this study reveal that acteoside at 50 μg/mL protected SH-SY5Y cells against Aβ 1-42-induced neural damage and inhibited Aβ 1-42 oligomerization, are revealed by ThT fluorescent staining. These results were consistent with earlier reports that acteoside protects Aβ 25-35-induced neural damage in SH-SY5Y and PC12 cells [10,18] and inhibits the fibril formation of Aβ 1-42 in vitro [19]. Acteoside was further found herein to reduce extracellular but not intracellular levels of Aβ 1-40, which was produced by amyloidogenic pathway in SH-SY5Y cells, and promoted Aβ 1-40 degradation in vitro. Hence, acteoside reduced extracellular Aβ 1-40 levels mainly by promoting Aβ 1-40 degradation. Isoacteoside possessed the same pharmacological potential to inhibit the amyloid cascade. The inhibiting by isoacteoside of amyloidogenesis and amyloid oligomerization exceeded that by acteoside. Some reports have indicated that acteoside has cytoprotective effects against Aβ 25-35, glutamate, okadaic acid, and MPP$^+$ in vitro, and this effect may be mediated by their antioxidant and antiapoptotic activities by maintaining mitochondrial function and the activities of antioxidative enzymes, decreasing intracellular oxidative stress and Bax/Bcl-2 ratio, and inhibiting caspase-3 activity [9,10,18,35,36]. Other reports have indicated that acteoside protects Aβ 25-35-induced neuronal damage by inducting heme oxygenase-1 (HO-1) and the activation of transcription factor NF-E2-related factor 2 (Nrf2) by extracellular signal–regulated kinases (ERKs) and phosphatidylinositol 3-kinase/protein kinase B (PI3K/Akt) signaling [18], and restored the expression of neurotrophins including nerve growth factor (NGF), neurotrophin 3 (NT-3), and tropomyosin receptor kinase A (TrkA) in a D-galactose or D-galactose plus AlCl$_3$-induced mouse senescence model [15,17]. NGF and NT-3 exhibited a neuroprotective function with therapeutic potential against neurodegenerative diseases [37]. NGF is synthesized by cortical and hippocampal neurons and retrogradely transported to BF cholinergic neurons through cholinergic projections that bearing the TrkA and low-affinity p75 neurotrophin receptor [38]. NGF maintains the survival of BF cholinergic neurons and enhances cholinergic neurotransmission through acute neurotransmitter-like and classical trophic mechanisms [39]. Accumulating evidence indicates that NGF improves the survival of cholinergic neurons and reduces cognitive decline in humans with mild AD [40]. Therefore, the linkage of NGF and ERK/Akt-Nrf2 signaling pathway on the memory-improving and cytoprotective effects of acteoside against Aβ 1-42 must be clarified and the cytoprotective mechanism of isoacteoside against Aβ-induced neural damage shall be investigated in the future.

4. Materials and Methods

4.1. Animals

Male Sprague-Dawley rats (300–350 g) were obtained from BioLASCO Taiwan Co., Ltd. They were housed in groups of four, chosen at random, in wire-mesh cages (39 cm × 26 cm × 21 cm) in a temperature

(23 ± 1 °C) and humidity (60%) regulated environment with a 12 h–12 h light/dark cycle (light phase: 08:00 to 20:00). The Institutional Animal Care and Use Committee of China Medical University approved the experimental protocol (Protocol No. 99-127-B), and the animals were cared according to the Guiding Principles for the Care and Use of Laboratory Animals. After one week of acclimatization, the rats were used in the experiments that are described below.

4.2. Drugs

Acteoside and isoacteoside (with purities of greater than 98%) were kindly provided by Sinphar Pharmaceutical Co., Ltd. (I-Lan, Taiwan) and freshly dissolved in sterile distilled water. Synthesized human Aβ 1-42 and Aβ 1-40 were purchased from Tocris Bioscience (Ellisville, MO, USA). Aβ 1-42 was freshly dissolved with 35% acetonitrile/0.1% trifluoroacetic acid at a concentration of 250 pmol/μL and used to fill into mini-osmotic pump (Alzet 2002; Alza, Palo Alto, CA, USA) in vivo test. Aβ 1-42 and Aβ 1-40 were prepared with sterile phosphate buffer saline (PBS) in vitro test.

4.3. In Vivo Aβ 1-42-Infused Model

An Aβ 1-42-infused rat model was developed by infusing Aβ 1-42 into the cerebral ventricle via a mini-osmotic pump, as described elsewhere [23]. Briefly, rats were anesthetized with phenobarbital (45 mg/kg, i.p.) and placed in a David Kopf stereotaxic instrument. An infusion cannula was implanted into the left cerebral ventricle (AP-1.5, ML + 0.9, V-3.6 from Bregma), and a continual infusion of Aβ 1-42 (300 pmol/day) was maintained for at least two weeks by attaching an infusion cannula to the mini-osmotic pump. Sham group was infused with 35% acetonitrile/0.1% trifluoroacetic acid.

4.3.1. Schedule of Aβ 1-42-Infused Model

Surgery, drug treatment, and behavioral tests were scheduled as in our previous report [24]. After implantation, Aβ 1-42 infusion began on a day that was designated as Day 0. On the next day (Day 1), the rats were orally administered with vehicle, acteoside or isoacteoside (2.5, 5.0 mg/kg) throughout Aβ 1-42 infusion period. The behavioral tests were carried out from Day 7 to Day 14 after Aβ 1-42 infusion, in the order, locomotor and exploratory tests (Day 7), passive avoidance test (Days 8–9), spatial performance test in MWM (Days 10–13), and probe test in MWM (Day 14). On Day 15 after Aβ 1-42 infusion, the rats were killed 1 h after their final treatment with acteoside or isoacteoside to measure AChE and MAO activities, levels of neurotransmitters and the metabolites in the brain.

4.3.2. Behavioral Tests

The behavioral tests were performed as described in our previous report [23]. On Day 7, locomotor and exploratory tests were simultaneously performed with open-field task (Coulbourn Instruments L.L.C., Holliston, MA, USA). Each rat was observed for 10 min to record the movement time, distance and velocity (locomotor activity), the number of entries it made into the hole and the time spent (exploratory activity) using TruScan software v 2.07 (Coulbourn Instruments L.L.C.) [24]. On Day 8, the training trial of passive avoidance test was performed with passive avoidance apparatus (Coulbourn Instruments L.L.C.). When the rat entered the dark compartment from the light compartment, the door was closed and an inescapable foot shock (0.8 mA for 2 s) was delivered through the grid floor. On the following day (Day 9), the retention trial of passive avoidance test was conducted. The rat was again placed in the light compartment and the latency was recorded [24]. An upper cut-off time of 300 s was set. On Days 10–13, the spatial performance in MWM was tested using a black circular stainless pool (with a diameter of 165 cm and a height of 60 cm) that was filled with water at 23 ± 1 °C to a depth of 35 cm. Each rat underwent eight training sessions over four consecutive days to find the Plexiglass hidden platform (with a diameter of 10 cm) that was submerged 1.0 cm below the surface of the water. The swim path and escape latency to the platform of a white rat in the black pool were recorded using a video camera and an automated video tracking system device equipped with EthoVision XT software (Noldus Information Technology, Leesburg, VA, USA) [24]. On the following day (Day 14), the probe test was performed to measure the reference memory. The platform was removed and the parameters, including the time spent and distance moved in each quadrant while searching for the platform [24].

4.3.3. Assessment of Aβ 1-42 Deposition, Neurotransmitter Levels, and AChE and MAO Activity in Brain

The rats in each group were separated into three groups: one for assaying Aβ 1-42 deposition, one for measuring neurotransmitter levels, one for measuring biochemical activities. To assay Aβ 1-42 deposition in the rat brain, the paraffin brain slices of rat were prepared and cut into sections (10 μm) using a microtome (Leica 2030 Biocut, Nussloch, Germany). The sections were labeled with a mouse anti-human amyloid β protein 17–24 monoclonal antibody (1:300, Dakopatts A/C; Glostrup, Denmark) and developed with 0.05% diaminobenzidine using a Vectastain kit (Vector Laboratories, Burlingame, CA, USA). The Aβ 1-42 labeled plaques at least 20 fields of each brain section were counted under 40× magnification using an image analyzer (Leica, Q500MC, Nussloch, Germany). The ratio of Aβ 1-42 deposition was obtained from Aβ 1-42 labeled plaques for each brain section. To measure the neurotransmitter levels, all rats were sacrificed and their brains were separated into cortex and hippocampus, which were placed on ice, according to the protocol of Glowinski and Iversen [41]. The supernatants of the brain tissues were prepared through homogenization, filtration and centrifugation, and then the neurotransmitter (and their metabolite) concentrations of brain supernatants were measured by high-performance liquid chromatography with electrochemical detection (EICOM HTEC-500, Kyoto, Japan). To measure brain AChE and MAO activities, all brains also were cut into cortex and hippocampus, and then the brain supernatants were prepared by homogenization and centrifugation. The brain supernatants and recombinant AChE enzyme were incubated with 5,5′-dithiobis(2-nitrobenzoic acid), and the absorbance at 412 nm was measured following the addition of acetylthiocholine. AChE activity was expressed as U AChE per mg protein. Brain homogenates were incubated with 5 U/mL horseradish peroxidase, 100 μM amplex red, and the substrate (5 mM serotonin for MAO-A or 5 mM benzylamine for MAO-B) at 25 °C for 60 min. The fluorescence intensity was measured, and MAO-A and MAO-B activities were expressed as percentage of the corresponding values for sham rats [24]. The protein content of brain supernatants was quantified using Bio-Rad protein assay kit.

4.4. In Vitro Test on Amyloid Cacade

Human SH-SY5Y neuroblastoma cells were cultured in DMEM that was supplemented with 10% fetal bovine serum, 100 units/mL penicillin and 100 μg/mL streptomycin in a water-saturated atmosphere with 5% CO_2 at 37 °C. Experiments were performed 24 h after the cells were seeded in 96- or 24-well sterile clear-bottom plates. For cytoprotective and amyloidogenic-inhibiting tests, acteoside or isoacteoside (25, 50 μg/mL) was dissolved with culture medium. For amyloid degradation and oligomerization, acteoside or isoacteoside (50 μg/mL) was dissolved with sterile phosphate buffer saline.

4.4.1. Assessment of Cytoprotective Effect in SH-SY5Y Cells

Acteoside or isoacteoside was treated 1 h before Aβ 1-42 (20 μM). The reduction of 3-[4,5-dimethylthiazol-2-yl]-2,5-diphenyl-tetrazolium bromide (MTT) to insoluble formazan was used to evaluate cell viability. Briefly, 24 h after exposure to Aβ 1-42 in 96-well plate, the medium was replaced and MTT (0.5 mg/mL) was added to each well. After incubation for 2 h at 37 °C, the cells were washed with PBS, and DMSO was added. The absorbance at 570 nm was measured using an ELISA reader. Cell viability was expressed as a percentage of corresponding value for untreated cells, which served as the control group (designated 100% viable). Each of four independent experiments was performed in triplicate.

4.4.2. Assessment of Intracellular and Extracellular Aβ 1-40 Levels in SH-SY5Y Cells

Following treatment with acteoside or isoacteoside, culture media and SH-SY5Y cells were collected separately and the levels of Aβ 1-40 therein were determined using human Aβ 1-40 immunoassay kits (Invitrogen, Carlsbad, CA, USA). Experiments were performed according to the protocol of the manufacturer of the kits.

4.4.3. Assessment of Cell-Free Aβ 1-40 Degradation In Vitro

The culture medium that contained the proteases to degrade Aβ was collected and used for the cell-free assay of Aβ degradation. Ten nanograms of Aβ 1-40 was added to 300 μL culture medium that contained acteoside or isoacteoside, and incubated at 37 °C for 24 h. The remaining Aβ was then quantified using human Aβ 1-40 immunoassay kits.

4.4.4. Assessment of Aβ 1-42 Oligomerization In Vitro

Aβ 1-42 (100 μM) was dissolved in F-12 medium that contained acteoside or isoacteoside, and incubated at 4 °C for 24 h to accelerate Aβ oligomerization. The reaction solution was mixed with 5 μM ThT, and was then incubated for 30 min. The intensity of fluorescence at an emission wavelength of 485 nm was measured under excitation at a wavelength of 450 nm.

4.5. Statistical Analysis

The data of passive avoidance response were analyzed by performing a Kruskal-Wallis non-parametric one-way analysis of variance, followed by Dunn's test. One-way analysis of variance (ANOVA) and then Scheff's test were applied to data concerning spatial performance, probe test, the ratio of amyloid deposition, the activities of AChE and MAO, the levels of central neurotransmitters and their metabolites, and cell viability and amyloidogenic test. Significant differences in all statistical evaluations were calculated using SPSS software (version 22, IBM, Armonk, NY, USA) and p-values < 0.05 were considered significance.

5. Conclusions

Based on our results and those presented elsewhere [13,16,18,19,35], we suggest that acteoside and isoacteoside are potential therapeutic phenylethanoid glycosides for AD. The memory-improving mechanism of acteoside and isoacteoside involves reducing Aβ deposition and Aβ cytotoxicity by inhibiting Aβ oligomerization through the catechol moiety [19] and promoting Aβ degradation, and then reversing cortical cholinergic dysfunction, which includes the inhibition of AChE activity. Isoacteoside is more effective than acteoside with respect to amyloidogenesis and amyloid oligomerization, and it exhibits a different behavioral-improving pattern against Aβ 1-42-induced behavioral dysfunction. In the future, molecular docking studies of acteoside and isoacteoside against amyloid protein should be conducted and the interaction between phenylethanoid glycosides and amyloid protein must be assayed.

Supplementary Materials: Supplementary materials can be found at www.mdpi.com/1422-0067/18/4/895/s1.

Acknowledgments: We would like to thank the financial support of Institute for Information Industry 96-EC-17-A-20-I1-0049. We also thank Sinphar Pharmaceutical Co., Ltd. (I-Lan, Taiwan) for providing acteoside and isoacteoside.

Author Contributions: Chi-Rei Wu performed the in vivo experiments and analyzed the data; Young-Ji Shiao and Muh-Hwan Su performed the in vitro experiments and analyzed the data; Hang-Ching Lin isolated and identified these compounds; Chi-Rei Wu conceived, designed, supervised the study, drafted and revised the manuscript.

Conflicts of Interest: The authors declare no conflict of interest.

Abbreviations

Aβ	Amyloid β peptide
Ach	Acetylcholine
AChE	Acetylcholinesterase
AD	Alzheimer's disease
APP	Amyloid precursor protein
BF	Basal forebrain
Ch	Choline
DA	Dopamine
MAO	Monoamine oxidase
MTT	3-[4,5-dimethylthiazol-2-yl]-2,5-diphenyl-tetrazolium bromide
MWM	Morris water maze
NE	Norepinephrine
PBS	Phosphate buffer saline
ThT	Thioflavin T

References

1. Hardy, J.; Selkoe, D.J. The amyloid hypothesis of Alzheimer's disease: Progress and problems on the road to therapeutics. *Science* **2002**, *297*, 353–356. [CrossRef] [PubMed]
2. Karran, E.; Mercken, M.; de Strooper, B. The amyloid cascade hypothesis for Alzheimer's disease: An appraisal for the development of therapeutics. *Nat. Rev. Drug Discov.* **2011**, *10*, 698–712. [CrossRef] [PubMed]
3. Querfurth, H.W.; LaFerla, F.M. Alzheimer's disease. *N. Engl. J. Med.* **2010**, *362*, 329–344. [CrossRef] [PubMed]
4. Li, L.; Tsao, R.; Liu, Z.; Liu, S.; Yang, R.; Young, J.C.; Zhu, H.; Deng, Z.; Xie, M.; Fu, Z. Isolation and purification of acteoside and isoacteoside from *Plantago psyllium* L. by high-speed counter-current chromatography. *J. Chromatogr. A* **2005**, *1063*, 161–169. [CrossRef] [PubMed]
5. Pettit, G.R.; Numata, A.; Takemura, T.; Ode, R.H.; Narula, A.S.; Schmidt, J.M.; Cragg, G.M.; Pase, C.P. Antineoplastic agents, 107. Isolation of acteoside and isoacteoside from *Castilleja linariaefolia*. *J. Nat. Prod.* **1990**, *53*, 456–458. [CrossRef] [PubMed]
6. Ohno, T.; Inoue, M.; Ogihara, Y.; Saracoglu, I. Antimetastatic activity of acteoside, a phenylethanoid glycoside. *Biol. Pharm Bull.* **2002**, *25*, 666–668. [CrossRef] [PubMed]
7. Chiou, W.F.; Lin, L.C.; Chen, C.F. Acteoside protects endothelial cells against free radical-induced oxidative stress. *J. Pharm Pharmacol.* **2004**, *56*, 743–748. [CrossRef] [PubMed]
8. Lee, K.J.; Woo, E.R.; Choi, C.Y.; Shin, D.W.; Lee, D.G.; You, H.J.; Jeong, H.G. Protective effect of acteoside on carbon tetrachloride-induced hepatotoxicity. *Life Sci.* **2004**, *74*, 1051–1064. [CrossRef] [PubMed]
9. Koo, K.A.; Kim, S.H.; Oh, T.H.; Kim, Y.C. Acteoside and its aglycones protect primary cultures of rat cortical cells from glutamate-induced excitotoxicity. *Life Sci.* **2006**, *79*, 709–716. [CrossRef] [PubMed]
10. Wang, H.; Xu, Y.; Yan, J.; Zhao, X.; Sun, X.; Zhang, Y.; Guo, J.; Zhu, C. Acteoside protects human neuroblastoma SH-SY5Y cells against β-amyloid-induced cell injury. *Brain Res.* **2009**, *1283*, 139–147. [CrossRef] [PubMed]
11. Yamada, P.; Iijima, R.; Han, J.; Shigemori, H.; Yokota, S.; Isoda, H. Inhibitory effect of acteoside isolated from *Cistanche tubulosa* on chemical mediator release and inflammatory cytokine production by RBL-2H3 and KU812 cells. *Planta Med.* **2010**, *76*, 1512–1518. [CrossRef] [PubMed]
12. Jing, W.; Chunhua, M.; Shumin, W. Effects of acteoside on lipopolysaccharide-induced inflammation in acute lung injury via regulation of NF-κB pathway in vivo and in vitro. *Toxicol. Appl. Pharmacol.* **2015**, *285*, 128–135. [CrossRef] [PubMed]
13. Lin, J.; Gao, L.; Huo, S.X.; Peng, X.M.; Wu, P.P.; Cai, L.M.; Yan, M. Effect of acteoside on learning and memory impairment induced by scopolamine in mice. *China J. Chin. Mater. Med.* **2012**, *37*, 2956–2959.
14. Xiong, L.; Mao, S.; Lu, B.; Yang, J.; Zhou, F.; Hu, Y.; Jiang, Y.; Shen, C.; Zhao, Y. *Osmanthus fragrans* flower extract and acteoside protect against D-galactose-induced aging in an ICR mouse model. *J. Med. Food* **2016**, *19*, 54–61. [CrossRef] [PubMed]

15. Gao, L.; Peng, X.; Huo, S.; He, Y.; Yan, M. Acteoside enhances expression of neurotrophin-3 in brain tissues of subacute aging mice induced by D-galactose combined with aluminum trichloride. *Chin. J. Cell. Mol. Immunol.* **2014**, *30*, 1022–1025.

16. Peng, X.M.; Gao, L.; Huo, S.X.; Liu, X.M.; Yan, M. The mechanism of memory enhancement of acteoside (Verbascoside) in the senescent mouse model induced by a combination of D-gal and AlCl$_3$. *Phytother. Res.* **2015**, *29*, 1137–1144. [CrossRef] [PubMed]

17. Gao, L.; Peng, X.M.; Huo, S.X.; Liu, X.M.; Yan, M. Memory enhancement of acteoside (Verbascoside) in a senescent mice model induced by a combination of D-gal and AlCl$_3$. *Phytother. Res.* **2015**, *29*, 1131–1136. [CrossRef] [PubMed]

18. Wang, H.Q.; Xu, Y.X.; Zhu, C.Q. Upregulation of heme oxygenase-1 by acteoside through ERK and PI3 K/Akt pathway confer neuroprotection against β-amyloid-induced neurotoxicity. *Neurotox. Res.* **2012**, *21*, 368–378. [CrossRef] [PubMed]

19. Kurisu, M.; Miyamae, Y.; Murakami, K.; Han, J.; Isoda, H.; Irie, K.; Shigemori, H. Inhibition of amyloid β aggregation by acteoside, a phenylethanoid glycoside. *Biosci. Biotechnol. Biochem.* **2013**, *77*, 1329–1332. [CrossRef] [PubMed]

20. Myhrer, T. Neurotransmitter systems involved in learning and memory in the rat: A meta-analysis based on studies of four behavioral tasks. *Brain Res. Rev.* **2003**, *41*, 268–287. [CrossRef]

21. Itoh, A.; Nitta, A.; Nadai, M.; Nishimura, K.; Hirose, M.; Hasegawa, T.; Nabeshima, T. Dysfunction of cholinergic and dopaminergic neuronal systems in β-amyloid protein—Infused rats. *J. Neurochem.* **1996**, *66*, 1113–1117. [CrossRef] [PubMed]

22. Tran, M.H.; Yamada, K.; Nabeshima, T. Amyloid β-peptide induces cholinergic dysfunction and cognitive deficits: A minireview. *Peptides* **2002**, *23*, 1271–1283. [CrossRef]

23. Yamada, K.; Tanaka, T.; Mamiya, T.; Shiotani, T.; Kameyama, T.; Nabeshima, T. Improvement by nefiracetam of β-amyloid-(1-42)-induced learning and memory impairments in rats. *Br. J. Pharmacol.* **1999**, *126*, 235–244. [CrossRef] [PubMed]

24. Wu, C.R.; Lin, H.C.; Su, M.H. Reversal by aqueous extracts of *Cistanche tubulosa* from behavioral deficits in Alzheimer's disease-like rat model: Relevance for amyloid deposition and central neurotransmitter function. *BMC Complement. Altern. Med.* **2014**, *14*, 202. [CrossRef] [PubMed]

25. Lee, K.Y.; Jeong, E.J.; Lee, H.S.; Kim, Y.C. Acteoside of Callicarpa dichotoma attenuates scopolamine-induced memory impairments. *Biol. Pharm. Bull.* **2006**, *29*, 71–74. [CrossRef] [PubMed]

26. Arai, H.; Ichimiya, Y.; Kosaka, K.; Moroji, T.; Iizuka, R. Neurotransmitter changes in early- and late-onset Alzheimer-type dementia. *Prog. Neuropsychopharmacol. Biol. Psychiatry* **1992**, *16*, 883–890. [PubMed]

27. Gulyas, B.; Pavlova, E.; Kasa, P.; Gulya, K.; Bakota, L.; Varszegi, S.; Keller, E.; Horvath, M.C.; Nag, S.; Hermecz, I.; et al. Activated MAO-B in the brain of Alzheimer patients, demonstrated by ^{11}C-L-deprenyl using whole hemisphere autoradiography. *Neurochem. Int.* **2011**, *58*, 60–68. [CrossRef] [PubMed]

28. Parihar, M.S.; Hemnani, T. Alzheimer's disease pathogenesis and therapeutic interventions. *J. Clin. Neurosci.* **2004**, *11*, 456–467. [CrossRef] [PubMed]

29. Kim, D.; Baik, S.H.; Kang, S.; Cho, S.W.; Bae, J.; Cha, M.Y.; Sailor, M.J.; Mook-Jung, I.; Ahn, K.H. Close correlation of monoamine oxidase activity with progress of Alzheimer's disease in mice, observed by in vivo two-photon imaging. *ACS Cent. Sci.* **2016**, *2*, 967–975. [CrossRef] [PubMed]

30. Nakamura, S.; Kawamata, T.; Akiguchi, I.; Kameyama, M.; Nakamura, N.; Kimura, H. Expression of monoamine oxidase B activity in astrocytes of senile plaques. *Acta Neuropathol.* **1990**, *80*, 419–425. [CrossRef] [PubMed]

31. Smale, G.; Nichols, N.R.; Brady, D.R.; Finch, C.E.; Horton, W.E., Jr. Evidence for apoptotic cell death in Alzheimer's disease. *Exp. Neurol.* **1995**, *133*, 225–230. [CrossRef] [PubMed]

32. Blanco, N.J.; Love, B.C.; Cooper, J.A.; McGeary, J.E.; Knopik, V.S.; Maddox, W.T. A frontal dopamine system for reflective exploratory behavior. *Neurobiol. Learn. Mem.* **2015**, *123*, 84–91. [CrossRef] [PubMed]

33. Young, J.W.; Kooistra, K.; Geyer, M.A. Dopamine receptor mediation of the exploratory/hyperactivity effects of modafinil. *Neuropsychopharmacology* **2011**, *36*, 1385–1396. [CrossRef] [PubMed]

34. Alttoa, A.; Seeman, P.; Koiv, K.; Eller, M.; Harro, J. Rats with persistently high exploratory activity have both higher extracellular dopamine levels and higher proportion of D2 (High) receptors in the striatum. *Synapse* **2009**, *63*, 443–446. [CrossRef] [PubMed]

Int. J. Mol. Sci. **2017**, *18*, 895

35. Bai, P.; Peng, X.M.; Gao, L.; Huo, S.X.; Zhao, P.P.; Yan, M. Study on protective effect of acteoside on cellular model of Alzheimer's disease induced by okadaic acid. *China J. Chin. Mater. Med.* **2013**, *38*, 1323–1326.

36. Pu, X.; Song, Z.; Li, Y.; Tu, P.; Li, H. Acteoside from *Cistanche salsa* inhibits apoptosis by 1-methyl-4-phenylpyridinium ion in cerebellar granule neurons. *Planta Med.* **2003**, *69*, 65–66. [CrossRef] [PubMed]

37. Skaper, S.D. The neurotrophin family of neurotrophic factors: An overview. *Methods Mol. Biol.* **2012**, *846*, 1–12. [PubMed]

38. Korsching, S.; Auburger, G.; Heumann, R.; Scott, J.; Thoenen, H. Levels of nerve growth factor and its mRNA in the central nervous system of the rat correlate with cholinergic innervation. *EMBO J.* **1985**, *4*, 1389–1393. [PubMed]

39. Auld, D.S.; Mennicken, F.; Quirion, R. Nerve growth factor rapidly induces prolonged acetylcholine release from cultured basal forebrain neurons: Differentiation between neuromodulatory and neurotrophic influences. *J. Neurosci.* **2001**, *21*, 3375–3382. [PubMed]

40. Mufson, E.J.; Counts, S.E.; Perez, S.E.; Ginsberg, S.D. Cholinergic system during the progression of Alzheimer's disease: Therapeutic implications. *Expert Rev. Neurother.* **2008**, *8*, 1703–1718. [CrossRef] [PubMed]

41. Glowinski, J.; Iversen, L.L. Regional studies of catecholamines in the rat brain. I. The disposition of [^{3}H]norepinephrine, [^{3}H]dopamine and [^{3}H]dopa in various regions of the brain. *J. Neurochem.* **1966**, *13*, 655–669. [CrossRef] [PubMed]

International Journal of
Molecular Sciences

MDPI

Article

Cultivar-Specific Changes in Primary and Secondary Metabolites in Pak Choi (*Brassica Rapa*, Chinensis Group) by Methyl Jasmonate

Moo Jung Kim [1], Yu-Chun Chiu [1], Na Kyung Kim [2], Hye Min Park [2], Choong Hwan Lee [2], John A. Juvik [3] and Kang-Mo Ku [1],*

[1] Division of Plant and Soil Sciences, West Virginia University, Morgantown, WV 26506, USA; mjkim@mail.wvu.edu (M.J.K); yuchiu@mix.wvu.edu (Y.-C.C.)
[2] Department of Bioscience and Biotechnology, Konkuk University, Seoul 143-701, Korea; god_1012@hanmail.net (N.K.K.); ramgee@naver.com (H.M.P.); chlee123@konkuk.ac.kr (C.H.L.)
[3] Department of Crop Sciences, University of Illinois at Urbana-Champaign, Urbana, IL 61801, USA; juvik@illinois.edu
* Correspondence: kangmo.ku@mail.wvu.edu; Tel.: +1-304-293-2549

Academic Editor: Toshio Morikawa
Received: 14 March 2017; Accepted: 2 May 2017; Published: 7 May 2017

Abstract: Glucosinolates, their hydrolysis products and primary metabolites were analyzed in five pak choi cultivars to determine the effect of methyl jasmonate (MeJA) on metabolite flux from primary metabolites to glucosinolates and their hydrolysis products. Among detected glucosinolates (total 14 glucosinolates; 9 aliphatic, 4 indole and 1 aromatic glucosinolates), indole glucosinolate concentrations (153–229%) and their hydrolysis products increased with MeJA treatment. Changes in the total isothiocyanates by MeJA were associated with epithiospecifier protein activity estimated as nitrile formation. Goitrin, a goitrogenic compound, significantly decreased by MeJA treatment in all cultivars. Changes in glucosinolates, especially aliphatic, significantly differed among cultivars. Primary metabolites including amino acids, organic acids and sugars also changed with MeJA treatment in a cultivar-specific manner. A decreased sugar level suggests that they might be a carbon source for secondary metabolite biosynthesis in MeJA-treated pak choi. The result of the present study suggests that MeJA can be an effective agent to elevate indole glucosinolates and their hydrolysis products and to reduce a goitrogenic compound in pak choi. The total glucosinolate concentration was the highest in "Chinese cabbage" in the control group (32.5 µmol/g DW), but indole glucosinolates increased the greatest in "Asian" when treated with MeJA.

Keywords: glucosinolate; isothiocyanate; methyl jasmonate; pak choi; *Brassica rapa*

1. Introduction

Pak choi (*Brassica rapa*, Chinensis group) is a cool-season crop similar to many other *Brassica* vegetables, such as kale and broccoli, and was domesticated in China [1]. Pak choi is a popularly-consumed vegetable in China and is showing an increase in consumption in Europe and North America, primarily due to its comparatively mild flavor [1,2]. In the U.S., its production is increasing for farmer's markets and community-supported agriculture [1]. As a *Brassica* vegetable, pak choi provides a number of phytonutrients, in particular glucosinolates (19.36–63.43 µmol/g DW according to Wiesner et al. [3]). Although its popularity is increasing, the nutritional quality of pak choi has not been thoroughly investigated, and only a few studies have reported glucosinolate profiles from pak choi [3,4].

Pak choi contains a number of glucosinolates including gluconapin, glucobrassicanapin, progoitrin, glucobrassicin and neoglucobrassicin [3,4]. Glucosinolates are nitrogen- and sulfur-containing secondary metabolites derived from amino acids. Depending on their structure and precursor amino acid, glucosinolates are classified into three major groups: aliphatic (from methionine), indole (from tryptophan) and aromatic (from phenylalanine or tyrosine) glucosinolates (Figure 1) [5]. The first two steps in the biosynthetic pathway facilitate chain elongation and core structure formation, converting the precursor amino acid into desulfoglucosinolates after which final products are formed through secondary structural modifications [6]. Once formed and stored in the vacuoles of cells, glucosinolates can be hydrolyzed by an endogenous enzyme myrosinase following cellular disruption [7]. After glucosinolate hydrolysis, a few forms of hydrolysis products including isothiocyanates, nitriles, epithionitriles, thiocyanates and indoles are formed. The concentrations of specific hydrolysis products are determined by a number of factors including the activity of epithiospecifier protein (ESP) and epithiospecifier protein modifier 1, pH and the presence and concentration of certain metal ions (Figure 1). The potential health benefits and bioactivity of glucosinolates are attributed to their hydrolysis products, not the parent glucosinolates. A number of cell culture and pre-clinical studies have reported that *Brassica* vegetables and glucosinolate hydrolysis products are beneficial against carcinogenesis [8–10]. However, it was also shown that glucosinolate concentration was positively correlated with quinone reductase-inducing activity, a biomarker for anti-carcinogenic activity, in arugula (*Eruca sativa*) and horseradish (*Armoracia rusticana*) [11,12]. Therefore, increasing glucosinolate concentration could be a good strategy to enhance the potential health benefits of *Brassica* vegetables including pak choi.

Glucosinolates are a secondary metabolite whose biosynthesis can be induced by various biotic and abiotic factors [13]. Since glucosinolates are involved in plant defense, insect damage or physical wounding can induce glucosinolate biosynthesis. Glucosinolate concentrations can also be regulated by sulfur and nitrogen fertilization. In addition, growing degree days, solar radiation, number of days after transplanting and precipitation have been shown to affect glucosinolate biosynthesis [14]. These studies suggest that cultural practices that are effective and economic to apply can be developed to increase glucosinolate concentration and to improve the potential health benefits of glucosinolate-containing crops. It has been suggested that aliphatic, indole and aromatic glucosinolates biosynthesis have different regulatory mechanisms. Brown et al. [15] have reported that in broccoli, aliphatic glucosinolates are primarily controlled by genetic factors (accounting for >60% of the total variance), whereas indole glucosinolates are inducible compounds that are primarily influenced by the environment and cultural conditions or insect damage [14]. This report found that indole glucosinolates are relatively easier to increase by manipulating cultural conditions compared to aliphatic glucosinolates. In addition, conventional production systems minimize plant biotic stress by applying fungicides and insecticides, resulting in lower concentrations of indole glucosinolates in *Brassica* crops.

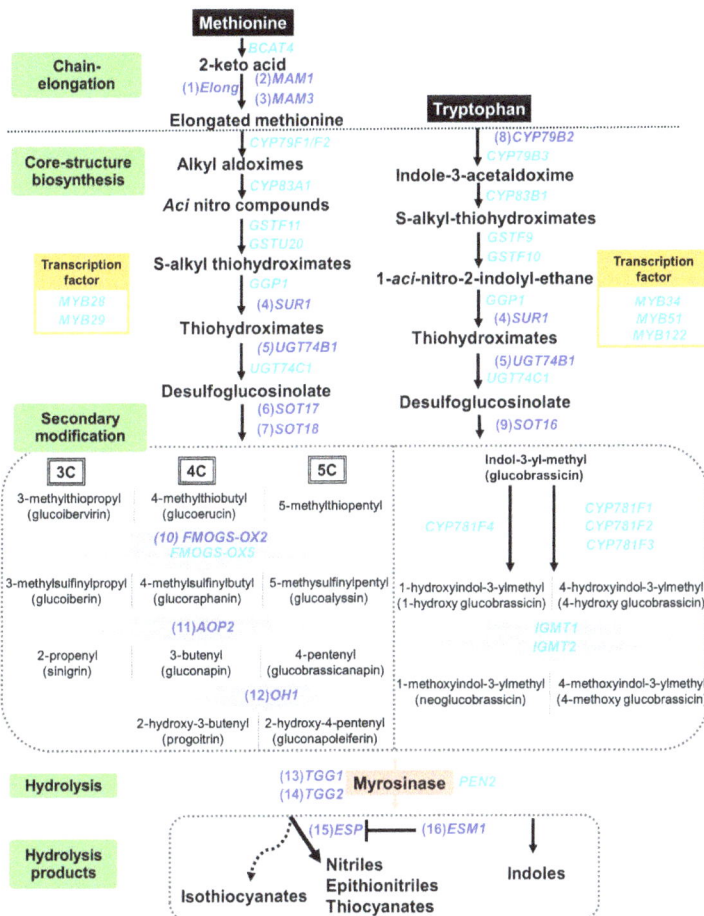

Figure 1. Schematic biosynthetic pathway of aliphatic and indole glucosinolates and involved genes (modified from Yi et al. [16]). Genes following numbers in brackets are analyzed in this study (Table S1). Gene names in light blue without a number were not analyzed in this study. The dashed line separates the biosynthesis pathway into three sections—chain elongation, core-structure biosynthesis, and secondary modification with addition of glucosinolate hydrolysis products. The dashed lines between compounds are to clarify and group the compounds from the same precursor or in the same group. Dotted arrow indicates another reaction pathway without involving of *ESP*. *ESM1* regulates *ESP* (bar and dash).

Previous studies have found that exogenous treatment of methyl jasmonate (MeJA) can induce indole glucosinolate biosynthesis in *Brassica* vegetables [14,17,18]. MeJA is registered in the Environmental Protection Agency (EPA) and is certified as a safe compound for all food commodities when applied preharvest [19]. Additionally, only 2–4 days are needed to elevate indole glucosinolate concentration prior to harvest in *Brassica oleracea* or *B. rapa*. [4,17] and, therefore, does not affect crop yield. Indole glucosinolates such as neoglucobrassicin are normally not accumulated at higher concentrations without severe herbivore or tissue damage, but can be regulated by MeJA. Ku et al. [18] have reported increased concentrations of glucobrassicin, an indole glucosinolate, by MeJA in the leaf tissues of the kale cultivars "Dwarf Blue Curled Vates" and "Red Winter" (98–166% of increase) in a two-year study while no significant change in aliphatic glucosinolate concentrations was observed. In the leaves of "Green Magic" and "VI-158" broccoli, indole glucosinolates also increased by MeJA treatment, whereas aliphatic glucosinolates were much less affected [17].

Although MeJA effects on pak choi glucosinolates have previously been reported [4,20], these studies only focused on major glucosinolates or analyzed one cultivar. Moreover, the profile of glucosinolate hydrolysis products, which are ultimately responsible for the bioactivity and potential health benefits, has not been investigated in pak choi. Considering its increasing popularity, pak choi merits further study on nutritional composition and the strategies to improve its health-promoting value. The objective of this study was to determine the metabolite changes in pak choi associated with foliar application of MeJA. Glucosinolates and their hydrolysis products, as well as the expression of genes in glucosinolate biosynthesis were analyzed in five pak choi cultivars to evaluate the effect of MeJA treatment on glucosinolate metabolism. Primary metabolites including amino acids, organic acids, sugars and sugar derivatives were analyzed to better understand how MeJA treatment changes metabolite flux from primary to secondary metabolites in pak choi.

2. Results and Discussions

2.1. Glucosinolate Concentrations

A total of 14 glucosinolates were detected with gluconapin and glucobrassicanapin as the predominant glucosinolates, representing 7–54% and 14–31% of the total glucosinolate concentration, respectively, depending on cultivar and treatment (Figure 2). Other aliphatic glucosinolates detected in pak choi included glucoiberin, progoitrin, glucoalyssin, gluconapoleiferin, glucoraphanin, sinigrin and glucoerucin (Table S2). The indole glucosinolates 4-hydroxyglucobrassicin, glucobrassicin, 4-methoxyglucobrassicin and neoglucobrassicin and aromatic glucosinolate gluconasturtiin were also present in the five pak choi cultivars. Our results are in agreement with a previous study that reported gluconapin and glucobrassicanapin as the major glucosinolates in pak choi [3]. Glucosinolate composition differed among cultivars and treatments. Gluconapin and glucobrassicanapin concentrations in the control ranged from 1.03 to 17.46 µmol/g DW and from 1.41 to 8.65 µmol/g DW, respectively (Figure 2). Sinigrin was only detected in control "Baby bok choy" (Supplementary Table S2). Glucoraphanin and glucoerucin were also detected in selected cultivars. These results agree with a previous report showing significant variation in glucosinolate concentrations among different pak choi cultivars [3,4]. Additionally, Wiesner et al. [3] reported that aliphatic glucosinolates represented 92–98% of the total glucosinolates, similar to our result (51–91% depending on cultivar and treatment). The control "Chinese cabbage" was found to contain the highest concentration of total glucosinolates, suggesting that this cultivar may possess the greatest glucosinolate-related health-promoting properties under standard production conditions among the five cultivars investigated in this study.

Glucosinolate composition changed in response to MeJA treatment differentially among cultivars (Figure 2). For example, gluconapin and glucobrassicanapin decreased in "Baby bok choy", "Chinese cabbage" and "Pak choi pechay", while increasing in "Asian" with MeJA treatment. Similarly, progoitrin increased in 'Baby bok choy' and 'Asian', but decreased in the other cultivars when treated with MeJA. However, neoglucobrassicin, an indole glucosinolate, increased in all cultivars after MeJA treatment by 153–229%. Glucobrassicin also increased in "Asian", "Col baby choi" and "Pak choi pechay", while a decreasing 4-methoxyglucobrassicin concentration was found in "Baby bok choy" and "Chinese cabbage". Our result is in agreement with the report by Zang et al. [4] who showed a significant increase in indole glucosinolate concentrations with MeJA treatment in four pak choi cultivars. However, they found increased aliphatic glucosinolates in only one cultivar, indicating that aliphatic glucosinolate concentrations are less affected by MeJA compared to indole glucosinolates, and their change could be cultivar specific, consistent with our result. MeJA's effect on glucosinolates has been reported in other *Brassica* vegetables, including broccoli, kale, cauliflower and Chinese cabbage [17,18,21,22]. These studies reported that indole glucosinolates, in particular neoglucobrassicin, increased while aliphatic glucosinolates were less influenced by MeJA. However, we found that aliphatic glucosinolates can also significantly change in response to MeJA in pak choi, with variation in their change depending on compound and cultivar, indicating that glucosinolate changes by MeJA treatment differ among species, crops and cultivars. When treated with MeJA, "Asian" showed the greatest increase in the total glucosinolate concentration (2.7-fold), indicating that this cultivar has the greatest sensitivity to MeJA treatment among the investigated cultivars, and thus, MeJA can be an effective agent to elevate glucosinolates of this pak choi cultivar. Additionally, changes in glucosinolate concentrations were partially associated with differential expression levels of genes involved in glucosinolate biosynthesis in response to MeJA treatment, depending on cultivar. In particular, the expression level of *OH1*, a gene converting gluconapin to progoitrin (Figure 1), increased by 14.4-, 49.8- and 7.3-fold in "Baby bok choy", "Asian" and "Pak choi pechay", respectively (Supplementary Table S1). The expression level of *OH1* was positively correlated with progoitrin ($r = 0.902$, $p = 0.0004$, $n = 10$). Gluconapoleiferin, where the same genes are responsible for its biosynthesis, increased in "Baby bok choy", "Asian" and "Pak choi pechay" with upregulation of those genes (Table S1). The expression level of *OH1* was also positively correlated with gluconapoleiferin ($r = 0.846$, $p = 0.0021$, $n = 10$). Although indole glucosinolates, in particular neoglucobrassicin, were more affected by MeJA than aliphatic glucosinolates, changes in indole glucosinolate concentration and related gene expression have been reported [17,20]. However, how aliphatic glucosinolates are affected by MeJA has been less elucidated.

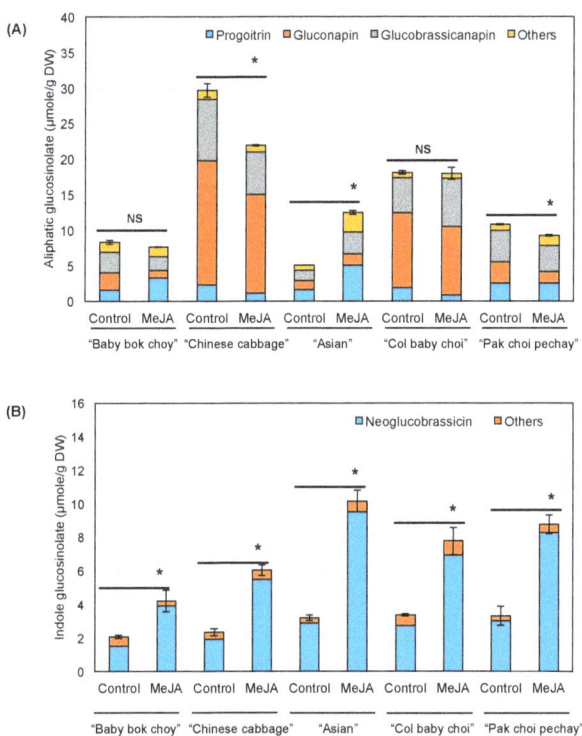

Figure 2. Aliphatic (**A**) and indole (**B**) glucosinolate concentration in control and MeJA-treated pak choi. Data are presented as the mean concentration ± the standard error of the total concentration ($n = 3$). Asterisks (*) above the bar indicate a significant difference of total concentration between treatments by Student's *t*-test at $p \leq 0.05$. MeJA, methyl jasmonate; DW, dry weight; NS, not-significant.

2.2. Glucosinolate Hydrolysis Products

We detected 14 hydrolysis products including isothiocyanates, nitriles, indoles and oxazolidine-thione (as shown by Rask et al. [23]) and their changing concentrations related to MeJA treatment depended on cultivar, similar to the glucosinolates (Table 1). The 3-butenyl isothiocyanate, 4-pentenyl isothiocyanate and phenethyl isothiocyanate, the hydrolysis product of gluconapin, glucobrassicanapin and gluconasturtiin, respectively, decreased with MeJA treatment in all cultivars, except for increased 4-pentenyl isothiocyanate and phenethyl isothiocyanate in "Asian". However, nitriles, including crambene (from progoitrin), 1-cyano-3,4-epithiobutane (from gluconapin), 1-cyano-4,5-epithiopentane (from glucobrassicanapin), 1-cyano-2-hydroxy-3,4-epithiobutane (from progoitrin) and an oxazolidine-thione goitrin (from progoitrin), changed differentially among cultivars with MeJA treatment. Most nitriles increased in "Asian"but goitrin decreased in all cultivars with MeJA treatment. Crambene increased in "Baby bok choy" and "Asian", but decreased in "Chinese cabbage" and "Col baby choi". In contrast to isothiocyanates and nitriles, 1-methoxyindole-3-carbinol, a hydrolysis product of neoglucobrassicin, and other compounds, including 1-methoxyindole-3-carboxaldehyde, 1-methoxyindole-3-acetonitrile and indole-3-acetonitrile, significantly increased in all five cultivars, except for no statistical change in 1-methoxyindole-3-carbinol in "Chinese cabbage".

Although the glucosinolate profile and hydrolysis products from indole glucosinolates have been investigated [3], a full profile of glucosinolate hydrolysis products has not been reported for pak choi. Ku et al. [18] observed differential hydrolysis product composition between "Dwarf Blue Curled Vates" and "Red Winter" kale and a significant increase in 1-methoxyindole-3-carbinol in both cultivars with MeJA treatment. Jasmonic acid also increased 1-methoxyindole-3-carbinol in "VI-158" and "Green Magic" broccoli, but depending on the concentration of MeJA [17]. Glucosinolate hydrolysis product concentrations clearly change with MeJA treatment, in particular the hydrolysis products of neoglucobrassicin.

In addition to an increased amount of the hydrolysis product of neoglucobrassicin, we also found that hydrolysis products from aliphatic glucosinolates changed with MeJA treatment. For instance, 3-butenyl and 4-pentenyl isothiocyanates decreased by MeJA in all cultivars except for increased 4-pentenyl isothiocyanate in "Asian" (Table 1). This is probably related to the activity of ESP, which enhances the formation of epithionitriles over isothiocyanates [24]. We indirectly measured the ESP activity by incubating pak choi protein extract with the extract of horseradish, which has a simple glucosinolate profile [11], and found that nitrile formation (%) increased in all cultivars when treated with MeJA except for "Chinese cabbage" and "Asian" for the hydrolysis of gluconasturtiin (Figure 3). Although nitrile formation (%) of MeJA-treated "Asian" was significantly higher than control according to allyl isothiocyanate (from sinigrin), there was no significant difference in phenethyl isothiocyanate (from gluconasturtiin) (Figure 3). There was a significant correlation ($r = 0.700$, $p = 0.0239$, $n = 10$) between nitrile formation from sinigrin and nitrile formation from gluconasturtiin. Additionally, we found a significant correlation ($r = 0.925$, $p < 0.0001$, $n = 10$) between nitrile formation from sinigrin and nitrile formation from neoglucobrassicin, where the nitrile formation (%) from neoglucobrassicin was determined as the percentage of 1-metoxyindole-3-acetonitrile to the total hydrolysis products (Table 1). Other compounds, such as crambene, 1-cyano-3,4-epithiobutane, 1-cyano-4,5-epithiopentane and 1-cyano-2-hydroxy-3,4-epithiobutane, also changed with MeJA treatment depending on cultivar. These results indicate that ESP activity in general increased with MeJA in most pak choi cultivars and thus partially explains why isothiocyanates tended to decrease with MeJA application. We also found that MeJA treatment significantly reduced myrosinase activity in "Baby bok choy", "Chinese cabbage" and "Pak choi pechay" cultivars, indicating a cultivar-specific response to MeJA (Supplementary Figure S1). Regardless of cultivar, goitrin, an oxazolidine-thione from progoitrin, decreased by MeJA. This is significant because goitrin is a goitrogenic compound that can disrupt hormone production in the thyroid gland by inhibiting uptake of iodine [25], and therefore, high level of goitrin intake could be a problem, especially under iodine malnutrition.

Table 1. GC-MS peak intensity changes in hydrolysis products by MeJA foliar spray treatment in pak choi.

Cultivar	Treatment	From Aliphatic Glucosinolates (Peak Intensity)[z]						
		3-Butenyl ITC [y]	4-Pentenyl ITC	Crambene	1-Cyano-3,4-Epithiobutane	1-Cyano-4,5-Epithiopentane	1-Cyano-2-Hydroxy-3,4-Epithiobutane	Goitrin
		C_5H_7NS	C_6H_9NS	C_5H_7NO [w]	C_5H_7NS	C_6H_9NS	C_5H_7NOS	C_5H_7NOS
Baby bok choy	Control	68.53a [x]	8.67a	ND [w]	104.19a	42.72a	20.47a	45.53a
	MeJA	10.76b	4.60b	18.01	20.49b	18.81b	21.48b	6.02b
Chinese cabbage	Control	868.29a	55.55a	7.70	242.24a	41.77a	10.46a	118.07
	MeJA	35.52b	3.33b	ND	23.85b	3.47b	0.96b	ND
Asian	Control	70.70a	10.82b	ND	27.01b	10.07b	5.42b	41.32a
	MeJA	49.19b	24.52a	24.49	35.04a	33.46a	32.52a	25.16b
Col baby choi	Control	501.80a	26.01a	9.00a	298.02a	39.28a	13.66a	81.26
	MeJA	113.48b	18.55b	6.31b	99.48b	28.84b	4.92b	ND
Pak choi pechay	Control	97.93a	15.37a	7.94a	58.26a	28.34a	15.77a	93.61
	MeJA	19.89b	10.91b	8.12a	11.64b	12.98b	6.55b	ND

		From Indole Glucosinolates (Peak Intensity)				
		1-MI3C	1-MI3Carx	1-MI3ACN	I3CA	I3A
		$C_{10}H_{11}$	$C_{10}H_9NO_2$	$C_{11}H_{10}N_2O$	C_9H_7NO	$C_{10}H_8N_2$
Baby bok choy	Control	26.40b	12.61b	5.13b	1.70	0b
	MeJA	72.16a	117.22a	280.24a	ND	27.02a
Chinese cabbage	Control	29.57a	32.24b	10.94b	4.17	0b
	MeJA	25.17a	97.89a	94.18a	ND	11.33a
Asian	Control	39.98b	32.47b	15.60b	3.44	1.42b
	MeJA	103.66a	129.81a	285.38a	ND	25.29a
Col baby choi	Control	39.11b	40.28b	12.05b	4.82	1.47b
	MeJA	90.46a	86.54a	134.25a	ND	13.88a
Pak choi pechay	Control	32.94b	24.43b	12.51b	ND	1.65b
	MeJA	93.14a	102.37a	208.84a	ND	17.86a

[z] Data represent peak count ($\times 10^3$) of each compound. [y] 3-butenyl ITC and 1-cyano-3,4-epithiobutane from glucobrassicanapin; 4-pentenyl ITC and 1-cyano-4,5-epithiopentane from gluconapin; crambene, goitrin and 1-cyano-2-hydroxy-3,4-epithiobutane from progoitrin; 1-MI3C, 1-MI3Carx and 1-MI3AC, from neoglucobrassicin; I3CA and I3A from glucobrassicin. [x] Means were separated by Student's *t*-test at $p \leq 0.05$ ($n = 3$). [w] ND, not detected.

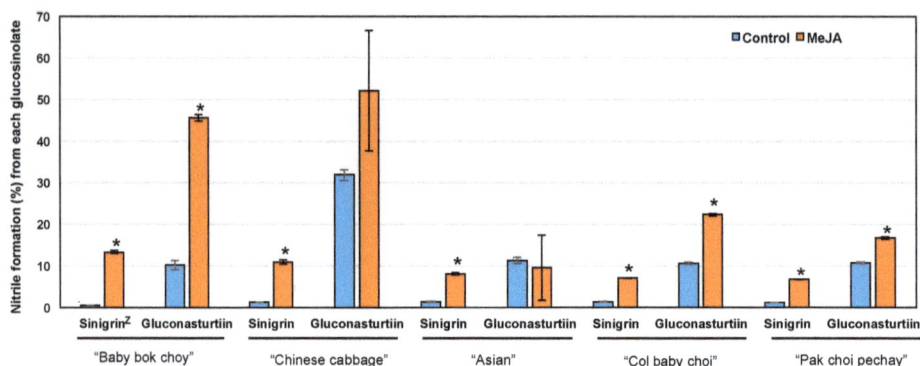

Figure 3. Nitrile formation (%) of control and MeJA-treated pak choi. Nitrile formation (%) is shown as the relative ratio of nitrile to the total hydrolysis product formed (sum of isothiocyanates and nitriles). Data are presented as the mean concentration $\pm$ standard error ($n = 3$). Asterisks (*) above the error bar indicate a significant difference between treatments within same glucosinolates substrate by Student's t-test at $p \leq 0.05$. ^{z} Substrate glucosinolate used in the estimation of nitrile formation activity [nitrile percentage (%) out of total hydrolysis products].

Hydrolysis products of glucosinolates have been considered to reduce the risk of degenerative diseases, especially against carcinogenesis. The 3-butenyl isothiocyanate, a hydrolysis product of gluconapin, has shown anti-proliferative activity against various cancer cell lines including human prostate, lung, cervical, liver and breast cancers, as well as human neuroblastoma and osteosarcoma cell lines [8]. In particular, prostate cancer PC-3 cells had the greatest inhibition by 3-butenyl isothiocyanate with IC$_{50}$ (50% inhibitory concentration) and IC$_{70}$ values of 0.041 and 0.060 μg/mL, respectively, compared to the positive control camptothecin (IC$_{50}$ of 121.60 μM; 42.36 μg/mL). Isothiocyanates have been generally accepted to possess a greater bioactivity compared to other forms of glucosinolate hydrolysis products, but some compounds in other forms have also been studied for their potential health benefits. Crambene, a nitrile from progoitrin, was reported to increase mRNA expression of quinone reductase, as well as the activity of quinone reductase and glutathione-S-transferase [26,27] in mouse hepatoma Hepa1c1c7 cells (0.1–10 mM) or in adult male CDF 344 rats (fed 50 mg/kg of crambene for seven days), indicating a potential anticarcinogenic property of crambene. Another compound, 1-methoxyindole-3-carbinol, a hydrolysis product of neoglucobrassicin, inhibited the growth of human colon cancer DLD-1 and HCT-116 cells in a dose-dependent manner at 10–100 μM [9]. However, the effect of 1-methoxyondole-3-carbinol on carcinogenesis is inconclusive. When applied to murine hepatoma Hepa1c1c7 cells at 50 μM or administered to Winster rat at 570 μmol/kg body weight, 1-methoxyindole-3-carbinol significantly increased cytochrome P-450 1A1, indicating that this compound may help carcinogenesis [28]. Another study reported that male transgenic C57BL/6-Tg(TRAMP)8247Ng/J mouse (transgenic adenocarcinoma of mouse prostate) fed a diet containing 10% of indole glucosinolate-elevated broccoli powder showed no difference from the mouse fed a diet containing normal broccoli powder in the reduction of prostate carcinogenesis [29]. Although not analyzed in this study, indole-3-carbinol, generated from glucobrassicin, has been reported for its potential anticarcinogenic activity in cell culture and preclinical studies, and it is sometimes reported that 1-methoxyindole-3-carbinol may possess a greater bioactivity compared to indole-3-carbinol [9,30–32]. It was suggested that methylation increases hydrophobicity of the compound and enhances cell membrane penetration, and therefore, 1-methoxyindole-3-carbinol may have a greater bioactivity [17]. The concentration of 1-methoxyindole-3-acetonitrile was significantly increased by MeJA treatment due to the upregulation of neoglucobrassicin synthesis and high ESP activity (indirectly estimated as nitrile formation), but there was no report on a health-promoting effect of this compound; therefore, further studies are needed.

This study revealed that MeJA not only increases total glucosinolates, but also changes hydrolysis product concentrations. Isothiocyanates were in general reduced, but 1-methoxyindole-3-acetonitrile increased in all five cultivars, and 1-methoxyindole-3-carbinol increased in four cultivars. Goitrin, a goitrogenic compound, decreased in all cultivars. In particular, "Asian" had increased 4-pentenyl isothiocyanate, crambene and 1-methoxyindole-3-carbinol, but decreased goitrin by MeJA. Our results suggest that MeJA can significantly change glucosinolate metabolites, and "Asian" is specifically responsive to MeJA. Based on the metabolite changes in "Asian" by MeJA, "Asian" might be an excellent choice to improve the health-promoting values of pak choi using MeJA.

2.3. Primary Metabolites

It has been suggested that exogenous MeJA can alter primary metabolites, such as sugars, organic acids and amino acids, and these changes may affect glucosinolate biosynthesis [33]. Therefore, analysis of these metabolites may help to understand their further transition to secondary metabolites. Although the MeJA effect on glucosinolate has been studied in a few *Brassica* crops, how MeJA affects primary metabolism has not been well reported. Primary metabolites with variable importance in projection (VIP) scores over 1.0 were selected based on the cut-off value for VIP advocated by Wold [34] to separate terms that do not make an important contribution to the dimensionality reduction involved in PLS (partial least squares) (VIP < 0.8) from those that might (VIP ≥ 0.8) (Table 2). A higher VIP score indicates a greater difference between treatments and, therefore, can be useful in selecting a biomarker that differs between treatments. According to the VIP scores, three amino acids (alanine, valine, glutamic acid), two organic acids (citric acid and cinnamic acid) and five sugars and sugar derivatives (glycerol, fructose, *myo*-inositol, galactose and maltose) among detected compounds were the most important primary metabolites that differentiate MeJA-treated pak choi from control plants (Table 2). A high VIP score of these compounds indicates that these compounds are more important biomarkers that describe the variation in the primary metabolites of pak choi.

Individual metabolites varied with MeJA treatment among cultivars (Table 2 and Table S3). Among the metabolites with VIP score >1.0, levels of the amino acids alanine, valine and glutamic acid were significantly higher when treated with MeJA in "Pak choi pechay" (Table 2). In contrast, glutamic acid decreased in "Chinese cabbage" and "Col baby choi". Organic acids also varied with MeJA treatment depending on cultivar. Citric acid increased in all cultivars with MeJA treatment. Cultivar-dependent changes with MeJA were also observed for sugar levels. MeJA treatment reduced fructose, maltose and galactose in "Asian", but fructose increased in "Col baby choi". Glycerol and *myo*-inositol increased in four cultivars. Among the selected metabolites, we found a general increase or no change in alanine and valine, increases in organic acids, decreases in mono- and di-saccharide sugars and elevated sugar alcohols. Although we observed increases in some amino acids depending on cultivar, Tytgat et al. [35] reported decreased amino acids in jasmonic acid-treated *B. oleracea* plants. They also reported reduced sugar concentration with jasmonic acid treatment, similar to our observations. Kim [33] reported reduced levels of hexose sugars and TCA cycle intermediates in MeJA-treated "Green Magic" broccoli leaves. These results and our data indicate that exogenous MeJA can decrease sugar levels and changes in secondary metabolism might partially be attributed to reduction of sugars, as sugars may provide the carbon skeleton for secondary metabolite biosynthesis (Figure 4). Since MeJA is synthesized from linolenic acid, MeJA may also affect fatty acid metabolism. When treated with MeJA, linolenic acid concentration, as well as fatty acid composition changed in mature green tomatoes [36]. When treated with MeJA, linolenic acid increased, while linoleic acid was reduced, suggesting a possible MeJA effect on fatty acids and their derivatives. In addition, changes in sugar metabolism could also have affected sugar alcohol biosynthesis, such as glycerol and *myo*-inositol.

Table 2. Fold change of primary metabolites in MeJA-treated pak choi compared to control pak choi.

Cultivar	Amino Acids			Organic Acids		Sugars and Sugar Derivatives				
	Alanine	Valine	Glutamic Acid	Citric Acid	Cinnamic Acid	Maltose	Fructose	Galactose	myo-Inositol	Glycerol
"Baby bok choy"	1.92 *,z	3.50 *	0.97	1.80 *	4.75 *	0.87	0.47 *	2.66 *	1.73 *	2.02 *
"Chinese cabbage"	2.00 *	1.53	0.83 *	2.40 *	1.31	1.33	0.67 *	6.87 *	1.03	3.33 *
"Asian"	1.14	2.00 *	2.85 *	1.20 *	2.32 *	0.24 *	0.65 *	0.54 *	2.26 *	2.17 *
"Col baby choi"	0.97	0.87	0.93 *	2.25 *	2.98 *	0.24 *	1.17 *	0.89	1.39 *	0.87
"Pak choi pechay"	2.23 *	9.36 *	2.68 *	1.50 *	3.27 *	0.79 *	0.86	4.75 *	1.85 *	2.68 *
Total change y	1.59	2.76	1.42	1.73	3.06	0.51	0.75	2.40	1.48	2.02
VIP score x	1.59	1.12	1.00	1.64	1.22	1.02	1.36	1.04	1.20	1.39

x Variable importance in projection. y Total change was calculated as the relative ratio of the total peak intensity of each metabolite in MeJA-treated plants to the total peak intensity of control plants (n = 3). z Values were calculated as the fold change compared to the control group. Asterisk (*) indicates a significant difference of the fold change compared to the control by Student's t-test at $p \leq 0.05$ based on peak intensity (n = 3).

To our knowledge, changes in primary metabolites in response to MeJA treatment have not been reported in pak choi. Additionally, information of primary metabolite changes by MeJA treatment in other *Brassica* crops is also lacking. Liang et al. [37] applied MeJA to turnip (*B. rapa* var. *rapa*) and found that most sugars and amino acids analyzed using nuclear magnetic resonance spectroscopy were reduced by MeJA, in contrast to our results. This difference from our result indicates that MeJA might affect primary metabolism differentially among crops. Moreover, our result shows that the MeJA effect varied among the five pak choi cultivars.

Many of the primary metabolites analyzed in this study play an important role in human diets. For instance, amino acids, in general, are involved in various biochemical mechanisms, such as protein synthesis, cell signaling, osmoregulation and metabolic regulation [38] with some amino acids also associated with mammalian immune systems [39]. Moreover, hexose sugars and organic acids are involved in the primary metabolisms such as the TCA cycle and are used as a precursor of amino acids and secondary metabolites (Figure 4). Therefore, understanding how these primary metabolites change by MeJA treatment in addition to secondary metabolites are important to improve the nutritional value of foods and for developing cultivation regimes to enhance the nutritional properties.

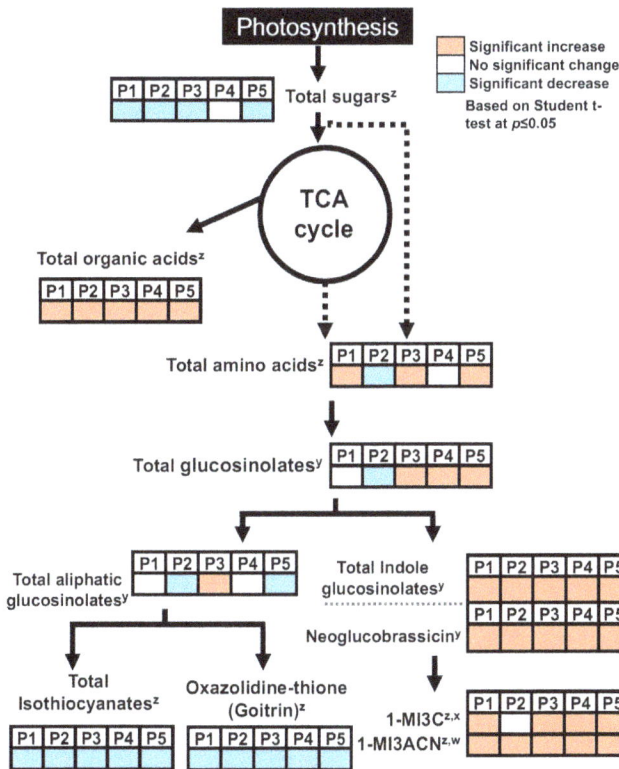

Figure 4. Summary of MeJA effect on primary and secondary metabolites. All data were compared to control by Student's *t*-test at $p \leq 0.05$. GS, glucosinolate; 1-MI3C, 1-methoxyindole-3-carbinol. P1, "Baby bok choy"; P2, "Chinese cabbage"; P3, "Asian"; P4, "Col baby choi"; P5, "Pak choi pechay"; [x] 1-methoxyindole-3-carbinol; [y] based on compound concentration; [w] 1-methoxyindole-3-acetonitrile; [z] Based on peak intensity. The dotted line indicates the conversion is simplified biosynthesis without intermediates.

3. Materials and Methods

3.1. Plant Materials

The pak choi cultivars used in this experiment were "Baby bok choy" (Lake Valley Inc., Boulder, CO, USA), "Chinese cabbage" (Heirloom; Lake Valley Inc., Boulder, CO, USA), "Asian" (Livingston Seed, Columbus, OH, USA), "Col baby choi" (Burpee Seeds, Warminster, PA, USA) and "Pak choi pechay" (Burpee Seeds, Warminster, PA, USA). Seeds of each pak choi cultivar were germinated in a 32-cell plug tray filled with Sunshine LC1 professional soil mix (Sun Gro Horticulture, Vancouver, BC, Canada). Plants were grown in a greenhouse at the University of Illinois at Urbana-Champaign under a 25/18 °C and 14/10 h: day/night temperature and light regimes with additional HID (high-intensity discharge) lighting provided for 14 h (from 06:00 to 20:00). Four weeks after germination, plants in the vegetative growth stage were transplanted to a 1-L pot in the greenhouse and grown under the same environmental conditions. After four weeks, all aerial parts of the pak choi cultivars were sprayed with 500 µM MeJA in 0.1% ethanol. Control plants were applied with 0.1% ethanol. Then, nine pak choi plants were harvested for each treatment with three plants in each of three biological replicates, two days after spray. All of the samples were freeze-dried, ground to a fine powder and stored at <-20 °C prior to extraction.

3.2. Analysis of Glucosinolates

Extraction and quantification of glucosinolates using HPLC were performed following a previously published protocol [18]. Freeze-dried pak choi powder (0.2 g) and 2 mL of 70% methanol were added to a 10-mL tube (Nalgene, Rochester, NY, USA) and heated on a heating block at 95 °C for 10 min. After cooling on ice, 0.5 mL of glucosinalbin (1 mM; purified from seeds of *Sinapis alba*; and the concentration was confirmed using sinigrin standard (Sigma-Aldrich, St. Louis, MO, USA)) were added as an internal standard, and the mixture was centrifuged at 8000× *g* for 5 min at 4 °C. The supernatant was collected, and the pellet was re-extracted with 2 mL of 70% methanol at 95 °C for 10 min. A subsample (1 mL) from each pooled extract was transferred to a 2-mL microcentrifuge tube (Fisher Scientific, Waltham, MA, USA). A mixture of 1 M lead acetate and 1 M barium acetate (1:1, *v/v*) (0.15 mL) was added to precipitate protein. After centrifugation at 12,000× *g* for 1 min, each sample was loaded onto a column containing DEAE Sephadex A-25 resin that was pre-charged with 1 M NaOH and 1 M pyridine acetate (GE Healthcare, Piscataway, NJ, USA). Samples were incubated for 18 h with *Helix pomatia* Type-1 arylsulfatase (Sigma-Aldrich, St. Louis, MO, USA) for desulfation, and the desulfo-glucosinolates were eluted with 3 mL of Millipore-filtered deionized distilled water. Samples (100 µL) were injected to a high performance liquid chromatography (Agilent 1100 HPLC system, Agilent Technologies, Santa Clara, CA, USA) equipped with a binary pump (G1311A, Agilent Technologies, St. Clara, CA, USA), a vacuum degasser (G1322A, Agilent Technologies), a thermostatic column compartment (G1316A, Agilent Technologies), a diode array detector (G1315B, Agilent Technologies) and an autosampler (HP 1100 series G1313A, Agilent Technologies, Santa Clara, CA, USA). An all-guard cartridge pre-column (Alltech, Lexington, KY, USA) and a Kromasil RP-C18 column (250 mm × 4.6 mm, 5-µm particle size, Supelco, Bellefonte, PA, USA) were used for glucosinolate separation. The flow rate was 1 mL/min with mobile Phase A (1 mM ammonium acetate containing 1% acetonitrile (*v/v*)) and B (100% acetonitrile), with the following elution profile: 0 min 0% B, 7 min 4% B, 20 min 20% B, 35 min 25% B, 36 min 80% B, 40 min 80% B, 41 min 0% B, and 50 min 0% B. The glucosinolates were detected at 229 nm. The UV response factor for each glucosinolate was used for quantification (Clarke, 2010). The identification of desulfo-glucosinolate profiles was validated using LC-tandem mass spectrometer (MS) (32 Q-Tof Ultima spectrometer, Waters Corp., Milford, MA, USA) coupled to HPLC (1525 HPLC system, Waters Corp.). The molecular ion and fragmentation patterns of individual desulfo-glucosinolates were compared to a previously published report [40].

3.3. Quantification of Glucosinolate Hydrolysis Products

Freeze-dried pak choi powder (50 mg) was suspended in 1 mL distilled water in a 2-mL microcentrifuge tube (Fisher Scientific, Waltham, MA, USA). Hydrolysis products were generated naturally by endogenous myrosinase in the absence of light at room temperature for 24 h. After adding 1 mL of dichloromethane, the samples were centrifuged at 12,000× g for 2 min, and the lower organic layer was collected. A gas chromatograph (GC) (6890N, Agilent Technologies) coupled to a MS detector (5975B, Agilent Technologies) equipped with an auto sampler (7683B, Agilent Technologies) and a capillary column (30 m × 0.32 mm × 0.25 μm J&W HP-5, Agilent Technologies) was used to determine glucosinolate hydrolysis products. A 1-μL sample of the dichloromethane extract was injected to the GC-MS with the split ratio of 1:1. After an initial temperature held at 40 °C for 2 min, the oven temperature was increased to 260 °C at 10 °C/min and held for 10 min [41]. Injector and detector temperatures were set at 200 and 280 °C, respectively. The flow rate of the helium carrier gas was set at 1.1 mL/min. Peaks were identified using standard compounds (goitrin, crambene, indole-3-acetonitrile and 3-butenyl isothiocyanate) or by comparing with the National Institute of Standards and Technology (NIST) library or previous publications [42,43] (Table S4).

3.4. Measurement of Nitrile Formation and Myrosinase Activity

Nitrile formation (%) was measured to estimate the ESP activity as ESP enhances the formation of nitriles over isothiocyanates. The nitrile formation of each pak choi cultivar was determined by incubating concentrated horseradish root extract with protein extract of pak choi. The horseradish extract was used as an exogenous substrate source of sinigrin and gluconasturtiin at the saturated level in order to minimize the reaction of pak choi protein with endogenous glucosinolates substrate. Subsequently, only hydrolysis products from sinigrin and gluconasturtiin were dominant compounds detected from GC-MS (Figure S2).

To measure the nitrile formation and myrosinase activity, freeze-dried pak choi powder (75 mg) was mixed with 1.5 mL of concentrated "1091" horseradish root extract [11] in 2-mL microcentrifuge tubes (10 g of horseradish were mixed with 100 mL of 70% methanol. This solution was centrifuged at 4000× g for 5 min. The supernatant of horseradish root extract was transferred to a beaker and boiled until all solvent was evaporated, then reconstituted with 50 mL of deionized water). The sinigrin and gluconasturtiin concentrations were 74 and 7 μmol/g DW, respectively. After centrifugation at 12,000× g for 2 min, 0.6 mL supernatant were transferred to a 1.5-mL Teflon centrifuge tube (Savillex Corporation, Eden Prairie, MN, USA), and then, 0.6 mL of dichloromethane were added. The tubes were placed upside down to minimize the loss of volatile compounds at room temperature for an hour. Then, tubes were vortexed and centrifuged at 12,000× g for 2 min. The dichloromethane organic layer was injected to GC-MS (Trace 1310 GC, Thermo Fisher Scientific, Waltham, MA, USA) coupled to a MS detector system (ISQ QD, Thermo Fisher Scientific, Waltham, MA, USA) and an auto sampler (Triplus RSH, Thermo Fisher ScientificA capillary column (DB-5MS, Agilent Technologies; 30 m × 0.25 mm × 0.25 μm capillary column) was used to determine glucosinolate hydrolysis products. After an initial temperature held at 40 °C for 2 min, the oven temperature was increased to 320 °C at 15 °C/min and held for 4 min. Injector and detector temperatures were set at 270 °C and 275 °C, respectively. The flow rate of the helium carrier gas was set at 1.2 mL/min. Standard curves of allyl isothiocyanate, 2-phenthyl isothiocyanate and 3-phenylpropionitrile (Sigma-Aldrich) were used for quantification, and the relative ratio of nitrile to the total hydrolysis products was calculated to determine ESP activity. The standard curve from allyl isothiocyanate was applied to quantify of 1-cyano-2,3-epithiopropane. Myrosinase activity was estimated as the total amount of hydrolysis products produced within 60 min [41]. One unit was defined as 1 μmole of the above four hydrolysis products released per min.

3.5. RNA Extraction and Quantitative Real-Time-PCR

Total RNA was isolated from control and MeJA-treated pak choi plants using RNeasy Mini Kit (QIAGEN, Hilden, Germany) following the manufacturer's instructions. The quantity of RNA was measured using a NanoDrop 3300 spectrophotometer (Thermo Scientific, Waltham, MA, USA). One microgram of RNA was reverse-transcribed with Superscript™ III First-Strand Synthesis SuperMix for qRT-PCR (Invitrogen, Carlsbad, CA, USA) according to the manufacturer's instructions. The qRT-PCR was carried out with the Power SYBR® Green RT-PCR Master Mix (QIAGEN) using an ABI 7900HT Fast Real-Time PCR System (Applied Biosystems, Foster City, CA, USA) following the manufacturer's instructions. The resulting cDNA samples were diluted to $1/10$ (v/v) for qRT-PCR. The primer sets of glucosinolate biosynthesis genes, hydrolysis genes and transcription factor genes were designed based on the sequences published in the online database (Available online: http://www.ocri-genomics.org/bolbase/index.html) [6]. The primers were synthesized by Integrated DNA Technologies (Coralville, IA, USA). A final list of the primers used, the gene models from which they were created and classifications of the genes can be found in Supplementary Table S5. The results were expressed after normalization to the *Brassica rapa* actin gene (*BrACT1*) [3,17]. The relative expression ratio was determined with the equation $2^{-\Delta\Delta Ct}$ using the *BrACT1* normalized ΔCt values generated by the ABI 7900HT Sequence Detection System Software 2.4 (Applied Biosystems).

3.6. Primary Metabolites Analysis by GC-Time-of-Flight-MS

Primary metabolites were analyzed following the method of Ku et al. [44] with minor modifications. Freeze-dried powder of pak choi (400 mg) was extracted with 10 mL of a mixture of methanol, deionized distilled water and chloroform (2.5:1:1, $v/v/v$) for 24 h using a Twist Shaker (Biofree, Seoul, Korea). The resulting mixture was centrifuged at 5000 rpm for 5 min (Universal 320, Hettich Zentrifugen, Tuttlingen, Germany). The supernatant of 900 μL was transferred to a 1.5-mL microcentrifuge tube. After adding of 400 μL distilled water and centrifugation at 5000 rpm for 5 min, 400 μL of the polar phase were transferred to another 1.5-mL tube and then concentrated using Modulspin 31 speed-vacuum concentrator (Biotron, Seoul, Korea). For the GC-MS analysis, the dried extract was oximated with 50 μL methoxyamine hydrochloride in pyridine (20 mg/mL) for 90 min at 30 °C using a thermomixer (Eppendoft, Hamburg, Germany), followed by silylation with 50 μL of *N*-methyl-*N*-(trimethylsilyl)-trifluoroacetamide (MSTFA) (Sigma-Aldrich) for 30 min at 37 °C using a thermomixer. The final concentration of the samples was 2.5 mg/mL for GC-time-of-flight (TOF)-MS analysis. The samples were then filtered through a 0.2-μm PTFE filter. Each biological replication ($n = 3$) from each pak choi cultivar sample (1 μL) was injected in triplicate with the split ratio of 1:5.

A GC system (7890A, Agilent Technologies), equipped with an autosampler (7693, Agilent Technologies) and a Pegasus HT TOF-MS (Leco Corporation, St. Joseph, MI, USA), and a capillary column (HP-5MS, 30 m × 0.25 mm × 0.25 μm, Agilent Technologies) were used for analysis. Chromatographic-grade helium with a constant flow of 1.0 mL/min was used as the carrier gas. The injector and transfer line temperatures were both set at 240 °C. The oven temperature was initially held at 75 °C for 2 min, then ramped to 300 °C at a rate of 15 °C/min and maintained at 300 °C for 3 min. The mass data collected in the electron ionization mode with 70 eV of ionization energy were used to conduct a mass scan ranges at m/z 50–1000. The average value from three analytical replications for each biological replication was used for statistical analysis. Identification of compounds was done by comparing with standard compounds or the NIST library (Table S3).

3.7. Statistical Analysis

All analyses were done with three biological replications (three plants per replication). Univariate analysis of variance (ANOVA) and Student's *t*-test were performed using JMP Pro 12 (SAS Institute, Cary, NC, USA) to determine the MeJA effect on primary and secondary metabolites and gene expression changes. For primary metabolite data, raw data files were converted to CDF format

Int. J. Mol. Sci. **2017**, *18*, 1004

(*.cdf) using the Leco ChromaTOF software program (Version 4.44, Leco Corp, Warrendale, PA, USA). After conversion, the MS data were processed using the metAlign software package (Available online: http://www.metalign.nl) to obtain a data matrix containing retention times, accurate masses and normalized peak intensities. The resulting data were exported to Excel (Microsoft, Redmond, WA, USA) for further analysis using MetaboAnalyst 3.0 (Available online: http://www.metaboanalyst.ca/faces/home.xhtml).

4. Conclusions

In the present study, the effects of foliar application of MeJA on glucosinolates, their hydrolysis products and primary metabolites were analyzed in five pak choi cultivars. The MeJA treatment significantly changed primary and secondary metabolite composition with the response to MeJA treatment differing among cultivars. In general, indole glucosinolates and their hydrolysis products were significantly increased by MeJA treatment, whereas other glucosinolates, in particular aliphatic glucosinolates, changed differentially depending on cultivar. Moreover, MeJA reduced goitrin, a goitrogenic compound produced from an aliphatic glucosinolate progoitrin, in all five cultivars. Amino acids, organic acids and sugar alcohols tended to increase, but mono- and di-saccharide sugars decreased with MeJA treatment, suggesting that the sugars may be a carbon source for secondary metabolite biosynthesis induced by MeJA treatment (Figure 4). Some gene expression data supported glucosinolate changes, for instance *OH1* for progoitrin biosynthesis. In conclusion, the results of this study suggest that MeJA treatment can act as an agent to regulate the primary and secondary metabolites in pak choi, and cultivar-specific responses to MeJA were found among the five pak choi cultivars. Additionally, these results can be used in developing cultural practice strategies to enhance nutritional value of pak choi and in breeding to develop a pak choi cultivar with improved nutritional properties. Specifically, "Asian" was found to be the most responsive to MeJA among investigated cultivars, with increase in glucosinolates and hydrolysis products that are potentially anticarcinogenic and a reduction in a goitrogen compound.

Supplementary Materials: Supplementary materials can be found at http://www.mdpi.com/1422-0067/18/5/1004/s1.

Acknowledgments: This research was supported by the West Virginia Agricultural and Forestry Experiment Station (WVA00698): Scientific Article No. 3310.

Author Contributions: Conceived of and designed the experiments: Kang-Mo Ku, Choong Hwan Lee and John A. Juvik; Performed the experiments: Kang-Mo Ku, M.K., Yu-Chun Chiu, Na Kyung Kim and Hye Min Park; Analyzed the data: Kang-Mo Ku, Moo Jung Kim and Yu-Chun Chiu; Wrote the paper: Kang-Mo Ku, Moo Jung Kim Yu-Chun Chiu and John A. Juvik. All authors have read and approve of the final manuscript.

Conflicts of Interest: The authors declare no conflict of interest.

References

1. Nair, A.; Irish, L. Commercial Production of Pak Choi. Available online: https://store.extension.iastate.edu/Product/Commercial-Production-of-Pak-Choi (accessed on 7 May 2017).
2. Wiesner, M.; Schreiner, M.; Glatt, H. High mutagenic activity of juice from pak choi (*Brassica rapa* ssp. *chinensis*) sprouts due to its content of 1-methoxy-3-indolylmethyl glucosinolate, and its enhancement by elicitation with methyl jasmonate. *Food Chem. Toxicol.* **2014**, *67*, 10–16. [PubMed]
3. Wiesner, M.; Zrenner, R.; Krumbein, A.; Glatt, H.; Schreiner, M. Genotypic variation of the glucosinolate profile in pak choi (*Brassica rapa* ssp. *chinensis*). *J. Agric. Food Chem.* **2013**, *61*, 1943–1953. [CrossRef] [PubMed]
4. Zang, Y.X.; Zhang, H.; Huang, L.H.; Wang, F.; Gao, F.; Lv, X.S.; Yang, J.; Zhu, B.; Hong, S.B.; Zhu, Z.J. Glucosinolate enhancement in leaves and roots of pak choi (*Brassica rapa* ssp. *chinensis*) by methyl jasmonate. *Hortic. Environ. Biotechnol.* **2015**, *56*, 830–840. [CrossRef]
5. Halkier, B.A.; Du, L. The biosynthesis of glucosinolates. *Trends Plant Sci.* **1997**, *2*, 425–431. [CrossRef]
6. Liu, S.; Liu, Y.; Yang, X.; Tong, C.; Edwards, D.; Parkin, I.A.P.; Zhao, M.; Ma, J.; Yu, J.; Huang, S.; et al. The *Brassica oleracea* genome reveals the asymmetrical evolution of polyploid genomes. *Nat. Commun.* **2014**, *5*, 3930. [CrossRef] [PubMed]

7. Bones, A.M.; Rossiter, J.T. The myrosinase-glucosinolate system, its organisation and biochemistry. *Physiol. Plant* **1996**, *97*, 194–208. [CrossRef]
8. Arora, R.; Kumar, R.; Mahajan, J.; Vig, A.P.; Singh, B.; Arora, B.S. 3-Butenyl isothiocyanate: A hydrolytic product of glucosinolate as a potential cytotoxic agent against human cancer cell lines. *J. Food Sci. Technol.* **2016**, *53*, 3437–3445. [CrossRef] [PubMed]
9. Neave, A.S.; Sarup, S.M.; Seidelin, M.; Duus, F.; Vang, O. Characterization of the *N*-methoxyindole-3-carbinol (NI3C)-induced cell cycle arrest in human colon cancer cell lines. *Toxicol. Sci.* **2005**, *83*, 126–135. [CrossRef] [PubMed]
10. Takada, Y.; Andreeff, M.; Aggarwal, B.B. Indole-3-carbinol suppresses NF-κB and IκBα kinase activation, causing inhibition of expression of NF-κB-regulated antiapoptotic and metastatic gene products and enhancement of apoptosis in myeloid and leukemia cells. *Blood* **2005**, *106*, 641–649. [CrossRef] [PubMed]
11. Ku, K.-M.; Jeffery, E.H.; Juvik, J.A.; Kushad, M.M. Correlation of quinone reductase activity and allyl isothiocyanate formation among different genotypes and grades of horseradish roots. *J. Agric. Food Chem.* **2015**, *63*, 2947–2955. [CrossRef] [PubMed]
12. Ku, K.-M.; Kim, M.J.; Jeffery, E.H.; Kang, Y.-H.; Juvik, J.A. Profiles of glucosinolates, their hydrolysis products, and quinone reductase inducing activity from 39 arugula (*Eruca Sativa* Mill.) accessions. *J. Agric. Food Chem.* **2016**, *64*, 6524–6532. [CrossRef] [PubMed]
13. Yan, X.; Chen, S. Regulation of plant glucosinolate metabolism. *Planta* **2007**, *226*, 1343–1352. [CrossRef] [PubMed]
14. Ku, K.M.; Jeffery, E.H.; Juvik, J.A. Influence of seasonal variation and methyl jasmonate mediated induction of glucosinolate biosynthesis on quinone reductase activity in broccoli florets. *J. Agric. Food Chem.* **2013**, *61*, 9623–9631. [CrossRef] [PubMed]
15. Brown, A.F.; Yousef, G.G.; Jeffery, E.H.; Klein, B.P.; Wallig, M.A.; Kushad, M.M.; Juvik, J.A. Glucosinolate profiles in broccoli: Variation in levels and implications in breeding for cancer chemoprotection. *J. Am. Soc. Hortic. Sci.* **2002**, *127*, 807–813.
16. Yi, G.-E.; Robin, A.H.K.; Yang, K.; Park, J.-I.; Kang, J.-G.; Yang, T.-J.; Nou, I.-S. Identification and expression analysis of glucosinolate biosynthetic genes and estimation of glucosinolate contents in edible organs of *Brassica oleracea* subspecies. *Molecules* **2015**, *20*, 13089–13111. [CrossRef] [PubMed]
17. Ku, K.-M.; Becker, T.M.; Juvik, J.A. Transcriptome and metabolome analyses of glucosinolates in two broccoli cultivars following jasmonate treatment for the Induction of glucosinolate defense to *Trichoplusia ni* (Hübner). *Int. J. Mol. Sci.* **2016**, *17*, 1135. [CrossRef] [PubMed]
18. Ku, K.-M.; Jeffery, E.H.; Juvik, J.A. Exogenous methyl jasmonate treatment increases glucosinolate biosynthesis and quinone reductase activity in kale leaf tissue. *PLoS ONE* **2014**, *9*, e103407. [CrossRef] [PubMed]
19. Bradbury, S. *Methyl Jasmonate; Exemption from the Requirement of a Tolerance*; EPA: Washington, DC, USA, 2013; pp. 22789–22794.
20. Wiesner, M.; Hanschen, F.S.; Schreiner, M.; Glatt, H.; Zrenner, R. Induced production of 1-methoxy-indol-3-ylmethyl glucosinolate by jasmonic acid and methyl jasmonate in sprouts and leaves of pak choi (*Brassica rapa* ssp. *chinensis*). *Int. J. Mol. Sci.* **2013**, *14*, 14996–15016. [CrossRef] [PubMed]
21. Ku, K.M.; Choi, J.-H.; Kushad, M.M.; Jeffery, E.H.; Juvik, J.A. Pre-harvest methyl jasmonate treatment enhances cauliflower chemoprotective attributes without a loss in postharvest quality. *Plant Foods Hum. Nutr.* **2013**, *68*, 113–117. [CrossRef] [PubMed]
22. Zang, Y.X.; Zheng, W.W.; He, Y.; Hong, S.B.; Zhu, Z.J. Global analysis of transcriptional response of Chinese cabbage to methyl jasmonate reveals JA signaling on enhancement of secondary metabolism pathways. *Sci. Hortic.* **2015**, *189*, 159–167. [CrossRef]
23. Rask, L.; Andréasson, E.; Ekbom, B.; Eriksson, S.; Pontoppidan, B.; Meijer, J. Myrosinase: Gene family evolution and herbivore defense in Brassicaceae. *Plant Mol. Biol.* **2000**, *42*, 93–114. [CrossRef] [PubMed]
24. Matusheski, N.V.; Juvik, J.A.; Jeffery, E.H. Heating decreases epithiospecifier protein activity and increases sulforaphane formation in broccoli. *Phytochemistry* **2004**, *65*, 1273–1281. [CrossRef] [PubMed]
25. Gaitan, E. Goitrogens in food and water. *Annu. Rev. Nutr.* **1990**, *10*, 21–39. [CrossRef] [PubMed]
26. Nho, C.W.; Jeffery, E. The synergistic upregulation of phase II detoxification enzymes by glucosinolate breakdown products in cruciferous vegetables. *Toxicol. Appl. Pharmacol.* **2001**, *174*, 146–152. [CrossRef] [PubMed]

27. Nho, C.W.; Jeffery, E. Crambene, a bioactive nitrile derived from glucosinolate hydrolysis, acts via the antioxidant response element to upregulate quinone reductase alone or synergistically with indole-3-carbinol. *Toxicol. Appl. Pharmacol.* **2004**, *198*, 40–48. [CrossRef] [PubMed]

28. Stephensen, P.U.; Bonnesen, C.; Schaldach, C.; Andersen, O.; Bjeldanes, L.F.; Vang, O. *N*-methoxyindole-3-carbinol is a more efficient inducer of cytochrome P-450 1A1 in cultured cells than indol-3-carbinol. *Nutr. Cancer* **2000**, *36*, 112–121. [CrossRef] [PubMed]

29. Liu, A.G.; Juvik, J.A.; Jeffery, E.H.; Berman-Booty, L.D.; Clinton, S.K.; John, W.; Erdman, J. Enhancement of broccoli indole glucosinolates by methyl jasmonate treatment and effects on prostate carcinogenesis. *J. Med. Food* **2014**, *17*, 1177–1182. [CrossRef] [PubMed]

30. Choi, Y.; Kim, Y.; Park, S.; Lee, K.W.; Park, T. Indole-3-carbinol prevents diet-induced obesity through modulation of multiple genes related to adipogenesis, thermogenesis or inflammation in the visceral adipose tissue of mice. *J. Nutr. Biochem.* **2012**, *23*, 1732–1739. [CrossRef] [PubMed]

31. Srivastava, B.H.; Shukla, Y. Antitumour promoting activity of indole-3-carbinol in mouse skin carcinogenesis. *Cancer Lett.* **1998**, *134*, 91–95. [CrossRef]

32. Wu, H.T.; Lin, S.H.; Chen, Y.H. Inhibition of cell proliferation and in vitro markers of angiogenesis by indole-3-carbinol, a major indole metabolite present in cruciferous vegetables. *J. Agric. Food Chem.* **2005**, *53*, 5164–5169. [CrossRef] [PubMed]

33. Kim, H.S. Functional Studies of Lignin Biosynthesis Genes and Putative Flowering Gene in *Miscanthus* × *Giganteus* and Studies on Indolyl Glucosinolate Biosynthesis and Translocation in *Brassica oleracea*. Ph.D. Dissertation, University of Illinois at Urbana-Champaign, Urbana, IL, USA, 2010.

34. Wold, S. Methods and principles in medicinal chemistry. In *Chemometric Methods in Molecular Design*; van de Waterbeemd, H., Ed.; VCH: Weinheim, Germany, 2008; Volume 2, pp. 195–218.

35. Tytgat, T.O.G.; Verhoeven, K.J.F.; Jansen, J.J.; Raaijmakers, C.E.; Bakx-Schotman, T.; McIntyre, L.M.; van der Putten, W.H.; Biere, A.; van Dam, N.M. Plants know where it hurts: Root and shoot jasmonic acid induction elicit differential responses in *Brassica oleracea*. *PLoS ONE* **2013**, *8*, e65502. [CrossRef]

36. Czapski, J.; Horbowicz, M.; Saniewski, M. The effect of methyl jasmonate on free fatty acids content in ripening tomato fruits. *Biol. Plant* **1992**, *34*, 71–76. [CrossRef]

37. Liang, Y.S.; Choi, Y.H.; Kim, H.K.; Linthorst, H.J.M.; Verpoorte, R. Metabolomic analysis of methyl jasmonate treated Brassica rapa leaves by 2-dimensional NMR spectroscopy. *Phytochemistry* **2006**, *67*, 2503–2511. [CrossRef] [PubMed]

38. Wu, G.Y. Functional amino acids in growth, reproduction, and health. *Adv. Nutr.* **2010**, *1*, 31–37. [CrossRef] [PubMed]

39. Li, P.; Yin, Y.L.; Li, D.; Kim, S.W.; Wu, G.Y. Amino acids and immune function. *Br. J. Nutr.* **2007**, *98*, 237–252. [CrossRef] [PubMed]

40. Kusznierewicz, B.; Iori, R.; Piekarska, A.; Namieśnik, J.; Bartoszek, A. Convenient identification of desulfoglucosinolates on the basis of mass spectra obtained during liquid chromatography-diode array-electrospray ionisation mass spectrometry analysis: Method verification for sprouts of different *Brassicaceae* species extracts. *J. Chromatogr. A* **2013**, *1278*, 108–115. [CrossRef] [PubMed]

41. Dosz, E.B.; Ku, K.M.; Juvik, J.A.; Jeffery, E.H. Total myrosinase activity estimates in brassica vegetable produce. *J. Agric. Food Chem.* **2014**, *62*, 8094–8100. [CrossRef] [PubMed]

42. Kjær, A.; Ohashi, M.; Wilson, J.M.; Djerassi, C. Mass spectra of isothiocyanates. *Acta Chem. Scand.* **1963**, *17*, 2143–2154. [CrossRef]

43. Spencer, G.F.; Daxenbichler, M.E. Gas chromatography-mass spectrometry of nitriles, isothiocyanates and oxazolidinethiones derived from cruciferous glucosinolates. *J. Sci. Food Agric.* **1980**, *31*, 359–367. [CrossRef]

44. Ku, K.M.; Choi, J.N.; Kim, J.; Kim, J.K.; Yoo, L.G.; Lee, S.J.; Hong, Y.S.; Lee, C.H. Metabolomics analysis reveals the compositional differences of shade grown tea (*Camellia sinensis* L.). *J. Agric. Food Chem.* **2010**, *58*, 418–426. [CrossRef] [PubMed]

International Journal of
Molecular Sciences

MDPI

Article

In Vitro Evaluation of the Antioxidant, Cytoprotective, and Antimicrobial Properties of Essential Oil from *Pistacia vera* L. Variety Bronte Hull

Antonella Smeriglio, Marcella Denaro, Davide Barreca *, Antonella Calderaro, Carlo Bisignano, Giovanna Ginestra, Ersilia Bellocco and Domenico Trombetta

Department of Chemical, Biological, Pharmaceutical and Environmental Sciences, University of Messina, Viale F. Stagno d'Alcontres 31, 98166 Messina, Italy; asmeriglio@unime.it (A.S.); denaromarcella.md@gmail.com (M.D.); anto.calderaro@gmail.com (A.C.); cbisignano@unime.it (C.B.); gginestra@unime.it (G.G.); ebellocco@unime.it (E.B.); dtrombetta@unime.it (D.T.)
* Correspondence: dbarreca@unime.it; Tel.: +39-090-676-5187

Academic Editor: Toshio Morikawa
Received: 10 May 2017; Accepted: 3 June 2017; Published: 6 June 2017

Abstract: Although the chemical composition and biological properties of some species of the genus *Pistacia* has been investigated, studies on hull essential oil of *Pistacia vera* L. variety Bronte (HEO) are currently lacking. In this work, we have carried out an in-depth phytochemical profile elucidation by Gas Chromatography-Mass Spectrometry (GC-MS) analysis, and an evaluation of antioxidant scavenging properties of HEO, using several different in vitro methods, checking also its cytoprotective potential on lymphocytes treated with tert-butyl hydroperoxide. Moreover, the antimicrobial activity against Gram-positive and Gram-negative strains, both American Type Culture Collection (ATCC) and clinical isolates, was also investigated. GC-MS analysis highlighted the richness of this complex matrix, with the identification of 40 derivatives. The major components identified were 4-Carene (31.743%), α-Pinene (23.584%), D-Limonene (8.002%), and 3-Carene (7.731%). The HEO showed a strong iron chelating activity and was found to be markedly active against hydroxyl radical, while scarce effects were found against 2,2-diphenyl-1-picrylhydrazyl (DPPH) radical. Moreover, pre-treatment with HEO was observed to significantly increase the cell viability, decreasing the lactate dehydrogenase (LDH) release. HEO was bactericidal against all the tested strains at the concentration of 7.11 mg/mL, with the exception of *Pseudomonas aeruginosa* ATCC 9027. The obtained results demonstrate the strong free-radical scavenging activity of HEO along with remarkable cytoprotective and antimicrobial properties.

Keywords: *Pistacia vera* L. variety Bronte; essential oil; antioxidant; cytoprotective activity; antimicrobial activity

1. Introduction

Plants producing essential oils represent a large part of natural flora and an important resource in various fields such as the pharmaceutical, food, and cosmetic industries, thanks to their flavour, fragrance, and biological activities [1]. Essential oils are a complex mixture of hydrocarbons and oxygenated hydrocarbons arising from the isoprenoid pathways and secreted by glandular trichomes disseminated mainly onto the surface of plant organs; therefore, they have a pivotal role in the growth and colonization of plants, conferring colour and scent to reproductive organs, attracting pollinators and favouring seed dispersal [2]. Furthermore, they seem to mediate the plant relationship with abiotic (e.g., light, temperature, and so on) and biotic factors, playing a defensive role against herbivores, harmful insects, and microbial pathogens [2].

More than 250 types of essential oils are commercialized annually on the international market, some of which are employed in aromatherapy and for the treatment of several diseases including cardiovascular and neurological diseases, diabetes, and cancer [1]. Moreover, the antimicrobial properties of essential oils have been widely recognized [3–6] thanks to several studies which showed a synergistic effect of several components of essential oils against various human pathogens [1]. Essential oils have been used since ancient times, in folk medicine throughout the world, to preserve a good health status [6–9]. Presently, several properties and therapeutic effects have been ascribed to essential oils such as antibiotic, rubefacient, anaesthetic, antispasmodic, balmy-expectorant, repellent, carminative, as well as beneficial effects on the central and peripheral nervous system [10–15].

Recently, antimicrobial drug resistance has brought researchers to evaluate novel antimicrobial lead molecules to treat various human pathogens. Synthetic drugs have often failed this goal, if not for a lack of efficacy, because of the greater risk to which the patient was exposed (acute and chronic toxicity and environmental hazard potential), leading the researchers to better explore natural remedies [1]. From this point of view, the study of essential oils has undergone a remarkable growth, and many new plants producing essential oil have been evaluated [16–20]. Among these, the authors decided to put their attention on *Pistacia vera* L., a nut tree that belongs to the Anacardiaceae family [21] cultivated in Iran, Turkey, the USA, Syria, Italy, Tunisia, and Greece.

In Italy, pistachio crops are typically Sicilian, mainly grown in the territory of Bronte (CT), situated on the eastern part of Sicily and characterized by lava-rich soils and very particular climatic conditions [21,22] which confer to the pistachio nuts superior organoleptic and nutritional characteristics. For this reason, the European Union recognized the variety Bronte as a D.O.P. (Protected Designation of Origin) product. Pistachio nuts are considered a rich source of many important biofunctional compounds that are useful for the human diet and known for their various pharmacological properties such as antimicrobial, anti-inflammatory, insecticidal, and anti-nociceptive activities [23]. Recently, we analysed and described the nutraceutical, antioxidant, and cytoprotective activity of crude phenols and anthocyanins-rich extracts derived from ripe pistachio hulls (*Pistacia vera* L., variety Bronte), a by-product of the pistachio industry [23,24]. These studies managed to identify this matrix as a promising source of healthy compounds. The results obtained highlighted antioxidant and cytoprotective properties directly correlated to the high total phenol content, in particular flavonols, phenolic acid, and flavan-3-ols and, among anthocyanins, to the high content of cyanidyn-3-*O*-galactoside [23,24]. The presence of these compounds makes pistachio hulls an attractive source of health-promoting compounds, which are potentially helpful against several oxidative stress-related diseases.

Despite the well-known ethnopharmacological relevance of some essential oils derived from some species of the genus *Pistacia* (*P. atlantica* Desf and *P. integerrima* J.L. Stewart ex Brandis) against several diseases, particularly those affecting the respiratory and the gastrointestinal tract [25–28], no literature data concerning the hull essential oil composition and biological properties of *Pistacia vera* L., variety Bronte, are today available.

Having considered the above, we have for the first time performed, a phytochemical characterization of ripe pistachio hull essential oil followed by the evaluation of antioxidant, cytoprotective, and antiperoxidative properties through cell-free and cell-based assays. The antimicrobial activity against some Gram-positive and Gram-negative standard and clinical bacterial has also been investigated.

2. Results and Discussion

2.1. Essential Oil Composition

The yield of hull essential oil of *Pistacia vera* L. variety Bronte (HEO) was 0.25% (*v*/*w* fresh material), well above others reported in literature (>3.50 times) [20], demonstrating how the Bronte variety possesses a much higher content of essential oil, probably due to characteristic pedoclimatic conditions of growth and probably to the extraction carried out immediately after harvest, which allowed for an increase in the extraction efficiency. This practice also led to a reduced formation of peroxidation, isomerization, or rearrangement of products due to temperature, light, and oxygen availability [29]. The essential oil composition, with retention indices and percentages, was reported in Table 1.

Table 1. Chemical composition of *Pistacia vera* L. variety Bronte hull essential oil.

#	KI [a]	Compound	Area [b] (%)
1	916	Bornylene	0.035
2	923	Tricyclene	0.709
3	935	α-Pinene	23.584
4	950	Camphene	4.133
5	978	β-Pinene	1.062
6	993	β-Myrcene	2.393
7	995	2-Carene	1.152
8	1006	α-Phellandrene	0.456
9	1011	δ-3-Carene	7.731
10	1018	α-Terpinene	2.195
11	1027	*p*-Cymene	1.621
12	1031	D-Limonene	8.002
13	1050	*trans*-β-Ocimene	0.509
14	1056	*cis*-β-Ocimene	0.412
15	1061	γ-Terpinene	0.582
16	1082	4-Carene	31.743
17	1096	α-Pinene oxide	0.787
18	1101	Linalol	0.278
19	1107	2-Fenchanol	0.385
20	1130	1,3,8-*p*-Menthatriene	0.225
21	1148	Camphor	0.236
22	1150	Menthone	0.031
23	1169	Borneol	0.831
24	1188	*p*-Cymen-8-ol	0.692
25	1194	α-Terpineol	4.036
26	1197	Myrtenal	0.011
27	1202	Myrtenol	0.082
28	1210	α-Methylcynnamaldehyde	0.016
29	1250	Piperitone	0.687
30	1232	Nerol	0.272
31	1285	Bornyl acetate	2.430
32	1365	Nerol acetate	0.136
33	1513	β-Bisabolene	0.010
34	1525	γ-Selinene	0.005
35	1530	δ-Cadinene	0.028
36	1568	*cis*-5-Dodecenoic acid	0.225
37	1810	1,13-Tetradecadiene	1.528
38	1880	1-Hexadecanol	0.160
39	1957	Palmitic acid	0.015
40	2549	1,15-Hexadecadiene	0.430
41	2106	Unknown	0.013
42	2111	Unknown	0.129
Monoterpene Hydrocarbons			86.591
Oxygenated Monoterpenes			8.235
Sesquiterpenes			0.056
Others			5.115

#: Components are listed in their elution order from HP-5 MS column; [a]: Retention index (KI) relative to standard mixture of *n*-alkanes on HP-5MS column; [b]: values (relative peak area percentage) represent averages of three determinations.

A total of 40 volatile constituents were identified, fully characterized, and grouped in four classes: monoterpene hydrocarbons, oxygenated monoterpenes, sesquiterpenes, and others (non-terpenoidic compounds). The major components of HEO were 4-Carene (31.743%), α-Pinene (23.584%), D-Limonene (8.002%), and 3-Carene (7.731%), all monoterpenes hydrocarbons that represented 86.591% of the HEO, followed by oxygenated monoterpenes, which presented an average value of 8.235% with α-Terpineol (4.036%), Borneol (0.831%), and α-Pinene oxide (0.787%) as the main compounds.

Although the monoterpenes hydrocarbons remained the most abundant family, followed by oxygenated monoterpenes (reflecting the results reported in the literature [20]), the characteristic absence of oxygenated sesquiterpenes and the very low presence of non-oxygenated sesquiterpenes, makes the essential oil of Bronte pistachio hulls easily recognized and distinguishable from essential oils derived from other pistachio hull varieties.

2.2. Antioxidant Activities

Reactive oxygen species are involved in the pathological development of many important human diseases, thus antioxidants play a crucial role in human health due to their unquestionable beneficial effects on living organisms that enable them to overcome oxidative injuries, and modulate biological pathways and membrane functionality [23]. In this article, the authors conducted the analysis of antioxidant and free radical scavenging properties of HEO by using several antioxidant assays (hydrogen atom transfer and electron transfer-based methods) in order to check their behaviour under different reaction environments and mechanism typologies. This aspect is often overlooked but is critical in order to establish the structure-scavenging activity relationship of the major components of the phytocomplex investigated [24]. Figure 1 shows the antioxidant and free radical-scavenging potential of HEO towards Fe^{3+}-TPTZ (A), $ABTS^{\bullet+}$ (B), and 2,2-diphenyl-1-picrylhydrazyl (DPPH)$^{\bullet}$ (C) as well as its iron chelating capacity (D). Moreover, the antiradical properties of HEO against two of the most common and dangerous reactive oxygen species produced in the organisms, superoxide anion ($O_2^{\bullet-}$) and hydroxyl radical ($^{\bullet}OH$), was investigated and reported in Figure 2A,B. As can be seen in Figures 1 and 2, the HEO showed a remarkable dose-dependent antioxidant and free-radical scavenging activity towards all assays performed, with the following order of potency, expressed as the half maximal inhibitory concentration (IC_{50}), $^{\bullet}OH$ (IC_{50} 0.003 mg/mL) > Ferric Reducing Antioxidant Power (FRAP; IC_{50} 0.063 mg/mL) > Trolox Equivalent Antioxidant Capacity (TEAC; IC_{50} 0.128 mg/mL) > DPPH (IC_{50} 0.878 mg/mL) as well as having showed a strong iron chelating capacity (IC_{50} 0.017 mg/mL). The strong antioxidant activity was also confirmed by Folin-Ciocalteu assay results (1278.44 ± 41.79 mg gallic acid equivalents (GAE)/100 g of HEO).

The ability to scavenge different reactive species makes the HEO an important source of antioxidants that are potentially useful in the detoxification mechanisms of living organism. Primary and relative weak antioxidants (such as $O_2^{\bullet-}$), in fact, can be the precursor or be combined with others (e.g., nitric oxide) to give rise to very dangerous reactive species (hydrogen peroxide, hydroxyl radical, singlet oxygen, peroxynitric radical, and so on). Furthermore, HEO, acting as a strong chelating agent, may reduce the availability of transition metals and inhibit the radical-mediated oxidative chain reactions in biological systems preserving the integrity and consequently the functionality of membranes. As showed in β-carotene bleaching assay (Figure 3), in fact, the presence of these antioxidant compounds in HEO, mainly p-Cymene, Borneol, and β-Myrcene, are able to form radical adduct with peroxyl radical [30]; the compounds exhibit antioxidant properties and are capable of preventing oxidative damage from free radical-mediated oxidation in a dose-dependent manner.

These activities are predominantly attributable to the high amount of monoterpenes hydrocarbons found in HEO. In fact, it has widely been demonstrated that monoterpene hydrocarbons are better antioxidant compounds in respect to sesquiterpenes and particularly those with strongly activated methylene groups in their structure, such as 4-Carene, α-Terpinene, and γ-Terpinene, were the most active. So, in agreement with the data reported in literature [31], among oxygenated monoterpenes, the following order of effectiveness in antioxidant assays can be also hypothesised in our work: monoterpene phenols > allylic alcohols > monoterpenes aldehydes and ketones.

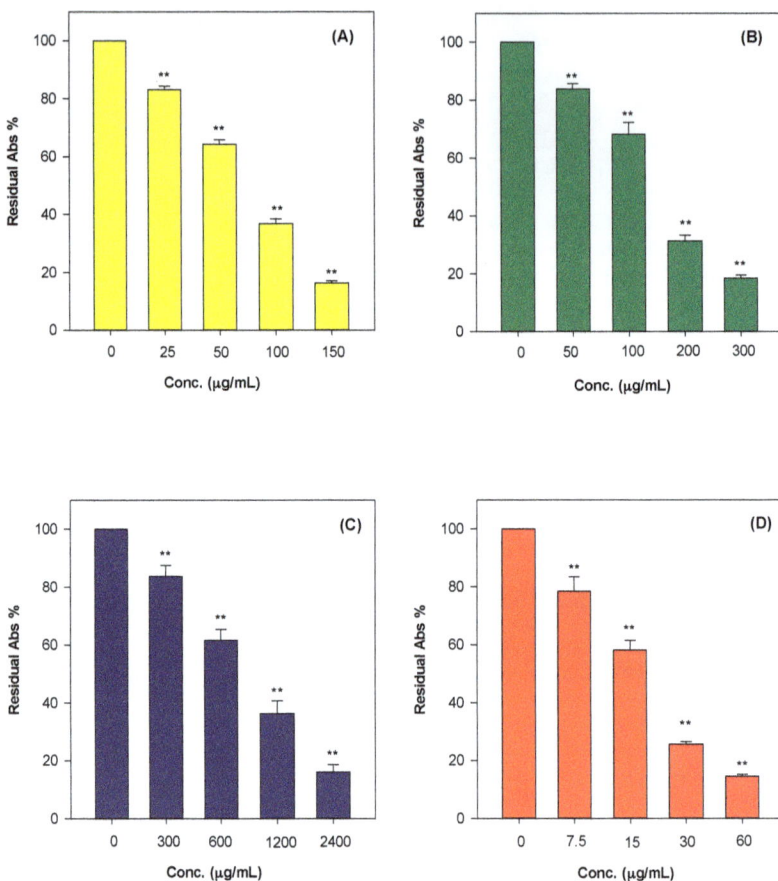

Figure 1. Antioxidant and free radical-scavenging activity of hull essential oil of *Pistacia vera* L. variety Bronte (HEO) towards Fe^{3+}-TPTZ (**A**); $ABTS^{\bullet+}$ (**B**); $DPPH^{\bullet}$ (**C**); and iron chelating capacity (**D**). The asterisks (**) indicate significant differences ($p < 0.05$).

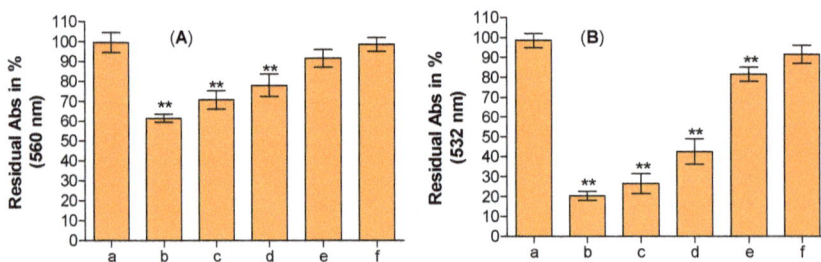

Figure 2. Superoxide anion (**A**) scavenging assay in the absence (a) or in the presence of 60, 30, 15, 7.5, and 3.7 µg/mL of HEO (b–f, respectively); Hydroxyl radical (**B**) scavenging assay in the absence (a) or in the presence of 15, 7.50, 3.75, 1.85, and 0.92 µg/mL of HEO (b–f, respectively). The asterisks (**) indicate significant differences ($p < 0.05$). Each value represents mean ± SD ($n = 3$).

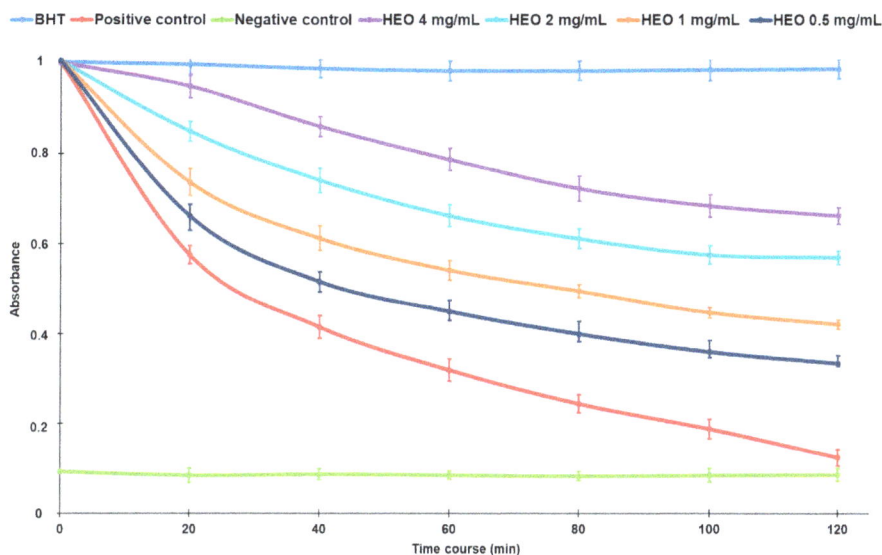

Figure 3. Beta-carotene bleaching curves of hull essential oil of *Pistacia vera* L. variety Bronte (HEO) at concentration range 0.5–4 mg/mL in respect to the reference compound butylated hydroxytoluene (BHT) (2 mg/mL). Results are expressed as mean (*n* = 3) of three independent experiments.

Concerning the sesquiterpene group, the radical scavenging properties of the hydrocarbons-type were quite low and lower than that of the monoterpene hydrocarbons group, whereas among the oxygenated type, mainly allylic alcohols showed good scavenging properties, similar to those of oxygenated monoterpenes [31], but they are completely lacking in this essential oil.

However, being a very complex matrix, the antioxidant properties of the essential oil do not always depend on the antioxidant activity of its main components and can be modulated by several mechanisms, so concepts of synergism and antagonism can be very relevant [30].

2.3. Cytoprotective Activity

The potential protective influences of the compounds present in HEO were analysed on lymphocytes treated with tert-butyl hydroperoxide (*t*-BOOH). A preliminary evaluation, obtained by incubating the cells with the essential oil at the concentration utilized in our work, shows no cytotoxicity effects of lymphocytes (data not shown). As can be seen in Figure 4, the treatment of lymphocytes with *t*-BOOH resulted in a net increase of cell mortality (~41% fall in cell viability vs. controls). The presence of 20, 17.5, 15, and 12.5 µg/mL of essential oil significantly increased cell viability by ~1.50, 1.40, 1.20, and 1.12-fold, respectively, compared cells treated only with *t*-BOOH (Figure 4A). Below this concentration, the effects were almost completely absent, with results superimposable with the ones obtained with lymphocytes treated with *t*-BOOH. The protective influences of the compounds present in the essential oil were also evident by monitoring lactate dehydrogenase (LDH) release in cell culture medium. In fact, this enzyme is commonly utilized as a marker of cell integrity and cytotoxicity, following oxidative burden, membrane damages, activation of apoptotic, and/or necrotic events. As we expected (Figure 4B), the incubation of lymphocytes with *t*-BOOH resulted in a remarkable increase of LDH release into the media (~4.2-fold higher than the control). The presence of 20 µg/mL of essential oil resulted in the complete absence of LDH release, confirming the data obtained with trypan blue coloration. The presence of 17.5, 15, 12, and 10 µg/mL of essential oil decreased enzyme release by ~66%, 52%, 31%, and 6%, respectively, in comparison to the sample treated with only *t*-BOOH, while concentrations below 10 µg/mL did not show a significant decrease of LDH release. Overall, the compounds present in the hull essential oil showed dose-response activity whereby they increased cell survival and prevented damages due to the presence of the strong antioxidant.

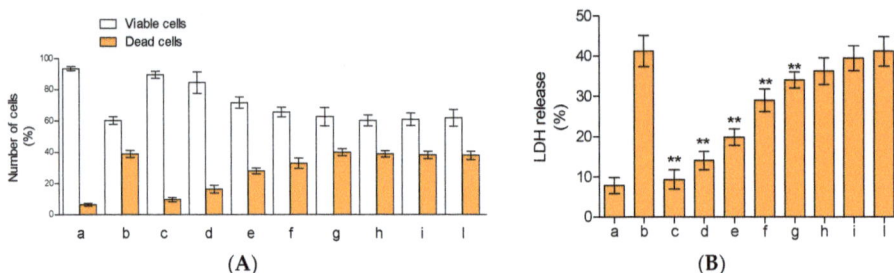

(A)   (B)

Figure 4. Cytoprotective effects of HEO on tert-butyl hydroperoxide (*t*-BOOH) treated lymphocytes. Lymphocytes plus 100 µM of *t*-BOOH were incubated for 24 h in the absence (b) or in the presence of 20, 17.5, 15, 12.5, 10, 7.5, 5, or 1 µg/mL of essential oil (c–l, respectively). Lymphocytes incubated under the same experimental condition without *t*-BOOH (a). Cell vitality and integrity were analysed by trypan blue staining (**A**) and lactate dehydrogenase (LDH) release (**B**). The asterisks (**) indicate significant differences ($p < 0.05$). Each value represents mean $\pm$ SD ($n = 3$).

2.4. Antimicrobial Activities

The minimal inhibitory concentrations (MICs) and minimal bactericidal concentrations (MBCs) values of HEO against the American Type Culture Collection (ATCC) and clinical isolates tested bacteria are reported in Table 2. Results of negative controls indicated the complete absence of inhibition of all the strains tested (data not shown). A concentration of 7.11 mg/mL inhibited the growth of all the tested strains with the exception of *Pseudomonas aeruginosa* ATCC 9027. The same concentration was found to be bactericidal against all strains. Several studies have previously demonstrated the antimicrobial activity of plant essential oils [32–34]. The compound 4-carene, also identified amongst the main constituents of the Iranian Cymbobogon Olivieri essential oil, has been implicated in the antimicrobial activity against Gram-positive bacteria, Gram-negative bacteria, and the yeast *Candida albicans* [35].

Table 2. Minimal inhibitory concentrations (MICs) and minimal bactericidal concentrations (MBCs) of HEO (expressed as mg/mL) against Gram-positive and Gram-negative bacteria.

Bacterial Strain	MIC	MBC
S. aureus ATCC 6538	7.11	7.11
E. coli ATCC 10,536	7.11	7.11
P. aeruginosa ATCC 9027	na	na
S. aureus ATCC 43,300 (MRSA)	7.11	7.11
S. aureus 74CCH	7.11	7.11
S. aureus 7786	7.11	7.11
S. aureus 815	7.11	7.11

HEO, hull essential oil; na, not active.

The seed oil from the Tunisian endemic *Ferula tunetana* Pomel ex Batt. containing α-pinene (39.8%) was also found active against *Salmonella typhimurium* LT2 DT104 and *Bacillus cereus* ATCC 14,579, with inhibition zones of 16.2 $\pm$ 1.0 mm and 15.8 $\pm$ 1.0 mm, respectively [36]. Although we have not performed any mechanistic investigation in the present study, we believe the polyphenols present in HEO damaged the cell wall or cell membrane.

3. Materials and Methods

3.1. Chemicals

Folin-Ciocalteu reagent, 2,2-diphenyl-1-picrylhydrazyl (DPPH), potassium peroxydisulfate, 2,2′-azino-bis(3-ethylbenzothiazoline-6-sulfonic acid) diammonium salt (ABTS$^{\bullet+}$), 2,4,6-Tris(2-pyridyl)-S-triazina (TPTZ), ethylenediaminetetraacetic acid (EDTA), iron sulphate heptahydrate, ferrozine, potassium peroxydisulfate, sodium phosphate dibasic, potassium phosphate monobasic, sodium acetate, iron(III) chloride hexahydrate, iron(II) chloride tetrahydrate, and C7–C40 saturated alkane standard were purchased from Sigma-Aldrich (Milan, Italy). α-Pinene, 2-Carene, 3-Carene, β-Myrcene, and D-limonene were purchased from Extrasynthese (Lyon, France). Dichloromethane was GC-grade and was purchased from Merck (Darmstadt, Germany). All other chemicals and solvents used were of analytical grade.

3.2. Plant Material and Isolation of Essential Oil

The ripe pistachio hulls (*P. vera* L., variety Bronte) were harvested on 3 October 2016 by a local farmer in Bronte (Catania, Italy) and immediately sent to the laboratory. Four hundred grams (400 g) of fresh pistachio hulls were manually separated from the inner woody shells and subjected to hydrodistillation, according to the current *European Pharmacopoeia* method, until no significant increase in the volume of the collected oil was observed (3 h). The HEOs were dried on Na_2SO_4 and stored in a sealed vial under N_2 until analysis.

3.3. GC/MS Analysis

Gas chromatographic (GC) analysis was performed on Agilent gas chromatograph, Model 7890A, equipped with a flame ionization detector (FID). Analytical conditions were HP-5MS capillary column (30 m × 0.25 mm coated with 5% phenyl methyl silicone, 95% dimethyl polysiloxane, 0.25 μm film thickness) and helium as the carrier gas (1 mL/min). Injection was done in split mode (1:60), injected volume was 1 μL (10% essential oil/CH_2Cl_2 *v/v*), and the injector and detector temperatures were 250 °C and 280 °C, respectively. The oven temperature was held at 50 °C for 1 min, then increased to 240 °C at 5 °C/min and held at 240 °C for 5 min. Percentages of compounds were determined from their peak areas in the GC-FID profiles.

Gas chromatography-mass spectrometry (GC-MS) analysis was carried out on the same instrument described above coupled with an Agilent 5975C mass detector (Santa Clara, CA, USA), with the same column and the same operative conditions used for the analytical GC. We adjusted the ionization voltage to 70 eV, the electron multiplier to 900 V, and the ion source temperature to 230 °C. Mass spectra data were acquired in the scan mode in an m/z range of 40–300. The compounds were identified based on: their GC retention index (relative to C9–C22 *n*-alkanes on the HP-5MS column), values reported in the literature, the computer matching of their mass spectral data with those of the MS library (NIST 08), the comparison of their MS fragmentation patterns with those reported in literature, and, whenever possible, the co-injection with authentic standards (α-Pinene, 2-Carene, 3-Carene, β-Myrcene and D-limonene).

3.4. Screening of Antioxidant and Free-Radical Scavenging Properties

3.4.1. Determination of Total Phenolic Compounds

The total phenols content was determined according to Smeriglio et al. [37] using gallic acid as the reference compound and the results were expressed as mg of gallic acid equivalents (GAE)/100 g of essential oil.

3.4.2. DPPH Assay

The DPPH free radical scavenging activity was evaluated according to Bellocco et al. [24]. Briefly, freshly DPPH methanol solution (10^{-4} M), was mixed with 37.5 μL of sample solution (range 2.4–0.300 mg/mL), and the mixture was vortexed for 10 s at room temperature (RT). The decrease in absorption at 517 nm, against blank, was measured after 20 min using an UV-VIS Spectrophotometer (Shimadzu UV-1601, Kyoto, Japan). The inhibition (%) of radical scavenging activity was calculated by the following equation:

$$\text{Inhibition (\%)} = \frac{A_0 - A_S}{A_0} \times 100 \tag{1}$$

where A_0 is absorbance of the control and A_S is absorbance of the sample after 20 min incubation.

3.4.3. Trolox Equivalent Antioxidant Capacity (TEAC) Assay

The antioxidant activity against ABTS$^{\bullet+}$ radical was carried out according to Smeriglio et al. [37]. Briefly, the reaction mixture (4.3 mM potassium persulfate and 1.7 mM ABTS solution 1:5, v/v) was incubated for 12–16 h in the dark at room temperature, and was diluted, before use, with phosphate buffer (pH 7.4) in order to obtain an absorbance at 734 nm of 0.7 ± 0.02. Fifty microliters (50 µL) of sample solution (range 300–50 µg/mL) was added to 1 mL of reaction mixture and incubated in the dark at room temperature for 6 min; the absorbance was then recorded at 734 nm using an UV-VIS Spectrophotometer (Shimadzu UV-1601). The inhibition (%) of radical scavenging activity was calculated using Equation (1).

3.4.4. Ferric Reducing Antioxidant Power (FRAP)

The free-radical scavenging capacity against TPTZ radical was performed according to Smeriglio et al. [37]. Twenty-five microliters (25 µL) of sample solution (range 150–25 µg/mL) was added to 1.5 mL of daily fresh FRAP reagent pre-warmed at 37 °C, and the absorbance was recorded at 593 nm by an UV-VIS Spectrophotometer (Shimadzu UV-1601) after an incubation time of 4 min at 20 °C, with the FRAP reagent used as blank. The inhibition (%) of radical scavenging activity was calculated using Equation (1).

3.4.5. Chelating Capacity on Fe^{2+}

Fe^{2+} chelating capacity was evaluated as described by Smeriglio et al. [37] with some modifications. Briefly, 25 µL of FeCl$_2$·4H$_2$O solution (1.8 mM) was added to 50 µL of sample solution (range 60–7.5 µg/mL) and incubated at RT for 5 min. Following, 50 µL of a ferrozine solution (4 mM) was added to the reaction mixture, and the sample volume was diluted to 1.5 mL with deionized water. After that, the mixture was vortexed and incubated for 10 min at room temperature. The absorbance was read at 562 nm using an UV/VIS Spectrophotometer (Shimadzu UV-1601). The inhibition (%) of Fe^{2+} chelating capacity was calculated using Equation (1).

3.4.6. β-Carotene Bleaching

The β-carotene bleaching assay was performed using an emulsion prepared according to Smeriglio et al. [37]. Aliquots of this emulsion (7.0 mL) were mixed with 0.28 mL of sample solution (range 4–0.5 mg/mL). An emulsion without β-carotene was used as control. The reaction mixture was initially recorded at the starting time ($t = 0$) at 470 nm and then incubated at 50 °C in a water bath for 120 min, with the absorbance recorded every 20 min. Butylated hydroxytoluene (BHT) was used as the reference compound and the results were expressed as inhibition (%) of β-carotene bleaching using Equation (1).

3.4.7. Superoxide Anion (O$_2$$^{\bullet-}$) Scavenging Assay

The superoxide anion scavenging activity of the sample was determined according to Barreca et al. [23]. The reaction mix was composed of 1.0 mL of nitroblue tetrazolium (NBT) solution (156 µM NBT in 100 mM phosphate buffer, pH 7.4), 1.0 mL of β-nicotinamide adenine dinucleotide (NADH) solution (468 µM in 100 mM phosphate buffer, pH 7.4), and 20 µL of sample solution (range 60–3.75 µg/mL). The reaction was started by adding 100 µL phenazine methosulphate (PMS) solution (60 µM PMS in 100 mM phosphate buffer, pH 7.4) to the mixture. The reaction mixture was incubated at 25 °C for 5 min, and absorbance at 560 nm was measured against blank samples with a Varian Cary 50 UV-VIS spectrophotometer. The inhibition (%) of radical scavenging activity was calculated using Equation (1).

3.4.8. Hydroxyl Radical (•OH) Scavenging Assay

The hydroxyl radical scavenging assay was performed as described by Tellone et al. [38]. This assay is based on quantification of the degradation product of 2-deoxyribose by condensation with thiobarbituric acid (TBA). The reaction mixture contained the following in the final volume of 1.0 mL: 2.8 mM 2-deoxy-2-ribose, 10 mM phosphate buffer pH 7.4, 25 µM $FeCl_3$, 100 µM EDTA, 2.8 mM H_2O_2, 100 µM ascorbic acid and sample solution in order to obtain the final concentration range of 15–0.92 µg/mL. The samples were incubated for 1 h at 37 ± 0.5 °C in a water bath. Then 1.0 mL of 1% (w/v) TBA was added to each mixture followed by the addition of 1 mL of 2.8% (w/v) trichloroacetic acid (TCA). The solutions were heated in a water bath at 100 °C for 15 min to develop the pink coloured malondialdehyde–thiobarbituric acid adduct. After cooling, the absorbance was measured at 532 nm against an appropriate blank solution. The inhibition (%) of radical scavenging activity was calculated using Equation (1).

3.5. Evaluation of Cytoprotective Properties

3.5.1. Lymphocyte Isolation

Lymphocytes were isolated from heparinized whole blood collected from healthy volunteers, who provided written medical histories through a standardized questionnaire and had not taken anti-inflammatory medication or nutritional supplements. Blood samples were diluted with equal volumes of balanced salt solution, layered over Histopaque-1077 in centrifuge tubes and centrifuged at $400 \times g$ for 30–40 min at 25 °C. The peripheral blood mononuclear cell (PBMC) layer was removed with a pipette and washed by centrifugation. The PBMCs were passed through a Percoll gradient according to Repnik et al. [39] to enrich the fraction in lymphocytes. Lymphocytes (viability > 90%) were counted on a haemocytometer and suspended in Roswell Park Memorial Institute (RPMI) 1640 medium supplemented with 10% fetal calf serum, 2 mM glutamine, 100 units/mL penicillin G and streptomycin. Cell concentration was adjusted to 1×10^5 cells/mL.

3.5.2. Cytotoxicity Assays

For the cytotoxicity assay, cells (1×10^5/mL) were incubated in complete medium without or with 20, 17.5, 15, 12.5, 10, 7.5, 5, or 1 µg/mL of sample solution for 24 h in the presence of 100 µM tert-butyl hydroperoxide (*t*-BOOH). The stock solution of HEO in dimethyl sulphoxide (DMSO) were conveniently diluted with phosphate saline buffer to maintain the concentration of the DMSO below 0.1% in the reaction mixture. Parallel controls were performed without *t*-BOOH and in the presence of the final concentrations of HEO. After incubation, cell viability was assessed by trypan blue staining. Briefly, an aliquot of the cell suspension was diluted 1:1 (v/v) with 0.4% trypan blue and the cells were counted using a haemocytometer. Results were expressed as the percentage of viable or dead cells (ratio of unstained or stained cells to the total number of cells, respectively). Cytotoxicity was also measured by lactate dehydrogenase (LDH) release from damaged cells into culture medium and expressed as a percentage of total cellular activity. LDH activity in the medium was determined using a commercially available kit from BioSystems S.A (Barcelona, Spain) and expressed as a function of the total amount of the enzyme present in the *t*-BOOH-no treated cells obtained after total cell lysis by sonication. At the concentration tested, HEO does not show any interference with the LDH assay.

3.6. Antimicrobial Activity

The following strains, obtained from the in-house culture collection of the University of Messina (Messina, Italy), were used for antimicrobial testing: *Staphylococcus aureus* ATCC 6538P, *Staphylococcus aureus* MRSA ATCC 43,300, three clinical isolates of *S. aureus* (of which two were methicillin resistant), *Escherichia coli* ATCC 10536, and *Pseudomonas aeruginosa* ATCC 9027.

Cultures for antimicrobial activity tests were grown in Mueller–Hinton Broth (MHB, Oxoid, CM0405) at 37 °C (24 h). For solid media, 1.5% (w/v) agar (Difco) was added.

The minimum inhibitory concentration (MIC) and the minimum bactericidal concentration (MBC) of HEO were determined by the broth microdilution method, according to CLSI (2009). MBCs were determined by seeding 20 mL from all clear MIC wells onto Mueller–Hinton agar (MHA, Oxoid) plates. The MBC was defined as the lowest extract concentration which killed 99.9% of the final inocula after 24 h incubation at 37 °C.

All experiments were performed in triplicate on three independent days. A number of positive and negative controls with selected antibiotics and solvent (DMSO) were included in each assay.

3.7. Statistical Analysis

Results (inhibition %) were expressed as mean ± standard deviation (S.D.) of three independent experiments in triplicate ($n = 3$) and as concentration inhibiting 50% of the initial activity (IC_{50}). Data were analysed by one-way analysis of variance (ANOVA). The significance of the difference from the respective controls for each experimental test condition was assayed by using Tukey's test for each paired experiment. Statistical significance was considered at $p < 0.05$. IC_{50} values were calculated by Graphpad Prism software (version 5.0, GraphPad Software Inc., La Jolla, CA, USA).

4. Conclusions

In the present study, we analysed, for the first time, the composition, as well as the antioxidant and biological potential of essential oil obtained from the hull of *Pistacia vera* L. variety Bronte. GC-MS analysis shed some light on its composition, allowing for the identification of 40 derivatives, characterized by the abundance of monoterpene hydrocarbons and oxygenated monoterpenes, both of which are some of the most studied and promising forms of secondary metabolites due to their remarkable and different biological properties. In fact, in the obtained results, the essential oil showed remarkable antioxidant and cytoprotective activities, which can also be attributed to its richness in monoterpene derivatives, which is dominated by the presence of 4-Carene, α-Pinene, D-Limonene, and 3-Carene. Therefore, this complex matrix could represent a suitable natural source of nutraceutical compounds to be employed in cosmetics, pharmaceutics, food preservation, and biotechnology applications.

Acknowledgments: The authors' research funds covered the cost of the study and open access fee.

Author Contributions: Antonella Smeriglio conceived the study, designed the experiments, and wrote the manuscript; Davide Barreca contributed to the design of the experiments and the interpretation of the data and wrote the manuscript; Marcella Denaro, Carlo Bisignano, Giovanna Ginestra, and Antonella Calderaro were responsible for performing the experiments; Ersilia Bellocco was responsible for data collection and data analysis; Domenico Trombetta is the principal investigator responsible for study design. All the authors read and approved the final manuscript.

Conflicts of Interest: The authors declare no conflict of interest.

References

1. Swamy, M.K.; Akhtar, M.S.; Sinniah, U.R. Antimicrobial Properties of Plant Essential Oils against Human Pathogens and Their Mode of Action: An Updated Review. *Evid. Based Complement. Altern. Med.* **2016**, *2016*, 3012462. [CrossRef] [PubMed]
2. Sharifi-Rad, J.; Sureda, A.; Tenore, G.C.; Daglia, M.; Sharifi-Rad, M.; Valussi, M.; Tundis, R.; Sharifi-Rad, M.; Loizzo, M.R.; Ademiluyi, A.O.; et al. Biological Activities of Essential Oils: From Plant Chemoecology to Traditional Healing Systems. *Molecules* **2017**, *22*, 70. [CrossRef] [PubMed]

3. García, C.C.; Acosta, E.G.; Carro, A.C.; Fernández Belmonte, M.C.; Bomben, R.; Duschatzky, C.B.; Perotti, M.; Schuff, C.; Damonte, E.B. Virucidal activity and chemical composition of essential oils from aromatic plants of central west Argentina. *Nat. Prod. Commun.* **2010**, *5*, 1307–1310. [PubMed]

4. Al-Mariri, A.; Safi, M. In Vitro Antibacterial Activity of Several Plant Extracts and Oils against Some Gram-Negative Bacteria. *Iran J. Med. Sci.* **2014**, *39*, 36–43. [PubMed]

5. Silva, F.; Domingues, F.C. Antimicrobial activity of coriander oil and its effectiveness as food preservative. *Crit. Rev. Food Sci. Nutr.* **2017**, *57*, 35–47. [CrossRef] [PubMed]

6. Jarić, S.; Mitrović, M.; Pavlović, P. Review of Ethnobotanical, Phytochemical, and Pharmacological Study of *Thymus serpyllum* L. *Evid. Based Complement. Altern. Med.* **2015**, *2015*, 101978. [CrossRef] [PubMed]

7. Shokri, H. A review on the inhibitory potential of *Nigella sativa* against pathogenic and toxigenic fungi. *Avicenna J. Phytomed.* **2016**, *6*, 21–33. [PubMed]

8. Arumugam, G.; Swamy, M.K.; Sinniah, U.R. *Plectranthus amboinicus* (Lour.) Spreng: Botanical, Phytochemical, Pharmacological and Nutritional Significance. *Molecules* **2016**, *21*, 369. [CrossRef] [PubMed]

9. Mahboubi, M. *Rosa damascena* as holy ancient herb with novel applications. *J. Tradit. Complement. Med.* **2015**, *6*, 10–16. [CrossRef] [PubMed]

10. Lillehei, A.S.; Halcon, L.L. A systematic review of the effect of inhaled essential oils on sleep. *J. Altern. Complement. Med.* **2014**, *20*, 441–451. [CrossRef] [PubMed]

11. Rehman, J.U.; Ali, A.; Khan, I.A. Plant based products: Use and development as repellents against mosquitoes: A review. *Fitoterapia* **2014**, *95*, 65–74. [CrossRef] [PubMed]

12. Sugawara, Y.; Shigetho, A.; Yoneda, M.; Tuchiya, T.; Matumura, T.; Hirano, M. Relationship between mood change, odour and its physiological effects in humans while inhaling the fragrances of essential oils as well as linalool and its enantiomers. *Molecules* **2013**, *18*, 3312–3338. [CrossRef] [PubMed]

13. Langeveld, W.T.; Veldhuizen, E.J.; Burt, S.A. Synergy between essential oil components and antibiotics: A review. *Crit. Rev. Microbiol.* **2014**, *40*, 76–94. [CrossRef] [PubMed]

14. Taga, I.; Lan, C.Q.; Altosaar, I. Plant essential oils and mastitis disease: Their potential inhibitory effects on pro-inflammatory cytokine production in response to bacteria related inflammation. *Nat. Prod. Commun.* **2012**, *7*, 675–682. [PubMed]

15. Solórzano-Santos, F.; Miranda-Novales, M.G. Essential oils from aromatic herbs as antimicrobial agents. *Curr. Opin. Biotechnol.* **2012**, *23*, 136–141. [CrossRef] [PubMed]

16. Božović, M.; Ragno, R. *Calamintha nepeta* (L.) Savi and its Main Essential Oil Constituent Pulegone: Biological Activities and Chemistry. *Molecules* **2017**, *22*, 290. [CrossRef] [PubMed]

17. Rodrigues Simões, R.; Dos Santos Coelho, I.; Célio Junqueira, S.; Regina Pigatto, G.; José Salvador, M.; Santos, A.R.; de Faria, F.M. Oral treatment with essential oil of *Hyptis spicigera* Lam. (Lamiaceae) reduces acute pain and inflammation in mice: Potential interactions with transient receptor potential (TRP) ion channels. *J. Ethnopharmacol.* **2017**, *200*, 8–15. [CrossRef] [PubMed]

18. Kumar, A.S.; Jeyaprakash, K.; Chellappan, D.R.; Murugan, R. Vasorelaxant and cardiovascular properties of the essential oil of *Pogostemon elsholtzioides*. *J. Ethnopharmacol.* **2017**, *199*, 86–90. [CrossRef] [PubMed]

19. Branquinho, L.S.; Santos, J.A.; Cardoso, C.A.; Mota, J.D.; Junior, U.L.; Kassuya, C.A.; Arena, A.C. Anti-inflammatory and toxicological evaluation of essential oil from *Piper glabratum* leaves. *J. Ethnopharmacol.* **2017**, *198*, 372–378. [CrossRef] [PubMed]

20. Rezaie, M.; Farhoosh, R.; Sharif, A.; Asili, J.; Iranshahi, M. Chemical composition, antioxidant and antibacterial properties of Bene (*Pistacia atlantica* subsp. mutica) hull essential oil. *J. Food Sci. Technol.* **2015**, *52*, 6784–6790. [CrossRef] [PubMed]

21. Martorana, M.; Arcoraci, T.; Rizza, L.; Cristani, M.; Bonina, F.P.; Saija, A.; Trombetta, D.; Tomaino, A. In vitro antioxidant and in vivo photoprotective effect of pistachio (*Pistacia vera* L., variety Bronte) seed and skin extracts. *Fitoterapia* **2013**, *85*, 41–48. [CrossRef] [PubMed]

22. Tomaino, A.; Martorana, M.; Arcoraci, T.; Monteleone, D.; Giovinazzo, C.; Saija, A. Antioxidant activity and phenolic profile of pistachio (*Pistacia vera* L., variety Bronte) seeds and skins. *Biochimie* **2010**, *92*, 1115–1122. [CrossRef] [PubMed]

23. Barreca, D.; Laganà, G.; Leuzzi, U.; Smeriglio, A.; Trombetta, D.; Bellocco, E. Evaluation of the nutraceutical, antioxidant and cytoprotective properties of ripe pistachio (*Pistachia vera* L. variety Bronte) hulls. *Food Chem.* **2016**, *196*, 493–502. [CrossRef] [PubMed]

24. Bellocco, E.; Barreca, D.; Laganà, G.; Calderaro, A.; El Lekhlifi, Z.; Chebaibi, S.; Smeriglio, A.; Trombetta, D. Cyanidin-3-*O*-galactoside in ripe pistachio (*Pistachia vera* L. variety Bronte) hulls: Identification and evaluation of its antioxidant and cytoprotective activities. *J. Funct. Foods* **2016**, *27*, 376–385. [CrossRef]

25. Kasabri, V.; Afifi, F.U.; Hamdan, I. In vitro and in vivo acute antihyperglycemic effects of five selected indigenous plants from Jordan used in traditional medicine. *J. Ethnopharmacol.* **2011**, *133*, 888–896. [CrossRef] [PubMed]

26. Mehla, K.; Balwani, S.; Kulshreshtha, A.; Nandi, D.; Jaisankar, P.; Ghosh, B. Ethyl gallate isolated from *Pistacia integerrima* Linn. inhibits cell adhesion molecules by blocking AP-1 transcription factor. *J. Ethnopharmacol.* **2011**, *137*, 1345–1352. [CrossRef] [PubMed]

27. Shirole, R.L.; Shirole, N.L.; Kshatriya, A.A.; Kulkarni, R.; Saraf, M.N. Investigation into the mechanism of action of essential oil of *Pistacia integerrima* for its antiasthmatic activity. *J. Ethnopharmacol.* **2014**, *153*, 541–551. [CrossRef] [PubMed]

28. Shirole, R.L.; Shirole, N.L.; Saraf, M.N. In vitro relaxant and spasmolytic effects of essential oil of *Pistacia integerrima* Stewart ex Brandis Galls. *J. Ethnopharmacol.* **2015**, *168*, 61–65. [CrossRef] [PubMed]

29. Ebadi, M.T.; Sefidkon, F.; Azizi, M.; Ahmadi, N. Packaging methods and storage duration affect essential oil content and composition of lemon verbena (*Lippia citriodora* Kunth.). *Food Sci. Nutr.* **2017**, *5*, 588–595. [CrossRef] [PubMed]

30. Dawidowicz, A.L.; Olszowy, M. Does antioxidant properties of the main component of essential oil reflect its antioxidant properties? The comparison of antioxidant properties of essential oils and their main components. *Nat. Prod. Res.* **2014**, *28*, 1952–1963. [CrossRef] [PubMed]

31. González-Burgos, E.; Gómez-Serranillos, M.P. Terpene compounds in nature: A review of their potential antioxidant activity. *Curr. Med. Chem.* **2012**, *19*, 5319–5341. [CrossRef] [PubMed]

32. Bisignano, C.; Filocamo, A.; Faulks, R.M.; Mandalari, G. In vitro antimicrobial activity of pistachio (*Pistacia vera* L.) polyphenols. *FEMS Microbiol. Lett.* **2013**, *341*, 62–67. [CrossRef] [PubMed]

33. Mandalari, G.; Bennett, R.N.; Bisignano, G.; Trombetta, D.; Saija, A.; Faulds, C.B.; Gasson, M.J.; Narbad, A. Antimicrobial activity of flavonoids extracted from bergamot (*Citrus bergamia* Risso) peel, a byproduct of the essential oil industry. *J. Appl. Microbiol.* **2007**, *103*, 2056–2064. [CrossRef] [PubMed]

34. Mandalari, G.; Bisignano, C.; D'Arrigo, M.; Ginestra, G.; Arena, A.; Tomaino, A.; Wickham, M.S. Antimicrobial potential of polyphenols extracted from almond skins. *Lett. Appl. Microbiol.* **2010**, *51*, 83–90. [CrossRef] [PubMed]

35. Mahboubi, M.; Kazempour, N. Biochemical Activities of Iranian *Cymbopogon olivieri* (Boiss) Bor. Essential Oil. *Indian J. Pharm. Sci.* **2012**, *74*, 356–360. [CrossRef] [PubMed]

36. Znati, M.; Filali, I.; Jabrane, A.; Casanova, J.; Bouajila, J.; Ben Jannet, H. Chemical Composition and In Vitro Evaluation of Antimicrobial, Antioxidant and Antigerminative Properties of the Seed Oil from the Tunisian Endemic *Ferula tunetana* Pomel ex Batt. *Chem. Biodivers.* **2017**, *14*. [CrossRef] [PubMed]

37. Smeriglio, A.; Galati, E.M.; Monforte, M.T.; Lanuzza, F.; D'Angelo, V.; Circosta, C. Polyphenolic Compounds and Antioxidant Activity of Cold-Pressed Seed Oil from Finola Cultivar of *Cannabis sativa* L. *Phytother. Res.* **2016**, *30*, 1298–1307.

38. Tellone, E.; Ficarra, S.; Russo, A.; Bellocco, E.; Barreca, D.; Laganà, G.; Leuzzi, U.; Pirolli, D.; De Rosa, M.C.; Giardina, B.; et al. Caffeine inhibits erythrocyte membrane derangement by antioxidant activity and by blocking caspase 3 activation. *Biochimie* **2012**, *94*, 393–402. [CrossRef] [PubMed]

39. Repnik, U.; Knezevic, M.; Jeras, M. Simple and costeffective isolation of monocytes from buffy coats. *J. Immunol. Methods* **2003**, *278*, 283–292. [CrossRef]

International Journal of
Molecular Sciences

MDPI

Review

Antibacterial and Antifungal Activities of Spices

Qing Liu [1], Xiao Meng [1], Ya Li [1], Cai-Ning Zhao [1], Guo-Yi Tang [1] and Hua-Bin Li [1,2,*]

[1]   Guangdong Provincial Key Laboratory of Food, Nutrition and Health, Department of Nutrition,
       School of Public Health, Sun Yat-sen University, Guangzhou 510080, China;
       liuq248@mail2.sysu.edu.cn (Q.L.); mengx7@mail2.sysu.edu.cn (X.M.); liya28@mail2.sysu.edu.cn (Y.L.);
       zhaocn@mail2.sysu.edu.cn (C.-N.Z.); tanggy5@mail2.sysu.edu.cn (G.-Y.T.)
[2]   South China Sea Bioresource Exploitation and Utilization Collaborative Innovation Center,
       Sun Yat-sen University, Guangzhou 510006, China
*   Correspondence: lihuabin@mail.sysu.edu.cn; Tel.: +86-20-8733-2391

Received: 16 May 2017; Accepted: 11 June 2017; Published: 16 June 2017

Abstract: Infectious diseases caused by pathogens and food poisoning caused by spoilage microorganisms are threatening human health all over the world. The efficacies of some antimicrobial agents, which are currently used to extend shelf-life and increase the safety of food products in food industry and to inhibit disease-causing microorganisms in medicine, have been weakened by microbial resistance. Therefore, new antimicrobial agents that could overcome this resistance need to be discovered. Many spices—such as clove, oregano, thyme, cinnamon, and cumin—possessed significant antibacterial and antifungal activities against food spoilage bacteria like *Bacillus subtilis* and *Pseudomonas fluorescens*, pathogens like *Staphylococcus aureus* and *Vibrio parahaemolyticus*, harmful fungi like *Aspergillus flavus*, even antibiotic resistant microorganisms such as methicillin resistant *Staphylococcus aureus*. Therefore, spices have a great potential to be developed as new and safe antimicrobial agents. This review summarizes scientific studies on the antibacterial and antifungal activities of several spices and their derivatives.

Keywords: spice; antibacterial activity; antifungal effect; antimicrobial property; essential oil; clove; oregano; thyme

---

1. Introduction

Microbial pathogens in food may cause spoilage and contribute to foodborne disease incidence, and the emergence of multidrug resistant and disinfectant resistant bacteria—such as *Staphylococcus aureus* (*S. aureus*), *Escherichia coli* (*E. coli*), and *Pseudomonas aeruginosa* (*P. aeruginosa*)—has increased rapidly, causing the increase of morbidity and mortality [1]. Weak acids such as benzoic and sorbic acids [2], which are commonly applied in food industry as chemical preservatives to increase the safety and stability of manufactured foods on its whole shelf-life by controlling pathogenic and spoilage food-related microorganisms [3], can result in the development of microbiological resistance [4]. Moreover, chemical preservatives cannot completely eliminate several pathogenic bacteria like *Listeria monocytogenes* (*L. monocytogenes*) in food products or delay the growth of spoilage microorganisms [5]. Natural products, as substitutes of synthetic chemical preservatives, are increasingly accepted because they are innately better tolerated in human body and have inherent superiorities for food industry [4]. The antimicrobial activities of natural products are necessary to be studied and applied in food industry.

Morbidity and mortality are mainly caused by infectious diseases all over the world. The World Health Organization reported that 55 million people died worldwide in 2011, with one-third of the deaths owing to infectious diseases [6]. Antibiotic resistant microorganisms can increase mortality rates because they can survive and recover through their ability to acquire and transmit resistance after exposure to antibiotic drugs, which are one of the therapies to infectious diseases [7]. Antibiotic resistant bacteria threaten the antibiotic effectiveness and limit the therapeutic options even for common infections [8]. The decline in research and development of new antibacterial agents, which are able to inhibit antibiotic resistant disease-causing microorganisms such as *S. aureus*, aggravates the emerging antibiotic resistance [9]. Therefore, much attention should be paid to natural products, which could be used as effective drugs to treat human diseases, with high efficacy against pathogens and negligible side effects.

Spices have been used as food and flavoring since ancient times [10], and as medicine and food preservatives in recent decades [11,12]. Many spices—such as clove, oregano, thyme, cinnamon, and cumin—have been applied to treat infectious diseases or protect food because they were experimentally proved to possess antimicrobial activities against pathogenic and spoilage fungi and bacteria [10,13,14]. Moreover, the secondary metabolites of these spices are known as antimicrobial agents, the majority of which are generally recognized as safe materials for food with insignificant adverse effects [11]. Therefore, spices could be candidates to discover and develop new antimicrobial agents against foodborne and human pathogens.

This review summarizes the scientific studies on the antibacterial and antifungal activities of spices and their derivatives, and some suggestions and prospects are offered for future studies.

2. Clove

Clove (*Eugenia caryohyllata*), belonging to family Myrtaceae, is widely used in medicine as antiseptic against infectious diseases like periodontal disease due to the antimicrobial activities against oral bacteria [15]. Clove is also commonly applied in food industry as a natural additive or antiseptic to increase shelf-life due to the effective antimicrobial activities against some foodborne pathogens [16]. The main active component of clove oil and extract was found, i.e., eugenol [15,17].

2.1. Antimicrobial Activities of Clove

Antimicrobial activities of clove water extract were studied in vitro and in vivo against pathogenic microorganisms (*S. aureus* and *E. coli*, in a model of pyelonephritis) [18]. An in vitro study was conducted with the agar well diffusion method, and the results suggested that clove water extract showed antibacterial activity against *S. aureus* (minimum inhibitory concentration (MIC): 2 mg/mL) and *E. coli* (MIC: 2.5 mg/mL). While in vivo, the study was conducted in 40 adult male albino rats, and the results confirmed the efficacy of clove extract as natural antimicrobials. The direct antimicrobial activities of ultra-fine powders of ball-milled cinnamon and clove were tested by Kuang et al. [19] against *E. coli*, *S. aureus*, *Brochothrix thermosphacta* (*B. thermosphacta*), *Lactobacillus rhamnosus* (*L. rhamnosus*), and *Pseudomonas fluorescens* (*P. fluorescens*) from meat, using broth dilution method. Clove powder showed strong inhibitory effects on five microorganisms tested with the MICs ranging from 1.0% w/v (*L. rhamnosus* and *B. thermosphacta*) to 2.0% w/v (*P. fluorescens*), and the inhibitory effects were positively associated with the concentrations of powder, which increased from 0.5% to 2.5% w/v.

Clove could destroy cell walls and membranes of microorganisms, and permeate the cytoplasmic membranes or enter the cells, then inhibit the normal synthesis of DNA and proteins [16]. Eugenol, the major component of clove, could inhibit the production of amylase and proteases in *Bacillus cereus* (*B. cereus*) and has the ability of cell wall deterioration and cell lysis [20].

2.2. Comparison of Antimicrobial Activities of Clove and Other Spices

Badei et al. [21] tested the antimicrobial activities of cardamom, cinnamon and clove essential oils (EOs) against nine Gram-positive bacterial strains, four Gram-negative bacterial strains, seven molds, and two yeasts, compared with phenol, using the disc diffusion method. Clove EO showed the highest antimicrobial activity, and the antimicrobial spectra (diameter of inhibition zones) of 10% clove EO was 1.48 times as that of 10% phenol. Schmidt et al. [22] evaluated the antifungal effects of eugenol-containing EOs of 4 spices on 38 *Candida albicans* (*C. albicans*) isolates, of which 12 were isolated from oropharynges, 16 from vaginas, and 10 from damaged skin, using the microdilution method. Clove EO possessed the strongest antifungal activities against all *C. albicans* strains among the tested spices. Pure eugenol alone exhibited weaker antifungal activities than clove leaf EO. Angienda et al. [23] investigated the antimicrobial activities of EOs of four spices against *Salmonella typhimurium* (*S. typhimurium*), *E. coli*, *B. cereus*, and *Listeria innocua* (*L. innocua*) by agar diffusion test. Clove EO showed the most effective inhibition against both Gram-positive bacteria and Gram-negative bacteria compared with three other EOs, with the MICs ranging from 1.25% *v/v* (*B. cereus*) to 2.50% *v/v* (*S. typhimurium* and *E. coli*). Lomarat et al. [17] reported the antimicrobial activities of EOs from nine spices against histamine-producing bacteria including *Morganella morganii* (*M. morganii*), by determining MICs and minimum bactericidal concentrations (MBCs) using the broth dilution assay, and also found the antibacterial compounds of EOs by bioautography-guided isolation. The results indicated that the clove EO was the most effective against *M. morganii* among nine tested spices with MIC 0.13% *v/v* and MBC 0.25% *v/v*. The eugenol was identified as the active component of clove EO by thin layer chromatography bioautography assay.

Shan et al. [24] tested the antibacterial activities of ethanol extracts from five spices and herbs against *L. monocytogenes*, *S. aureus*, and *Salmonella enterica* (*S. enterica*) in raw pork by counting bacterial enumeration. When treated with clove extract, raw pork samples were found with the fewest colonies of tested bacteria. Bayoub et al. [25] reported the antimicrobial activities of ethanol extracts of 13 plants including clove against *L. monocytogenes*, the MICs were determined by agar well diffusion test. The results showed that clove extract was the most effective inhibitor against *L. monocytogenes* compared with the other 12 selected plant ethanol extracts, with the MIC 0.24 mg/mL. Cui et al. [26] tested the antimicrobial activities of 90 plant extracts (water and 99.5% ethanol extracts) against *Clostridium* spp. Clove water extract was found with the greatest antimicrobial activity against *Clostridium botulinum* in trypticase peptone glucose yeast extract broth (pH = 7.0) among all the water extracts, and the MICs of clove extract ranged from 0.1% to 0.2% against *Clostridium* spp. Antimicrobial effects of 3 extracts (ethyl acetate, acetone, and methanol extracts) of 12 plants were tested on 2 fungi (*Kluyveromyces marxianus* (*K. marxianus*) and *Rhodotorula rubra* (*R. rubra*)) and 8 bacteria (*Klebsiella pneumoniae* (*K. pneumoniae*), *Bacillus megaterium* (*B. megaterium*), *P. aeruginosa*, *S. aureus*, *E. coli*, *Enterobacter cloacae* (*E. cloacae*), *Corynebacterium xerosis* (*C. xerosis*), and *Streptococcus faecalis* (*S. faecalis*)) by the disc diffusion method [27]. Clove exhibited the most effective inhibitory impacts. The methanol extract from clove showed inhibition against microorganisms (diameter of inhibition zones (DIZs): 8–24 mm) tested except *K. pneumoniae*. The acetone extract showed inhibition against microorganisms (DIZs: 8–18 mm) tested except *R. rubra* and *K. pneumoniae*. The ethyl acetate extract only showed antibacterial activity against *B. megaterium* (DIZ: 7 mm). Liang et al. [28] observed the antimicrobial activities of seven spices, and different concentrations of extracts and EOs in each spice were used to test the effects on the growth of spoilage microorganisms in apple cider by total plate counts. Clove products showed the strongest antimicrobial activities compared with other spices tested. Nearly seven log reduction of microorganisms was observed at 0.8% *v/v* in the cider for clove oil and 2% *w/w* for clove powder at room temperature. Badhe et al. [29] tested the antimicrobial activities of many spice and herb powders against *S. aureus*, *S. typhimurium*, *E. coli*, and *B. cereus* at refrigerated temperature (8 ± 2 °C) for intervals of 0, 3, 6, 12, 24, and 48 h. The results indicated that at the concentration of 2%, clove powder showed highest effect on *S. aureus* followed by *E.coli* and *S. typhimurium*, and at 24 h under refrigeration, clove powder led to a significant reduction of bacteria counting.

2.3. The Application of Clove as Antimicrobial Agents in Food Packaging

Clove EO and its functional extracts have been incorporated into films, the antimicrobial activities of which have been evaluated in some studies. In a study, chitosan at high, moderate and low molecular mass were elaborated with antimicrobial films which were incorporated with EOs and extracts from two spices [30]. Then the antimicrobial effects of the films were investigated on *E. coli*, *S. typhimurium*,

S. aureus, *B. cereus*, and *L. monocytogenes*. The films prepared by low molecular mass chitosan with 2% EO and ethyl heptanoate extract from clove showed antimicrobial activities against a majority of the tested strains. In another study, the researchers tested the antimicrobial activities of EOs and functional extracts of cumin, clove, and elecampane against *E. coli*, *S. typhimurium*, *B. cereus*, *S. aureus*, and *L. monocytogenes* by determining the MICs and MBCs [31]. They also evaluated the antibacterial activities of edible films prepared by EOs and functional extracts of spices based on chitosan polymeric structure against the same bacteria by determining the DIZs. Clove EO showed the best inhibitory effects with the MIC of 500 mg/L on all the bacteria tested, clove extracts showed very similar MICs to those of EO, except ethyl caproate extract of clove against *L. monocytogenes* (MIC of 750 mg/L) and ethyl heptanoate extract of clove against *B. cereus* (MIC of 250 mg/L). Among the chitosan films added with EOs, only clove showed inhibition zones of all tested bacteria except *L. monocytogenes*. The ethyl heptanoate extract of clove film also possessed antibacterial activities against all tested bacteria, weaker than those of clove EO though. Liu et al. [32] evaluated the antimicrobial activities of spice EOs against microbial populations in chilled pork stored in PE film antimicrobial package using the disk diffusion method to determine the DIZs and serial dilution assay to determine the MICs. Clove EO was the most effective against microorganisms tested among all the spice EOs tested. The MICs of clove EO were 0.10%, 0.10%, and 0.30% *v/v* against *Enterobacteriaceae*, *S. aureus*, and *Pseudomonas* sp., respectively. Spice EOs possessed the ability to decrease the number of spoilage populations, but not the species diversity of spoilage microbiota.

Collectively, clove EO and extracts could prevent against some food spoilage and foodborne pathogens (Table 1), especially Gram-positive bacteria. The MICs of clove were less than 2.5% against tested microorganisms like *P. fluorescens*, *S. typhimurium*, *E. coli*, *B. cereus*, and *L. innocua*. Generally speaking, the qualities of the papers cited are good and the results are reliable.

Int. J. Mol. Sci. **2017**, *18*, 1283

Table 1. Antibacterial and antifungal activities of clove.

Type of Samples	Bacteria and Fungi	Main Results	Reference
Clove and cinnamon water extracts	*Staphylococcus aureus* and *Escherichia coli*	Both in vivo and in vitro results confirmed the efficacy of clove extract as natural antimicrobials.	[18]
Ultra-fine powders of ball-milled clove	*E. coli, S. aureus, Brochothrix thermosphacta, Lactobacillus rhamnosus, Pseudomonas fluorescens*	Clove powder showed a strong inhibitory effect with the minimum inhibitory concentrations (MICs) ranging from 1.0% to 2.0% w/v.	[19]
Cardamom, cinnamon, clove essential oils (EOs) and phenol	13 bacterial strains, 7 molds and 2 yeasts	Clove EO possessed the highest antimicrobial activities.	[21]
4 spice EOs	*Candida albicans*	Clove EO possessed the strongest activities against all *C. albicans* strains.	[22]
4 spice EOs	*Salmonella typhimurium, E. coli, Bacillus cereus, Listeria innocua*	Clove EO showed the most effective inhibition with the MICs ranging from 1.25% to 2.50% v/v.	[23]
9 spice EOs	*Morganella morganii*	Clove EO was the most effective with MIC of 0.13% v/v.	[17]
Ethanol extracts from 5 spices and herbs	*Listeria monocytogenes, S. aureus, Salmonella enterica*	Clove extract was the most effective against bacteria tested.	[24]
13 plant ethanol extracts	*L. monocytogenes*	Clove extract was the most effective with the MIC of 0.24 mg/mL.	[25]
90 plant extracts	*Clostridium* spp.	Clove water extract was the most effective among all the water extracts with the MIC ranging from 0.1% to 0.2%.	[26]
Ethyl acetate, acetone and methanol extracts of 12 plant species	*Kluyveromyces marxianus, Rhodotorula rubra, Klebsiella pneumoniae, Bacillus megaterium, Pseudomonas aeruginosa, S. aureus, E. coli, Enterobacter cloacae, Corynebacterium xerosis, Streptococcus faecalis*	Clove possessed the most effective inhibitory effects.	[27]
7 spices, their extracts and EOs	Microorganisms in apple cider	Clove products had the strongest antimicrobial activities compared with other spices tested.	[28]
Many spice and herb powders	*S. aureus, S. typhimurium, E. coli, B. cereus*	2% level of clove powder was more effective against *S. aureus* followed by *E. coli* and *S. typhimurium* under refrigeration.	[29]
EOs and functional extracts of cumin and clove.	*E. coli, S. typhimurium, S. aureus, B. cereus, L. monocytogenes*	The films of low molecular weight chitosan with a concentration of 2% of EO of clove and clove ethyl heptanoate extract had antimicrobial activities against most strains tested.	[30]
EOs and functional extracts of cumin, clove, and elecampane	*E. coli, S. typhimurium, B. cereus, S. aureus, L. monocytogenes*	Chitosan films added with clove EO showed the best inhibitory effects with the MICs of 500 mg/L.	[31]
3 spice EOs	*Enterobacteriaceae, S. aureus, Pseudomonas* sp., Lactic acid bacteria, *Brocithrix thermosphacta*	Clove EO was the most effective against microorganisms tested.	[32]

3. Oregano

Oregano (*Origanum vulgare*), belonging to family Lamiaceae, has been used as food seasoning and flavoring for a long time. The major components associated with antimicrobial activities in oregano EO were proved to be carvacrol and thymol [33].

3.1. Antimicrobial Activities of Oregano

Babacan et al. [34] evaluated the antimicrobial activities of oregano extract against various *Salmonella* serotypes by evaluating the bacterial growth with disc diffusion method. The results showed that DIZs of oregano were 15, 19, and 16 mm for *Salmonella gallinarum* (*S. gallinarum*), *Salmonella enteritidis* (*S. enteritidis*), and *S. typhimurium*, respectively. Santoyo et al. [35] observed the antimicrobial activities of EO-rich fractions of oregano which were selectively precipitated in the second separator in different conditions against six microorganism strains (*S. aureus*, *Bacillus subtilis* (*B. subtilis*), *E. coli*, *P. aeruginosa*, *C. albicans*, and *Aspergillus niger* (*A. niger*)), using the disk diffusion and broth dilution methods. The results showed that all of the supercritical fluid extraction fractions exhibited antimicrobial effects on tested microorganisms, and the most efficient fraction was obtained with 7% ethanol at 150 bar and 40 °C. De Souza et al. [36] evaluated the effects of heating (at the temperatures of 60, 80, 100, and 120 °C, at a duration of 1 h for each) on the antimicrobial activities of oregano EO against 9 microorganism strains (*C. albicans*, *Candida krusei* (*C. krusei*), *Candida tropicalis* (*C. tropicalis*), *B. cereus*, *E. coli*, *S. aureus*, *Yersinia enterocolitica* (*Y. enterocolitica*), *S. enterica*, and *Serratia marcescens* (*S. marcescens*)), using the solid medium diffusion procedure. The results indicated that heating treatment showed no significant effects on the antimicrobial activities of EO, with the DIZs and MICs of heated EO close to those of EO kept at room temperature (MICs ranging from 10 to 40 μL/mL).

Oregano could bind to sterols in the fungal membranes of *C. albicans* strains [37], but the exact mechanisms of action on other microorganisms are to be further studied. Carvacrol, one of the major components of oregano, could interact with cell membranes through changing the permeability for small cations [38]. As the chemical compounds in EO and extracts of oregano are complex, they could inhibit microorganisms through different cell targets.

3.2. Comparison of Antimicrobial Activities of Oregano and Other Spices

Ozcan et al. [39] investigated the antifungal activities of four spice decoctions against six molds (*Fusarium oxysporum* f. sp. *phaseoli*, *Macrophomina phaseoli* (*M. phaseoli*), *Botrytis cinerea* (*B. cinerea*), *Rhizoctonia solani* (*R. solani*), *Alternaria solani* (*A. solani*), and *Alternaria parasiticus* (*A. parasiticus*)). The results showed that the mycelial growth were 100% inhibited by 10% oregano decoction in culture medium. Ai-Turki et al. [40] tested the antimicrobial activities of aqueous extracts of four plants against *E. coli* and *B. subtilis* using the disc diffusion method. Oregano extract showed the best antibacterial effects on two bacteria compared with three other spice extracts, and *B. subtilis* showed more sensitivity than *E. coli*. Marques et al. [41] assessed the antimicrobial activities of the EOs of oregano and marjoram against *S. aureus* isolated from poultry meat using the disk diffusion method, and the MICs and MBCs were tested using the microdilution technique. All the *S. aureus* strains were susceptible to oregano EO with the MICs ranging from 6.25 to 25 μL/mL, but four of the isolates were resistant to ampicillin and one was resistant to tetracycline. Bozin et al. [42] investigated the antimicrobial activities of 3 spice EOs against 13 bacterial strains using the hole-plate agar diffusion method and 6 fungi by the microdilution technique. The results indicated that the most effective antibacterial activities were expressed by oregano EO, even on multiresistant strains of *P. aeruginosa* and *E. coli*. Viuda-Martos et al. [43] studied the antimicrobial activities of EOs from six spices against six bacteria (*Lactobacillus curvatus* (*L. curvatus*), *Lactobacillus sakei* (*L. sakei*), *Staphylococcus carnosus* (*S. carnosus*), *Staphylococcus xylosus* (*S. xylosus*), *Enterobacter gergoviae* (*E. gergoviae*) and *Enterobacter amnigenus* (*E. amnigenus*)), using the disc diffusion method. Oregano EO was the most effective against bacteria tested, with DIZs ranging from 35.29 mm (*S. xylosus*) to 57.90 mm (*E. amnigenus*). Santurio et al. [44] reported the antimicrobial activities of

EOs of eight spices against *E. coli* strains isolated from poultry and cattle faeces by determining the MICs using the broth microdilution technique. The results showed that the most effective against all *E. coli* strains in the study was oregano EO. Khosravi et al. [45] investigated the antifungal activities of *Artemisia sieberi* and oregano EOs against *Candida glabrata* (*C. glabrata*) isolated from patients with vulvovaginal candidiasis by determining the MICs and minimal fungicidal concentrations (MFCs), using the broth macrodilution method. The results indicated that the EOs inhibited all tested *C. glabrata* isolates concentration-dependently, with the MICs ranging from 0.5 to 1100 µg/mL (mean: 340.2 µg/mL) for oregano. Dal Pozzo et al. [46] studied the antimicrobial activities of 7 spice EOs, and some majority constituents of these spices such as carvacrol, thymol, cinnamaldehyde, and cineole against 33 *Staphylococcus* spp. isolates from herds of dairy goats, by determining the MICs and MBCs using the broth microdilution method. Oregano and thyme possessed equally strong antimicrobial activities among EOs. Santos et al. [47] evaluated the antimicrobial activities of four spices against several bacteria like *S. aureus* and *E. coli* isolated from vongole and bacteria standard ATCC (American Type Culture Collection): *E. coli*, *S. aureus*, and *Salmonella choleraesuis* (*S. choleraesuis*), by determining the MICs using diffusion test. Oregano and clove EOs presented antimicrobial activities against all tested bacteria, but oregano presented larger DIZs of 26.7 mm (*E. coli*) and 29.3 mm (*S. aureus*). Hyun et al. [48] tested the antibacterial effects of various spice EOs including oregano on total mesophilic microorganisms in products (fresh leaf lettuce and radish sprouts) using the dipping method. One species of oregano (in the USA) EO showed the best effects on maintaining reduced levels of total mesophilic microorganisms in fresh leaf lettuce and radish sprouts compared with the control.

3.3. The Application of Oregano as Antimicrobial Agents in Food Packaging

The antimicrobial effects of pure EOs of four spices and chitosan-EOs films on *L. monocytogenes* and *E. coli* were evaluated in vitro by agar diffusion test [49]. The antimicrobial activities of EOs alone and incorporated in the films were similar following the order: oregano >> coriander > basil > anise. When used in inoculated bologna samples at 10 °C and stored for five days, pure chitosan films led to 2 log reduction of *L. monocytogenes*, 3.6–4 log reduction of *L. monocytogenes*, and 3 log reduction of *E. coli* were observed in films incorporated with 1% and 2% oregano EO.

All the above studies are of good quality, and oregano showed strong antimicrobial activities against microorganism strains such as *Staphylococcus* spp. and *S. aureus* isolates with larger DIZs and lower MICs, MBCs, and MFCs compared with several other spices (Table 2). Future studies could focus on the application of oregano and its EO in food industry, and also the possible mode of action.

Table 2. Antibacterial and antifungal activities of oregano.

Type of Study	Bacteria and Fungi	Main Results	Reference
Oregano extract	*Salmonella gallinarum, Salmonella enteritidis, S. typhimurium*	Oregano extract had antibacterial effects on *Salmonella* serotypes.	[34]
EO-rich fractions of oregano	*S. aureus, B. subtilis, E. coli, P. aeruginosa, C. albicans, Aspergillus niger*	All of the supercritical fluid extraction fractions showed antimicrobial activities against all tested microorganisms.	[35]
Oregano EO	*C. albicans, Candida krusei, Candida tropicalis, B. cereus, E. coli, S. aureus, Yersinia enterocolitica, S. enterica, Serratia marcescens*	Heating treatment showed no significant effects on the antimicrobial activities of EO.	[36]
4 spice decoctions	*F. oxysporum* f. sp. *phaseoli, Macrophomina phaseoli, Botrytis cinerea, Rhizoctonia solani, Alternaria solani, Alternaria parasiticus*	The 10% level of oregano decoction was 100% inhibitive to mycelial growth in the culture medium.	[39]
4 plant aqueous extracts		Oregano extract had the highest antibacterial activities against all tested bacteria.	[40]
Oregano and marjoram EOs	*E. coli* and *B. subtilis*	All the isolates tested were sensitive to EO of oregano.	[41]
3 spice EOs	*S. aureus* isolated from poultry meat.	Oregano EO showed the most effective antibacterial activities.	[42]
6 spice EOs	13 bacterial strains and 6 fungi	Oregano EO was the most effective.	[43]
8 spice EOs	*Staphylococcus xylosus, Staphylococcus carnosus, Lactobacillus sakei, Lactobacillus curvatus, Enterobacter gergoviae, Enterobacter amnigenus*	Oregano EO was the most effective against *E. coli*.	[44]
Oregano and *A. sieberi* EOs	*E. coli* strains isolated from poultry and cattle faeces.	The MICs of oregano EO ranged from 0.5 to 1100 μg/mL for all tested *C. glabrata* isolates.	[45]
7 spice EOs and the majority constituents	*Candida glabrata* isolated from patients with vulvovaginal candidiasis.	Oregano and thyme EOs possessed the equal and strongest antimicrobial activities among EOs.	[46]
4 spice EOs	33 *Staphylococcus* spp. isolates	Oregano presented antimicrobial activities against all tested bacteria.	[47]
Various spice EOs	*S. aureus* and *E. coli* isolated from vongole and bacteria standard ATCC: *E. coli, S. aureus, Salmonella choleraesuis*	Oregano-2 (in the USA) oil was the most effective at maintaining the reduced levels of total mesophilic microorganisms.	[48]
Pure EOs of 4 spices and chitosan–EOs films	Microorganisms in fresh leaf lettuce and radish sprouts.	Both oregano EO alone and incorporated in the films possessed the best antimicrobial activities.	[49]
	L. monocytogenes and *E. coli*		

4. Thyme

Thyme (*Thymus vulgaris*), belonging to family Lamiaceae, is a subshrub native to the western Mediterranean region. Thyme is widely used as a spice to add special flavor to foods. In recent studies, thyme was found to possess efficient antimicrobial activities and was used in some foods to extend the shelf-life [50].

4.1. Antimicrobial Activities of Thyme

A study evaluated the antimicrobial activities of thyme EO against bacteria (*B. subtilis*, *S. aureus*, *Staphylococcus epidermidis* (*S. epidermidis*), *P. aeruginosa*, *E. coli*, and *Mycobacterium smegmatis* (*M. smegmatis*)) and fungal strains (*C. albicans* and *Candida vaginalis*) [51]. Thyme EO showed effective bactericidal and antifungal activities against tested microorganism strains with MICs ranging from 75 to 1100 µg/mL for bacteria, and from 80 to 97 µg/mL for fungi. In another study, EOs obtained from thyme harvested at four ontogenetic stages were tested for their antibacterial activities against nine strains of Gram-negative bacteria and six strains of Gram-positive bacteria using the bioimpedance method to test the bacteriostatic activities and plate counting technique to study the inhibitory effects by direct contact [52]. The results indicated that all the thyme EOs had significant bacteriostatic activities against the microorganisms tested. Furthermore, the antimicrobial activities of EOs of four *Thymus* species (*T. vulgaris*, *T. serpyllum*, *T. pulegioides*, and *T. glabrescens*) were determined by agar diffusion method [53]. *T. vulgaris* and *T. serpyllum* EOs were the most efficient as they inhibited all the tested bacteria (*P. aeruginosa*, *Cronobacter sakazakii* (*C. sakazakii*), *L. innocua*, and *Streptococcus pyogenes* (*S. pyogenes*)) and yeasts (*C. albicans* and *Saccharomyces cerevisiae* (*S. cerevisiae*)) both at original and half-diluted concentrations. *P. aeruginosa*, *L. innocua*, and *S. pyogenes* were highly and equally sensitive to the *Thymus* oils, while *C. sakazakii* exhibited limited sensitivity, and the sensitivity of the two yeast strains were similar to that of *C. sakazakii*, but *S. cerevisiae* was a little more sensitive than *C. albicans*.

The major active compound of thyme is thymol, which exerted its antimicrobial action through binding to membrane proteins by hydrophobic bonding and hydrogen bonding, and then changing the permeability of the membranes [20]. Thymol also decreased intracellular adenosine triphosphate (ATP) content of *E. coli* and increased extracellular ATP, which could disrupt the function of plasma membranes [54]. As thymol was proved to act differently against Gram-positive and Gram-negative bacteria [20], the exact mechanisms of antimicrobial action should be further studied.

4.2. Comparison of Antimicrobial Activities of Thyme and Other Spices

Al-Turki et al. [55] reported the antibacterial activities of thyme, peppermint, sage, black pepper and garlic hydrosols against *B. subtilis* and *S. enteritidis*, using the agar disk diffusion method. Thyme hydrosol demonstrated more significant inhibitory effects on *B. subtilis* and *S. enteritidis* than sage, peppermint, and black pepper hydrosols, with the mean DIZs 20 mm for *B. subtilis* and 15 mm for *S. enteritidis*. According to another study, the antimicrobial effects of the six plant hydrosols on *S. aureus*, *E. coli*, *S. typhimurium*, *P. aerugenosa*, and *C. albicans* were tested by determining the microbial growth zones on hydrosol agar plates and control agar plates [56]. The results showed that at 15% thyme hydrosol completely inhibited *E. coli* and *S. typhimurium*, but *C. albicans* was inactive to the tested hydrosols. Girova et al. [57] assessed the antimicrobial activities of five plant EOs against psychrotrophic microorganisms (*P. fluorescens*, *Pseudomonas putida* (*P. putida*), *P. fragi*, *B. thermosphacta*, and *C. albicans*) isolated from spoiled chilled meat products and some reference strains (*P. fluorescens* ATCC 17397, *P. putida* NBIMCC (National Bank for Industrial Microorganisms and Cell Cultures) 561, *P. aeruginosa* ATCC 9027, and *C. albicans* ATCC 10231) using the method of disc diffusion and serial broth dilution. The results indicated that the antimicrobial effects of the EOs were equal at 37 °C and 4 °C. Thyme EO exhibited the highest antimicrobial activities with the MICs ranging from 0.05% to 0.8% *w*/*v*. Hajlaoui et al. [58] observed the anti-*Vibrio alginolyticus* (*V. alginolyticus*) activities of five aromatic plant EOs using agar well diffusion test, and the MICs and MBCs were examined using the broth microdilution susceptibility method. Thyme EO was proved to be the most efficient against 13

V. alginolyticus strains compared with 4 other EOs, with the MICs ranges of 0.078–0.31 mg/mL and MBCs ranges of 0.31–1.25 mg/mL. Also, Viuda-Martos et al. [59] assessed the growth inhibition of some indicators of spoilage bacteria strains (*L. innocua, S. marcescens,* and *P. fluorescens*) and the concentration effects of five spice EOs using the agar disc diffusion method. Only the EO of thyme showed inhibitive effects on all tested bacteria at all added doses (100%, 50%, 25%, 12.5%, and 5%). Aliakbarlu et al. [60] evaluated the antibacterial activities of EOs from thyme, *Thymus kotschyanus, Ziziphora tenuior,* and *Ziziphora clinopodioides,* against two Gram-positive bacteria (*B. cereus* and *L. monocytogenes*) and two Gram-negative bacteria (*S. typhimurium* and *E. coli*), using the agar disc diffusion and micro-well dilution assay. The EO of thyme showed the highest antibacterial activities, with the widest inhibition zones and the lowest MICs (0.312–1.25 μL/mL), and *B. cereus* was the most sensitive bacterium tested. Hyun et al. [48] investigated the antibacterial effects of several EOs on 18 pathogenic bacteria and 15 spoilage bacteria by agar disc diffusion test. The results showed that thyme-1 (*T. vulgaris*) EO and thyme-2 (*T. vulgaris* ct linalool) EO exerted the highest antibacterial activities against 18 pathogenic bacteria strains compared with other spices, except for *P. aeruginosa.* Thyme-1 EO also demonstrated the best antibacterial effects on spoilage bacteria. In addition, the antimicrobial effects of 17 spices and herbs against *Shigella* strains were tested in another study [61]. The MICs were determined by the agar dilution method with dried ground spices and herbs added to the broth and agar, and the results showed that MICs of thyme were 0.5–1% *w/v* for the *Shigella* strains. The study also used various combinations of temperatures (12, 22, and 37 °C), pH values (5.0, 5.5, and 6.0), and NaCl concentrations (1%, 2%, 3%, and 4% *w/v*), and the inclusion or exclusion of thyme or basil at 1% *w/v* in a Mueller–Hinton agar model system to test the inhibitory effects of thyme and basil. In the presence of thyme, *Shigella flexneri* (*S. flexneri*) did not develop Colony-Forming Units (CFU) during the seven-day incubation period for 16 of the 18 tested combinations.

Some studies compared the antimicrobial activities of different extracts of thyme. Martins et al. [62] evaluated and compared the antimicrobial activities of the infusion, decoction, and hydroalcoholic extracts prepared from thyme against *S. aureus, S. epidermidis, E. coli, Klebsiella* spp., *P. aeruginosa, Enterobacter aerogenes* (*E. aerogenes*), *Proteus vulgaris* (*P. vulgaris*), and *Enterobacter sakazakii* (*E. sakazakii*) using the disc diffusion halo test. For Gram-positive species, thyme extracts only presented activity against *S. epidermidis,* and hydroalcoholic extract showed a lower antibacterial activity than decoction and infusion extracts, which showed the similar activities. For Gram-negative species, thyme extracts showed antimicrobial activities in the order of *E. coli* > *P. vulgaris, P. aeruginosa* > *E. aerogenes* = *E. sakazakii;* decoction and hydroalcoholic extracts had similar effects against the bacteria except *P. aeruginosa,* while the lowest activity was observed in infusion extracts. Moreover, the antifungal effects of thyme EO, hydrosol and propolis extracts on natural mycobiota on the surface of sucuk were evaluated [63]. The results showed that potassium sorbate (15% *w/v,* in water), thyme EO (10 mg/mL, in dimethyl sulfoxide), and propolis extract (50 mg/mL, in dimethyl sulfoxide) reduced by 4.88, 2.45, and 2.05 log CFU/g in yeast-mold counting compared with sterile water, respectively.

Aman et al. [64] analyzed the polyphenolic fractions and oil fractions of oilseeds from 4 spices, including thyme, for their antimicrobial activities against 35 bacterial strains. The results showed that oil fractions of all spice oilseeds were more active than their polyphenolic fractions, and thyme oil fraction had the highest antibacterial activities compared with other spice oilseeds. Aznar et al. [65] studied the growth of *Candida lusitaniae* (*C. lusitaniae*) on different concentrations of nisin (0.1–3 mmol/L), thymol (0.02–1.5 mmol/L), carvacrol (0.02–1 mmol/L), or cymene (0.02–3 mmol/L) in broths (pH = 5, 25 °C), and also evaluated the inhibitory activity of thymol against *C. lusitaniae* in tomato juice. Thymol, carvacrol, and cymene totally inhibited the yeast growth for more than 21 days at 25 °C when the concentrations were higher than 1 mmol/L. Compared with the control without thymol, the activity of thymol against *C. lusitaniae* in tomato juice was significant.

In conclusion, the results obtained from a number of investigations with good quality indicated that thyme possessed effective antimicrobial activities against several pathogenic and spoilage bacteria and fungi, like *S. aureus* and *E. coli,* with low MICs (≤1100 μL/mL) (Table 3).

Table 3. Antibacterial and antifungal activities of thyme.

Type of Samples	Tested Bacteria and Fungi	Main Results	Reference
Thyme EO	B. subtilis, S. aureus, Staphylococcus epidermidis, P. aeruginosa, E. coli, Mycobacterium smegmatis, C. albicans, Candida vaginalis	MICs ranged from 75 to 1100 µg/mL for bacteria, and from 80 to 97 µg/mL for fungi.	[51]
Thyme EOs of 4 ontogenetic stages	E. coli, Proteus mirabilis, Proteus vulgaris, S. typhimurium, S. marcescens, Y. enterocolitica, P. fluorescens, Pseudomonas putida, Micrococcus spp., S. flava, S. aureus, Bacillus licheniformis, Bacillus thuringiensis, L. innocua	All the thyme EOs had significant antibacterial activities against the microorganisms tested.	[52]
4 *Thymus* species EOs	P. aeruginosa, Cronobacter sakazakii, L. innocua, Streptococcus pyogenes, C.albicans, Saccharomyces cerevisiae	Thyme EO was the most efficient on all the tested bacteria and yeast both in original and half-diluted concentrations.	[53]
5 spice hydrosols	B. subtilis and S. enteritidis	Thyme hydrosol was more effective than sage, peppermint, and black pepper.	[55]
6 plant hydrosols	S. aureus, E. coli, S. typhimurium, P. aeruginosa, C. albicans	15% hydrosol concentration of thyme completely inhibited E. coli and S. typhimurium.	[56]
5 plant EOs	P. fluorescens, P. putida, Pseudomonas fragi, Brochothrix thermosphacta C. albicans, P. aeruginosa	Thyme EO showed the highest antimicrobial activities with MICs ranging from 0.05% to 0.8% w/v.	[57]
5 aromatic plant EOs	13 Vibrio alginolyticus strains	The MICs of thyme EO ranged from 0.078 to 0.31 mg/mL, and MBCs ranged from 0.31 to 1.25 mg/mL.	[58]
5 spice EOs	L. innocua, S. marcescens, P. fluorescens	Only the thyme EO showed inhibition effects on all tested bacteria at all added doses.	[59]
4 spice EOs	B. cereus, L. monocytogenes, S. typhimurium, E. coli	MICs of thyme EO ranged from 0.312 to 1.25 µL/mL.	[60]
Various EOs	18 pathogens and 15 spoilage bacteria	Thyme EO showed the strongest antibacterial activities against spoilage bacteria.	[48]
17 spices and herbs	Shigella sonnei and Shigella flexneri	MICs of thyme ranged from 0.5% to 1% (w/v) depending on the Shigella strains used.	[61]
Thyme infusion, decoction and hydroalcoholic extracts	S. aureus, S. epidermidis, E. coli, Klebsiella spp., P. aeruginosa, Enterobacter aerogenes, P. vulgaris, Enterobacter sakazakii	Decoction presented the most pronounced effects.	[62]
Thyme EO, hydrosol and propolis extracts	Natural mycobiota on the surface of sucuk	Thyme EO and propolis extract provided reductions of 2.45 and 2.05 log CFU/g in yeast-mold counts respectively.	[63]
Polyphenolic fractions and oil fractions from 4 spice oilseeds	35 bacterial strains	Thyme oil fraction had the highest antibacterial activities comparing with other spices oilseeds.	[64]
Thymol, nisin, carvacrol, cymene	Candida lusitaniae	Thymol completely inhibited the yeast growth at concentrations over 1 mmol/L for at least 21 days at 25 °C.	[65]

5. Cinnamon

Cinnamon (*Cinnamomum zeylanicum*), belonging to family Lauraceae, is widely applied in savory dishes, pickles, and soups [66]. Cinnamaldehyde, cinnamyl acetate, and cinnamyl alcohol are the three main compounds of cinnamon [67]. Due to its antimicrobial activities, cinnamon is also used in cosmetics or food products [11], and also used as health-promoting agents to treat diseases like inflammation, gastrointestinal disorders, and urinary infections [68,69].

5.1. Antimicrobial Activities of Cinnamon

The antimicrobial activities of cinnamon were evaluated in some studies. Gupta et al. [70] compared the antimicrobial activities of cinnamon extract (50% ethanol) and EO against 10 bacteria and 7 fungi by the agar well diffusion method. Cinnamon EO was more effective than cinnamon extract against tested microorganisms, with the MICs ranging from 1.25% to 5% *v*/*v*. Cinnamon EO exerted the strongest effect on *B. cereus* among bacteria, and *Rhizomucor* sp. among fungi. Cinnamon extract showed the highest activities against *B. cereus* among bacteria, and *Penicillium* sp. among fungi. Ceylan et al. [71] tested the antibacterial effects of cinnamon, sodium benzoate, potassium sorbate, and their combinations on *E. coli* at 8 and 25 °C in apple juice. The results showed that 0.3% *w*/*v* cinnamon provided 1.6 log CFU/mL reduction on *E. coli* at 8 °C and 2.0 log CFU/mL reduction at 25 °C. Cinnamon had synergistic effects with sodium benzoate and potassium sorbate on *E. coli* at 8 and 25 °C. Recently, the anti-biofilm effects of cinnamon EO and liposome-encapsulated cinnamon EO on methicillin resistant *S. aureus* (MRSA) were evaluated in a study by scanning electron microscopy and laser scanning confocal microscopy analyses [72]. Cinnamon EO possessed effective antibacterial activity and prominent anti-biofilm activity against MRSA. In the presence of liposomes, the stability and the acting time of cinnamon EO were further improved.

The major component of cinnamon, cinnamaldehyde, possesses antimicrobial effects on microorganisms, as it inhibited cell wall biosynthesis, membrane function, and specific enzyme activities. More specific cellular targets of cinnamaldehyde are still required to be studied in detail [73].

5.2. Comparison of Antimicrobial Activities of Cinnamon and Other Spices

Mvuemba et al. [74] evaluated the inhibitory effects of aqueous extracts of four spices (cinnamon, ginger, nutmeg, and horseradish) on the growth of mycelial of various spoilage pathogens (*A. niger*, *Fusarium sambucinum* (*F. sambucinum*), *Pythium sulcatum* (*P. sulcatum*), and *Rhizopus stolonifera* (*R. stolonifera*)). At the concentration of 0.05 g/mL, cinnamon extract totally inhibited *A. niger* and *P. sulcatum*, while at the level of 0.10 and 0.15 g/mL *F. sambucinum* and *R. stolonifer* were completely inhibited, respectively. Another study conducted by Wang et al. [75] tested the antibacterial effects of five plant aqueous extracts on five bacteria (*S. aureus*, *Lactobacillus* sp., *B. thermosphacta*, *Pseudomonas* spp., and *E. coli*) by the aerobic plate count method and disc diffusion. Cinnamon aqueous extract was the only one to inhibit all the tested microorganisms at the concentration of 1% *w*/*v*. The inhibitory effects were stronger with the increase of extract concentrations from 1% to 5% *w*/*v*. In the same way, the antimicrobial activities of the hydrosols of six spices (basil, clove, cardamom, cinnamon, mustard, and thyme) against five microorganisms (*S. aureus*, *E. coli*, *S. typhimurium*, *P. aeruginosa*, and *C. albicans*) were tested [56]. The inhibition percentage of cinnamon hydrosol was 10–33.8% at 5% *v*/*v* hydrosol, 10–66.5% at 10% *v*/*v* hydrosol, and 10–100% at 15% *v*/*v* hydrosol against microorganisms tested except *C. albicans*. Moreover, *S. aureus* was the most sensitive strain to cinnamon hydrosol, while *P. aeruginosa* was the least sensitive strain. Agaoglu et al. [76] examined the antimicrobial activities of diethyl ether extracts of six spices used in meat products against eight strains of bacteria (*S. aureus*, *K. pneumoniae*, *P. aeruginosa*, *E. coli*, *Enterococcus faecalis* (*E. faecalis*), *M. smegmatis*, *Micrococcus luteus* (*M. luteus*), and *C. albicans*), by the disc diffusion. Among all the spices tested, only cinnamon exerted inhibitory activities against all the tested microorganisms. *S. aureus* and *C. albicans* were the most sensitive to cinnamon, while *E. coli* was the least. Keskin et al. [27] investigated the antimicrobial effects of the ethyl acetate, acetone, and methanol extracts of 12 plant species on 8 bacterial and 2 fungi species using the disc assay. Cinnamon methanol extract exerted antimicrobial effects on all tested microorganisms, while the ethyl acetate extract showed inhibition against tested microorganisms except *P. aeruginosa* and *R. rubra*, and the acetone extract showed inhibition against tested microorganisms except *R. rubra*. Revati et al. [77] explored the antimicrobial activities of seven Indian spice ethanol extracts against *Enterococci* (including 215 enterococcal strains) isolated from human clinical samples with the agar well diffusion method. Crude ethanol extract of cinnamon was the most effective against all the clinical isolates of *Enterococci*, with the DIZs ranging from 31 to 34 mm. Moreover, the antimicrobial activities of 8 spice EOs against 6 bacterial species and 10 fungal species were tested in a study using the disk diffusion assay and MICs were determined using the agar dilution test [78]. Cinnamon EO possessed the strongest inhibition effects on all tested microorganisms among all spices examined with the MICs ranges of 0.015–2.0 mg/mL. Compared with bacteria, fungi were more sensitive to cinnamon EO.

Collectively, all the mentioned studies with good quality demonstrated that cinnamon showed antimicrobial activities covering a wide range of species, such as MRSA and *A. niger*, at low MICs (Table 4), indicating that cinnamon had great potential to provide health benefits through application in food industry.

Table 4. Antibacterial and antifungal activities of cinnamon.

Type of Samples	Tested Bacteria and Fungi	Mian Results	Reference
Cinnamon extract and oil	7 Gram-positive bacteria, 3 Gram-negative bacteria, and 7 fungi	Cinnamon oil was more effective than cinnamon extract with MICs ranging from 1.25% to 5% v/v.	[70]
Cinnamon, sodium benzoate, potassium sorbate	E. coli	E. coli was reduced by 1.6 log CFU/mL at 8 °C and 2.0 log CFU/mL at 25 °C by 0.3% cinnamon.	[71]
Cinnamon oil and liposome-encapsulated cinnamon oil	Methicillin resistant Staphylococcus aureus (MRSA)	Cinnamon oil possessed effective antibacterial activity and prominent anti-biofilm activity against MRSA.	[72]
4 spice aqueous extracts	A. niger, Fusarium sambucinum, Pythium sulcatum, Rhizopus stolonifer	0.05 g/mL of cinnamon extract completely inhibited A. niger and P. sulcatum, 0.10 g/mL of cinnamon extract completely inhibited F. sambucinum.	[74]
5 plant aqueous extracts	S. aureus, Lactobacillus sp., B. thermosphacta, Pseudomonas spp., E. coli	Cinnamon aqueous extract inhibited all the tested microorganisms at the concentration of 1%.	[75]
6 spice hydrosols	S. aureus, E. coli, S. typhimurium, P. aeruginosa, C. albicans	The percent inhibition ranged from 10% to 33.8% at 5% hydrosol of cinnamon.	[56]
6 spice diethyl ether extracts	S. aureus, K. pneumoniae, P. aeruginosa, E. coli, Enterococcus faecalis, M. smegmatis, Micrococcus luteus, C. albicans	Cinnamon possessed inhibitory activities against all the tested microorganisms.	[76]
Ethyl acetate, acetone, and methanol extracts from 12 plants	K. pneumonia, B. megaterium, P. aeruginosa, S. aureus, E. coli, E. cloacae, C. xerosis, S. faecalis, K. marxianus, R. rubra	The methanol extract of cinnamon showed antibacterial activities against all the microorganisms tested.	[27]
7 Indian spice ethanol extracts	215 enterococcal strains	Crude ethanol extract of cinnamon was the most effective against all the clinical isolates.	[77]
8 spice EOs	B. cereus, E. coli, L. monocytogenes, S. rissen, P. fluorescens, S. aureus, Candida lipolytica, Hanseniaspora uvarum, Pichia membranaefaciens, Rhodotorula glutinis, Schizosaccharomyces pombe, Zygosaccharomyces rouxii, A. flavus, Aspergillus versicolor, A. parasiticus, Fusarium moniliforme	Cinnamon EO possessed the strongest inhibition effects with the MICs ranging from 0.015 to 2.0 mg/mL.	[78]
10 spice EOs	B. cereus, B. subtilis, E. coli, K. pneumoniae, L. monocytogenes, P. aeruginosa, S. aureus, S. enterica, S. marcencens, Y. enterocolitica	Cinnamon EO was efficient in inhibiting all tested bacterial strains.	[79]

6. Cumin

Cumin (*Cuminum cyminum*) is an aromatic plant belonging to the Apiaceae family. Cumin has been used since ancient time as an ingredient in foods in Middle East, and cumin seeds have long been used as antiseptic and disinfectant in India [80]. Cuminaldehyde, cymene, and terpenoids are the major bioactive constituents of cumin EOs [81].

6.1. Antimicrobial Activities of Cumin

In a study, the antimicrobial activities of cumin EO against *E. coli*, *S. aureus*, *S. faecalis*, *P. aeruginosa*, and *K. pneumoniae* were investigated by agar diffusion and dilution methods [81]. *E. coli*, *S. aureus*, and *S. faecalis* were susceptive to various cumin EO dilutions while *P. aeruginosa* and *K. pneumoniae* were resistant. In another study, the antifungal activities of cumin seeds EO against 1230 fungi isolated from food samples were tested [82]. The EO was fungicidal to most of the fungal species, and exerted a broad spectrum of fungal toxicity at MIC (0.6 μL/mL) against all 19 foodborne fungi strains except *R. stolonifer*. Furthermore, Abd El Mageed et al. [83] explored the effects of microwaves on EO of cumin seeds on its antimicrobial activities against *E. coli*, *S. aureus*, *P. aeruginosa*, *A. niger*, *A. parasiticus*, and *C. albicans* using the disk diffusion method. Both microwave and conventionally (oven) roasted cumin oils had similar antimicrobial effects on microorganisms tested and were more effective than those of raw oils. Reza et al. [80] studied the effects of ã-irradiation (10 and 25 kGy) on the antibacterial activities of cumin against *E. coli*, *P. aeruginosa*, *B. cereus*, and *S. aureus*, by the agar well diffusion method and disk diffusion method. The results indicated that cumin EO exerted antibacterial effects on bacteria tested, and ã-irradiation (10 and 25 kGy) to cumin seeds had no significant effects on the antimicrobial activities of cumin.

6.2. Comparison of Antimicrobial Activities of Cumin and Other Spices

Chaudhry et al. [84] determined the antibacterial effects of aqueous infusions and aqueous decoctions of 3 spices on 188 bacteria from 11 genera isolated from oral cavity of apparently healthy individuals, by the disc diffusion test. Aqueous decoction of cumin possessed the highest antimicrobial activities for it showed inhibitory effects on 73% of the bacteria strains tested. Cumin EO was also more effective than some spice EOs as reported. Iacobellis et al. [85] evaluated the antimicrobial activities of EOs of cumin and *Carum carvi* L. against *E. coli* and the genera *Pseudomonas*, *Clavibacter*, *Curtobacterium*, *Rhodococcus*, *Erwinia*, *Xanthomonas*, *Ralstonia*, and *Agrobacterium* using the agar diffusion test. Cumin EO showed antibacterial effects on both Gram-positive and Gram-negative bacteria except *Pseudomonas viridiflava*, which was resistant to 8 μL EO, the highest level tested. Ozcan et al. [86] examined the antimicrobial activities of nine spice EOs at three concentrations (1%, 10%, and 15% *v*/*v*) against *S. typhimurium*, *B. cereus*, *S. aureus*, *E. faecalis*, *E. coli*. *Y. enterocolitica*, *S. cerevisiae*, *Candida rugosa*, *Rhizopus oryzae*, and *A. niger*. The results showed that cumin EO was effective against all tested bacterial species as well as *S. cerevisiae* and *Candida rugosa* among fungi. Stefanini et al. [87] analyzed the antimicrobial activities of EOs of spices (fennel seeds, dill, cumin, and coriander) by determining the DIZs. The results indicated that cumin was effective against *E. coli*, *P. aeruginosa*, and *Salmonella* sp. with DIZs of 18, 10, and 23 mm, respectively. In another study, the antimicrobial activities of EOs of six spices against *L. curvatus*, *L. sakei*, *S. carnosus*, *S. xylosus*, *E. gergoviae*, and *E. amnigenus* were assessed using the agar disc diffusion method [43]. Cumin EO was the second effective against bacteria tested with DIZs ranging from 31.23 mm (*L. sakei*) to 38.17 mm (*E. gergoviae*). Moreover, another study evaluated the antimicrobial activities of EOs of five spices against different microorganism species by the disc diffusion method and discussed the possible effects in vitro between plants and antibiotics [88]. Cumin inhibited all tested bacteria and fungi. The application of cumin with gentamicin, cephalothin, and ceftriaxone showed synergistic effects against *Pseudomonas pyocyaneus* (*P. pyocyaneus*) and *Aeromonas hydrophila* (*A. hydrophila*), but showed antagonistic effects against other bacteria tested. Similarly, the possible synergistic interactions of some spice EOs on antibacterial activities against six foodborne bacteria—*B. cereus, L. monocytogenes, M. luteus, S. aureus, E. coli,* and *S. typhimurium*—were evaluated by micro broth dilution, checkerboard titration, and time-kill methods [89]. The results showed that coriander and cumin seed oil combination exhibited synergistic interactions on antibacterial activities.

Consequently, cumin had antimicrobial effects on several microorganisms like *E. coli, S. aureus,* and *S. faecalis* at low concentrations (Table 5). In the future, the mechanisms of antimicrobial action of cumin and its major components—cuminaldehyde and cymene—on other microorganisms should be further studied.

Table 5. Antibacterial and antifungal activities of cumin.

Type of Study	Bacteria and Fungi	Main Results	Reference
Cumin EO	*E. coli, S. aureus, S. faecalis, P. aeruginosa, K. pneumoniae*	*E. coli, S. aureus*, and *S. faecalis* were sensitive to various cumin EO dilutions.	[81]
Cumin seeds EO	1230 fungal isolates obtained from food samples	The EO was fungicidal against most of the fungal species at MIC of 0.6 μL/mL.	[82]
Cumin seeds EOs	*E. coli, S. aureus, P. aeruginosa, A. niger, A. parasiticus, C. albicans*	Both microwave and conventionally (oven) roasted cumin oils showed higher effects than raw oils.	[83]
Cumin EO	*E. coli, P. aeruginosa, B. cereus, S. aureus*	ã-Irradiation to cumin seeds at 10 and 25 kGy had no significant effects on the antibacterial effects.	[80]
Aqueous infusions and aqueous decoctions from kalonji, cumin and poppy seed	188 bacterial isolates isolated from oral cavity of apparently healthy individuals	Aqueous decoction of cumin inhibited 73% of the tested bacteria.	[84]
Cumin and *C. carvi* EOs	*E. coli*, the genera *Pseudomonas, Clavibacter, Curtobacterium, Rhodococcus, Erwinia, Xanthomonas, Ralstonia, Agrobacterium*	Cumin EO showed antibacterial activities against all tested bacteria except *Pseudomonas viridiflava*.	[85]
9 spice EOs	*S. typhimurium, B. cereus, S. aureus, E. faecalis, E. coli. Y. enterocolitica, S. cerevisiae, Candida rugosa, Rhizopus oryzae, A. niger*	Cumin EO was effective against all bacterial species and two fungi (*S. cerevisiae* and *Candida rugosa*).	[86]
4 spice EOs	*S. aureus, Salmonella* sp., *E. coli, P. aeruginosa*, etc.	Cumin EO was effective against *E. coli, P. aeruginosa* and *Salmonella* sp.	[87]
6 spice EOs	*L. curvatus, L. sakei, S. carnosus, S. xylosus, E. gergoviae, E. amnigenus*	Cumin EO was the second effective among tested spices.	[43]
5 spice EOs	*M. luteus, B. megaterium, Brevibacillus brevis, E. faecalis, Pseudomonas pyocyaneus, M. smegmatis, E. coli, Aeromonas hydrophila, Y. enterocolitica, S. aureus, S. faecalis, S. cerevisiae, Kluvyeromyces fragilis*	Cumin inhibited all tested bacteria and fungi and showed synergistic and antagonistic effect with antibiotics.	[88]
EOs of 9 spices in combination	*B. cereus, L. monocytogenes, M. luteus, S. aureus, E. coli, S. typhimurium*	Coriander/cumin seed oil combination showed synergistic interactions on antibacterial activities.	[89]

7. Rosemary

Rosemary (*Rosmarinus officinalis*), belonging to the Lamiaceae family, is a perennial shrub with pleasant smell and grows all over the world. Rosemary has been used in pharmaceutical products and traditional medicine, and also used as a flavoring agent in food products due to its desirable flavor, antioxidant activities, and antimicrobial activities [90,91].

7.1. Antimicrobial Activities of Rosemary

Tavassoli et al. [91] reported rosemary EO suppressed *Leuconostoc mesenteroides*, *Lactobacillus delbruekii*, *S. cerevisia*, and *C. krusei*. The results indicated that rosemary EO showed higher inhibitory effects on bacteria (MICs: 0.5–1.5 mg/mL) tested than on yeasts. Bozin et al. [92] identified the antimicrobial activities of EOs of rosemary and sage against 13 bacterial strains and 6 fungi by the microdilution technique. Compared with bifonazole, rosemary EO showed better antifungal activities especially against *C. albicans*, *Trichophyton tonsurans* (*T. tonsurans*), and *Trichophyton rubrum* at lower MICs (15.0–30.2 μL). Rosemary EO also expressed important antibacterial activities on *E. coli*, *S. typhimurium*, *S. enteritidis*, and *Shigella sonei*. Weerakkody et al. [93] compared the antibacterial effects of extracts from seven spices and herbs on *E. coli*, *S. typhimurium*, *L. monocytogenes*, and *S. aureus* by the agar disc diffusion and broth dilution assay. The results of both methods indicated that hexane extract of rosemary exhibited significantly higher antibacterial activities than ethanol and water extracts against all bacteria tested except *S. typhimurium* with the MICs ranging from 1.25 to 5.0 mg/mL.

7.2. Comparison of Antimicrobial Activities of Rosemary and Other Spices

Additionally, Krajcova et al. [94] observed the antimicrobial activities of five plant ethanol extracts against *B. cereus*, *E. coli*, *P. aeruginosa*, *S. aureus*, and *L. monocytogenes* using the dilution method and the description of growth curves of the tested bacteria. Rosemary extract was proved to be the most effective at all concentrations (0.1%, 0.05%, 0.02%, and 0.01% *w/w*). At the concentration of 0.01% *w/w*, rosemary extract only inhibited *P. aeruginosa* and *E. coli*, while the higher extract concentrations were effective against all other bacteria. Zhang et al. [95] examined the antimicrobial effects of 14 spice ethanol extracts and their mixtures on *L. monocytogenes*, *E. coli*, *P. fluorescens*, and *L. sake* using the well diffusion test. Individual extract of rosemary showed strong antimicrobial activities, and the combination of rosemary and liquorice extracts showed the best inhibitory effects on all tested microorganisms. Kozlowska et al. [96] tested the antimicrobial activities of aqueous extracts from 5 spices against 8 Gram-positive bacteria and 12 Gram-negative bacteria by the disc diffusion assay. Rosemary exhibited its inhibitory effects with a broader spectrum than the other four spices, as the MICs were 0.125–0.5 mg/mL for all the tested Gram-positive bacteria and 0.25–0.5 mg/mL for four Gram-negative bacteria. Weerakkody et al. [97] studied the antimicrobial activities of two extract combinations against *L. monocytogenes* and *S. aureus* and naturally spoilage microflora on instant shrimp stored for 16 days at 4 or 8 °C. Both combinations (galangal, rosemary, and lemon; galangal and rosemary) significantly decreased the levels of aerobic bacteria and lactic acid bacteria, but showed no effects on *L. monocytogenes* or *S. aureus*. Azizkhani et al. [90] evaluated the antimicrobial effects of rosemary, mint, and a mixture of tocopherols against microorganisms from the sausages. The application of rosemary significantly inhibited the growth of microorganisms and the lowest microbial counts were obtained in samples containing both rosemary and mint, indicating the possible synergistic effects. Toroglu [88] evaluated the antimicrobial activities of five spice EOs by the disc diffusion method and discussed possible effects of plants and antibiotics. Rosemary had antimicrobial effects on all tested fungi and bacteria. The combination of rosemary EO and cephalothin antibiotics showed synergic effects on *S. aureus*, while the combination of rosemary EO and ceftriaxone antibiotics showed no effect.

Above all, the papers cited are of good quality and indicated that rosemary EO and extracts were found antimicrobial at low MICs against some bacteria and fungi, especially Gram-positive bacteria such as *S. aureus* (Table 6). Some studies indicated that rosemary showed synergic effects with some spices and antibiotics such as galangal and cephalothin. The mechanisms of antimicrobial action of both rosemary and its major components should be further studied.

Table 6. Antibacterial and antifungal activities of rosemary.

Type of Study	Bacteria and Fungi	Main Results	Reference
Rosemary EO	*Leuconostoc mesenteroides, Lactobacillus delbruekii, S. cerevisiae, C. krusei*	Rosemary EO showed higher effects against bacteria tested than yeasts.	[91]
Rosemary and sage EOs	*C. albicans, Trichophyton mentagrophytes, Trichophyton tonsurans, Trichophyton rubrum, Epidermophyton floccosum, Microsporum canis, P. aeruginosa, E. coli, S. typhimurium, S. enteritidis, Shigella sonei, Micrococcus flavus, Sarcina lutea, S. aureus, S. epidermidis, B. subtilis*	The EO of rosemary showed significant antifungal activities and antibacterial activities.	[92]
7 spice and herb extracts	*E. coli, S. typhimurium, L. monocytogenes, S. aureus*	The hexane extract of rosemary exhibited significantly higher antibacterial activities than ethanol and water extracts.	[93]
5 plant ethanol extracts	*B. cereus, E. coli, P. aeruginosa, S. aureus, L. monocytogenes*	Rosemary extract was the most effective against all the tested microorganisms.	[94]
14 spice ethanol extracts and their mixture	*L. monocytogenes, E. coli, P. fluorescens, L. sake*	The mixture of rosemary and liquorice extracts was the most effective against all tested bacteria.	[95]
5 spice aqueous extracts	8 Gram-positive bacteria and 12 Gram-negative bacteria	MICs ranged from 0.125 to 0.5 mg/mL for Gram-positive bacteria and 0.25–0.5 mg/mL for Gram-negative bacteria.	[96]
2 spice and herb extract combinations	*L. monocytogenes, S. aureus* and naturally present spoilage microflora on cooked ready-to-eat shrimp stored for 16 days at 4 or 8 °C	Both combination of galangal, rosemary, and lemon and combination of galangal and rosemary significantly reduced levels of aerobic bacteria and lactic acid bacteria.	[97]
Rosemary, mint and a mixture of tocopherols	Microorganisms from the sausages	The addition of rosemary resulted in significant inhibition of microbial growth and showed possible synergistic effects with mint.	[90]
5 spice EOs	*M. luteus, B. megaterium, B. brevis, E. faecalis, P. pyocyaneus, M. smegmatis, E. coli, A. hydrophila, Y. enterocolitica, S. aureus, S. faecalis, S. cerevisiae, K. fragilis*	Rosemary EO showed synergic effects with cephalothin.	[88]

8. Garlic

Garlic (*Allium sativum*) belongs to the Liliaceae family [98]. The antimicrobial activities of garlic have been recognized for many years, and the active component was identified as allicin, a diallyl thiosulfinate (2-propenyl-2-propenethiol sulfonate) [99].

8.1. Antimicrobial Activities of Garlic

In a study, Sallam et al. [100] examined the antimicrobial effects of fresh garlic, garlic powder, and garlic oil on microorganisms in raw chicken sausage by aerobic plate count. Garlic materials showed antimicrobial activities in such an order: fresh garlic > garlic powder > garlic oil > butylated hydroxyanisole. Another study also assessed the antimicrobial activities of dried garlic powders made by different drying methods against *S. aureus*, *E. coli*, *S. typhimurium*, *B. cereus* and a mixed lactic culture containing *Lactobacillus delbrueckii* subsp. bulgaricus and *Streptococcus thermophilus* [99]. Fresh garlic exhibited the highest activities followed by freeze-dried powder. The retaining of active components responsible for antimicrobial activities was mainly affected by both drying temperature and time.

Chopped garlic at concentrations from 0% to 10% were investigated for the antimicrobial effects in ground beef (stored at refrigerator and ambient temperatures) and raw meatballs (stored at room temperature) by determining the colony counts of total aerobic mesophilic bacteria, yeast, and molds at 2, 6, 12, and 24 h after storage [101]. The results indicated that chopped garlic delayed the growth of microorganisms in ground meat, which depended on the garlic concentrations. The addition of garlic (5% or 10%) to the raw meatball mix reduced the microorganism counting, in terms of total aerobic mesophilic bacteria, yeast, and mold counts.

Garlic EO penetrated the cellular membranes and even the menbranes of organelles like mitochondria, damaged organelles, and resulted in the death of *C. albicans* [102]. Furthermore, garlic EO induced differential expression of several critical genes including those involved in oxidation-reduction processes, and cellular response to drugs and starvation.

8.2. Comparison of Antimicrobial Activities of Garlic and Other Spices

Some studies compared the antimicrobial activities of different spices. Indu et al. [103] studied the antimicrobial effects of 5 spice extracts on 20 serogroups of *E. coli*, 8 serotypes of *Salmonella*, *L. monocytogenes* and *A. hydrophila* using the agar well method and filter paper method. Garlic extract exhibited significant antibacterial activities at all concentrations (100%, 75%, 50%, and 25%) against all test microorganisms except *L. monocytogenes*, and the activity against *E. coli* was linearly dependent with concentration. Joe et al. [104] reported the antimicrobial effects of garlic, ginger, and pepper ethanol extracts on *K. pneumoniae*, *S. aureus*, *M. morgani*, *C. albicans*, *E. coli*, and *P. vulgaris* using the filter paper assay. Garlic extract exerted superior antibacterial activities at all concentrations (1000, 1500, and 2000 ppm), especially against *P. vulgaris* and *M. morgani*, and the activities were a linear function of concentrations. Geremew et al. [105] examined the antimicrobial activities of six spice crude extracts (acetone, ethanol, and hexane extracts) against *E. coli*, *S. aureus*, *S. flexneri*, and *Streptococcus pneumoniae* by the agar well diffusion method. Garlic was the most effective against all tested pathogens except *S. flexneri*. Among different solvent extracts used, garlic acetone extract exhibited the highest antibacterial activities. Touba et al. [106] tested the antimicrobial activities of crude extracts of seven spices against three Roselle pathogens by poisoned food technique. The results indicated that the cold water extract of garlic exhibited good antifungal activities against all three tested fungi, and hot water extract of garlic showed the best antifungal activities. Nejad et al. [98] reported the antibacterial effect of garlic aqueous extract on *S. aureus* in hamburger. Samples treated with garlic aqueous extract were kept in refrigerator for one and two weeks, and were frozen for one, two, and three months, before being tested by the microbial counts. The first- and second-week samples were significantly reduced by all the 1, 2, and 3-mL extracts, which were added to 100 g hamburger samples, respectively, showing 2 and 3-mL extracts were more effective. In treatment of one, two, and three-month samples, the growth of *S. aureus* was significantly decreased by the 2 and 3-mL extracts. Al-Turki [55] explored the antimicrobial activities of five spice hydrosols (thyme, peppermint, sage, black pepper, and garlic) against *B. subtilis* and *S. enteritidis* using the agar disk diffusion method. Garlic hydrosol exhibited stronger antibacterial activities against *B. subtilis* and *S. enteritidis* compared with thyme, peppermint, sage, and black pepper hydrosols.

In conclusion, garlic showed great antimicrobial activities at low concentrations against several pathogenic microorganisms like *E. coli* and *S. aureus* (Table 7). Fresh garlic was found to possess higher antimicrobial activities than garlic powder and oil.

Table 7. Antibacterial and antifungal activities of garlic.

Type of Study	Bacteria and Fungi	Main Results	Reference
Fresh garlic, garlic powder, garlic oil	Microorganisms in raw chicken sausage	The order of antimicrobial activities were fresh garlic > garlic powder > garlic oil > butylated hydroxyanisole.	[100]
Garlic powder	*S. aureus, E. coli, S. typhimurium, B. cereus,* and a mixed lactic culture consisting of *Lactobacillus delbrueckii* subsp. bulgaricus and *Streptococcus thermophilus*	Fresh garlic produced the greatest inhibition followed by freeze-dried powder.	[99]
Chopped garlic	Microorganisms in ground beef and raw meatball	Chopped garlic had slowing-down effects on microbiological growth.	[101]
5 spice extracts	20 serogroups of *E. coli,* 8 serotypes of *Salmonella, L. monocytogenes* and *A. hydrophila*	Garlic extract exhibited significant activities against microorganisms except *L. monocytogenes* at all concentrations.	[103]
3 ethanol extracts	*K. pneumoniae, S. aureus, M. morganii, C. albicans, E. coli, P. vulgaris*	Garlic extract exerted superior antibacterial activities at all concentrations	[104]
6 spice crude ethanol, hexane and acetone extracts	*E. coli, S. aureus, S. flexneri; Streptococcus pneumoniae*	Garlic was the most effective against all the tested pathogens except *S. flexneri.*	[105]
7 spice crude extracts	*Phoma exigua, Fusarium nygamai, R. solani*	Cold water extract of garlic exhibited good antifungal activities against all three tested fungi.	[106]
Garlic aqueous extract	*S. aureus*	The first and second week samples were significantly decreased by all the 1, 2, and 3-mL garlic extracts.	[98]
5 spice hydrosols	*B. subtilis* and *S. enteritidis*	Garlic hydrosols demonstrated stronger antibacterial activities than other spices hydrosols.	[54]

9. Ginger

Ginger (*Zingiber officinale*), belonging to the family of Zingiberaceae [107], is widely used as an ingredient in food, pharmaceutical, cosmetic, and other industries. Some volatile compounds which are responsible for antimicrobial activities in ginger were á-pinene, borneol, camphene, and linalool [108].

9.1. Antimicrobial Activities of Ginger

Ginger was proved to possess antimicrobial activities in several studies. Singh et al. [109] determined the antifungal activities of EO and oleoresin of ginger against *Aspergillus terrus, A. niger, Aspergillus flavus (A. flavus), Trichothecium roseum (T. roseum), Fusarium graminearum (F. graminearum), F. oxysporum, Fusarium oxysporum (F. monoliforme)*, and *Curvularia palliscens*, by food poison and inverted petri-plate technique. The results showed that the EO 100% inhibited *F. oxysporum*, while the oleoresin 100% inhibited *A. niger*. Park et al. [107] compared the ethanol and *n*-hexane extracts of ginger and five ginger constituents against three anaerobic Gram-negative bacteria, *Porphyromonas gingivalis (P. gingivalis), Porphyromonas endodontalis*, and *Prevotella intermedia*. The results indicated that ginger extracts exhibited antibacterial activities against three tested bacteria. Two highly alkylated gingerols showed significant inhibition against the growth of these oral pathogens with the MICs ranging from 6 to 30 µg/mL, and also killed the oral pathogens at a MBC range of 4–20 µg/mL. Sa-Nguanpuag et al. [108] evaluated the in vitro and in vivo antimicrobial activities of ginger oils which were obtained by hydrodistillation and solvent extraction method. The results showed that the oils extracted by both methods possessed antimicrobial activities against *B. subtilis, Bacillus nutto, P. aerugenosa, Rhodoturola* sp., *Samonella newport, S. enteritidis*, and *Fusarium* sp.; except *E. coli, Campylobactor coli*, and *Campylobactor jejuni (C. jejuni)* in vitro. In the case of shredded green papaya, when the package was added with 5 and 10 µL ginger oils the growth of microorganisms was inhibited well, while with 15 µL ginger oil a reduction in growth rate was observed.

9.2. Comparison of Antimicrobial Activities of Ginger and Other Spices

Yoo et al. [110] investigated the antibacterial activities of EOs from ginger and mustard against *Vibrio* species at various temperatures. The results indicated that EOs from ginger and mustard could inhibit the growth of *Vibrio parahaemolyticus* and *Vibrio vulnificus* at 5 °C of storage. Indu et al. [103] tested the antibacterial activities of 5 spice extracts against 20 serogroups of *E. coli*, 8 serotypes of *Salmonella, L. monocytogenes*, and *A. hydrophila* by the agar well method and filter paper method. The results indicated that ginger extract possessed inhibitory effects on two serogroups of *E. coli*. Mvuemba et al. [74] assessed the antimicrobial activities of four spice water extracts against the mycelial growth of *A. niger, F. sambucinum, P. sulcatum*, or *R. stolonifera*. The results demonstrated that ginger extract significantly suppressed the mycelial growth of tested microorganisms, and *P. sulcatum* was 100% inhibited by 0.05 g/mL of ginger extract. Touba et al. [106] tested the antifungal activities of crude extracts of seven spices made by cold water and hot water against *Phoma exigua (P. exigua), Fusarium nygamai (F. nygamai)*, and *R. solani* by poisoned food technique. The results showed that hot water extracts from garlic and ginger possessed the best antifungal activities. Cold water extracts were commonly more effective than hot water extracts on tested pathogens. In another study, the antibacterial activities of 7 ethanol extracts of spices against 215 high levels gentamicin resistant enterococcal strains isolated from clinical samples were evaluated by the well diffusion method [77]. The results indicated that only cinnamon and ginger extracts were found to have activities against all the isolates, with the DIZs of ginger ranged from 27 to 30 mm.

Collectively, ginger was proved to possess significant antimicrobial activities against some common microorganisms such as *P. aerugenosa* both in vivo and in vitro at low concentrations (Table 8). Ginger could also inhibit pathgens like *P. gingivalis* and enterococcal isolates with low MICs and MBCs. The exact mechanisms of action of ginger on bacteria and fungi were rarely studied and need futher exploration.

Int. J. Mol. Sci. **2017**, *18*, 1283

Table 8. Antibacterial and antifungal activities of ginger.

Type of Sample	Bacteria and Fungi	Main Results	Reference
3 spice extracts and EOs	5 strains of *L. monocytogenes*, 4 strains of *S. typhimurium* DT104	Commercial EOs of ginger inhibited all *L. monocytogenes* at ≤ 0.6	[111]
Ginger EO and oleoresin	*Aspergillus terrus, A. niger, A. flavus, Trichothecium roseum, Fusarium graminearum, Fusarium oxysporum, Fusarium monoliforme, Curvularia palliscens*	EO and oleoresin of ginger were 100% antifungal against *F. oxysporum* and *A. niger*, respectively.	[109]
Ginger ethanol and *n*-hexane extracts	*Porphyromonas gingivalis, Porphyromonas endodontialis, Prevotella intermedia*	Only [10]-gingerol and [12]-gingerol effectively inhibited the growth of tested bacteria at a MIC range of 6–30 μg/mL.	[107]
Ginger oil extracted by hydrodistillation and solvent extraction method	*B. subtilis, Bacillus nutto, P. aerugenosa, Rhodoturola sp., Samonella nexport, S. enteritidis, Fusarium sp.*	Extracts obtained by both extraction methods inhibited listed microorganisms.	[108]
Ginger and mustard EOs	*Vibrio* species	Ginger and mustard EOs inhibited the growth of *Vibrio parahaemolyticus* and *Vibrio vulnificus*.	[110]
5 spice extracts	20 serogroups of *E. coli*, 8 serotypes of *Salmonella*, *L. monocytogenes* and *A. hydrophila*.	Ginger extract possessed inhibitory effects on two serogroups of *E. coli*.	[102]
4 spice water extracts	*A. niger, F. sambucinum, P. sulcatum, R. stolonifera*	Ginger extract significantly inhibited the mycelial growth of tested microorganisms.	[74]
7 spice crude extracts	*P. exigua, F. nygamai, R. solani*	In the case of the hot water extracts, garlic and ginger showed the best antifungal activities.	[106]
7 spice ethanol extracts	215 enterococcal strains isolated from clinical samples	Ginger was found to have antibacterial activities against all the isolates.	[76]

10. Basil

Basil (*Ocimum basilicum*) is one of the oldest spices, which is widely used in the flavoring of confectionary, baked goods, condiments, etc. Basil oil was also used in perfumery, as well as in dental and oral products [112]. Basil is a natural spice which possesses antimicrobial activities as many studies have reported.

10.1. Antimicrobial Activities of Basil

In a study, the antimicrobial activities of EOs from aerial parts of basil (collected at full flowering stage during summer, autumn, winter, and spring) against *S. aureus*, *E. coli*, *B. subtilis*, and *Pasteurella multocida*, as well as pathogenic fungi *A. niger*, *Mucor mucedo*, *Fusarium solani* (*F. solani*), *Botryodiplodia theobromae*, and *R. solani* were assessed by the disc diffusion method and the MICs were determined by a microdilution broth susceptibility assay [113]. The results indicated that basil EOs possessed antimicrobial activities against all tested microorganisms. Antimicrobial activities of the EOs varied significantly as seasons changed, and EOs from winter and autumn crops exhibited greater antimicrobial activities. In another study, the antimicrobial activities of chloroform, acetone and 2 different concentrations of methanol extracts of basil against 10 bacteria and 4 yeasts were determined by the disc diffusion assay [114]. Methanol extracts provided inhibition zones on *P. aeruginosa*, *Shigella* sp., *L. monocytogenes*, *S. aureus*, and two strains of *E. coli*, but the chloroform and acetone extracts exhibited no effects. Kocic-Tanackov et al. [115] reported the antifungal effects of basil extract on *Fusarium* species (*Fusarium oxysporum*, *Fusarium proliferatum*, *Fusarium subglutinans*, and *Fusarium verticillioides* isolated from cakes), by the agar plate test. Basil extract showed significant activities against *F. proliferatum* and *F. subglutinans* at the concentration of 0.35 and 0.70% *v/v*, but showed lower activities against other tested *Fusarium* species. Basil extract 100% inhibited aerial mycelium of all tested *Fusarium* spp. at 1.50% *v/v*. Beatovic et al. [116] investigated the antimicrobial activities of EOs of 12 basil cultivars against 8 bacterial species (*B. cereus*, *Micrococcus flavus*, *S. aureus* and *E. faecalis*, *E. coli*, *P. aeruginosa*, *S. typhimurium*, and *L. monocytogenes*) and 7 fungi (*Aspergillus fumigatus* (*A. fumigatus*), *A. niger*, *Aspergillus versicolor* (*A. versicolor*), *Aspergillus ochraceus* (*A. ochraceus*), *Penicillium funiculosum*, *Penicillium ochrochloron*, and *Trichoderma viride*) by a modified microdilution technique. All basil EOs tested showed significant antimicrobial activities, with MICs ranging from 0.009 to 23.48 μg/mL for bacteria and 0.08–5.00 μg/mL for fungi. All the EOs showed 100-fold higher antibacterial activities than ampicillin for some bacteria, and 10- to 100-fold higher antifungal activities than the commercial antifungal agents, e.g., ketoconazole and bifonazole.

10.2. Comparison of Antimicrobial Activities of Basil and Other Spices

El-Habib [117] investigated the antifungal activities of seven spice EOs against *A. flavus* and aflatoxin producted by *A. flavus* strain. The results showed that basil EO delayed the growth of *A. flavus*. At 150 μL/100 mL, basil EO completely inhibited *A. flavus*, and effectively controlled the aflatoxin B1 production. Lomarat et al. [17] tested the antibacterial activities of eight EOs against *M. morganii*, a histidine decarboxylase producing bacteria, by microdilution assay. Basil EO possessed the antibacterial activity against *M. morganii* (MIC: 2.39 mg/mL, MBC: 4.77 mg/mL), and the active compound of basil oil was methyl chavicol.

Generally, basil has been proved to possess effects of inhibiting some microorganisms at low MICs especially fungi like *A. flavus* (Table 9), but the mechanisms of action have been rarely explored. Therefore, future studies are needed.

Table 9. Antibacterial and antifungal activities of basil.

Type of Samples	Bacteria and Fungi	Main Results	Reference
EO from aerial parts of basil	S. aureus, E. coli, B. subtilis, Pasteurella multocida, A. niger, Mucor mucedo, F. solani, Botryodiplodia theobromae, R. solani	All the tested microorganisms were sensitive to EOs of basil.	[113]
Chloroform, acetone and methanol extracts of basil	E. gallinarum, E. faecalis, B. subtilis, E. coli, Shigella sp., S. pyogenes, S. aureus, L. monocytogenes, P. aeruginosa, S. cerevisiae, C. albicans, C. crusei	The methanol extract inhibited P. aeruginosa, Shigella sp., L. monocytogenes, S. aureus, and two strains of E. coli.	[114]
Basil extracts	Fusarium oxysporum, Fusarium proliferatum, Fusarium subglutinans, Fusarium verticillioides	At the concentration of 1.50% v/v, basil extract completely inhibited Fusarium spp. tested.	[115]
EOs from 12 cumin cultivars	B. cereus, M. flavus, S. aureus, E. faecalis, E. coli, P. aeruginosa, S. typhimurium, L. monocytogenes, 7 fungi, Aspergillus fumigatus, A. niger, A. versicolor, Aspergillus ochraceus, Penicillium funiculosum, Penicillium ochrochloron, Trichoderma viride	MICs of basil EOs ranged from 0.009 to 23.48 µg/mL for bacteria and 0.08–5.00 µg/mL for fungi.	[116]
7 spice EOs	A. flavus	Basil EO completely inhibited A. flavus at 150 µL/100 mL.	[117]
8 spice EOs	Histamine-producing bacteria including M. morganii	Basil EO inhibited M. morganii with the MIC of 2.39 mg/mL.	[17]

11. Fennel

Fennel (*Foeniculum vulgare*), belonging to family Umbellifarae [118], is widely planted in temperate zones and the tropical belt for its aromatic fruits, and is used as an ingredient in the cooking [119]. The EO of fennel seeds has been reported with significant antifungal activities and antibacterial activities.

11.1. Antimicrobial Activities of Fennel

In a study, the antibacterial activities of fennel seeds EO against *Streptococcus mutans* (*S. mutans*) strains were tested [120]. The results showed that growths of all *S. mutans* strains tested were completely inhibited by fennel seeds EOs at concentrations higher than 80 ppm. Diao et al. [119] also determined the antibacterial activities of EO from fennel seeds against several foodborne pathogens by the kill-time curve assay method. The results showed that fennel seeds EO exerted antibacterial effects on *Streptomyces albus* (*S. albus*), *B. subtilis*, *S. typhimurium*, *Shigella dysenteriae* (*S. dysenteriae*). and *E. coli*, among which *S. dysenteriae* was the most sensitive with the lowest MIC (0.125 mg/mL) and MBC (0.25 mg/mL). In another study, the antimicrobial activities of crude extract of fennel was determined using the agar diffusion method against *E. coli*, *S. blanc*, *P. merabilis*, *P. vulgaris*, *S. epidemidis*, *S. saprophyticus*, *A. versicolor*, *A. fumigates*, and *Penicilium camemberti* [121]. The results indicated that the crude extract of fennel had a great potential as an antimicrobial material against all the nine microorganisms tested, especially fungal strains. Some studies also tested the methanol, ethanol, and acetone extracts of fennel. In a study, the antifungal activities of EO and acetone extract of fennel against 10 fungi were assessed by the inverted petriplate method [118]. The results showed that fennel EO completely inhibited *A. niger*, *A. flavus*, *F. graminearum*, and *Fusarium moniliforme* (*F. moniliforme*) at 6 μL (in 20 mL culture medium), and it was effective on *A. niger* even at 4 μL.

Fennel seed EO could break the permeability of cell membrane of *S. dysenteriae* and result in the leakage of electrolytes, losses of proteins, reducing sugars, etc., and eventually lead to the decomposition and death of cells [119].

11.2. Comparison of Antimicrobial Activities of Fennel and Other Spices

The antimicrobial activities of cumin and fennel EOs on *S. typhimurium* and *E. coli* were compared by the disc diffusion method and dilution method [122]. Fennel EO was more effective than cumin EO, with the lowest MICs of 0.031% and 0.062% *v/v* against *S. typhimurium* and *E. coli*, respectively. Nguyen et al. [123] evaluated the antimicrobial activities of methanol and ethanol extracts of eight spices against *B. subtilis*, *E. faecalis*, *L. innocua*, *E. coli*, *P. putida*, *Providencia stuartii*, and *Acetobacter calcoaceticus* (*A. calcoaceticus*) by the Kirby-Bauer disc diffusion method. Methanol and ethanol extracts from fennel seeds exhibited the best antimicrobial effects with the largest DIZs on six out of the seven bacteria except *E. coli*.

Fennel EO and extracts were effective against several foodborne pathogens with low MICs and MBCs such as *S. dysenteriae*, *S. typhimurium*, and *E. coli* (Table 10). The mechanisms of fennel and its major components need further studies.

Table 10. Antibacterial and antifungal activities of fennel.

Type of Sample	Bacteria and Fungi	Main Results	Reference
Fennel seeds EO	*Streptococcus mutans*	MICs: 80 ppm	[120]
Fennel seeds EO	*Streptomyces albus, B. subtilis, S. typhimurium, P. aeruginosa, Shigella dysenteriae, E. coli*	EO of fennel seeds inhibited several foodborne pathogens with lowest MIC of 0.125 mg/mL.	[119]
Fennel crude extract	*E. coli, S. blanc, P. merabilis, P. vulgaris, S. epidemidis, Staphylococcus saprophyticus, A. versicolor, A. fumigates, Penicilium camemberti*	Fennel crude extract had antimicrobial activities against all nine microorganisms, especially fungi.	[121]
Fennel EO and acetone extract	*A. niger, A. flavus, Aspergillus oryzae, A. ochraceus, F. graminearum, F. moniliforme, P. ctrium, Penicillium viridicatum, Penicillium madriti, Curvularia lunata*	Fennel EO showed complete zone inhibition against several strains at 6 μL dose.	[118]
Cumin and fennel EOs	*S. typhimurium* and *E. coli*	The MICs of fennel EO was 0.031% *v/v* against *S. typhimurium* and 0.062% *v/v* E. coli.	[122]
8 spice methanol and ethanol extracts	*B. subtilis, E. faecalis, L. innocua, E. coli, P. putida, Providencia stuartii, Acetobacter calcoaceticus*	Fennel seeds extracts showed the largest zones of inhibitions in six out of the seven bacteria.	[123]

12. Coriander

Coriander (*Coriandrum sativum*), belonging to family Umbelliferae, is a native plant of the Mediterranean region and is widely cultivated in India, Russia, Central Europe, Asia, and Morocco. Coriander was widely applied in producing chutneys and sauces, flavoring pastry, cookies, buns, and tobacco products, and extensively employed for preparation of curry powder, pickling spices, sausages, seasonings, and food preservatives [4,118].

12.1. Antimicrobial Activities of Coriander

Duarte et al. [124] investigated the antimicrobial activities of coriander EO and its major compound, linalool, against *C. jejuni* and *C. coli* strains by the disc diffusion test, vapor-phase method and microdilution method. The MICs of coriander EO and linalool against *C. jejuni* and *C. coli* strains ranged between 0.5 and 1 μL/mL. Coriander EO also showed inhibitory effects on the biofilm formation of *Campylobacter* spp. Also, the antimicrobial activities of coriander EO against multidrug resistant pathogen, *Acinetobacter baumannii* (*A. baumannii*), were tested [125]. The MICs and MBCs were determined by a microdilution broth susceptibility assay. The MICs and MBCs of coriander EO against *A. baumannii* strains both ranged between 1 and 4 μL/mL. Another study investigated the synergistic antibacterial effects of coriander EO and six antibacterial drugs (cefoperazone, chloramphenicol, ciprofloxacin, gentamicin, tetracycline, and piperacillin) against two *A. baumannii* strains [126]. The results indicated that coriander EO showed synergistic action with chloramphenicol, ciprofloxacin, and tetracycline, and contributed to resensitizing *A. baumannii* to the action of chloramphenicol. Freires et al. [127] investigated the antifungal activities of EO from coriander leaves against *Candida* spp. The results showed that the MICs ranged from 15.6 to 31.2 μg/mL, and MFCs ranged from 31.2 to 62.5 μg/mL against *Candida* spp. for coriander EO. Sliva et al. [128] assessed the bacterial activities of coriander EO against 12 bacterial strains by microdilution broth susceptibility assay. The results indicated that coriander EO showed antimicrobial activities against all tested bacteria and showed bactericidal activities against bacteria except *B. cereus* and *E. faecalis*. The MICs of coriander against all tested bacteria ranged from 0.1% to 1.6% *v/v*, and MBCs ranged from 0.1% to 3.2% *v/v* except *B. cereus* and *E. faecalis*. Acimovic et al. [129] assessed the antifungal activities of EOs of six coriander accessions of different origins against *Colletotrichum acutatum* and *Colletotrichum gloeosporioides* using the inverted petriplate method. The results indicated that coriander EOs could inhibit *Colletotrichum* genus at higher application rates ($\geq$0.16 µL/mL of air).

Singh et al. [130] reported the antifungal effects of coriander EO and oleoresin on eight fungi by the inverted petriplate and food poison techniques. The results of the former method showed that EO was highly active against *Curpularia palliscens*, *F. oxysporum*, *Fusarium monitiforme*, and *Aspergillus terreus* (*A. terreus*), and the oleoresin inhibited more than 50% mycelial zones for *F. oxysporum*, *A. niger*, and *A. terreus*. The results of the latter method indicated that EO 100% inhibited the growth of *A. terreus*, *A. niger*, *F. graminearum*, and *F. oxysporum*, but the oleoresin exhibited weaker fungitoxic activities, which only 100% inhibited the growth of *F. oxysporum*. In another study, the antimicrobial activities of ethanol and aqueous-ethanol extracts of coriander were investigated against *B. subtilis*, *S. aureus*, *P. vulgaris*, *E. coil*, *P. aeruginosa*, *K. peunomonia*, *L. monocytogenes*, and *C. albicans* [131]. Ethanol extract revealed the elevated antimicrobial activities against *P. vulgaris* and *C. albicans*, and was more potent against tested microorganisms. Besides, aqueous-ethanol extract exhibited the highest activities against *B. subtilis* and *L. monocytogenes*. Furthermore, the effect of microwaves on EO of coriander on its antimicrobial activities was also tested [83]. The antimicrobial effects against microorganisms of both microwave and conventionally roasted oils were similar and more effective than those of raw oils.

Coriander EO permeated the cell membranes, resulting in the loss of all cellular functions [4]. The mechanisms of antibacterial action of coriander EO on Gram-positive and Gram-negative bacteria are different and need further exploring. Coriander EO was found to bind to membrane ergosterol and increase ionic permeability, ultimately causing cell death of *C. albicans* [127].

12.2. Comparison of Antimicrobial Activities of Coriander and Other Spices

The antimicrobial activities of four spice EOs against isolated clinical specimens were compared using the diffusion method [87], and the results showed that coriander oil was active only against *Salmonella* sp. Dimic et al. [132] tested the antifungal activities of lemon EO, coriander extract and cinnamon extract against five molds (*A. parasiticus*, *Cladosporium cladosporioides* (*C. cladosporioides*), *Eurotium herbariorum*, *Penicillium chrysogenum*, and *Aspergillus carbonarius*) by the agar dilution method and vapor phase method. The results indicated that coriander extract had the best antifungal activities in the vapor phase as it completely inhibited *A. parasiticus*, *C. cladosporioides*, *E. herbariorum*, and *P. chrysogenum* at 4.17 µL/mL.

The papers cited are of high quality and indicated that coriander possessed significant antimicrobial activities at low concentrations against several pathogens such as *A. baumannii*, *Campylobacter* spp. at low MICs, MBCs, and MFCs (Table 11).

Table 11. Antibacterial and antifungal activities of coriander.

Type of Sample	Bacteria and Fungi	Main Results	Reference
Coriander EO and linalool	*Campylobactor jejuni* and *Campylobactor coli* strains	MICs ranged between 0.5 and 1 mL/mL.	[124]
Coriander EO	*A. baumannii* strains	MICs and MBCs ranged between 1 and 4 μL/mL.	[125]
Coriander EO and 6 antibacterial drugs	*A. baumannii* strains	Coriander EO showed synergistic action with chloramphenicol, ciprofloxacin, and tetracycline.	[126]
Coriander leaves EO	*Candida* spp.	MICs ranged from 15.6 to 31.2 μg/mL, and MFCs ranged from 31.2 to 62.5 μg/mL.	[127]
Coriander EO	12 bacterial strians	MICs of coriander against all tsted bacteria ranged from 0.1% to 1.6%, *v/v*.	[128]
EOs of 6 coriander accessions	*Colletotrichum acutatum* and *Colletotrichum gloeosporioides*	Coriander EOs could inhibit *Colletotrichum* genus at higher application rates.	[129]
Coriander EO and oleoresin	*Aspergillus terreus, A. niger, F. graminearum, F. oxysporum*	Both EO and oleoresin of coriander were effective against tested fungi.	[130]
Coriander ethanol and aqueous-ethanol extracts	*B. subtilis, S. aureus, P. vulgaris, E. coil, P. aeruginosa, K. peumomonia, L. monocytogenes, C. albicans*	The ethanol extract showed clear difference and more potent against tested microorganisms in comparison with the aqueous-ethanol extract.	[131]
Coriander EO	-	Microwave and conventionally roasted oils exhibit similar antimicrobial effects but were higher effect than raw oils.	[83]
4 spice EOs	Microorganisms isolated from clinical specimens of patients	Coriander oil was active only against *Salmonella* sp.	[87]
Lemon EO, coriander extract and cinnamon extract	*A. parasiticus, Cladosporium cladosporioides, Eurotium herbariorum, Penicillium chrysogenum and Aspergillus carbonarius*	Coriander extract had the best antifungal activities in the vapor phase	[132]

13. Galangal

Galangal (*Alpinia galangal*) (Table 12) has been used as a food additive in Thailand and other Asian countries since ancient time [133]. In a study, the antimicrobial activities of extracts of seven spices and herbs against *E. coli*, *S. typhimurium*, *L. monocytogenes*, and *S. aureus* were compared by the agar disc diffusion and broth dilution assays [93]. The hexane and ethanol extracts of galangal had strong antimicrobial activities against *S. aureus* (MIC < 0.625 mg/mL) and *L. monocytogenes* (MIC < 0.625 mg/mL at 24 h and 1.25 mg/mL at 48 h). Moreover, the synergistic antimicrobial effects of extract combination (galangal, rosemary, and lemon iron bark) on *S. aureus*, *L. monocytogenes*, *E. coli*, *S. typhimurium*, and *Clostridium perfringens* were evaluated [134]. Galangal and rosemary showed synergistic activities against *S. aureus* and *L. monocytogenes*, while galangal and lemon iron bark showed synergistic activities against *E. coli* and *S. typhimurium*. Additionally, Rao et al. [133] tested the antibacterial activities of galangal methanol, acetone, and diethyl ether extracts against *B. subtilis*, *E. aerogenes*, *E. cloacae*, *E. faecalis*, *E. coli*, *K. pneumoniae*, *P. aeruginosa*, *S. typhimurium*, *S. aureus*, and *S. epidermis* using agar well diffusion method and macrodilution method. Among the three solvents used, the activities of methanol extract at pH 5.5 were excellent against all the pathogens (MIC: 0.04–1.28 mg/mL, MBCs: 0.08–2.56 mg/mL). Another study also evaluated the antimicrobial activities of methanol extracts of four *Alpinia* strains against six strains of bacteria and four strains of fungi, using the disc diffusion assay [135]. The results demonstrated that galangal flower possessed the best effects on *M. luteus* and only the extract from galangal rhizome showed antifungal activity toward *A. niger*. The mechanisms of action of galangal have been rarely explored up till now.

Table 12. Antibacterial and antifungal activities of galangal.

Type of Sample	Bacteria and Fungi	Main Results	Reference
7 spice and herb extracts	*E. coli*, *S. typhimurium*, *L. monocytogenes*, *S. aureus*	Galangal hexane and ethanol extracts had strong antimicrobial activities against *S. aureus* and *L. monocytogenes*.	[93]
Combination of extracts from galangal, rosemary and lemon iron bark	*S. aureus*, *L. monocytogenes*, *E. coli*, *S. typhimurium*, *Clostridium perfringens*	Galangal showed synergistic activities against tested microorganisms with rosemary and lemon iron bark.	[134]
Galangal methanol, acetone and diethyl ether extracts	*B. subtilis*, *E. aerogenes*, *E. cloacae*, *E. faecalis*, *E. coli*, *K. pneumoniae*, *P. aeruginosa*, *S. typhimurium*, *S. aureus*, *S. epidermis*	The activities of methanol extract at pH 5.5 were excellent with MICs ranging from 0.04 to 1.28 mg/mL.	[133]
4 *Alpinia* strains methanol extracts	6 strains of bacteria and 4 strains of fungi	Galangal flower possessed the highest activity against *M. luteus* and only the extract from galangal rhizome showed antifungal activity toward *A. niger*.	[135]

14. Black Pepper

Black pepper (*Piper nigrum*) (Table 13) is largely used as a flavoring agent in foods. The antifungal effects of EO and acetone extract of black pepper on various pathogenic fungi were tested by the inverted petriplate technique and food poisoning technique [136]. The results showed that the EO was 100% controlled the mycelial growth of *F. graminearum*, while the acetone extract 100% inhibited mycelial growth of *Penicillium viridcatum* and *A. ochraceus*. In another study, the bacterial effects of EOs and acetone extracts of four spices on *S. aureus*, *B. cereus*, *B. subtilis*, *E. coli*, *S. typhi*, and *P. aeruginosa* were studied using the disk diffusion and poison food assay [137]. The results showed that black pepper extracts completely reduced colonies of *S. aureus*, *B. cereus*, and *B. subtilis* at 5 and 10 µL levels using the poison food method. Zarai et al. [138] evaluated the antimicrobial effects of various solvent extracts, piperine, and piperic acid from pepper against *E. coli*, *K. pneumonia*, *S. enterica*, *S. aureus*, *S. epidermidis*, *E. faecalis*, and *B. subtilis* by the agar diffusion assay and micro-well dilution assay. The results showed that the ethanol extract was the most effective to the tested bacteria with the MICs ranging from 156.25 µg/mL (*S. aureus* and *B. subtilus*) to 1250 µg/mL (*E. coli* and *K. pneumonia*).

Table 13. Antibacterial and antifungal activities of black pepper

Type of Samples	Bacteria and Fungi	Main Results	Reference
Black pepper EO and acetone extract	*A. flavus*, *A. ochraceus*, *A.oryzae*, *A. niger*, *F. moniliforme*, *F. graminearum*, *Penicillium citrinum*, *Penicillium viridcatum*, *P. madriti*, *Curvularia lunata*	The EO was effective against *F. graminearum*, while the acetone extract was effective against *P. viridcatum* and *A. ochraceus*.	[136]
4 spice EOs and acetone extracts	*S. aureus*, *B. cereus*, *B. subtilis*, *E. coli*, *S. typhimurium*, *P. aeruginosa*	Black pepper extracts showed complete reduction of colonies against tested bacterial strains at 5 and 10 µL levels.	[137]
Various solvent extracts, piperine and piperic acid from pepper	*E. coli*, *K. pneumonia*, *S. enterica*, *S. aureus*, *S. epidermidis*, *E. faecalis*, *B. subtilis*	The ethanol extract was the most effective with the MICs ranging from 156.25 to 1250 µg/mL.	[138]

Black pepper EO could cause physical and morphological alterations in the cell walls and membranes of *E. coli*, and then result in the leakage of electrolytes, ATP, proteins, and DNA materials [139]. Chemical components of black pepper and its mechanisms of antimicrobial action need further exploration.

15. Other Spices

The antimicrobial activities of the spices mentioned above against several common microorganisms are summarized in Table 14. Other spices—such as *Allium roseum* L., *Cinnamomum verum*, *Laurus nobilis*, *Myristica fragrans*, and *Pimpinella anisum*—were also proved to possess significant antifungal and antibacterial activities (Table 15).

Table 14. Antimicrobial activities of spices against several common microorganisms.

Bacteria and Fungi	Spices	Type of Samples	Reference
E. coli	Clove	Aqueous extract	[18]
		Acetone extract	[27]
		EO	[23,30,31]
		Ethyl acetate extract	[27]
		Ethyl heptanoate extract	[30]
		Methanol extract	[27]
		Powder	[19,29]
	Oregano	Aqueous extract	[40]
		EO	[36,44,47,49]
		EO-rich fractions	[35]
	Thyme	Decoction	[62]
		EO	[51,52,60]
		Hydroalcoholic extract	[62]
		Hydrosol	[56]
		Infusion	[62]
	Cinnamon	Acetone extract	[27]
		Aqueous extract	[75]
		Diethyl ether extract	[76]
		EO	[78,79]
		Ethyl acetate extract	[27]
		Hydrosol	[56]
		Methanol extract	[27]
		Powder	[71]
	Cumin	EO	[80,81,83,85–89]
	Rosemary	Aqueous extract	[93]
		EO	[88,92]
		Ethanol extract	[93–95]
		Hexane extract	[93]
	Garlic	Acetone extract	[105]
		Aqueous extract	[103]
		Ethanol extract	[104,105]
		Hexane extract	[105]
		Powder	[99]
	Ginger	Aqueous extract	[103]
	Basil	Acetone extract	[114]
		Chloroform extract	[114]
		EO	[113,116]
		Methanol extract	[114]
	Fennel	Crude extract	[121]
		EO	[121,122]
		Ethanol extract	[123]
		Methanol extract	[123]
	Coriander	Aqueous-ethanol extract	[131]
		Ethanol extract	[131]
	Galangal	Acetone extract	[133]
		Diethyl ether extract	[133]
		Hexane extract	[134]
		Methanol extract	[133]
	Black pepper	Acetone extract	[137]
		EO	[137]
		Ethanol extract	[138]

Table 14. *Cont.*

Bacteria and Fungi	Spices	Type of Samples	Reference
S. aureus	Clove	Acetone extract	[27]
		Aqueous extract	[18]
		EO	[30–32]
		Ethanol extract	[24]
		Ethyl heptanoate extract	[30,31]
		Methanol extract	[27]
		Powder	[19,29]
	Oregano	EO	[36,41,47]
		EO-rich fractions	[35]
	Thyme	Decoction	[62]
		EO	[51,52]
		Hydroalcoholic extract	[62]
		Infusion	[62]
	Cinnamon	Acetone extract	[27]
		Aqueous extract	[75]
		Diethyl ether extract	[76]
		EO	[72,78,79]
		Ethyl acetate extract	[27]
		Hydrosol	[56]
		Methanol extract	[27]
	Cumin	EO	[80,81,83,86,88,89]
	Rosemary	EO	[88,92]
		Ethanol extract	[94,97]
		Hexane extract	[93]
	Garlic	Aqueous extract	[98]
		Acetone extract	[105]
		Ethanol extract	[104,105]
		Hexane extract	[105]
		Powder	[99]
	Basil	Acetone extract	[114]
		Chloroform extract	[114]
		EO	[113,116]
		Methanol extract	[114]
	Coriander	Aqueous-ethanol extract	[131]
		Ethanol extract	[131]
	Galangal	Acetone extract	[133]
		Diethyl ether extract	[133]
		Ethanol extract	[93]
		Hexane extract	[93,134]
		Methanol	[133]
	Black pepper	Acetone extract	[137]
		EO	[137]
		Ethanol extract	[138]

Table 14. *Cont.*

Bacteria and Fungi	Spices	Type of Samples	Reference
L. monocytogenes	Clove	Ethanol extract	[24,25]
		Ethyl heptanoate extract	[30,31]
		EO	[30,31]
	Oregano	EO	[49]
	Thyme	EO	[60]
	Cinnamon	EO	[78,79]
	Cumin	EO	[89]
	Rosemary	Aqueous extract	[93]
		Ethanol extract	[93–95,97]
		Hexane extract	[93]
	Garlic	Aqueous extract	[103]
	Ginger	EO	[111]
	Basil	Acetone extract	[114]
		Chloroform	[114]
		EO	[116]
		Methanol extract	[114]
	Coriander	Aqueous-ethanol extract	[131]
		Ethanol extract	[131]
	Galangal	Ethanol extract	[93]
		Hexane extract	[93,134]
S. typhimurium	Clove	EO	[23,30,31]
		Ethyl heptanoate extract	[30]
		Powder	[29]
	Oregano	Extract	[34]
	Thyme	EO	[52,60]
		Hydrosol	[56]
	Cumin	EO	[86,89]
	Rosemary	Aqueous extract	[93]
		EO	[92]
		Ethanol extract	[93]
		Hexane extract	[93]
	Garlic	Powder	[99]
	Basil	EO	[116]
	Fennel	EO	[121,122]
	Galangal	Acetone extract	[133]
		Diethyl ether extract	[133]
		Hexane extract	[134]
		Methanol extract	[133]
	Black pepper	Acetone extract	[137]
		EO	[137]
P. aeruginosa	Clove	Acetone extract	[27]
		EO	[32]
		Ethyl acetate extract	[27]
		Methanol extract	[27]
	Oregano	EO-rich fractions	[35]
	Thyme	Decoction	[62]
		EO	[51,53,57]
		Hydroalcoholic extract	[62]
		Infusion	[62]
	Cinnamon	Diethyl ether extract	[76]
		EO	[79]
		Hydrosol	[56]
		Methanol extract	[27]

Table 14. *Cont.*

Bacteria and Fungi	Spices	Type of Samples	Reference
P. aeruginosa	Cumin	EO	[83,87]
	Rosemary	EO	[92]
		Ethanol extract	[94]
	Ginger	EO	[108]
	Basil	Acetone extract	[114]
		Chloroform	[114]
		EO	[116]
		Methanol extract	[114]
	Fennel	EO	[121]
	Coriander	Aqueous-ethanol extract	[131]
		Ethanol extract	[131]
	Galangal	Acetone extract	[133]
		Diethyl ether extract	[133]
		Methanol extract	[133]
	Black pepper	Acetone extract	[137]
		EO	[137]
B. subtilis	Oregano	Aqueous extract	[40]
		EO-rich fractions	[35]
	Thyme	EO	[51]
		Hydrosol	[55]
	Cinnamon	EO	[79]
	Rosemary	EO	[92]
	Garlic	Hydrosol	[55]
	Ginger	EO	[108]
	Basil	EO	[113]
	Fennel	EO	[121]
		Ethanol extract	[123]
		Methanol extract	[123]
	Coriander	Aqueous-ethanol extract	[131]
		Ethanol extract	[131]
	Galangal	Acetone extract	[133]
		Diethyl ether extract	[133]
		Methanol extract	[133]
	Black pepper	Acetone extract	[137]
		EO	[137]
		Ethanol extract	[138]
B. cereus	Clove	EO	[23,30,31]
		Ethyl heptanoate extract	[30]
	Oregano	EO	[36]
	Thyme	EO	[60]
	Cinnamon	EO	[78,79]
	Cumin	EO	[80,86,89]
	Rosemary	Ethanol extract	[94]
	Garlic	Powder	[99]
	Basil	EO	[116]
	Black pepper	Acetone extract	[137]
		EO	[137]

Int. J. Mol. Sci. **2017**, *18*, 1283

Table 14. *Cont.*

Bacteria and Fungi	Spices	Type of Samples	Reference
E. faecalis	Cinnamon	Diethyl ether extract	[76]
	Cumin	EO	[86,88]
	Rosemary	EO	[88]
	Basil	EO	[116]
	Fennel	Ethanol extract	[123]
		Methanol extract	[123]
	Galangal	Acetone extract	[133]
		Diethyl ether extract	[133]
		Methanol extract	[133]
	Black pepper	Ethanol extract	[138]
E. faecalis	Cinnamon	Diethyl ether extract	[76]
	Cumin	EO	[86,88]
	Basil	EO	[116]
	Fennel	Ethanol extract	[123]
		Methanol extract	[123]
	Galangal	Acetone extract	[133]
		Diethyl ether extract	[133]
		Methanol extract	[133]
	Black pepper	Ethanol extract	[138]
K. pneumoniae	Clove	Acetone extract	[27]
		Ethyl acetate	[27]
		Methanol extract	[27]
	Cinnamon	Acetone extract	[27]
		Diethyl ether extract	[76]
		EO	[79]
		Ethyl acetate	[27]
		Methanol extract	[27]
	Garlic	Ethanol extract	[104]
	Galangal	Acetone extract	[133]
		Diethyl ether extract	[133]
		Methanol extract	[133]
	Black pepper	Ethanol extract	[138]
P. vulgaris	Thyme	Decoction	[62]
		EO	[52]
		Hydroalcoholic extract	[62]
		Infusion	[62]
	Garlic	Ethanol extract	[104]
	Fennel	Crude extract	[121]
	Coriander	Aqueous-ethanol extract	[131]
		Ethanol extract	[131]
P. fluorescens	Clove	Powder	[19]
	Thyme	EO	[52,57,59]
	Cinnamon	EO	[78]
	Rosemary	Ethanol extract	[95]
L. innocua	Clove	EO	[23]
	Thyme	EO	[52,52,59]
	Fennel	Ethanol extract	[123]
		Methanol extract	[123]

Table 14. *Cont.*

Bacteria and Fungi	Spices	Type of Samples	Reference
S. faecalis	Clove	Acetone extract	[27]
		Ethyl acetate	[27]
		Methanol extract	[27]
	Cinnamon	Acetone extract	[27]
		Ethyl acetate	[27]
		Methanol extract	[27]
	Cumin	Decoctions	[81]
		EO	[88]
		Infusions	[81]
S. enteritidis	Thyme	Hydrosol	[55]
	Garlic	Hydrosol	[55]
	Ginger	EO	[108]
M. luteus	Cinnamon	Diethyl ether extract	[76]
	Cumin	EO	[88]
	Galangal	Methanol extract	[135]
B. megaterium	Clove	Acetone extract	[27]
		Ethyl acetate	[27]
		Methanol extract	[27]
	Cinnamon	Acetone extract	[27]
		Ethyl acetate	[27]
		Methanol extract	[27]
	Cumin	EO	[88]
A. hydrophila	Cumin	EO	[88]
	Garlic	Aqueous extract	[103]
S. epidermidis	Thyme	Decoction	[62]
		EO	[51]
		Hydroalcoholic extract	[62]
		Infusion	[62]
	Black pepper	Ethanol extract	[138]
C. albicans	Clove	EO	[22]
	Oregano	EO	[35,36]
	Thyme	EO	[51,57]
	Cinnamon	Hydrosol	[56,76]
		Diethyl ether extract	[76]
	Cumin	EO	[83]
	Rosemary	EO	[92]
	Garlic	Ethanol extract	[104]
	Coriander	Aqueous-ethanol extract	[131]
		EO	[127]
		Ethanol extract	[131]

Table 14. *Cont.*

Bacteria and Fungi	Spices	Type of Samples	Reference
A. niger	Oregano	EO-rich fraction	[35]
	Cinnamon	Aqueous extract	[74]
	Cumin	EO	[83]
	Ginger	Aqueous extract	[74,109]
		EO	[109]
		Oleoresin	[109]
	Basil	EO	[113,116]
	Fennel	Acetone extract	[118]
		EO	[118]
	Coriander	EO	[130]
	Galangal	Methanol extract	[135]
A. flavus	Basil	EO	[117]
	Fennel	Acetone extract	[118]
		EO	[118]
F. oxysporum	oregano	Decoction	[39]
	ginger	EO	[109]
	basil	Extract	[115]
	coriander	EO	[130]
		Oleoresin	[130]
F. graminearum	Fennel	Acetone extract	[118]
		EO	[118]
	Coriander	EO	[130]
		Oleoresin	[130]
	Black pepper	Acetone extract	[136]
		EO	[136]
S. cerevisiae	Thyme	EO	[52]
	Cumin	EO	[86]

Table 15. Antibacterial and antifungal activities of other spices.

Spices	Type of Samples	Bacteria and Fungi	Main Results	Reference
Achillea species	Ethanol extract	*K. pneumoniae*, *E. cloacae*, *S. typhimurium*, *S. epidermis*, *E. coli*, *E. aerogenes*, *S. aureus*, *Klebsiella oxytoca*, *S. pyogenes*, *P. aeruginosa*, *C. albicans*	*Achillea* species showed a broad spectrum of strong antibacterial activities against all tested microorganisms.	[140]
Achillea millefolium	Ethanol extract	*S. aureus*, *S. enteritidis*, *E. coli*, *S. pneumoniae*, *K. pneumoniae*, *P. aeruginosa*, *E. aerogenes*, *P. mirabilis*, *A. niger*, *C. albicans*	The antibacterial activities of *A. millefolium* were greater or similar to other penicillin derivatives but lesser than Ampicillin.	[141]
Aframomum corrorima	Seeds, pods, leaves and rhizomes extract	*A. flavus* and *Penicillum expansum*	*A. corrorima* crude seed extract was the most active against *A. flavus* and *P. expansum* at concentration of 0.4 mg/mL.	[142]
Allium hirtifolium Boiss.	Hydromethanol extract	MRSA, *S. epidermidis*, *S. pneumoniae*, *E. coli*, *S. typhimurium*, *P. mirabilis*, *K. pneumoniae*	*A. hirtifolium* extract was effective against 10 species of pathogenic bacteria with MICs ranging from 1.88 to 7.50 mg/mL.	[143]
Allium roseum L.	Extracts of bulbs, leaves, flowers and seeds by 3 extraction methods	*S. aureus*, *S. epidermidis*, *M. luteus*, *B. cereus*, *B. subtilis*, *E. faecalis*, *S. typhimurium*, *E. coli*, *P. aeruginosa*, *C. albicans*	*A. roseum* extract showed very significant antimicrobial activities to strains such as *C. albicans* (MICs: 1.00–3.44 µg/µL) and *E. coli* (MICs: 2.00–3.44 µg/µL).	[144]
Allium ursinum L.	Pressurized-liquid extract	*S. aureus* and *A. niger*	*A. ursinum* extract showed antimicrobial activities against *S. aureus* with DIZs of 12 and 10 mm (two parallel determinations) and *A. niger* of 6 mm.	[145]
Amomum kravanh	EO	Different foodborne pathogens	*A. kravanh* EO exhibited the best antibacterial activities against *B. subtilis* and *E. coli*.	[146]
Anethum graveolens L.	EO and acetone extract	*P. citrinum*, *A. niger*, *S. aureus*, *B. cereus*, *P. aeruginosa*	EO and extract showed different but both effective activities against tested microorganisms.	[147]
Anethum graveolens L.	diethyl-ether extract	*P. aeruginosa*, *E. coli*, *K. pneumoniae*, *M. luteus*, *E. faecalis*, *B. megaterium*, *S. aureus*	*A. graveolens* extract affected all of the bacteria tested.	[148]
Anethum graveolens L.	EO	*A. flavus*	*A. graveolens* EO is the most effective against aflatoxin production.	[117]
Brassica jancea	EO	*Vibrio parahaemolyticus* and *Vibrio vulnificus*	*B. jancea* EO could inhibit *V. parahaemolyticus* and *Vibrio vulnificus* inoculated sliced raw flatfish at 5 °C of storage.	[110]
Brassica jancea	Water extract	*E. coli*, *S. aureus*, *B. cereus*	*B. jancea* extract showed good inhibitory action at 1% concentration.	[149]

350

Table 15. Cont.

Spices	Type of Samples	Bacteria and Fungi	Main Results	Reference
Bunium persicum	Volatile compounds	F. oxysporum	B. persicum showed the strongest effect compared with other 51 spices and herbs.	[150]
Caesulia axillaris Roxb.	EO	A. flavus	C. axillaris EO showed complete inhibition against A. flavus at 1.0 μg/mL.	[151]
Capsicum froutescens	Ethanol extract	S. aureus	C. froutescens extract showed the highest activity.	[152]
Capsicum frutescens L.	n-hexane, chloroform, ethyl acetate, acetone, and methanol extracts of dried seeds	B. cereus, S. aureus, MRSA, E. coli, S. typhimurium, P. aeruginosa, K. pneumoniae, P. vulgaris, C. albicans, C. krusei	Microwave assisted solvent extracts showed significant activities and n-hexane extract was effective against P. aeruginosa and C. albicans, while ethyl acetate extract was effective against C. krusei.	[153]
Carum capticum	EO	Corynebacterium diphtheriae, S. aureus, Staphylococcus haemolyticus, B. subtilis, P. aeruginosa, E. coli, Klebsiella species, P. vulgaris	C. capticum was very effective against all tested bacteria.	[154]
Carum copticum	EO	S. aureus, B. cereus, E. coli, S. enteritidis, L. monocytogenes	C. copticum EO was the most effective against tested bacteria with MICs of 0.03–0.5 mg/mL compared with two other spices.	[155]
Cinnamomum burmannii	Methanol crude extract	B. cereus, L. monocytogenes, S. aureus, E. coli, Salmonella anatum	MIC and MBC for B. cereus were 625 and 2500 μg/mL, respectively, for four other bacteria were more than 2500 μg/mL.	[156]
Cinnamomum cassia	Ultra-fine powder	E. coli, S. aureus, P. fluorescens, L. rhamnosus, B. thermosphacta	C. cassia powder significantly reduced the microorganisms tested at the concentration ≤2.5% w/v and the inhibitory effects were positive correlated with concentrations.	[19]
Cinnamomum tamala	Leaves EO	C. albicans, A. niger, A. fumigatus, R. stolonifer, Penicillium spp.	The MFCs of EO against all the tested fungi were 230 μg/mL.	[157]
Cinnamomum verum	Bark and leaf extracts and EO	Bacteria isolated from urine samples, and A. niger	C. verum oil possessed stronger antimicrobial activities than extracts. A. niger showed no growth in the presence of oil.	[158]
Cinnamomum verum	EO	E. coli, S. typhimurium, S. aureus, B. subtilis, A. flavus, C. albicans	C. verum EO treated group showed significant decrease in viable bacterial counts.	[159]
Cinnamomum verum	EO	S. typhimurium, S. paratyphi, E. coli, S. aureus, P. fluorescens, B. licheniformis	C. verum bark EO showed the best antibacterial activities with mean MICs ranging from 2.9 to 4.8 mg/mL.	[160]

Table 15. *Cont.*

Spices	Type of Samples	Bacteria and Fungi	Main Results	Reference
Citrus aurantium L.	Ethanol extract	*E. coli, P. aeruginosa, S. aureus, B. cereus*	*C. aurantium* showed strong antimicrobial activities against tested bacteria.	[161]
Clinopodium ascendens	EO	*S. aureus, S. faecium, S. mutans, Agrobacterium tumefasciens, E. coli, B. cinerea, C. albicans*	*C. ascendens* exhibited remarkable activity against *E. coli* and was active against *A. tumefasciens, S. aureus,* and *B. cinerea.*	[162]
Corydothymus capitatus	EO	*P. putida*	*C. capitatus* EO was the most active with a MIC of 0.025% *w/v* and a MTC of 0.006% *w/v.*	[163]
Cotoneaster nummularioides	Leaves EO	*B. cereus, S. aureus, Salmonella entrica, E. coli*	The extract of *C. nummularioides* showed strong effects on two Gram-positive microorganisms tested with higher sensitivity for *B. cereus* (MIC: 3.125 mg/mL).	[164]
Croton hirtus	EO	*E. coli, S. aureus*	*C. hirtus* EO was effective against *S. aureus* with MIC of 512 µg/mL.	[165]
Cuminum nigrum L.	Polyphenolic compounds	*B. subtilis, B. cereus, Enterobacter* spp., *E. coli, L. monocytogenes, S. aureus, Y. enterocolitica*	*C. nigrum* extract possessed significantly inhibitory effects on *B. subtilis, B. cereus,* and *S. aureus.*	[166]
Curcuma longa	Curcumin	*S. aureus*	Antibacterial activity of curcumin against *S. aureus* was enhanced with the increase of the concentration.	[167]
Cunila galioides	EO from aerial parts	15 bacterial species including *Bacillus* sp., *L. monocytogenes, S. aureus, A. hydrophila, E. faecalis* etc.	The oil of *C. galioides* citral efficiently controlled some microorganisms, showing both contact and gaseous activity.	[168]
Dichrostachys glomerata	Methanol extract	*Providencia stuartii, P. aeruginosa, K.pneumoniae, E. coli, E. aerogenes, E. cloacae*	*D. glomerata* extract inhibited the growth of all the 29 tested bacteria with MICs ≤ 1024 µg/mL.	[169]
Echinops giganteus	Methanol extract	*Mycobacterium tuberculosis* H(37)Rv, *Mycobacterium tuberculosis* H37Ra	The extract of *E. giganteus* was the most effective with MICs of 32 µg/mL and 16 µg/mL, respectively against H37Ra and H(37)Rv, compared with other 19 spices.	[170]
Elettaria cardamomum	Ethanol extract	4 strains of Gram-positive bacteria and 12 strains of Gram-negative bacteria	*E. cardamomum* extract was effective against a majority of the pathogens, MICs ranged from 9.4 to 18.75 mg/mL except *E. coli, B. cereus,* and *E. cloacae* which had a great sensitivity to the spice extract (MICs < 2.34 mg/mL).	[171]

Table 15. *Cont.*

Spices	Type of Samples	Bacteria and Fungi	Main Results	Reference
Elettaria cardamomum	EO and various oleoresins	*S. aureus, B. cereus, E. coli, S. typhimurium, A. terreus, Penicillium purpurogenum, F. graminearum, Penicillium madriti*	The EO showed strong effects against bacteria tested at 3000 ppm, and the methanol and ethanol oleoresins gave the best results against *A. terreus* at 3000 ppm.	[172]
Eucalyptus globulus	Hydrodistillated extract	*S. aureus, B. subtilis, L. innocua, E. coli, P. aeruginosa*	*E. globulus* extract showed an inhibition effects against all the tested bacteria with MIC of 3 and 4 mg/mL.	[173]
Eucalyptus largiflorens	EO	*A. flavus, A. parasiticus, A. niger, Penicillium chryzogenum, P. citrinum*	The leaf oil of *E. largiflorens* showed higher antifungal activities than four other *Eucalyptus* spices.	[174]
Eucalyptus radiata	EO	*P. aeruginosa, E. coli , K. pneumoniae, S. typhimurium, Acinetobacter baumannii, P. aeruginosa, K. pneumoniae*	*E. radiate* showed better antibacterial activities with MICs ranging from 8 to 32 μL/mL.	[175]
Eugenia caryophyllum Bullock and Harrison	Aqueous extract	*S. aureus, S. typhimurium, E. coli, S. epidermidis, L. plantarum, P. vulgaris*	The MICs and MBCs against all tested bacteria ranged from 1 to 4 g/L and 2 to 8 g/L, respectively.	[176]
Foeniculum vulgare ssp. piperitum	EO	*A. alternate, F. oxysporum, R. solani*	100% fungistatic effects were observed with 40 ppm doses of *F. vulgare* oils.	[177]
Glaucium elegans	Methanol extract	*E. coli, S. aureus, S. enteritidis, Bacillus anthracis, Proteus*	*G. elegans* methanol extract had significant antibacterial effects.	[178]
Gloriosa superba Linn	Methanol extract and fractions in different solvent systems	*C. albicans, Candida glaberata, Trichophyton longifusus, M. canis, S. aureus, E. coli, B. subtilis, K. pneumonae, S. flexneri, S. typhimurium*	The n-butanol fraction of *G. superba* showed excellent antifungal activities and chloroform fraction showed the highest antibacterial activity against *S. aureus*.	[179]
Helichrysum species	Methanol extracts	13 bacteria and 2 yeasts	All the extracts showed significant antimicrobial activities against all tested microorganisms.	[180]
4 *Helichrysum* Mill. plants	Methanol extracts	*A. hydrophila, Bacillus brevis, B. cereus, K. pneumoniae, P. aeruginosa, S. aureus, E. coli, M. morganii, M. smegmatis, P. mirabilis, Y. enterocolitica, S. cerevisiae*	The methanol extracts had antibacterial activities against the first six microorganisms listed.	[181]
horseradish	Aqueous extract	*S. aureus*	Horseradish water extract showed a higher biological activity.	[182]

Table 15. Cont.

Spices	Type of Samples	Bacteria and Fungi	Main Results	Reference
Hyssopus officinalis L.	EO	A. niger, A. ochraceus, A. versicolor, A. fumigatus, Cladosporium cladosporioides, Cladosporium fulvum, Penicillium funiculosum, Penicillium ochrochloron, Trichoderma viride, C. albicans	All tested EO and deodorized extracts showed activities with the MICs ranging from 4 to 16 mg/mL.	[183]
Laser trilobum L.	Methanol extract	S. aureus, P. vulgaris, P. mirabilis, B. cereus, A. hydrophila, E. faecalis, K. pneumoniae, S. typhimurium, E. aerogenes, E. coli	The fruit extract had significant antimicrobial effects on pathogen bacteria.	[184]
Laurus nobilis	Ethanol extract	4 Gram-positive bacteria and 12 Gram-negative bacteria	L. nobilis extract was effective in inhibiting a majority of the pathogens, MICs ranged from 4.7 to 9.4 mg/mL.	[185]
Laurus nobilis L.	EO and leaves ethanol, water and hot water extract	B. thermosphacta, E. coli, L. innocua, L. monocytogenes, P. putida, S. typhimurium, Shewanella putrefaciens	L. nobilis EO exhibited strong antibacterial activities against all tested bacteria.	[186]
Laurus nobilis L.	Aqueous, ethanol, ethyl acetate and hexane extracts	B. cereus, S. aureus, E. coli, K. pneumoniae, C. albicans	Only aqueous extract of L. nobilis showed anticandidal activities among the tested 8 plants.	[187]
Lavandula officinalis	EO	L. innocua and P. fluorescens	L. officinalis EO showed the highest activity against L. innocua.	[188]
Lichen Xanthoria parietina	Acetone extract	S. aureus, E. faecalis, P. vulgaris, P. mirabilis, S. typhimurium, E. cloacae, E. aerogenes, P. aeruginosa, K. pneumoniae, R. solani, Botridis cinerea, C. albicans	X. parietina acetone extract and parietin showed similar activities on the nine bacteria tested, but less active than parietin on the three fungi tested.	[189]
Lippia grandis Schauer.	EO	E. coli, P. aeruginosa, K. pneumoniae, S. aureus, E. faecalis	The EO was effective against 75% of the microorganisms analyzed especially S. aureus, E. faecalis, and E. coli.	[190]
Lippia javanica	Acetone and aqueous extracts	S. aureus, L. monocytogenes, S. typhimurium, E. coli, A. fumigatus, A. niger, M. canis, Microsporum gypseum, T. tonsurans, T. rubrum, T. mucoides, Penicillium aurantiogriseum, Penicillium chrysogenum	The aqueous and acetone extracts were active against the bacterial strains, and the acetone extract exhibited the antifungal activities higher than even the reference drugs.	[191]
Lippia origanoides H.B.K.	EO	C. albicans, Candida parapsilosis, Candida guilliermondii, Cryptococcus neoformans, Trichophyton rubrum, Fonsecaea pedrosoi, S. aureus, Lactobacillus casei, S. mutans	L. origanoides EO showed highly significant inhibition zones for all microorganisms tested.	[192]

Table 15. *Cont.*

Spices	Type of Samples	Bacteria and Fungi	Main Results	Reference
Litsea cubeba	EO	*E. coli*	The MIC and MBC of *L. cubeba* against *E. coli* were both 0.125% v/v.	[193]
Melissa officinalis L.	Ethanol, ethyl acetate and aqueous extracts	*Agrobacterium tumefaciens, Bacillus mycoides, B. subtilis, E. cloaceae, Erainia carotovora, E. coli, Proteus* sp., *P. fluorescens, S. aureus*	*M. officinalis* ethanol, ethyl acetate, and aqueous extracts significantly enhanced the effectiveness of tested preservatives (sodium benzoate, sodium nitrite, and potassium sorbate).	[194]
Mentha piperita L.	EO	*T. rubrum, T. tonsurans, T. schoenleinii, T. mentagrophytes, M. canis, M. fulcum*	For effective concentration of *M. piperita* oil against tested antropophilic dermatophytes, and MICs ranged from 0.1 to 1.5 μL/mL.	[195]
Mentha spicata L.	hexane, chloroform, ethyl acetate, and aqueous fractions of ethanol extract	*Salmonella paratyphi, Shigella boydii, S. aureus, E. coli, Vibrio cholera, P. aeruginosa, E. faecalis, S. typhimurium, P. vulgaris, K. pneumoniae*	*M. spicata* ethanol extract and its solvent fractions effectively inhibited half of the microorganism growth.	[196]
Myristica argentea	Water extract	*E. coli* and *S. aureus*	*M. argentea* were more effective against *E. coli* (MIC of 9.80 mg/mL) and *S. aureus* (MIC of 6.20 mg/mL).	[197]
Myristica fragrans	-	20 different serogroups of *E. coli*, 8 serotypes of *Salmonella, L. monocytogenes, A. hydrophila*	*M. fragrans* showed good anti-listerial activity, although activities against *E. coli* and *Salmonella* were serotype dependent.	[103]
Myristica fragrans	Ethyl acetate and ethanol extracts of flesh, mace and seed	*S. mutans, Streptococcus mitis, Streptococcus salivarius, Aggregatibacter actinomycetemcomitans, P. gingivalis, Fusobacterium nucleatum*	Flesh ethyl acetate extract had the highest effects against tested bacteria with mean MICs ranging from 0.625 to 1.25 mg/mL among all tested extracts.	[198]
Myrtus communis	EO	*P. aeruginosa, S. typhimurium, E. coli, A. hydrophila, L. monocytogenes, C. albicans*	*M. communis* EO exhibited antimicrobial activities against all tested microorganisms, especially Gram-negative bacteria.	[199]
Myrtus communis L.	Methanol, ethyl acetate, acetone extracts	*S. aureus, P. vulgaris, P. mirabilis*	The most effective extract was the methanol extract from *M. communis* leaves against *S. aureus*.	[200]
Myrica gale L.	EO	*A. flavus, Cladosporium cladosporioides, Penicillium expansum*	A complete antifungal activity was observed at 1000 ppm of *M. gale* EO against *Cladosporium cladosporioides*.	[201]
Nepeta alpina	EO	*Bacillus pumilus, E. coli, Kocuria varians, L. monocytogenes, P. aeruginosa, S. typhimurium, A. niger, A. flavus, C. glabrata*	The EO was active against *L. monocytogenes* with MIC of 32 μg/mL.	[202]

Table 15. *Cont.*

Spices	Type of Samples	Bacteria and Fungi	Main Results	Reference
Nigella sativa L.	Aqueous extracts	*Uromyces appendiculatus*	*N. sativa* extract was effective against *U. appendiculatus* and controlled rust similar to mancozeb fungicide at 2 and 3% concentrations.	[203]
Nigella sativa L.	n-hexan extract	24 pathogenic, spoilage and lactic acid bacteria	*N. sativa* oil showed antibacterial activities against all the bacteria at all concentrations (0.5%, 1.0% and 2.0%) tested.	[204]
Ocimum canum	EO	*B. subtilis, E. coli, K. pneumoniae, M. luteus, P. aeruginosa, Raoultella planticola, S. typhimurium, S. mutans*	MICs of *O. canum* ranged from 0.43 to 2.08 µL/mL against 7 out of 10 bacteria tested.	[205]
Ocimum gratissimum L.	EO	*A. flavus, A. niger, Aspergillus fumigatus, Aspergillus terreus, Aspergillus sydowi, Aspergillus alternate, Penicillium italicum, Fusarium nivale, C. lunata, Cladosporium spp.*	The EO exhibited antifungal activities against fungal isolates from some spices and showed better efficacy as fungi toxicant than prevalent fungicide Wettasul-80.	[206]
Ocimum sanctum	EO	*A. flavus*	MIC: 0.3 µL/mL.	[207]
Ocimum sanctum L.	EO	*A. flavus, Aspergillus fumigatus, Aspergillus clavatus, Aspergillus orizae S. aureus, E. faecalis, E. coli, enterohemorrhagic E. coli, P. aeruginosa, S. flexneri*	*O. sanctum* EO exhibited antimicrobial activities against all tested pathogens at concentrations of 0.125–32 µL/mL except *P. aeruginosa*.	[208]
Ocimum suave	EO	*S. aureus, S. epidermidis, S. mutans, S. viridans, E. coli, E. cloacae, K. pneumoniae, P. aeruginosa, C. albicans, C. tropicalis, C. glabrata*	*O. suave* EO showed the strongest antibacterial activities with MICs ranging from 0.05 to 1.37 mg/mL.	[209]
Olea europaea L.	Methanol extract	*S. aureus, S. epidermidis, S. pyogenes, Streptococcus agalactiae, S. enterica serovar Typhi, P. aeruginosa, Acetobacter calcoaceticus, C. albicans, P. vulgaris, S. faecalis, S. dysenteriae, K. pneumoniae, E. coli, V. cholera, C. xerosis*	*O. europaea* methanol extract showed strong antibacterial activities against *S. aureus, S. epidermidis,* and *S. pyogenes* at MICs range of 31.25–62.5 ug/mL.	[210]
Origanum marjorana	Water extract	*Vibrio parahaemolyticus*	*O. marjorana* showed the lowest MICs against *V. parahaemolyticus* both in a nutrient rich and poor medium.	[211]
Origanum minutiflorum	EO	*E. coli, S. aureus, S. enteritidis, L. monocytogenes, L. plantarum*	Whey protein based edible films incorporated with *O. minutiflorum* EO was the most effective at 2% level.	[212]
Orthosiphon stamineus Benth.	Methanol and aqueous extracts	*V. parahaemolyticus*	*V. parahaemolyticus* was more susceptible to 50–100% methanol extracts of *O. stamineus*.	[213]

Table 15. *Cont.*

Spices	Type of Samples	Bacteria and Fungi	Main Results	Reference
Peganum harmala L.	Methanol extract	*S. aureus, S. epidermidis, S. pyogenes, S. agalactiae, S. enterica serovar Typhi, P. aeruginosa, Acetobacter calcoaceticus, C. albicans, P. vulgaris, S. faecalis, S. dysenteriae, K. pneumoniae, E. coli, V. cholera, C. xerosis*	*P. harmala* seed showed MICs of 31.25–62.5, 250, 125–250, and 31.25–250 µg/mL, respectively for *S. aureus, S. enterica serovar Typhi, Acetobacter calcoacticus,* and *C. albicans.*	[210]
Pimenta dioica L.	Alcoholic and hexane extracts	*P. fluorescens, B. megaterium, A. niger, Penicillium sp.*	Alcoholic and hexane extracts of *P. dioica* exerted significant inhibitory effects on both the bacteria and fungi.	[214]
Pimpinella anisum L.	EO of fruit	*A. alternate, A. niger, A. parasiticus*	The most sensitive fungus for *P. anisum* oil was *A. parasiticus.*	[215]
Pimpinella anisum L.	EO	16 microorganisms	*P. anisum* EO exhibited strong antifungal activities against *R. glutinis, A. ochraceus,* and *F. moniliforme.*	[78]
Pimpinella anisum L.	EO	*C. lipolytica, H. uvarum, Pichia membranaefaciens, R. glutinis, S. pombe, Z. rouxii, A. flavus, A. ochraceus, A. parasiticus, F. moniliforme*	*P. anisum* EO completely inhibited the growth of tested fungi.	[78]
Piper capense	EO	*S. aureus, E. faecalis, C. albicans*	*P. capense* showed moderate activities against tested microorganisms.	[216]
Piper guineense	powder	*B. cereus, Bacillus coagulans, B. enterobacter sp., A. niger, R. stolonifer*	*P. guineense* inhibited *R. stolonifer* at concentrations above 0.5%.	[217]
Phlomis oppositiflora	Methanol, ethanol, ethyl acetate extracts and EO	*E. coli, S. aureus, K. pneumonia, M. smegmatis, P. aeruginosa, E. cloacae, B. megaterium, M. luteus, R. rubra, C. albicans, K. marxianus*	*P. oppositiflora* contains antimicrobial components against various microorganisms.	[218]
Ramalina species	Acetone, methanol and ethanol extracts	*E. coli* and *S. aureus*	The MICs of all extracts ranged from 64 to 512 g/mL for all bacterial strains tested.	[219]
Rhus coriaria L.	80% (v/v) aqueous alcohol extract	*S. aureus, B. cereus, E. coli, S. typhimurium, P. vulgaris, S. flexneri*	The MICs of *R. coriaria* extract against the tested bacteria ranged from 0.04% to 0.2%.	[220]
Rhus coriaria	Water extract	*B. cereus, L. monocytogenes, E. coli, S. typhimurium*	*R. coriaria* extract was the most effective against the four bacteria tested.	[221]
Salvia officinalis L.	EO	13 bacterial strains and 6 fungi	Sage EO was more effective against *E. coli, S. typhimurium, S. enteritidis,* and *S. sonei.*	[92]
Salvia officinalis L. (sage)	80% ethanol extract	*Campylobacter coli, E. coli, Streptococcus infantis, B. cereus, L. monocytogenes, S. aureus*	Sage extract showed the best antibacterial activities compared with four other plants, especially against Gram-positive bacteria and *C. coli.*	[222]

Table 15. *Cont.*

Spices	Type of Samples	Bacteria and Fungi	Main Results	Reference
Salvia officinalis L.	EO	*E. coli, P. aeruginosa, Enterobacter* sp., *S. aureus*	Microwave-EO of *S. officinalis* possessed good antibacterial activities than the hydrodistilled oil.	[223]
Salvia leriifolia	Methanol extract	*S. aureus*	*S. leriifolia* extract exhibited antimicrobial activity against *S. aureus*.	[224]
Santolina chamaecyparissus L.	EO	*K. pneumonia* and *C. albicans*	*S. chamaecyparissus* EO was very active against the two microorganisms listed.	[225]
Satureja cuneifolia Ten.	EO	*E. coli, Campylobacter jejuni, S. sonnei, S. aureus, L. monocytogenes, B. cereus, P. aeruginosa, S. enteritidis*	MICs of *S. cuneifolia* EO for tested bacteria were in the range of 600–1400 µg/mL.	[226]
Satureja kitaibelii	EO	30 pathogenic microorganisms	*S. kitaibelii* EO showed significant activities against foodborne microbes (MIC: 0.18–25.5 µg/mL), multiresistant bacterial isolates (MIC: 6.25–50.0 µg/mL), and dermatophyte strains (MIC: 12.5–50.0 µg/mL).	[227]
Satureja wiedemanniana	EO	37 *Bacillus* strains	Both *S. wiedemanniana* EO and its main component p-cymene exhibited strong antimicrobial activities against some *Bacillus* strains.	[228]
Satureja species	EOs	*A. niger, Penicillium digitatum, B. cinerea, R. stolonifer*	The EOs exhibited fungicidal activities against *P. digitatum, B. cinereal,* and *R. stolonifer.*	[229]
Silene laxa	Ethyl acetate, chloroform, methanol, ethanol and acetone extract	*P. aeruginosa, E. cloacae, B. megaterium, E. cloacae, S. aureus*	*S. laxa* leaves ethanol extract showed the best activities against *P. aeruginosa, E. cloacae, B. megaterium,* while the methanol extracts of *S. laxa* fruits showed the best antibacterial activity against *B.megaterium.*	[230]
Summer savory	-	*A. niger, A. alternate, A. parasiticus*	0.5% summer savory extract showed 100% inhibition till the seventh day of incubation.	[231]
Syzygium aromaticum L.	Water extract	*S. aureus, S. epidermidis, S. pyogenes, S. agalactiae, S. enterica serovar Typhi, P. aeruginosa, Acetobacter calcoaceticus, C. albicans, P. vulgaris, S. faecalis, S. dysenteriae, K. pneumoniae, E. coli, V. cholera, C. xerosis*	*S. aromaticum* water extract showed antibacterial activities with MICs in the range of 31.25–250 µg/mL for *S. aureus, S. epidermidis, S. pyogenes, S. enterica serovar Typhi, Acetobacter calcoacticus,* and *P. aeruginosa.*	[210]
Thymbra spicata L.	Decoction	*F. oxysporum f. sp. phaseoli, M. phaseoli, B. cinerea, R. solani, A. solani, A. parasiticus*	*T. spicata* completely inhibited the mycelial growth of fungi and showed a complete fungicidal effect on molds.	[39]

Table 15. *Cont.*

Spices	Type of Samples	Bacteria and Fungi	Main Results	Reference
Thymus capitata	EO	*L. monocytogenes*	MICs ranged from 0.32 to 20 mg/mL.	[232]
Thymus capitatus	EO	*L. innocua, S. marcescens, P. fragi, P. fluorescens, A. hydrophila, Shewanella putrefaciens, Achromobacter denitrificans, E. amnigenus, E. gergoviae, Alcaligenes faecalis, Leuconostoc carnosum*	*T. capitatus* EOs showed inhibitory effects on the 10 tested bacteria with MICs ranging from 1.87 to 7.5 μL/mL.	[233]
Thymus cappadocicus Boiss.	EO	13 bacteria and 2 yeasts	*T. cappadocicus* EO showed great antimicrobial activities against microorganisms tested.	[234]
Thymus eigii	EO	*M. luteus, B. megaterium, B. brevis, E. faecalis, P. pyocyaneus, M. smegmatis, E. coli, A. hydrophila, Y. enterocolitica, S. aureus, S. faecalis, S. cerevisiae, K. fragilis*	*T. eigii* EO showed the highest antimicrobial activities compared with two other plants.	[235]
Thymus piperella	EO	*L. innocua, S. marcescens , P. fragi, P. fluorescens, A. hydrophila, S. putrefaciens, A. denitrificans, E. amnigenus, E. gergoviae, A. faecalis, L. carnosum*	*T. piperella* EO had inhibitory effects on 5 of the 11 bacteria tested.	[236]
Thymus serpyllum	EO	*Penicillium* sp., *Alternaria* sp., *Aureobasidium* sp.	8 mg/disc EO of *T. serpyllum* has a good efficiency by inhibiting the germination of spores from 80% to 100%.	[237]
Trachyspermum ammi L.	EO	*A. niger, A. flavus, A. oryzae, A. ochraceus, F. monoliforme, F. graminearum, Pencillium citrium, P. viridicatum, P. madriti, C. lunata*	*T. ammi* EO exhibited a broad spectrum of fungi toxic behavior against all tested fungi.	[238]
Xylopia aethiopica	-	*Sclerotium rolfsii*	*X. aethiopica* extract was the most effective against *S. rolfsii* compared with four other spices.	[239]
Zanthoxylum piperitum	Polymeric procyanidin	*S. aureus*	A polymeric proanthocyanidin purified from the fruit of *Z. piperitum*, noticeably decreased the MICs of β-lactam antibiotics for MRSA.	[240]
Zanthoxylum schinifolium	EO	*S. aureus, S. epidermidis, B. subtilis, S. typhimurium, P. aeruginosa, S. dysenteriae, E. coli*	*Z. schinifolium* EO was particularly strong against *S. epidermidis*, with MIC 2.5 mg/mL.	[241]
Zataria multiflora Boiss.	80% (v/v) aqueous alcohol extract	*S. aureus, B. cereus, E. coli, S. typhimurium, P. vulgaris, S. flexneri*	The MICs of *Z. multiflora* against the tested bacteria ranged from 0.4% to 0.8%.	[220]

16. Conclusions

The antibacterial and antifungal activities of commonly used spices have been summarized. Several spices—such as clove, oregano, thyme, cinnamon, and cumin—have exhibited significant antimicrobial activities against food spoilage bacteria like *B. subtilis* and *P. fluorescens*; pathogens like *S. aureus*, *V. parahaemolyticus*, and *S. typhimurium*; harmful fungi like *A. flavus* and *A. niger*; and even antibiotic resistant microorganisms such as MRSA. Therefore, these spices could be used to decrease the possibility of food poisoning and spoilage, to increase the food safety and shelf-life of products, and to treat some infectious diseases. In the future, as the combinations of several spices were proven to possess higher inhibitory effects on specific bacteria than those of individual spices, the interactions of more spices should be studied and evaluated to inhibit different microorganisms in different food products. Additionally, spices could be used in food packaging as published, but more studies are required to take the other aspects into consideration, such as how to prevent odor/flavor transferring from packages containing natural spice extracts to the packaged foods. Furthermore, spice products may be considered as an alternative to common antibiotics to treat infectious diseases. As the majority of the studies focused on the in vitro activities of spices against human pathogenic bacteria, in vivo studies and clinical trials are needed to be conducted in future. The mechanisms of antimicrobial action of spices remain to be clarified in order to make the best use of spices. Furthermore, the potential toxicity of spices on humans should be evaluated.

Acknowledgments: This work was supported by the National Natural Science Foundation of China (No. 81372976), Key Project of Guangdong Provincial Science and Technology Program (No. 2014B020205002), and the Hundred-Talents Scheme of Sun Yat-Sen University.

Author Contributions: Qing Liu and Hua-Bin Li conceived this paper; Qing Liu, Xiao Meng, Ya Li, Cai-Ning Zhao, and Guo-Yi Tang wrote this paper; Hua-Bin Li revised the paper.

Conflicts of Interest: The authors declare no conflict of interest.

References

1. Miladi, H.; Zmantar, T.; Chaabouni, Y.; Fedhila, K.; Bakhrouf, A.; Mandouani, K.; Chaieb, K. Antibacterial and efflux pump inhibitors of thymol and carvacrol against food-borne pathogens. *Microb. Pathog.* **2016**, *99*, 95–100. [CrossRef] [PubMed]

2. Brul, S.; Coote, P. Preservative agents in foods. Mode of action and microbial resistance mechanisms. *Int. J. Food Microbiol.* **1999**, *50*, 1–17. [CrossRef]

3. De Souza, E.L.; Stamford, T.; Lima, E.D.; Trajano, V.N.; Barbosa, J.M. Antimicrobial effectiveness of spices: An approach for use in food conservation systems. *Braz. Arch. Biol. Technol.* **2005**, *48*, 549–558. [CrossRef]

4. Silva, F.; Domingues, F.C. Antimicrobial activity of coriander oil and its effectiveness as food preservative. *Crit. Rev. Food Sci. Nutr.* **2017**, *57*, 35–47. [CrossRef] [PubMed]

5. Tajkarimi, M.M.; Ibrahim, S.A.; Cliver, D.O. Antimicrobial herb and spice compounds in food. *Food Control* **2010**, *21*, 1199–1218. [CrossRef]

6. Nabavi, S.M.; Marchese, A.; Izadi, M.; Curti, V.; Daglia, M.; Nabavi, S.F. Plants belonging to the genus *Thymus* as antibacterial agents: From farm to pharmacy. *Food Chem.* **2015**, *173*, 339–347. [CrossRef] [PubMed]

7. Marchese, A.; Barbieri, R.; Sanches-Silva, A.; Daglia, M.; Nabavi, S.F.; Jafari, N.J.; Izadi, M.; Ajami, M.; Nabavi, S.M. Antifungal and antibacterial activities of allicin: A review. *Trends Food Sci. Technol.* **2016**, *52*, 49–56. [CrossRef]

8. Paphitou, N.I. Antimicrobial resistance: Action to combat the rising microbial challenges. *Int. J. Antimicrob. Agents* **2013**, *42*, S25–S28. [CrossRef] [PubMed]

9. Högberg, L.D.; Heddini, A.; Cars, O. The global need for effective antibiotics: Challenges and recent advances. *Trends Pharmacol. Sci.* **2010**, *31*, 509–515. [CrossRef] [PubMed]

10. Lai, P.K.; Roy, J. Antimicrobial and chemopreventive properties of herbs and spices. *Curr. Med. Chem.* **2004**, *11*, 1451–1460. [CrossRef] [PubMed]

11. Nabavi, S.F.; di Lorenzo, A.; Izadi, M.; Sobarzo-Sánchez, E.; Daglia, M.; Nabavi, S.M. Antibacterial effects of cinnamon: From farm to food, cosmetic and pharmaceutical industries. *Nutrients* **2015**, *7*, 7729–7748. [CrossRef] [PubMed]

12. Zheng, J.; Zhou, Y.; Li, Y.; Xu, D.P.; Li, S.; Li, H.B. Spices for prevention and treatment of cancers. *Nutrients* **2016**, *8*, 495. [CrossRef] [PubMed]

13. De, M.; De, A.K.; Banerjee, A.B. Antimicrobial screening of some Indian spices. *Phytother. Res.* **1999**, *13*, 616–618. [CrossRef]

14. Arora, D.S.; Kaur, J. Antimicrobial activity of spices. *Int. J. Antimicrob. Agents* **1999**, *12*, 257–262. [CrossRef]

15. Chaieb, K.; Hajlaoui, H.; Zmantar, T.; Kahla-Nakbi, A.B.; Rouabhia, M.; Mahdouani, K.; Bakhrouf, A. The chemical composition and biological activity of clove essential oil, *Eugenia caryophyllata* (*Syzigium aromaticum* L. Myrtaceae): A short review. *Phytother. Res.* **2007**, *21*, 501–506. [CrossRef] [PubMed]

16. Xu, J.G.; Liu, T.; Hu, Q.P.; Cao, X.M. Chemical composition, antibacterial properties and mechanism of action of essential oil from clove buds against *Staphylococcus aureus*. *Molecules* **2016**, *21*, 1194. [CrossRef] [PubMed]

17. Lomarat, P.; Phanthong, P.; Wongsariya, K.; Chomnawang, M.T.; Bunyapraphatsara, N. Bioautography-guided isolation of antibacterial compounds of essential oils from Thai spices against histamine-producing bacteria. *Pak. J. Pharm. Sci.* **2013**, *26*, 473–477. [PubMed]

18. Nassan, M.A.; Mohamed, E.H.; Abdelhafez, S.; Ismail, T.A. Effect of clove and cinnamon extracts on experimental model of acute hematogenous pyelonephritis in albino rats: Immunopathological and antimicrobial study. *Int. J. Immunopathol. Pharmacol.* **2015**, *28*, 60–68. [CrossRef] [PubMed]

19. Kuang, X.; Li, B.; Kuang, R.; Zheng, X.D.; Zhu, B.; Xu, B.L.; Ma, M.H. Granularity and antibacterial activities of ultra-fine cinnamon and clove powders. *J. Food Saf.* **2011**, *31*, 291–296. [CrossRef]

20. Burt, S. Essential oils: Their antibacterial properties and potential applications in foods—A review. *Int. J. Food Microbiol.* **2004**, *94*, 223–253. [CrossRef] [PubMed]

21. Badei, A.; Faheld, S.; El-Akel, A.; Mahmoud, B. Application of some spices in flavoring and preservation of cookies: 2-Antimicrobial and sensory properties of cardamom, cinnamon and clove. *Dtsch. Lebensm. Rundsch.* **2002**, *98*, 261–265.

22. Schmidt, E.; Jirovetz, L.; Wlcek, K.; Buchbauer, G.; Gochev, V.; Girova, T.; Stoyanova, A.; Geissler, M. Antifungal activity of eugenol and various eugenol-containing essential oils against 38 clinical isolates of *Candida albicans*. *J. Essent. Oil Bear. Plants* **2007**, *10*, 421–429. [CrossRef]

23. Angienda, P.O.; Onyango, D.M.; Hill, D.J. Potential application of plant essential oils at sub-lethal concentrations under extrinsic conditions that enhance their antimicrobial effectiveness against pathogenic bacteria. *Afr. J. Microbiol. Res.* **2010**, *4*, 1678–1684.

24. Shan, B.; Cai, Y.Z.; Brooks, J.D.; Corke, H. Antibacterial and antioxidant effects of five spice and herb extracts as natural preservatives of raw pork. *J. Sci. Food Agric.* **2009**, *89*, 1879–1885. [CrossRef]

25. Bayoub, K.; Baibai, T.; Mountassif, D.; Retmane, A.; Soukri, A. Antibacterial activities of the crude ethanol extracts of medicinal plants against *Listeria monocytogenes* and some other pathogenic strains. *Afr. J. Biotechnol.* **2010**, *9*, 4251–4258.

26. Cui, H.Y.; Gabriel, A.A.; Nakano, H. Antimicrobial efficacies of plant extracts and sodium nitrite against *Clostridium botulinum*. *Food Control* **2010**, *21*, 1030–1036. [CrossRef]

27. Keskin, D.; Toroglu, S. Studies on antimicrobial activities of solvent extracts of different spices. *J. Environ. Biol.* **2011**, *32*, 251–256. [PubMed]

28. Liang, Z.W.; Cheng, Z.H.; Mittal, G.S. Inactivation of microorganisms in apple cider using spice powders, extracts and oils as antimicrobials with and without low-energy pulsed electric field. *J. Food Agric. Environ.* **2003**, *1*, 28–33.

29. Badhe, S.R.; Fairoze, M.N. Antibacterial efficacy of clove powder of chicken legs spiked with pathogenic reference strains under refrigeration temperature (8 ± 2 °C). *Indian J. Anim. Res.* **2012**, *46*, 371–375.

30. Hernández-Ochoa, L.; Gonzales-Gonzales, A.; Gutierrez-Mendez, N.; Munoz-Castellanos, L.N.; Quintero-Ramos, A. Study of the antibacterial activity of chitosan-based films prepared with different molecular weights including spices essential oils and functional extracts as antimicrobial agents. *Rev. Mex. Ing. Quim.* **2011**, *10*, 455–463.

31. Hernández-Ochoa, L.; Macías-Castañeda, C.A.; Nevárez-Moorillón, G.V.; Salas-Muñoz, E.; Sandoval-Salas, F. Antimicrobial activity of chitosan-based films including spices' essential oils and functional extracts. *CyTA J. Food* **2012**, *10*, 85–91. [CrossRef]

32.   Liu, G.Q.; Zhang, L.L.; Zong, K.; Wang, A.M.; Yu, X.F. Effects of spices essential oils on the spoilage-related microbiota in chilled pork stored in antimicrobial pack. *Food Sci. Technol. Res.* **2012**, *18*, 695–704. [CrossRef]
33.   Rodriguez-Garcia, I.; Silva-Espinoza, B.A.; Ortega-Ramirez, L.A.; Leyva, J.M.; Siddiqui, M.W.; Cruz-Valenzuela, M.R.; Gonzalez-Aguilar, G.A.; Ayala-Zavala, J.F. Oregano essential oil as an antimicrobial and antioxidant additive in food products. *Crit. Rev. Food Sci. Nutr.* **2016**, *56*, 1717–1727. [CrossRef] [PubMed]
34.   Babacan, O.; Cengiz, S.; Akan, M. Detection of antibacterial effect of oregano plant on various *Salmonella* serotypes. *Ank. Univ. Vet. Fak. Derg.* **2012**, *59*, 103–106.
35.   Santoyo, S.; Cavero, S.; Jaime, L.; Ibanez, E.; Senorans, F.J.; Reglero, G. Supercritical carbon dioxide extraction of compounds with antimicrobial activity from *Origanum vulgare* L.: Determination of optimal extraction parameters. *J. Food Prot.* **2006**, *69*, 369–375. [CrossRef] [PubMed]
36.   De Souza, E.L.; Stamford, T.; Lima, E.; Barbosa, J.M.; Marques, M. Interference of heating on the antimicrobial activity and chemical composition of *Origanum vulgare* L. (Lamiaceae) essential oil. *Cienc. Tecnol. Aliment.* **2008**, *28*, 418–422. [CrossRef]
37.   Lima, I.O.; Pereira, F.D.O.; de Oliveira, W.A.; Lima, E.D.O.; Menezes, E.A.; Cunha, F.A.; Formiga Melo Diniz, M.D.F. Antifungal activity and mode of action of carvacrol against *Candida albicans* strains. *J. Essent. Oil Res.* **2013**, *25*, 138–142. [CrossRef]
38.   Ultee, A.; Kets, E.; Smid, E.J. Mechanisms of action of carvacrol on the food-borne pathogen *Bacillus cereus*. *Appl. Environ. Microb.* **1999**, *65*, 4606–4610.
39.   Ozcan, M.; Boyraz, N. Antifungal properties of some herb decoctions. *Eur. Food Res. Technol.* **2000**, *212*, 86–88.
40.   Ai-Turki, A.I.; Ei-Ziney, M.G.; Abdel-Salam, A.M. Chemical and anti-bacterial characterization of aqueous extracts of oregano, marjoram, sage and licorice and their application in milk and labneh. *J. Food Agric. Environ.* **2008**, *6*, 39–44.
41.   Marques, J.D.; Volcao, L.M.; Funck, G.D.; Kroning, I.S.; da Silva, W.P.; Fiorentini, A.M.; Ribeiro, G.A. Antimicrobial activity of essential oils of *Origanum vulgare* L. and *Origanum majorana* L. against *Staphylococcus aureus* isolated from poultry meat. *Ind. Crop. Prod.* **2015**, *77*, 444–450. [CrossRef]
42.   Bozin, B.; Mimica-Dukic, N.; Simin, N.; Anackov, G. Characterization of the volatile composition of essential oils of some Lamiaceae spices and the antimicrobial and antioxidant activities of the entire oils. *J. Agric. Food Chem.* **2006**, *54*, 1822–1828. [CrossRef] [PubMed]
43.   Viuda-Martos, M.; Ruiz-Navajas, Y.; Fernández-López, J.; Pérez-Álvarez, J.A. Antibacterial activity of different essential oils obtained from spices widely used in Mediterranean diet. *Int. J. Food Sci. Technol.* **2008**, *43*, 526–531. [CrossRef]
44.   Santurio, D.F.; da Costa, M.M.; Maboni, G.; Cavalheiro, C.P.; de Sá, M.F.; Dal Pozzo, M.; Alves, S.H.; Fries, L. Antimicrobial activity of spice essential oils against *Escherichia coli* strains isolated from poultry and cattle. *Cienc. Rural* **2011**, *41*, 1051–1056. [CrossRef]
45.   Khosravi, A.R.; Shokri, H.; Kermani, S.; Dakhili, M.; Madani, M.; Parsa, S. Antifungal properties of *Artemisia sieberi* and *Origanum vulgare* essential oils against *Candida glabrata* isolates obtained from patients with vulvovaginal candidiasis. *J. Mycol. Med.* **2011**, *21*, 93–99. [CrossRef]
46.   Dal Pozzo, M.; Viégas, J.; Santurio, D.F.; Rossatto, L.; Soares, I.H.; Alves, S.H.; da Costa, M.M. Antimicrobial activities of essential oils extracted from spices against *Staphylococcus* spp. isolated from goat mastitis. *Cienc. Rural* **2011**, *41*, 667–672. [CrossRef]
47.   Santos, J.C.; Carvalho, C.D.; Barros, T.F.; Guimaraes, A.G. In vitro antimicrobial activity of essential oils from oregano, garlic, clove and lemon against pathogenic bacteria isolated from *Anomalocardia brasiliana*. *Semin. Cienc. Agrar.* **2011**, *32*, 1557–1564.
48.   Hyun, J.E.; Bae, Y.M.; Song, H.; Yoon, J.H.; Lee, S.Y. Antibacterial effect of various essential oils against pathogens and spoilage microorganisms in fresh produce. *J. Food Saf.* **2015**, *35*, 206–219. [CrossRef]
49.   Zivanovic, S.; Chi, S.; Draughon, A.F. Antimicrobial activity of chitosan films enriched with essential oils. *J. Food Sci.* **2005**, *70*, M45–M51. [CrossRef]
50.   Assiri, A.M.A.; Elbanna, K.; Abulreesh, H.H.; Ramadan, M.F. Bioactive compounds of cold-pressed thyme (*Thymus vulgaris*) oil with antioxidant and antimicrobial properties. *J. Oleo Sci.* **2016**, *65*, 629–640. [CrossRef] [PubMed]

51. Al Maqtari, M.; Alghalibi, S.M.; Alhamzy, E.H. Chemical composition and antimicrobial activity of essential oil of *Thymus vulgaris* from Yemen. *Turk. J. Biochem.* **2011**, *36*, 342–349.

52. Marino, M.; Bersani, C.; Comi, G. Antimicrobial activity of the essential oils of *Thymus vulgaris* L. measured using a bioimpedometric method. *J. Food Prot.* **1999**, *62*, 1017–1023. [CrossRef] [PubMed]

53. Varga, E.; Bardocz, A.; Belak, A.; Maraz, A.; Boros, B.; Felinger, A.; Boszormenyi, A.; Horvath, G. Antimicrobial activity and chemical composition of thyme essential oils and the polyphenolic content of different thymus extracts. *Farmacia* **2015**, *63*, 357–361.

54. Tiwari, B.K.; Valdramidis, V.P.; O'Donnell, C.P.; Muthukumarappan, K.; Bourke, P.; Cullen, P.J. Application of natural antimicrobials for food preservation. *J. Agric. Food Chem.* **2009**, *57*, 5987–6000. [CrossRef] [PubMed]

55. Al-Turki, A.I. Antibacterial effect of thyme, peppermint, sage, black pepper and garlic hydrosols against *Bacillus subtilis* and *Salmonella enteritidis*. *J. Food Agric. Environ.* **2007**, *5*, 92–94.

56. Hussien, J.; Teshale, C.; Mohammed, J. Assessment of the antimicrobial effects of some ethiopian aromatic spice and herb hydrosols. *Int. J. Pharmacol.* **2011**, *7*, 635–640. [CrossRef]

57. Girova, T.; Gochev, V.; Jirovetz, L.; Buchbauer, G.; Schmidt, E.; Stoyanova, A. Antimicrobial activity of essential oils from spices against psychrotrophic food spoilage microorganisms. *Biotechnol. Biotechnol. Equip.* **2010**, *24*, 547–552. [CrossRef]

58. Hajlaoui, H.; Snousi, M.; Noumi, E.; Zaneti, S.; Ksouri, R.; Bakhrouf, A. Chemical composition, antioxidant and antibacterial activities of the essential oils of five Tunisian aromatic plants. *Ital. J. Food Sci.* **2010**, *22*, 320–329.

59. Viuda-Martos, M.; Mohamady, M.A.; Fernández-López, J.; Abd ElRazik, K.A.; Omer, E.A.; Perez-Alvarez, J.A.; Sendra, E. In vitro antioxidant and antibacterial activities of essentials oils obtained from Egyptian aromatic plants. *Food Control* **2011**, *22*, 1715–1722. [CrossRef]

60. Aliakbarlu, J.; Shameli, F. In vitro antioxidant and antibacterial properties and total phenolic contents of essential oils from *Thymus vulgaris*, *T. Kotschyanus*, *Ziziphora tenuior* and *Z. Clinopodioides*. *Turk. J. Biochem.* **2013**, *38*, 425–431. [CrossRef]

61. Bagamboula, C.F.; Uyttendaele, M.; Debevere, J. Antimicrobial effect of spices and herbs on *Shigella sonnei* and *Shigella flexneri*. *J. Food Prot.* **2003**, *66*, 668–673. [CrossRef] [PubMed]

62. Martins, N.; Barros, L.; Santos-Buelga, C.; Silva, S.; Henriques, M.; Ferreira, I. Decoction, infusion and hydroalcoholic extract of cultivated thyme: Antioxidant and antibacterial activities, and phenolic characterisation. *Food Chem.* **2015**, *167*, 131–137. [CrossRef] [PubMed]

63. Ozturk, I. Antifungal activity of propolis, thyme essential oil and hydrosol on natural mycobiota of sucuk, a turkish fermented sausage: Monitoring of their effects on microbiological, color and aroma properties. *J. Food Process Preserv.* **2015**, *39*, 1148–1158. [CrossRef]

64. Aman, S.; Naim, A.; Siddiqi, R.; Naz, S. Antimicrobial polyphenols from small tropical fruits, tea and spice oilseeds. *Food Sci. Technol. Int.* **2014**, *20*, 241–251. [CrossRef] [PubMed]

65. Aznar, A.; Fernandez, P.S.; Periago, P.M.; Palop, A. Antimicrobial activity of nisin, thymol, carvacrol and cymene against growth of *Candida lusitaniae*. *Food Sci. Technol. Int.* **2015**, *21*, 72–79. [CrossRef] [PubMed]

66. Ranasinghe, P.; Jayawardana, R.; Galappaththy, P.; Constantine, G.R.; de Vas Gunawardana, N.; Katulanda, P. Efficacy and safety of "true" cinnamon (*Cinnamomum zeylanicum*) as a pharmaceutical agent in diabetes: A systematic review and meta-analysis. *Diabet. Med.* **2012**, *29*, 1480–1492. [CrossRef] [PubMed]

67. Khasnavis, S.; Pahan, K. Sodium benzoate, a metabolite of cinnamon and a food additive, upregulates neuroprotective Parkinson disease protein DJ-1 in astrocytes and neurons. *J. Neuroimmune Pharm.* **2012**, *7*, 424–435. [CrossRef] [PubMed]

68. Brierley, S.M.; Kelber, O. Use of natural products in gastrointestinal therapies. *Curr. Opin. Pharmacol.* **2011**, *11*, 604–611. [CrossRef] [PubMed]

69. Al-Jiffri, O.; El-Sayed, Z.M.F.; Al-Sharif, F.M. Urinary tract infection with *Esherichia coli* and antibacterial activity of some plants extracts. *Int. J. Microbiol. Res.* **2011**, *2*, 1–7.

70. Gupta, C.; Garg, A.P.; Uniyal, R.C.; Kumari, A. Comparative analysis of the antimicrobial activity of cinnamon oil and cinnamon extract on some food-borne microbes. *Afr. J. Microbiol. Res.* **2008**, *2*, 247–251.

71. Ceylan, E.; Fung, D.; Sabah, J.R. Antimicrobial activity and synergistic effect of cinnamon with sodium benzoate or potassium sorbate in controlling *Escherichia coli* O157: H7 in apple juice. *J. Food Sci.* **2004**, *69*, M102–M106. [CrossRef]

72. Cui, H.Y.; Li, W.; Li, C.Z.; Vittayapadung, S.; Lin, L. Liposome containing cinnamon oil with antibacterial activity against methicillin-resistant *Staphylococcus aureus* biofilm. *Biofouling* **2016**, *32*, 215–225. [CrossRef] [PubMed]

73. Shreaz, S.; Wani, W.A.; Behbehani, J.M.; Raja, V.; Irshad, M.; Karched, M.; Ali, I.; Siddiqi, W.A.; Hun, L.T. Cinnamaldehyde and its derivatives, a novel class of antifungal agents. *Fitoterapia* **2016**, *112*, 116–131. [CrossRef] [PubMed]

74. Mvuemba, H.N.; Green, S.E.; Tsoomo, A.; Avis, T.J. Antimicrobial efficacy of cinnamon, ginger, horseradish and nutmeg extracts against spoilage pathogens. *Phytoprotection* **2009**, *90*, 65–70. [CrossRef]

75. Wang, Q.; Ou, Z.B.; Lei, H.W.; Zeng, X.H.; Ying, Y.; Bai, W.D. Antimicrobial activities of a new formula of spice water extracts against foodborne bacteria. *J. Food Process Preserv.* **2012**, *36*, 374–381. [CrossRef]

76. Agaoglu, S.; Dostbil, N.; Alemdar, S. Antimicrobial activity of some spices used in the meat industry. *Bull. Vet. Inst. Pulawy* **2007**, *51*, 53–57.

77. Revati, S.; Bipin, C.; Chitra, P.; Minakshi, B. In vitro antibacterial activity of seven Indian spices against high level gentamicin resistant strains of enterococci. *Arch. Med. Sci.* **2015**, *11*, 863–868. [CrossRef] [PubMed]

78. Nanasombat, S.; Wimuttigosol, P. Antimicrobial and antioxidant activity of spice essential oils. *Food Sci. Biotechnol.* **2011**, *20*, 45–53. [CrossRef]

79. Trajano, V.N.; Lima, E.D.; de Souza, E.L.; Travassos, A. Antibacterial property of spice essential oils on food contaminating bacteria. *Ciencia Tecnol. Aliment.* **2009**, *29*, 542–545. [CrossRef]

80. Reza, Z.M.; Atefeh, J.Y.; Faezeh, F. Effect of ā-irradiation on the antibacterial activities of *Cuminum cyminum* L. essential oils in vitro and in vivo systems. *J. Essent. Oil Bear. Plants* **2015**, *18*, 582–591. [CrossRef]

81. Allahghadri, T.; Rasooli, I.; Owlia, P.; Nadooshan, M.J.; Ghazanfari, T.; Taghizadeh, M.; Astaneh, S. Antimicrobial property, antioxidant capacity, and cytotoxicity of essential oil from cumin produced in Iran. *J. Food Sci.* **2010**, *75*, H54–H61. [CrossRef] [PubMed]

82. Kedia, A.; Prakash, B.; Mishra, P.K.; Dubey, N.K. Antifungal and antiaflatoxigenic properties of *Cuminum cyminum* (L.) seed essential oil and its efficacy as a preservative in stored commodities. *Int. J. Food Microbiol.* **2014**, *168*, 1–7. [CrossRef] [PubMed]

83. Abd El Mageed, M.A.; Mansour, A.F.; El Massry, K.F.; Ramadan, M.M.; Shaheen, M.S.; Shaaban, H. Effect of microwaves on essential oils of coriander and cumin seeds and on their antioxidant and antimicrobial activities. *J. Essent. Oil Bear. Plants* **2012**, *15*, 614–627. [CrossRef]

84. Chaudhry, N.; Tariq, P. In vitro antibacterial activities of kalonji, cumin and poppy seed. *Pak. J. Bot.* **2008**, *40*, 461–467.

85. Iacobellis, N.S.; Lo Cantore, P.; Capasso, F.; Senatore, F. Antibacterial activity of *Cuminum cyminum* L. and *Carum carvi* L. essential oils. *J. Agric. Food Chem.* **2005**, *53*, 57–61. [CrossRef] [PubMed]

86. Ozcan, M.; Erkmen, O. Antimicrobial activity of the essential oils of Turkish plant spices. *Eur. Food Res. Technol.* **2001**, *212*, 658–660.

87. Stefanini, M.B.; Figueiredo, R.O.; Ming, L.C.; Junior, A.F. Antimicrobial activity of the essential oils of some spice herbs. In Proceedings of the International Conference on Medicinal and Aromatic Plants Possibilities and Limitations of Medicinal and Aromatic Plant Production in the 21st Century, Budapest, Hungary, 8–11 July 2001; Szoke, E., Mathe, I., Blunden, G., Kery, A., Eds.; International Society for Horticultural Science: Leuven, Belgium, 2003; pp. 215–216.

88. Toroglu, S. In vitro antimicrobial activity and synergistic/antagonistic effect of interactions between antibiotics and some spice essential oils. *J. Environ. Biol.* **2011**, *32*, 23–29. [PubMed]

89. Bag, A.; Chattopadhyay, R.R. Evaluation of synergistic antibacterial and antioxidant efficacy of essential oils of spices and herbs in combination. *PLoS ONE* **2015**, *10*, e0131321. [CrossRef] [PubMed]

90. Azizkhani, M.; Tooryan, F. Antioxidant and antimicrobial activities of rosemary extract, mint extract and a mixture of tocopherols in beef sausage during storage at 4C. *J. Food Saf.* **2015**, *35*, 128–136. [CrossRef]

91. Tavassoli, S.K.; Mousavi, S.M.; Emam-Djomeh, Z.; Razavi, S.H. Chemical composition and evaluation of antimicrobial properties of *Rosmarinus officinalis* L. essential oil. *Afr. J. Biotechnol.* **2011**, *10*, 13895–13899.

92. Bozin, B.; Mlmica-Dukic, N.; Samojlik, I.; Jovin, E. Antimicrobial and antioxidant properties of rosemary and sage (*Rosmarinus officinalis* L. and *Salvia officinalis* L., Lamiaceae) essential oils. *J. Agric. Food Chem.* **2007**, *55*, 7879–7885. [CrossRef] [PubMed]

93. Weerakkody, N.S.; Caffin, N.; Turner, M.S.; Dykes, G.A. In vitro antimicrobial activity of less-utilized spice and herb extracts against selected food-borne bacteria. *Food Control* **2010**, *21*, 1408–1414. [CrossRef]

94. Krajcova, E.; Greifova, M.; Schmidt, S. Study of antimicrobial activity of selected plant extracts against bacterial food contaminants. *J. Food Nutr. Res.* **2008**, *47*, 125–130.

95. Zhang, H.Y.; Kong, B.H.; Xiong, Y.; Sun, X. Antimicrobial activities of spice extracts against pathogenic and spoilage bacteria in modified atmosphere packaged fresh pork and vacuum packaged ham slices stored at 4 °C. *Meat Sci.* **2009**, *81*, 686–692. [CrossRef] [PubMed]

96. Kozlowska, M.; Laudy, A.E.; Przybyl, J.; Ziarno, M.; Majewska, E. Chemical composition and antibacterial activity of some medicinal plants from Lamiaceae family. *ACTA Pol. Pharm.* **2015**, *72*, 757–767. [PubMed]

97. Weerakkody, N.S.; Caffin, N.; Dykes, G.A.; Turner, M.S. Effect of antimicrobial spice and herb extract combinations on *Listeria monocytogenes*, *Staphylococcus aureus*, and *Spoilage Microflora* growth on cooked ready-to-eat vacuum-packaged shrimp. *J. Food Prot.* **2011**, *74*, 1119–1125. [CrossRef] [PubMed]

98. Nejad, A.; Shabani, S.; Bayat, M.; Hosseini, S.E. Antibacterial effect of garlic aqueous extract on *Staphylococcus aureus* in hamburger. *Jundishapur J. Microbiol.* **2014**, *7*, e13134.

99. Rahman, M.S.; Al-Sheibani, H.I.; Al-Riziqi, M.H.; Mothershaw, A.; Guizani, N.; Bengtsson, G. Assessment of the anti-microbial activity of dried garlic powders produced by different methods of drying. *Int. J. Food Prop.* **2006**, *9*, 503–513. [CrossRef]

100. Sallam, K.I.; Ishloroshi, M.; Samejima, K. Antioxidant and antimicrobial effects of garlic in chicken sausage. *LWT-Food Sci. Technol.* **2004**, *37*, 849–855. [CrossRef] [PubMed]

101. Aydin, A.; Bostan, K.; Erkan, M.E.; Bingol, B. The antimicrobial effects of chopped garlic in ground beef and raw meatball (Çið Köfte). *J. Med. Food* **2007**, *10*, 203–207. [CrossRef] [PubMed]

102. Li, W.; Shi, Q.; Dai, H.; Liang, Q.; Xie, X.; Huang, X.; Zhao, G.; Zhang, L. Antifungal activity, kinetics and molecular mechanism of action of garlic oil against *Candida albicans*. *Sci. Rep. UK* **2016**, *6*, 22805. [CrossRef] [PubMed]

103. Indu, M.N.; Hatha, A.; Abirosh, C.; Harsha, U.; Vivekanandan, G. Antimicrobial activity of some of the south-Indian spices against serotypes of *Escherichia coli*, *Salmonella*, *Listeria monocytogenes* and *Aeromonas hydrophila*. *Braz. J. Microbiol.* **2006**, *37*, 153–158. [CrossRef]

104. Joe, M.M.; Jayachitra, J.; Vijayapriya, M. Antimicrobial activity of some common spices against certain human pathogens. *J. Med. Plants Res.* **2009**, *3*, 1134–1136.

105. Geremew, T.; Kebede, A.; Andualem, B. The role of spices and lactic acid bacteria as antimicrobial agent to extend the shelf life of *metata ayib* (traditional Ethiopian spiced fermented cottage cheese). *J. Food Sci. Technol. Mysore* **2015**, *52*, 5661–5670. [CrossRef] [PubMed]

106. Touba, E.P.; Zakaria, M.; Tahereh, E. Anti-fungal activity of cold and hot water extracts of spices against fungal pathogens of roselle (*Hibiscus sabdariffa*) in vitro. *Microb. Pathog.* **2012**, *52*, 125–129. [CrossRef] [PubMed]

107. Park, M.; Bae, J.; Lee, D.S. Antibacterial activity of [10]-gingerol and [12]-gingerol isolated from ginger rhizome against periodontal bacteria. *Phytother. Res.* **2008**, *22*, 1446–1449. [CrossRef] [PubMed]

108. Sa-Nguanpuag, K.; Kanlayanarat, S.; Srilaong, V.; Tanprasert, K.; Techavuthiporn, C. Ginger (*Zingiber officinale*) oil as an antimicrobial agent for minimally processed produce: A case study in shredded green papaya. *Int. J. Agric. Biol.* **2011**, *13*, 895–901.

109. Singh, G.; Maurya, S.; Catalan, C.; de Lampasona, M.P. Studies on essential oils, Part 42: Chemical, antifungal, antioxidant and sprout suppressant studies on ginger essential oil and its oleoresin. *Flavour Frag. J.* **2005**, *20*, 1–6. [CrossRef]

110. Yoo, M.J.; Kim, Y.S.; Shin, D.H. Antibacterial effects of natural essential oils from ginger and mustard against *Vibrio* species inoculated on sliced raw flatfish. *Food Sci. Biotechnol.* **2006**, *15*, 462–465.

111. Thongson, C.; Davidson, P.M.; Mahakarnchanakul, W.; Vibulsresth, P. Antimicrobial effect of Thai spices against *Listeria monocytogenes* and *Salmonella typhimurium* DT104. *J. Food Prot.* **2005**, *68*, 2054–2058. [CrossRef] [PubMed]

112. Suppakul, P.; Miltz, J.; Sonneveld, K.; Bigger, S.W. Antimicrobial properties of basil and its possible application in food packaging. *J. Agric. Food Chem.* **2003**, *51*, 3197–3207. [CrossRef] [PubMed]

113. Hussain, A.I.; Anwar, F.; Sherazi, S.; Przybylski, R. Chemical composition, antioxidant and antimicrobial activities of basil (*Ocimum basilicum*) essential oils depends on seasonal variations. *Food Chem.* **2008**, *108*, 986–995. [CrossRef] [PubMed]

114. Kaya, I.; Yigit, N.; Benli, M. Antimicrobial activity of various extracts of *Ocimum basilicum* L. and observation of the inhibition effect on bacterial cells by use of scanning electron microscopy. *Afr. J. Tradit. Complement. Altern. Med.* **2008**, *5*, 363–369. [CrossRef] [PubMed]

115. Kocic-Tanackov, S.; Dimic, G.; Levic, J.; Tanackov, I.; Tuco, D. Antifungal activities of basil (*Ocimum basilicum* L.) extract on *Fusarium* species. *Afr. J. Biotechnol.* **2011**, *10*, 10188–10195.

116. Beatovic, D.; Krstic-Milosevic, D.; Trifunovic, S.; Siljegovic, J.; Glamoclija, J.; Ristic, M.; Jelacic, S. Chemical composition, antioxidant and antimicrobial activities of the essential oils of twelve *Ocimum basilicum* L. cultivars grown in Serbia. *Rec. Nat. Prod.* **2015**, *9*, 62–75.

117. El-Habib, R. Antifungal activity of some essential oils on *Aspergillus flavus* growth and aflatoxin production. *J. Food Agric. Environ.* **2012**, *10*, 274–279.

118. Singh, G.; Maurya, S.; de Lampasona, M.P.; Catalan, C. Chemical constituents, antifungal and antioxidative potential of *Foeniculum vulgare* volatile oil and its acetone extract. *Food Control* **2006**, *17*, 745–752. [CrossRef]

119. Diao, W.R.; Hu, Q.P.; Zhang, H.; Xu, J.G. Chemical composition, antibacterial activity and mechanism of action of essential oil from seeds of fennel (*Foeniculum vulgare* Mill.). *Food Control* **2014**, *35*, 109–116. [CrossRef]

120. Park, J.S.; Baek, H.H.; Bai, D.H.; Oh, T.K.; Lee, C.H. Antibacterial activity of fennel (*Foeniculum vulgare* Mill.) seed essential oil against the growth of *Streptococcus mutans*. *Food Sci. Biotechnol.* **2004**, *13*, 581–585.

121. Zellagui, A.; Gherraf, N.; Elkhateeb, A.; Hegazy, M.; Mohamed, T.A.; Touil, A.; Shahat, A.A.; Rhouati, S. Chemical constituents from Algerian *Foeniculum Vulgare* aerial parts and evaluation of antimicrobial activity. *J. Chil. Chem. Soc.* **2011**, *56*, 759–763. [CrossRef]

122. Bisht, D.S.; Menon, K.; Singhal, M.K. Comparative antimicrobial activity of essential oils of *Cuminum cyminum* L. and *Foeniculum vulgare* Mill seeds against *Salmonella typhimurium* and *Escherichia coli*. *J. Essent. Oil Bear. Plants* **2014**, *17*, 617–622. [CrossRef]

123. Nguyen, S.; Huang, H.; Foster, B.C.; Tam, T.W.; Xing, T.; Smith, M.L.; Arnason, J.T.; Akhtar, H. Antimicrobial and P450 inhibitory properties of common functional foods. *J. Pharm. Pharm. Sci.* **2014**, *17*, 254–265. [CrossRef]

124. Duarte, A.; Luis, A.; Oleastro, M.; Domingues, F.C. Antioxidant properties of coriander essential oil and linalool and their potential to control *Campylobacter* spp. *Food Control* **2016**, *61*, 115–122. [CrossRef]

125. Duarte, A.F.; Ferreira, S.; Oliveira, R.; Domingues, F.C. Effect of coriander oil (*Coriandrum sativum*) on planktonic and biofilm cells of *Acinetobacter baumannii*. *Nat. Prod. Commun.* **2013**, *8*, 673–678.

126. Duarte, A.; Ferreira, S.; Silva, F.; Domingues, F.C. Synergistic activity of coriander oil and conventional antibiotics against *Acinetobacter baumannii*. *Phytomedicine* **2012**, *19*, 236–238. [CrossRef] [PubMed]

127. Freires, I.D.; Murata, R.M.; Furletti, V.F.; Sartoratto, A.; de Alencar, S.M.; Figueira, G.M.; Rodrigues, J.; Duarte, M.; Rosalen, P.L. *Coriandrum sativum* L. (coriander) essential oil: Antifungal activity and mode of action on *Candida* spp., and molecular targets affected in human whole-genome expression. *PLoS ONE* **2014**, *9*, e99086.

128. Silva, F.; Ferreira, S.; Queiroz, J.A.; Domingues, F.C. Coriander (*Coriandrum sativum* L.) essential oil: Its antibacterial activity and mode of action evaluated by flow cytometry. *J. Med. Microbiol.* **2011**, *60*, 1479–1486. [CrossRef] [PubMed]

129. Acimovic, M.G.; Grahovac, M.S.; Stankovic, J.M.; Cvetkovic, M.T.; Masirevic, S.N. Essential oil composition of different coriander (*Coriandrum sativum* L.) accessions and their influence on mycelial growth of *Colletotrichum* spp. *Acta Sci. Pol. Hortorum Cultus* **2016**, *15*, 35–44.

130. Singh, G.; Maurya, S.; de Lampasona, M.P.; Catalan, C. Studies on essential oils, Part 41. Chemical composition, antifungal, antioxidant and sprout suppressant activities of coriander (*Coriandrum sativum*) essential oil and its oleoresin. *Flavour Frag. J.* **2006**, *21*, 472–479. [CrossRef]

131. Yakout, S.M.; Abd-Alrahman, S.H.; Mostafa, A.; Salem-Bekhit, M.M. Antimicrobial effect of seed ethanolic extract of coriander. *J. Pure Appl. Microbiol.* **2013**, *7*, 459–463.

132. Dimic, G.; Kocic-Tanackov, S.; Mojovic, L.; Pejin, J. Antifungal activity of lemon essential oil, coriander and cinnamon extracts on foodborne molds in direct contact and the vapor phase. *J. Food Process. Pres.* **2015**, *39*, 1778–1787. [CrossRef]

133. Rao, K.; Ch, B.; Narasu, L.M.; Giri, A. Antibacterial activity of *Alpinia galanga* (L.) willd crude extracts. *Appl. Biochem. Biotechnol.* **2010**, *162*, 871–884. [CrossRef] [PubMed]

134. Weerakkody, N.S.; Caffin, N.; Lambert, L.K.; Turner, M.S.; Dykes, G.A. Synergistic antimicrobial activity of galangal (*Alpinia galanga*), rosemary (*Rosmarinus officinalis*) and lemon iron bark (*Eucalyptus staigerana*) extracts. *J. Sci. Food Agric.* **2011**, *91*, 461–468. [CrossRef] [PubMed]

135. Wong, L.F.; Lim, Y.Y.; Omar, M. Antioxidant and antimicrobial activities of some *Alpina* species. *J. Food Biochem.* **2009**, *33*, 835–851. [CrossRef]

136. Singh, G.; Marimuthu, P.; Catalan, C.; DeLampasona, M.P. Chemical, antioxidant and antifungal activities of volatile oil of black pepper and its acetone extract. *J. Sci. Food Agric.* **2004**, *84*, 1878–1884. [CrossRef]

137. Singh, G.; Marimuthu, P.; Murali, H.S.; Bawa, A.S. Antioxidative and antibacterial potentials of essential oils and extracts isolated from various spice materials—Part 48. *J. Food Saf.* **2005**, *25*, 130–145. [CrossRef]

138. Zarai, Z.; Boujelbene, E.; Ben Salem, N.; Gargouri, Y.; Sayari, A. Antioxidant and antimicrobial activities of various solvent extracts, piperine and piperic acid from *Piper nigrum*. *LWT-Food Sci. Technol.* **2013**, *50*, 634–641. [CrossRef]

139. Zhang, J.; Ye, K.; Zhang, X.; Pan, D.; Sun, Y.; Cao, J. Antibacterial activity and mechanism of action of black pepper essential oil on meat-borne *Escherichia coli*. *Front. Microbiol.* **2017**, *7*. [CrossRef] [PubMed]

140. Baris, D.; Kizil, M.; Aytekin, C.; Kizil, G.; Yavuz, M.; Ceken, B.; Ertekin, A.S. In vitro antimicrobial and antioxidant activity of ethanol extract of three *Hypericum* and three *Achillea* species from Turkey. *Int. J. Food Prop.* **2011**, *14*, 339–355. [CrossRef]

141. Tajik, H.; Jalali, F.; Sobhani, A.; Shahbazi, Y.; Zadeh, M.S. In vitro assessment of antimicrobial efficacy of alcoholic extract of *Achillea millefolium* in comparison with penicillin derivatives. *J. Anim. Vet. Adv.* **2008**, *7*, 508–511.

142. Eyob, S.; Martinsen, B.K.; Tsegaye, A.; Appelgren, M.; Skrede, G. Antioxidant and antimicrobial activities of extract and essential oil of korarima (*Aframomum corrorima* (Braun) P.C.M. Jansen). *Afr. J. Biotechnol.* **2008**, *7*, 2585–2592.

143. Ismail, S.; Jalilian, F.A.; Talebpour, A.H.; Zargar, M.; Shameli, K.; Sekawi, Z.; Jahanshiri, F. Chemical composition and antibacterial and cytotoxic activities of *Allium hirtifolium* Boiss. *BioMed Res. Int.* **2013**, *2013*. [CrossRef] [PubMed]

144. Najjaa, H.; Ammar, E.; Neffati, M. Antimicrobial activities of protenic extracts of *Allium roseum* L., a wild edible species in North Africa. *J. Food Agric. Environ.* **2009**, *7*, 150–154.

145. Mihaylova, D.S.; Lante, A.; Tinello, F.; Krastanov, A.I. Study on the antioxidant and antimicrobial activities of *Allium ursinum* L. pressurised-liquid extract. *Nat. Prod. Res.* **2014**, *28*, 2000–2005. [CrossRef] [PubMed]

146. Diao, W.R.; Zhang, L.L.; Feng, S.S.; Xu, J.G. Chemical composition, antibacterial activity, and mechanism of action of the essential oil from *Amomum kravanh*. *J. Food Prot.* **2014**, *77*, 1740–1746. [CrossRef]

147. Singh, G.; Maurya, S.; de Lampasona, M.P.; Catalan, C. Chemical constituents, antimicrobial investigations, and antioxidative potentials of *Anethum graveolens* L. essential oil and acetone extract: Part 52. *J. Food Sci.* **2005**, *70*, M208–M215. [CrossRef]

148. Akkoyun, H.T.; Dostbil, N. Antibacterial activity of some species of *Umbelliferae*. *Asian J. Chem.* **2007**, *19*, 4862–4866.

149. Sofia, P.K.; Prasad, R.; Vijay, V.K.; Srivastava, A.K. Evaluation of antibacterial activity of Indian spices against common foodborne pathogens. *Int. J. Food Sci. Technol.* **2007**, *42*, 910–915. [CrossRef]

150. Sekine, T.; Sugano, M.; Majid, A.; Fujii, Y. Antifungal effects of volatile compounds from black zira (*Bunium persicum*) and other spices and herbs. *J. Chem. Ecol.* **2007**, *33*, 2123–2132. [CrossRef] [PubMed]

151. Mishra, P.K.; Shukla, R.; Singh, P.; Prakash, B.; Dubey, N.K. Antifungal and antiaflatoxigenic efficacy of *Caesulia axillaris* Roxb essential oil against fungi deteriorating some herbal raw materials, and its antioxidant activity. *Ind. Crop. Prod.* **2012**, *36*, 74–80. [CrossRef]

152. Wahba, N.M.; Ahmed, A.S.; Ebraheim, Z.Z. Antimicrobial effects of pepper, parsley, and dill and their roles in the microbiological quality enhancement of traditional Egyptian kareish cheese. *Foodborne Pathog. Dis.* **2010**, *7*, 411–418. [CrossRef] [PubMed]

153. Gurnani, N.; Gupta, M.; Shrivastava, R.; Mehta, D.; Mehta, B.K. Effect of extraction methods on yield, phytochemical constituents, antibacterial and antifungal activity of *Capsicum frutescens* L. *Indian J. Nat. Prod. Resour.* **2016**, *7*, 32–39.

154. Singh, G.; Kapoor, I.; Pandey, S.K.; Singh, U.K.; Singh, R.K. Studies on essential oils: Part 10; Antibacterial activity of volatile oils of some spices. *Phytother. Res.* **2002**, *16*, 680–682. [CrossRef] [PubMed]

155. Oroojalian, F.; Kasra-Kermanshahi, R.; Azizi, M.; Bassami, M.R. Phytochemical composition of the essential oils from three *Apiaceae* species and their antibacterial effects on food-borne pathogens. *Food Chem.* **2010**, *120*, 765–770. [CrossRef]

156. Shan, B.; Cai, Y.Z.; Brooks, J.D.; Corke, H. Antibacterial properties and major bioactive components of cinnamon stick (*Cinnamomum burmannii*): Activity against foodborne pathogenic bacteria. *J. Agric. Food Chem.* **2007**, *55*, 5484–5490. [CrossRef] [PubMed]

157. Pandey, A.K.; Mishra, A.K.; Mishra, A. Antifungal and antioxidative potential of oil and extracts derived from leaves of Indian spice plant *Cinnamomum tamala. Cell. Mol. Biol.* **2012**, *58*, 142–147. [PubMed]

158. Shreya, A.; Manisha, D.; Sonali, J. Phytochemical screening and anti-microbial activity of cinnamon spice against urinary tract infection and fungal pathogens. *Int. J. Life Sci. Pharma Res.* **2015**, *5*, P30–P38.

159. Naveed, R.; Hussain, I.; Mahmood, M.S.; Akhtar, M. In vitro and in vivo evaluation of antimicrobial activities of essential oils extracted from some indigenous spices. *Pak. Vet. J.* **2013**, *33*, 413–417.

160. Naveed, R.; Hussain, I.; Tawab, A.; Tariq, M.; Rahman, M.; Hameed, S.; Mahmood, M.S.; Siddique, A.; Iqbal, M. Antimicrobial activity of the bioactive components of essential oils from Pakistani spices against *Salmonella* and other multi-drug resistant bacteria. *BMC Complement. Altern. Med.* **2013**, *13*, 265. [CrossRef] [PubMed]

161. Hashemi, S.; Amininezhad, R.; Shirzadinezhad, E.; Farahani, M.; Yousefabad, S. The antimicrobial and antioxidant effects of *Citrus aurantium* L. flowers (Bahar narang) extract in traditional yoghurt stew during refrigerated storage. *J. Food Saf.* **2016**, *36*, 153–161. [CrossRef]

162. Castilho, P.; Liu, K.; Rodrigues, A.I.; Feio, S.; Tomi, F.; Casanova, J. Composition and antimicrobial activity of the essential oil of *Clinopodium ascendens* (Jordan) Sampaio from Madeira. *Flavour Frag. J.* **2007**, *22*, 139–144. [CrossRef]

163. Oussalah, M.; Caillet, S.; Saucier, L.; Lacroix, M. Antimicrobial effects of selected plant essential oils on the growth of a *Pseudomonas putida* strain isolated from meat. *Meat Sci.* **2005**, *73*, 236–244. [CrossRef] [PubMed]

164. Sani, A.M.; Yaghooti, F. Antibacterial effects and chemical composition of essential oil from *Cotoneaster nummarioides* leaves extract on typical food-borne pathogens. *J. Essent. Oil Bear. Plants* **2016**, *19*, 290–296. [CrossRef]

165. Daouda, T.; Prevost, K.; Gustave, B.; Joseph, D.A.; Nathalie, G.; Raphael, O.; Rubens, D.; Claude, C.J.; Mireille, D.; Felix, T. Terpenes, antibacterial and modulatory antibiotic activity of essential oils from *Croton hirtus* L' Hér. (Euphorbiaceae) from Ivory Coast. *J. Essent. Oil Bear. Plants* **2014**, *17*, 607–616. [CrossRef]

166. Ani, V.; Varadaraj, M.C.; Naidu, K.A. Antioxidant and antibacterial activities of polyphenolic compounds from bitter cumin (*Cuminum nigrum* L.). *Eur. Food Res. Technol.* **2006**, *224*, 109–115. [CrossRef]

167. Li, X.F.; Jin, C.; He, J.; Zhou, J.; Wang, H.T.; Dai, B.; Yan, D.; Wang, J.B.; Zhao, Y.L.; Xiao, X.H. Microcalorimetric investigation of the antibacterial activity of curcumin on *Staphylococcus aureus* coupled with multivariate analysis. *J. Therm. Anal. Calorim.* **2012**, *109*, 395–402. [CrossRef]

168. Sandri, I.G.; Zacaria, J.; Fracaro, F.; Delamare, A.; Echeverrigaray, S. Antimicrobial activity of the essential oils of Brazilian species of the genus *Cunila* against foodborne pathogens and spoiling bacteria. *Food Chem.* **2007**, *103*, 823–828. [CrossRef]

169. Fankam, A.G.; Kuete, V.; Voukeng, I.K.; Kuiate, J.R.; Pages, J.M. Antibacterial activities of selected Cameroonian spices and their synergistic effects with antibiotics against multidrug-resistant phenotypes. *BMC Complement. Altern. Med.* **2011**, *11*, 104. [CrossRef] [PubMed]

170. Tekwu, E.M.; Askun, T.; Kuete, V.; Nkengfack, A.E.; Nyasse, B.; Etoa, F.X.; Beng, V.P. Antibacterial activity of selected Cameroonian dietary spices ethno-medically used against strains of *Mycobacterium tuberculosis*. *J. Ethnopharmacol.* **2012**, *142*, 374–382. [CrossRef] [PubMed]

171. El Malti, J.; Mountassif, D.; Amarouch, H. Antimicrobial activity of *Elettaria cardamomum*: Toxicity, biochemical and histological studies. *Food Chem.* **2007**, *104*, 1560–1568. [CrossRef]

172. Singh, G.; Kim, S.; Marimuthu, P.; Isidorov, V.; Vinogorova, V. Antioxidant and antimicrobial activities of essential oil and various oleoresins of *Elettaria cardamomum* (seeds and pods). *J. Sci. Food Agric.* **2008**, *88*, 280–289. [CrossRef]

173. Bey-Ould Si Said, Z.; Haddadi-Guemghar, H.; Boulekbache-Makhlouf, L.; Rigou, P.; Remini, H.; Adjaoud, A.; Khoudja, N.K.; Madani, K. Essential oils composition, antibacterial and antioxidant activities of hydrodistillated extract of *Eucalyptus globulus* fruits. *Ind. Crop. Prod.* **2016**, *89*, 167–175. [CrossRef]

174. Nikbakht, M.R.; Rahimi-Nasrabadi, M.; Ahmadi, F.; Gandomi, H.; Abbaszadeh, S.; Batooli, H. The chemical composition and in vitro antifungal activities of essential oils of five *Eucalyptus* species. *J. Essent. Oil Bear. Plants* **2015**, *18*, 666–677. [CrossRef]

175. Luís, Â.; Duarte, A.; Gominho, J.; Domingues, F.; Duarte, A.P. Chemical composition, antioxidant, antibacterial and anti-quorum sensing activities of *Eucalyptus globulus* and *Eucalyptus radiata* essential oils. *Ind. Crop. Prod.* **2016**, *79*, 274–282. [CrossRef]

176. Puangpronpitag, D.; Niamsa, N.; Sittiwet, C. Anti-microbial properties of clove (*Eugenia caryophyllum* Bullock and Harrison) aqueous extract against food-borne pathogen bacteria. *Int. J. Pharmacol.* **2009**, *5*, 281–284.

177. Ozcan, M.M.; Chalchat, J.C.; Arslan, D.; Ates, A.; Unver, A. Comparative essential oil composition and antifungal effect of bitter fennel (*Foeniculum vulgare* ssp. *piperitum*) fruit oils obtained during different vegetation. *J. Med. Food* **2006**, *9*, 552–561. [PubMed]

178. Soureshjan, E.H.; Heidari, M. In vitro variation in antibacterial activity plant extracts on *Glaucium elegans* and saffron (*Crocus sativus*). *Bangladesh J. Pharmacol.* **2014**, *9*, 275–278.

179. Khan, H.; Khan, M.A.; Mahmood, T.; Choudhary, M.I. Antimicrobial activities of *Gloriosa superba* Linn (Colchicaceae) extracts. *J. Enzym. Inhib. Med. Chem.* **2008**, *23*, 855–859. [CrossRef] [PubMed]

180. Albayrak, S.; Aksoy, A.; Sagdic, O.; Hamzaoglu, E. Compositions, antioxidant and antimicrobial activities of *Helichrysum* (Asteraceae) species collected from Turkey. *Food Chem.* **2010**, *119*, 114–122. [CrossRef]

181. Albayrak, S.; Aksoy, A.; Sagdic, O.; Budak, U. Phenolic compounds and antioxidant and antimicrobial properties of *Helichrysum* species collected from eastern Anatolia, Turkey. *Turk. J. Biol.* **2010**, *34*, 463–473.

182. Czapska, A.; Balasinska, B.; Szczawinski, J. Antimicrobial and antioxidant properties of aqueous extracts from selected spice plants. *Med. Weter.* **2006**, *62*, 302–305.

183. Džamiæ, A.M.; Sokoviæ, M.D.; Novakoviæ, M.; Jadranin, M.; Ristiæ, M.S.; Teševiæ, V.; Marin, P.D. Composition, antifungal and antioxidant properties of *Hyssopus officinalis* L. subsp *pilifer* (Pant.) Murb essential oil and deodorized extracts. *Ind. Crop. Prod.* **2013**, *51*, 401–407. [CrossRef]

184. Parlatan, A.; Saricoban, C.; Ozcan, M.M. Chemical composition and antimicrobial activity of the extracts of Kefe cumin (*Laser trilobum* L.) fruits from different regions. *Int. J. Food Sci. Nutr.* **2009**, *60*, 606–617. [CrossRef] [PubMed]

185. El Malti, J.; Amarouch, H. Antibacterial effect, histological impact and oxidative stress studies *Fromlaurus nobilis* extract. *J. Food Qual.* **2009**, *32*, 190–208. [CrossRef]

186. Ramos, C.; Teixeira, B.; Batista, I.; Matos, O.; Serrano, C.; Neng, N.R.; Nogueira, J.; Nunes, M.L.; Marques, A. Antioxidant and antibacterial activity of essential oil and extracts of bay laurel *Laurus nobilis* Linnaeus (Lauraceae) from Portugal. *Nat. Prod. Res.* **2012**, *26*, 518–529. [CrossRef] [PubMed]

187. Ceyhan, N.; Keskin, D.; Ugur, A. Antimicrobial activities of different extracts of eight plant species from four different family against some pathogenic microorganisms. *J. Food Agric. Environ.* **2012**, *10*, 193–197.

188. Marín, I.; Sayas-Barberá, E.; Viuda-Martos, M.; Navarro, C.; Sendra, E. Chemical composition, antioxidant and antimicrobial activity of essential oils from organic fennel, parsley, and lavender from Spain. *Foods* **2016**, *5*, 18. [CrossRef]

189. Basile, A.; Rigano, D.; Loppi, S.; Di Santi, A.; Nebbioso, A.; Sorbo, S.; Conte, B.; Paoli, L.; de Ruberto, F.; Molinari, A.M. Antiproliferative, antibacterial and antifungal activity of the lichen *Xanthoria parietina* and its secondary metabolite parietin. *Int. J. Mol. Sci.* **2015**, *16*, 7861–7875. [CrossRef] [PubMed]

190. Sarrazin, S.; Oliveira, R.B.; Barata, L.; Mourao, R. Chemical composition and antimicrobial activity of the essential oil of *Lippia grandis* Schauer (Verbenaceae) from the western Amazon. *Food Chem.* **2012**, *134*, 1474–1478. [CrossRef] [PubMed]

191. Asowata-Ayodele, A.M.; Otunola, G.A.; Afolayan, A.J. Assessment of the polyphenolic content, free radical scavenging, anti-inflammatory, and antimicrobial activities of acetone and aqueous extracts of *Lippia javanica* (Burm.F.) Spreng. *Pharmacogn. Mag.* **2016**, *12*, S353–S362. [PubMed]

192. Oliveira, D.R.; Leitao, G.G.; Bizzo, H.R.; Lopes, D.; Alviano, D.S.; Alviano, C.S.; Leitao, S.G. Chemical and antimicrobial analyses of essential oil of *Lippia origanoides* HBK. *Food Chem.* **2007**, *101*, 236–240. [CrossRef]

193. Li, W.R.; Shi, Q.S.; Liang, Q.; Xie, X.B.; Huang, X.M.; Chen, Y.B. Antibacterial activity and kinetics of *Litsea cubeba* oil on *Escherichia coli*. *PLoS ONE* **2014**, *9*, e110983. [CrossRef] [PubMed]

194. Stanojevic, D.; Comic, L.; Stefanovic, O.; Sukdolak, S.S. In vitro synergistic antibacterial activity of *Melissa officinalis* L. and some preservatives. *Span. J. Agric. Res.* **2010**, *8*, 109–115. [CrossRef]

195. Sharma, M.; Sharma, M. Antimicrobial potential of essential oil from *Mentha piperita* L. against anthropophilic dermatophytes. *J. Essent. Oil Bear. Plants* **2012**, *15*, 263–269. [CrossRef]

196. Arumugam, P.; Murugan, R.; Subathra, M.; Ramesh, A. Superoxide radical scavenging and antibacterial activities of different fractions of ethanol extract of *Mentha spicata* (L.). *Med. Chem. Res.* **2010**, *19*, 664–673. [CrossRef]

197. Esekhiagbe, M.; Agatemor, M.; Agatemor, C. Phenolic content and antimicrobial potentials of *Xylopia aethiopica* and *Myristica argentea*. *Maced. J. Chem. Chem. Eng.* **2009**, *28*, 159–162.

198. Shafiei, Z.; Shuhairi, N.N.; Yap, N.; Sibungkil, C.; Latip, J. Antibacterial activity of *Myristica fragrans* against oral pathogens. *Evid. Based Complement. Altern.* **2012**, *2012*. [CrossRef]

199. Mhamdi, B.; Abbassi, F.; Marzouki, L. Antimicrobial Activities effects of the essential oil of spice food *Myrtus communis* leaves vr. Italica. *J. Essent. Oil Bear. Plants* **2014**, *17*, 1361–1366. [CrossRef]

200. Ozcan, M.M.; Uyar, B.; Unver, A. Antibacterial effect of myrtle (*Myrtus communis* L.) leaves extract on microorganisms. *J. Food Saf. Food Qual.* **2015**, *66*, 18–21.

201. Popovici, J.; Bertrand, C.; Bagnarol, E.; Fernandez, M.P.; Comte, G. Chemical composition of essential oil and headspace-solid microextracts from fruits of *Myrica gale* L. and antifungal activity. *Nat. Prod. Res.* **2008**, *22*, 1024–1032. [CrossRef] [PubMed]

202. Aboee-Mehrizi, F.; Rustaiyan, A.; Zare, M. Chemical composition and antimicrobial activity of the essential oils of *Nepeta alpina* growing wild in Iran. *J. Essent. Oil Bear. Plants* **2016**, *19*, 236–240. [CrossRef]

203. Arslan, U.; Ilhan, K.; Karabulut, O.A. Antifungal activity of aqueous extracts of spices against bean rust (*Uromyces appendiculatus*). *Allelopath. J.* **2009**, *24*, 207–213.

204. Arici, M.; Sagdic, O.; Gecgel, U. Antibacterial effect of Turkish black cumin (*Nigella sativa* L.) oils. *Grasas Aceites* **2005**, *56*, 259–262. [CrossRef]

205. Wouatsa, N.; Misra, L.; Kumar, R.V. Antibacterial activity of essential oils of edible spices, *Ocimum canum* and *Xylopia aethiopica*. *J. Food Sci.* **2014**, *79*, M972–M977. [CrossRef] [PubMed]

206. Prakash, B.; Shukla, R.; Singh, P.; Mishra, P.K.; Dubey, N.K.; Kharwar, R.N. Efficacy of chemically characterized *Ocimum gratissimum* L. essential oil as an antioxidant and a safe plant based antimicrobial against fungal and aflatoxin B_1 contamination of spices. *Food Res. Int.* **2011**, *44*, 385–390. [CrossRef]

207. Kumar, A.; Dubey, N.K.; Srivastaya, S. Antifungal evaluation of *Ocimum sanctum* essential oil against fungal deterioration of raw materials of *Rauvolfia serpentina* during storage. *Ind. Crop. Prod.* **2013**, *45*, 30–35. [CrossRef]

208. Saharkhiz, M.J.; Kamyab, A.A.; Kazerani, N.K.; Zomorodian, K.; Pakshir, K.; Rahimi, M.J. Chemical compositions and antimicrobial activities of *Ocimum sanctum* L. essential oils at different harvest stages. *Jundishapur J. Microbiol.* **2015**, *8*, e13720. [CrossRef] [PubMed]

209. Runyoro, D.; Ngassapa, O.; Vagionas, K.; Aligiannis, N.; Graikou, K.; Chinou, I. Chemical composition and antimicrobial activity of the essential oils of four *Ocimum* species growing in Tanzania. *Food Chem.* **2010**, *119*, 311–316. [CrossRef]

210. Ali, N.H.; Faizi, S.; Kazmi, S.U. Antibacterial activity in spices and local medicinal plants against clinical isolates of Karachi, Pakistan. *Pharm. Biol.* **2011**, *49*, 833–839. [CrossRef] [PubMed]

211. Yano, Y.; Satomi, M.; Oikawa, H. Antimicrobial effect of spices and herbs on *Vibrio parahaemolyticus*. *Int. J. Food Microbiol.* **2006**, *111*, 6–11. [CrossRef] [PubMed]

212. Seydim, A.C.; Sarikus, G. Antimicrobial activity of whey protein based edible films incorporated with oregano, rosemary and garlic essential oils. *Food Res. Int.* **2006**, *39*, 639–644. [CrossRef]

213. Ho, C.H.; Noryati, I.; Sulaiman, S.F.; Rosma, A. In vitro antibacterial and antioxidant activities of *Orthosiphon stamineus* Benth extracts against food-borne bacteria. *Food Chem.* **2010**, *122*, 1168–1172. [CrossRef]

214. Boyd, F.; Benkeblia, N. In vitro evaluation of antimicrobial activity of crude extracts of *Pimenta dioica* L. (Merr.). In Proceedings of the 3rd International Conference on Postharvest and Quality Management of Horticultural Products of Interest for Tropical Regions, Port of Spain, Trinid & Tobago, 1–5 July 2013; Mohammed, M., Francis, J.A., Eds.; International Society for Horticultural Science: Leuven, Belgium, 2014; pp. 199–205.

215. Ozcan, M.M.; Chalchat, J.C. Chemical composition and antifungal effect of anise (*Pimpinella anisum* L.) fruit oil at ripening stage. *Ann. Microbiol.* **2006**, *56*, 353–358. [CrossRef]

216. Woguem, V.; Maggi, F.; Fogang, H.; Tapondjou, L.A.; Womeni, H.M.; Quassinti, L.; Bramucci, M.; Vitali, L.A.; Petrelli, D.; Lupidi, G.; et al. Antioxidant, antiproliferative and antimicrobial activities of the volatile oil from the wild pepper piper capense used in Cameroon as a culinary spice. *Nat. Prod. Commun.* **2013**, *8*, 1791–1796. [PubMed]

217. Eruteya, O.C.; Odunfa, S.A. Antimicrobial properties of three spices used in the preparation of suya condiment against organisms isolated from formulated samples and individual ingredients. *Afr. J. Biotechnol.* **2009**, *8*, 2316–2320.

218. Toroglu, S.; Cenet, M. Comparison of antimicrobial activities of essential oil and solvent extracts of endemic *Phlomis oppositiflora* Boiss. & Hausskn. from Turkey. *Pak. J. Zool.* **2013**, *45*, 475–482.

219. Sahin, S.; Oran, S.; Sahinturk, P.; Demir, C.; Ozturk, S. *Ramalina* lichens and their major metabolites as possible natural antioxidant and antimicrobial agents. *J. Food Biochem.* **2015**, *39*, 471–477. [CrossRef]

220. Fazeli, M.R.; Amin, G.; Attari, M.; Ashtiani, H.; Jamalifar, H.; Samadi, N. Antimicrobial activities of Iranian sumac and avishan-e shirazi. (*Zataria multiflora*) against some food-borne bacteria. *Food Control* **2007**, *18*, 646–649. [CrossRef]

221. Aliakbarlu, J.; Mohammadi, S.; Khalili, S. A study on antioxidant potency and antibacterial activity of water extracts of some spices widely consumed in Iranian diet. *J. Food Biochem.* **2014**, *38*, 159–166. [CrossRef]

222. Mekinic, I.G.; Skroza, D.; Ljubenkov, I.; Simat, V.; Mozina, S.S.; Katalinic, V. In vitro antioxidant and antibacterial activity of Lamiaceae phenolic extracts: A correlation study. *Food Technol. Biotechnol.* **2014**, *52*, 119–127.

223. Bouajaj, S.; Benyamna, A.; Bouamama, H.; Romane, A.; Falconieri, D.; Piras, A.; Marongiu, B. Antibacterial, allelopathic and antioxidant activities of essential oil of *Salvia officinalis* L. growing wild in the Atlas Mountains of Morocco. *Nat. Prod. Res.* **2013**, *27*, 1673–1676. [CrossRef] [PubMed]

224. Mehr, H.M.; Hosseini, Z.; Khodaparast, M.; Edalatian, M.R. Study on the antimicrobial effect of *Salvia leriifolia* (nowroozak) leaf extract powder on the growth of *Staphylococcus aureus* in hamburger. *J. Food Saf.* **2010**, *30*, 941–953. [CrossRef]

225. Djeddi, S.; Djebile, K.; Hadjbourega, G.; Achour, Z.; Argyropoulou, C.; Skaltsa, H. In vitro Antimicrobial properties and chemical composition of *Santolina chamaecyparissus* essential oil from Algeria. *Nat. Prod. Commun.* **2012**, *7*, 937–940. [PubMed]

226. Oke, F.; Aslim, B.; Ozturk, S.; Altundag, S. Essential oil composition, antimicrobial and antioxidant activities of *Satureja cuneifolia* Ten. *Food Chem.* **2009**, *112*, 874–879. [CrossRef]

227. Mihajilov-Krstev, T.; Kitic, D.; Radnovic, D.; Ristic, M.; Mihajlovic-Ukropina, M.; Zlatkovic, B. Chemical composition and antimicrobial activity of *Satureja kitaibelii* essential oil against pathogenic microbial strains. *Nat. Prod. Commun.* **2011**, *6*, 1167–1172. [PubMed]

228. Yucel, N.; Aslim, B. Antibacterial activity of the essential oil of *Satureja wiedemanniana* against *Bacillus* species isolated from chicken meat. *Foodborne Pathog. Dis.* **2011**, *8*, 71–76. [CrossRef] [PubMed]

229. Farzaneh, M.; Kiani, H.; Sharifi, R.; Reisi, M.; Hadian, J. Chemical composition and antifungal effects of three species of *Satureja* (*S. hortensis*, *S. spicigera* and *S. khuzistanica*) essential oils on the main pathogens of strawberry fruit. *Postharvest Biol. Technol.* **2015**, *109*, 145–151.

230. Toroglu, S.; Keskin, D.; Dadandi, M.Y.; Yildiz, K. Comparision of antimicrobial activity of *Silene laxa* Boiss. & Kotschy and *Silene caramanica* Boiss. & Heldr different extracts from Turkey. *J. Pure Appl. Microbiol.* **2013**, *7*, 1763–1768.

231. Ozcan, M.M.; Al Juhaimi, F.Y. Antioxidant and antifungal activity of some aromatic plant extracts. *J. Med. Plants Res.* **2011**, *5*, 1361–1366.

232. El Abed, N.; Kaabi, B.; Smaali, M.I.; Chabbouh, M.; Habibi, K.; Mejri, M.; Marzouki, M.N.; Ahmed, S.B. Chemical composition, antioxidant and antimicrobial activities of *Thymus capitata* essential oil with its preservative effect against *Listeria monocytogenes* inoculated in minced beef meat. *Evid. Based Complement. Altern.* **2014**, *2014*. [CrossRef]

233. Ballester-Costa, C.; Sendra, E.; Fernandez-Lopez, J.; Perez-Alvarez, J.A.; Viuda-Martos, M. Chemical composition and in vitro antibacterial properties of essential oils of four *Thymus* species from organic growth. *Ind. Crop. Prod.* **2013**, *50*, 304–311. [CrossRef]

234. Albayrak, S.; Aksoy, A. Essential oil composition and in vitro antioxidant and antimicrobial activities of *Thymus cappadocicus* Boiss. *J. Food Process Preserv.* **2013**, *37*, 605–614. [CrossRef]

235. Toroglu, S. In vitro antimicrobial activity and antagonistic effect of essential oils from plant species. *J. Environ. Biol.* **2007**, *28*, 551–559. [PubMed]

236. Ruiz-Navajas, Y.; Viuda-Martos, M.; Sendra, E.; Perez-Alvarez, J.A.; Fernández-López, J. Chemical characterization and antibacterial activity of *Thymus moroderi* and *Thymus piperella* essential oils, two *Thymus* endemic species from southeast of Spain. *Food Control* **2012**, *27*, 294–299. [CrossRef]

237. Georgescu, C.; Mironescu, M. Obtaining, characterisation and screening of the antifungal activity of the volatile oil extracted from *Thymus serpyllum*. *J. Environ. Prot. Ecol.* **2011**, *12*, 2294–2302.

238. Singh, G.; Maurya, S.; Catalan, C.; de Lampasona, M.P. Chemical constituents, antifungal and antioxidative effects of ajwain essential oil and its acetone extract. *J. Agric. Food Chem.* **2004**, *52*, 3292–3296. [CrossRef] [PubMed]

239. Adesegun, E.A.; Adebayo, O.S.; Akintokun, A.K. Antifungal activity of spices extracts against *Sclerotium rolfsii*. In Proceedings of the 28th International Horticultural Congress on Science and Horticulture for People (IHC)/International Symposium on Organic Horticulture—Productivity and Sustainability, Lisbon, Portugal, 22–27 August 2010; Mourao, I., Aksoy, U., Eds.; International Society for Horticultural Science: Leuven, Belgium, 2012; pp. 415–419.

240. Kusuda, M.; Inada, K.; Ogawa, T.O.; Yoshida, T.; Shiota, S.; Tsuchiya, T.; Hatano, T. Polyphenolic constituent structures of *Zanthoxylum piperitum* fruit and the antibacterial effects of its polymeric procyanidin on methicillin-resistant *Staphylococcus aureus*. *Biosci. Biotechnol. Biochem.* **2006**, *70*, 1423–1431. [CrossRef] [PubMed]

241. Diao, W.R.; Hu, Q.P.; Feng, S.S.; Li, W.Q.; Xu, J.G. Chemical composition and antibacterial activity of the essential oil from green huajiao (*Zanthoxylum schinifolium*) against selected foodborne pathogens. *J. Agric. Food Chem.* **2013**, *61*, 6044–6049. [CrossRef] [PubMed]

International Journal of
Molecular Sciences

MDPI

Review

Fucaceae: A Source of Bioactive Phlorotannins

Marcelo D. Catarino, Artur M. S. Silva and Susana M. Cardoso *

Department of Chemistry & Organic Chemistry, Natural Products and Food Stuffs Research Unit (QOPNA), University of Aveiro, Aveiro 3810-193, Portugal; mcatarino@ua.pt (M.D.C.); artur.silva@ua.pt (A.M.S.S.)
* Correspondence: susanacardoso@ua.pt; Tel.: +351-234-370-360; Fax: +351-234-370-084

Received: 29 April 2017; Accepted: 15 June 2017; Published: 21 June 2017

Abstract: Fucaceae is the most dominant algae family along the intertidal areas of the Northern Hemisphere shorelines, being part of human customs for centuries with applications as a food source either for humans or animals, in agriculture and as remedies in folk medicine. These macroalgae are endowed with several phytochemicals of great industrial interest from which phlorotannins, a class of marine-exclusive polyphenols, have gathered much attention during the last few years due to their numerous possible therapeutic properties. These compounds are very abundant in brown seaweeds such as Fucaceae and have been demonstrated to possess numerous health-promoting properties, including antioxidant effects through scavenging of reactive oxygen species (ROS) or enhancement of intracellular antioxidant defenses, antidiabetic properties through their acarbose-like activity, stimulation of adipocytes glucose uptake and protection of β-pancreatic cells against high-glucose oxidative stress; anti-inflammatory effects through inhibition of several pro-inflammatory mediators; antitumor properties by activation of apoptosis on cancerous cells and metastasis inhibition, among others. These multiple health properties render phlorotannins great potential for application in numerous therapeutical approaches. This review addresses the major contribution of phlototannins for the biological effects that have been described for seaweeds from Fucaceae. In addition, the bioavailability of this group of phenolic compounds is discussed.

Keywords: seaweeds; algae; Fucaceae; phlorotannins; bioactivities; antioxidant; antidiabetes; anti-inflammatory; antitumor; bioavailability

---

1. Introduction

Fucaceae is a family of brown algae containing five subordinate taxa currently recognized, including *Ascophyllum*, *Fucus*, *Pelvetia*, *Pelvetiopsis* and *Silvetia* (Figure 1), which dominate the biomass in the intertidal areas of many cold and warm temperate regions in the Northern Hemisphere, being distributed along the Northeast-Atlantic coastlines, from the White Sea to the south of the Canary Islands, and the Northwest-Atlantic, from south Greenland to North Carolina, as well as along the Northeast-Pacific coastline, extending from Alaska to California [1,2]. *Fucus* is undoubtedly the most prominent genus from this family. It currently comprises 66 taxonomically accepted species, which are characterized by a greenish brown trisected thallus, i.e., a structure consisting of a holdfast, a small stipe and flattened dichotomously-branched blades with terminal receptacles that swell during the reproductive season. The blades usually have a central-thickened area called the midrib, and in some species, such as *F. vesiculosus*, air bladders can be found to keep them floating in a vertical position when submerged [3]. This is also the most widely-distributed genus from Fucaceae being scattered throughout all of the regions covered by this family [4,5]. The most well-known species of this genus is *F. vesiculosus*, that commonly dominates the shallow macroalgae communities growing on high salinity waters from 0.5–4 m in depth and forming large belts that constitute the habitats for species-rich epiphytic and epibenthic communities [6,7].

Figure 1. Type species of each genus composing the Fucaceae family. (**A**) *Fucus vesiculosus* L., photo by Emőke Dénes licenced by CC BY-SA/resized from the original; (**B**) *Ascophyllum nodosum* (L.) Le Jolis, photo by Anne Burgess licensed by CC BY-SA/resized from the original; (**C**) *Pelvetia canaliculata* (L.) Decaisne & Thuret, photo by Tom Corser licensed by CC BY-SA/resized from the original; (**D**) *Silvetia compressa* (J. Agardh) E. Serrão, T.O. Cho, S.M. Boo & Brawley, photo by Plocamium licensed by CC BY-NC/resized from the original; and (**E**) *Pelvetiopsis limitata* (Setchell) N.L. Gardner, photo by Peter D. Tillman licensed by CC BY/resized from the original.

Ascophyllum and *Pelvetia* are two monotypic genera, i.e., each comprises solely one species, namely *A. nodosum* and *P. canaliculata*, respectively, and are both exclusive to the North-Atlantic, although the latter is only found in the European coastlines [1,5,8–10]. As the most tolerant species to the exposure conditions, *P. canaliculata* forms a zone at the upper region of the shore, sometimes growing among coarse grass and other longshore angiosperms [11].

On the other hand, *Pelvetiopsis* and *Silvetia* are two genera from Fucaceae that are exclusive to the North-Pacific, the former distributed from south Canada to north California, while the latter covers the west coast of North America and has also been reported to occur in the Japan, China and Korea coastlines [10,12,13]. *Silvetia* species were originally classified as members of *Pelvetia*; however, owing to differences in oogonium structures and rDNA sequences, the new genus was created in 1999 [14]. Due to the lack of scientific interest in these two genera, little is known about them.

The use of Fucaceae, alongside other seaweeds, has long been part of human activities with applications in the most varied fields. Historically, *Ascophyllum*, *Fucus*, *Pelvetia* and *Silvetia* have been harvested and used as a food source for humans, typically in countries from Far East Asia, where seaweed consumption is part of their culture. Furthermore, although with less incidence, some *Fucus* species have also been consumed as foods in coastal countries of Western Europe and Alaska [15]. In the Azores Islands, the swollen receptacles of *F. spiralis* are a popular delicacy, known as sea lupines and eaten fresh [16].

Besides being used as food, *Ascophyllum* sp. and *Fucus* spp. have been used for distinct purposes over the centuries. Note that these seaweeds are often known as kelp, which is the name of the alkaline ashes produced from brown algae and used as an alkali agent for soap, paper and glass production, dying and in linen bleaching during the eighteenth–nineteenth centuries [17,18]. Later in the 1940s, *A. nodosum* was the most important feedstock for the business of alginate production in countries such as Ireland, Scotland and Norway, which were the principal suppliers of this phycocolloid [18,19]. However, because this species is relatively costly to harvest and it has a lower extract quality compared to other species, its use for this purpose has dramatically decreased during the recent years and been replaced by more attractive and versatile seaweeds including *Laminaria hyperborea* and *Lessonia* spp. [20]. Nevertheless, due to their combination of macro- and micro-nutrients, as well as the presence of natural plant growth hormones and other biostimulants, *A. nodosum* and, to a lesser extent, *Fucus* spp. continue to be used as biofertilizers [21–24], animal nutrition [25–27] and pest control [28–31]. Indeed, nowadays, *A. nodosum* has found its major application in the fertilizers, animal feed and phytopharmaceuticals industries, it being possible to find a series of *A. nodosum*-based products, such Acadian®, Agri-Gro Ultra, Alg-A-Mic, Maxicrop, Nitrozime, Soluble Seaweed Extract, Stimplex®, Tasco® and several others currently available on the market.

In turn, the current most popular application of *Fucus* spp. is for the treatment of goiter, i.e., the swelling of thyroid, and thyroid-related complications caused by iodine deficits. In fact, *F. vesiculosus* along with *Laminaria* sp. were the original sources of iodine, found in 1811 by Bernard Courtois [32]. This element was further described by Moro and Basile [33] as the most important active principle of *F. vesiculosus*, since it is essential for the production of thyroid hormones, which in turn are responsible for the increase of the metabolism in most tissues and consequently raise the basal metabolic rate [32]. Because of that, *F. vesiculosus* supplements are commonly used not only for the treatment of goiter, but also for treating obesity [34]. *F. vesiculosus* has also been commonly used for the treatment of rheumatoid arthritis, asthma, atherosclerosis, psoriasis and skin diseases, as well as several other complications [35–37]. Likewise, *Ascophyllum* and *Pelvetia* are endowed with several medicinal properties including antioxidant [38,39], anticoagulant [40,41], anti-inflammatory [41,42], antitumor [43,44] and antidiabetic [45–47], among others. In addition, these Fucaceae can be currently found in the ingredient labels of a dozen cosmetic products used as antiaging, anti-wrinkle, anti-photoaging, slimming, moisturizing and skin-whitening agents [48,49].

Among the various Fucaceae secondary metabolites, one can detach the importance of the phlorotannins, i.e., a class of phenolic compounds that is found exclusively in marine organisms, particularly in brown macroalgae [50]. These are very hydrophilic compounds, consisting of dehydro-oligomers or dehydro-polymers formed through C–C and/or C–O–C oxidative coupling of phloroglucinol (1,3,5-trihydroxybenzene) monomeric units, which are biosynthesized through the acetate–malonate pathway [51]. Phlorotannins may be found in a wide range of molecular sizes, comprised between 126 Da and 650 kDa [52], and according to the number of hydroxyl groups and nature of the structural linkages between phloroglucinol units, they can be characterized into four different subclasses: fuhalols and phlorethols (possessing an ether linkage), fucols (possessing an aryl linkage), fucophlorethols (possessing an ether and aryl linkage) and eckols and carmalols (possessing a dibenzodioxin linkage) [53]. From these subclasses, the most commonly found in Fucaceae are fucols and fucophlorethols (Figure 2).

Phlorotannins have been suggested to be multifunctional in brown seaweeds, with putative roles as primary cell wall components also involved in its biosynthesis and defensive mediators against natural enemies, working as herbivore deterrents, digestive inhibitors and as antibacterial and antifouling agents [54–56]. Besides, they also contribute to the protection of algae against ultraviolet radiation (UV-B) and may act as chelators of metal ions [57–59]. These compounds are known to accumulate mainly in the cell cytoplasm in specialized membrane-bound vesicles named physodes, representing up to 25% of seaweed´s dry weight (DW) [54].

During the last few years, an increasing interest has been paid to these algal metabolites since they have been demonstrated to exert numerous biological activities with potential application in food, pharmaceutical and cosmetic industries, among others. Because of their high abundance in phlorotannins, most studies involving the bioactivities of these phenolic compounds have been performed mainly with Laminariales, particularly those belonging to the Lessoniaceae family including *Ecklonia* spp. and *Eisenia* spp., while other algae families, such as Sargassaceae or Fucaceae, which could also represent a good source of these compounds, remain virtually unexploited. In this context, Fucaceae is of particular interest since, contrary to Sargassaceae (with some exceptions), most of the species from this family are considered edible. Therefore, and because of Fucaceae's abundance in phlorotannins, these seaweeds might be of particular economic interest as they own great potential to be used as natural raw ingredients in foods, nutraceutical or pharmaceutical industries [60]. In this context, this manuscript revises the major biological properties described so far for the Fucaceae family, with special focus on their phlorotannin composition and importance for such effects, hoping to contribute to boost their industrial interest and utilization.

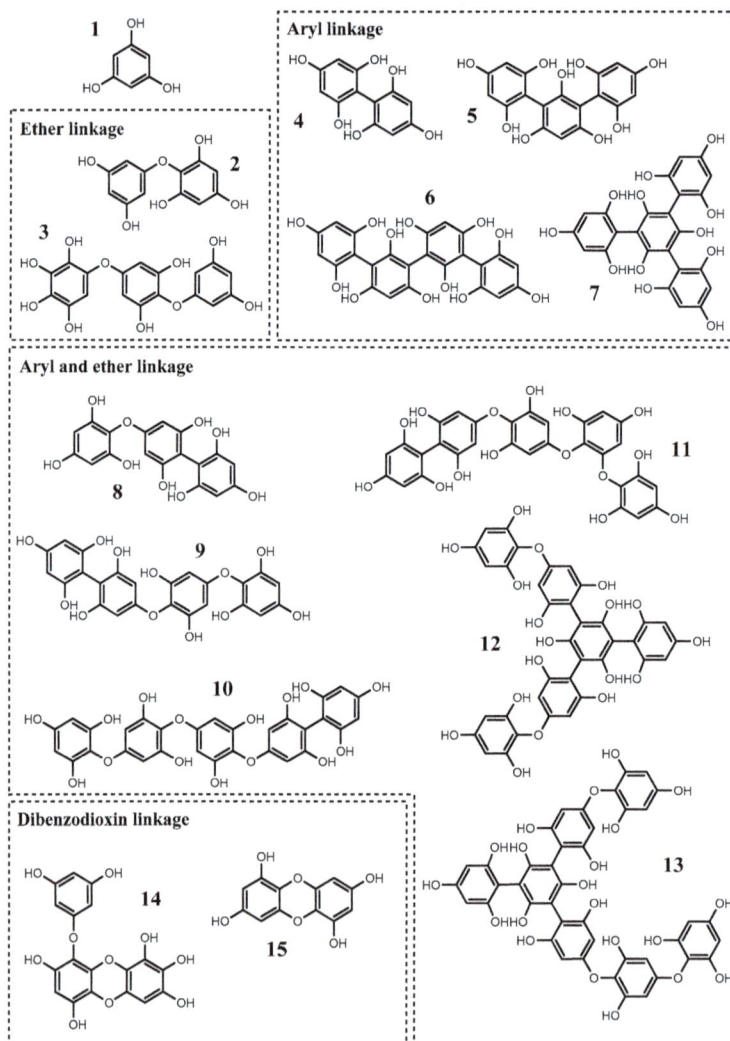

Figure 2. Structure of phlorotannins isolated from algae belonging to Fucaceae: (**1**) phloroglucinol; (**2**) diphlorethol; (**3**) hydroxytrifuhalol; (**4**) difucol; (**5**) trifucol; (**6**) tetrafucol A; (**7**) tetrafucol B; (**8**) fucophlorethol; (**9**) fucodiphlorethol; (**10**) fucotriphlorethol A; (**11**) fucotriphlorethol E; (**12**) trifucodiphlorethol A; (**13**) trifucotriphlorethol A; (**14**) 7-hydroxyeckol; and (**15**) phloroglucinol C–O–C dimer.

2. Phlorotannins from Fucaceae

Although some authors have reported the presence of phenolic acids and flavonoids in brown algae [61,62], phlorotannins represent their major phenolic constituents and, therefore, also in Fucaceae [63]. In fact, phlorotannins have been reported as the only phenolic compounds in *F. vesiculosus* [64,65]. According to what was revised by Holdt et al. [60], the highest phlorotannin contents registered for *A. nodosum* and *Fucus* sp. are 14% and 12% dry weight, respectively. Nevertheless, Fucaceae phlorotannins are very susceptible to inter-species variations. Connan et al. [66] observed that species such as *A. nodosum* and *F. vesiculosus* growing in the mid-tide zone have the highest content in phenolics (about 5.8% dry weight), while those growing in the lower intertidal level, such as *F. serratus*, have a lower phenolic content (4.3% dry weight), and the species growing at the upper level of the intertidal zone, such as *F. spiralis* and *P. canaliculata*, contain the lowest phenolic content (3.9% and 3.4% dry weight, respectively). Moreover, phlorotannins are also subject to significant intra-species variability depending on several factors such as algae size, age, tissue type and environmental factors, including nutrients, light, salinity, water depth and season [67]. The seasonal variations observed by Ragan and Jensen for *A. nodosum* and *F. vesiculosus* indicated that the polyphenols content was minimum (approximately 9–10% and 8–10% of dry matter, respectively) at the end of spring, during the period of fertility, and maximum (approximately 12–14% and 11–13% of dry matter, respectively) during the winter [68]. However, contradictory results were later reported, revealing that the phlorotannin peak of these Fucaceae occurs during the summer, matching with the higher solar exposure period and thus agreeing with the UV-protective functions invoked for these compounds [66]. Likewise, in Pavia and Toth's [67] experiments, the authors observed that the thalli from *A. nodosum* and *F. vesiculosus* that had been exposed to sunlight contained higher phlorotannins than the shaded ones. Observations of maximum phlorotannin content in the summer, when the irradiance is highest, has been described for other Fucales as well [69–71], and current evidence suggests that the production of phlorotannins by seaweeds is tightly correlated with UV radiation [72–74].

Salinity is another parameter considered determinant for phlorotannin concentrations in seaweeds since, according to Pedersen [75], the phenolic content of *A. nodosum* and *F. vesiculosus* increases with increasing salinity in their habitats. Further research confirmed that the decrease of the salinity coincided with high exudation of *A. nodosum* and *F. vesiculosus* phenolics into the surrounding water, thus resulting in a significant reduction of the phenolic content of these two species [58].

Identification and characterization of phlorotannins from brown algae has been a challenging subject since, in addition to their high susceptibility to oxidation and lack of commercially available standards, the large size and complexity, structural similarity and reactivity with other compounds make them very difficult to isolate and purify from such polymeric mixtures as crude seaweed extracts [76,77]. Therefore, the exact characterization of phlorotannins commonly requires the combination of ultra-performance liquid chromatography (UPLC) (equipped with column technologies capable of resolving extremely polar complex polymer mixtures) with mass spectrometry (MS) and nuclear magnetic resonance (NMR) techniques [78,79]. Nevertheless, a few works have already focused on the phlorotannin profile from *Fucus* spp., *Pelvetia canaliculata* and *Ascophyllum nodosum* in terms of degree of polymerization (DP). In *A. nodosum* and *P. canaliculata*, phlorotannins of DP 6–13 were found to be predominant, while *F. spiralis* was particularly rich in compounds with lower DP (4–6) [79]. Similarly, Steevensz et al. [80] reported that higher DP phlorotannins were observed with more abundance in *P. canaliculata* > *F. vesiculosus* > *A. nodosum* > *F. spiralis*. Interestingly, phloroglucinol monomers up to 39 units were detected in all of these seaweeds, except *P. canaliculata*, which contained phlorotannins composing up to 49 monomeric units. This fact has been hypothesized by the authors to be correlated with the higher exposure of this species to extreme conditions, consequently requiring more complex phlorotannin structures for their protection.

In addition to DP studies, some authors were also able to isolate and identify phlorotannins from *F. vesiculosus* including phloroglucinol (1 in Figure 2), difucol, trifucol and tetrafucols A and B; fucophlorethol, fucodiphlorethol and fucotriphlorethols A and E; trifucodiphlorethol A and trifucotriphlorethol A (4–13, respectively, in Figure 2) [77,81–84]. More recently, compounds such as hydroxyfuhalol A, difucol/diphlorethol, tetrafucol, fucodiphlorethol, 7-hydroxyeckol and the C–O–C dimer of phloroglucinol (2–4, 6, 9, 14 and 15, respectively, in Figure 2) were identified in *A. nodosum* extracts, as well [85].

3. Biological Activities

Although phlorotannins have been subject to thorough research focusing on their numerous potential biological activities, the majority of these studies have been performed with extracts from *Ecklonia* spp. or *Eisenia bicyclis* [52]. Still, interesting results focusing on phlorotannins extracted from brown algae belonging to Fucaceae, mostly from *Fucus* spp. and *Ascophyllum nodosum*, have arisen. Phenolic extracts from these algae have been demonstrated to exhibit various biological activities including antioxidant, anti-inflammatory, antimicrobial, antidiabetic and several others that could be of great interest for the development of new functional and/or therapeutic agents with high value for the food and pharmaceutical industries, thus strengthening the commercial exploitation of such macroalgae.

3.1. Antioxidant Activity

As phenolic compounds, the most characteristic biological effect of phlorotannins is the antioxidant activity. Among four species of brown algae, including *Cystoseira nodicaulis*, *Himanthalia elongata*, *F. serratus* and *F. vesiculosus*, the latter exhibited a total phenolic content (TPC) of 232.0 μg phloroglucinol equivalents (PE)/mg ethanolic extract, corresponding to the extract with the highest phenolic abundance, and the strongest activity in ferric reducing antioxidant power (FRAP) and 1,1-diphenyl-2-picrylhydrazyl radical (DPPH$^\bullet$) assays (307.3 μg trolox equivalents (TE)/mg extract and IC_{50} = 4 μg/mL, respectively). In turn, *C. nodicaulis* ethanolic extract, which yields the lowest phenolic content (89.1 μg PE/mg extract), tendentially revealed the lowest antioxidant activity in these two methods (101.4 μg TE/mg extract and IC_{50} = 28.0 μg/mL, respectively) [86]. A similar study performed with ten species belonging either to green, red or brown algae revealed that the group of Fucaceae (*F. vesiculosus*, *F. serratus* and *A. nodosum*) gave origin to the richest acetone (70%, *v:v*) extracts in terms of phenolic content, representing 24.2, 24.0 and 15.9 g PE/100 g extract, respectively. Likewise, these three were the most active antioxidant extracts, revealing EC_{50} values of 10.7, 11.0 and 18.5 μg/mL, respectively, in DPPH$^\bullet$ and oxygen radical antioxidant capacity (ORAC) values of 2.57, 2.55 and 1.42, respectively, against the >25.8 μg/mL and >0.98 mmol TE/g extract observed for the remaining extracts [87]. Nevertheless, although this evidence indicates a strong correlation between antioxidant activity and total phenolic content, this might not always be true in every case. In fact, according to O'Sullivan and co-workers [88], despite that the total phenolic content of the methanolic extract (60%, *v:v*) from *F. vesiculosus* only accounted for 2.5 mg gallic acid equivalents (GAE)/g DW, this exhibited an overall antioxidant activity in the FRAP (109.8 μM ascorbic acid/g DW), DPPH$^\bullet$ (31.2% radical scavenging) and β-carotene bleaching inhibition (71.2% protection) assays, which was better than the equivalent extracts from *A. nodosum* (81.0 μM ascorbic acid/g DW, 25.6 and 76.3%, respectively), *P. canaliculata* (71.5 μM ascorbic acid/g DW, 7.3 and 53.9%, respectively) and *F. serratus* (113.5 μM ascorbic acid/g DW, 5.5 and 62.2%, respectively), all containing approximately 4 mg GAE/g DW. Although this fact could result from the contribution of non-phenolic compounds present in the extract, it may also suggest that more important than the total phenolics content in the extract is the nature of such compounds. Indeed, when considering the specific activity of phlorotannins, Breton et al. [89] observed that the oligophenols fraction (<2 kDa) from *A. nodosum* methanol 100% extract revealed an antioxidant index (AI_{50}) values (i.e., the amount of phenols in μg contained in the fraction necessary to obtain 50% of inhibition in the DPPH$^\bullet$ assay) below 20 μg, whereas the fraction of >50 kDa phenols exhibited AI_{50} values of 34 μg, thus evidencing the importance of the molecular weight for the physiological roles and putative function of phlorotannins. Through an electrochemical approach, the specific activity was also found stronger for subfractions of 2–50 kDa and <2 kDa isolated from *A. nodosum* methanol 100% extracts, rather than subfractions over 50 kDa (AI_{50} = 0.24, 0.78 and 1.24×10^3 μM PE/L, respectively), and 1–4-times more active than the corresponding subfractions obtained from the crude methanol 50% extract, indicating differences on phlorotannins activity based on polarity [90]. However, when testing the relationships between the degree of polymerization, molecular size and antioxidant activity of different molecular weight subfractions obtained from *F. vesiculosus* ethanol 80% extracts, no clear correlations were found, except for the Fe^{2+} chelating ability, which was greater for the 100–300 and >300 kDa subfractions (47.6 and 45.1%, respectively) than for the 30–100, 5–30 and <5 kDa subfractions (36.6, 33.7 and 25.1%, respectively) [91].

Cérantola et al. [92] showed that fucol and fucophlorethol polymers, both isolated from *F. spiralis*, presented identical Q_{50} values (approximately 33 μg), i.e., the amount of compound in μg necessary to obtain 50% of inhibition in DPPH$^{\bullet}$ assay, which in turn were lower than those obtained for ascorbic acid and phloroglucinol (38.2 and 41.7 μg, respectively), thus evidencing higher antioxidant activity than these two reference compounds. More recently, positive results were observed for three phlorotannins isolated from *F. vesiculosus*, namely trifucodiphlorethol A, trifucotriphlorethol A and fucotriphlorethol A, which revealed good DPPH$^{\bullet}$ scavenging activity (IC_{50} = 14.4, 13.8 and 10.0 μg/mL, respectively) comparable to that of phloroglucinol (IC_{50} = 13.2 μg/mL), as well as a potential for scavenging peroxyl radical three-times more strongly than that of trolox in the ORAC assay. Additionally, moderate inhibitory effects towards xanthine oxidase activity were observed for trifucotriphlorethol A [84].

Because of these promising antioxidant effects, Fucaceae seaweeds are endowed with a great potential for the development of novel antioxidant products with high commercial interest for pharmaceutical, nutraceutical, cosmetic and especially food industries. Indeed, the introduction of Fucaceae phlorotannin extracts in food matrixes has already been demonstrated to effectively act as rancidification inhibitors/retarders, thus contributing for the enhancement of their shelf-lives and standing out as good candidates for exploitation as natural food additives. In this context, Honold et al. [93] found that introducing 1.5–2 g/kg of *F. vesiculosus* ethanol 80% or acetone 70% extracts in fish-oil-mayonnaise resulted in a significant enhancement of the product's oxidative stability by reducing the hydroperoxides' formation and lipid oxidation reactions. In a similar study conducted with fish muscle, the addition of 300 mg/kg muscle of oligomeric purified phlorotannin subfractions from *F. vesiculosus* was capable of inhibiting the lipid peroxidation of the product, demonstrating an effectiveness comparable to that of 100 mg/kg propyl gallate, one of the most potent antioxidant additives in food systems [94]. O'Sullivan et al. [95] also observed that the introduction of 0.5% (w/w) of *F. vesiculosus* ethanolic extracts into raw milk had promising effects against lipid oxidation of this dairy product, although it was not well accepted from a sensorial perspective due to the green/yellowish color and fishy taste. Further studies conducted by the same research group showed that the incorporation of 0.5% (w/w) *A. nodosum* or *F. vesiculosus* extracts (ethanolic 80% and 60%, respectively) into yogurts resulted in good inhibitory effects against lipid oxidation, without affecting the product's acidity, microbiology or whey separation parameters. Once again, introduction of *F. vesiculosus* extract was sensorially rejected, while yogurts with *A. nodosum* extract were generally well accepted by the panelists [96]. The promising antioxidant effects of Fucaceae phenolic extracts have also been demonstrated in cellular models (Table 1).

In Wang et al. work [91], five phlorotannin subfractions from *F. vesiculosus* (separated by dialysis according to their different molecular weights) produced a decrease in ROS production inversely proportional to the compounds molecular weight in phorbol-12-myristate-13-acetate (PMA)-induced human mononuclear cell primary cultures. Indeed, incubation of Raw 264.7 macrophages with two *F. vesiculosus* ethanolic extracts (Ext1, 35%, and Ext2, 70%) resulted in the reduction of PMA or lipopolysaccharide (LPS)-stimulated $O_2^{\bullet-}$ production, the former showing IC_{50} values of approximately 38 μg/mL in both assays, while the latter was more effective towards PMA rather than LPS stimulation (IC_{50} = 31 and 68 μg/mL, respectively) [97]. *A. nodosum* phlorotannin extract at 0.2% was also shown to significantly reduce the *tert*-butyl hydroperoxide (*t*-BHP)-induced ROS production in epithelial cells to levels close to the negative control [38].

Table 1. Selected studies of antioxidant activity of phlorotannin extracts of some Fucaceae, as measured by in vitro and in vivo biological models.

Extraction Method	Model	Treatment Conditions	Effect	References
F. vesiculosus				
EtOH 80% → fractionation with n-Hex and EtOAc → subfractionation of EtOAc in Sephadex LH-20	PMA-treated mononuclear cells from human blood	10 µM PMA + 1.5 µg/mL of 6 different EtOAc sub-fractions	All sub-fractions (except the 4th) ↓ ROS levels below 65%	[91]
MeOH 60%	Caco-2 cells	100 µg/mL of extract for 24 h	↑ GSH levels by 31.9%	[88]
MeOH 60%	H_2O_2-induced Caco-2 cells	24 h pre-treatment with 100 µg/mL of extract for 24 h + 200 µM H_2O_2	Restored SOD levels from 64.9 to 89% and ↓ 9.5% of the DNA damage	[39,88]
MeOH 60%	t-BHP-induced Caco-2 cells	100 µg/mL of extract for 24 h + 200 µM t-BHP	↓ apx. 12% DNA damage in t-BHP-induced cells	[98]
Ext1: EtOH 35% Ext2: EtOH 70%	In vitro: PMA or LPS-induced Raw 264.7 cells In vivo: Sprague–Dawley rats	In vitro: 100 ng/mL PMA or LPS + different concentrations of extracts In vivo: oral treatment with 200 mg/kg/day during 4 weeks	In vitro: Ext2: ↓ of $O_2^{\bullet-}$ in PMA-induced cells (IC_{50} = 31 µg/mL), Ext1: ↓ of $O_2^{\bullet-}$ in both cell models (IC_{50} = 38 and 39 µg/mL, respectively); In vivo: Ext2: ↑ reducing power, PON-1 activity and $O_2^{\bullet-}$ scavenging activity in the blood plasma (29%, 33% and 25%, respectively)	[97]
F. serratus				
MeOH 60%	Caco-2 cells	100 µg/mL of extract for 24 h	↑ GSH levels by 37.4%	[88]
MeOH 60%	t-BHP or H_2O_2-induced Caco-2 cells	100 µg/mL of extract for 24 h + 1 mM t-BHP or 200 µM H_2O_2	Restored SOD levels in both t-BHP and H_2O_2-induced cells from 73.9–108% and 64.9–89.5%, respectively, and ↓ 13.2% of the H_2O_2-induced DNA damage	[39,88]
Ext1: H_2O Ext2: EtOH 80%	t-BHP-induced Caco-2 cells	100 µg/mL of extracts for 24 h + 1 mM t-BHP	Both extracts ↓ apx. 13% DNA damage in t-BHP-induced cells	[98]

Int. J. Mol. Sci. **2017**, *18*, 1327

Table 1. *Cont.*

Extraction Method	Model	Treatment Conditions	Effect	References
		A. nodosum		
Extract with 18% phlorotannins	*t*-BHP-induced ARPE-19 and WKD cells	0.1–0.5% extract for 20 min + 500 μM *t*-BHP	↓ ROS production close to the negative control on cells treated with 0.2% extract	[38]
MeOH 60% → digestion with pepsin at 37 °C and pH 2 → digestion with pancreatin/bile extract at 37 °C pH 6.9 → dialysis with cutoff at 1 kDa	*t*-BHP-induced HepG-2 cells	0.5–50 μg/mL of extract for 20 h + 400 μM *t*-BHP	↓ ROS and lipid, restored GSH levels to apx. 75% and regulated the activity of GSH-px, GSH-red GSH-tr	[99]
MeOH 60%	Caco-2 cells	100 μg/mL of extract for 24 h	↑ GSH levels by 35.5%	[88]
MeOH 60%	H$_2$O$_2$-induced Caco-2 cells	100 μg/mL of extract for 24 h + 200 μM H$_2$O$_2$	Restored SOD levels from 64.9–89.5%	[88]
Ext1: H$_2$O Ext2: EtOH 60% Ext3: EtOH 80%	*t*-BHP or H$_2$O$_2$-induced Caco-2 cells	100 μg/mL of extracts for 24 h + 1 mM *t*-BHP or 200 μM H$_2$O$_2$	Ext1: ↓ 20% H$_2$O$_2$-induced DNA damage; Ext2: ↓ apx. 15% *t*-BHP -induced DNA damage, Ext3: ↓ apx. 13% DNA damage in both models	[98]
		P. canaliculata		
MeOH 60%	Caco-2 cells	100 μg/mL of extract for 24 h	↑ GSH levels by 38.7%	[88]
MeOH 60%	*t*-BHP or H$_2$O$_2$-induced Caco-2 cells	100 μg/mL of extract for 24 h + 1 mM *t*-BHP or 200 μM H$_2$O$_2$	Restored SOD levels from 73.9–97% and 64.9–97.4%, respectively	[39,88]

apx., approximately; EtOAc, ethyl acetate; EtOH, ethanol; Ext, extraction; GSH, glutathione; GSH-px, glutathione peroxidase; GSH-red, glutathione reductase; GSH-tr, glutathione transferase; LPS, lipopolysaccharide; MeOH, methanol; *n*-Hex, *n*-hexane; PON-1, paraoxonase 1; SOD, superoxide dismutase; PMA, phorbol-12-myristate-13-acetate; ROS, reactive oxygen species; *t*-BHP, and *tert*-butyl hydroperoxide. Cell lines: ARPE-19, human retinal pigment epithelium; Caco-2, human epithelial colorectal adenocarcinoma; HepG-2, liver hepatocellular carcinoma; Raw 264.7, murine macrophages; and WKD, human conjunctival cells.

Quéguineur et al. [99] further observed that a digested-dialyzed phlorotannin extract (rich in compounds over 1 kDa) from *A. nodosum* not only caused the reduction of intracellular ROS and lipid peroxidation in *t*-BHP-induced HepG-2 cells, as also enhanced their endogenous antioxidant defenses by increasing the levels of glutathione (GSH) and the enzyme activities of GSH-peroxidase (GSH-px), GSH-reductase (GSH-red) and GSH-S-transferase (GSH-tr). Augmented levels of GSH (32–39% higher than the control) were observed as well on Caco-2 cells incubated not only in the presence of *A. nodosum* hydromethanolic extract at 100 μg/mL, but also with those of *P. canaliculata*, *F. vesiculosus* and *F. serratus*. Furthermore, all of these extracts, particularly that of *P. canaliculata*, could almost completely restore the H_2O_2-induced depletion of superoxide dismutase (SOD) activity, although only two Fucaceae extracts, namely *F. serratus* followed by *F. vesiculosus*, exhibited a reduction of H_2O_2-induced oxidative damage to DNA (from 63% in control to 53% and 50%, respectively) [88]. The same *F. serratus* and *F. vesiculosus* methanolic 50% extracts were posteriorly confirmed to exhibit DNA protective effects in Caco-2 cells treated with H_2O_2, but not with *t*-BHP, although both *F. serratus* alongside with *P. canaliculata* extracts completely restored the SOD activity that was impaired by the *t*-BHP stimulation [39]. However, when testing extracts obtained by different procedures, instead of observing DNA protective effect against H_2O_2-induced oxidative damage, *F. serratus* (aqueous and ethanolic 80% extracts) and *F. vesiculosus* (methanolic 60% extract) revealed a decrease of the *t*-BHP-induced DNA damage of approximately 50% compared to the non-treated Caco-2 cells, most likely due to the extraction of phlorotannins with different polarities. In turn, *A. nodosum* ethanolic 80% extracts exhibited DNA protective effects either in the presence of H_2O_2 or *t*-BHP, while the ethanol 60% extract was only active against H_2O_2 and the aqueous extract was effective against *t*-BHP [98]. Hence, overall, the above-mentioned works point out the promising effects of phlorotannins of Fucaceae origin towards distinct oxidative stress events. Still, it is relevant to note a common flaw in all of these studies, i.e., the lack of comparison of the antioxidant activities of these seaweed extracts and/or phlorotannin compounds with that of well-known compounds. The gathering of this information would be helpful to achieve a better comprehension of the actual potential of these compounds.

In vivo experiments conducted by Zaragozá et al. [97] revealed that the feeding of Sprague–Dawley rats with *F. vesiculosus* phenolic extracts resulted in an increased blood plasma antioxidant activity slightly better than that of phloroglucinol, which is commonly used as a standard compound. In more detail, after a four-week oral treatment of 200 mg/kg body weight/day of *F. vesiculosus* ethanol 70% extract, the reducing power, paraoxonase 1 (PON-1) activity and $O_2^{\bullet-}$ scavenging activity in the plasma were increased by 29%, 33% and 25%, respectively. Phloroglucinol administered in the same conditions also produced positive, although slightly lower effects in these parameters, causing a 31% and 12% increase of reducing power and PON-1 activity, respectively, and no activity against $O_2^{\bullet-}$. The fact that thiobarbituric acid reactive substances (TBARS) were also reduced by 17% in the *F. vesiculosus*-treated rats and 12% in phloroglucinol-treated group might be a consequence not only the plasma's increased ability to scavenge free radicals, but also PON-1's greater hydrolytic activity. Particularly, in the phloroglucinol-treated group, in which no effects were seen on $O_2^{\bullet-}$, the increase of this enzyme activity might be the major cause for TBARS reduction since this enzyme is known for protecting low-density lipoproteins from oxidative modification by ROS and contributing for the degradation of hydrogen peroxide (peroxidase activity) [100].

In several studies, phloroglucinol was proven to display a very pleiotropic role in oxidative stress events. Indeed, this compound was shown to reduce several oxidative stress hallmarks in numerous cell lines, stimulate the intracellular antioxidant defenses including the activation of nuclear factor (erythroid-derived 2)-like 2 (Nrf2) [101–106] and even positively contribute for photoprotective effects on skin [107] and improvement of motor functions and oxidative damage in the brain of animal models of Parkinson's disease [105,108].

3.2. Antidiabetic Activity

In 2012, diabetes mellitus was the direct cause of 1.5 million deaths, reaching an estimated prevalence of approximately 9% among the worldwide adult population in 2014. Moreover, in 2030, it is projected that this disease will be the 7th main cause of death in the world [109]. In the specific case of type 2 diabetes mellitus, the most common therapeutic targets are α-amylase and α-glucosidase, two enzymes responsible for the starch hydrolysis releasing the glucose monomers for subsequent absorption by the small intestine. Therefore, the inhibition of these enzymes reduces the availability of free glucose monomers and consequently decreases the postprandial peak of blood glucose levels [110].

In this context, phenolic extracts from *Fucus* spp. and particularly from *A. nodosum* have demonstrated promising effects against these enzymes (Table 2). Per Zhang et al. [45], the inhibitory effects of different fractions from *A. nodosum* ethanol 50% extracts towards α-glucosidase activity was highly correlated with their phlorotannin content, as the lowest IC_{50} value (24.0 µg/mL) was observed for the C18 purified ethyl acetate fraction (TPC = 70.2% PE), followed by non-purified ethyl acetate fraction (IC_{50} = 38.0 µg/mL; TPC = 39.8% PE) and crude ethanol extract (IC_{50} = 77.0 µg/mL; TPC = 22.5% PE). When comparing the α-glucosidase inhibitory activities of ethanol 96% and acetone 70% extracts from *A. nodosum* and *F. vesiculosus*, both rich in phlorotannins, to that of acarbose (i.e., a well-known inhibitor of α-glucosidase and α-amylase currently used as an antidiabetic drug), IC_{50} values of 8.9 and 0.72 µg/mL, respectively, for the former, and 4.4 and 0.34 µg/mL, respectively, for the latter were obtained, corresponding to an inhibitory activity 160–2000-times stronger than that of acarbose (IC_{50} = 720 µg/mL) [111]. The methanolic extract of *P. siliquosa* (currently *S. siliquosa*), also rich in phlorotannins, was shown to be an effective inhibitor of α-glucosidase, as well [112]. It should be noted that the biological effects of Fucaceae algae towards this enzyme can be significantly affected depending on the harvesting season. In the specific case of *A. nodosum*, the highest inhibitory activity against α-glucosidase was observed during the summer, more precisely in July, when the authors found the highest phlorotannin accumulation for this species [113].

In addition to the strong inhibitory effect against α-glucosidase, *A. nodosum* extracts were also proven to display inhibition towards α-amylase [46]. Indeed, an acetonitrile 50% extract from *A. nodosum* purified in a solid-phase extraction (SPE) column was shown to exert higher inhibitory activity on α-amylase rather than on α-glucosidase, with an IC_{50} value eight-times lower than that of acarbose (0.8 µg/mL) [114]. Similar results were described for phlorotannin-purified fractions of an extract from *F. distichus*, which were capable of reducing the activity of both α-glucosidase and α-amylase 126- and 10-times more effectively than the above-mentioned pharmaceutical drug [115]. The aqueous and ethanolic 80% extracts of *A. nodosum*, *F. vesiculosus*, *F. serratus*, *F. spiralis* and *P. canaliculata* presented inhibitory properties against these two enzymes comparable to that of acarbose as well, although depending on the extract procedure, some differences could be observed. In particular, for the aqueous extracts, the strongest α-amylase inhibitor was *A. nodosum* (IC_{50} = 53.6 µg/mL), followed by *F. vesiculosus* > *P. canaliculata* > *F. serratus* > *F. spiralis*, while for the ethanol extracts, *A. nodosum* (IC_{50} = 44.7 µg/mL) still exhibited the best activity, but the *P. canaliculata* extract was more active than that of *F. vesiculosus*. These differences were more evident in the case of α-glucosidase. The aqueous extracts from *F. vesiculosus* and *P. canaliculata* exhibited similar inhibitory activity (IC_{50} approximately 0.3 µg/mL), followed by *A. nodosum* > *F. serratus* > *F. spiralis*, while for the ethanol extract, *F. vesiculosus* maintained the strongest activity (IC_{50} = 0.49 µg/mL), but *P. canaliculata* activity was only followed by that of *F. spiralis*. It is worth noting that overall, the ethanol extracts were more effective against α-amylase, while the opposite was observable for α-glucosidase. Nevertheless, with the exception of *F. spiralis*, α-amylase inhibitory profiles of all aqueous and ethanolic extracts were very similar to that of acarbose. Notably, both *F. vesiculosus* extracts exhibited stronger α-glucosidase inhibitory activity than that of the pharmaceutical drug [47].

Table 2. Selected studies of the anti-diabetic activity of phlorotannin extracts of some Fucaceae, as measured in vitro and in vivo.

Extraction Method	Model	Test Conditions	Effect	References
F. vesiculosus				
Sequential extraction with $CHCl_3 \rightarrow$ EtOH 96% $\rightarrow$ Ac 70%	Measurement of α-glucosidase activity	Crescent concentrations of extracts	EtOH and Ac extracts had the highest inhibitory activity (IC_{50} = 4.4 and 0.34 µg/mL, respectively)	[111]
Ext1: H_2O / Ext2: EtOH	Measurement of α-glucosidase and α-amylase activities	0.1–1000 µg/mL of extracts	↓ enzymatic activity (α-glucosidase: IC_{50} = 0.32 and 0.49 µg/mL, respectively; α-amylase: IC_{50} = 59.1 and 63.5 µg/mL, respectively)	[47]
Ac 70% $\rightarrow$ fractionation with DCM, EtOAc and But $\rightarrow$ subfractionation of EtOAc in Sephadex LH-20 (F1–F4)	BSA-methylglyoxal and BSA-glucose assay	Crescent concentrations of fractions or sub-fractions	Strong ↓ BSA glycation by subfractions, (EC_{50} apx. 0.16 mg/mL for F1–F4 in BSA-methylglyoxal and 0.05 mg/mL for F1 and F2 in BSA-glucose)	[116]
F. distichus				
EtOH 80% $\rightarrow$ Fractionation with *n*-hex, EtOAc, 1-But $\rightarrow$ subfractionation of EtOAc in Sephadex LH-20	Measurement of α-glucosidase and α-amylase activities	1.5–200 µg/mL of subfractions	Subfraction 22 showed ↑ inhibitory activity (IC_{50} = 0.89 and 13.98 µg/mL, respectively)	[115]
A. nodosum				
EtOH 50% at 80 °C $\rightarrow$ Fractionation with EtOAc and 1-But $\rightarrow$ purification in C18 column	Measurement of α-glucosidase activity	Crescent concentrations of fractions	Purified fraction showed ↑ inhibitory activity (IC_{50} = 24 µg/mL)	[117]
Sequential extraction with $CHCl_3 \rightarrow$ EtOH 96% $\rightarrow$ Ac 70%	Measurement of α-glucosidase activity	Crescent concentrations of extracts	Ac extracts showed ↑ inhibitory activity (IC_{50} = 0.72 µg/mL)	[111]
H_2O at 80 °C from algae collected at different seasons	Measurement of α-glucosidase activity	0.05–0.5 µg/mL of extract	Summer extracts have ↑ inhibitory activity (IC_{70} = 2.23 µg/mL)	[113]
Ext1: H_2O / Ext2: EtOH	Measurement of α-glucosidase and α-amylase activities	0.1–1000 µg/mL of extracts	↓ enzymatic activity (α-glucosidase: IC_{50} = n.d.; α-amylase: IC_{50} = 44.7 and 53.6 µg/mL, respectively)	[47]
EtOH 50%	2-deoxyglucose-cultured 3T3-L1 cells	50–400 µg/mL of extract for 20 min + 1 µCi/mL 2-deoxyglucose	↑ basal glucose uptake by 3-fold at 400 µg/mL	[117]
ACN:0.2% CH_2O_2 (1:1) $\rightarrow$ purification in SPE column $\rightarrow$ fractionation in Sephadex LH-20	Measurement of α-glucosidase and α-amylase activities in absence or presence of acarbose	Phlorotannin fraction: 2.5–100 µg GAE/mL for α-glucosidase and 50–400 µg GAE/mL for α-amylase; acarbose + phlorotannin fraction: 1 µg/mL + 0.1 µg/GAE –0.25 µg/mL + 0.025 µg/GAE	↓ enzymatic activity (α-glucosidase: IC_{50} = 10 µg GAE/mL, α-amylase: IC_{50} = 0.15 µg GAE/mL). ↓ acarbose concentration needed for an effective enzymatic inhibition (from 1–0.5 µg/mL)	[118]

Table 2. *Cont.*

Extraction Method	Model	Test Conditions	Effect	References
		P. canaliculata		
MeOH 70%	In vitro: measurement of sucrase and maltase activities	In vitro: 0–16.7 mg/mL extract	In vitro: ↓ enzymatic activity (IC_{50} = 2.24 and 2.84 mg/mL, respectively)	[112]
	In vivo: sucrose-fed Wistar rats	In vivo: oral administration of 1 mg/kg of extract + 0.5 mg/kg of sucrose	In vivo: ↓ postprandial blood glucose levels	
		A. nodosum combined with *F. vesiculosus*		
Commercial hot water extract InSea2™ (10% polyphenol content in CAE)	In vitro: measurement of α-glucosidase and α-amylase activities	In vitro: 1.25–25 μg/mL of InSea2™	In vitro: ↓ enzymatic activity (IC_{50} = 2.8 and 5 μg/mL, respectively)	[119]
	In vivo: Wistar rats fed with corn starch + safflower oil	In vivo: oral administration of 7.5 mg/kg of InSea2™ + 2 mL/kg of starch and oil (1:1)	In vivo: ↓ 90% postprandial blood glucose and ↓ 40% insulin peak	
Commercial hot water extract InSea2™ (10% polyphenol content in CAE)	Human trial	Oral administration of two capsules (500 mg) 30 min prior to carbohydrate ingestion	↓ insulin incremental area of the curve by 12.1% and ↑ insulin sensitivity by 7.9%	[120]

Ac, acetone; apx., approximately; BSA, bovine albumin serum; But, butanol; CAE, chlorogenic acid equivalents; DCM, dichloromethane; EtOAc, ethyl acetate; EtOH, ethanol; Ext, extraction; HCl, chloridric acid; GAE, gallic acid equivalents; SPE column, solid-phase extraction column; and ROS, reactive oxygen species; Cell lines: INS-1, rat pancreatic β-cells; and 3T3-L1, preadipocytes.

In Roy et al. [119], the incubation of α-amylase and α-glucosidase with a commercial mixture of *A. nodosum* and *F. vesiculosus* phlorotannin extract resulted in inhibitions of approximately 100% at concentrations below 0.2 μM. Furthermore, the immediate postprandial blood glucose levels of Wistar rats orally treated with this mixture (7.5 mg/kg) were decreased by 90%, and the peak increase of insulin secretion was reduced by 22%. In addition, in a previous study, the oral administration of 200 mg/kg/day of two different *A. nodosum* phlorotannin extracts (crude ethanol 50% extract and a HP-20 column purified ethanol 50% extract) to streptozotocin-diabetic mice fed with sucrose during four weeks was shown to improve the fasting serum glucose levels and lower the postprandial blood glucose level at the 14th day by 27% and 25% comparing to the diabetic controls [45]. Identical results were described for *P. siliquosa* (currently *S. siliquosa*) methanolic 70% extracts, which not only suppressed the enzymatic activities of sucrase and maltase in vitro (IC_{50} = 2.24 and 2.84 mg/mL), but also reduced the postprandial blood glucose levels in vivo on sucrose-fed Wistar rats orally treated with 1 g/kg body weight of this extract [112].

All these evidences suggest that the extracts from *A. nodosum* and/or *Fucus* spp. have great potential to be used either as an anti-diabetic therapeutic approach targeting α-glucosidase and α-amylase and/or as a co-ingredient of already existent pharmaceutical drugs. Indeed, Pantidos et al. [118] demonstrated that the combination of acarbose with a purified phlorotannin-rich fraction from *A. nodosum* exerted a synergistic inhibitory effect towards these two enzymes, thus allowing one to reduce the concentration of acarbose necessary for obtaining an effective inhibitory activity from 1.0–0.5 μg/mL. Moreover, in a human clinical trial, the single ingestion of 500 mg of a commercial extract mixture from *A. nodosum* and *F. vesiculosus* 30 min prior to the consumption of 50 g of carbohydrates was associated with a 12.1% reduction in the insulin incremental area of the curve and a 7.9% increase in insulin sensitivity [120].

Although the Fucaceae phenolic extracts have been mainly described for their acarbose-like effects when evaluating their anti-diabetic effects, other possible mechanisms were also reported. For example, the phenolic-rich ethanol extract from *A. nodosum* was shown to stimulate the basal glucose uptake into 3T3-L1 adipocytes, thus contributing to the reduction of blood glucose levels and the amelioration of hyperglycemia [45]. Furthermore, phloroglucinol alongside with four purified phlorotannin (from *F. vesiculosus* acetone 70% extract) fractions demonstrated very effective inhibitory activities against the bovine serum albumin (BSA)-methylglyoxal assay (IC_{50} = 58 μg/mL for phloroglucinol and approximately 160 μg/mL for algal fractions) and the BSA-glucose assay (IC_{50} = 68 μg/mL for phloroglucinol and 45–1526 μg/mL for algal fractions) and, therefore, a promising anti-advanced glycated end-products (AGEs) formation activity, i.e., a class of compounds generated by the exposure of proteins and other endogenous molecules to reducing sugars [116]. Due to the high blood glucose levels on diabetic patients, AGEs are produced in concentrations beyond the normal levels, thus leading to pathological consequences that are on the basis of the diabetic complications like retinopathy, nephropathy, neuropathy and cardiomyopathy [121]. Therefore, the ability of phloroglucinol and *F. vesiculosus* phlorotannins to prevent their formation indicate that they may contribute to the protection against the diabetic-related pathologies. Other studies with phloroglucinol demonstrated that it has the capacity to protect pancreatic β-cells from high glucose-induced oxidative stress and consequent apoptosis [122].

3.3. Anti-Inflammatory Activity

In addition to the previously mentioned biological activities, phlorotannins have also been closely related to the targeting of numerous inflammatory events. Note that inflammation is a complex and coordinated immunological response of the organism to harmful stimuli, consisting of a tightly regulated signaling cascade that is orchestrated by a series of pro-inflammatory mediators including cytokines, chemokines, adhesion molecules, enzymes and others [123]. Among these mediators, one can highlight the importance of tumor necrosis factor-α (TNF-α), whose main function is the activation of nuclear factor-κB [124], which in turn is responsible for the transcription of several genes encoding other pro-inflammatory mediators including TNF-α itself, interleukins (ILs), chemokines, adhesion molecules and key inflammatory enzymes including cyclooxygenase-2 and inducible nitric oxide synthase (COX-2 and iNOS, respectively), which further disseminate the pro-inflammatory stimuli [125]. Therefore, the described anti-inflammatory activities of Fucaceae phlorotannins are based on the screening of their ability to target one or multiple of these mediators (Table 3).

According to Zaragozá et al. [97], the production of NO• (i.e., a pivotal free radical involved in the signaling and pathogenesis of inflammation) in PMA-stimulated RAW 264.7 cells was inhibited in a dose-dependent fashion by a phlorotannin-rich *F. vesiculosus* ethanol 35% extract (IC$_{50}$ of 37 µg/mL). Likewise, a phlorotannin extract from *A. nodosum* was shown to dose-dependently decrease the LPS-induced expression of TNF-α and IL-6 in U937 macrophages [38]. Similar results were later observed in an identical cellular model, thus endorsing the hypothesis that phenolic extracts of this species could act as anti-inflammatory agents by blocking the propagation of the pro-inflammatory stimuli [126]. Bahar and co-workers [42] reported that the treatment of porcine colonic tissues ex vivo either with *A. nodosum* ethanol 80% or *F. serratus* aqueous extracts, caused a significant downregulation of the LPS-induced pro-inflammatory genes including *IL6*, *IL8* and *TNFA* (encoding for the cytokines IL-6, IL-8 and TNF-α, respectively), comparable to that of dexamethasone (i.e., a corticosteroid medication used for the treatment of inflammation and autoimmune diseases). More recently, this research group also observed that the treatment of TNF-α-challenged Caco-2 cells with an ethanol 80% extract of *A. nodosum* significantly suppressed the expression of several pro-inflammatory genes encoding cytokines (*IL8*, *TNFA*, *IL1B*, *IL18* and *CSF1*), chemokines (*CXCL10*, *CCL5*), components of the NF-κB pathway (*NFKB2* and *IKBKB*) and other mediators (*PTGS2* and *MIF*) by more than two-fold compared to the negative control. Further experiments in LPS-stimulated porcine colonic tissue ex vivo revealed that this *A. nodosum* extract caused the downregulation of immune-related genes, including *LYZ*, *IL8*, *PTGS2*, *TLR6*, *CXCL10*, *IL6*, *CXCL11*, *ICAM*, *NFKB1* and *CXCL2* [127]. Identical results were also reported for a cold water extract of *F. vesiculosus*, which inhibited the expressions (>2-fold) of the genes *IL17A* and *IL8* (encoding for cytokines), *CCL2*, *CXCL2*, *CXCL10* and *CXCL11* (encoding for chemokines), *ICAM1* and *VCAM1* (encoding for cell adhesion molecules), *TLR4* and *TLR7* (encoding for Toll-like receptors), *NFKB1* and *RELB* (encoding for NF-κB components), *MAP3K8* and *CJUN* (encoding for mitogen activated protein kinases and activator protein-1 components, respectively) and *PTGS2*, *C5* and *LYZ* (encoding for other pro-inflammatory mediators), in the same ex vivo model. Notably, Toll-like receptor 4 (TLR-4) was identified in this study as an important target for the anti-inflammatory effect of this extract [128]. It is interesting to note that dexamethasone (as shown in pig or in rat colonic tissue models) does not seem to interfere with the expression of TLR-4 [129,130]. However, further investigations are still needed in order to understand whether the *A. nodosum* anti-inflammatory bioactivity mediated through inhibition of TLR-4 expression has any distinct advantage over the inflammatory immune diseases treatments based on dexamethasone.

Such a broad spectrum of anti-inflammatory bioactivity of *A. nodosum* and *F. vesiculosus* suggests that there is a great potential for future exploitation of these seaweeds as therapeutic agents for the treatment of inflammatory conditions, particularly those related with mammalian intestine diseases, although further studies, namely in vivo, would be necessary to better evaluate the feasibility of these results.

Table 3. Selected studies of the anti-inflammatory activity of phlorotannin extracts of some Fucaceae, as measured in in vitro and ex vivo biological models.

Extraction method	Model	Test Conditions	Effect	References
F. vesiculosus				
H₂O	LPS-induced porcine colonic tissue ex vivo	1 mg/mL extract + 10 µg/mL LPS	↓ expression of the genes *IL17A, IL8, CCL2, CXCL2, CXCL10, CXCL11, ICAM1, VCAM1, TLR4, TLR7, NFKB1, RELB, MAP3K8, CJUN, PTGS2, C5* and *LYZ* >2× compared to the control	[128]
EtOH 35%	PMA-stimulated RAW 264.7	100 ng/mL PMA + different concentrations of extracts	↓ production of NO• (IC₅₀ = 37 µg/mL)	[97]
F. serratus				
H₂O	LPS-induced porcine colonic tissue ex vivo	1 mg/mL extract + 10 µg/mL LPS	↓ expression of the genes *IL8, IL6* and *TNFA* below 0.70, 0.69 and 1.15× compared to LPS control, respectively	[42]
F. distichus				
MeOH 80% → fractionation with *n*-hex, EtOAc and 1-But → subfractionation of EtOAc in flash chromatography	LPS-induced RAW 264.7 cells	12.5–50 µg/mL, a subfraction rich in fucophlorethols for 1 h + 1 µg/mL LPS	↓ expression of IL-1β, IL-6, IL-17, TNF-α, MCP-1, iNOS, COX-2, ICAM-1, TLR-4 and TLR-9 in a dose-dependent manner	[131]
A. nodosum				
Extract with 18% phlorotannins	LPS-induced U937 cells	0.05–0.2% of extract for 2 h + 0.5 µg/mL LPS	↓ levels of TNF-α and IL-6 close to control	[38]
H₂O → alginate precipitation → ultrafiltration	LPS-induced U937 cells	0.1 µg extract for 2 h + 0.5 µg/mL LPS	↓ levels of TNF-α by 94% and IL-6 by 84%	[126]
EtOH 80%	LPS-induced porcine colonic tissue ex vivo	1 mg/mL extract + 10 µg/mL LPS	↓ expression of the genes *IL8, IL6* and *TNFA* below 0.99, 0.75 and 1.01× compared to LPS control, respectively	[42]
EtOH 80%	TNF-α-induced Caco-2 cells	0.1–1 mg/mL extract + 10 ng/mL TNF-α	↓ expression of the genes *IL8, TNFA, IL1B, IL18, CSF1, CXCL10, CCL5, NFKB2, IKBKB, PTGS2* and *MIF* by >2×	[127]
EtOH 80% → dialysis fractionation into three M_w fractions (<3.5 kDa, 3.5–100 kDa, >100 kDa)	LPS-induced porcine colonic tissue ex vivo	1 mg/mL extract or M_w fractions + 10 µg/mL LPS	↓ expression of the genes *LYZ, IL8, PTGS2, TLR6, CXCL10, IL6, CXCL11, ICAM, NFKB1* and *CXCL2* by >2× either by the crude extract or the three M_w fractions	[127]

COX-2, cyclooxygenase-2; EtOAc, ethyl acetate; IL, interleukin; iNOS, inducible nitric oxide synthase; LPS, lipopolysaccharide; M_w, molecular weight; NO•, nitric oxide; PMA, phorbol-12-myristate-13-acetate; and TNF-α, tumor necrosis factor-α. Cell lines: RAW 264.7, murine macrophages; U937, human leukemic monocytes; and Caco-2, human colon epithelium.

F. distichus is another example of a Fucaceae with promising anti-inflammatory properties, comparable to those of dexamethasone. Kellogg et al. [131] reported that the fucophlorethols-rich fraction isolated from a methanolic 80% extract of this seaweed was remarkably effective against the expression of an array of inflammatory markers triggered by LPS-stimulation of RAW 264.7 macrophages, showing particular high activity towards COX-2, iNOS, IL-1β, IL-6, TNF-α, intercellular adhesion molecule-1 and TLR-4, reducing their expression to below 10% at 50 µg/mL when comparing to the LPS control. Monocyte chemoattractant protein-1, IL-17 and TLR-9 were also found strongly inhibited, below 60%, for the same concentration. Based on this data and on the fact that fucophlorethols are one of the most abundant phlorotannin groups in Fucaceae, it is possible to suggest that these compounds might be important contributors for the anti-inflammatory activity that has been observed for the phenolic extracts from this family.

3.4. Antitumor Activity

Both oxidative stress and inflammation have long been associated with the development of cancer. The production of ROS, including hydroxyl radical (OH$^\bullet$) and superoxide (O$_2^{\bullet-}$), and reactive nitrogen species (RNS), such as nitric oxide (NO$^\bullet$) and peroxynitrite (ONOO-), associated with chronic inflammatory states may lead to environments that foster genomic lesions and tumor initiation [132]. In this field, reported data suggest that Fucaceae phlorotannins can exert important chemopreventive and antiproliferative effects against some cancer cell lines (Table 4).

According to Nwosu et al. [114], a purified phlorotannin extract from *A. nodosum* origin was shown to strongly inhibit the proliferation of colon cancer cells in a dose-responsive manner, with IC$_{50}$ values of 33 µg/mL. Likewise, an HPLC fraction obtained from an *F. vesiculosus* acetone extract was reported to have potent anti-proliferative effects on different pancreatic cancer cell lines, showing EC$_{50}$ values between 17.4 and 28.9 µg/mL. The authors also mentioned that this extract affected only proliferating, but not resting cells through stimulation of cell cycle arrest, which is comparable to the effects of common chemotherapeutic drugs clinically used, such as gemcitabine [133]. Further studies from this research group concluded that the multistep fractionation of *F. vesiculosus* acetone extract through precipitation, normal phase HPLC and reversed phase HPLC could result in the obtainment of two active fractions (F15/16 and F36/37) against human pancreatic cancer (Panc89) (EC$_{50}$ of approximately 16 and 47 µg/mL for F15/16 and F36/37, respectively) and human pancreatic cancer PancTu1 (EC$_{50}$ of approximately 17 and 80 µg/mL for F15/16 and F36/37, respectively), despite that their anti-proliferative effects were far from those of the chemotherapeutic gemcitabine (EC$_{50}$ = 3.5 ng/mL and 14 ng/mL against Panc89 and PancTu1 cells, respectively) commonly used as a first line treatment for pancreatic cancer [134]. Antitumor activity against HeLa cells was reported for an *F. spiralis* dichloromethane extract, which reduced their proliferation by 50% at 10.7 µg/mL. However, the phenolic content of this extract was only 13 µg GAE/mg extract, which makes phlorotannins unlikely to contribute for these results [135]. Nevertheless, a phlorotannin extract from this species, particularly abundant in fucophlorethols, was shown to inhibit the activity of hyaluronidase, an enzyme overexpressed in breast cancer, revealing an IC$_{50}$ of 0.73 mg/mL DW, which was 2–4-times lower than the results observed for three other Fucales, namely *Cystoseira nodicaulis*, *C. usneoides* and *C. tamariscifolia* [136]. Still, one should note that these inhibitory effects are considerably lower when comparing with the IC$_{50}$ reported for other compounds such as catechin (0.18 mg/mL), epigallocatechin gallate (0.09 mg/mL) or sodium cromoglycate (0.14 mg/mL), known as good inhibitors of this enzyme [137].

Focusing three fucophlorethols isolated from *F. vesiculosus*, namely trifucodiphlorethol A, trifucotriphlorethol A and fucotriphlorethol A, Parys et al. [84] reported that the good chemopreventive properties of these compounds were due to their capacity to inhibit the activity of aromatase (an enzyme also involved in the carcinogenesis from breast and other estrogen-related cancers) and CYP1A, which is an enzyme belonging to the cytochrome P450 family and known to be involved in carcinogen activation of mutagens derived from cooked food.

Table 4. Selected studies of the antitumor activity of phlorotannin extracts of some Fucaceae, as measured in in vitro biological models.

Extraction Method	Model	Test Conditions	Effect	References
		F. vesiculosus		
Acetone 99.5% → purification by HPLC	PancTu1, Panc89, Panc1 and Colo357 cells	12.5–100 µg/mL of purified extract	↓ cell proliferation, ↑ cell cycle inhibitors ($IC_{50} = 17.35$ µg/mL, 17.5 µg/mL, 19.23 µg/mL and 28.9 µg/mL, for each cell line, respectively)	[133]
H_2O → precipitation → normal phase HPLC → reversed phase HPLC → F15 + F16	Panc89 and PancTu1 cells	0.2–200 µg/mL of fractions	↓ cell proliferation (F15: $IC_{50} = 15.2$ and 18.3 µg/mL, respectively; F16: $IC_{50} = 16.4$ and 16.2 µg/mL)	[134]
		F. spiralis		
Ext1: DCM Ext2: MeOH 100% Ext3: *n*-hex fraction of Ext2	HeLa cells	Crescent concentrations of dichloromethane extract	↑ apoptosis, with Ext1 showing highest activity ($IC_{50} = 10.7$ µg/mL)	[135]
Ac 70% → purification with cellulose	Hyaluronidase activity measurement	0.5–2.25 mg/mL of extract	↓ enzymatic activity ($IC_{50} = 0.73$ mg/mL dry weight)	[136]
		A. nodosum		
ACN:0.2% CH_2O_2 (1:1) → purification in SPE columns	Caco-2 cells	15–42.5 µg/mL of extract	↓ cell proliferation ($IC_{50} = 33$ µg/mL)	[114]

Ac, acetone; ACN, acetonitrile; DCM, dichloromethane; Ext, extraction; F15, fraction 15; F16, fraction 16; HPLC, high performance liquid chromatography; MeOH, methanol; SPE column, and solid-phase extraction column. Cell lines: Caco-2, human colon cancer; HeLa, human cervix carcinoma; PancTu1, human pancreatic cancer; Colo357, human pancreatic adenosquamous carcinoma; Panc89, human pancreatic cancer; and Panc1, pancreatic carcinoma.

These data suggest that *A. nodosum* and *Fucus* spp. phenolic compounds could represent possible new agents with therapeutic applications on the treatment of pancreatic and colon cancer, the former being one of the most aggressive cancer entities and the latter one of the most incident cancers worldwide [138]. Still, much work needs to be carried out in order to prove both the efficacy and safety of these agents in vivo.

3.5. Other Biological Activities

The typical phlorotannin profile from brown algal with antimicrobial activity mainly consists of phloroglucinol, eckol and dieckol [139,140]. Fucaceae seaweeds are, however, more prevalent in fucols and fucophlorethols. Yet, some positive results in this field have already been reported. Indeed, Sandsdalen et al. [141] have shown that a fucophlorethol derivative isolated from *F. vesiculosus* was a potent bactericidal agent against both Gram-positive (*Staphylococcus aureus*, *Staphylococcus epidermidis*) and Gram-negative (*Escherichia coli*, *Proteus mirabilis*, *Pseudomonas aeruginosa*) bacteria, reducing their growth by 85% compared to the controls. Likewise, the phlorotannins purified from *F. spiralis* acetone 70% extract showed antibacterial effects against Gram-positive bacteria, exhibiting minimum inhibitory concentrations of 2 mg/mL for *Micrococcus luteus*, 2 mg/mL for *S. epidermidis*, 7.8 mg/mL for *S. aureus* and *Bacillus cereus* and 15.6 mg/mL for *Enterococcus faecalis*, while no activity was observed for the Gram-negative ones [142]. Identical results were observed for an acetone extract from *A. nodosum*, which also produced more effective inhibition towards Gram-positive (MIC of 0.25 and 0.2 mg/mL for *M. luteus* and *S. aureus*, respectively) than Gram-negative (MIC of 0.4 and 0.5 mg/mL for *E. coli* and *Enterococcus aerogenes*, respectively) bacteria. Notably, the inhibitory effectiveness of this extract towards Gram-positive and Gram-negative bacteria was respectively 25–30- and 12–15-times stronger than those of ethylparaben, sodium benzoate and potassium sorbate, which are three important bactericides used as food preservatives [143]. On the other hand, Wang et al. [144] observed that a purified phlorotannin extract of *A. nodosum* origin exhibited strong bactericidal activity against *E. coli* O157:H7, and a complete eradication of this microorganism was observed after 6 h treatment with extract at 50 µg/mL. Furthermore, in combination with a silver-zeolite, *A. nodosum* aqueous phenolic-rich extract obtained by alginate precipitation and a series of filtrations resulted in a complete inhibition of the film formation of *Streptococcus gordonii* alone and altered the film formation of co-cultured *Porphyromonas gingivalis* and *S. gordonii*, thus indicating a possible therapeutic approach for preventing and/or treating periodontal diseases [126]. Fungicidal properties were described for *F. spiralis* phlorotannin-purified acetone 70% extract, which exhibited particular inhibitory effects against the growth of several dermatophytes, revealing MICs of 3.9–31.3 mg/mL DW for *Trichophyton rubrum* > *Epidermophyton floccosum* > *Trichophyton mentagrophytes* = *Microsporum canis* > *Microsporum gypseum*, thus being of some interest for the development of skincare products for the treatment of dermatophytosis [145]. In addition, *F. vesiculosus* was shown to be a rich source of polysaccharides and polyphenols with the capacity to inhibit both HIV-induced syncytium formation and reverse transcriptase activity [146].

Photoprotective activity is another biological property that has been described for phenolics from Fucaceae as well. The phenolic-rich water-soluble fraction from *F. vesiculosus* and *A. nodosum* acetone 70% extracts revealed moderate photoprotective effects in vivo, as they could prevent the UV-B-exposed zebrafish embryos from dying, although a big percentage of these embryos presented low-level malformations. Notwithstanding, the number of normal embryos was higher in the presence 0.4 mg PE/mL of *F. vesiculosus* extract (17%) than in the presence of the same concentration of *A. nodosum* (8.3%), which is very likely due to the differences in the phenolic profile between species and consequently their different radical scavenging activities [147].

Consumption of Fucaceae seaweeds may also have a significant impact on the control of hypertensive conditions. In this context, a methanolic extract from *F. spiralis* was reported for its capacity to inhibit angiotensin I-converting enzyme, a key player in the control of blood pressure, by approximately 80%. In addition, the fractionation of this extract according to their molecular weight

(<1 kDa, 1–3 kDa and >3 kDa) resulted in distinct inhibition abilities towards the enzyme, being the strongest (almost 90% inhibition) observed for the >3 kDa fraction at 200 μg/mL, which was almost as effective as captopril (i.e., a pharmaceutical anti-hypertensive agent) that caused 97% of inhibition for the same concentration [148]. *S. siliquosa* ethanol 95% extract has shown good inhibitory effects towards this enzyme as well (45% at 164 μg/mL), although this was considerably less effective than captopril (33% inhibition at 1.6 ng/mL) [149].

Recently, Kellogg et al. [131] observed that the treatment of 3T3-L1 adipocytes with 100 μg/mL of an ethyl acetate fraction that was obtained from a *F. distichus* crude methanolic extract resulted in a reduction of cellular lipid accumulation to 77.5%. Moreover, the authors further observed that, at 50 μg/mL, a fucophlorethol-rich subfraction obtained by an addition purifying step onto a Sephadex LH-20 column not only reduced lipid accumulation in the same cellular model down to 45.9%, but also had the capacity to inhibit leptin mRNA expression close to 0% and enhance that of adiponectin in 20% compared to the untreated control. These results were even more pronounced than those obtained for dexamethasone, which only caused a 20% reduction of the leptin and negatively interfered with the adiponectin expression, reducing it by approximately 50%. Therefore, these data suggest that *F. distichus* phlorotannins may exert anti-obesity effects through regulation of lipid metabolism.

It should be noted that phlorotannins are endowed with several other biological properties such as neuroprotective, cardioprotective, antiallergic, anti-arthritis and many others. However, it must also be emphasized that these properties are very frequently reported for phlorotannins from brown algae species such as *Ecklonia* spp., *Eisenia* spp. and *Ishige okamurae*, which are abundant in phlorotannins (such as eckol, dieckol and several derivatives) that have different structural features when compared to the typical phlorotannins found in Fucaceae [52,150,151]. Hence, further studies are necessary in order to clarify the possible targeting of Fucaceae typical phlorotannins in such mechanisms.

4. Bioavailability

Dietary habits are the major source of polyphenols. However, the biological activity of these compounds in vivo is critically influenced by their bioaccessibility, absorption and metabolism [152]. Since very few data concerning the bioavailability of phlorotannins are currently available, it is common to consider that this group of compounds follow an identical behavior to that of plant polyphenols, which are better absorbed in the large intestine after undergoing an extensive transformation by enzymatic activity or colon microbial fermentation [153,154].

Until recently, the bioavailability of phlorotannins was still an unexplored subject. However, the earliest studies in this field are already emerging. A preliminary approach was carried out by Bangoura et al. [155,156], who observed that the phlorotannin concentration in the flesh of abalones was raised after feeding them either with *Ecklonia cava* or *Ecklonia stolonifera*, i.e., two brown algae species rich in these compounds.

More recently, Corona and co-workers [85] conducted a study aiming to determine the gastrointestinal stability and bioavailability of a food-grade phlorotannin extract from *A. nodosum*. In more detail, this extract was submitted to an in vitro gastric and ileal digestion followed by colonic bacteria fermentation and, ultimately, a dialysis filtration to simulate the absorption into the circulation. Through HPLC-MS, the authors could identify 11 compounds in the dialysate, four of them corresponding to hydroxytrifuhalol A, a C–O–C dimer of phloroglucinol, diphlorethol/difucol and 7-hydroxyeckol, which had been previously identified in the crude extract of *A. nodosum*, and seven new uncharacterized compounds that corresponded to in vitro-absorbed metabolites. Some of these compounds were further detected on urine samples of human volunteers who were administered with a single capsule containing 100 mg of the *A. nodosum* extract, thus confirming that they were absorbed into the blood circulation in vivo. In plasma, the total level of phlorotannins/metabolites detected varied between 0.011 and 7.757 μg/mL, while in urine, the values ranged between 0.15–33.52 μg/mL, and although some metabolites were found in samples collected at 2–4 h after capsule ingestion, the majority were detected at late time points, indicating that the high molecular weight phlorotannins

Int. J. Mol. Sci. **2017**, *18*, 1327

were poorly absorbed in the upper tract and went through colonic fermentation, which resulted in the formation of lower molecular weight derivatives that were more likely to be absorbed. During the passage through the digestive tract, phenolic compounds are known to undergo extensive modification by glucosidase enzymes, phase I enzymes, including cytochrome P450, and phase II enzymes (glucuronosyltransferases, sulfotransferases) found both in the small intestine and the liver [157]. Indeed, in this work, the authors found that some metabolites were only detectable in blood or urine samples after an enzymatic treatment with glucuronidase or sulfatase, while others were only observable in untreated samples, indicating that these compounds corresponded to the conjugated metabolites. On the other hand, some compounds were detected in samples either with or without enzymatic treatment, which means that these were the unconjugated metabolites [85].

Based on these data, it seems that, similarly to what happens to plant polyphenols, phlorotannins may undergo different modifications during their transit in the gastrointestinal tract, and the resultant metabolites might represent active forms that will pass through the gut barrier and exert their physiological and biological functions in the organism [158,159].

5. Concluding Remarks

In conclusion, Fucaceae seaweeds are a valuable source of phlorotannins, which have drawn much attention during recent years due to their numerous possible therapeutic properties. Common features of phlorotannin extracts from Fucaceae include antioxidant effects through scavenging of ROS or enhancement of intracellular antioxidant defenses, antidiabetic properties through their acarbose-like activity and capacity to increase adipocytes glucose uptake and β-pancreatic cells resistance to high-glucose oxidative stress, anti-inflammatory effects through inhibition of several pro-inflammatory mediators and antitumor properties through activation of apoptosis on cancerous cells and inhibition of metastasis. Other important biological activities have been demonstrated, such as antimicrobial, anti-hypertensive, anti-obesity and photoprotective activities. Besides, the bioavailability of phlorotannins is presently suggested to resemble that of plant tannins, with the majority of these compounds being modified by the gut microflora and the resultant metabolites possibly representing true bioactive forms. In sum, it can be suggested that Fucaceae phlorotannins present powerful and versatile bioactivities that grant them great potential for exploitation as renewable feedstocks for the development of new nutraceutical, cosmetic and pharmaceutical products.

Acknowledgments: Thanks are due to University of Aveiro, Science and Technology Foundation/Ministry of Education and Science (FCT/MEC) for the financial support to the Organic Chemistry, Natural Products and Food Stuffs Research Unit (QOPNA) research Unit (FCT UID/QUI/00062/2013), through national funds and, where applicable, co-financed by the European Regional Development Fund (FEDER), within the Portugal 2020 (PT2020) Partnership Agreement. The authors also acknowledge the funding through the Project Seaweed for Healthier Traditional Products (SHARP), Research & Development Co-promotion No. 3419, supported by European structural and investment funds (FEEI) under the Program "Portugal 2020". Marcelo D. Catarino and Susana M. Cardoso acknowledge FCT for financial support (PD/BD/114577/2016 and SFRH/BPD/113080/2015 fellowships, respectively).

Conflicts of Interest: The authors declare no conflict of interest.

Abbreviations

AGEs	Advanced Glycated End-Products
BSA	Bovine Serum Albumin
CAT	Catalase
COX-2	Cyclooxygenase-2
DP	Degree of Polymerization
DPPH$^\bullet$	1,1-Diphenyl-2-Picrylhydrazyl Radical

FRAP	Ferric Reducing Antioxidant Power
GAE	Gallic Acid Equivalents
GSH	Glutathione
GSH-px	Glutathione Peroxidase
GSH-red	Glutathione Reductase
GSH-tr	Glutathione Transferase
HPLC	High Performance Liquid Chromatography
IL	Interleukin
iNOS	Inducible Nitric Oxide Synthase
LPS	Lipopolysaccharide
MIC	Minimum Inhibitory Concentration
MS	Mass Spectrometry
NF-κB	Nuclear Factor-κB
NMR	Nuclear Magnetic Spectroscopy
Nrf2	Nuclear Factor (Erythroid-Derived 2)-Like 2
ORAC	Oxygen Radical Absorbance Capacity
PE	Phloroglucinol Equivalents
PMA	Phorbol-12-Myristate-13-Acetate
PON-1	Paraoxonase-1
RNS	Reactive Nitrogen Species
ROS	Reactive Oxygen Species
SOD	Superoxide Dismutase
SPE	Solid Phase Extraction
t-BHP	*tert*-Butyl Hydroperoxide
TBARS	Thiobarbituric Acid Reactive Substances
TE	Trolox Equivalents
TLR	Toll-Like Receptor
TNF-α	Tumor Necrosis Factor-α
TPC	Total Phenolic Content
UV	Ultraviolet
UPLC	Ultra-Performance Liquid Chromatography

References

1. Guiry, M.D. AlgaeBase. World-wide electronic publication, National University of Ireland, Galway. Available online: http://www.algaebase.org (accessed on 31 March 2017).
2. GBIF Secretariat. GBIF Backbone Taxonomy—Fucaceae. Available online: http://www.gbif.org/species/9641 (accessed on 3 April 2017).
3. Kucera, H.; Saunders, G.W. Assigning morphological variants of *Fucus* (Fucales, Phaeophyceae) in Canadian waters to recognized species using DNA barcoding. *Botany* **2008**, *86*, 1065–1079. [CrossRef]
4. GBIF Secretariat. GBIF Backbone Taxonomy—*Fucus* L. Available online: http://www.gbif.org/species/7832266 (accessed on 3 April 2017).
5. Jueterbock, A.; Tyberghein, L.; Verbruggen, H.; Coyer, J.A.; Olsen, J.L.; Hoarau, G. Climate change impact on seaweed meadow distribution in the North Atlantic rocky intertidal. *Ecol. Evol.* **2013**, *3*, 1356–1373. [CrossRef] [PubMed]
6. Torn, K.; Krause-Jensen, D.; Martin, G. Present and past depth distribution of bladderwrack (*Fucus vesiculosus*) in the Baltic Sea. *Aquat. Bot.* **2006**, *84*, 53–62. [CrossRef]
7. Malm, T.; Kautsky, L.; Engkvist, R. Reproduction, recruitment and geographical distribution of *Fucus serratus* L. in the Baltic Sea. *Bot. Mar.* **2001**, *44*, 101–108. [CrossRef]
8. GBIF Secretariat. GBIF Backbone Taxonomy—*Pelvetia* Decaisne & Thuret. Available online: http://www.gbif.org/species/3196494 (accessed on 3 April 2017).
9. GBIF Secretariat. GBIF Backbone Taxonomy—*Ascophyllum* Stackhouse. Available online: http://www.gbif.org/species/3196523 (accessed on 3 April 2017).

10. Lee, Y.K.; Yoon, H.S.; Motomura, T.; Kim, Y.J.; Boo, S.M. Phylogenetic relationships between *Pelvetia* and *Pelvetiopsis* (Fucaceae, Phaeophyta) inferred from sequences of the RuBisCo spacer region. *Eur. J. Phycol.* **1999**, *34*, 205–211. [CrossRef]

11. Lewis, J.R. *The Ecology of Rocky Shores*; English Universities Press: London, UK, 1964.

12. GBIF Secretariat. GBIF Backbone Taxonomy—Silvetia E. Serrão, T.O. Cho, S.M. Boo & S.H. Brawley. Available online: http://www.gbif.org/species/3196480 (accessed on 3 April 2017).

13. GBIF Secretariat. GBIF Backbone Taxonomy—Pelvetiopsis N.L. Gardener. Available online: http://www.gbif.org/species/3196508 (accessed on 3 April 2017).

14. Serrão, E.A.; Alice, L.A.; Brawley, S.H. Evolution of the Fucaceae (Phaeophyceae) inferred from nrDNA-ITS. *J. Phycol.* **1999**, *35*, 382–394. [CrossRef]

15. Pereira, L. *Edible Seaweeds of the World*; CRC Press: Boca Raton, FL, USA, 2016.

16. Patarra, R.F.; Paiva, L.; Neto, A.I.; Lima, E.; Baptista, J. Nutritional value of selected macroalgae. *J. Appl. Phycol.* **2011**, *23*, 205–208. [CrossRef]

17. Stansbury, J.; Saunders, P.; Winston, D. Promoting healthy thyroid function with iodine, bladderwrack, guggul and iris. *J. Restor. Med.* **2012**, *1*, 83–90. [CrossRef]

18. Guiry, M.D.; Morrison, L. The sustainable harvesting of *Ascophyllum nodosum* (Fucaceae, Phaeophyceae) in Ireland, with notes on the collection and use of some other brown algae. *J. Appl. Phycol.* **2013**, *25*, 1823–1830. [CrossRef]

19. Guiry, M.D.; Garbary, D.J. Geographical and taxonomic guide to European seaweeds of economic importance. In *Seaweed Resources in Europe: Uses and Potential*; Guiry, M.D., Blunden, G., Eds.; John Wiley & Sons: Chichester, UK, 1991.

20. Bixler, H.J.; Porse, H. A decade of change in the seaweed hydrocolloids industry. *J. Appl. Phycol.* **2011**, *23*, 321–335. [CrossRef]

21. Colapietra, M.; Alexander, A. Effect of foliar fertilization on yield and quality of table grapes. *Acta Hortic.* **2006**, *721*, 213–218. [CrossRef]

22. Bozorgi, H.; Bidarigh, S.; Bakhshi, D. Marine brown alga extract (*Ascophyllum nodosum*) under foliar spraying of methanol and iron fertilizers on flower tube length of saffron (*Crocus sativus* L.) in North of Iran. *Int. J. Agric. Crop Sci.* **2012**, *4*, 1512–1518.

23. Sharma, S.H.S.; Lyons, G.; McRoberts, C.; McCall, D.; Carmichael, E.; Andrews, F.; Swan, R.; McCormack, R.; Mellon, R. Biostimulant activity of brown seaweed species from Strangford Lough: Compositional analyses of polysaccharides and bioassay of extracts using mung bean (*Vigno mungo* L.) and pak choi (*Brassica rapa chinensis* L.). *J. Appl. Phycol.* **2012**, *24*, 1081–1091. [CrossRef]

24. Shah, M.T.; Zodape, S.T.; Chaudhary, D.R.; Eswaran, K.; Chikara, J. Seaweed sap as an alternative liquid fertilizer for yield and quality improvement of wheat. *J. Plant Nutr.* **2013**, *36*, 192–200. [CrossRef]

25. Anderson, M.J.; Blanton, J.R.; Gleghorn, J.; Kim, S.W.; Johnson, J.W. *Ascophyllum nodosum* supplementation strategies that improve overall carcass merit of implanted english crossbred cattle. *Asian-Australas. J. Anim. Sci.* **2006**, *19*, 1514–1518. [CrossRef]

26. Evans, F.D.; Critchley, A.T. Seaweeds for animal production use. *J. Appl. Phycol.* **2014**, *26*, 891–899. [CrossRef]

27. Turner, J.L.; Dritz, S.S.; Higgins, J.J.; Minton, J.E. Effects of *Ascophyllum nodosum* extract on growth performance and immune function of young pigs challenged with *Salmonella typhimurium*. *J. Anim. Sci.* **2002**, *80*, 1947–1953. [CrossRef] [PubMed]

28. Hankins, S.D.; Hockey, H.P. The effect of a liquid seaweed extract from *Ascophyllum nodosum* (Fucales, Phaeophyta) on the two-spotted red spider mite *Tetranychus urticae*. *Hydrobiologia* **1990**, *204*, 555–559. [CrossRef]

29. Jayaraman, J.; Norrie, J.; Punja, Z.K. Commercial extract from the brown seaweed *Ascophyllum nodosum* reduces fungal diseases in greenhouse cucumber. *J. Appl. Phycol.* **2011**, *23*, 353–361. [CrossRef]

30. Radwan, M.A.; Farrag, S.A.A.; Abu-Elamayem, M.M.; Ahmed, N.S. Biological control of the root-knot nematode, *Meloidogyne incognita* on tomato using bioproducts of microbial origin. *Appl. Soil Ecol.* **2012**, *56*, 58–62. [CrossRef]

31. Sultana, V.; Baloch, G.N.; Ara, J.; Ehteshamul-Haque, S.; Tariq, R.M.; Athar, M. Seaweeds as an alternative to chemical pesticides for the management of root diseases of sunflower and tomato. *J. Appl. Bot. Food Qual.* **2011**, *84*, 162–168.

32. Küpper, F.C.; Feiters, M.C.; Olofsson, B.; Kaiho, T.; Yanagida, S.; Zimmermann, M.B.; Carpenter, L.J.; Luther, G.W.; Lu, Z.; Jonsson, M.; et al. Commemorating two centuries of iodine research: An interdisciplinary overview of current research. *Angew. Chem. Int. Ed. Engl.* **2011**, *50*, 11598–11620. [CrossRef] [PubMed]
33. Moro, C.O.; Basile, G. Obesity and medicinal plants. *Fitoterapia* **2000**, *71*, S73–S82. [CrossRef]
34. Balch, P.A.; Bell, S. *Prescription for Herbal Healing*, 2nd ed.; Penguin Group Inc.: London, UK, 2012.
35. Pereira, L. A review of the nutrient composition of selected edible seaweeds. In *Seaweed: Ecology, Nutrient Composition and Medicinal Uses*; Pomin, V.H., Ed.; Nova Science Publishers, Inc.: New York, USA, 2011; pp. 15–47.
36. Ale, M.T.; Mikkelsen, J.D.; Meyer, A.S. Important determinants for fucoidan bioactivity: A critical review of structure-function relations and extraction methods for fucose-containing sulfated polysaccharides from brown seaweeds. *Mar. Drugs* **2011**, *9*, 2106–2130. [CrossRef] [PubMed]
37. Laekeman, G. *Assessment Report on Fucus vesiculosus L., Thallus*; EMA/HMPC/313675/2012; European Medicines Agency: London, UK, 2014.
38. Dutot, M.; Fagon, R.; Hemon, M.; Rat, P. Antioxidant, anti-inflammatory, and anti-senescence activities of a phlorotannin-rich natural extract from brown seaweed *Ascophyllum nodosum*. *Appl. Biochem. Biotechnol.* **2012**, *167*, 2234–2240. [CrossRef] [PubMed]
39. O'Sullivan, A.M.; O'Callaghan, Y.C.; O'Grady, M.N.; Queguineur, B.; Hanniffy, D.; Troy, D.J.; Kerry, J.P.; O'Brien, N.M. Assessment of the ability of seaweed extracts to protect against hydrogen peroxide and *tert*-butyl hydroperoxide induced cellular damage in Caco-2 cells. *Food Chem.* **2012**, *134*, 1137–1140. [CrossRef] [PubMed]
40. Colliec, S.; Boisson-vidal, C.; Jozefonvicz, J. A low molecular weight fucoidan fraction from the brown seaweed *Pelvetia canaliculata*. *Phytochemistry* **1994**, *35*, 697–700. [CrossRef]
41. Cumashi, A.; Ushakova, N.A.; Preobrazhenskaya, M.E.; D'Incecco, A.; Piccoli, A.; Totani, L.; Tinari, N.; Morozevich, G.E.; Berman, A.E.; Bilan, M.I.; et al. A comparative study of the anti-inflammatory, anticoagulant, antiangiogenic, and antiadhesive activities of nine different fucoidans from brown seaweeds. *Glycobiology* **2007**, *17*, 541–552. [CrossRef] [PubMed]
42. Bahar, B.; O'Doherty, J.V.; Hayes, M.; Sweeney, T. Extracts of brown seaweeds can attenuate the bacterial lipopolysaccharide-induced pro-inflammatory response in the porcine colon ex vivo. *J. Anim. Sci.* **2012**, *90*, 46–48. [CrossRef] [PubMed]
43. Nakayasu, S.; Soegima, R.; Yamaguchi, K.; Oda, T. Biological activities of fucose-containing polysaccharide ascophyllan isolated from the brown alga *Ascophyllum nodosum*. *Biosci. Biotechnol. Biochem.* **2009**, *73*, 961–964. [CrossRef] [PubMed]
44. Abu, R.; Jiang, Z.; Ueno, M.; Isaka, S.; Nakazono, S.; Okimura, T.; Cho, K.; Yamaguchi, K.; Kim, D.; Oda, T. Anti-metastatic effects of the sulfated polysaccharide ascophyllan isolated from *Ascophyllum nodosum* on B16 melanoma. *Biochem. Biophys. Res. Commun.* **2015**, *458*, 727–732. [CrossRef] [PubMed]
45. Zhang, J.; Tiller, C.; Shen, J.; Wang, C.; Girouard, G.S.; Dennis, D.; Barrow, C.J.; Miao, M.; Ewart, H.S. Antidiabetic properties of polysaccharide- and polyphenolic-enriched fractions from the brown seaweed *Ascophyllum nodosum*. *Can. J. Physiol. Pharmacol.* **2007**, *85*, 1116–1123. [CrossRef] [PubMed]
46. Apostolidis, E.; Lee, C.M. In vitro potential of *Ascophyllum nodosum* phenolic antioxidant-mediated α-glucosidase and α-amylase inhibition. *J. Food Sci.* **2010**, *75*, H97–H102. [CrossRef] [PubMed]
47. Lordan, S.; Smyth, T.J.; Soler-Vila, A.; Stanton, C.; Paul Ross, R. The α-amylase and α-glucosidase inhibitory effects of Irish seaweed extracts. *Food Chem.* **2013**, *141*, 2170–2176. [CrossRef] [PubMed]
48. Bedoux, G.; Hardouin, K.; Burlot, A.S.; Bourgougnon, N. Bioactive components from seaweeds: Cosmetic applications and future development. In *Advances in Botanical Research*; Bourgougnon, N., Ed.; Academic Press: London, UK, 2014; Volume 71, pp. 345–378.
49. European Commission. CosIng Database. Available online: https://ec.europa.eu/growth/sectors/cosmetics/cosing (accessed on 5 April 2017).
50. Lee, S.H.; Jeon, Y.J. Anti-diabetic effects of brown algae derived phlorotannins, marine polyphenols through diverse mechanisms. *Fitoterapia* **2013**, *86*, 129–136. [CrossRef] [PubMed]
51. Isaza Martínez, J.H.; Torres Castañeda, H.G.; Martinez, J.H.I.; Castaneda, H.G.T. Preparation and chromatographic analysis of phlorotannins. *J. Chromatogr. Sci.* **2013**, *51*, 825–838. [CrossRef] [PubMed]
52. Li, Y.-X.; Wijesekara, I.; Li, Y.; Kim, S.-K. Phlorotannins as bioactive agents from brown algae. *Process Biochem.* **2011**, *46*, 2219–2224. [CrossRef]

53. Pal Singh, I.; Bharate, S.B. Phloroglucinol compounds of natural origin. *Nat. Prod. Rep.* **2006**, *23*, 558–591. [CrossRef] [PubMed]

54. Koivikko, R.; Loponen, J.; Honkanen, T.; Jormalainen, V. Contents of soluble, cell-wall-bound and exuded phlorotannins in the brown alga *Fucus vesiculosus*, with implications on their ecological functions. *J. Chem. Ecol.* **2005**, *31*, 195–212. [CrossRef] [PubMed]

55. Wikström, S.A.; Pavia, H. Chemical settlement inhibition versus post-settlement mortality as an explanation for differential fouling of two congeneric seaweeds. *Oecologia* **2004**, *138*, 223–230. [CrossRef] [PubMed]

56. Targett, N.M.; Arnold, T.M. Predicting the effects of brown algal phlorotannins on marine herbivores in tropical and temperate oceans. *J. Phycol.* **1998**, *34*, 195–205. [CrossRef]

57. Pavia, H.; Cervin, G.; Lindgren, A.; Aberg, P. Effects of UV-B radiation and simulated herbivory on phlorotannins in the brown alga *Ascophyllum nodosum*. *Mar. Ecol. Prog. Ser.* **1997**, *157*, 139–146. [CrossRef]

58. Connan, S.; Stengel, D.B. Impacts of ambient salinity and copper on brown algae: 2. Interactive effects on phenolic pool and assessment of metal binding capacity of phlorotannin. *Aquat. Toxicol.* **2011**, *104*, 1–13. [CrossRef] [PubMed]

59. Gómez, I.; Huovinen, P. Induction of phlorotannins during UV exposure mitigates inhibition of photosynthesis and DNA damage in the kelp *Lessonia nigrescens*. *Photochem. Photobiol.* **2010**, *86*, 1056–1063. [CrossRef] [PubMed]

60. Holdt, S.L.; Kraan, S. Bioactive compounds in seaweed: Functional food applications and legislation. *J. Appl. Phycol.* **2011**, *23*, 543–597. [CrossRef]

61. Sabeena Farvin, K.H.; Jacobsen, C. Phenolic compounds and antioxidant activities of selected species of seaweeds from Danish coast. *Food Chem.* **2013**, *138*, 1670–1681. [CrossRef] [PubMed]

62. Yoshie-Stark, Y.; Hsieh, Y.; Suzuki, T. Distribution of flavonoids and related compounds from seaweeds in Japan. *Tokyo Univ. Fish.* **2003**, *89*, 1–6.

63. Kim, S.K.; Himaya, S.W.A. Medicinal effects of phlorotannins from marine brown algae. In *Advances in Food and Nutrition Research*; Kim, S., Ed.; Academic Press: San Diego, CA, USA, 2011; Volume 64, pp. 97–109.

64. Koivikko, R.; Loponen, J.; Pihlaja, K.; Jormalainen, V. High-performance liquid chromatographic analysis of phlorotannins from the brown alga *Fucus vesiculosus*. *Phytochem. Anal.* **2007**, *18*, 326–332. [CrossRef] [PubMed]

65. Jormalainen, V.; Honkanen, T. Variation in natural selection for growth and phlorotannins in the brown alga *Fucus vesiculosus*. *J. Evol. Biol.* **2004**, *17*, 807–820. [CrossRef] [PubMed]

66. Connan, S.; Goulard, F.; Stiger, V.; Deslandes, E.; Gall, E.A. Interspecific and temporal variation in phlorotannin levels in an assemblage of brown algae. *Bot. Mar.* **2004**, *47*, 410–416. [CrossRef]

67. Pavia, H.; Toth, G.B. Influence of light and nitrogen on the phlorotannin content of the brown seaweeds *Ascophyllum nodosum* and *Fucus vesiculosus*. *Hydrobiology* **2000**, *440*, 299–305. [CrossRef]

68. Ragan, M.A.; Jensen, A. Quantitative studies on brown algal phenols. II. Seasonal variation in polyphenol content of *Ascophyllum nodosum* (L.) Le Jol. and *Fucus vesiculosus* (L.). *J. Exp. Mar. Biol. Ecol.* **1978**, *34*, 245–258. [CrossRef]

69. Stiger, V.; Deslandes, E.; Payri, C.E. Phenolic contents of two brown algae, *Turbinaria ornata* and *Sargassum mangarevense* on Tahiti (French Polynesia): Interspecific, ontogenic and spatio-temporal variations. *Bot. Mar.* **2004**, *47*, 402–409. [CrossRef]

70. Peckol, P.; Krane, J.M.; Yates, J.L. Interactive effects of inducible defense and resource availability on phlorotannins in the North Atlantic brown alga *Fucus vesiculosus*. *Mar. Ecol. Prog. Ser.* **1996**, *138*, 209–217. [CrossRef]

71. Parys, S.; Kehraus, S.; Pete, R.; Küpper, F.C.; Glombitza, K.-W.; König, G.M. Seasonal variation of polyphenolics in *Ascophyllum nodosum* (Phaeophyceae). *Eur. J. Phycol.* **2009**, *44*, 331–338. [CrossRef]

72. Kamiya, M.; Nishio, T.; Yokoyama, A.; Yatsuya, K.; Nishigaki, T.; Yoshikawa, S.; Ohki, K. Seasonal variation of phlorotannin in sargassacean species from the coast of the Sea of Japan. *Phycol. Res.* **2010**, *58*, 53–61. [CrossRef]

73. Pavia, H.; Brock, E. Extrinsic factors influencing phlorotannin production in the brown alga. *Mar. Ecol. Prog. Ser.* **2000**, *193*, 285–294. [CrossRef]

74. Roleda, M.Y.; Wiencke, C.; Lüder, U.H. Impact of ultraviolet radiation on cell structure, UV-absorbing compounds, photosynthesis, DNA damage, and germination in zoospores of Arctic *Saccorhiza dermatodea*. *J. Exp. Bot.* **2006**, *57*, 3847–3856. [CrossRef] [PubMed]

75. Pedersen, A. Studies on phenol content and heavy metal uptake in fucoids. In *Eleventh International Seaweed Symposium*; Bird, C.J., Ragan, M.A., Eds.; Springer: Dordrecht, The Netherlands, 1984; pp. 498–504.

76. Stern, J.L.; Hagerman, A.E.; Steinberg, P.D.; Winter, F.C.; Estes, J.A. A new assay for quantifying brown algal phlorotannins and comparisons to previous methods. *J. Chem. Ecol.* **1996**, *22*, 1273–1293. [CrossRef] [PubMed]

77. Parys, S.; Rosenbaum, A.; Kehraus, S.; Reher, G.; Glombitza, K.W.; König, G.M. Evaluation of quantitative methods for the determination of polyphenols in algal extracts. *J. Nat. Prod.* **2007**, *70*, 1865–1870. [CrossRef] [PubMed]

78. Li, Y.; Fu, X.; Duan, D.; Liu, X.; Xu, J.; Gao, X. Extraction and identification of phlorotannins from the brown alga, *Sargassum fusiforme* Setchell, Harvey. *Mar. Drugs* **2017**, *15*, 49. [CrossRef] [PubMed]

79. Tierney, M.S.; Soler-Vila, A.; Rai, D.K.; Croft, A.K.; Brunton, N.P.; Smyth, T.J. UPLC-MS profiling of low molecular weight phlorotannin polymers in *Ascophyllum nodosum*, *Pelvetia canaliculata* and *Fucus spiralis*. *Metabolomics* **2014**, *10*, 524–535. [CrossRef]

80. Steevensz, A.J.; MacKinnon, S.L.; Hankinson, R.; Craft, C.; Connan, S.; Stengel, D.B.; Melanson, J.E. Profiling phlorotannins in brown macroalgae by liquid chromatography-high resolution mass spectrometry. *Phytochem. Anal.* **2012**, *23*, 547–553. [CrossRef] [PubMed]

81. Glombitza, K.W.; Rauwald, H.W.; Eckhardt, G. Fucole, polyhydroxyoligophenyle aus *Fucus vesiculosus*. *Phytochemistry* **1975**, *14*, 1403–1405. [CrossRef]

82. Craige, J.S.; McInnes, A.G.; Ragan, M.A.; Walter, J.A. Chemical constituents of the physodes of brown algae. Characterization by ^{1}H and ^{13}C nuclear magnetic resonance spectroscopy of oligomers of phloroglucinol from *Fucus vesiculosus* (L.). *Can. J. Chem.* **1977**, *55*, 1575–1582. [CrossRef]

83. Glombitza, K.; Rauwald, H.; Eckhardt, G. Fucophloretholes, Polyhydroxyoligophenyl ethers from *Fucus vesiculosus*. *Planta Med.* **1977**, *32*, 33–45. [CrossRef]

84. Parys, S.; Kehraus, S.; Krick, A.; Glombitza, K.W.; Carmeli, S.; Klimo, K.; Gerhäuser, C.; König, G.M. *In vitro* chemopreventive potential of fucophlorethols from the brown alga *Fucus vesiculosus* L. by anti-oxidant activity and inhibition of selected cytochrome P450 enzymes. *Phytochemistry* **2010**, *71*, 221–229. [CrossRef] [PubMed]

85. Corona, G.; Ji, Y.; Anegboonlap, P.; Hotchkiss, S.; Gill, C.; Yaqoob, P.; Spencer, J.P.E.; Rowland, I. Gastrointestinal modifications and bioavailability of brown seaweed phlorotannins and effects on inflammatory markers. *Br. J. Nutr.* **2016**, *15*, 1–14. [CrossRef] [PubMed]

86. Heffernan, N.; Brunton, N.P.; FitzGerald, R.J.; Smyth, T.J. Profiling of the molecular weight and structural isomer abundance of macroalgae-derived phlorotannins. *Mar. Drugs* **2015**, *13*, 509–528. [CrossRef] [PubMed]

87. Wang, T.; Jónsdóttir, R.; Ólafsdóttir, G. Total phenolic compounds, radical scavenging and metal chelation of extracts from Icelandic seaweeds. *Food Chem.* **2009**, *116*, 240–248. [CrossRef]

88. O'Sullivan, A.M.; O'Callaghan, Y.C.; O'Grady, M.N.; Queguineur, B.; Hanniffy, D.; Troy, D.J.; Kerry, J.P.; O'Brien, N.M. In vitro and cellular antioxidant activities of seaweed extracts prepared from five brown seaweeds harvested in spring from the west coast of Ireland. *Food Chem.* **2011**, *126*, 1064–1070. [CrossRef]

89. Breton, F.; Cérantola, S.; Gall, E.A. Distribution and radical scavenging activity of phenols in *Ascophyllum nodosum* (Phaeophyceae). *J. Exp. Mar. Biol. Ecol.* **2011**, *399*, 167–172. [CrossRef]

90. Blanc, N.; Hauchard, D.; Audibert, L.; Ar Gall, E. Radical-scavenging capacity of phenol fractions in the brown seaweed *Ascophyllum nodosum*: An electrochemical approach. *Talanta* **2011**, *84*, 513–518. [CrossRef] [PubMed]

91. Wang, T.; Jónsdóttir, R.; Liu, H.; Gu, L.; Kristinsson, H.G.; Raghavan, S.; Ólafsdóttir, G. Antioxidant capacities of phlorotannins extracted from the brown algae *Fucus vesiculosus*. *J. Agric. Food Chem.* **2012**, *60*, 5874–5883. [CrossRef] [PubMed]

92. Cérantola, S.; Breton, F.; Gall, E.A.; Deslandes, E. Co-occurrence and antioxidant activities of fucol and fucophlorethol classes of polymeric phenols in *Fucus spiralis*. *Bot. Mar.* **2006**, *49*, 347–351. [CrossRef]

93. Honold, P.J.; Jacobsen, C.; Jónsdóttir, R.; Kristinsson, H.G.; Hermund, D.B. Potential seaweed-based food ingredients to inhibit lipid oxidation in fish-oil-enriched mayonnaise. *Eur. Food Res. Technol.* **2016**, *242*, 571–584. [CrossRef]

94. Wang, T.; Jónsdóttir, R.; Kristinsson, H.G.; Thorkelsson, G.; Jacobsen, C.; Hamaguchi, P.Y.; Ólafsdóttir, G. Inhibition of haemoglobin-mediated lipid oxidation in washed cod muscle and cod protein isolates by *Fucus vesiculosus* extract and fractions. *Food Chem.* **2010**, *123*, 321–330. [CrossRef]

95. O'Sullivan, A.M.; O'Callaghan, Y.C.; O'Grady, M.N.; Waldron, D.S.; Smyth, T.J.; O'Brien, N.M.; Kerry, J.P. An examination of the potential of seaweed extracts as functional ingredients in milk. *Int. J. Dairy Technol.* **2014**, *67*, 182–193. [CrossRef]

96. O'Sullivan, A.M.; O'Grady, M.N.; O'Callaghan, Y.C.; Smyth, T.J.; O'Brien, N.M.; Kerry, J.P. Seaweed extracts as potential functional ingredients in yogurt. *Innov. Food Sci. Emerg. Technol.* **2015**, *37*, 293–299. [CrossRef]

97. Zaragozá, M.C.; López, D.; Sáiz, M.P.; Poquet, M.; Pérez, J.; Puig-Parellada, P.; Màrmol, F.; Simonetti, P.; Gardana, C.; Lerat, Y.; et al. Toxicity and antioxidant activity in vitro and in vivo of two *Fucus vesiculosus* extracts. *J. Agric. Food Chem.* **2008**, *56*, 7773–7780. [CrossRef] [PubMed]

98. O'Sullivan, A.M.; O'Callaghan, Y.C.; O'Grady, M.N.; Hayes, M.; Kerry, J.P.; O'Brien, N.M. The effect of solvents on the antioxidant activity in Caco-2 cells of Irish brown seaweed extracts prepared using accelerated solvent extraction (ASE®). *J. Funct. Foods* **2013**, *5*, 940–948. [CrossRef]

99. Quéguineur, B.; Goya, L.; Ramos, S.; Martín, M.A.; Mateos, R.; Guiry, M.D.; Bravo, L. Effect of phlorotannin-rich extracts of *Ascophyllum nodosum* and *Himanthalia elongata* (Phaeophyceae) on cellular oxidative markers in human HepG2 cells. *J. Appl. Phycol.* **2013**, *25*, 1–11. [CrossRef]

100. Aviram, M.; Hardak, E.; Vaya, J.; Mahmood, S.; Milo, S.; Hoffman, A.; Billicke, S.; Draganov, D.; Rosenblat, M. Human serum paraoxonases (PON1) Q and R selectively decrease lipid peroxides in human coronary and carotid atherosclerotic lesions: PON1 esterase and peroxidase-like activities. *Circulation* **2000**, *101*, 2510–2517. [CrossRef] [PubMed]

101. Ahn, G.-N.; Kim, K.-N.; Cha, S.-H.; Song, C.-B.; Lee, J.; Heo, M.-S.; Yeo, I.-K.; Lee, N.-H.; Jee, Y.-H.; Kim, J.-S.; et al. Antioxidant activities of phlorotannins purified from *Ecklonia cava* on free radical scavenging using ESR and H_2O_2-mediated DNA damage. *Eur. Food Res. Technol.* **2007**, *226*, 71–79. [CrossRef]

102. Kang, K.A.; Lee, K.H.; Chae, S.; Zhang, R.; Jung, M.S.; Ham, Y.M.; Baik, J.S.; Lee, N.H.; Hyun, J.W. Cytoprotective effect of phloroglucinol on oxidative stress induced cell damage via catalase activation. *J. Cell. Biochem.* **2006**, *97*, 609–620. [CrossRef] [PubMed]

103. Kim, H.S.; Lee, K.; Kang, K.A.; Lee, N.H.; Hyun, J.W. Phloroglucinol exerts protective effects against oxidative stress-induced cell damage in SH-SY5Y cells. *J. Pharmacol. Sci.* **2012**, *119*, 186–192. [CrossRef] [PubMed]

104. Kim, M.M.; Kim, S.K. Effect of phloroglucinol on oxidative stress and inflammation. *Food Chem. Toxicol.* **2010**, *48*, 2925–2933. [CrossRef] [PubMed]

105. Ryu, J.; Zhang, R.; Hong, B.H.; Yang, E.J.; Kang, K.A.; Choi, M.; Kim, K.C.; Noh, S.J.; Kim, H.S.; Lee, N.H.; et al. Phloroglucinol attenuates motor functional deficits in an animal model of Parkinson's disease by enhancing Nrf2 activity. *PLoS ONE* **2013**, *8*, e71178. [CrossRef] [PubMed]

106. Kang, M.-C.; Cha, S.H.; Wijesinghe, W.A.; Kang, S.-M.; Lee, S.-H.; Kim, E.-A.; Song, C.B.; Jeon, Y.-J. Protective effect of marine algae phlorotannins against AAPH-induced oxidative stress in zebrafish embryo. *Food Chem.* **2013**, *138*, 950–955. [CrossRef] [PubMed]

107. Kim, K.C.; Piao, M.J.; Cho, S.J.; Lee, N.H.; Hyun, J.W. Phloroglucinol protects human keratinocytes from ultraviolet B radiation by attenuating oxidative stress. *Photodermatol. Photoimmunol. Photomed.* **2012**, *28*, 322–331. [CrossRef] [PubMed]

108. Yang, E.-J.; Ahn, S.; Ryu, J.; Choi, M.-S.; Choi, S.; Chong, Y.H.; Hyun, J.-W.; Chang, M.-J.; Kim, H.-S. Phloroglucinol attenuates the cognitive deficits of the 5XFAD mouse model of Alzheimer's Disease. *PLoS ONE* **2015**, *10*, e0135686. [CrossRef] [PubMed]

109. World Health Organization. *Global Report on Diabetes*; World Health Organization: Geneva, Switzerland, 2016; Volume 978.

110. Thilagam, E.; Parimaladevi, B.; Kumarappan, C.; Mandal, S.C. α-Glucosidase and α-amylase inhibitory activity of *Senna surattensis*. *J. Acupunct. Meridian Stud.* **2013**, *6*, 24–30. [CrossRef] [PubMed]

111. Liu, B.; Kongstad, K.T.; Wiese, S.; Jäger, A.K.; Staerk, D. Edible seaweed as future functional food: Identification of α-glucosidase inhibitors by combined use of high-resolution α-glucosidase inhibition profiling and HPLC-HRMS-SPE-NMR. *Food Chem.* **2016**, *203*, 16–22. [CrossRef] [PubMed]

112. Ohta, T.; Sasaki, S.; Oohori, T.; Yoshikawa, S. α-Glucosidase inhibitory activity of a 70% methanol extract from ezoishige (*Pelvetia babingtonii* de toni) and its effect on the elevation of blood glucose level in rats. *Biosci. Biotechnol. Biochem.* **2002**, *66*, 1552–1554. [CrossRef] [PubMed]

113. Apostolidis, E.; Karayannakidis, P.D.; Kwon, Y.I.; Lee, C.M.; Seeram, N.P. Seasonal variation of phenolic antioxidant-mediated α-glucosidase inhibition of *Ascophyllum nodosum*. *Plant Foods Hum. Nutr.* **2011**, *66*, 313–319. [CrossRef] [PubMed]

114. Nwosu, F.; Morris, J.; Lund, V.A.; Stewart, D.; Ross, H.A.; McDougall, G.J. Anti-proliferative and potential anti-diabetic effects of phenolic-rich extracts from edible marine algae. *Food Chem.* **2011**, *126*, 1006–1012. [CrossRef]

115. Kellogg, J.; Grace, M.H.; Lila, M.A. Phlorotannins from alaskan seaweed inhibit carbolytic enzyme activity. *Mar. Drugs* **2014**, *12*, 5277–5294. [CrossRef] [PubMed]

116. Liu, H.; Gu, L. Phlorotannins from brown algae (*Fucus vesiculosus*) inhibited the formation of advanced glycation endproducts by scavenging reactive carbonyls. *J. Agric. Food Chem.* **2012**, *60*, 1326–1334. [CrossRef] [PubMed]

117. Kitts, D.D.; Popovich, D.G.; Hu, C. Characterizing the mechanism for ginsenoside-induced cytotoxicity in cultured leukemia (THP-1) cells. *Can. J. Physiol. Pharmacol.* **2007**, *85*, 1173–1183. [CrossRef] [PubMed]

118. Pantidos, N.; Boath, A.; Lund, V.; Conner, S.; McDougall, G.J. Phenolic-rich extracts from the edible seaweed, *Ascophyllum nodosum*, inhibit α-amylase and α-glucosidase: Potential anti-hyperglycemic effects. *J. Funct. Foods* **2014**, *10*, 201–209. [CrossRef]

119. Roy, M.C.; Anguenot, R.; Fillion, C.; Beaulieu, M.; Bérubé, J.; Richard, D. Effect of a commercially-available algal phlorotannins extract on digestive enzymes and carbohydrate absorption in vivo. *Food Res. Int.* **2011**, *44*, 3026–3029. [CrossRef]

120. Paradis, M.E.; Couture, P.; Lamarche, B. A randomised crossover placebo-controlled trial investigating the effect of brown seaweed (*Ascophyllum nodosum* and *Fucus vesiculosus*) on postchallenge plasma glucose and insulin in men and women. *Appl. Physiol. Nutr. Metab.* **2011**, *36*, 913–919. [CrossRef] [PubMed]

121. Singh, V.P.; Bali, A.; Singh, N.; Jaggi, A.S. Advanced glycation end products and diabetic complications. *Korean J. Physiol. Pharmacol.* **2014**, *18*, 1–14. [CrossRef] [PubMed]

122. Park, M.H.; Han, J.S. Phloroglucinol protects INS-1 pancreatic β-cells against glucotoxicity-induced apoptosis. *Phytother. Res.* **2015**, *29*, 1700–1706. [CrossRef] [PubMed]

123. Ashley, N.T.; Weil, Z.M.; Nelson, R.J. Inflammation: Mechanisms, costs, and natural variation. *Annu. Rev. Ecol. Evol. Syst.* **2012**, *43*, 385–406. [CrossRef]

124. Catarino, M.D.; Alves-Silva, J.M.; Pereira, O.R.; Cardoso, S.M. Mediaterranean diet: A precious tool for fighting inflammatory diseases. In *Polyphenols: Food Sources, Bioactive Properties and Antioxidant Effects*; Cobb, D.T., Ed.; Nova Science Publishers: New York, NY, USA, 2014; pp. 87–112.

125. Catarino, M.D.; Talhi, O.; Rabahi, A.; Silva, A.M.S.; Cardoso, S.M. The anti-inflammatory potential of flavonoids: Mechanistic aspects. In *Studies in Natural Products Chemistry*; Atta-ur-Rahman, Ed.; Elsevier: Amsterdam, The Netherlands, 2016; Volume 48, pp. 65–99.

126. Cillard, J.; Bonnaure-mallet, M. Silver-zeolite combined to polyphenol-rich extracts of *Ascophyllum nodosum*: Potential active role in prevention of periodontal diseases. *PLoS ONE* **2014**, *9*, e105475.

127. Bahar, B.; O'Doherty, J.V.; Smyth, T.J.; Sweeney, T. A comparison of the effects of an *Ascophyllum nodosum* ethanol extract and its molecular weight fractions on the inflammatory immune gene expression in vitro and ex vivo. *Innov. Food Sci. Emerg. Technol.* **2016**, *37*, 276–285. [CrossRef]

128. Bahar, B.; Doherty, J.V.O.; Smyth, T.J.; Ahmed, A.M.; Sweeney, T. A cold water extract of *Fucus vesiculosus* inhibits lipopolysaccharide (LPS) induced pro-inflammatory responses in the porcine colon ex vivo model. *Innov. Food Sci. Emerg. Technol.* **2016**, *37*, 229–236. [CrossRef]

129. Liu, L.; Yuan, S.; Long, Y.; Guo, Z.; Sun, Y.; Li, Y.; Niu, Y.; Li, C.; Mei, Q. Immunomodulation of Rheum tanguticum polysaccharide (RTP) on the immunosuppressive effects of dexamethasone (DEX) on the treatment of colitis in rats induced by 2,4,6-trinitrobenzene sulfonic acid. *Int. Immunopharmacol.* **2009**, *9*, 1568–1577. [CrossRef] [PubMed]

130. Bahar, B.; O'Doherty, J.V.; Vigors, S.; Sweeney, T. Activation of inflammatory immune gene cascades by lipopolysaccharide (LPS) in the porcine colonic tissue ex vivo model. *Clin. Exp. Immunol.* **2016**, *186*, 266–276. [CrossRef] [PubMed]

131. Kellogg, J.; Esposito, D.; Grace, M.H.; Komarnytsky, S.; Lila, M.A. Alaskan seaweeds lower inflammation in RAW 264.7 macrophages and decrease lipid accumulation in 3T3-L1 adipocytes. *J. Funct. Foods* **2015**, *15*, 396–407. [CrossRef]

132. Rakoff-Nahoum, S. Why cancer and inflammation? *Yale J. Biol. Med.* **2006**, *79*, 123–130. [PubMed]

133. Geisen, U.; Zenthoefer, M.; Peipp, M.; Kerber, J.; Plenge, J.; Managò, A.; Fuhrmann, M.; Geyer, R.; Hennig, S.; Adam, D.; et al. Molecular mechanisms by which a *Fucus vesiculosus* extract mediates cell cycle inhibition and cell death in pancreatic cancer cells. *Mar. Drugs* **2015**, *13*, 4470–4491. [CrossRef] [PubMed]

134. Zenthoefer, M.; Geisen, U.; Hofmann-peiker, K.; Fuhrmann, M.; Geyer, R.; Piker, L.; Kalthoff, H.; Fuhrmann, M.; Kerber, J.; Kirchhöfer, R.; et al. Isolation of polyphenols with anticancer activity from the Baltic Sea brown seaweed *Fucus vesiculosus* using bioassay-guided fractionation. *J. Appl. Phycol.* **2017**, *148*, 1–17. [CrossRef]

135. Barreto, M.D.C.; Mendonça, E.A.; Gouveia, V.F.; Anjos, C.; Medeiros, J.S.; Seca, A.M.L.; Neto, A.I. Macroalgae from S. Miguel Island as a potential source of antiproliferative and antioxidant products. *Arquipel. Life Mar. Sci.* **2012**, *29*, 53–58.

136. Ferreres, F.; Lopes, G.; Gil-Izquierdo, A.; Andrade, P.B.; Sousa, C.; Mouga, T.; Valentão, P. Phlorotannin extracts from fucales characterized by HPLC-DAD-ESI-MS^n: Approaches to hyaluronidase inhibitory capacity and antioxidant properties. *Mar. Drugs* **2012**, *10*, 2766–2781. [CrossRef] [PubMed]

137. Shibata, T.; Fujimoto, K.; Nagayama, K.; Yamaguchi, K.; Nakamura, T. Inhibitory activity of brown algal phlorotannins against hyaluronidase. *Int. J. Food Sci. Technol.* **2002**, *37*, 703–709. [CrossRef]

138. Siegel, R.L.; Miller, K.D.; Jemal, A. Cancer statistics, 2016. *CA Cancer J. Clin.* **2016**, *66*, 7–30. [CrossRef] [PubMed]

139. Pérez, M.J.; Falqué, E.; Domínguez, H. Antimicrobial action of compounds from marine seaweed. *Mar. Drugs* **2016**, *14*, 52. [CrossRef] [PubMed]

140. Lee, D.-S.; Kang, M.-S.; Hwang, H.-J.; Eom, S.-H.; Yang, J.-Y.; Lee, M.-S.; Lee, W.-J.; Jeon, Y.-J.; Choi, J.-S.; Kim, Y.-M. Synergistic effect between dieckol from *Ecklonia stolonifera* and β-lactams against methicillin-resistant *Staphylococcus aureus*. *Biotechnol. Bioprocess Eng.* **2008**, *13*, 758–764. [CrossRef]

141. Sandsdalen, E.; Haug, T.; Stensvag, K.; Styrvold, O.B. The antibacterial effect of a polyhydroxylated fucophlorethol from the marine brown alga, *Fucus vesiculosus*. *World J. Microbiol. Biotechnol.* **2003**, *19*, 777–782. [CrossRef]

142. Lopes, G.; Sousa, C.; Silva, L.R.; Pinto, E.; Andrade, P.B.; Bernardo, J.; Mouga, T.; Valentão, P. Can phlorotannins purified extracts constitute a novel pharmacological alternative for microbial infections with associated inflammatory conditions? *PLoS ONE* **2012**, *7*, e31145. [CrossRef] [PubMed]

143. Jiménez, J.T.; O'Connell, S.; Lyons, H.; Bradley, B.; Hall, M. Antioxidant, antimicrobial, and tyrosinase inhibition activities of acetone extract of *Ascophyllum nodosum*. *Chem. Pap.* **2010**, *64*, 434–442. [CrossRef]

144. Wang, Y.; Xu, Z.; Bach, S.J.; McAllister, T.A. Sensitivity of *Escherichia coli* to seaweed (*Ascophyllum nodosum*) phlorotannins and terrestrial tannins. *Asian-Australas. J. Anim. Sci.* **2009**, *22*, 238–245. [CrossRef]

145. Lopes, G.; Pinto, E.; Andrade, P.B.; Valentão, P. Antifungal activity of phlorotannins against dermatophytes and yeasts: Approaches to the mechanism of action and influence on *Candida albicans* virulence factor. *PLoS ONE* **2013**, *8*, e72203. [CrossRef] [PubMed]

146. Béress, A.; Wassermann, O.; Tahhan, S.; Bruhn, T.; Béress, L.; Kraiselburd, E.N.; Gonzalez, L.V.; de Motta, G.E.; Chavez, P.I. A new procedure for the isolation of anti-HIV compounds (polysaccharides and polyphenols) from the marine alga *Fucus vesiculosus*. *J. Nat. Prod.* **1993**, *56*, 478–488.

147. Guinea, M.; Franco, V.; Araujo-Bazán, L.; Rodríguez-Martín, I.; González, S. In vivo UVB-photoprotective activity of extracts from commercial marine macroalgae. *Food Chem. Toxicol.* **2012**, *50*, 1109–1117. [CrossRef] [PubMed]

148. Paiva, L.; Lima, E.; Neto, A.I.; Baptista, J. Angiotensin I-converting enzyme (ACE) inhibitory activity of *Fucus spiralis* macroalgae and influence of the extracts storage temperature—A short report. *J. Pharm. Biomed. Anal.* **2016**, *131*, 503–507. [CrossRef] [PubMed]

149. Jung, H.A.; Hyun, S.K.; Kim, H.R.; Choi, J.S. Angiotensin-converting enzyme I inhibitory activity of phlorotannins from *Ecklonia stolonifera*. *Fish. Sci.* **2006**, *72*, 1292–1299. [CrossRef]

150. Wijesekara, I.; Yoon, N.Y.; Kim, S.K. Phlorotannins from *Ecklonia cava* (Phaeophyceae): Biological activities and potential health benefits. *BioFactors* **2010**, *36*, 408–414. [CrossRef] [PubMed]

151. Cardoso, S.; Pereira, O.; Seca, A.; Pinto, D.; Silva, A. Seaweeds as preventive agents for cardiovascular diseases: From nutrients to functional foods. *Mar. Drugs* **2015**, *13*, 6838–6865. [CrossRef] [PubMed]

152. Balasundram, N.; Sundram, K.; Samman, S. Phenolic compounds in plants and agri-industrial by-products: Antioxidant activity, occurrence, and potential uses. *Food Chem.* **2006**, *99*, 191–203. [CrossRef]

153. Crozier, A.; Del Rio, D.; Clifford, M.N. Bioavailability of dietary flavonoids and phenolic compounds. *Mol. Asp. Med.* **2010**, *31*, 446–467. [CrossRef] [PubMed]

154. D'Archivio, M.; Filesi, C.; Varì, R.; Scazzocchio, B.; Masella, R. Bioavailability of the polyphenols: Status and controversies. *Int. J. Mol. Sci.* **2010**, *11*, 1321–1342. [CrossRef] [PubMed]

155. Bangoura, I.; Chowdhury, M.T.H.; Kang, J.Y.; Cho, J.Y.; Jun, J.C.; Hong, Y.K. Accumulation of phlorotannins in the abalone *Haliotis discus hannai* after feeding the brown seaweed *Ecklonia cava*. *J. Appl. Phycol.* **2013**, *26*, 967–972. [CrossRef]

156. Bangoura, I.; Hong, Y.-K. Dietary intake and accumulation of phlorotannins in abalone after feeding the phaeophyte *Ecklonia stolonifera*. *J. Life Sci.* **2015**, *25*, 780–785. [CrossRef]

157. Manach, C.; Williamson, G.; Morand, C.; Scalbert, A.; Rémésy, C. Bioavailability and bioefficacy of polyphenols in humans. I. Review of 97 intervention studies. *Am. J. Clin. Nutr.* **2005**, *81*, 243S–255S.

158. Bohn, T. Dietary factors affecting polyphenol bioavailability. *Nutr. Rev.* **2014**, *72*, 429–452. [CrossRef] [PubMed]

159. Scalbert, A.; Williamson, G. Dietary intake and bioavailability of polyphenols. *J. Med. Food* **2000**, *3*, 121–125.

International Journal of
Molecular Sciences

MDPI

Article

Hepatoprotective Role of *Hydrangea macrophylla* against Sodium Arsenite-Induced Mitochondrial-Dependent Oxidative Stress via the Inhibition of MAPK/Caspase-3 Pathways

Md Rashedunnabi Akanda [1,2], Hyun-Jin Tae [1], In-Shik Kim [1], Dongchoon Ahn [1], Weishun Tian [1], Anowarul Islam [1], Hyeon-Hwa Nam [3], Byung-Kil Choo [3] and Byung-Yong Park [1,*]

1   College of Veterinary Medicine and Biosafety Research Institute, Chonbuk National University, Iksan 54596, Korea; rashed.mvd@gmail.com (M.R.A.); hjtae@jbnu.ac.kr (H.-J.T.); iskim@jbnu.ac.kr (I.-S.K.); ahndc@jbnu.ac.kr (D.A.); tianws0502@126.com (W.T.); anowarul.vet@gmail.com (A.I.)
2   Department of Pharmacology and Toxicology, Sylhet Agricultural University, Sylhet 3100, Bangladesh
3   Department of Crop Science and Biotechnology, Chonbuk National University, Jeonju 54896, Korea; hh_hh@jbnu.ac.kr (H.-H.N.); bkchoo@jbnu.ac.kr (B.-K.C.)
*   Corresponding: parkb@jbnu.ac.kr; Tel.: +82-63-850-0961; Fax: +82-63-850-0910

Received: 26 May 2017; Accepted: 5 July 2017; Published: 10 July 2017

Abstract: Sodium arsenite ($NaAsO_2$) has been recognized as a worldwide health concern. *Hydrangea macrophylla* (HM) is used as traditional Chinese medicine possessing antioxidant activities. The study was performed to investigate the therapeutic role and underlying molecular mechanism of HM on $NaAsO_2$-induced toxicity in human liver cancer (HepG2) cells and liver in mice. The hepatoprotective role of HM in HepG2 cells was assessed by using 3-(4,5-dimethylthiazol-2-Yl)-2,5-diphenyltetrazolium bromide (MTT), reactive oxygen species (ROS), and lactate dehydrogenase (LDH) assays. Histopathology, lipid peroxidation, serum biochemistry, quantitative real-time polymerase chain reaction (qPCR) and Western blot analyses were performed to determine the protective role of HM against $NaAsO_2$ intoxication in liver tissue. In this study, we found that co-treatment with HM significantly attenuated the $NaAsO_2$-induced cell viability loss, intracellular ROS, and LDH release in HepG2 cells in a dose-dependent manner. Hepatic histopathology, lipid peroxidation, and the serum biochemical parameters alanine aminotransferase (ALT) and aspartate aminotransferase (AST) were notably improved by HM. HM effectively downregulated the both gene and protein expression level of the mitogen-activated protein kinase (MAPK) cascade. Moreover, HM well-regulated the Bcl-2-associated X protein (Bax)/B-cell lymphoma-2 (Bcl-2) ratio, remarkably suppressed the release of cytochrome *c*, and blocked the expression of the post-apoptotic transcription factor caspase-3. Therefore, our study provides new insights into the hepatoprotective role of HM through its reduction in apoptosis, which likely involves in the modulation of MAPK/caspase-3 signaling pathways.

Keywords: hepatoprotection; *Hydrangea macrophylla*; $NaAsO_2$; mitogen-activated protein kinase (MAPK); caspase-3

1. Introduction

Inorganic arsenic compounds are heavy metal toxicants recognized as human carcinogens [1,2]. Among them, sodium arsenite is the most hazardous inorganic arsenic compound for human and animal health [3]. Arsenic is found in the environment surrounding the industrial and natural sources, raising eco-friendly public health concerns due to modern globalization [4]. Trivalent arsenical (arsenite) in ground water is the foremost source of arsenicosis, affecting more than 140 million people globally, particularly in India, Bangladesh, and neighboring countries [5]. Epidemiological investigations reveal links between arsenicosis and pathogenesis of various adverse health effects such as liver disorders, vascular diseases, diabetes, and cancer [6,7]. Studies at the molecular and cellular level show that arsenicosis enhances the production of reactive oxygen species (ROS) that causes DNA methylation, lipid peroxidation, increase oxidation of protein and disrupt enzymes [8,9]. Among them, oxidative stress due to the excessive release of free radicals has been implicated in $NaAsO_2$-mediated damage in liver, kidney, heart, brain, skin and other tissues [10].

Oxidative stress is a fairly new but widely standard theory of $NaAsO_2$-induced hepatotoxicity [11]. The increasing oxidative stress and thereby reduction of the endogenous antioxidant system during arsenic intoxication assists as a crucial factor in liver disorders. Particularly, it may lead to the several pathological conditions such as hepatic degeneration, alteration of lipid status and progressive fibrosis [12]. Generation of intracellular ROS leads to disruption in the matrix metalloproteinase (MMP), lipid peroxidation, carbonylation of protein, an imbalance in the Bcl-2-associated X protein (Bax)/B-cell lymphoma-2 (Bcl-2) ratio, and releases of cytochrome *c* following stimulation of the mitogen-activated protein kinase (MAPK) cascade [13]. Arsenic-mediated oxidative stress stimulates the MAPK cascade and induces apoptosis in hepatocytes via the mitochondria-dependent caspase signaling pathway [14,15].

Antioxidants have been recognized as favorable for mitigating arsenic-mediated oxidative stress in liver [16]. Naturally-found phytomedicine and their active ingredients have received significant attention as antioxidant agents and might offer some protection against oxidative stress, thus having a potential role in reducing the toxicity of trivalent arsenite [17]. Traditional herbal phytomedicine has received much attention as effective and alternative remedies for liver diseases [18]. In this study, we emphasize for the first time a simple and competent process of obtaining an extract from *Hydrangea macrophylla* (HM) seeds that have strong efficacy on $NaAsO_2$-induced hepatotoxicity in vitro and in vivo. HM is a Hydrangeaceae plant native to the Korean mountains known as "soogook", and is traditionally used as a folk medicine to treat many diseases such as diabetes and liver disorders. The major components of the HM extract such as phyllodulcin, hydrangenol, and hydrangeic acid has been determined by the high-performance liquid chromatography method [19]. The biological properties of HM and its active compounds have been reported with respect to antioxidant [20] anti-diabetic [21] and anti-malarial [22] activities. However, to our knowledge, HM has not been previously reported for its hepatoprotective effect. Therefore, based on the traditional uses and pharmacological actions of the active component of HM, our study investigates the hepatoprotective activities of HM and underlying molecular mechanisms involved in the action of $NaAsO_2$-induced oxidative stress in liver.

2. Results

2.1. Analysis of Total Phenolic and Flavonoid Content of HM

Phenolic and flavonoid contents are the secondary metabolites of the plant, which exhibits a series of biological activities, and certainly, has antioxidant properties. The total phenolic and flavonoid contents of HM were investigated and are presented in Table 1.

Table 1. Total phenolics, flavonoids and extraction yield of *Hydrangea macrophylla* (HM).

Plant Extract	Total Phenolics (mg Gallic Acid Equivalent/g Extract)	Total Flavonoids (mg Rutin/g Extract)	Total Yield (%)
HM extract	92.358 ± 0.342	220.725 ± 3.263	26.9

2.2. HM Reduced NaAsO₂-Induced Oxidative Stress in Human Liver Cancer (HepG2) Cells

We performed the 3-(4,5-dimethylthiazol-2-Yl)-2,5-diphenyltetrazolium bromide (MTT) assay to evaluate the hepatoprotective effects of HM against $NaAsO_2$-induced cytotoxicity in HepG2 cells. We first determined the optimum concentration of the HM extract and found that cell viability was more than 90% after 24 h of incubation with various concentrations of HM (5, 10, 20, and 30 µg/mL) (Figure 1a). After that, NaAsO₂ markedly ($p < 0.05$) decreased cell viability as compared to the control. We found that HM (10, 20 and 30 µg/mL) significantly ($p < 0.05$) protected HepG2 cells against NaAsO₂-mediated oxidative damage in a dose-dependent manner (Figure 1b). Likewise, the morphology of HepG2 cells was improved by co-treatment of the HM (Figure 1c). Additionally, HM (30 µg/mL) did not show any detrimental effect on HepG2 cells viability and morphology.

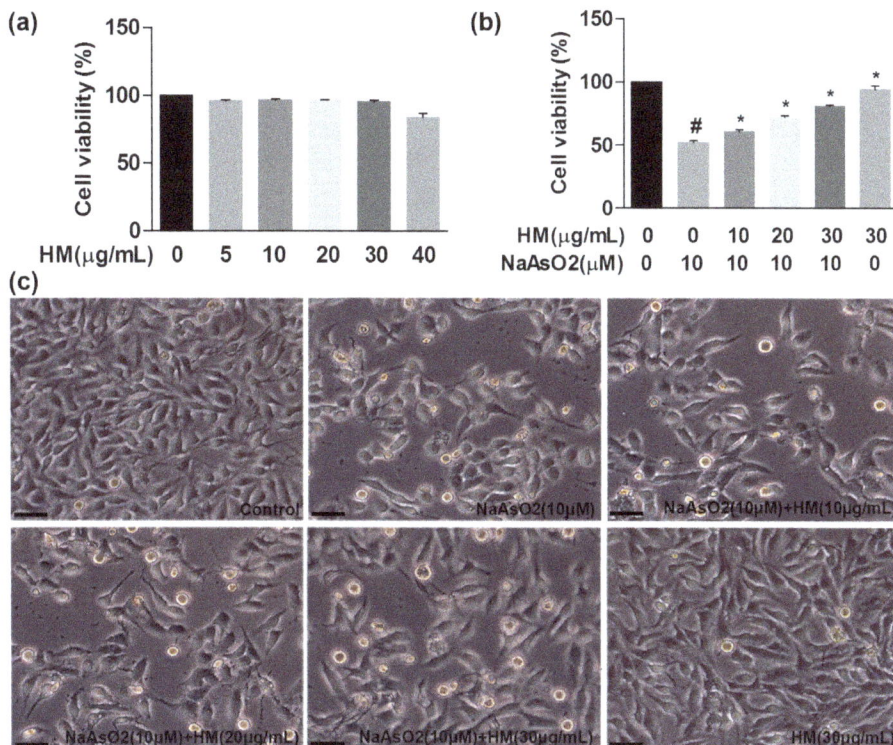

Figure 1. Hepatoprotective effects of HM against NaAsO₂-induced oxidative stress in human liver cancer (HepG2) cells. (**a**) Cytotoxicity and (**b**) cell viability were measured by the 3-(4,5-dimethylthiazol-2-Yl)-2,5-diphenyltetrazolium bromide (MTT) assay; (**c**) observation of HepG2 cell morphology. Cells were pretreated with different concentrations of HM for 1 h, followed by co-treatment with NaAsO₂ for 24 h. Scale bar: 200 µM. Data are expressed as mean ± standard error mean (SEM) of three independent experiments. # $p < 0.05$ compared with the control and NaAsO₂ group, and * $p < 0.05$ compared with the NaAsO₂ and HM extract-treated groups.

2.3. HM Decreased the Intracellular ROS Generation

Intracellular ROS generation followed by accumulation of free radicals is supposed to be an important marker for understanding NaAsO₂-induced hepatic cell death. To investigate the role of HM on NaAsO₂-induced ROS generation, HepG2 cells were pretreated with HM (10, 20, and 30 µg/mL) for 1 h and subsequently exposed to HM and NaAsO₂ (10 µM) for another 24 h. NaAsO₂ markedly ($p < 0.05$) increased the generation of intercellular ROS as compared to the control. Conversely, co-treatment with the HM (20 and 30 µg/mL) significantly ($p < 0.05$) and dose-dependently reduced ROS generation (Figure 2a). Also, HM (30 µg/mL) did not show any effects on the intracellular ROS generation. These data showed that HM protects hepatic cells from the oxidative damaging effect caused by NaAsO$_2$.

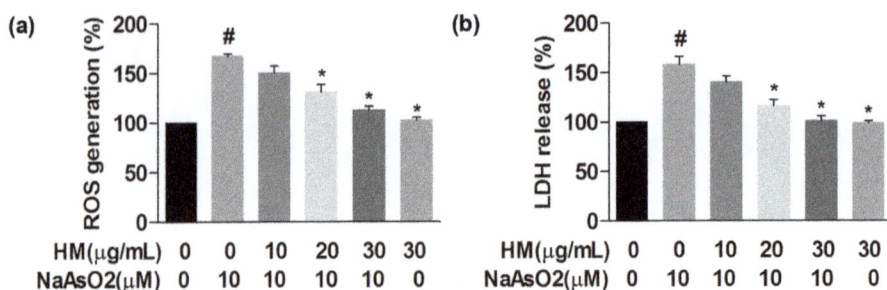

Figure 2. HM inhibited the reactive oxygen species (ROS) and lactate dehydrogenase (LDH) release in HepG2 cells. (**a**) Intracellular ROS and (**b**) LDH levels were measured. Cells were pretreated with different concentrations of HM (10, 20, and 30 μg/mL) for 1 h, followed by co-treatment with 10 μM NaAsO$_2$ for another 24 h. Data are expressed as mean ± standard error mean (SEM) of three independent experiments. # $p < 0.05$ compared with the control and NaAsO$_2$ group, and * $p < 0.05$ compared with the NaAsO$_2$ and HM extract-treated groups.

2.4. HM Inhibited the Lactate Dehydrogenase (LDH) Release

In vitro hepatoprotective effect of HM was determined by performing LDH assay using HepG2 cells culture supernatant. The LDH level was significantly ($p < 0.05$) increased in NaAsO$_2$ exposed cells compared to the control; however, co-treatment with the HM (20 and 30 μg/mL) significantly ($p < 0.05$) and dose-dependently reduced the LDH release (Figure 2b). Besides, HM (30 μg/mL) alone did not show any effect on the LDH release. Our data revealed that HM protects the HepG2 cells from cytotoxicity caused by NaAsO$_2$.

2.5. HM Improved the Liver Histopathology and Body Weight

NaAsO$_2$-mediated liver damage and its protection by HM treatment in mice were confirmed by microscopic evaluation of histopathological changes. Microscopic analysis of the liver indicated a normal structure of hepatocytes arranged around the portal vein in the control mice liver (Figure 3a); however, the NaAsO$_2$-exposed group showed damage in the hepatic lobules surrounding the hepatic artery, degenerated nuclei, a dilated portal vein, and blurred cytoplasm (Figure 3b). The histopathological changes induced by NaAsO$_2$ were considerably improved by co-treatment with the HM (Figure 3c). The dilated portal vein diameter was also markedly ($p < 0.05$) reduced by HM co-treatment (Figure 3d). Moreover, arsenic-exposed mice showed a significant ($p < 0.05$) decline in body weight compared to normal control mice, whereas co-administration of HM and NaAsO$_2$ effectively ($p < 0.05$) increased body weight compared to treatment with NaAsO$_2$ alone (Figure 3e).

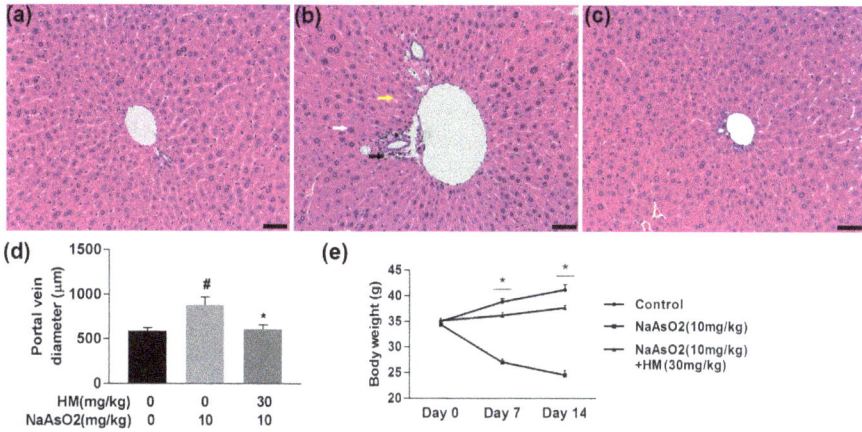

Figure 3. HM improved the liver histology and body weight in experimental mice. Untreated mice were used as a control to compare histological changes induced by $NaAsO_2$. (**a**) Normal control; (**b**) $NaAsO_2$ (10 mg/kg); (**c**) Co-treatment with HM (30 mg/kg) + $NaAsO_2$ (10 mg/kg); (**d**) portal vein diameter; and (**e**) Body weight. In the $NaAsO_2$ group, the white arrow indicates the degenerative nucleus, the yellow arrow indicates the blurred cytoplasm and the black arrow indicates the damaged hepatic lobule surrounding hepatic artery. In contrast, co-treatment with HM improved histological changes compared to $NaAsO_2$ alone. Portal vein diameter and body weight also significantly decreases and increased in HM-treated mice, respectively. Data are expressed as mean ± standard error mean (SEM) of three independent experiments. Scale bar: 200 μM. # $p < 0.05$ compared with the control and $NaAsO_2$ group, and * $p < 0.05$ compared with the $NaAsO_2$ and HM extract-treated group.

2.6. HM Regulated the Serum Biochemical Parameters

Alanine aminotransferase (ALT) and aspartate aminotransferase (AST) levels were investigated as biomarkers of liver cell integrity. The levels of the serum cytosolic enzymes ALT and AST were considerably ($p < 0.05$) higher in $NaAsO_2$-intoxicated mice than in normal control. We found that co-treatment with HM significantly ($p < 0.05$) improved liver physiology by reducing the level of ALT and AST as compared to $NaAsO_2$ alone (Figure 4a,b). These data indicate that HM improves the liver physiology in $NaAsO_2$-mediated hepatotoxicity.

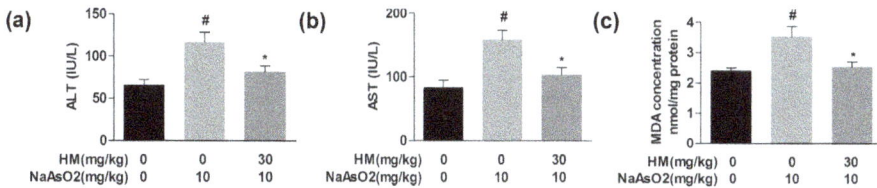

Figure 4. Effect of HM on serum markers and lipid peroxidation in experimental mice. (**a**) Serum alanine aminotransferase (ALT); (**b**) serum aspartate aminotransferase (AST), and (**c**) Malondialdehyde (MDA) levels in liver tissue. ALT, AST and MDA levels were increased in $NaAsO_2$-intoxicated mice. Co-treatment with HM significantly decreased the serum ALT and AST and tissue MDA levels as compared to $NaAsO_2$ alone. Data are expressed as mean ± standard error mean (SEM) of three independent experiments. # $p < 0.05$ compared with the control and $NaAsO_2$ group, and * $p < 0.05$ compared with the $NaAsO_2$ and HM extract-treated group.

2.7. HM Controlled the Lipid Peroxidation Production

Thiobarbituric acid reactive substances (TBARS) concentration was evaluated to determine the level of malondialdehyde (MDA) in liver samples, which is the end-product of lipid peroxidation in oxidative stress. The MDA level in the $NaAsO_2$ group was significantly ($p < 0.05$) higher than in the control. The elevated level of MDA was markedly ($p < 0.05$) reduced after co-treatment with the HM (Figure 4c). This result suggested that HM might have anti-oxidative effect against $NaAsO_2$-induced hepatic damage in mice.

2.8. HM Suppressed the Gene Expression of MAPKs (Extracellular Signal-Regulated Kinases (ERK), C-Jun N-Terminal Kinases (JNK), and p38)

To reveal the possible molecular pathways of hepatoprotection by HM, we evaluated gene expression in liver homogenates by quantitative real-time polymerase chain reaction (qPCR) analysis. Treatment with $NaAsO_2$ markedly ($p < 0.05$) increased the gene expression level of *ERK*, *JNK*, and *p38* compared to the control, and these higher gene expression level were significantly ($p < 0.05$) attenuated by co-treatment with HM compared with $NaAsO_2$ alone (Figure 5). These effects suggest that HM significantly inhibits the gene expression of MAPKs, thereby reducing liver damage caused by $NaAsO_2$.

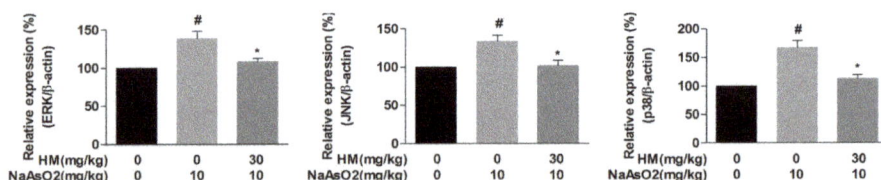

Figure 5. HM attenuated the gene expression of mitogen-activated protein kinase (MAPK) (extracellular signal-regulated kinases (*ERK*), C-Jun N-terminal kinases (*JNK*), and *p38*) in liver tissue. The expression level of *MAPK* genes was significantly upregulated in $NaAsO_2$-exposed liver tissue, but co-treatment with HM effectively downregulated the gene expression. Data are expressed as mean ± standard error mean (SEM) of three independent experiments. # $p < 0.05$ compared with the control and $NaAsO_2$ group, and * $p < 0.05$ compared with the $NaAsO_2$ and HM extract-treated group.

2.9. HM Mitigated NaAsO2-Mediated Hepatotoxicity by Regulating Anti-Apoptotic Signaling Pathways

Mitochondrial damage is an important marker of apoptotic cell death and is executed through ROS-mediated oxidative stress [23]. Here, we evaluated the involvement of the mitochondrial pathway of hepatic apoptosis by western blot analysis. Our results revealed that $NaAsO_2$ markedly ($p < 0.05$) upregulated phosphorylation of the MAPK cascade (pERK1/2, pJNK, and pp38), whereas co-treatment with HM effectively ($p < 0.05$) downregulated the phosphorylation level (Figure 6). The expression ratio of the Bax/Bcl-2 family protein was also significantly ($p < 0.05$) controlled by HM co-treatment. Meanwhile, the expression of cytochrome c and activated caspase-3 was also remarkably ($p < 0.05$) increased with $NaAsO_2$ treatment. However, HM co-treatment significantly ($p < 0.05$) attenuated the cytochrome c and activated caspase-3 expression (Figure 7). Together, these results supported that HM considerably regulates the mitochondrial-dependent hepatic damage caused by $NaAsO_2$.

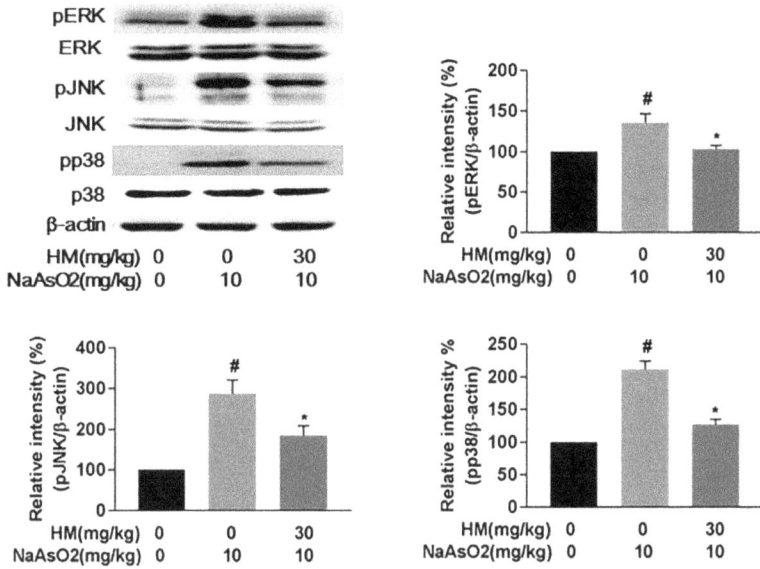

Figure 6. Effect of HM on the MAPK (pERK1/2, pJNK, and pp38) signaling pathway in liver tissue. The expression of pERK, pJNK, and pp38 in liver tissue was analyzed by Western blot. Co-treatment with HM significantly downregulated the expression level of pERK1/2, pJNK, and pp38. Data are expressed as mean ± standard error mean (SEM) of three independent experiments. # $p < 0.05$ compared with the control and NaAsO$_2$ group, and * $p < 0.05$ compared with the NaAsO$_2$- and HM extract-treated group.

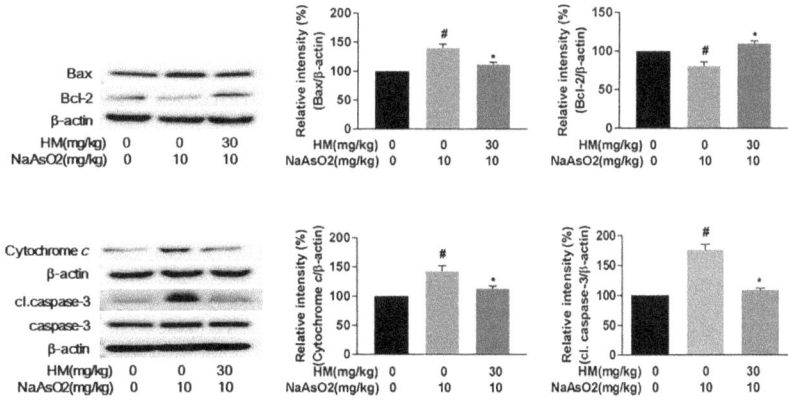

Figure 7. Effect of HM on the apoptotic signaling pathways in response to NaAsO$_2$ and HM exposure. The expression of Bcl-2-associated X protein (Bax), B-cell lymphoma-2 (Bcl-2), cytochrome *c*, and caspase-3 in liver tissue was analyzed by Western blot analysis. Co-treatment with HM effectively regulated the Bax/Bcl-2 ratio and notably downregulated the expression of cytochrome *c*, and cleaved caspase-3. Data are expressed as mean ± standard error mean (SEM) of three independent experiments. # $p < 0.05$ compared with the control and NaAsO$_2$ group, and * $p < 0.05$ compared with the NaAsO$_2$ and HM extract-treated group.

3. Discussion

Sodium arsenite (NaAsO$_2$) is a ubiquitous environmental stressor that has become a danger to human and animal health [24]. Long-term exposure to arsenic compounds has been directly related to major health disorders such as hepatitis, hepatic cancer, diabetes, coronary disease, stroke, peripheral vascular disease, and skin disease [25,26]. Among them, the liver is the most target site for arsenic toxicity due to its physiology, particularly for biochemical alteration of arsenic metabolites. Oxidative injury plays a vital role in such kinds of alteration-related pathophysiology. Treatment preventing the hepatic damage may lead to prospective therapeutic strategies against the hepatic disorders and HM extract may provide a novel therapeutic candidate. Evidence has stated that extract from HM has potential antioxidative properties [21]. In this study, we demonstrated that HM can be used as a novel indigenous phytomedicine due to its strong hepatoprotective effects against NaAsO$_2$-mediated oxidative stress in vitro and in vivo.

Phenolics and flavonoids are the most important plant secondary metabolites and have the strong antioxidant capacity [27,28]. Their antioxidant ability is mainly due to their redox properties, which allow them to act as reducing agents, oxygen scavengers and transition metal ions chelator [29]. In our study, we found a considerable amount of phenolic and flavonoid content in HM extract that may be the major contributor for the antioxidant role against oxidative stress-induced hepatic damage. NaAsO$_2$-mediated cytotoxicity is mainly associated with the generation of ROS, increase in lipid peroxidation, DNA dysfunction, cell cycle disruption, and apoptotic cell death [30,31]. An in vitro hepatocellular model, HepG2 cells were exposed to NaAsO$_2$ and led to cell death by oxidative stress. However, with pretreatment with HM, the cell viability was restored, indicating its hepatoprotective role. Intracellular ROS release and LDH production have established a mechanism that is associated with hepatic cell death [5]. Excessive accumulation of intracellular ROS and LDH production may accelerate unstable cellular homeostasis that leads to mitochondrial membrane dysfunction [32]. Natural phytomedicine, with the capability for scavenging free radicals, may reduce conditions correlated to oxidative stress. We found that exposure to NaAsO$_2$ in HepG2 cells considerably increased the ROS and LDH release that boosted the oxidative stress and prompted cell apoptosis. However, dose-dependent and significant inhibition of ROS generation and LDH leakage in HepG2 cells were observed after HM treatment. Therefore, HM may protect the cells against NaAsO$_2$-induced oxidative stress via its antioxidant capacity.

To evaluate either HM exhibits the same defensive role in vivo, NaAsO$_2$-intoxicated mice were studied. In an attempt to assess the internal hepatotoxicity by NaAsO$_2$, the histopathological changes were evaluated. We observed that arsenic caused hepatic tissue destruction, degenerated the nucleus, dilated the portal vein, and blurred cytoplasm, perhaps due to the accumulation of free radicals and following lipid peroxidation. Such findings are related to the previous study [33]. Co-treatment with HM during arsenic exposure effectively improved the hepatic histological architecture. The serum biochemical indicators ALT and AST were positively correlated with hepatic histopathology [34]. The activities of ALT and AST are frequently used as a diagnostic marker of liver damage since they are linked with liver physiology [35]. Increased levels of ALT and AST in the bloodstream damage the hepatic cell integrity. We found HM was effective in restoring the serum enzyme biomarkers. MDA is the end product of lipid peroxidation and is a well-established and standard mechanism of cellular injury used as an indicator of oxidative stress in cells and tissues [36]. We found the significant increase of MDA level in arsenic-exposed liver suggested higher lipid peroxidation, leading to tissue damage. However, HM treatment effectively attenuated the MDA that plays an important role in inhibiting oxidative stress [37]. At the end of the experiment, a significant increase in body weight was recorded in HM-treated mice compared to the NaAsO$_2$ group alone.

Multiple biological mechanisms and molecular signaling pathways are involved in apoptotic cell death [38]. Mitochondria play a crucial role in the induction of cellular apoptosis [39]. Cytotoxic ROS activate mitochondrial-dependent apoptosis via stimulation of the MAPK cascade and subsequent modulation of the pro-apoptotic protein Bax and anti-apoptotic protein Bcl-2, followed by cytochrome *c* release, and finally executes hepatic cell death by activation of the caspase cascade pathway [25,40]. The relevant observation was confirmed in this study. We observed that $NaAsO_2$ markedly increased the gene expression and phosphorylation of the MAPK cascade that ultimately leads to apoptosis, while HM effectively reduced both the gene expression and phosphorylation of the MAPK cascade in arsenic-exposed liver tissue. Likewise, HM notably decreased the expression of Bax and increased expression of the Bcl-2 protein, subsequently controlling the release of cytosolic cytochrome *c*. These findings suggest that HM positively regulates the $NaAsO_2$-induced mitochondrial-dependent apoptotic signaling pathway. It is well established that Bax/Bcl-2 proteins control permeabilization of the mitochondrial membrane and thus control the release of apoptotic factors from the intermembrane space of mitochondria [25].

An imbalance in the Bax/Bcl-2 ratio leads to the production of cytochrome *c* from mitochondria and subsequently activates the apoptotic protein caspase to trigger cellular apoptosis [41,42]. Cytochrome *c* promotes the ATP-dependent formation of the apoptosome, resulting in activation of the caspase cascade via a mitochondrial-dependent pathway [43]. Predominantly, overactivation of caspase-3 indicates pathogenesis in hepatic cell death [5,44]. We found that $NaAsO_2$ markedly upregulated the expression of caspase-3, whereas notable attenuation of caspase-3 was found after co-treatment with HM. This result was supported by earlier investigation [45]. Therefore, our data revealed that HM could protect the liver from $NaAsO_2$-induced oxidative stress through the mitochondria-dependent pathway, indicating that the radical scavenging potency of HM is responsible for its anti-apoptotic activity.

4. Materials and Methods

4.1. Chemicals and Antibodies

The highest analytical grades of all chemicals were used. Sodium arsenite ($NaAsO_2$), 3-(4,5-dimethylthiazol-2-Yl)-2,5-diphenyltetrazolium bromide (MTT), penicillin/streptomycin, gallic acid, rutin, hematoxylin, eosin, and protease inhibitor were purchased from Sigma-Aldrich (St. Louis, MO, USA). Fetal bovine serum (FBS), Dulbecco's Modified Eagle's Medium (DMEM), and other cell culture reagents were obtained from Gibco (Carlsbad, CA, USA). Dimethyl sulfoxide (DMSO) was obtained from Bioshop (Burlington, ON, Canada). RNA extraction kits were purchased from RiboEx and Hybrid-R (Gene All, Seoul, South Korea). Tissue protein extraction reagent (T-PER), complementary DNA (cDNA) synthesis (ReverTra Ace® qPCR RT Kit, Toyobo, Osaka, Japan), and bicinchoninic acid (BCA) protein assay kits were purchased from Thermo Scientific (Waltham, MA, USA). The SYBR Green qPCR kit was purchased from Toyobo (Osaka, Japan). Primary antibodies (pERK1/2, pJNK, pp38, tERK1/2, tJNK, tp38, Bax, Bcl-2, cytochrome *c*, cleaved caspase-3, and caspase-3) and β-actin were purchased from Cell Signaling (Danvers, MA, USA). The goat anti-rabbit immunoglobulin G horseradish peroxidase (IgG-HRP) secondary antibody was purchased from Santa Cruz (Santa Cruz, CA, USA). The WESTSAVE Gold Enhanced Chemiluminescence (ECL) detection kit was acquired from Abfrontire (Seoul, Korea), and the ALT and AST kits were from ASAN (Hwaseong, Korea). LDH cytotoxicity assay kit was obtained from TAKARA (Tokyo, Japan), and the ROS-Glo H_2O_2 assay kit was from Promega (Madison, WI, USA). Zoletil 50 was supplied by Virbac S.A (Carros, France).

4.2. Preparation of Hydrangea Macrophylla Seed Extract

The seeds of *Hydrangea macrophylla* (HM) plant were collected from Jirisan located in the southern part of South Korea and authenticated based on its microscopic and macroscopic features by the Korea Institute of Oriental Medicine. We prepared HM seeds extract according to the previously described method with minor modifications [46]. Briefly, the plant seeds were sliced and dried completely. The extract was prepared by maceration of seeds sample with 70% ethanol (twice for 2 h reflux), and then filtered extract was concentrated under vacuum centrifuge and dehydrated with a lyophilizer.

The powder extract was liquefied in DMSO and sterilized using a 0.22-μm syringe filter. The dried extract was kept at −20 °C. The study was conducted using a single batch of plant extract to avoid batch-to-batch variation and maximize the product constancy.

4.3. Determination of Total Phenolic and Flavonoid Content

Total phenolic and flavonoid content of HM extract was measured according to the previously described method [47].

4.4. Cell Culture

HepG2 cells were maintained at 37 °C in a 5% CO_2 humidified incubator. Cells were cultured in DMEM supplemented with 10% FBS and 1% penicillin and streptomycin. The cell culture medium was changed for every 2 days, and the cells were subcultured when they reached about 90% confluency in the culture flask.

4.5. Assessment of Cell Viability

MTT assay was used to measure cell viability. HepG2 cells were seeded (1×10^4 cells/well in 96-well plates) and cultured in a 37 °C incubator overnight. For evaluating the cytotoxicity of HM, cells were treated with HM (5, 10, 20, 30 and 40 μg/mL) for 24 h. In contrast, measuring the cell viability, cells were pretreated for 1 h with different concentrations of HM (10, 20 and 30 μg/mL) and then co-incubated with HM and $NaAsO_2$ (10 μM) for an additional 24 h. The medium was replaced with 0.5 mg/mL of the MTT working solution and incubated for 2 h. The blue formazan crystals were solubilized by DMSO. Optical density was measured at 570 nm absorbance by a tunable versa max microplate reader (Molecular Devices, Sunnyvale, CA, USA). Similarly, for observation of HepG2 cell morphology, the image of the cell was captured by an inverted microscope (Olympus, CKX41, Tokyo, Japan) at fixed 100× magnification.

4.6. Measurement of Reactive Oxygen Species (ROS) Generation

To evaluate the level of intracellular ROS, HepG2 cells (1×10^4 cells/well) were cultured in 96-well plates overnight. After adherence, cells were pretreated for 1 h with different concentrations of HM (10, 20 and 30 μg/mL) and then co-incubated with HM and $NaAsO_2$ (10 μM) for an additional 24 h. The intracellular ROS level was measured according to the manufacturers' procedure for the kit, and the absorbance was measured at 490 nm using a tunable versa max microplate reader.

4.7. Determination of Lactate Dehydrogenase (LDH) Release

To measure the level of extracellular LDH release, HepG2 cells (1×10^4 cells/well) were cultured in 96-well plates overnight. After adherence, cells were pretreated for 1 h with different concentrations of HM (10, 20 and 30 μg/mL) and then co-incubated with HM and $NaAsO_2$ (10 μM) for an additional 24 h. The LDH level was measured according to the manufacturers' procedure of kit, and the absorbance was measured at 490 nm using a tunable versa max microplate reader.

4.8. Mice Management and Experimental Design

Male ICR mice (6 weeks old) were maintained in accordance with the animal welfare regulations of the Institutional Animal Care and Use Committee (IACUC; CBNU 2016-68), Chonbuk National University Laboratory Animal Centre, South Korea. Mice were kept in standard mouse cages with an ad libitum supply of food and distilled water. Ideal conditions for temperature (23 ± 2 °C), humidity (35–60%), and photoperiod cycle (12 h light and 12 h dark) were maintained over the experimental period. Mice were adapted to the laboratory conditions for 1 week before starting the experiment. A total of 36 mice were randomly divided into three groups: (1) normal control mice were treated with saline; (2) toxic control mice were treated with $NaAsO_2$ once daily (10 mg/kg body weight, (per os/orally) p.o., for 10 days); and (3) experimental mice were treated with HM once daily (30 mg/kg body weight, p.o., for 15 days) prior to treatment with $NaAsO_2$ (10 mg/kg body weight, p.o., for 10 days). After the experimental period, mice were fasted overnight and anesthetized with Zoletil 50.

4.9. Histopathological Study of the Liver

Mice liver was collected for histopathological examination. Liver samples were immediately fixed in 10% neutral buffered formalin (NBF) and processed in an auto processor (Excelsior ES, Thermo Scientific, Waltham, MA, USA). After embedding in paraffin, 5-μm sections were stained with hematoxylin and eosin and mounted on glass slides. Digital images were obtained using a Leica DM2500 microscope (Leica Microsystems, Wetzlar, Germany) at a fixed 200× magnification. The diameter of the portal vein was measured using image measurement software (v 22.1., iSolution DTM, Vancouver, BC, Canada).

4.10. Serum Biochemical Analysis

The levels of the basic liver function biomarker enzymes serum ALT and AST was examined. Blood samples were collected from the mouse and incubated for 30 min at room temperature. Blood was centrifuged at 3000 rpm for 15 min at 4 °C to collect the serum. ALT and AST levels were analyzed according to the manufacturer's recommendation.

4.11. Lipid Peroxidation Assay

The level of MDA is an important marker of oxidative stress condition. MDA concentration was measured in the liver tissue. The samples homogenized in a ratio of 1/10 in 1.15% (w/v) ice-cold KCl solution with the aid of the thiobarbituric acid (TBA) established method [48]. The standards of 2.5, 5, 10 and 20 nmol/mL tetra ethoxy propane (TEP) were used. The results were expressed as nmol MDA/mg protein.

4.12. Quantitative Real-Time Polymerase Chain Reaction (qPCR) Analysis

Total RNA was isolated from mouse liver tissue according to the manufacturer's instructions, and RNA concentration was quantified using a BioSpec-nano spectrophotometer (Shimadzu Biotech, Tokyo, Japan) at a 260/280 nm ratio. For cDNA synthesis, 3 μg of total RNA was used, and the cDNA synthesis procedure was performed according to the manufacturer's instructions. qPCR was performed using the SYBR Green Real-Time PCR master mix with the Roche LightCyclerTM, and the conditions were maintained according to the manufacturer's instructions. β-Actin was used as the housekeeping gene. Relative expression of target genes was normalized with reference gene (β-actin). The nucleotide sequences of the primers are presented in Table 2 [49].

Table 2. Nucleotide sequences of the primers for qPCR

Gene	Primers Sequence (5'–3')	Size (bp)	Genebank Accession No.
ERK	TCAGAGGCAGGTGGATCTCT ACGGGGAGGACTCTGTTTTT	109	NM_011949.3
JNK	CGGAACACCTTGTCCTGAAT CACATCGGGGAACAGTTTCT	93	NM_016700.4
p38	AGCCAATTCCAGTGTTGGAC TTCTGGGCTCCAAATGATTC	120	NM_011951.3
β-actin	AGAAGATCTGGCACCACACC TACGACCAGAGGCATACAGG	195	NM_007393.5

4.13. Western Blot Analysis

Liver lysates were prepared in ice-cold lysis buffer containing tissue protein extraction reagent (T-PER), phenylmethanesulfonyl (PMSF), Na_3VO_4 (sodium orthovanadate), and protease inhibitor cocktail. The lysate was centrifuged at 12,000 rpm for 20 min at 4 °C, and the supernatant was collected. The total protein concentration of the lysate supernatant was measured using the BCA protein assay kit. An equal amount of protein was separated by 12% sodium dodecyl sulfate-polyacrylamide gel electrophoresis (SDS-PAGE) and transferred to a nitrocellulose membrane. The membrane was blocked with a regular blocking solution (5% non-fat skim milk in Tris-buffered saline (TBST)) for 1 h at room temperature, followed by incubation with primary antibodies against pERK, pJNK, pp38, Bax, Bcl-2, cytochrome *c*, caspase-3, cleaved caspase-3, and β-actin overnight at 4 °C. The blot was washed and then incubated with anti-rabbit secondary antibodies for 1 h at room temperature. Protein band was detected using an ECL detection kit, and images were obtained using an imaging system (LAS-400 image system, GE Healthcare, UK). β-actin was used as the control.

4.14. Statistical Analysis

Data were analyzed with Graph Pad Prism 7.0 (La Jolla, CA, USA) and expressed as mean ± standard error mean (SEM). Group comparisons were performed using analysis of variance (ANOVA), followed by Tukey's multiple comparisons tests. The minimum statistical significance was considered $p < 0.05$ for all analyses.

5. Conclusions

In conclusion, both in vitro and in vivo findings offer evidence of the hepatoprotective potency of HM on $NaAsO_2$-mediated oxidative damage via attenuation of free radical generation, restoration of hepatic physiology, and reduction in mitochondrial-dependent apoptosis. Thus, HM is a natural phytomedicine that seems to be a promising therapeutic agent for treatment of hepatic disorders by targeting oxidative stress.

Acknowledgments: The present research was supported by the Basic Science Research Program through the National Research Foundation of South Korea, funded by the Ministry of Education (2017R1D1A1B03035765) and by funds for newly appointed professors of Chonbuk National University in 2016.

Author Contributions: Md Rashedunnabi Akanda and Hyun-Jin Tae contributed equally to this work. Md Rashedunnabi Akanda and Hyun-Jin Tae: conceptualization, methodology, data analysis and writing original draft. In-Shik Kim; Dongchoon Ahn; Weishun Tian; Anowarul Islam and Byung-Kil Choo: review and editing. Hyeon-Hwa Nam: methodology (extract preparation). Byung-Yong Park: conceptualization, supervision, review and editing.

Conflicts of Interest: The authors declare no conflict of interest.

References

1. Jan, A.T.; Azam, M.; Siddiqui, K.; Ali, A.; Choi, I.; Haq, Q.M. Heavy metals and human health: Mechanistic insight into toxicity and counter defense system of antioxidants. *Int. J. Mol. Sci.* **2015**, *16*, 29592–29630. [CrossRef] [PubMed]
2. Palmieri, M.A.; Molinari, B.L. Effect of sodium arsenite on mouse skin carcinogenesis. *Toxicol. Pathol.* **2015**, *43*, 704–714. [CrossRef] [PubMed]
3. Chung, J.Y.; Yu, S.D.; Hong, Y.S. Environmental source of arsenic exposure. *J. Prev. Med. Public Health* **2014**, *47*, 253–257. [CrossRef] [PubMed]
4. Srivastava, P.; Yadav, R.S.; Chandravanshi, L.P.; Shukla, R.K.; Dhuriya, Y.K.; Chauhan, L.K.; Dwivedi, H.N.; Pant, A.B.; Khanna, V.K. Unraveling the mechanism of neuroprotection of curcumin in arsenic induced cholinergic dysfunctions in rats. *Toxicol. Appl. Pharmacol.* **2014**, *279*, 428–440. [CrossRef] [PubMed]
5. Saha, S.; Rashid, K.; Sadhukhan, P.; Agarwal, N.; Sil, P.C. Attenuative role of mangiferin in oxidative stress-mediated liver dysfunction in arsenic-intoxicated murines. *Biofactors* **2016**, *42*, 515–532. [CrossRef] [PubMed]

6. Kharroubi, W.; Dhibi, M.; Haouas, Z.; Chreif, I.; Neffati, F.; Hammami, M.; Sakly, R. Effects of sodium arsenate exposure on liver fatty acid profiles and oxidative stress in rats. *Environ Sci. Pollut. Res. Int.* **2014**, *21*, 1648–1657. [CrossRef] [PubMed]

7. States, J.C.; Srivastava, S.; Chen, Y.; Barchowsky, A. Arsenic and cardiovascular disease. *Toxicol. Sci.* **2009**, *107*, 312–323. [CrossRef] [PubMed]

8. Jomova, K.; Jenisova, Z.; Feszterova, M.; Baros, S.; Liska, J.; Hudecova, D.; Rhodes, C.J.; Valko, M. Arsenic: Toxicity, oxidative stress and human disease. *J. Appl. Toxicol.* **2011**, *31*, 95–107. [CrossRef] [PubMed]

9. Yamauchi, H.; Aminaka, Y.; Yoshida, K.; Sun, G.F.; Pi, J.B.; Waalkes, M.P. Evaluation of DNA damage in patients with arsenic poisoning: Urinary 8-hydroxydeoxyguanine. *Toxicol. Appl. Pharmacol.* **2004**, *198*, 291–296. [CrossRef] [PubMed]

10. Das, J.; Ghosh, J.; Manna, P.; Sinha, M.; Sil, P.C. Taurine protects rat testes against NaAsO$_2$-induced oxidative stress and apoptosis via mitochondrial dependent and independent pathways. *Toxicol. Lett.* **2009**, *187*, 201–210. [CrossRef] [PubMed]

11. Kitchin, K.T.; Conolly, R. Arsenic-induced carcinogenesis—Oxidative stress as a possible mode of action and future research needs for more biologically based risk assessment. *Chem. Res. Toxicol.* **2010**, *23*, 327–335. [CrossRef] [PubMed]

12. Ghatak, S.; Biswas, A.; Dhali, G.K.; Chowdhury, A.; Boyer, J.L.; Santra, A. Oxidative stress and hepatic stellate cell activation are key events in arsenic induced liver fibrosis in mice. *Toxicol. Appl. Pharmacol.* **2011**, *251*, 59–69. [CrossRef] [PubMed]

13. Jing, Y.; Dai, J.; Chalmers-Redman, R.M.; Tatton, W.G.; Waxman, S. Arsenic trioxide selectively induces acute promyelocytic leukemia cell apoptosis via a hydrogen peroxide-dependent pathway. *Blood* **1999**, *94*, 2102–2111. [PubMed]

14. Flora, S.J.; Mehta, A.; Gupta, R. Prevention of arsenic-induced hepatic apoptosis by concomitant administration of garlic extracts in mice. *Chem. Biol. Interact.* **2009**, *177*, 227–233. [CrossRef] [PubMed]

15. Suzuki, T.; Tsukamoto, I. Arsenite induces apoptosis in hepatocytes through an enhancement of the activation of Jun N-terminal kinase and p38 mitogen-activated protein kinase caused by partial hepatectomy. *Toxicol. Lett.* **2006**, *165*, 257–264. [CrossRef] [PubMed]

16. Chang, S.I.; Jin, B.; Youn, P.; Park, C.; Park, J.D.; Ryu, D.Y. Arsenic-induced toxicity and the protective role of ascorbic acid in mouse testis. *Toxicol. Appl. Pharmacol.* **2007**, *218*, 196–203. [CrossRef] [PubMed]

17. Chen, C.; Jiang, X.; Zhao, W.; Zhang, Z. Dual role of resveratrol in modulation of genotoxicity induced by sodium arsenite via oxidative stress and apoptosis. *Food Chem. Toxicol.* **2013**, *59*, 8–17. [CrossRef] [PubMed]

18. Ma, M.; Liu, G.H.; Yu, Z.H.; Chen, G.; Zhang, X. Effect of the lycium barbarum polysaccharides administration on blood lipid metabolism and oxidative stress of mice fed high-fat diet in vivo. *Food Chem.* **2009**, *113*, 872–877.

19. Jung, C.H.; Kim, Y.; Kim, M.S.; Lee, S.; Yoo, S.H. The establishment of efficient bioconversion, extraction, and isolation processes for the production of phyllodulcin, a potential high intensity sweetener, from sweet hydrangea leaves (hydrangea macrophylla thunbergii). *Phytochem. Anal.* **2016**, *27*, 140–147. [CrossRef] [PubMed]

20. Lorentz, C.; Dulac, A.; Pencreac'h, G.; Ergan, F.; Richomme, P.; Soultani-Vigneron, S. Lipase-catalyzed synthesis of two new antioxidants: 4-O- and 3-O-palmitoyl chlorogenic acids. *Biotechnol. Lett.* **2010**, *32*, 1955–1960. [CrossRef] [PubMed]

21. Zhang, H.L.; Matsuda, H.; Yamashita, C.; Nakamura, S.; Yoshikawa, M. Hydrangeic acid from the processed leaves of hydrangea macrophylla var. Thunbergii as a new type of anti-diabetic compound. *Eur. J. Pharmacol.* **2009**, *606*, 255–261. [CrossRef] [PubMed]

22. Ishih, A.; Miyase, T.; Terada, M. Comparison of antimalarial activity of the alkaloidal fraction of hydrangea macrophylla var. Otaksa leaves with the hot-water extract in ICR mice infected with plasmodium yoelii 17 XL. *Phytother. Res.* **2003**, *17*, 633–639. [CrossRef] [PubMed]

23. Dua, T.K.; Dewanjee, S.; Gangopadhyay, M.; Khanra, R.; Zia-Ul-Haq, M.; De Feo, V. Ameliorative effect of water spinach, ipomea aquatica (convolvulaceae), against experimentally induced arsenic toxicity. *J. Transl. Med.* **2015**, *13*, 81. [CrossRef] [PubMed]

24. Abdul, K.S.; Jayasinghe, S.S.; Chandana, E.P.; Jayasumana, C.; de Silva, P.M. Arsenic and human health effects: A review. *Environ Toxicol. Pharmacol.* **2015**, *40*, 828–846. [CrossRef] [PubMed]

25. Rashid, K.; Sinha, K.; Sil, P.C. An update on oxidative stress-mediated organ pathophysiology. *Food Chem. Toxicol.* **2013**, *62*, 584–600. [CrossRef] [PubMed]
26. Byass, P. The global burden of liver disease: A challenge for methods and for public health. *BMC Med.* **2014**, *12*, 159. [CrossRef] [PubMed]
27. Lin, D.; Xiao, M.; Zhao, J.; Li, Z.; Xing, B.; Li, X.; Kong, M.; Li, L.; Zhang, Q.; Liu, Y.; et al. An overview of plant phenolic compounds and their importance in human nutrition and management of type 2 diabetes. *Molecules* **2016**, *21*. [CrossRef] [PubMed]
28. Mierziak, J.; Kostyn, K.; Kulma, A. Flavonoids as important molecules of plant interactions with the environment. *Molecules* **2014**, *19*, 16240–16265. [CrossRef] [PubMed]
29. Dzoyem, J.P.; Eloff, J.N. Anti-inflammatory, anticholinesterase and antioxidant activity of leaf extracts of twelve plants used traditionally to alleviate pain and inflammation in South Africa. *J. Ethnopharmacol.* **2015**, *160*, 194–201. [CrossRef] [PubMed]
30. Zhang, C.; Jia, X.; Bao, J.; Chen, S.; Wang, K.; Zhang, Y.; Li, P.; Wan, J.B.; Su, H.; Wang, Y.; et al. Polyphyllin vii induces apoptosis in HepG2 cells through Ros-mediated mitochondrial dysfunction and MAPK pathways. *BMC Complement. Altern. Med.* **2016**, *16*, 58. [CrossRef] [PubMed]
31. Su, M.; Yu, T.; Zhang, H.; Wu, Y.; Wang, X.; Li, G. The antiapoptosis effect of glycyrrhizate on HepG2 cells induced by hydrogen peroxide. *Oxid. Med. Cell Longev.* **2016**, *2016*, 6849758. [CrossRef] [PubMed]
32. Arciello, M.; Gori, M.; Balsano, C. Mitochondrial dysfunctions and altered metals homeostasis: New weapons to counteract HCV-related oxidative stress. *Oxid. Med. Cell Longev.* **2013**, *2013*, 971024. [CrossRef] [PubMed]
33. Messarah, M.; Klibet, F.; Boumendjel, A.; Abdennour, C.; Bouzerna, N.; Boulakoud, M.S.; El Feki, A. Hepatoprotective role and antioxidant capacity of selenium on arsenic-induced liver injury in rats. *Exp. Toxicol. Pathol.* **2012**, *64*, 167–174. [CrossRef] [PubMed]
34. Liu, Y.; Chen, H.; Wang, J.; Zhou, W.; Sun, R.; Xia, M. Association of serum retinoic acid with hepatic steatosis and liver injury in nonalcoholic fatty liver disease. *Am. J. Clin. Nutr.* **2015**, *102*, 130–137. [CrossRef] [PubMed]
35. Kim, W.R.; Flamm, S.L.; Di Bisceglie, A.M.; Bodenheimer, H.C. Public Policy Committee of the American Association for the Study of Liver Disease. Serum activity of alanine aminotransferase (ALT) as an indicator of health and disease. *Hepatology* **2008**, *47*, 1363–1370. [CrossRef] [PubMed]
36. Zhang, Z.; Gao, L.; Cheng, Y.; Jiang, J.; Chen, Y.; Jiang, H.; Yu, H.; Shan, A.; Cheng, B. Resveratrol, a natural antioxidant, has a protective effect on liver injury induced by inorganic arsenic exposure. *Biomed. Res. Int.* **2014**, *2014*, 617202. [CrossRef] [PubMed]
37. Laouar, A.; Klibet, F.; Bourogaa, E.; Benamara, A.; Boumendjel, A.; Chefrour, A.; Messarah, M. Potential antioxidant properties and hepatoprotective effects of juniperus phoenicea berries against ccl4 induced hepatic damage in rats. *Asian Pac. J. Trop. Med.* **2017**, *10*, 263–269. [CrossRef] [PubMed]
38. Wang, K. Molecular mechanisms of hepatic apoptosis. *Cell Death Dis.* **2014**, *5*, e996. [CrossRef] [PubMed]
39. Guicciardi, M.E.; Malhi, H.; Mott, J.L.; Gores, G.J. Apoptosis and necrosis in the liver. *Compr. Physiol.* **2013**, *3*, 977–1010. [PubMed]
40. Pearson, G.; Robinson, F.; Gibson, T.B.; Xu, B.E.; Karandikar, M.; Berman, K.; Cobb, M.H. Mitogen-activated protein (MAP) kinase pathways: Regulation and physiological functions. *Endocrine Rev.* **2001**, *22*, 153–183. [CrossRef]
41. Gogvadze, V.; Orrenius, S.; Zhivotovsky, B. Multiple pathways of cytochrome c release from mitochondria in apoptosis. *Biochim. Biophys. Acta* **2006**, *1757*, 639–647. [CrossRef] [PubMed]
42. Huttemann, M.; Pecina, P.; Rainbolt, M.; Sanderson, T.H.; Kagan, V.E.; Samavati, L.; Doan, J.W.; Lee, I. The multiple functions of cytochrome *c* and their regulation in life and death decisions of the mammalian cell: From respiration to apoptosis. *Mitochondrion* **2011**, *11*, 369–381. [CrossRef] [PubMed]
43. Green, D.R.; Reed, J.C. Mitochondria and apoptosis. *Science* **1998**, *281*, 1309–1312. [CrossRef] [PubMed]
44. McIlwain, D.R.; Berger, T.; Mak, T.W. Caspase functions in cell death and disease. *Cold Spring Harb. Perspect. Biol.* **2013**, *5*, a008656. [CrossRef] [PubMed]
45. Santra, A.; Chowdhury, A.; Ghatak, S.; Biswas, A.; Dhali, G.K. Arsenic induces apoptosis in mouse liver is mitochondria dependent and is abrogated by N-acetylcysteine. *Toxicol. Appl. Pharmacol.* **2007**, *220*, 146–155. [CrossRef] [PubMed]
46. Lee, D.Y.; Choi, G.; Yoon, T.; Cheon, M.S.; Choo, B.K.; Kim, H.K. Anti-inflammatory activity of chrysanthemum indicum extract in acute and chronic cutaneous inflammation. *J. Ethnopharmacol.* **2009**, *123*, 149–154. [CrossRef] [PubMed]

47. Jing, L.; Ma, H.; Fan, P.; Gao, R.; Jia, Z. Antioxidant potential, total phenolic and total flavonoid contents of rhododendron anthopogonoides and its protective effect on hypoxia-induced injury in PC12 cells. *BMC Complement. Altern. Med.* **2015**, *15*, 287. [CrossRef] [PubMed]
48. Mihara, M.; Uchiyama, M. Determination of malonaldehyde precursor in tissues by thiobarbituric acid test. *Anal. Biochem.* **1978**, *86*, 271–278. [PubMed]
49. Akanda, M.R.; Kim, M.J.; Kim, I.S.; Ahn, D.; Tae, H.J.; Rahman, M.M.; Park, Y.G.; Seol, J.W.; Nam, H.H.; Choo, B.K.; et al. Neuroprotective effects of Sigesbeckia pubescens extract on glutamate-induced oxidative stress in HT22 cells via downregulation of MAPK/Caspase-3 pathways. *Cell. Mol. Neurobiol.* **2017**. [CrossRef] [PubMed]

International Journal of
Molecular Sciences

MDPI

Article

The *Alternaria alternata* Mycotoxin Alternariol Suppresses Lipopolysaccharide-Induced Inflammation

Shivani Grover and Christopher B. Lawrence *

Department of Biological Sciences, Virginia Tech, Blacksburg, VA 24061, USA; shgrover@vt.edu
* Correspondence: cblawren@vt.edu; Tel.: +1-540-231-9708

Received: 1 May 2017; Accepted: 8 July 2017; Published: 20 July 2017

Abstract: The *Alternaria* mycotoxins alternariol (AOH) and alternariol monomethyl ether (AME) have been shown to possess genotoxic and cytotoxic properties. In this study, the ability of AOH and AME to modulate innate immunity in the human bronchial epithelial cell line (BEAS-2B) and mouse macrophage cell line (RAW264.7) were investigated. During these studies, it was discovered that AOH and to a lesser extent AME potently suppressed lipopolysaccharide (LPS)-induced innate immune responses in a dose-dependent manner. Treatment of BEAS-2B cells with AOH resulted in morphological changes including a detached pattern of growth as well as elongated arms. AOH/AME-related immune suppression and morphological changes were linked to the ability of these mycotoxins to cause cell cycle arrest at the G2/M phase. This model was also used to investigate the AOH/AME mechanism of immune suppression in relation to aryl hydrocarbon receptor (AhR). AhR was not found to be important for the immunosuppressive properties of AOH/AME, but appeared important for the low levels of cell death observed in BEAS-2B cells.

Keywords: *Alternaria alternata*; alternariol; innate immunity; immunosuppression

1. Introduction

The fungal genus *Alternaria* harbors many plant and human pathogens, saprophytes, and allergenic species, and has been shown to be a prolific producer of secondary metabolites [1]. *Alternaria* spores are ubiquitous, and exposure has been clinically associated with the development, onset, and exacerbation of allergic diseases such as allergic rhinitis, asthma, and chronic rhinosinusitis (CRS) [2–7]. Indeed, up to 70% of mold allergy patients have skin test reactivity to *Alternaria*. Over 10 allergen proteins have been described from *Alternaria*, however, the secreted major allergen, Alt a 1, produces a prolonged and intense IgE-mediated reaction in sensitized patients [3–6]. Despite the well-documented clinical relevance of proteinaceous allergens, no small molecules (secondary metabolites) from *Alternaria* have been studied in regard to lung epithelium, inflammation, and immune responses.

Fungal mycotoxins are products of their secondary metabolism that can often cause deleterious effects in vertebrates. These secondary metabolites belong to different chemical classes such as steroids quinones, pyrones, peptides, phenolics, and the fumonisin-like toxins. These toxins can enter the body through skin, mucous, airways, and ingestion. Constant exposure can lead to hypersensitivity and mycotoxicosis, leading to a potentially compromised immune system and the onset of other illnesses and infections (HIV, kidney and liver damage) [7,8]. However, of all the mycotoxins known, only a few are subject to regular monitoring of contamination and level intake such as aflatoxins from *Aspergillus* spp., and fumonisins, deoxyivalenol, zearlenone, and ochratoxin-A from other fungi like *Fusarium* spp. Legal authorities from both food and feed industries acknowledge the importance of detecting and quantifying mycotoxin levels and identifying the effects of their contamination [9].

Besides producing deleterious mycotoxins, fungi are also an important resource of potential beneficial compounds with therapeutic properties. Ever since the discovery of penicillin from *Penicillium notatum* in 1929, the importance of elucidating the potential of fungal secondary metabolites has been beyond question [10]. *Alternaria* metabolites have exhibited a variety of therapeutic and biological properties such as phytotoxicity, cytotoxicity, anti-HIV, anti-cancer, and anti-microbial properties, to name a few, all of which have generated considerable research interest worldwide. For example, porritoxin from *Alternaria porri* is a likely cancer chemo-preventive agent, and depudecin from *Alternaria brassicicola* is an inhibitor of histone deacetylase [11–14].

The most well-studied deleterious *Alternaria* mycotoxins alternariol (AOH) and alternariol methyl ether (AME) have been detected in most foods and grains, often at high concentrations (Figure 1) [15]. Foods such as apples, apple products, mandarins, olives, pepper, tomatoes, oilseed, sunflower seeds, sorghum, wheat, edible oils, citrus fruits, melons, pears, prune nectar, raspberries, red currant, carrots, barley, oats, red wine, and lentils are known to be frequently contaminated with AOH/AME. The maximum levels reported are in the range of 1–103 µg/kg, with higher levels found in food products visibly rotted with *Alternaria* [15]. However, as of yet, no data concerning tissue levels of AOH exists in animals and humans [13,16].

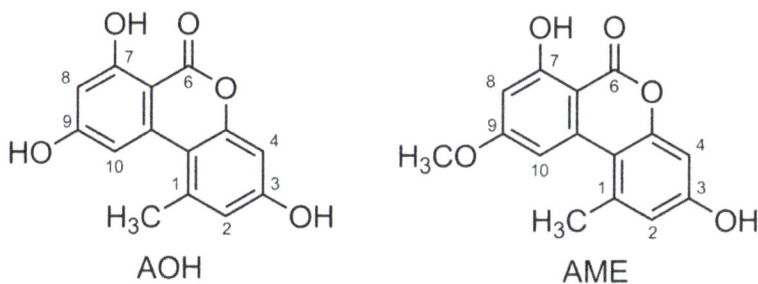

Figure 1. Chemical structure of alternariol (AOH) and alternariol monomethyl ether (AME).

AOH was shown to cause mutagenicity and cytotoxicity in Chinese hamster V79 cells [17,18]. AOH is also known to cause the formation of micronuclei (MN) in V79 and human endometrial adenocarcinoma cell line (Ishikawa cells) [17,18]. Treatment of AOH on murine macrophage cell line RAW 264.7 showed cytotoxicity and DNA strand breakage, as well as oxidative damage, cellular stress, and cell cycle arrest in the G2/M phase [18]. Human adenocarcinoma cells (HT29) treated with AOH indicated that the toxin modulates levels of reactive oxygen species (ROS) [19].

The aryl hydrocarbon receptor (AhR) is a ligand-activated transcription factor that often binds to environmental toxins and subsequently modulates the downstream expression of genes involved in detoxification and transport [20]. It has been studied in relation to various environmental contaminants such as the xenobiotic compound TCDD (2,3,7,8-tetrachlorodibenzo-p-dioxin) [21]. Binding of the AhR to the ligand causes the translocation of the complex to the nucleus and binding with AhR nucleus translocator (ARNT) [20]. The AhR-ARNT complex then binds to various xenobiotic response elements (XREs) and modulates the induction of genes, for instance the cytochrome P450 family [20]. AhR is a potential receptor for AOH and AME. The CYP450 family of genes, including the highly expressed CYP1A1, are a major target of the AhR-ARNT complex and often mediate toxin hydroxylation and further metabolism [21,22]. AOH has a planar structure that is similar to other AhR ligands and may be a substrate of CYP1A1. Further evidence of this was substantiated by the treatment of AOH on murine hepatoma cells, resulting in the differential expression of CYP1A1 in the presence and absence of activated and inactivated AhR [21,22]. This study is the first attempt at providing an experimental framework to investigate the immune-modulatory properties and potential clinical importance of *Alternaria* secondary metabolites, the mycotoxins alternariol (AOH) and alternariol methyl ether (AME).

2. Results

2.1. Alternariol Suppresses Innate Immune Responses in Human Lung Epithelial and Mouse Macrophage Cell Lines

We first sought to characterize the response of the mammalian innate immune system by quantifying cytokine and chemokine inflammatory markers upon AOH treatment on airway epithelial cells. We profiled cytokine interleukin-6 (IL6) in addition to chemokines interleukin-8 (IL8) and monocyte chemoattractant protein-1 (MCP-1/CCL2), which have been shown to be highly induced in many inflammatory diseases including sepsis [23].

We hypothesized initially that AOH/AME would have proinflammatory effects on cells. We evaluated the protein and mRNA levels of IL6 and IL8 after the treatment of bronchial epithelial cells (BEAS-2B) with AOH. Surprisingly, we found that AOH did not cause an increase in the protein levels of IL6 and IL8, but resulted in downregulation at the mRNA level after 6, 12, and 24 h of incubation. This data is summarized in Appendix A, Table A1. Because AOH may not be able to induce the primary inflammation markers (IL6, IL8, MCP-1/CCL2); we also examined other cytokine and chemokine markers that might be stimulated by AOH. We subsequently analyzed the protein levels of thymic stromal lymphopoietin (TSLP), tumor necrosis factor α (TNF-α), interleukin-1 β (IL-1β), interleukin-10 (IL-10), and transforming growth factor β (TGF-β), but no induction was observed. Primer sequences for gene expression analyses performed throughout the study are shown in Appendix A, Table A2.

Upon discovering that AOH treatment resulted in the downregulation of *IL6* and *IL8* genes at the mRNA level, we next used the proinflammatory bacterial cell wall lipopolysaccharide (LPS) as an inducer of innate immunity. In the presence of 10 μM AOH and 10 μg/mL of LPS at 24 h, the levels of IL6, IL8, and MCP-1/CCL2 were reduced by several fold (Figure 2). In AME (10 μM)-treated cells, basal and LPS-induced cytokine and chemokine levels were reduced approximately half as much as compared to AOH-treated cells, leading to the conclusion that both AOH and AME have immunosuppressive properties, although AOH is the far more potent molecule of the two. We repeated this experimental design with mouse macrophage RAW264.7 cell line and observed similar results. In macrophages, LPS-induced IL6 was completely suppressed at a dose of 10 μM AOH (Figure 2).

Quantitative real-time reverse transcription-polymerase chain reaction (qRT-PCR) was used to detect changes in mRNA abundance as a measure of gene expression induced by AOH, in the presence and absence of LPS. Chemokine and cytokine gene expression profiles after normalization to the control housekeeping gene GAPDH were generated in this study for the investigation of the AOH (10 μM dose) phenotype response after a 24-h treatment in the presence and absence of 10 μg LPS. LPS-induced IL6 mRNA levels were reduced two-fold in presence of AOH. LPS-induced chemokines IL8 and MCP-1/CCL2 showed a four-fold decrease in the presence of AOH. While MCP-1/CCL2 qRT-PCR analysis showed a similar decrease of LPS-induced inflammation, it showed downregulation of the gene in the presence of AOH alone. Furthermore, we analyzed caspase 1. Caspase 1 aids in the formation of mature peptides for inflammatory cytokines interleukin-1β and interleukin-18, and is also involved in cell death and inflammasome (NLRP1 multi-molecular complex) formation [24,25]. An AOH dose of 10 μM upregulated caspase-1 but downregulated LPS-induced caspase 1 by almost five-fold, suggesting a complex mode of regulation (Figure 3). Interestingly, we observed that AOH induced cytochrome P450 CYP1A1 gene expression and partially prevented LPS-induced downregulation.

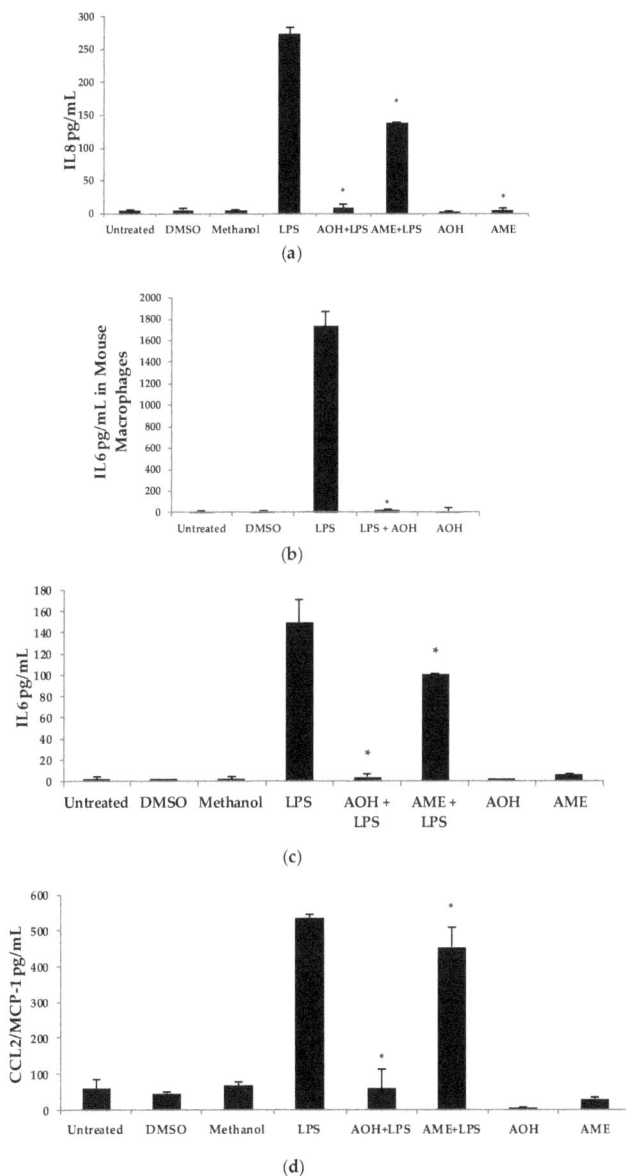

Figure 2. Alternariol (AOH) and alternariol monomethyl ether (AME) suppress lipopolysaccharide (LPS)-induced innate immunity. BEAS-2B airway epithelial cells (panels a, c, and d) and RAW 264.7 mouse macrophages (panel b) at a density of 5×10^5 cells/well were treated with 10 μM of AOH and 10 μM of AME in the presence and absence of 10 μg of LPS and incubated for 24 h under normal conditions at 37 °C, 5% CO_2. Supernatants were subsequently analyzed using enzyme-linked immunosorbent assay (ELISA) (**a**) IL8 BEAS-2B cells, (**b**) IL6 in RAW264.7 cells, (**c**) IL6 BEAS-2B cells, and (**d**) CCL2/MCP-1 BEAS-2B cells. An * indicates $p < 0.05$ according to Student's t-test when comparing AOH/AME + LPS to LPS-induced controls. DMSO, dimethyl sulfoxide.

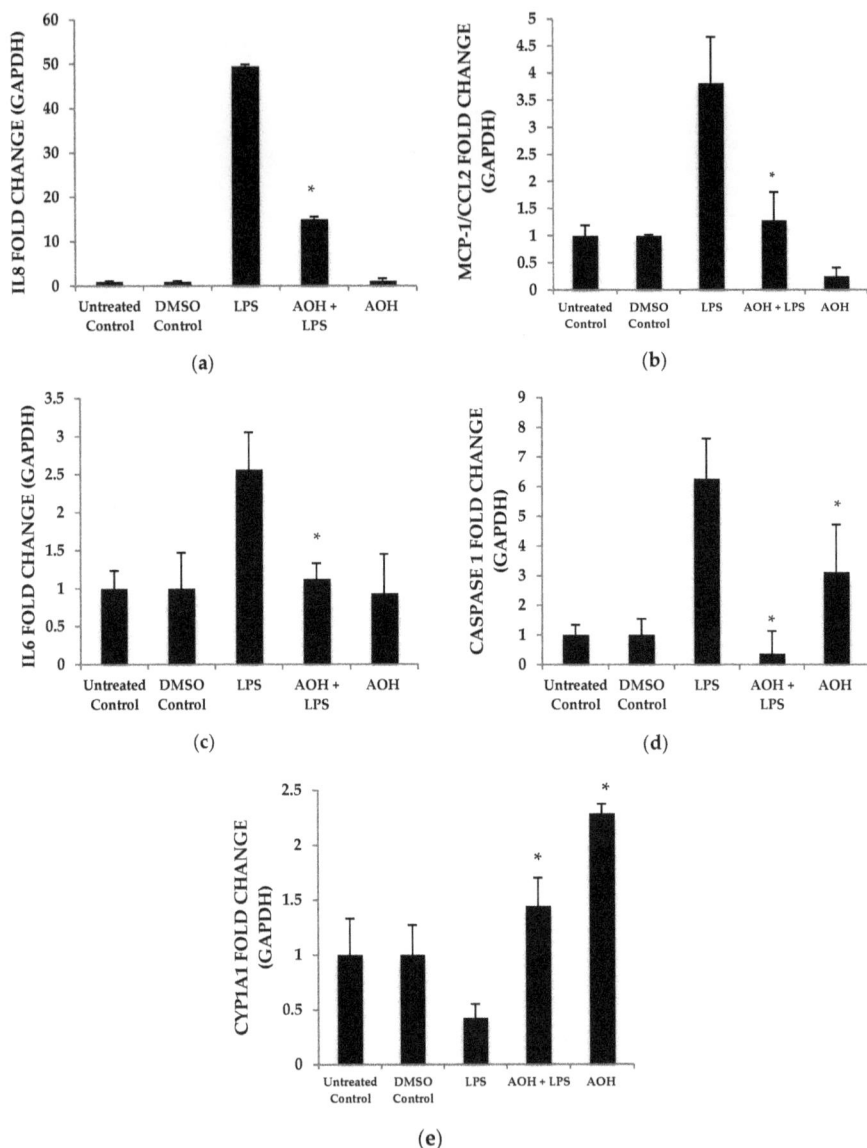

Figure 3. Airway epithelium treated with alternariol (AOH) and alternariol monomethyl ether (AME) results in the downregulation of LPS-induced mRNAs. BEAS-2B cells seeded at a density of 5×10^5 cells/well were treated with 10 μM AOH and 10 μg LPS for 24 h. The resulting RNA was harvested and measured with quantitative real-time reverse transcription-polymerase chain reaction (qRT-PCR). Each graph here demonstrates the upregulation and downregulation (fold change) of gene expression by normalization with the control GAPDH. (**a**) IL8, (**b**) CCL2, (**c**) IL6, (**d**) Caspase 1, and (**e**) CYP1A1 fold change. An * indicates $p < 0.05$ according to Student's *t*-test when comparing AOH to dimethyl sulfoxide (DMSO) control and AOH + LPS to LPS corresponding control.

2.2. Dose-Dependent Analysis of Alternariol (AOH) and Lipopolysaccharide (LPS)

We next evaluated varying doses of AOH and LPS in order to determine minimum concentrations of AOH with immunosuppressive activity. We found that AOH is highly immunosuppressive in a dose-dependent manner in both BEAS-2B and RAW264.7 cells (Figure 4). IL8 protein levels were measured in BEAS-2B cells. BEAS-2B supernatants following AOH doses of 10, 100 nM, 1, 5, and 10 µM were analyzed by enzyme-linked immunosorbent assay (ELISA). As expected, we observed an AOH dose-dependent decrease in LPS (10 µg)-induced inflammation in BEAS-2B cells. Although in all the above-mentioned doses IL8 was not detected with AOH alone, significant LPS-induced IL8 suppression was observed starting at the 5 µM dosage. A dose of 10 µM showed the highest amount of IL8 suppression. IL8 levels in LPS-induced cells treated with 10 µM AOH were equivalent or less than the levels in untreated cells (Figure 4). Similar results were observed in experiments using RAW264.7 macrophages (see Appendix A, Figure A3).

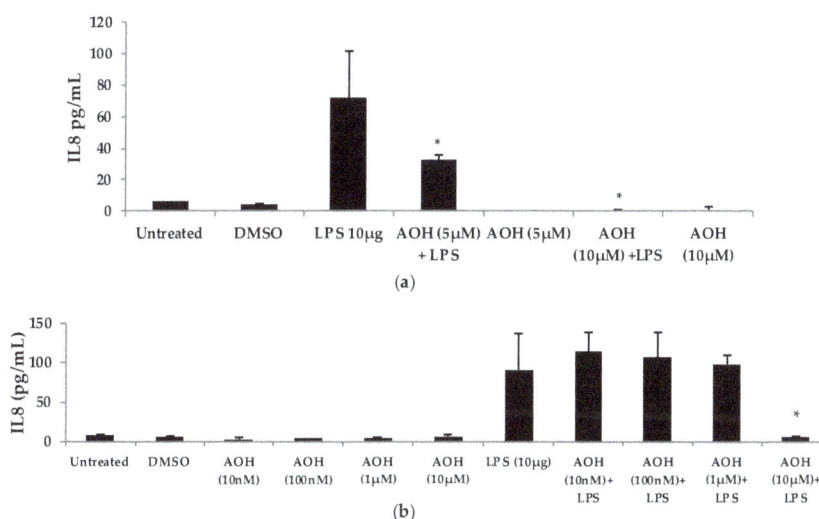

Figure 4. Dose-dependent response of airway epithelium cells (BEAS-2B) after treatment with alternariol (AOH) and lipopolysaccharide (LPS). (**a**) BEAS-2B cells were treated with (5–10 µM) of AOH in the presence and absence of 10 µg of LPS to measure IL8 levels released. Cell densities were 5×10^5 cells/well and were incubated for 24 h under normal conditions at 37 °C, 5% CO_2 after treatment; (**b**) BEAS-2B cells were treated with (10 nM–10 µM) of AOH in the presence and absence of 10 µg of LPS to measure IL8 levels released in supernatants using enzyme-linked immunosorbent assay (ELISA). Cell densities were 5×10^5 cells/well and were incubated for 24 h under normal conditions at 37 °C, 5% CO_2 after treatment. An * indicates $p < 0.05$ according to Student's *t*-test when comparing AOH + LPS treatments to LPS-induced control.

To further investigate the dose-dependent response of AOH, we conducted an experiment to investigate LPS doses on bronchial lung epithelial cells (BEAS-2B). Our previous experimental design of a 24-h cell treatment was applied to evaluate the protein levels of IL6 and IL8 using ELISA. We tested LPS doses including 10 ng, 50 ng, 100 ng, 500 ng, 1 µg, 5 µg, and 10 µg (Figure 5). A dose of 10 µg of LPS resulted in the induction of 132 pg/mL of IL6 and 221 pg/mL of IL8. With these results, we validated the doses of 10 µg of LPS and 10 µM of AOH as sufficiently substantiated for further experiments. Similar results were observed in experiments with RAW264.7 macrophages (see Appendix A, Figure A3).

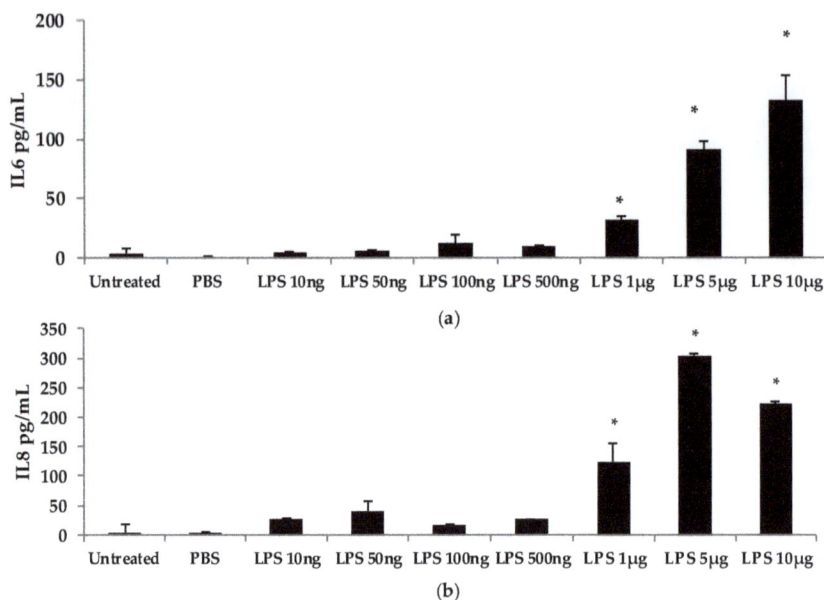

Figure 5. Dose-dependent response of bronchial epithelial BEAS2-B cells to LPS. LPS was added to BEAS-2B cells at a density of 500,000 cells/well for 24 h. (**a**) IL6 measured by enzyme-linked immunosorbent assay (ELISA); (**b**) IL8 measured by ELISA. An * indicates $p < 0.05$ according to Student's *t*-test comparing individual treatments to phospho-buffered saline (PBS) control.

To test whether AOH-induced immune suppression is dependent on the timing of LPS addition, we treated BEAS-2B cells with AOH and added LPS two hours later. A range of doses of AOH (10–100 µM) and AME (1–30 µM) were tested on BEAS-2B cells. All treatments resulted in a marked decrease in IL6, IL8, and MCP-1/CCL2 in the presence of LPS, indicating that AOH can prevent LPS-induced innate immune response when LPS is added after several hours (see Appendix A, Figure A1).

2.3. Cell Morphology Alterations in Response to AOH

Up until this time, no microscopic studies have been performed on AOH-treated mammalian lung epithelial cells. Multiple morphological changes were observed in the cells after treatment with AOH. BEAS-2B cells treated with AOH showed a marked change after 24 h (Figure 6). The cells demonstrated a more detached and spread out pattern of growth as well as elongated arms. This indicates that the cell cycle arrest at the G2/M phase reported in earlier studies with other cell types could be an underlying cause of cell stress observed in BEAS-2B cells, and therefore could be responsible for the change in morphology.

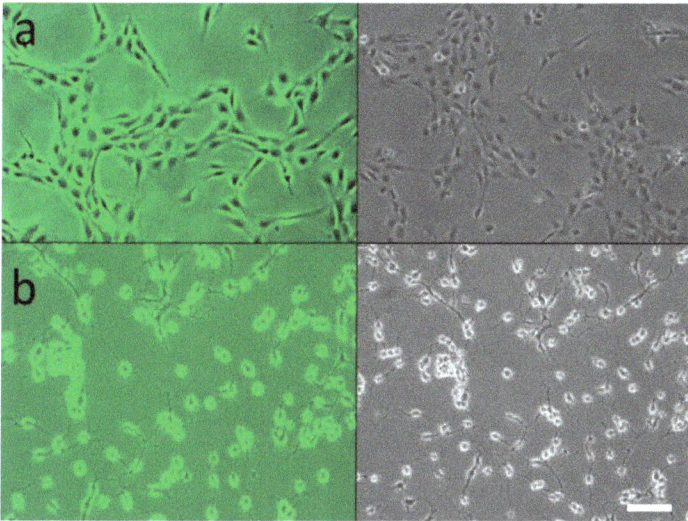

Figure 6. Human airway epithelial cells in the presence of alternariol (AOH). BEAS-2B cells were incubated with 10 μM of AOH for 24 h under normal conditions at 37 °C, 5% CO_2. The images were taken with confocal microscopy with a cell density of 5×10^5 cells/well (magnification 200×). (**a**) Untreated BEAS-2B cells at 24 h in color (upper left panel) and grey-scale (upper right panel); (**b**) BEAS-2B with 10 μM AOH at 24 h in color (lower left panel) and grey-scale (lower right panel). Scale bar = 100 μm.

2.4. AOH Inhibits Cell Proliferation and Has Minimal Effects on Cell Death in BEAS-2B Cells

Previous studies have emphasized the ability of AOH to cause cell death and cell cycle arrest [16–19]. Hence, a colorimetric MTT assay was performed in order to first investigate cell proliferation with doses ranging from 1 to 100 μM of AOH. In the MTT assay, the yellow MTT 3-(4,5-dimethylthiazol-2-yl)-2,5-diphenyltetrazolium bromide is reduced to purple formazan in the mitochondria of living cells. At a concentration of 10 μM of AOH, proliferation was 56% compared to control cells. It was further reduced to 23% at 20 μM and at 100 μM; only 12% of the cells were proliferating compared to controls (Figure 7).

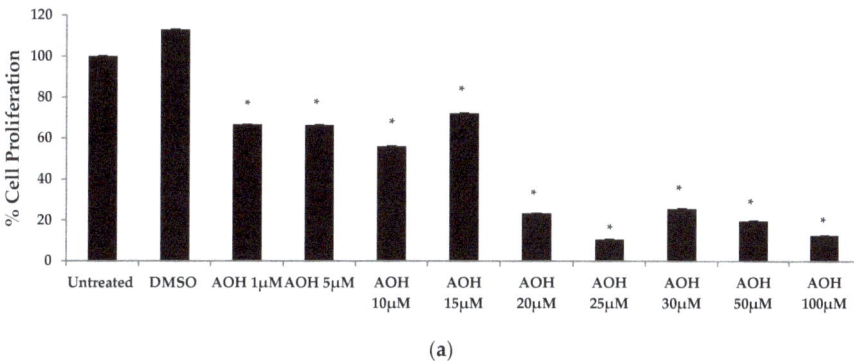

(**a**)

Figure 7. *Cont.*

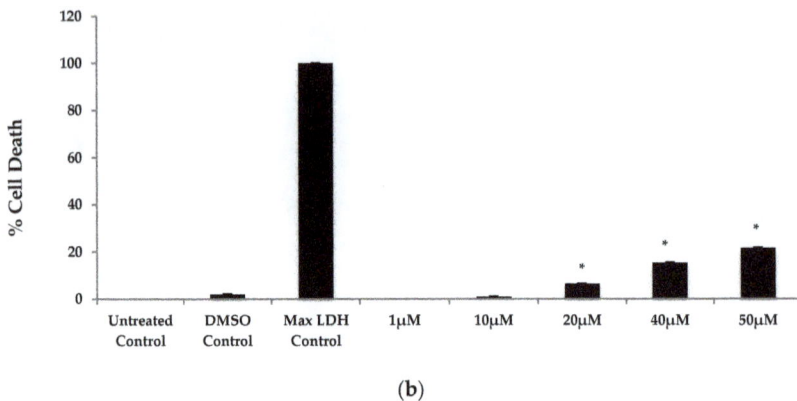

(b)

Figure 7. Cell proliferation and cell death analysis of BEAS-2B cells treated with alternariol (AOH). (a) A dose-dependent analysis of cell proliferation of BEAS-2B cells after treatment with AOH was performed by MTT assay. Cells were seeded at a density of 500,000 cells/well for 24 h; (b) A dose curve of lactate dehydrogenase (LDH) assay to measure the amount of LDH released by dead cells upon treatment with AOH for 24 h at a cell density of 20,000 cells/well. An * indicates $p < 0.05$ according to Student's *t*-test for AOH treatments compared to dimethyl sulfoxide (DMSO)-treated controls.

Next, we employed a lactate dehydrogenase (LDH) assay to measure cell death. Less than 10% cell death was detected at 10 μM AOH-treated cells compared to control cells. Collectively, the MTT and LDH assay results suggest that cell cycle arrest, not cell death, is most likely responsible for the ability of AOH to suppress LPS-induced innate immune responses.

2.5. AOH Causes Cell Cycle Arrest in Lung Epithelium

The reduction in cell proliferation may be an intrinsic property of AOH, caused by the short and yet reversible arrest in the G2/M phase of the cell cycle. To further investigate whether or not cell cycle arrest may be the cause of immune suppression observed in our studies, we used the compound RO-3306, a selective ATP-competitive inhibitor of the cyclin dependent kinase, CDK1 [26]. 1 CDK1 is a serine/threonine kinase that controls the progression of cell cycle through each checkpoint (courtesy of the Cimini Lab, Virginia Tech, Blacksburg, VA, USA). RO-3306 has been identified to cause cell cycle arrest at the G2/M phase, similar to what has been reported for AOH at a dose of 10 μM [18,26]. Hence, we treated BEAS-2B cells with 10 μM AOH and 10 μM RO-3306 in the presence and absence of 10 μg LPS. We profiled the IL8 protein levels using ELISA. The data showed that RO-3306 exhibited similar immune suppressive properties as AOH (revealing a reduction of LPS-induced IL8), but was less potent. No IL8 induction was seen in cells treated with RO-3306 alone (Figure 8). These data suggest that the ability of AOH to suppress IL8 may be at least partially dependent upon its ability to cause cell cycle arrest.

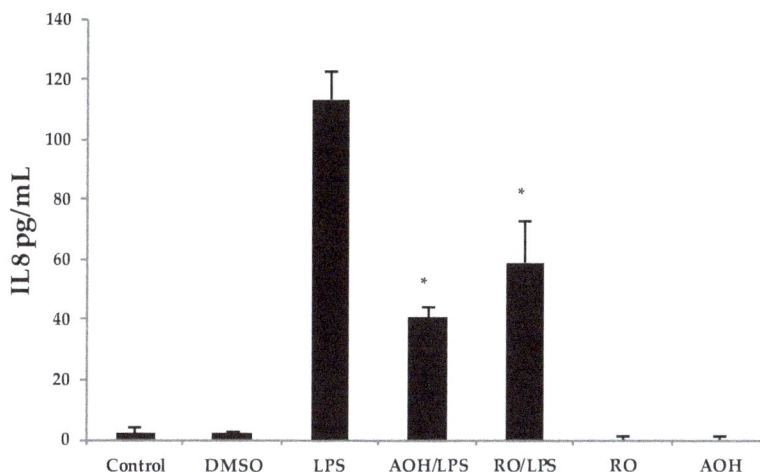

Figure 8. Treatment of airway epithelium cells by alternariol (AOH), RO-3306, and lipopolysaccharide (LPS). BEAS-2B cells seeded at a density of 5×10^5 cells/well were treated with 10 μM of AOH or 10 μM of RO-3306 in the presence and absence of 10 μg of LPS and incubated for 24h. Supernatants were analyzed using enzyme-linked immunosorbent assay (ELISA). An * indicates $p < 0.05$ according to Student's *t*-test for AOH/LPS and RO/LPS treatments compared to the LPS treatment alone.

2.6. Aryl Hydrocarbon Receptor Analysis and Mechanism of AOH-Induced Immune Suppression

Next, we hypothesized that the aryl hydrocarbon receptor (AhR) is the target receptor for AOH that triggers downstream signaling related to its immunosuppressive properties. RNA silencing was used to knockdown the gene encoding AhR in BEAS-2B cells. Following optimization, we typically obtained a minimum of 70% knockdown of AhR using gene-specific siRNAs (see Appendix A, Figure A2). After AhR knockdown using gene-specific siRNAs, BEAS-2B cells were treated with AOH in the presence and absence of LPS for 24 h, and supernatants were then subject to ELISA. No change was observed that correlated with the silenced AhR gene, suggesting that this receptor may not be important for the ability of AOH to suppress LPS-induced immunity in BEAS-2B cells (Figure 9).

Finally, we attempted to determine whether the modest cell death-inducing property of AOH is dependent upon AhR in BEAS-2B cells. Significant differences were detected in cell death after AOH treatments when comparing AhR silenced cells to scrambled controls, indicating that AhR may be important in causing cell death in BEAS-2B cells (Figure 10).

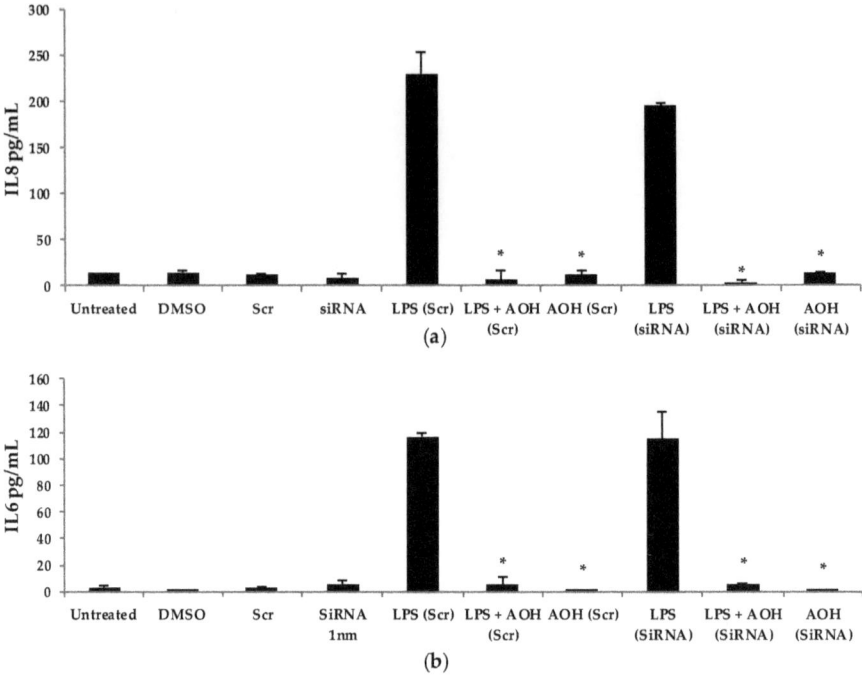

Figure 9. RNA silencing of the aryl hydrocarbon receptor (AhR) gene followed by treatment with lipopolysaccharide (LPS) and alternariol (AOH) in BEAS-2B cells. Cells were seeded at a density of 150,000 cells/well. Cells were treated with AhR siRNA for 24 h twice to successfully knockdown AhR. (**a**) IL8, and (**b**) IL6 released upon treatment with 10 μM AOH and 10 μg LPS for 24 h as measured by enzyme-linked immunosorbent assay (ELISA). An * indicates $p < 0.05$ according to Student's *t*-test for AOH treatments (Scr or SiRNA) compared to LPS (Scr) or LPS (siRNA) controls, respectively.

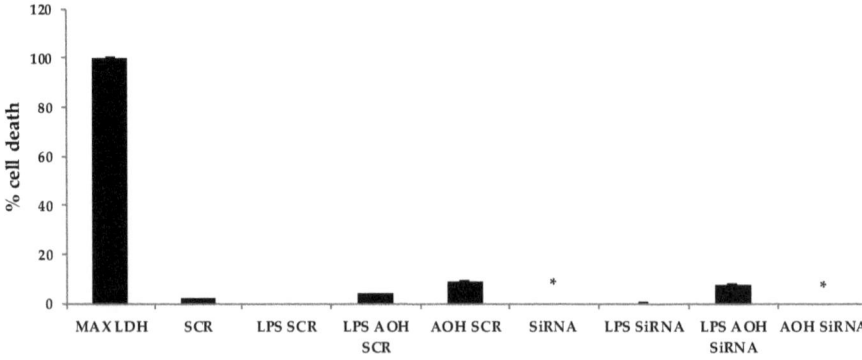

Figure 10. Alternariol (AOH) induced cell death is dependent upon the aryl hydrocarbon receptor (AhR). An LDH assay was performed on BEAS-2B cells with silenced AhR, 10 μM AOH, and 10 μg lipopolysaccharide (LPS). An * indicates $p < 0.05$ according to Student's *t*-test when comparing AhR gene-specific (siRNA) treatments to their appropriate scrambled siRNA (SCR) controls.

3. Discussion

This is the first study to investigate the immunomodulatory effects of AOH/AME on mammalian cells. Collectively, our data shows that AOH/AME caused a decrease in inflammatory responses in BEAS-2B bronchial epithelial cells and murine RAW264.7 macrophages when stimulated with LPS. Our data implicated that the immunosuppressive property of AOH/AME may be associated with cell cycle arrest at the G2 phase. The cell proliferation and cell death assays conducted in this study raised our understanding of the cytotoxic effects of this compound at various doses in lung epithelial cells. For example, the MTT assay results in BEAS-2B cells suggest that AOH decreased cell proliferation by almost 50% at a 10 μM dose. Along with our experiments using the G2 phase cell cycle arresting agent RO-3306, this may further implicate that the immunosuppressive properties of AOH may be related to its ability to cause cell cycle arrest. The cell death assay showed that AOH is cytotoxic to lung epithelial cells, primarily at a dose of 20 μM or higher. Results of our experiments using the 10 μM dose provides evidence that the cell death at this concentration is minimal and thus has little effect on the immunosuppressive properties of AOH. Our data using an siRNA knockdown approach also indicates that the modest cell death caused by AOH but not immunosuppression in BEAS-2B cells is most likely dependent on the AhR receptor.

In the context of allergic disease, our results suggest that AOH/AME are not plausible targets for designing therapeutics for reducing *Alternaria*-induced inflammation. In fact, the opposite may be true. Further supporting the results of this study, preliminary experiments utilizing AOH/AME-deficient *Alternaria* mutant spores have shown that they cause dramatically increased innate immune responses in BEAS-2B cells compared to wild-type fungal strains that produce normal levels of AOH/AME [27].

4. Materials and Methods

4.1. Materials

Alternariol (AOH) (Cayman Chemical, Ann Arbor, MI, USA) was reconstituted at 1 mg/mL in DMSO (Sigma-Aldrich, St. Louis, MO, USA). Alternariol monomethyl ether (AME) (Sigma-Aldrich) was reconstituted at 1 mg/mL in methanol. Ultrapure bacterial endotoxin Lipopolysaccharide (Sigma-Aldrich), cell culture grade, was reconstituted to a final concentration of 1 mg/mL in phosphate buffered saline (PBS) (Fisher Scientific, Pittsburgh, PA, USA). RO-3306 (Sigma-Aldrich) was reconstituted at 1 mg/mL in DMSO. The stock solutions were stored at $-20\,^\circ$C until further use.

4.2. Cell Culture and Cell Lines

BEAS-2B, a human bronchial lung epithelial cell line and mouse macrophage raw 264.7 cell lines were maintained in RPMI-1640 culture medium (Fisher Scientific) with 10% heat-inactivated fetal bovine serum (FBS) (Fisher Scientific) and 1% penicillin-streptomycin (ThermoFisher, Pittsburgh, PA, USA) in round bottom tissue culture-treated plates (Fisher Scientific). BEAS-2B and macrophages were incubated in 5% CO_2 at 37 $^\circ$C. Both above-mentioned cell lines were starved for 2 h before treatment with secondary metabolites in RPMI-1640 and 1% penicillin-streptomycin. BEAS-2B and macrophage cells were seeded at a density of 500,000 cells/well in 6-well tissue culture plates prior to treatment, unless indicated otherwise.

4.3. Quantification of Protein Levels of Cytokines and Chemokines

The cells in 6-well plates were seeded in 1.5 mL RPMI-1640 media and cells in 12-well plates were seeded in 1 mL RPMI-1640 media. BEAS-2B cells were seeded on the plates in triplicates and, after an overnight incubation at 37 $^\circ$C and 5% CO_2, washed with Dulbecco's phosphate-buffered saline (DPBS) (Fisher Scientific). The cells were then placed in the starve media for 2 h, after which they were washed again with DPBS before being placed in fresh RPMI-1640 media. AOH, AME, and LPS were then added to the media. BEAS-2B cells were incubated for 24 h. The resulting supernatant and genetic material were collected with trypsin (Sigma-Aldrich) and stored at $-80\,^\circ$C. The protein levels in the cells were analyzed with enzyme-linked immunosorbent assay (ELISA) kits (Biolegend, San Diego, CA, USA) and (ThermoFisher, Pittsburgh, PA USA) following the instructions of the manufacturers. The absorbance was recorded with a microplate reader at 450 nm.

4.4. Quantitatification of Gene Expression

The RNA samples were isolated by applying trypsin (Sigma-Aldrich) to the AOH/AME/LPS-treated BEAS-2B or RAW264.7 cells and RNA was extracted using mammalian RNA extraction kits and manufacturer protocols (Qiagen, Valencia, CA, USA). The samples were processed into cDNA following the manufacturer's instructions from the Bioline Tetro cDNA synthesis kit, and then stored at −20 °C (Bioline, London, UK). All the qRT-PCR reactions for the biological triplicates were performed as technical duplicates using the cDNA as a template. GAPDH was used as a control housekeeping gene for all experiments, as it has a continuous expression in mammalian cell lines. A BIO-RAD Iq5 Multicolor Real-Time PCR Detection System machine was used to conduct the qRT-PCR reaction (Bio-Rad, Hercules, CA, USA). All reactions were carried out at 20 μL volume with SYBR Green (Bioline) as the fluorescent reporter molecule. Relative fold change in gene expression was calculated using the 2(-Delta Delta C(T)) method and Pfaffl equation by normalization to GAPDH, as suggested by Bio-Rad protocols.

4.5. Analysis of Cell Death and Proliferation

The 3-(4,5-dimethylthiazol-2-yl) 2,5-diphenyltetrazolium bromide (MTT) solution was added to 50 μL of RPMI-1640 starve media harvested from cells treated with AOH at the 24-h time point. The plates were incubated at 37 °C for 4 h for the reduction of MTT formazan. Subsequently, 100 μL of DMSO was used to stop the reaction. Absorbance was measured at a wavelength of 570 nm using a microplate reader. The lactate dehydrogenase (LDH) assay was performed using the Pierce™ LDH Cytotoxicity Assay Kit (ThermoFisher). Cells were seeded at a density of 10,000 cells/well in 100 μL RPMI-1640 media, in 96-well flat bottom plates and incubated overnight at 37 °C and 5% CO_2. After a 24-h treatment, 50 μL of media from each well was transferred to a new plate and 50 μL of LDH reaction mixture was added. To measure LDH activity, the reaction was stopped after a 30-min incubation and the absorbance measured at 490 nm was subtracted from the absorbance measured at 680 nm. All treatments were performed in biological triplicates and technical replicates. Percent cytotoxicity was calculated using the formula:

$$\% \: Cytotoxicity = \frac{Compound-treated \: LDH \: activity - Spontaneous \: LDH \: activity \: DH \: activity}{Maximum \: LDH \: activity - Spontaneous \: L} \times 100$$

4.6. Microscopy

The surface morphology of bronchial lung epithelial cells was imaged by a Nikon Eclipse TE2000-U Inverted Microscope (Nikon, Tokyo, Japan). Cells were seeded at a density of 500,000 cells/wells and treated with 10 μM AOH for 24 h before imaging.

4.7. RNA Silencing

Cells were seeded at a density of 125,000 cells/well. Silencing of AhR receptor was performed using a target-specific 19–25 nucleotide siRNA (Santa Cruz Biotechnology, Santa Cruz, CA, USA) designed to knockdown its gene expression. The siRNA reagent was mixed with HiPerFect Transfection Reagent (Qiagen) for transfection. BEAS-2B cells were treated with two consecutive doses of 1 nM of siRNA for 24 h to achieve a knockdown efficiency of >70%. Cells were then treated with AOH, AME, and LPS for 24 h. All experiments were performed in biological triplicates, along with a scrambled siRNA control (Santa Cruz Biotechnology). Primer pair human AHR_F and human AHR_R were used for determining knockdown efficiency.

4.8. Statistical Analysis

All tests were performed as biological triplicates and technical triplicates. Standard deviation was calculated from among biological replicates. The difference between individual treatment groups was validated using an unpaired Student's *t* test for independent samples, including LPS alone and LPS stimulation in the presence of AOH. A *p*-value < 0.05 was regarded as statistically significant.

5. Conclusions

In conclusion, this is the first study to demonstrate that the mycotoxins AOH and AME from the fungus *Alternaria* are capable of preventing LPS-induced inflammatory responses in both a human bronchial epithelial cell line (BEAS-2B) and a mouse macrophage cell line (RAW264.7). It will be interesting in the future to further dissect the role of AOH/AME in the context of allergic inflammation in *Alternaria*-mouse models of allergy and asthma. From a broad perspective, AOH/AME may have potential and serve as the structural basis for designing new anti-inflammatory therapeutics.

Acknowledgments: We would like to thank Liwu Li and Daniela Cimini for many helpful discussions. No funds were received for covering the costs to publish in open access.

Author Contributions: Shivani Grover and Christopher B. Lawrence conceived and designed the experiments; Shivani Grover performed the experiments; Shivani Grover analyzed the data; Shivani Grover and Christopher B. Lawrence wrote the paper.

Conflicts of Interest: The authors declare no conflict of interest.

Appendix A

Table A1. Summary of alternariol (AOH) dosage and treatment conditions for BEAS-2B cells. No IL6 and IL8 protein level production was observed. IL6 and IL8 mRNAs were found to be downregulated upon stimulation with AOH in BEAS-2B cells.

AOH Dose	Time	IL6 (Protein)	*IL6* (Gene)	IL8 (Protein)	*IL8* (Gene)
25, 50, 100 μM	6 h	No Induction	Downregulation	No Induction	Downregulation
25, 50, 100 μM	12 h	No Induction	Downregulation	No Induction	Downregulation
25, 50, 100 μM	24 h	No Induction	Downregulation	No Induction	Downregulation

Table A2. Primers used in this study.

Gene Primer Name	Primer Sequence
Human AHR_F	5′ TGGTTGTGATGCCAAAGGAAG 3′
Human AHR_R	5′ GACCCAAGTCCATCGGTTGTT 3′
Human CYP1A1_F	5′ GAACCTTCCCTGATCCTTGTG 3′
Human CYP1A1_R	5′ CCCTGATTACCCAGAATACCAG 3′
Human Caspase1_F	5′ GTTCCTGGTGTTCATGTCTCA 3′
Human Caspase1_R	5′ CCTACTGAATCTTTAAACCACACC 3′
Human IL6_F	5′ GACAGCCACTCACCTCTT 3′
Human IL6_R	5′ TGTTTTCTGCCAGTGCC 3′
Human IL8_F	5′ TCCTGATTTCTGCAGCTCTG 3′
Human IL8_R	5′ GTCCACTCTCAATCACTCTCAG 3′
Human MCP-1/CCL2_F	5′ TGTCCCAAAGAAGCTGTGATC 3′
Human MCP-1/CCL2_R	5′ ATTCTTGGGTTGTGGAGTGAG 3′

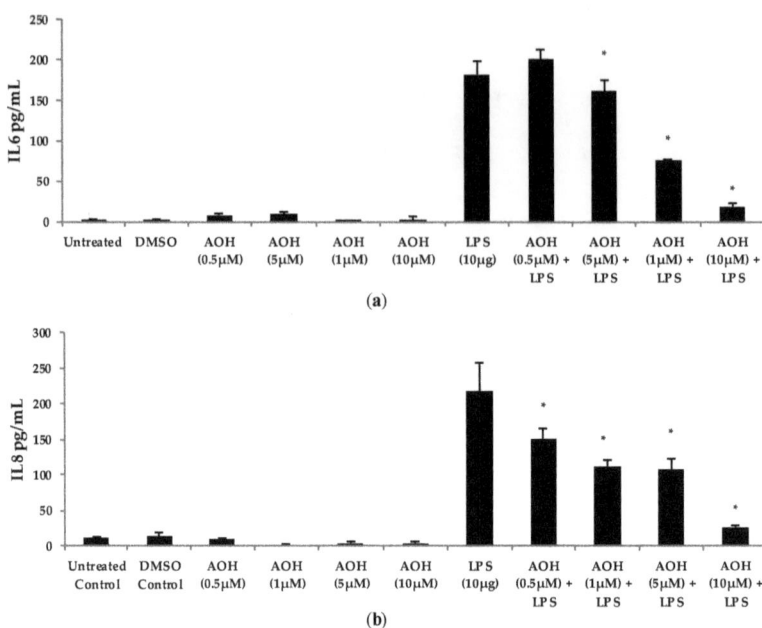

Figure A1. Dose-dependent analysis of the alternariol (AOH) response. Lipopolysaccharide (LPS) was added 2 h after AOH. BEAS-2B cells were seeded at a density of 500,000 cells/well. Cells were treated with 0.5–10 μM of AOH in the presence and absence of 10 μg of LPS to measure the cytokine levels released. Cell densities were 5×10^5 cells/well, and they were incubated for 24 h under normal conditions at 37 °C, 5% CO_2. Cells treated with AOH showed a marked suppression of cytokines both in the presence and absence of LPS. (**a**) IL6 measured by enzyme-linked immunosorbent assay (ELISA); (**b**) IL8 measured by ELISA. An * indicates $p < 0.05$ according to Student's t-test when comparing LPS + AOH treatments to LPS-induced control.

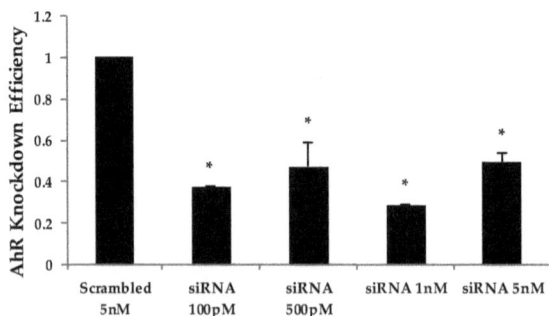

Figure A2. Quantitiatve real-time reverse transcription polymerase chain reaction (qRT-PCR) analysis of doses for aryl hydrocarbon receptor (AhR) silencing. Cells were seeded with a density of 150,000 cells/well. Cells were treated with AhR siRNA for 24 h. A dose curve was performed to elucidate the concentration required for AhR silencing in lung epithelial cells BEAS-2B. An * indicates $p < 0.05$ according to Student's t-test for AhR-specific siRNA treatments compared to the scrambled siRNA control.

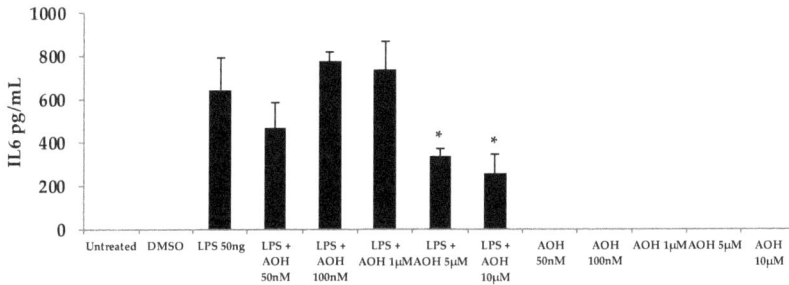

Figure A3. Dose-dependent response of mouse macrophages (RAW264.7) after treatment with alternariol (AOH) and lipopolysaccharide (LPS). Cells were treated with 50–10μM of AOH in the presence and absence of 50 ng of LPS to measure the IL6 levels released by enzyme-linked immunosorbent assay (ELISA). Cell densities were 5×10^5 cells/well, and they were incubated for 24 h under normal conditions at 37 °C, 5% CO_2 after treatment. An * indicates $p < 0.05$ according to Student's *t*-test when comparing LPS + AOH treatments to LPS-induced control.

Abbreviations

AhR	Aryl hydrocarbon Receptor
ARNT	Aryl hydrocarbon Receptor Nuclear Translocator
AOH	Alternariol
AME	Alternariol Monomethyl ether
BEAS-2B	Bronchial Lung Epithelial Cells
CCL2	Chemokine (C–C motif) Ligand 2
cDNA	Complementary DNA
CDK1	Cyclin-Dependent Kinase 1
DMSO	Dimethyl Sulfoxide
DPBS	Dulbecco's Phosphate-Buffered Saline
DNA	Deoxyribonucleic Acid
ELISA	Enzyme-Linked Immunosorbent Assay
FBS	Fetal Bovine Serum
GAPDH	Glyceraldehyde 3-Phosphate Dehydrogenase
HT29	Human Adenocarcinoma Cells
IL-1β	Interleukin-1 β
IL6	Interleukin 6
IL8	Interleukin 8
IgE	Immunoglobulin E
LDH	Lactate Dehydrogenase
LPS	Lipopolysaccharide
MTT	3-(4,5-dimethylthiazol-2-yl)-2,5-Diphenyltetrazolium Bromide
MN	Micronucleus Assay
MLC	Mouse Lymphoma Cell Line
PBS	Phosphate-Buffered Saline
RNA	Ribonucleic Acid
ROS	Reactive Oxygen Species
siRNA	Small Interfering RNA
TCDD	2,3,7,8-Tetrachlorodibenzo-p-Dioxin
TGF-β	Transforming Growth Factor β
TNF-α	Tumor Necrosis Factor α
qRT-PCR	Quantitative Real-Time Reverse Transcription-Polymerase Chain Reaction
XRE	Xenobiotic Response Element

References

1. Dang, H.X.; Pryor, B.; Peever, T.; Lawrence, C.B. The Alternaria genomes database: A comprehensive resource for a fungal genus comprised of saprophytes, plant pathogens, and allergenic species. *BMC Genom.* **2015**, *16*, 239. [CrossRef] [PubMed]
2. Van Leeuwen, W.S. Bronchial asthma in relation to climate. *Proc. R. Soc. Med.* **1924**, *17*, 19–26. [PubMed]
3. Sanchez, H.; Bush, R.K. A review of Alternaria alternata sensitivity. *Rev. Iberoam. Micol.* **2001**, *18*, 56–59. [PubMed]
4. Hedayati, M.T.; Arabzadehmoghadam, A.; Hajheydari, Z. Specific IgE against *Alternaria alternata* in atopic dermatitis and asthma patients. *Eur. Rev. Med. Pharmacol. Sci.* **2009**, *13*, 187–191. [PubMed]
5. Hong, S.G.; Cramer, R.A.; Lawrence, C.B.; Pryor, B.M. Alt a 1 allergen homologs from Alternaria and related taxa: Analysis of phylogenetic content and secondary structure. *Fungal Genet. Biol.* **2005**, *42*, 119–129. [CrossRef] [PubMed]
6. Kobayashi, T.; Iijima, K.; Radhakrishnan, S.; Mehta, V.; Vassallo, R.; Lawrence, C.B.; Cyong, J.C.; Pease, L.R.; Oguchi, K.; Kita, H. Asthma-related environmental fungus, Alternaria, activates dendritic cells and produces potent Th2 adjuvant activity. *J. Immunol.* **2009**, *182*, 2502–2510. [CrossRef] [PubMed]
7. Auger, P.L.; Gourdeau, P.; Miller, J.D. Clinical experience with patients suffering from a chronic fatigue–like syndrome and repeated upper respiratory infections in relation to airborne molds. *Am. J. Ind. Med.* **1994**, *25*, 41–42. [CrossRef] [PubMed]
8. Corrier, D.E. Mycotoxicosis: Mechanisms of immunosuppression. *Vet. Immunol. Immunopathol.* **1991**, *30*, 73–87. [CrossRef]
9. Streit, E.; Schwab, C.; Sulyok, M.; Naehrer, K.; Krska, R.; Schatzmayr, G. Multi-mycotoxin screening reveals the occurrence of 139 different secondary metabolites in feed and feed ingredients. *Toxins* **2013**, *5*, 504–523. [CrossRef] [PubMed]
10. Deshmukh, S.K.; Verekar, S.A.; Bhave, S.V. Endophytic fungi: A reservoir of antibacterials. *Front. Microbiol.* **2014**, *5*, 715. [CrossRef] [PubMed]
11. Lou, J.; Fu, L.; Peng, Y.; Zhou, L. Metabolites from Alternaria fungi and their bioactivities. *Molecules* **2013**, *18*, 5891–5935. [CrossRef] [PubMed]
12. Panel, E.; Chain, F. Scientific opinion on the risks for animal and public health related to the presence of Alternaria toxins in feed and food. *EFSA J.* **2011**, *9*, 2407. [CrossRef]
13. Fleck, S.C.; Burkhardt, B.; Pfeiffer, E.; Metzler, M. Alternaria toxins: Altertoxin II is a much stronger mutagen and DNA strand breaking mycotoxin than alternariol and its methyl ether in cultured mammalian cells. *Toxicol. Lett.* **2012**, *214*, 27–32. [CrossRef] [PubMed]
14. Bashyal, B.P.; Wellensiek, B.P.; Ramakrishnan, R.; Faeth, S.H.; Ahmad, N.; Gunatilaka, L. Altertoxins with potent anti-HIV activity from *Alternaria tenuissima* QUE1Se, a fungal endophyte of *Quercus emoryi*. *Bioorg. Med. Chem.* **2014**, *22*, 6112–6116. [CrossRef] [PubMed]
15. Ostry, V. Alternaria mycotoxins: An overview of chemical characterization, producers, toxicity, analysis and occurrence in foodstuffs. *World Mycotoxin J.* **2008**, *1*, 175–188. [CrossRef]
16. Solhaug, A.; Holme, J.A.; Haglund, K.; Dendele, B.; Sergent, O.; Pestka, J.; Lagadic-Gossmann, D.; Eriksen, G.S. Alternariol induces abnormal nuclear morphology and cell cycle arrest in murine RAW 264.7 macrophages. *Toxicol. Lett.* **2013**, *219*, 8–17. [CrossRef] [PubMed]
17. Lehmann, L.; Wagner, J.; Metzler, M. Estrogenic and clastogenic potential of the mycotoxin alternariol in cultured mammalian cells. *Food Chem. Toxicol.* **2005**, *44*, 398–408. [CrossRef] [PubMed]
18. Solhaug, A.; Vines, L.L.; Ivanova, L.; Spilsberg, B.; Holme, J.A.; Pestka, J.; Collins, A.; Eriksen, G.S. Mechanisms involved in alternariol-induced cell cycle arrest. *Mutat. Res.* **2012**, *738*, 1–11. [CrossRef] [PubMed]
19. Tiessen, C.; Fehr, M.; Schwarz, C.; Baechler, S.; Domnanich, K.; Böttler, U.; Pahlke, G.; Marko, D. Modulation of the cellular redox status by the Alternaria toxins alternariol and alternariol monomethyl ether. *Toxicol. Lett.* **2013**, *216*, 23–30. [CrossRef] [PubMed]
20. Nguyen, N.T.; Hanieh, H.; Nakahama, T.; Kishimoto, T. The roles of aryl hydrocarbon receptor in immune responses. *Int. Immunol.* **2013**, *25*, 335–343. [CrossRef] [PubMed]

21. Schreck, I.; Deigendesch, U.; Burkhardt, B.; Marko, D.; Weiss, C. The Alternaria mycotoxins alternariol and alternariol methyl ether induce cytochrome P450 1A1 and apoptosis in murine hepatoma cells dependent on the aryl hydrocarbon receptor. *Arch. Toxicol.* **2012**, *86*, 625–632. [CrossRef] [PubMed]
22. Beischlag, T.V.; Luis, M.J.; Hollingshead, B.D.; Perdew, G.H. The aryl hydrocarbon receptor complex and the control of gene expression. *Crit. Rev. Eukaryot. Gene Expr.* **2008**, *18*, 207–250. [CrossRef] [PubMed]
23. Opal, S.M. Endotoxins and other sepsis triggers. *Contrib. Nephrol.* **2010**, *167*, 14–24. [CrossRef] [PubMed]
24. Martinon, F.; Burns, K.; Tschopp, J. The inflammasome: A molecular platform triggering activation of inflammatory caspases and processing of proIL-β. *Mol. Cell.* **2002**, *10*, 417–426. [CrossRef]
25. Mariathasan, S.; Newton, K.; Monack, D.M.; Vucic, D. Differential activation of the inflammasome by caspase-1 adaptors ASC and Ipaf. *Nature* **2004**, *430*, 213–218. [CrossRef] [PubMed]
26. Vassilev, L.T.; Tovar, C.; Chen, S.; Knezevic, D.; Zhao, X.; Sun, H.; Heimbrook, D.C.; Chen, L. Selective small-molecule inhibitor reveals critical mitotic functions of human CDK1. *Proc. Natl. Acad. Sci. USA* **2006**, *103*, 10660–10665. [CrossRef] [PubMed]
27. Grover, S. The Role of the Alternaria Secondary Metabolite Alternariol in Inflammation. Master's Thesis, Virginia Tech University, Blacksburg, VA, USA, 2017.

International Journal of
Molecular Sciences

MDPI

Article

Immuno-Modulatory and Anti-Inflammatory Effects of Dihydrogracilin A, a Terpene Derived from the Marine Sponge *Dendrilla membranosa*

Elena Ciaglia [1,†], Anna Maria Malfitano [2,†], Chiara Laezza [3,4], Angelo Fontana [5], Genoveffa Nuzzo [5], Adele Cutignano [5], Mario Abate [1], Marco Pelin [6], Silvio Sosa [6], Maurizio Bifulco [1,7,*,‡] and Patrizia Gazzerro [2,*]

[1]   Department of Medicine, Surgery and Dentistry "Scuola Medica Salernitana", University of Salerno, Via Salvatore Allende, 84081 Baronissi Salerno, Italy; eciaglia@unisa.it (E.C.); m.abate23@studenti.unisa.it (M.A.)
[2]   Department of Pharmacy, University of Salerno, Via Giovanni Paolo II, 84084 Fisciano, Salerno, Italy; amalfitano@unisa.it
[3]   Department of Biology and Cellular and Molecular Pathology, University of Naples Federico II, Via Pansini, 80131 Naples, Italy; chiara.laezza@cnr.it
[4]   Institute of Endocrinology and Experimental Oncology, IEOS CNR, Via Pansini 5, 80131 Naples, Italy
[5]   Bio-Organic Chemistry Unit, Institute of Biomolecular Chemistry-CNR, Via Campi Flegrei 34, Pozzuoli, 80131 Naples; Italy, afontana@icb.cnr.it (A.F.); nuzzo.genoveffa@icb.cnr.it (G.N.); acutignano@icb.cnr.it (A.C.)
[6]   Department of Life Sciences, University of Trieste, 34127 Trieste, Italy; mpelin@units.it (M.P.), ssosa@units.it (S.S.)
[7]   CORPOREA-Fondazione Idis-Città della Scienza, via Coroglio 104 e 57, 80124 Naples, Italy
*   Correspondence: mbifulco@unisa.it (M.B.); pgazzerro@unisa.it (P.G.); Tel.: +39-089-969452 (P.G.); Fax: +39-089-969602 (P.G.)
†   These authors contributed equally to this work.
‡   M.B. is considered co-last author.

Received: 1 May 2017; Accepted: 23 June 2017; Published: 28 July 2017

Abstract: We assessed the immunomodulatory and anti-inflammatory effects of 9,11-dihydrogracilin A (DHG), a molecule derived from the Antarctic marine sponge *Dendrilla membranosa*. We used in vitro and in vivo approaches to establish DHG properties. Human peripheral blood mononuclear cells (PBMC) and human keratinocytes cell line (HaCaT cells) were used as in vitro system, whereas a model of murine cutaneous irritation was adopted for in vivo studies. We observed that DHG reduces dose dependently the proliferative response and viability of mitogen stimulated PBMC. In addition, DHG induces apoptosis as revealed by AnnexinV staining and downregulates the phosphorylation of nuclear factor kappa-light-chain-enhancer of activated B cells (NF-κB), signal transducer and activator of transcription (STAT) and extracellular signal–regulated kinase (ERK) at late time points. These effects were accompanied by down-regulation of interleukin 6 (IL-6) production, slight decrease of IL-10 and no inhibition of tumor necrosis factor-alpha (TNF-α) secretion. To assess potential properties of DHG in epidermal inflammation we used HaCaT cells; this compound reduces cell growth, viability and migration. Finally, we adopted for the in vivo study the croton oil-induced ear dermatitis murine model of inflammation. Of note, topical use of DHG significantly decreased mouse ear edema. These results suggest that DHG exerts anti-inflammatory effects and its anti-edema activity in vivo strongly supports its potential therapeutic application in inflammatory cutaneous diseases.

Keywords: marine sponge; natural compound; inflammation; lymphocytes

---

Int. J. Mol. Sci. **2017**, *18*, 1643

1. Introduction

Sponges, seaweeds, snails and soft corals are marine organisms representing an unexploited source of novel compounds with promising application to human wellbeing [1,2]. These metabolites exhibit various bioactivities and potential pharmacological properties, and many of them are currently on the market or clinical trials as anti-cancer, analgesic, immunomodulatory or anti-inflammatory agents [3–11]. Among the natural products, one of the most studied groups of molecules is represented by terpenes [12], secondary metabolites formed by repetitions of C5 isoprene units that derive from two distinct biochemical routes named mevalonate or non-mevalonate pathways [13–16]. Terpenes are generally classified into hemi, mono, sesqui, di, sester, or tri based according to the number of the isoprene units. These compounds have been found largely in higher plants [17,18] and in lower invertebrates including marine organisms [19–24]. Marine diterpenoids embrace a diverse and promising class of molecules exhibiting a range of effects including antiviral, antibacterial, antiparasite, anticancer, and anti-inflammatory activity [19,25–29]. Studies suggested that briarane-or cembrane-type diterpenes exert anti-inflammatory activity as they inhibit pro-inflammatory enzymes, such as cyclooxygenase (COX-2) and inducible nitric oxide synthase (iNOS) in murine macrophages activated with lipopolysaccharide (LPS) [30–33]. Cembranoids such as gibberosenes, grandilobatin, sarcocrassocolides, querciformolides, crassarines, crassumolides, sinularolides, durumolides and columnariols have shown capacity to block the expression of iNOS and/or COX-2 by LPS-activated RAW264.7 cells [34]. Some cembranoids have been identified as modulators of nuclear factor kappa-light-chain-enhancer of activated B cells (NF-κB) signaling pathway [35–37]. Indeed, in recent years, anti-inflammatory activity for eunicellin-based diterpenoids isolated from soft corals, has been described [38,39]. Marine diterpene glycosides are characterized by a diterpene aglycone core and a carbohydrate moiety. Among these compounds, glycosides, eleutherobin, fuscosides, and pseudopterosins are the most studied [40]. Eleutherobin is a microtubule-targeted agent currently used in preclinical studies [41,42]. The pseudopterosins have been described as molecules with important anti-inflammatory and analgesic properties [43,44]. Fuscosides A and B when topically applied, decrease phorbol myristate acetate (PMA)-induced edema in mouse ears by blocking neutrophil infiltration. Fuscoside B blocks the synthesis of leukotriene C4 in calcium ionophore-activated murine macrophages [45,46]. Another family of diterpenoids, verticillane-based diterpenoids, have recently demonstrated anti-inflammatory properties, members of this family inhibit iNOS in LPS-stimulated RAW264.7 cells [47]. Furthermore, a tricyclic brominated diterpenoid, the neorogioltriol isolated from algae, inhibited the activation of NF-κB and the secretion of tumor necrosis factor-alpha (TNF-α), nitric oxide (NO), and COX-2 in LPS stimulated macrophages. In an animal model of carrageenan-induced local inflammation, neorogioltriol decreased edema formation [48]. Mollusc derived dolabellane diterpenoids, also isolated from plants, exert antiprotozoa [49], antiviral [50], and antibacterial [51] activities [52]. Dolabelladienetriol has been suggested to have anti-inflammatory properties as downregulates the secretion of TNF-α and NO through the inhibition of NF-κB activation in *Leishmania amazonensis* infected and uninfected macrophages [53].

In a recent work about a novel platform of drug discovery based on an automatic fractionation of marine samples by a simple and versatile protocol of solid-phase extraction [54], we reported that the diterpenoid 9,11-dihydrogracilin (DHG) [55] from the Antarctic sponge *Dendrilla membranosa* stimulated the response of human peripheral blood mononuclear cells (PBMCs). DHG belongs to the spongiane family, a very large class of sponge-derived natural products showing several promising activities [56]. Here, we investigate the immunomodulatory and anti-inflammatory properties of DHG by in vitro and in vivo models. The effects exhibited by this compound support its potential as novel anti-inflammatory drug.

2. Results

2.1. 9,11-Dihydrogracilin A (DHG)-Mediated Inhibition of Human Peripheral Blood Mononuclear Cells (PBMC) Proliferation and Viability

First we asked if the immuno-modulatory potential of DHG could affect mitosis of CD3 monoclonal antibody (OKT3)- and Phytohemagglutinin (PHA)-activated healthy-PBMC. Proliferation of PBMC was determined after 4 or 6 days of stimulation with OKT3 (1 μg/mL) and PHA (1.5%) respectively, by measuring [^{3}H]-thymidine incorporation. As shown in Figure 1A,B, all mitogenic stimuli induced a significant proliferation of PBMC. The co-treatment with DHG at selected concentrations, ranging from 0.3 to 10 μM, resulted in a dose-response inhibition of mitosis of PHA and to a more extent of OKT3-stimulated PBMC. A better dose–response profile was observed using PHA as stimulus, thus for further experiments we used only PHA.

In order to assess whether besides inhibition of DNA synthesis, DHG could affect cell viability of PBMC, we counted the cells after the staining with trypan blue. DHG decreased the number of viable cells in a concentration-dependent manner (Figure 1C), specifically, at 10 μM, it significantly reduced viable cell number of 73 ± 2.4%. Of note the viability of DHG-treated resting cells was not significantly affected, thus excluding its possible toxic effect. Then, to better characterize the nature of cytotoxic effects mediated by DHG in activated PBMC, we next performed cell death assays by Annexin-V and propidium iodide double staining (Supplementary Figure S1). Here, we registered a dose-dependent induction of apoptosis, resulting in the death of 43.1 ± 2.4% of cells already after 48h exposure at the highest dose of 10 μM DHG (Figure 1D).

2.2. DHG Effects on Signaling Pathways

Since signal transducer and activator of transcription 5 (STAT5), extracellular signal–regulated kinase (ERK), and NF-κB signaling pathways are critical for PBMC activation following stimulation with PHA, we moved to investigate whether and in which way these signaling events were affected by increasing doses of DHG at early time points. As reported in Figure 2A, ERK was phosphorylated in response to 30 min-PHA stimulation. However, DHG 10 μM led to significantly greater levels of phospho-extracellular signal–regulated kinase (p-ERK) compared with the effect observed in response to the mitogen alone. On the other hand, phospho-nuclear factor kappa-light-chain-enhancer of activated B cells (p-NF-κB) was not affected by DHG treatment. Moreover, no signals were observed in the activation of STAT5 pathway at this early time point. On the contrary, as expected, after in vitro stimulation for 120 min, PHA enhanced tyrosine phosphorylation of STAT5, which is instead significantly inhibited by DHG co-treatment. Similarly, DHG at the highest dose decreased ERK and NF-κB activity, compared to control PHA-activated cells (Figure 2B). Kinetic studies (Supplementary Figure S2) revealed that inhibition of NF-κB phosphorylation by DHG co-treatment at all the concentrations used was still observed after 5 days of stimulation, suggesting a long lasting effect of the compound.

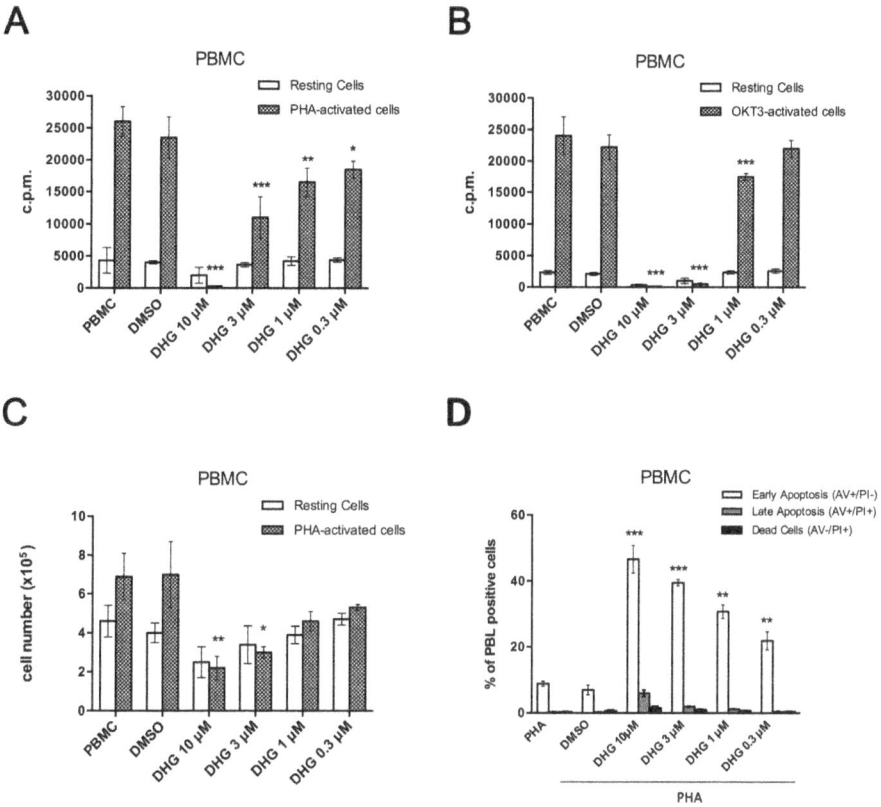

Figure 1. 9,11-Dihydrogracilin A (DHG) inhibits Peripheral Blood Mononuclear Cells (PBMC) proliferation and viability and induces apoptosis. (**A**) Unstimulated PBMC and phytohemagglutinin (PHA)-activated PBMC from healthy donors were treated with DHG at the indicated concentrations. Proliferation was measured after 18h of ^{3}H-thymidine incorporation (1 µCi). The counts per minutes (c.p.m.) ± the SD of the triplicates of five independent experiments are shown. (ANOVA * $p < 0.05$, *** $p < 0.001$, ** $p < 0.01$ versus PHA-treated PBMC); (**B**) Unstimulated PBMC and CD3 monoclonal antibody (OKT3)-activated PBMC of healthy donors were treated with DHG at the indicated concentrations. Proliferation was measured after 18h of ^{3}H-thymidine incorporation (1 µCi). The c.p.m. ± the SD of the triplicates of five independent experiments are shown. (ANOVA * $p < 0.05$, *** $p < 0.001$, ** $p < 0.01$ versus OKT3-treated PBMC); (**C**) Unstimulated PBMC and PHA-activated PBMC from healthy donors were treated with DHG, cultured for 6 days and stained with trypan blue. Cell viability was compared to that observed in PHA-activated PBMC (ANOVA * $p < 0.05$, ** $p < 0.01$). The histogram reported show the percent of live PBMC; (**D**) Induction of apoptosis was measured by annexin V and propidium iodide (PI) double staining through fluorescence-activated cell sorting (FACS) analysis in DHG-treated healthy donor PBMC, after 48 h. The panel reporting representative dot plots of 4 different experiments performed with similar results is included in the supplementary section (Supplementary Figure S1). Histograms in **D** indicate total percentage of early (Annexin V-positive cells/PI-negative cells) and late apoptotic events (Annexin V/PI-double positive cells) as well as necrotic cells (Annexin V-negative cells/PI-positive cells). Results are representative of 4 independent experiments and expressed as mean ± SD (ANOVA, *** $p < 0.001$, ** $p < 0.01$). DMSO, dimethyl sulfoxide.

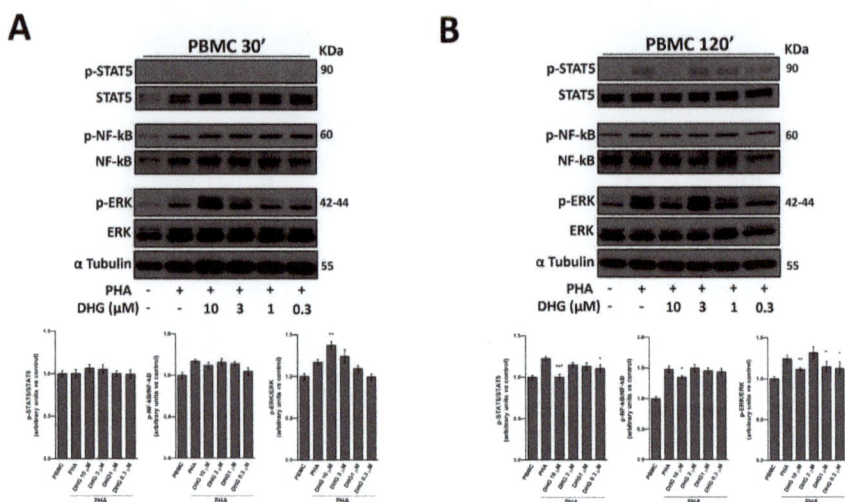

Figure 2. DHG effects on NF-κB, Signal Transducer and Activator of Transcription 5 (STAT5) and Extracellular Signal–regulated Kinase (ERK) phosphorylation. Western blot analysis performed on whole cell extracts from 30 min (**A**) and 120 min (**B**) of culture in the presence and in the absence of DHG at the indicated concentrations. α-tubulin was used as control of protein loading. Panels show representative results from 3 different experiments performed independently. Histograms below represent mean ± SD in densitometry units of scanned immunoblots from the 3 different experiments (ANOVA, *** $p < 0.001$, ** $p < 0.01$, * $p < 0.05$).

2.3. Cytokine Production by DHG-Treated PBMC upon Phytohemagglutinin (PHA) Stimulation

After proper activation, PBMC can secrete numerous cytokines, through which they coordinate immune response, such as interleukin 6 (IL-6), TNF-α and IL-10 [57,58]. After 24h of incubation, we then assessed the capacity of DHG-treated PBMC to produce these soluble factors. As expected, stimulation of PBMC by PHA induced secretion of all soluble factors tested. Notably, the treatment with DHG, at the concentration of 3 μM significantly inhibited the production of IL-6 (Figure 3A), without interfering with the levels of the other pro-inflammatory factor TNF-α (Figure 3B). Finally, a'slight decrease in the level of the contro-regulatory action of IL-10 was also documented (Figure 3C).

2.4. DHG Effects on Activation Marker Surface Expression in T and Natural Killer (NK) Cell Compartments

All these results suggest an anti-inflammatory action of DHG. So, we asked which particular lymphocyte cell subset could be affected by DHG. In particular, we focused both on T cells (CD3+/CD56−) and natural killer (NK) cell (CD3−/CD56+) compartments. It is well known that in activation state, CD25 and CD69 are induced in lymphocytes as classical markers to monitor T and NK cell reactivity. Therefore, by fluorescence-activated cell sorting (FACS) analysis we determined the level of PBMC activation from healthy donors treated with DHG at a selected concentration (3 μM) following PHA stimulation. In samples treated with DHG compared to the control, we documented a significant decrease of the percent of CD25+ NK cells (Figure 4A) but not in T cell population. At the same way, we reported no significant difference in the surface level of CD69 in regard to the percentage of CD3+ T cells expressing the activation marker following DHG treatment, while the same compound significantly reduces the number of CD69+ NK cells (Figure 4B), suggesting a possible specific inhibition of NK cell activation that needs to be underpinned in the near future.

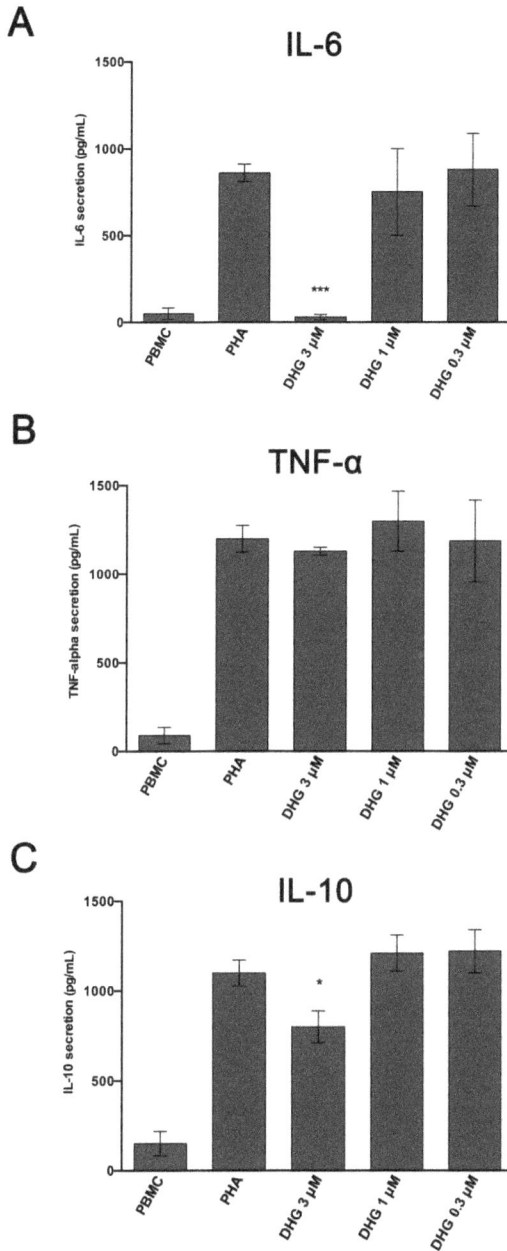

Figure 3. Cytokine secretion profile of DHG-treated PBMC. PBMC of healthy donors (*n* = 4) were activated with PHA (1.5%) for 24 h in the presence and in the absence of DHG at the indicated concentrations. Unstimulated cells are included as control (PBMC) in the figure. Supernatants were harvested and the concentrations of interleukin 6 (IL-6) (**A**), tumor necrosis factor-alpha (TNF-α) (**B**) and IL-10 (**C**) determined by ELISA immunoassay. Values reported refer to mean ± SD of four different donors. Statistical analyses are reported (ANOVA; * $p < 0.05$; *** $p < 0.001$).

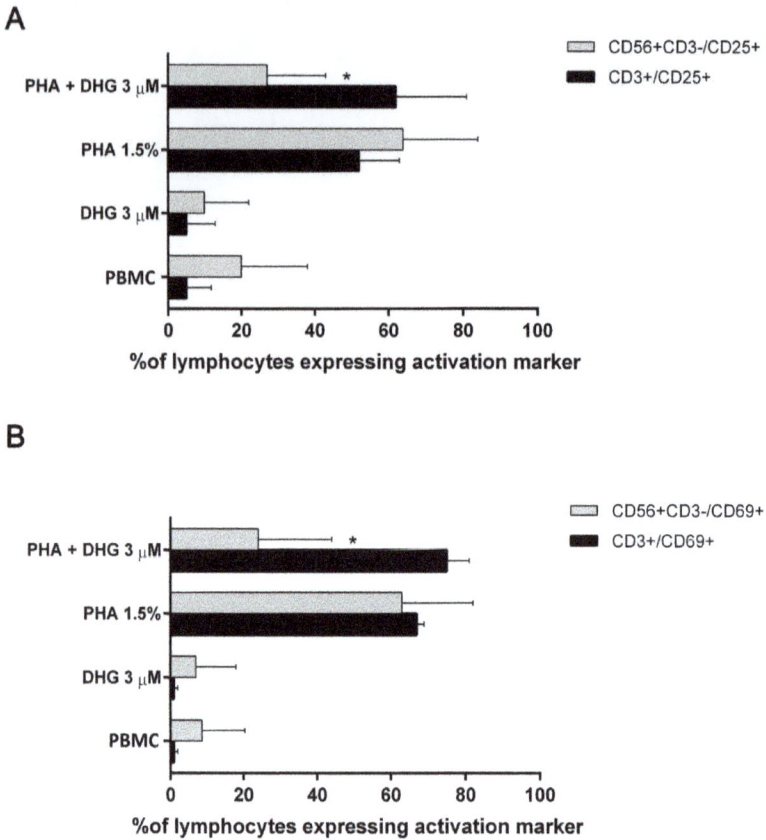

Figure 4. Flow cytometric analysis of CD25 and CD69 surface expression on DHG-treated PBMC. PBMC from healthy donors ($n = 7$) were stimulated with PHA (1.5%) in the presence and in the absence of DHG. Unstimulated cells are included as control (PBMC) in the figure. Following 24 h of activation, CD3+/CD56− (**black bars**) and CD3−/CD56+ (**gray bars**) populations were analyzed for CD25+ expression (**A**) and CD69+ expression (**B**) and compared by ANOVA (* $p < 0.05$, compared with untreated PHA-activated cells). Bar graphs report mean values ± SD.

2.5. Mitogenic and Migratory Capacity of Human Keratinocytes Cell Line (HaCaT) Exposed to DHG

Then we asked if the DHG action was confined to immune compartment or if it might interfere also with proliferation and function of other cell lines. To this end, we moved to gain insight into DHG effects on keratinocyte activity, by testing its ability to modulate the viability and growth of the spontaneously immortalized human keratinocytes cell line (HaCaT) cell line, a well-established keratinocyte model, easy to propagate and near to a normal phenotype.

Cell counting using Trypan Blue viability dye and sulforhodamine B assay revealed respectively a reduced viability (Figure 5A) and cell growth rate (Figure 5B) of HaCaT cells exposed to 48 h treatment with increasing concentrations of DHG.

Then, to evaluate the potential interference of DHG with the migratory function of HaCaT cells, we assessed a scratch wound assay using a micropipette tip. After 24 h of cell culture, whereas serum 10% favored narrowing of the scratch wound, in the presence of DHG, the wounded cells resulted in a less enhancement of wound healing to all doses tested (Figure 5C,D).

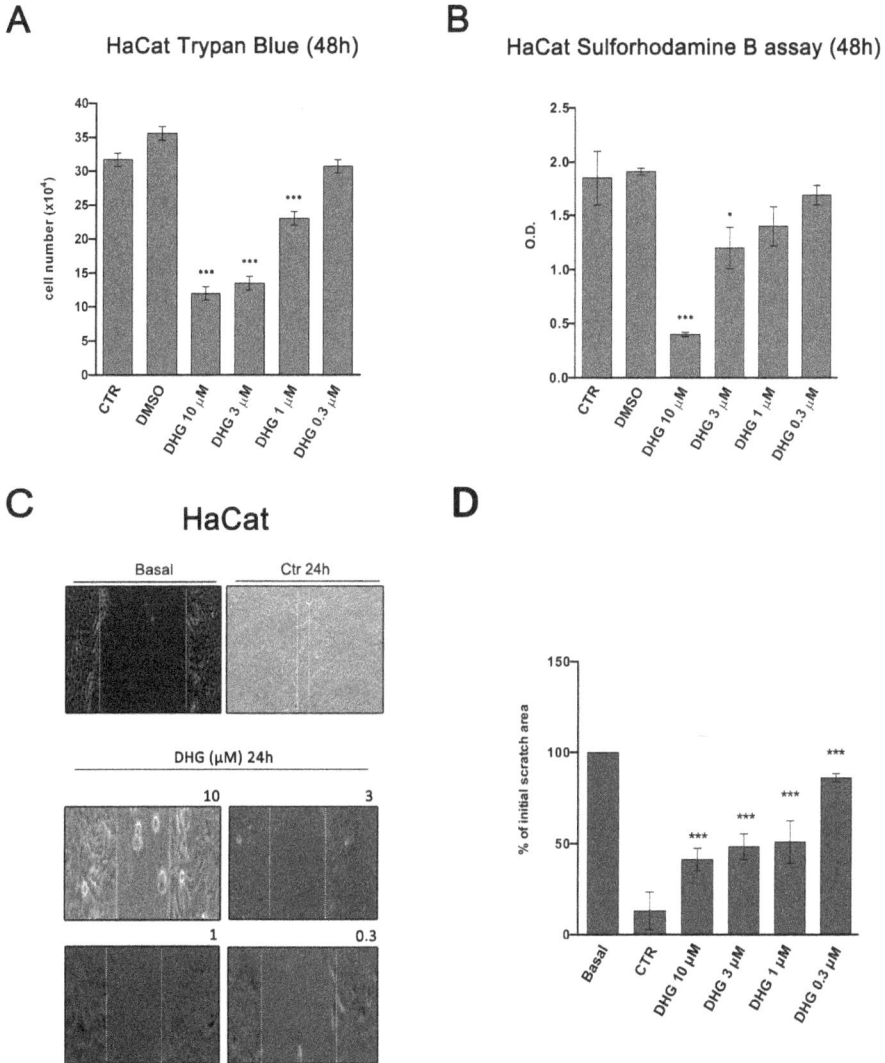

Figure 5. DHG effects on cell vitality and migration of HaCaT cells. (**A**) HaCaT cells were treated with vehicle alone (DMSO) or DHG at the reported concentrations, cultured for 48 h and stained with trypan blue as described in material and methods. Control cells without vehicle are also included in the figure (CTR). Cells were counted and the percent of cell viability was calculated compared to untreated cells (ANOVA * $p < 0.05$; *** $p < 0.001$); (**B**) HaCaT cells were treated with DMSO or DHG at the indicated concentrations, cultured for 48 h and next the sulforhodamine B assay was performed as described in material and methods (ANOVA * $p < 0.05$ and *** $p < 0.001$ versus DMSO-treated cells); (**C**) Wound healing assay performed in HaCaT cells treated for 24 h with vehicle (CTR) or DHG (0.3–10 μM) in complete medium. Representative light microscope images from three independent experiments are shown. Dotted white lines indicate the wounded area from the initial scratch. Magnification, × 20. Basal bar = 348.5 μm; (**D**) Histograms represent the mean scratch area observed in HaCaT cells expressed as percent of initial area. The measurement was made in three different experiments. Results are presented as mean ± standard error (ANOVA *** $p < 0.001$).

2.6. Topical Anti-Inflammatory Action of DHG In Vivo

In view of the immunomodulatory potential of DHG along with its effect on hyper-activation of keratinocytes, we finally tested the marine compound in vivo in a murine model of inflammation, the croton oil ear test to evaluate its topical anti-inflammatory effect at the non-toxic dose of 1 μmol/cm^2. This dose was selected on the basis of our experience on doses range of natural compounds active as anti-inflammatory agents in this in vivo model of skin inflammation. Interestingly, after 6 h, when oedema is already formed [59,60], DHG treatment was able to induce its significant reduction (Table 1). In particular, the oedema was decreased to about 58%, comparable to the activity of indomethacin, the reference non-steroidal anti-inflammatory drug (NSAID).

Table 1. Anti-inflammatory activity of DHG in a murine model of dermatitis. Dose-dependent anti-oedema activity of topically administered DHG (1 μmol/cm^2) and indomethacin (0.3 μmol/cm^2) in croton oil-induced ear dermatitis after 6 h. * $p < 0.001$ at the analysis of variance, as compared to controls.

Substance	Dose (μmol/cm^2)	Number of Animals	Edema (mg)	Reduction (%)	p
Controls	-	10	8.5 ± 0.2	-	-
DHG	1.0	10	3.6 ± 0.2 *	58	0.001
Indomethacin	0.3	10	3.7 ± 0.3 *	56	0.001

3. Discussion

Inflammation plays a crucial role in many physio/pathological states, and different cell populations are involved in all phases of inflammatory process, including neutrophils, dendritic cells, monocytes/macrophages, and lymphocytes. Previous findings suggested that marine diterpenoids elicit among numerous activities, anti-inflammatory effects on murine macrophages [30–37,47,53]. In this study, starting from a previous drug discovery study aimed to fractionate marine samples, we successfully demonstrated that the sponge metabolite DHG possesses promising anti-inflammatory properties in vitro and in vivo. Firstly, we assessed the ability of this natural product to reduce cell proliferation induced by different mitogens. We described efficacy of DHG to reduce dose dependently PBMC proliferation and viability (Figure 1A,B). To further investigate cellular effects of DHG, we evaluated if cell death might be caused by apoptosis. As reported in Figure 1D, we ascertained that DHG induced apoptosis in PBMC, and such effect was dose dependent as more apoptotic cells were detected at the highest concentrations of DHG. These effects are accompanied by DHG down-regulation of NF-κB, STAT and ERK phosphorylation at later time points. It is of note that the enhancement of ERK activation by 30 min treatment of DHG confirms the early apoptotic events observed (Figure 1D) and may reflect the contro-regulatory actions between these signaling events in lymphocytes biology [59]. Indeed, a prolonged and intense activation of the ERK pathway, as that in response to strong T-cell receptor (TCR) signals results in transient inhibition of IL-2-mediated activation of STAT5. In contrast, when cells receive weak signals, as that achieved by DHG low doses, the degree of activation of the ERK pathway is not strong enough to block STAT5 activation in response to small amounts of IL-2 secreted by T cells in an autocrine fashion. Our data are also in agreement with previous studies showing modulation of NF-κB, a key regulator in inflammatory processes, by several classes of marine diterpenoids in murine macrophages [35–37,53]. Our assays provided an initial evidence of the anti-inflammatory properties of this compound. It is well established that activated PBMC produce soluble factors that play relevant role in inflammation. Thus, cytokine secretion was assessed and the results obtained corroborate our findings. We selected three cytokines IL-6, TNF-α and IL-10. In particular, IL-6 and TNF-α are inflammatory cytokines, while IL-10 is an anti-inflammatory cytokine. The results obtained show that DHG down-regulates the expression of IL-6 that is a cytokine known to be active during inflammation, it does not affect TNF-α, thus suggesting a specific effect on particular cytokines like IL-6 and it slightly affected IL-10 secretion. The reduction of IL-10 is unexpected since it is an anti-inflammatory cytokine, however, the inhibition is very weak compared to the strong effect observed on IL-6. To establish if the effects observed on cell proliferation might be due to a reduced activation state of PBMC induced by DHG, we examined the expression of CD69, a well-known T and NKT cell early activation marker. We found that NK cell (CD3−/CD56+) compartment was affected by DHG since a reduced expression of this marker was observed. This finding is of particular interest because in agreement with previous reports describing that NK cells are protagonists of inflammatory skin diseases like psoriasis [60,61], a chronic relapsing-remitting inflammatory skin pathology characterized by thickened epidermis, as the result of keratinocyte hyper-proliferation and abnormal differentiation, increased vascularity and accumulation of inflammatory infiltrates.

Our findings are in agreement with previous studies that, in the past years, have offered suggestions that compounds from marine organisms, in particular diterpenoids, might found application in inflammatory disorders. In marine species, the ability to produce some compounds is evolutionarily selected as a substantial characteristic of defense from natural competitors. In recent years, the various properties of some natural substances, mainly of terpenoids, have made these compounds a good source of products potentially exploitable in several pharmacological applications. Published studies mainly described the effects of these molecules in murine macrophages. Here, we provided the first evidence that a diterpenoid affects also human lymphocytes. In order to investigate potential application of DHG in dermatological inflammation we analyzed its effect on immortalized keratinocytes. In these cells, we confirmed that DHG inhibits cell viability and it is also able to precociously decrease cell migration without affecting cell survival. Finally, we used an in vivo model of skin inflammation to establish a potential anti-inflammatory activity of DHG related to epidermal dysfunction. As expected, in the murine model of acute inflammation used [60,62], we observed that DHG elicited a significant anti-edema effect comparable to that of indomethacin after the induction of dermatitis.

In conclusion, our findings show that DHG reduces lymphocyte and keratinocyte proliferation and viability. In PBMC, DHG induces apoptosis reducing cell activation in a specific cell population, interferes with cytokine secretion and inhibits inflammatory pathways. In keratinocytes DHG reduces cell migration and croton oil-induced ear dermatitis in mice. Overall our results suggest a potential therapeutic use of DHG as a topical anti-inflammatory agent.

4. Materials and Methods

4.1. Reagents and Antibodies (Abs)

DHG was isolated from *D. membranosa* according to the recently described solid phase extraction (HRX-SPE) method [54]. The product (2.2 mg) was solubilized in dimethyl sulfoxide (DMSO) (0.01% in our assays) and added to cell cultures at the reported concentrations. Phytohemagglutinin (PHA) and OKT3 monoclonal antibody were from Sigma-Aldrich (St. Louis, MO, USA) and used at 1.5% and 1 µg/mL respectively.

For western blot analysis, rabbit polyclonal anti-human β-actin was purchased from Abcam (Cambridge, UK); rabbit monoclonal anti-human p-NF-κB, rabbit monoclonal anti-human NF-κB and the secondary HRP-linked antibodies were purchased from Cell Signaling Technology (Danvers, MA, USA), rabbit monoclonal anti-human p-STAT5 (Tyr 694), rabbit polyclonal anti-human STAT5, rabbit monoclonal anti-human p-p44/42 MAPK (p-ERK, Thr202/Tyr 204), rabbit monoclonal anti-human p44/42 MAPK were purchased from Cell Signaling Technology (Danvers, MA, USA); mouse monoclonal anti-human α-tubulin from Sigma-Aldrich Inc. (St. Louis, MO, USA).

The following mAbs were used for immunostaining or as blocking Abs: anti-CD56/PerCP/Cy5.5, anti-CD69/PE, anti-AnnexinV/FITC, anti-CD25 from BioLegend (San Diego, CA, USA); anti-CD3/FITC from BD Pharmingen (San Jose, CA, USA). FACSCalibur flow cytometer (BD Biosciences, San Jose, CA, USA) was used for data collection. For data analysis Cell Quest Pro program (BD Biosciences, San Jose, CA, USA) was used. Data are reported as logarithmic values of fluorescence intensity.

4.2. Cells

Healthy peripheral blood mononuclear cells were separated over Ficoll-Hypaque gradients (MP Biomedicals, Aurora, OH, USA). PBMC were grown in RPMI 1640 (Invitrogen, San Diego, CA, USA), supplemented with 2 mM L-glutamine, 50 ng/mL, streptomycin, 50 units/mL penicillin, and 10% heat-inactivated fetal bovine serum (Hyclone Laboratories, Logan, UT, USA). All volunteers provided written informed consent in agreement with the Declaration of Helsinki to the use of their residual buffy coats for research aims with approval from the University Hospital of Salerno Review Board.

Human immortalized keratinocytes (HaCaT) were grown in Dulbecco's modified Eagle's medium (DMEM, GIBCO, Grand Island, NY, USA) supplemented with 2 mM L-glutamine, 50 ng/mL, streptomycin, 50 units/mL penicillin, and 10% heat-inactivated fetal bovine serum (Hyclone Laboratories, Logan, UT, USA). HaCaT cells were kindly provided by Giuseppe Monfrecola (Department of Experimental Dermatology, University of Naples, Naples, Italy).

4.3. Proliferation, Cell Viability and Sulforhodamine B Assays

PBMC isolated from ten healthy donors (2×10^5 cells per well) were cultured in triplicate in round bottom 96-well plates in a final volume of 200 μL of RPMI 10% FBS. Cells were activated with OKT3 or PHA. DHG was then added to the cells at the indicated concentration and its effects on proliferation were measured by the procedure described in detail elsewhere [63]. Viability of healthy PBMC (2×10^5 cells per well) activated with PHA and cultured in 96-well plates in the presence and in the absence of DHG was determined by trypan blue staining and haemocytometer counting after 6 days of incubation. Unstimulated PBMC were used as control, a further control was the solvent DMSO in the presence of our stimuli (OKT3 and/or PHA). Cell viability of HaCaT cells (4×10^3/well) incubated with increasing doses of DHG for 48 h was also determined by trypan blue staining and haemocytometer counting. In particular, cells were plated in 48-well plates after 24h to let them to adhere to the plastic, we added DHG at the doses indicated and after further 48h cells were detached with trypsin, stained with trypan blue and counted.

For the Sulforhodamine B assays HaCaT cells were plated in 96-well plates at a density of 2000 cells per well and after 24 h of incubation, to allow cells to adhere to the plate, DHG was added to the cells at shown concentrations. After 48 h, the supernatant was eliminated, cells were washed with PBS and finally, trichloroacetic acid (TCA) at 10% was added for one hour under stirring at 4 °C. After incubation, TCA was deleted and several washes with water were made. After drying the plates, the sulforhodamine B was added and the cells were incubated at RT for 30 min. After having eliminated the dye, washes were made with 1% acetic acid, until the removal of the unbound dye and the plates were left to dry. In the next step, the dye was solubilized with a solution of TRIS HCL 10 mM. The reading was performed at 570 nm using the spectrophotometer.

4.4. Scratch Wound Healing Assay

To evaluate the effect of DHG on HaCaT cell migration, the cells were plated in 6-well plates at a density of 5×10^3 cells/well. After 6 days, the confluent cells formed a homogeneous carpet and a vertical wound in the wells was performed using a 200 μL tip. After a wash to remove the cells detached from the plate, culture medium containing DHG (10–0.3 μM) or the vehicle alone was added to the wells. The wound area was recorded immediately and after 24 h through microscope analysis.

4.5. Apoptosis Analysis

The determination of apoptosis of PBMC was conducted by Annexin V (BioLegend, San Diego, CA, USA) and PI staining. PBMC isolated from ten healthy donors (2×10^5 cells per well) were cultured in a final volume of 200 μL of RPMI 10% FBS in triplicate in round bottom 96-well plates. Cells were activated with PHA (1.5%) in the presence or in the absence of increasing concentrations of

Int. J. Mol. Sci. **2017**, *18*, 1643

DHG. After 48 h of incubation, PBMC were washed in PBS and subjected to apoptosis determination by the procedure described in detail elsewhere [64].

4.6. Flow Cytometric Assay

PBMC were cultured in medium, activated with PHA in the presence and in the absence of DHG 3 μM, in U-bottom 96-well plates. After 24 h, cells were washed with PBS 2% FBS, stained with anti-CD3 FITC (BD Biosciences) and anti-CD56 PE-Cy5 (BD Biosciences). The cells were acquired by flow cytometer and analyzed by Cell-Quest Pro software (BD Biosciences, San Jose, CA, USA). Results are reported as logarithmic values of fluorescence intensity.

4.7. Cytokine Secretion Measurement

PBMC isolated from ten healthy donors (2×10^5 cells per well) were cultured in a final volume of 200 μL of complete medium in triplicate in round bottomed 96-well plates. PBMC were incubated with DHG for 24 h. Thereafter, supernatant was collected and the concentrations of TNF-α, IL-6 and IL-10 evaluated by enzyme-linked immunosorbent assay (ELISA), according to the manufacturers' instruction (R&D Systems, Minneapolis, MN, USA; and BioSource International, Camarillo, CA, USA). Of note, the remaining cell pellets were used to analyze the surface expression of the CD3, CD69 and CD56.

4.8. Western Blot (WB) Analysis

PBMC (1×10^6) were serum starved for 4h in T25 flasks and pre-treated with DHG at different concentrations and then activated with 1.5% PHA in sterile eppendorf for 30 min and 120 min in RPMI free medium. Then cells were centrifuged, cell pellets were lysed in ice-cold lysis buffer (20% SDS, 50% Tris, HCl 1 M, pH 6.8, 5%-ME, 25% glycerol, bromophenol blue) and then assayed for WB by the procedure described in detail elsewhere [60].

4.9. Topical Anti-Inflammatory Activity

DHG topical anti-inflammatory effect was reported as block of the croton oil-induced ear dermatitis in mice by the procedure described in detail elsewhere [59,60]. All the in vivo assays complied with the Italian Decree n. 116/1992 (as well as the EU Directive 2010/63/EU) and the European Convention ETS 123.

4.10. Statistical Analysis

In all experiments statistical analysis was performed by GraphPad prism 6.0 software for Windows (GraphPad Software Inc., La Jolla, CA, USA). Results from multiple experiments are calculated as mean ± SD and analyzed for statistical significance using the 2-tailed Student *t*-test, for independent groups, or ANOVA followed by Bonferroni correction for multiple comparisons. *p* values less than 0.05 were considered significant.

Supplementary Materials: Supplementary materials can be found at www.mdpi.com/1422-0067/18/8/1643/s1.

Acknowledgments: This study was supported by: National project PON01_02093 from the Italian Ministry of Education, University and Research; Associazione Italiana Ricerca sul Cancro (AIRC; IG 13312 and IG18999 to Maurizio Bifulco); Elena Ciaglia was supported by a fellowship from Fondazione Umberto Veronesi (FUV 2017, cod.1072); Anna Maria Malfitano was supported by fellowships from the Italian Ministry of Education, University and Research (PON01_02093).

Author Contributions: Conception and design: Patrizia Gazzerro and Maurizio Bifulco. Development of methodology: Elena Ciaglia, Anna Maria Malfitano, Angelo Fontana, Genoveffa Nuzzo, Adele Cutignano, Marco Pelin. Acquisition of data: Elena Ciaglia, Anna Maria Malfitano, Angelo Fontana, Genoveffa Nuzzo, Adele Cutignano, Marco Pelin. Analysis and interpretation of data: Elena Ciaglia, Anna Maria Malfitano, Adele Fontana, Genoveffa Nuzzo, Adele Cutignano, Marco Pelin, Silvio Sosa, Patrizia Gazzerro. Writing, review, and/or revision of the manuscript: Elena Ciaglia, Anna Maria Malfitano, Chiara Laezza, Angelo Fontana,

Int. J. Mol. Sci. **2017**, *18*, 1643

Genoveffa Nuzzo, Adele Cutignano, Mario Abate, Marco Pelin, Silvio Sosa, Maurizio Bifulco, Patrizia Gazzerro. Study supervision: Patrizia Gazzerro, Maurizio Bifulco.

Conflicts of Interest: The authors declare no conflict of interest.

References

1. Blunt, J.W.; Copp, B.R.; Munro, M.H.G.; Northcote, P.T.; Prinsep, M.R. Marine natural products. *Nat. Prod. Rep.* **2016**, *33*, 382–431. [CrossRef] [PubMed]
2. La Barre, S.; Kornprobst, J.M. *Outstanding Marine Molecules: Chemistry, Biology, Analysis*, 2014th ed.; Barre, S.L., Kornprobst, J.-M., Eds.; Wiley-VCH Verlag GmbH & Co.: Weinheim, Germany, 2014.
3. Bhatnagar, I.; Kim, S.K. Marine antitumor drugs: Status, shortfalls and strategies. *Mar. Drugs* **2010**, *8*, 2702–2720. [CrossRef] [PubMed]
4. Fusetani, N. Marine natural products. In *Natural Products in Chemical Biology*, 1st ed.; Civjan, N., Ed.; Wiley & Sons, Inc.: Hoboken, NJ, USA, 2012; pp. 31–64.
5. Kingston, D.G.I.; Newman, D.J. Natural products as anticancer agents. In *Natural Products in Chemical Biology*, 1st ed.; Civjan, N., Ed.; Wiley & Sons, Inc.: Hoboken, NJ, USA, 2012; pp. 325–349.
6. Mayer, M.S.; Glaser, K.B.; Cuevas, C.; Jacobs, R.S.; Kem, W.; Little, R.D.; McIntosh, J.M.; Newman, D.J.; Potts, B.C.; Shuster, D.E. The odyssey of marine pharmaceuticals: A current pipeline perspective. *Trends Pharmacol. Sci.* **2010**, *31*, 255–265. [CrossRef] [PubMed]
7. Molinski, T.F.; Dalisay, D.S.; Lievens, S.L.; Saludes, J.P. Drug development from marine natural products. *Nat. Rev. Drug Discov.* **2009**, *8*, 69–85. [CrossRef] [PubMed]
8. Mondal, S.; Bandyopadhyay, S.; Ghosh, M.K.; Mukhopadhyay, S.; Roy, S.; Mandal, C. Natural products: Promising resources for cancer drug discovery. *Anticancer Agents Med. Chem.* **2012**, *12*, 49–75. [CrossRef] [PubMed]
9. Proksch, P.; Putz, A.; Ortlepp, S.; Kjer, J.; Bayer, M. Bioactive natural products from marine sponges and fungal endophytes. *Phytochem. Rev.* **2010**, *9*, 475–489. [CrossRef]
10. Sakai, R.; Swanson, G.T. Recent progress in neuroactive marine natural products. *Nat. Prod. Rep.* **2014**, *31*, 273–309. [CrossRef] [PubMed]
11. Williams, P.; Sorribas, A.; Howes, M.J.R. Natural products as a source of Alzheimer's drug leads. *Nat. Prod. Rep.* **2011**, *28*, 48–77. [CrossRef] [PubMed]
12. Gershenzon, J.; Dudareva, N. The functiFon of terpene natural products in the natural world. *Nat. Chem. Biol.* **2007**, *3*, 408–414. [CrossRef] [PubMed]
13. Dewick, P.M. The Mevalonate and Methylerythritol Phosphate Pathways: Terpenoids and Steroids. Medicinal Natural Products. In *Medicinal Natural Products: A Biosynthetic Approach*, 3rd, ed.; John Wiley & Sons, Ltd.: Hoboken, NJ, USA, 2009. [CrossRef]
14. Fontana, A.; Manzo, E.; Ciavatta, M.L.; Cutignano, A.; Gavagnin, M.; Cimino, G. Biosynthetic Studies Through Feeding Experiments in Marine Organisms. In *Handbook of Marine Natural Products*; Springer: Dordrecht, The Netherlands, 2012; pp. 895–946.
15. Ro, D.K. Terpenoid biosynthesis. *Plant Metab. Biotechnol.* **2011**, 217–240. [CrossRef]
16. Rohmer, M. Mevalonate-independent methylerythritol phosphate pathway for isoprenoid biosynthesis. Elucidation and distribution. *Pure Appl. Chem.* **2003**, *75*, 375–388. [CrossRef]
17. Singh, B.; Sharma, R. Plant terpenes: Defense responses, phylogenetic analysis, regulation and clinical applications. *Biotech* **2015**, *5*, 129–151. [CrossRef] [PubMed]
18. Tholl, D. Biosynthesis and biological functions of terpenoids in plants. *Adv. Biochem. Eng. Biotechnol.* **2015**, *148*, 63–106. [CrossRef] [PubMed]
19. Gavagnin, M.; Fontana, A. Diterpenes from marine opisthobranch molluscs. *Curr. Org. Chem.* **2000**, *4*, 1201–1248. [CrossRef]
20. Gordaliza, M. Cytotoxic terpene quinones from marine sponges. *Mar. Drugs* **2010**, *8*, 2849–2870. [CrossRef] [PubMed]
21. Liang, L.F.; Guo, Y.W. Terpenes from the soft corals of the genus Sarcophyton: Chemistry and biological activities. *Chem. Biodivers.* **2013**, *10*, 2161–2196. [CrossRef] [PubMed]
22. Mehbub, M.F.; Lei, J.; Franco, C.; Zhang, W. Marine sponge derived natural products between 2001 and 2010: Trends and opportunities for discovery of bioactives. *Mar. Drugs* **2014**, *12*, 4539–4577. [CrossRef] [PubMed]

23. Sipkema, D.; Franssen, M.C.R.; Osinga, R.; Tramper, J.; Wijffels, R.H. Marine sponges as pharmacy. *Mar. Biotechnol.* **2005**, *7*, 142–162. [CrossRef] [PubMed]

24. Wang, K. Terpenoids form the sea: Chemical diversity and bioactivity. *Curr. Org. Chem.* **2014**, *18*, 840–866. [CrossRef]

25. Guella, G.; Callone, E.; Mancini, I.; Dini, F.; di Giuseppe, G. Diterpenoids from marine ciliates: Chemical polymorphism of Euplotes rariseta. *Eur. J. Org. Chem.* **2012**, *27*, 5208–5216. [CrossRef]

26. Hanson, J.R. Diterpenoids of terrestrial origin. *Nat. Prod. Rep.* **2013**, *30*, 1346–1356. [CrossRef] [PubMed]

27. Keyzers, R.A.; Northcote, P.T.; Davies-Coleman, M.T. Spongian diterpenoids from marine sponges. *Nat. Prod. Rep.* **2006**, *23*, 321–334. [CrossRef] [PubMed]

28. Putra, M.Y.; Murniasih, T. Marine soft corals as source of lead compounds for anti-inflammatories. *J. Coast. Life Med.* **2016**, *4*, 73–77. [CrossRef]

29. De las Heras, B.; Hortelano, S. Molecular basis of the anti-inflammatory effects of terpenoids. *Inflamm. Allergy Drug Targets* **2009**, *8*, 28–39. [CrossRef] [PubMed]

30. Chen, W.F.; Chakraborty, C.; Sung, C.S.; Feng, C.W.; Jean, Y.H.; Lin, Y.Y.; Hung, H.C.; Huang, T.Y.; Huang, S.Y.; Su, T.M.; et al. Neuroprotection by marine-derived compound, 11-dehydrosinulariolide, in an in vitro Parkinson's model: A promising candidate for the treatment of Parkinson's disease. *Naunyn Schmiedebergs Arch. Pharmacol.* **2012**, *385*, 265–275. [CrossRef] [PubMed]

31. Lin, Y.Y.; Jean, Y.H.; Lee, H.P.; Chen, W.F.; Sun, Y.M.; Su, J.H.; Lu, Y.; Huang, S.Y.; Hung, H.C.; Sung, P.J.; et al. A soft coral-derived compound, 11-epi-sinulariolide acetate suppresses inflammatory response and bone destruction in adjuvant-induced arthritis. *PLoS ONE* **2013**, *8*, e62926. [CrossRef] [PubMed]

32. Sheu, J.H.; Chen, Y.H.; Chen, Y.H.; Su, Y.D.; Chang, Y.C.; Su, J.H.; Weng, C.F.; Lee, C.H.; Fang, L.S.; Wang, W.H.; et al. Briarane diterpenoids isolated from gorgonian corals between 2011 and 2013. *Mar. Drugs* **2014**, *12*, 2164–2181. [CrossRef] [PubMed]

33. Su, Y.D.; Su, T.R.; Wen, Z.H.; Hwang, T.L.; Fang, L.S.; Chen, J.J.; Wu, Y.C.; Sheu, J.H.; Sung, P.J. Briarenolides K and L, new anti-inflammatory briarane diterpenoids from an octocoral *Briareum* sp. (Briareidae). *Mar. Drugs* **2015**, *13*, 1037–1050. [CrossRef] [PubMed]

34. Hsiao, T.H.; Sung, C.S.; Lan, Y.H.; Wang, Y.C.; Lu, M.C.; Wen, Z.H.; Wu, Y.C.; Sung, P.J. New anti-inflammatory cembranes from the cultured soft coral *Nephthea columnaris*. *Mar. Drugs* **2015**, *13*, 3443–3453. [CrossRef] [PubMed]

35. Cuong, N.X.; Thao, N.P.; Luyenet, B.T.; Ngan, N.T.; Thuy, D.T.; Song, S.B.; Nam, N.H.; Kiem, P.V.; Kim, Y.H.; Minh, C.V. Cembranoid diterpenes from the soft coral *Lobophytum crassum* and their anti-inflammatory activities. *Chem. Pharm. Bull.* **2014**, *62*, 203–208. [CrossRef] [PubMed]

36. Thao, N.P.; Luyen, B.T.; Ngan, N.T.; Song, S.B.; Cuong, N.X.; Nam, N.H.; Kiem, P.V.; Kim, Y.H.; Minh, C.V. New anti-inflammatory cembranoid diterpenoids from the Vietnamese soft coral *Lobophytum crassum*. *Bioorg. Med. Chem. Lett.* **2014**, *24*, 228–232. [CrossRef] [PubMed]

37. Thao, N.P.; Nam, N.H.; Cuong, N.X.; Luyen, B.T.T.; Tai, B.H.; Kim, J.E.; Song, S.B.; Kiem, P.V.; Minh, C.V.; Kim, Y.H. Inhibition of NF-κB transcriptional activation in HepG2 cells by diterpenoids from the soft coral *Sinularia maxima*. *Arch. Pharm. Res.* **2014**, *37*, 706–712. [CrossRef] [PubMed]

38. Tai, C.J.; Su, J.H.; Huang, C.Y.; Huang, M.S.; Wen, Z.H.; Dai, C.F.; Sheu, J.H. Cytotoxic and anti-inflammatory eunicellin-based diterpenoids from the soft coral *Cladiella krempfi*. *Mar. Drugs* **2013**, *11*, 788–799. [CrossRef] [PubMed]

39. Chen, B.W.; Chao, C.H.; Su, J.H.; Wen, Z.H.; Sung, P.J.; Sheu, J.H. Anti-inflammatory eunicellin-based diterpenoids from the cultured soft coral *Klyxum simplex*. *Org. Biomol. Chem.* **2010**, *8*, 2363–2366. [CrossRef] [PubMed]

40. Berrué, F.; McCulloch, M.W.; Kerr, R.G. Marine diterpene glycosides. *Bioorg. Med. Chem.* **2011**, *19*, 6702–6719. [CrossRef] [PubMed]

41. Dumontet, C.; Jordan, M.A. Microtubule-binding agents: A dynamic field of cancer therapeutics. *Nat. Rev. Drug Discov.* **2010**, *9*, 790–803. [CrossRef] [PubMed]

42. Long, B.H.; Carboni, J.M.; Wasserman, A.J.; Cornell, L.A.; Casazza, A.M.; Jensen, P.R.; Lindel, T.; Fenical, W.; Fairchild, C.R. Eleutherobin, a novel cytotoxic agent that induces tubulin polymerization, is similar to paclitaxel (Taxol). *Cancer Res.* **1998**, *58*, 1111–1115. [PubMed]

43. Look, S.A.; Fenical, W.; Matsumoto, G.K.; Clardy, J. The pseudopterosins: A new class of antiinflammatory and analgesic diterpene pentosides from the marine sea whip *Pseudopterogorgia elisabethae* (Octocorallia). *J. Org. Chem.* **1986**, *51*, 5140–5145. [CrossRef]

44. Rodrìguez, I.I.; Shi, Y.P.; Garcìa, O.J.; Rodríguez, A.D.; Mayer, A.M.S.; Sánchez, J.A.; Ortega-Barria, E.; González, J. New pseudopterosin and seco-pseudopterosin diterpene glycosides from two Colombian isolates of *Pseudopterogorgia elisabethae* and their diverse biological activities. *J. Nat. Prod.* **2004**, *67*, 1672–1680. [CrossRef] [PubMed]

45. Shin, J.; Fenical, W. Fuscosides A–D: Anti-inflammatory diterpenoid glycosides of new structural classes from the Caribbean gorgonian *Eunicea fusca*. *J. Org. Chem.* **1991**, *56*, 3153–3158. [CrossRef]

46. Jacobson, P.B.; Jacobs, R.S. Fuscoside: An anti-inflammatory marine natural product which selectively inhibits 5-lipoxygenase. Part I: Physiological and biochemical studies in murine inflammatory models. *J. Pharmacol. Exp. Ther.* **1992**, *262*, 866–873. [PubMed]

47. Cheng, S.Y.; Lin, E.H.; Wen, Z.H.; Chiang, M.Y.; Duh, C.Y. Two new verticillane-type diterpenoids from the formosan soft coral *Cespitularia hypotentaculata*. *Chem. Pharm. Bull.* **2010**, *58*, 848–851. [CrossRef] [PubMed]

48. Chatter, R.; Othman, R.B.; Rabhi, S.; Kladi, M.; Tarhouni, S.; Vagias, C.; Roussis, V.; Guizani-Tabbane, L.; Kharrat, R. In vivo and in vitro anti-inflammatory activity of neorogioltriol, a new diterpene extracted from the red algae *Laurencia glandulifera*. *Mar. Drugs* **2011**, *9*, 1293–1306. [CrossRef] [PubMed]

49. Wei, X.; Rodríguez, A. D.; Baran, P.; Raptis, R.G. Dolabellane-type diterpenoids with antiprotozoan activity from a southwestern Caribbean gorgonian octocoral of the genus *Eunicea*. *J. Nat. Prod.* **2010**, *73*, 925–934. [CrossRef] [PubMed]

50. Cirne-Santos, C.C.; Souza, T.M.; Teixeira, V.L.; Fontes, C.F.; Rebello, M.A.; Castello-Branco, L.R.; Abreu, C.M.; Tanuri, A.; Frugulhetti, I.C.; Bou-Habib, D.C. The dolabellane diterpene Dolabelladienetriol is a typical noncompetitive inhibitor of HIV-1 reverse transcriptase enzyme. *Antiviral. Res.* **2008**, *77*, 64–71. [CrossRef] [PubMed]

51. Ioannou, E.; Quesada, A.; Rahman, M.M.; Gibbons, S.; Vagias, C.; Roussis, V. Dolabellanes with antibacterial activity from the brown alga *Dilophus spiralis*. *J. Nat. Prod.* **2011**, *74*, 213–222. [CrossRef] [PubMed]

52. Barbosa, J.P.; Pereira, R.C.; Abrantes, J.L.; Cirne dos Santos, C.C.; Rebello, M.A.; de Palmer Paixão Frugulhetti, I.C.; Laneuville Teixeira, V. In vitro antiviral diterpenes from the Brazilian brown alga *Dictyota pfaffii*. *Planta Med.* **2004**, *70*, 856–860. [CrossRef] [PubMed]

53. Soares, D.C.; Calegari-Silva, T.C.; Lopes, U.G.; Teixeira, V.L.; de Palmer Paixão, I.C.N.; Cirne-Santos, C.; Bou-Habib, D.C.; Saraiva, E.M. Dolabelladienetriol, a compound from *Dictyota pfaffii* algae, inhibits the infection by *Leishmania amazonensis*. *PLoS Negl. Trop. Dis.* **2012**, *6*, e1787. [CrossRef] [PubMed]

54. Cutignano, A.; Nuzzo, G.; Ianora, A.; Luongo, E.; Romano, G.; Gallo, C.; Sansone, C.; Aprea, S.; Mancini, F.; D'Oro, U.; et al. Development and application of a novel SPE-method for bioassay-guided fractionation of marine extracts. *Mar. Drugs* **2015**, *13*, 5736–5749. [CrossRef] [PubMed]

55. Puliti, R.; Fontana, A.; Cimino, G.; Mattia, C.A.; Mazzarella, L. Structure of a keto derivative of 9,11-dihydrogracilin A. *Acta Cryst. Sect. C* **1993**, *49*, 1373–1376. [CrossRef]

56. Maximo, P.; Ferreira, L.M.; Branco, P.; Lima, P.; Lourenco, A. The role of *Spongia* sp. in the discovery of marine lead compounds. *Mar. Drugs* **2016**, *14*, 139. [CrossRef] [PubMed]

57. Fan, J.; Nishanian, P.; Breen, E.C.; McDonald, M.; Fahey, J.L. Cytokine gene expression in normal human lymphocytes in response to stimulation. *Clin. Diagn. Lab. Immunol.* **1998**, *5*, 335–340. [PubMed]

58. Ciaglia, E.; Pisanti, S.; Picardi, P.; Laezza, C.; Malfitano, A.M.; D'Alessandro, A.; Gazzerro, P.; Vitale, M.; Carbone, E.; Bifulco, M. N6-isopentenyladenosine, an endogenous isoprenoid end product, directly affects cytotoxic and regulatory functions of human NK cells through FDPS modulation. *J. Leukoc. Biol.* **2013**, *94*, 1207–1219. [CrossRef] [PubMed]

59. Yamane, H.; Paul, W.E. Cytokines of the γ_c family control CD4+ T cell differentiation and function. *Nat. Immunol.* **2012**, *13*, 1037–1044. [CrossRef] [PubMed]

60. Ciaglia, E.; Pisanti, S.; Picardi, P.; Laezza, C.; Sosa, S.; Tubaro, A.; Vitale, M.; Gazzerro, P.; Malfitano, A.M.; Bifulco, M. N6-isopentenyladenosisne affects cytotoxic activity and cytokines production by IL-2 activated NK and exerts topical anti-inflammatory activity in mice. *Pharmacol. Res.* **2014**, *89*, 1–10. [CrossRef] [PubMed]

61. Ottaviani, C.; Nasorri, F.; Bedini, C.; de Pità, O.; Girolomoni, G.; Cavani, A. CD56^{bright}CD16$^-$ NK cells accumulate in psoriatic skin in response to CXCL10 and CCL5 and exacerbate skin inflammation. *Eur. J. Immunol.* **2006**, *36*, 118–128. [CrossRef] [PubMed]

62. Tubaro, A.; Dri, P.; Delbello, G.; Zilli, C.; Della Loggia, R. The croton oil ear test revisited. *Agents Actions* **1986**, *17*, 347–349. [CrossRef] [PubMed]

63. Manera, C.; Malfitano, A.M.; Parkkari, T.; Lucchesi, V.; Carpi, S.; Fogli, S.; Bertini, S.; Laezza, C.; Ligresti, A.; Saccomanni, G.; et al. New quinolone- and 1,8-naphthyridine-3-carboxamides as selective CB2 receptor agonists with anticancer and immuno-modulatory activity. *Eur. J. Med. Chem.* **2015**, *97*, 10–18. [CrossRef] [PubMed]

64. Ciaglia, E.; Torelli, G.; Pisanti, S.; Picardi, P.; D'Alessandro, A.; Laezza, C.; Malfitano, A.M.; Fiore, D.; Pagano Zottola, A.C.; Proto, M.C.; et al. Cannabinoid receptor CB1 regulates STAT3 activity and its expression dictates the responsiveness to SR141716 treatment in human glioma patients' cells. *Oncotarget* **2015**, *17*, 15464–15481. [CrossRef] [PubMed]

MDPI

St. Alban-Anlage 66

4052 Basel

Switzerland

Tel. +41 61 683 77 34

Fax +41 61 302 89 18

www.mdpi.com

International Journal of Molecular Sciences Editorial Office

E-mail: ijms@mdpi.com

www.mdpi.com/journal/ijms